Post-Transcriptional Control of Gene Expression

NATO ASI Series

Advanced Science Institutes Series

A series presenting the results of activities sponsored by the NATO Science Committee, which aims at the dissemination of advanced scientific and technological knowledge, with a view to strengthening links between scientific communities.

The Series is published by an international board of publishers in conjunction with the NATO Scientific Affairs Division

A Life Sciences B Physics	Plenum Publishing Corporation London and New York
C Mathematical and Physical Sciences D Behavioural and Social Sciences E Applied Sciences	Kluwer Academic Publishers Dordrecht, Boston and London
F Computer and Systems Sciences G Ecological Sciences H Cell Biology	Springer-Verlag Berlin Heidelberg New York London Paris Tokyo Hong Kong Barcelona

Series H: Cell Biology Vol. 49

Post-Transcriptional Control of Gene Expression

Edited by

John E. G. McCarthy

Gesellschaft für Biotechnologische Forschung mbH
W-3300 Braunschweig, FRG

Mick F. Tuite

Biological Laboratory, University of Kent
Canterbury, Kent CT2 7NJ, England

Springer-Verlag Berlin Heidelberg New York
London Paris Tokyo Hong Kong Barcelona
Published in cooperation with NATO Scientific Affairs Division

Proceedings of the NATO/CEC Advanced Research Workshop on "Post-Transcriptional Control of Gene Expression", held in Goslar, FRG, April 6–12, 1990

ISBN-13:978-3-642-75141-7 e-ISBN-13:978-3-642-75139-4
DOI: 10.1007/978-3-642-75139-4

Softcover reprint of the hardcover 1st edition 1990

2131/3140-543210 – Printed on acid-free-paper

PREFACE

The last ten years have witnessed a remarkable increase in our awareness of the importance of events subsequent to transcriptional initiation in terms of the regulation and control of gene expression. In particular, the development of recombinant DNA techniques that began in the 1970s provided powerful new tools with which to study the molecular basis of control and regulation at all levels. The resulting investigations revealed a diversity of post-transcriptional mechanisms in both prokaryotes and eukaryotes. Scientists working on translation, mRNA stability, transcriptional (anti)termination or other aspects of gene expression will often have met at specialist meetings for their own research area. However, only rarely do workers in different areas of post-transcriptional control/ regulation have the opportunity to meet under one roof. We therefore thought it was time to bring together leading representatives of most of the relevant areas in a small workshop intended to encourage interaction across the usual borders of research, both in terms of the processes studied, and with respect to the evolutionary division prokaryotes/eukaryotes.

Given the breadth of topics covered and the restrictions in size imposed by the NATO workshop format, it was an extraordinarily difficult task to choose the participants. However, we regarded this first attempt as an experiment on a small scale, intended to explore the possibilities of a meeting of this kind. Judging by the response of the participants during and after the workshop, the effort had been worthwhile. The general impression was that most participants had been impressed by the high level of the contributions and discussions, and also felt that they had learnt something new from the meeting. This encourages us to look forward optimistically to organising a second meeting with a similar philosophy and related themes, but on a larger scale, in 1992.

The present volume contains contributions from almost all of the main speakers at the Goslar workshop, and therefore provides a representative and up-to-date cross-section of the activities of leading groups in the respective fields of research. The contributions are in general review-oriented, and thus offer the reader a series of overviews covering most of the key aspects of post-transcriptional control/regulation in prokaryotes and eukaryotes.

We are grateful to the NATO Scientific Affairs Division, the Scientific Commission of the European Communities, and the Gesellschaft für Biotechnologische Forschung (GBF) in Braunschweig, West Germany, for financial support. We would also like to thank the participants of the workshop, the members of the McCarthy group, and the staff of the Hotel Kaiserworth for helping to make this workshop both a productive and an enjoyable experience.

August 1990

John E.G. McCarthy, Braunschweig, FRG
Mick F. Tuite, Canterbury, UK

CONTENTS

mRNA Stability

Chapter 1: **Ribonucleases: diversity and regulation** 1
Murray P. Deutscher and J. Zhang

Chapter 2: **Degradation of *puf* mRNA in *Rhodobacter capsulatus* and its role in the regulation of gene expression** 13
Gabriele Klug and S.N. Cohen

Chapter 3: **Mutational analysis of a RNase E dependent cleavage site from a bacteriophage T4 mRNA** 21
C. Ehretsmann, A.J. Carpousis and Henry M. Krisch

Chapter 4: **The role of a novel site-specific endoribonuclease in the regulated decay of *E.coli* mRNA - a model for growth-stage dependent mRNA stability in bacteria** 31
Alexander von Gabain, D. Georgellis, U. Lundberg, Ö. Melefors, L. Melin and O. Resnekov

Chapter 5: **Regulation of mRNA stability in yeast** 45
Allan Jacobson, A.H. Brown, J.L. Donahue, D. Herrick, R. Parker and S.W. Peltz

Chapter 6: **Mutations involved in mRNA stability and in the length of their poly(A) tails in the yeast *Saccharomyces cerevisiae*** 55
L. Minvielle-Sebastia, A. Petitjean, B. Winsor, N. Bonneaud and Francois Lacroute

Chapter 7: **Rapid degradation of the c-*fos* proto-oncogene transcript** 65
Joel G. Belasco, A.-B. Shyu and M.E. Greenberg

Chapter 8: **Post-transcriptional control of gene expression in chloroplasts** 73
David B. Stern, C.H.-Ching, C.C. Adams and K.L. Kindle

Chapter 9: **Differential mRNA stability: a regulatory strategy for Hsp70 synthesis** 83
Robert B. Petersen and S. Lindquist

Antisense Regulation

Chapter 10: **Post-transcriptional control of IS10 transposase expression: antisense RNA binding and other conformational changes affecting messenger RNA stability and translation** 93
C. Ma, J.E. Gonzalez, C.C. Case, T. Sonnabend, J. Rayner and Robert W. Simons

Chapter 11: **The antisense approach and early *Xenopus* development** 103
Alan Colman, C. Baker and J. Shuttleworth

Transcriptional (Anti-)Termination

Chapter 12: **The role of *boxA* in transcription of ribosomal RNA operons of *Escherichia coli*: changes in the processivity of RNA polymerase** 113
R.J. Gaudino and Edward A. Morgan

Chapter 13: **Regulation of Lambda N-gene expression** 125
L. Kameyama, L. Fernandez, G. Guarneros and Donald L. Court

Chapter 14: **Messenger RNA 3'end formation in *E.coli* and *S. cerevisiae*** 135
Terry Platt, C.A. Brennan, J.S. Butler, D.A. Campbell, P.P. Sadhale, P. Spear, E.J. Steinmetz, S.-Y. Wu and F.M. Zalatan

Translational Control

Chapter 15: **Culture conditions affect differently the translation of individual *Escherichia coli* mRNAs** 145
N. Jacques, M. Chevrier-Miller, J. Guillerez and Marc Dreyfus

Chapter 16: **Post-transcriptional control in *Escherichia coli*: the translation and degradation of mRNA** 157
John E.G. McCarthy, B. Gerstel, K. Hellmuth, V. Lang, B. Surin and P. Ziemke

Chapter 17: **Control of translational initiation by mRNA secondary structure: a quantitative analysis** 169
M.H. de Smit and Jan van Duin

Chapter 18: **The role of ribosomal RNA in the control of gene expression** 185
M.A. Firpo and Albert E. Dahlberg

Chapter 19: **The phage f1 gene VII start site and its mutants reveal that translational coupling can confer function to inherently inactive initiation sites** 197
Deborah A. Steege and M. Ivey-Hoyle

Chapter 20: **Measurement of translation rates in vivo at individual codons and implication of these rate differences for gene expression** 207
M.J. Sorensen, K.F. Jensen and Steen Pedersen

Chapter 21: **Regulation of gene expression by minor codons in Escherichia coli: minor codon modulator hypothesis** 217
Masayori Inouye and G. Chen

Chapter 22: **A short review of scanning** 227
Marilyn Kozak

Chapter 23: **Yeast mRNA structure and translational efficiency** 237
Hendrik A. Raue, J.J. van den Heuvel and R.J. Planta

Chapter 24: **Human fetal Gγ- and Aγ- globin mRNA and adult β-globin exhibit distinct translation efficiency and affinity for eukaryotic initiation factor 2** 249
S. Marsh, Y. Banai, M. Na'amad and Raymond Kaempfer

Chapter 25: **Effects of the 5'-leader sequence of tobacco mosaic virus RNA, or derivatives thereof, on foreign mRNA and native viral gene expression** 261
T. Michael Wilson, K. Saunders, M.J. Dowson-Day, D.E. Sleat, H. Trachsel and K.W. Mundry

Translational Regulatory Circuits

Chapter 26: **Translational control of the cIII gene of bacteriophage lambda** 277
Amos B. Oppenheim, S. Altuvia, D. Kornitzer and D. Teff

Chapter 27: **The mom operon of bacteriophage Mu is regulated by a combination of transcriptional and translational controls** 285
F. Gregory Wulczyn, F. Schmidt, A. Seiler and R. Kahmann

Chapter 28: **Mechanisms of ribosomal protein translational autoregulation** 299
David E. Draper

Chapter 29: **Translational feedback control in E.coli: the role of $tRNA^{Thr}$ and $tRNA^{Thr}$-like structures in the operator of the gene for threonyl-tRNA synthetase** 309
Matthias Springer, P. Lesage, M. Graffe, J. Dondon and M. Grunberg-Manago

Chapter 30: **Translational control of the transcriptional activator GCN4 involves upstream open reading frames, a general initiation factor and a protein kinase** 325
Alan G. Hinnebusch, J.-P. Abastado, E.M. Hannig, B.M. Jackson, P.F. Miller, M. Ramirez, R.C. Wek and N.P. Williams

Chapter 31: **Translational control by arginine of yeast gene CPA1** 337
Michel Werner, A. Feller, P. Delbecq and A. Pierard

Chapter 32: **Expression from polycistronic cauliflower mosaic virus pregenomic RNA** 347
J. Fütterer, J.-M. Bonneville, K. Gordon, M. deTapia, S. Karlsson and Thomas Hohn

Chapter 33: **Polyadenylation of cauliflower mosaic virus RNA is controlled by promoter proximity** 359
H. Sanfacon, P. Brodman and Thomas Hohn

Chapter 34: **The HIV-1 Rev trans-activator is a sequence specific RNA binding protein** 367
Michael H. Malim, D.F. McCarn, L.S. Tiley and B.R. Cullen

Chapter 35: **Control of protein synthesis by RNA regulators** 377
Michael B. Mathews, S. Gunnery, L. Manche, K.H. Mellits and T. Pe'ery

Chapter 36: **Untranslated leader sequences and enhanced messenger RNA translational efficiency** 389
Lee Gehrke and S.A. Jobling

Chapter 37: **Coordinate post-transcriptional regulation of ferritin and transferrin receptor expression: the role of regulated RNA-protein interaction** 399
Joe B. Harford and R.D. Klausner

Chapter 38: **Translation in yeast mitochondria: a review of general features and a case of mRNA-specific positive control** 411
Thomas D. Fox, T.W. McMullin, P. Haffter and L.S. Folley

Chapter 39: **The yeast pyruvate kinase gene is regulated at multiple levels** 421
P.A. Moore, A.J.E. Bettany and Alistair J.P. Brown

Biochemistry and Genetics of Translation

Chapter 40: **Initiation of protein synthesis in E.coli: the two crucial steps** 433
Larry Gold and D. Hartz

Chapter 41: **The function of initiation factors in relation to mRNA-ribosome interaction and regulation of gene expression** 443
A. La Teana, M. Falconi, R.T. Pawlik, R. Spurio, C.L. Pon and Claudio Gualerzi

Chapter 42: **Alternative translation and functional diversity of release factor 2 and lysyl-tRNA synthetase** 455
Yoshikazu Nakamura, K. Kawakami and O. Mikuni

Chapter 43: **Gene products that mediate translation initiation in yeast** 465
H. Yoon, A.M. Cigan and Thomas F. Donahue

Chapter 44: **Structure/function of mammalian initiation factors** 475
John W.B. Hershey, J. Schnier, Z. Smit-McBride, S.S. Milburn, N. Gaspar, S.-Y. Choi and M. Himbelin

Chapter 45: **Initiation factors involved in mRNA binding to ribosomes in Saccharomyces cerevisiae** 487
Peter P. Müller, S. Blum, M. Altmann, S. Lanker and H. Trachsel

Chapter 46: **The eukaryotic mRNA cap binding protein (eIF-4E): phosphorylation and regulation of cell growth** 497
R. Frederickson, A. Lazaris-Karatzas and Nahum Sonenberg

Chapter 47: **Peptide chain initiation in animal cells: mechanism and regulation** 511
Naba K. Gupta, B. Datta, A.L. Roy and M.K. Ray

Chapter 48: **New insights into an old problem: ternary complex (Met-tRNA$_f$·eIF-2·GTP) formation in animal cells** 521
Naba K. Gupta, A.L. Roy, M.K. Nag, T.G. Kinzy, S. MacMillan, R.E. Hileman, T.E. Dever, S. Wu, W.C. Merrick and J.W.B. Hershey

Chapter 49: **Phosphorylation of initiation and elongation factors and the control of translation** 527
Christopher G. Proud, N.T. Redpath and N.T. Price

Chapter 50: **Regulation of protein synthesis in mammalian systems by the guanine nucleotide exchange factor** 539
J.N. Dholakia and Albert J. Wahba

Chapter 51: **Isolation of the lsd mutations in Saccharomyces cerevisiae** 549
Alan B. Sachs

Chapter 52: **Functional role and biochemical properties of yeast peptide elongation factor 3 (EF-3)** 557
Masazumi Miyazaki, M. Uritani, Y. Kitaoka, K. Ogawa and H. Kagiyama

Translational Accuracy/Frameshifting

Chapter 53: **Translational Movements** 567
Charles G. Kurland

Chapter 54: **Ribosomal frameshift and frame-jump sites as control points during elongation** 579
Robert B. Weiss

Chapter 55: **Frameshifting in the expression of the *trpR* gene of *Escherichia coli*** 591
I. Benhar, C. Miller and Hanna Engelberg-Kulka

Chapter 56: **The ribosomal frame-shift signal of infectious bronchitis virus** 603
Stephen C. Inglis, N. Rolley and I. Brierley

Chapter 57: **Control of translational accuracy in yeast: the role of the Sal4 (Sup45) protein** 611
Mick F. Tuite, Akhmaloka, M. Firoozan, J.A.B. Duarte and C.M. Grant

Chapter 58: **HIV *pol* expression via a ribosomal frameshift** 623
Alan J. Kingsman, W. Wilson and S.M. Kingsman

Post-translational events

Chapter 59: **Protein modifications and mitochondrial import of yeast cytochrome c: an overview** 637
Fred Sherman

Subject index 647

PARTICIPANTS AND CONTRIBUTORS

Belasco, Dr. Joel G., Dept. of Microbiology and Molecular Genetics, Harvard Medical School, 25 Shattuck Street, Boston Massachusetts 02115 USA

Berkhout, Dr. Benjamin, Dept. of Health & Human Services, National Institutes of Health, Bldg. 4, Room 307, Bethesda, MD 20892 USA

Boni, Dr. Irina V., M. Shemyakin Institute f Bioorg. Chemistry, USSR Academy of Sciences, Ul. Miklucho-Maklay 16/10, 11781 GSP Moscow USSR

Brody, Dr. Edward, Centre de Génetique Moleculaire, 91190 Gif-Sur-Yvette France

Brown, Dr. Al, Dept. of Genetics & Microbiology, University of Aberdeen, Marischal College, Aberdeen AB9 1AS UK

Cleveland, Dr. Don W., Dept. of Biological Chemistry, John Hopkins University, 725 North Wolfe Street, Baltimore, Maryland 21205 USA

Colman, Dr. Alan, School of Biochemistry, University of Birmingham, P.O. Box 363, Birmingham B15 277 UK

Court, Dr. Donald L., Laboratory of Chromosome Biology, Frederick Cancer Research Facility, P.O.Box B, Bldg. 539, Frederick, Maryland 21701 USA

Dahlberg, Dr. Albert, Biochemistry Box G, Brown University, , Providence, R.I., 029212 USA

de Lorenzo, Victor, Dr., Consejo Superior de Investigaciones Cientificas, Centro de Investigaciones Biologicas, C/Velazquez 144, 28006 Madrid Espana

Deutscher, Dr. Murray, Dept. of Biochemistry, Univ. of Connecticut, Health Center, Farmington, Connecticut 06032 USA

Donahue, Dr. Tom, Dept. of Biology, Jordan Hall, Room A318, Indiana University, Bloomington, IN 47405 USA

Draper, Dr. David E., Dept. of Chemistry, John Hopkins University, , Baltimore, MD 21218 USA

Dreyfus, Dr. Marc, Laboratoire de Genetique Moleculaire, C.N.R.S. U.A. 238, 46 rue d'Ulm, 75230 Paris Cedex 05 France

Engelberg-Kulka, Dr. Hanna, Dept. of Molecular Biology, Hebrew-University, P.O. Box 1172, 91010 Jerusalem Israel

Fox, Dr. Thomas D., Section of Genetics and Development, Cornell University, Biotechnology Bldg., Ithaca, NY 14853-4835 USA

Gallie, Dr. Daniel, Dept. of Biological Sciences, Stanford University, Stanford, California 94305-5020 USA

Gehrke, Dr. Lee, Mass. Inst. of Technology, Bldg. E25-545, Cambridge, MA 02139 USA

Gerstel, Birgit, Gesellschaft für Biotechnologische Forschung, GBF, Mascheroder Weg 1, 3300 Braunschweig FRG

Gold, Dr. Larry, Dept. of Molec., Cell. and Developm. Biology, University of Colorado, Campus Box 347, Boulder, CO 80309 USA

Gross, Dr. Gerhard, Dept. of Genetics, GBF, Mascheroder Weg 1, 3300 Braunschweig FRG

Gualerzi, Dr. Claudio, Lab. of Genetics, Dept. of Cell Biology, University of Camerino, 62032 Camerino Italy

Gupta, Dr. Naba K., University of Nebraska, Dept. of Chemistry, Hamilton Hall, Lincoln, NE 68588-0304 USA

Haenni, Dr. Anne-Lise, Institut Jacques Monod, Universite Paris VII - CNRS, 2, Place Jussieu - Tour 43, 75251 Paris Cedex 05 France

Harford, Dr. Joe, Cell Biology and Metabolism Branch, National Institutes of Health, Bldg. 18, Bethesda, MD 20892 USA

Hartley, Alan, Biological Laboratory of Kent, Canterbury, Kent CT2 7NJ UK

Hauser, Dr. Hansjörg, GBF, Mascheroder Weg 1, 3300 Braunschweig, FRG

Hellmuth, Karsten, GBF, Mascheroder Weg 1, 3300 Braunschweig FRG

Hentze, Dr. Matthias, EMBL, Meyerhofstr. 1, 6900 Heidelberg FRG

Hershey, Dr. John W.B., Dept. of Biological Chemistry, School of Medicine, University of California, Davis, CA 95616 USA

Higgins, Dr. Chris F., ICRF Labs., Institute of Mol. Medicine, University of Oxford, John Radcliffe Hospital, Oxford OX3 9DU UK

Hinnebusch, Dr. Alan, NIH, Bldg. 6, Room 320, Bethesda, MD 20892 USA

Hohn, Dr. Thomas, Friedrich Miescher Institut, P.O. Box 2543, 4002 Basel Switzerland

Inglis, Dr. Stephen, Dept. of Pathology, University of Cambridge, Tennis Court Road, Cambridge CB2 1QP UK

Inouye, Dr. Masayori, Dept. of Biochemistry, University of New Jersey, 675 Hoes Lane, Piscataway, New Jersey 08854-5635 USA

Jackson, Dr. Richard, University of Cambridge, Dept. of Biochemistry, Tennis Court Road, Cambridge CB2 1QW UK

Jacobson, Dr. Allan, Dept. of Mol. Genetics and Microbiology, University of Massachusetts Medical School, 55 Lake Avenue North, Worcester, MA 01655 USA

Kaempfer, Dr. Raymond, Dept. of Molecular Virology, The Hebrew University of Jerusalem, Hadassah Medical School, 91010 Jerusalem USA

Kingsman, Dr. Alan J., Dept. of Biochemistry, Oxford University, South Parks Road, Oxford OX1 3QU UK

Klug, Dr. Gabriele, ZMBH, Postfach 106 249, 6900 Heidelberg FRG

Kozak, Dr. Marilyn, Dept. of Biochemistry, Rbt. Wood Johnson Medical School, 675 Hoes Lane, Piscataway, New Jersey 08854 USA

Krisch, Dr. Henry M., Dept. of Mol. Biology, 30, Quai E.-Ansermet, 1211 Geneva 4 Suisse

Kurland, Dr. Charles G., Dept. of Molecular Biology, Biomedicum Box 590, Husargatan 3, S-751 24 Uppsala Sweden

Lacroute, Dr. Francois, Centre Genetique Moleculaire Du CNRS, , Gif sur Yvette, 91190 France

Laird-Offringa, Dr. Ite A., Dept. of Medical Biochemistry, University of Leiden, P.O. Box 9503, 2300 Leiden The Netherlands

Lang, Volker, GBF, Mascheroder Weg 1, 3300 Braunschweig FRG

Malim, Dr. Michael H., Howard Hughes Medical Institute, Duke University Medical Center, Box 3025, Durham, North Carolina 27710 USA

Mathews, Dr. Michael B., C.S.H. Laboratory, P.O. Box 100, Cold Spring Harbor, NY 11724 USA

McCarthy, Dr. John E.G., GBF, Mascheroder Weg 1, 3300 Braunschweig FRG

Miyazaki, Dr. Masazumi, Dept. of Molecular Biology, School of Science, Nagoya University, Chikusa-ku, Nagoya 464-01 Japan

Morgan, Dr. Edward A., Dept. of Experimental Biology, Roswell Park Memorial Institute, 666 Elm Street, Buffalo, N.Y. 14263 USA

Müller, Dr. Peter P., Inst. f. Biochemie und Molekularbiologie, Universität Bern, Bühlstr. 28, CH-3012 Bern Switzerland

Nakamura, Dr. Yoshikazu, Institute of Medical Science, University of Tokyo, 4-6-1, Shirokanedai Minato-ku, Tokyo 108 Japan

Oberbäumer, Dr. Ilse, Max-Planck-Institut für Biochemie, 8003 Martinsried FRG

Oppenheim, Dr. Amos B., Dept. of Molecular Genetics, The Hebrew University-Haddassah Medical School, P.O. Box 1172, Jerusalem 91010 Israel

Pedersen, Dr. Steen, Institute of Microbiology, University of Copenhagen, Oster Farimagsgade 2A, 1353 Copenhagen K Denmark

Petersen, Dr. Bob, HHMI-N339, Box 391, 5841 S. Maryland, Chicago, FL 60637 USA

Platt, Dr. Terry, Dept. of Biochemistry, Univ. of Rochester Medical Center, 601 Elmwood Avenue, Rochester, NY 14642 USA

Proud, Dr. Chris, Dept. of Biochemistry, University of Bristol, University Walk, Bristol, BS8 1TD UK

Proudfoot, Dr. Nick J., Sir William Dunn School of Pathology, University of Oxford, South Parks Road, Oxford OX1 3RE UK

Raué, Dr. Dick, Dept. of Biochemistry, Free University, De Boelelaan 1083, 1081 Amsterdam The Netherlands

Régnier, Dr. Philippe, Institut de Biologie Physico-Chimique, 13, Rue Pierre et Marie Curie, 75005 Paris France

Sachs, Dr. Alan, Whitehead Institute for Biomedical Research, Cambridge, MA 02142 USA

Saggliocco, Dr. Francis, Dept. of Genetics & Microbiology, University of Aberdeen, Marischal College, Aberdeen AB9 1AS UK

Shatsky, Dr. Ivan N., A.N. Belozersky Laboratory, Moscow State University, Leninsky Hills, Bldg. "A", 11899 Moscow USSR

Sherman, Dr. Fred, Dept. of Biochemistry, University of Rochester, 601 Elmwood Avenue, Rochester, New York 14642 USA

Simons, Dr. Robert W., Dept. of Microbiology, University of California at Los Angeles, 405 Hilgard Avenue, Los Angeles, California 900042 USA

Sonenberg, Dr. Nahum, Dept. of Biochemistry, McGill University , 3655 Drummond Street, Montreal, Quebec, H3G 1Y6 Canada

Steege, Dr. Deborah, Dept. of Biochemistry, Duke University Medical Center, Durham, NC 27710 USA

Stern, Dr. David B., Boyce Thompson Institute of Plant Research, Tower Road, Ithaca, New York 14853-1801 USA

Thireos, Dr. George, Foundation for Research and Technology, Institute of Molecular Biology, P.O. Box 1527, Heraklion 71110, Crete Greece

Tuite, Dr. Mick, Biological Laboratory, The University of Canterbury, Canterbury, Kent CT2 7NJ UK

van Duin, Dr. Jan, Dept. of Chemistry, Leiden University, Einsteinweg 5, 2300 RA Leiden The Netherlands

Vega Laso, Dr. Charo, GBF, Mascheroder Weg 1. 3300 Braunschweig FRG

von Gabain, Dr. Alexander, Dept. of Bacteriology, Karolinska Institute, Box 60400, 10401 Stockholm Sweden

Voorma, Dr. Harry O., Dept. of Molec. Cell Biology, Transitorium 3, Padualaan 8, NL-3508 TB Utrecht The Netherlands

Wahba, Dr. Albert J., Dept. of Biochemistry, University of Mississippi, North State Street, Jackson, Mississippi 39216-4505 USA

Walter, Dr. Friedrich, Akademie der Wissenschaften der DDR, Zentralinstitut für Mikrobiologie und Experimentelle Therapie, Beutenbergstr. 11, 6900 Jena GDR

Walter, Dr. Peter, Dept. of Biochem. Biophysics, University of California, San Francisco, CA 94143-0448 USA

Weiss, Dr. Robert B., Dept. of Human Genetics, University of Utah , Medical Center, Salt Lake City, Utah 84132 USA

Werner, Dr. Michel, Service de Biochimie, Bat. 142, Cen Saclay, 91191 Gif-sur-Yvette CEDEX France

Wilson, Dr. T. Michael A., Center for Agric. Mol. Biology, Cook College, Rutgers University, P.O. Box 231, New Brunswick, New Jersey 08903-0231 USA

Wulczyn, Dr. F. Gregory, Inst. für Genbiologische Forschung Berlin GmbH, Ihnestr. 63, 1000 Berlin 33 FRG

Zhu, Dr. Delin, Biological Laboratory, University of Kent, Canterbury, Kent CT2 7NJ UK

1. L. Gold; 2. N. Sonenberg; 3. B. Gerstel; 4. M. Mathews; 5. D. Stern; 6. C. Vega Laso; 7. T. Platt; 8. M. Malim; 9. A. Haenni; 10. M. Deutscher; 11. S. Inglis; 12. A. Hinnebusch; 13. I. Laird-Offringa; 14. A. von Gabain; 15. J. Belasco; 16. G. Wulczyn; 17. A. Kingsman; 18. A. Sachs; 19. B. Berkhout; 20. D. Gallie; 21. M. Werner; 22. M. Tuite; 23. A. Dahlberg; 24. H.J. Hauser; 25. H. Voorma; 26. H. Krisch; 27. D.Steege; 28. D. Raué; 29. D. Draper; 30. J. Harford; 31. E. Morgan; 32. H. Engelberg-Kulka; 33. D. Cleveland; 34. F. Walter; 35. R. Jackson; 36. G. Thireos; 37. C. Kurland; 38. V. Lang; 39. S. Pedersen; 40. E. Brody; 41. P. Walter, 42. J. McCarthy; 43. Y. Nakamura; 44. D. Court; 45. R. Simons; 46. R. Petersen; 47. G. Gross; 48. F. Sherman; 49. C. Higgins; 50. T. Fox; 51. N. Gupta; 52. R. Kaempfer; 53. F. Hoeren (projectionist); 54. G. Klug; 55. A. Wahba; 56. D. Zhu; 57. J. van Duin; 58. A. Brown; 59. C. Proud; 60. M. Wilson; 61. A. Hartley

62. T. Donahue; **63**. F. Lacroute; **64**. A. Jacobson; **65**. A. Oppenheim; **66**. T. Hohn; **67**. J. Hershey; **68**. L. Gehrke; **69**. M. Kozak; **70**. C. Gualerzi; **71**. M. Miyazaki; **72**. M. Dreyfus; **73**. I. Oberbäumer; **74**. R. Weiss; **75**. P. Müller

RIBONUCLEASES: DIVERSITY AND REGULATION

Murray P. Deutscher and Jiren Zhang
Department of Biochemistry
University of Connecticut Health Center
Farmington, Connecticut 06032
U.S.A.

Introduction

The complexity of RNA metabolism has become much more apparent in recent years. First of all, it is now clear that there are many more types of RNA molecules present in cells than the original classes of rRNA, tRNA and mRNA. Secondly, most, if not all, of these RNA molecules are initially synthesized as precursors that must be processed to generate the mature, functional species. In addition, some of these functional RNAs undergo other turnover or modification reactions that further alter their structure. Finally, RNA molecules are ultimately degraded, and these degradative reactions proceed at different rates among classes of RNA molecules and even among members of the same class. These latter findings add an additional level of complexity to RNA metabolism because they imply that regulatory processes may be involved in the differential stability of RNA molecules.

Most of the reactions of RNA metabolism require a ribonuclease (RNase). Much work has gone into trying to identify and characterize these enzymes, but this field is still in its infancy (Deutscher, 1985, 1988). What is clear is that cells contain many more distinct RNases than might have been imagined just a few years ago, that in many cases alternate enzymes and pathways may exist for the metabolism of certain RNAs, and that the levels of the RNases themselves may be subject to regulation.

The first part of this article will provide some general information on the RNases already identified. The primary emphasis will be on information obtained from studies with Escherichia coli since it is only in this organism that a detailed picture is beginning to emerge. The second part of the article will focus on the question of RNase regulation using as an example the enzyme, RNase D, and its structural gene, rnd.

RNA Metabolism

The reactions that make up RNA metabolism, and for which RNase involvement is expected, can be divided into several different groups based on

NATO ASI Series, Vol. H 49
Post-Transcriptional Control of Gene Expression
Edited by J. E. G. McCarthy and M. F. Tuite

their ultimate effects on RNA structure (Table 1). In one group, termed "processing" reactions are those reactions that convert one RNA molecule to another which is either a functional RNA or is on the pathway to a functional RNA. Invariably, these reactions lead to an RNA of smaller size. Included in this group are the reactions that mature the 5' and 3' termini of various classes of RNAs, those that remove introns from various RNA precursors, and reactions that separate individual RNAs from complex transcripts (e.g., mRNA-mRNA, tRNA-tRNA, tRNA-rRNA, tRNA-mRNA) (Deutscher, 1984; King et al., 1986). From what is already known about RNA metabolism, this first group would encompass a large number of very different types of reactions because of the variety of potential substrates and the varied positions of cleavage. Consequently, it might be expected that many different RNases with differing specificities and modes of action would be required to carry out all of these reactions.

A second group of reactions encompasses those in which there is "turnover" of part of an RNA molecule, i.e., the RNA is altered in some manner and subsequently restored to its original, or to an almost original, functional form. Several examples of this type of reaction are already known, and others may exist as well. These reactions include the end-turnover of the -C-C-A sequence of tRNA (Deutscher, 1984), the breakage and re-joining of the anticodon loop of certain host tRNAs upon phage infection (Amitsur et al., 1989), and the modulation of the length of the poly(A) tract of mRNA (Bernstein and Ross, 1989). Highly-specific RNases are already known to participate in the first two processes (Deutscher et al., 1985; Amitsur et al., 1989).

The third division of RNA metabolism is the "degradative" reactions, and this group includes reactions which initiate the process as well as those which complete degradation to mononucleotides. Various types of RNA molecules are destined for removal by degradative reactions. Amongst these are the RNA fragments that are the by-products of the many RNA processing reactions, and RNA molecules that may have been synthesized incorrectly or inadvertantly damaged. These RNAs potentially could compete with functioning RNAs, and consequently, need to be eliminated. Another major class of RNA degradative reactions involves the normal cellular turnover of mRNA molecules. These reactions serve to readjust the mRNA pool to the needs of the cell, and play an important role in the overall regulation of gene expression (Belasco and Higgins, 1988).

In addition to the aforementioned reactions, cells also have mechanisms for degrading the usually stable RNA molecules, rRNA and tRNA, in response to specific physiological stresses. Thus, starvation for various nutrients leads to degradation of stable RNAs (Apirion, 1974), undoubtedly to provide nucleotides for synthetic processes. Likewise, exposure of cells to Hg^{2+} (Beppu and Arima, 1969) and elevated temperature are associated with degradation of stable RNA (Tomlins and Ordal, 1971; Ito and Ohnishi, 1983). Bacterial mutants have also been isolated which display increased degradation of stable RNA under certain physiological conditions (Ohnishi and Schlessinger, 1972; Lennette et al., 1972).

Finally, many cells have mechanisms for degrading extracellular RNA in their environment, possibly to provide nutrients or for some protective function. The enzymes involved in this process generally are secreted from cells and often are regulated by an intracellular inhibitor (e.g. Hartley, 1989).

Table 1. Summary of RNase involvement in RNA metabolism

Type of reaction	Specific Process
Processing	Maturation of termini of RNA precursors Removal of introns Separation of RNAs in polycistronic transcripts
Turnover	End-turnover of CCA sequence of tRNA Cleavage and rejoining of tRNA anticodon Modulation of poly(A) length on mRNA
Degradation	Removal of processing by-products Removal of "bad" RNAs Turnover of mRNAs Digestion of stable RNAs during stress Digestion of extracellular RNA

The total number of RNases that might be involved in these various degradative reactions is not clear, but it need not be large. A small number of non-specific RNases could accomplish the task if the mechanism of degradation simply involved attack on any RNA molecule that was not protected in some way. In such circumstances a combination of a non-specific endo- and exoribonuclease could rapidly degrade RNA to the mononucleotide level. On the other hand, if the initiation of degradation of some RNA molecules involved very specific cleavages, a larger number of RNases would be required. In fact, recent information suggests that specific cleavages do occur (e.g., Uzan et al., 1988; Nilsson et al., 1988).

Catalogue of RNases

The number of known RNase activities has expanded greatly in recent years as investigators have developed many ingenious substrates for their identification (Deutscher, 1985; 1988). Unfortunately, as work in this area has proliferated there has also been more confusion because it has not been clear whether all the identified activities represent distinct enzymes. This problem will be corrected as more of these activities are purified and as their in vivo functions are ascertained.

The information in Table 2 is a compilation of our current understanding of E. coli RNases. Some of the activities have been grouped together when there is the possibility that they may be manifestations of the same enzyme. Some other reported activities have not been listed if it is probable that they are mixtures of enzymes or another name for a more well-characterized activity. Other RNases from diverse organisms with specificities and modes of action different from any yet observed in E. coli are known, but are not listed here.

Our knowledge of the repertoire of RNases is still at an early stage of development. Nevertheless, several important conclusions can already be drawn from an examination of the Table:

1) A single cell, such as E. coli, can contain a large number of distinct RNases. Based on substrate specificity, mode of action, and the study of mutants, the number of clearly different enzymes in E. coli is approaching twenty, and there are still many functions for which RNases have not yet been identified.

2) RNases exist with several different modes of action. Thus, both exoribonucleases and endoribonucleases are known. Among the exoribonucleases, most of the known enzymes act in the 3'→5' direction, releasing 5' mononucleotides; however, eukaryotes also contain enzymes that degrade 5'→3' and release 5' mononucleotides (e.g., Stevens and Maupin, 1987). The absence so far of a 5' exoribonuclease in prokaryotes is particularly puzzling since functions are known for which such an activity might be expected (e.g. Bechhofer and Zen, 1989). Among the endoribonucleases are those, such as RNase I, which cleave RNAs non-specifically, and others such as RNases P, E and K which have highly specific cleavage sites. Also, there are those enzymes such as RNase III and RNase H which recognize general structural features of the substrate, double-strandedness and RNA-DNA hybrids, in this instance, but which may cleave somewhat less specifically.

3) Exoribonucleases can also be highly specific. Until a few years ago, the only exoribonucleases known were the non-specific enzymes, such as snake venom and spleen phosphodiesterases and RNase II. Exoribonucleases are now known which display high degrees of substrate specificity. For example, RNases D, BN and T all act on tRNA molecules, but with very different rates depending on whether the 3' terminus has an intact CCA sequence, nucleotide changes within the CCA sequence, or nucleotide residues following the CCA sequence (Deutscher, 1984).

4) RNases can have overlapping specificities. This is a particular problem because mutants defective in some RNases may not display any phenotype due to other enzymes that take over their function in vivo. A prime example of this type of compensation involves RNase II and polynucleotide phosphorylase (PNPase).

Table 2. Escherichia coli RNases

RNase	Structural properties	Mutant available	Cation requirements	Substrates/properties/function
E. coli endoribonucleases				
RNase I	~26 kDa	+	none	most RNAs, periplasmic, 3'P oligo products
RNase III	50 kDa, α_2 dimer	+	di, mono	double-stranded region, prerRNA and mRNA processing
RNase P	377nuc. RNA; 13.8 kDa protein	+	di, mono	tRNA and 4.5 S RNA 5' processing
RNase E	~ 70 kDa	+	di, mono	5S and tRNA processing
RNase H	17.5 kDa	+	di	RNA-DNA hybrids, DNA replication
RNase N	120 kDa	–	none	most RNAs, 5'P mono and oligo products
RNase K	55 kDa	–	di	mRNA, site-specific cleavage
RNases P2, O, PC	~ 40 kDa	–	di	multimeric pre-tRNAs, may be same enzyme
RNases IV, F, M	26-31 kDa	–	none	various RNAs at pyr-A sites, 3' P oligo products, may be same enzyme
E. coli exoribonucleases (all 3'→5')				
RNase II	70-80 kDa	+	di, mono	unstructured RNAs, mRNA degradation
PNPase	α_3, 250 kDa or $\alpha_3\beta_2$, 360 kDa	+	di	unstructured RNAs, phosphorolytic, mRNA degradation
RNase D	42.7 kDa	+	di	tRNA precursors, tRNAs lacking CCA
RNase BN	~ 65 kDa	+	di	phage tRNA 3' processing
RNase T	50 kDa, α_2 dimer	+	di	end-turnover of tRNA
RNase R	~ 80 kDa	+	di, mono	rRNA, mRNA, homopolymers, 5' P mono products
RNase PH	~ 45 kDa	+	di, mono	tRNA precursors, phospholytic
oligoribonuclease	38 kDa	–	di	oligonucleotides, 5' P mono products

References can be found in Deutscher (1985, 1987) except RNase M (Cannistraro and Kennell, 1989), RNase K (Lundberg et al., 1990), RNase PH (Deutscher et al., 1988).

Mutant strains lacking either enzyme are viable, but strains cannot survive when both nucleases are absent (Donovan and Kushner, 1986).

5) Exoribonucleases can be hydrolytic or phosphorolytic. Two phosphorolytic exoribonucleases are now known, PNPase and RNase PH. Interestingly, the substrate specificity of PNPase is similar to that of RNase II, and RNase PH is similar to RNase D. It seems that with these enzymes cells can use either a hydrolytic or phosphorolytic mode of degradation for a particular substrate. This raises the possibility that the choice of enzyme could be dictated by the energy state of the cell.

Importance of RNA Structure

In addition to the availability of particular RNases, it is becoming increasingly clear that RNA structure also plays an important role in determining RNA metabolic events. For example, it was shown early on that extended double-stranded stems are important for the maturation of 16S and 23S rRNAs by RNase III (King et al., 1986). Numerous examples have now been reported in which stem-loop structures are determinants of mRNA stability. These include structures near the 3' end of a message which can help to stabilize it (Gilson et al., 1987; Hayashi and Hayashi, 1985) by preventing 3' exonuclease digestion, and similar structures which can destabilize part of a mRNA by acting as sites for an initial endonucleolytic attack (Plunkett and Echols, 1989). Stem-loop structures in intercistronic regions also can act to give differential stability to different parts of a polycistronic mRNA (Newbury et al., 1987; Chen et al., 1988). Finally, stem-loop structures at the 5' end of a message can influence its stability since RNase III cleavages in this region serve to initiate degradation of the RNA (Takata et al., 1987; Portier et al., 1987). In fact, in the case of the rnc gene such a site actually functions in autoregulation of RNase III expression (Bardwell et al., 1989).

Since stem-loop structures often act as sites for initiation of RNA degradation, it is possible that degradation can be regulated by the binding to this region of proteins that prevent the initial cleavage step. Such a situation may be responsible for the iron-dependent stability of transferrin receptor mRNA (Mullner et al., 1989). Likewise, the AU sequences in the 3' untranslated region of certain mRNAs that mediate mRNA breakdown (Shaw and Kamen, 1986) may be modulated by a recently-discovered RNA binding protein (Malter, 1989).

The importance of RNA structure is further emphasized by some examples from tRNA processing. $tRNA_1^{Tyr}$ contains 3 consecutive CCA sequences at its 3' terminus, yet the processing machinery stops at the correct one (Schedl et al., 1976). Likewise, RNase D, which removes extra residues following the normal

CCA sequence, will rapidly remove a second CCA sequence, but will stop at the normal 3' terminus (Cudny *et al*., 1981). Thus, it is not the CCA sequence *per se* that is important, but rather its location in the overall three-dimensional structure of the tRNA molecule.

What's Next for RNases?

Our increased understanding of RNA metabolism has made it possible to know what types of substrates to design to look for new RNases. As a consequence, the number of RNases identified, and purified, will continue to increase. The continuing success with the *E. coli* system makes it likely that within the not too distant future every RNase in this cell will be known. Progress will be slower in eukaryotic systems because of the lack of appropriate mutants to confirm function and to eliminate interfering activities. For this reason, a concerted effort to study yeast RNases, where the possibility of ultimately using genetics exists, would prove worthwhile. As the repertoire of RNases continues to expand, studies of the mechanism of action of these enzymes will become more prominent, as will efforts to understand the regulation of their expression and activity. With the appreciation that RNases play a central role in gene expression, there is no doubt that their study will continue to be an important area of investigation.

Regulation of RNase D Expression

RNase D is a random exonuclease whose mode of action and substrate specificity suggest that it might be involved in the 3' processing of tRNA precursors (Deutscher, 1984). *In vitro* RNase D is highly specific for tRNA substrates. It removes extra 3' residues following the CCA sequence, and slows down dramatically upon reaching the mature terminus. As a consequence it can generate functionally active tRNAs from precursor molecules. On the other hand, it has not been possible to demonstrate a role for this enzyme *in vivo*. Mutant strains devoid of RNase D grow normally and process tRNA normally, suggesting that if it does participate in tRNA processing, another enzyme can take over in its absence. However, elevated levels of RNase D are deleterious to the cell (Zhang and Deutscher, 1988a); in particular, engineered strains with elevated RNase D activity and lacking tRNA nucleotidyltransferase grow extremely poorly and contain a large amount of defective tRNA (Zhang and Deutscher, 1988b). These latter findings indicate that RNase D can utilize tRNA as a substrate *in vivo*.

RNase D is encoded by the *rnd* gene located at 40 min on the *E. coli* genetic map (Deutscher, 1984). Cloning and sequencing of the *rnd* gene (Zhang and

Deutscher, 1988b,c) indicated that its coding region extends for 1128 nucleotides beginning at an unusual UUG codon and encoding a protein of about 42,600 Da. A single promoter, determined by deletion analysis (Zhang and Deutscher, 1989), is located upstream of the coding region; primer extension and nuclease S1 mapping indicates that transcription starts at residue 99 of the cloned fragment, 70 residues upstream of the initiation codon (Fig. 1). Interestingly, a hairpin structure, with features of a rho-independent terminator (GC-rich stem followed by 8 T residues), is located between the transcription and translation start sites (Fig. 1).

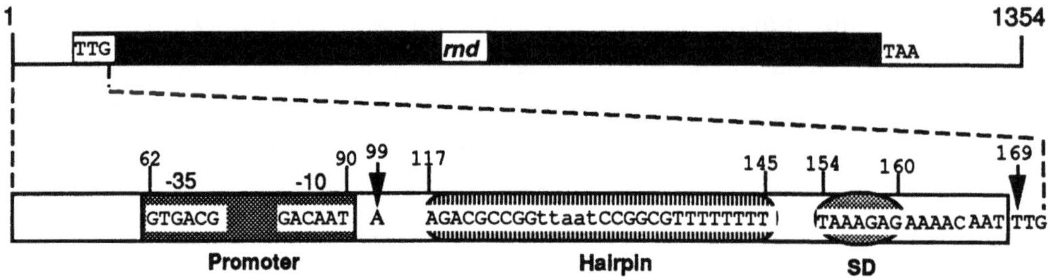

Fig. 1. Upstream region of the E. coli rnd gene. The promoter is located between residues 62 and 90 relative to the beginning of the cloned fragment, a hairpin structure with a GC-rich stem followed by 8 T residues is found between residues 117 and 145, and a ribosome-binding site (SD) is at residues 154 to 160. Transcription starts at the A residue at position 99 and translation begins at the TTG sequence at residue 169. Taken from Zhang and Deutscher (1989).

In order to ascertain the functional significance of the UUG initiation codon and the hairpin structure, the upstream region of the rnd gene was examined using deletion analysis and site-directed mutagenesis; expression of the rnd gene was monitored by dot-blots, primer extension, RNase D assays and immunoblotting. Conversion of the UUG codon to the more common AUG led to a greater than 10-fold increase in RNase D activity in cells transformed with a single-copy plasmid (Table 3). It was not possible to see the increase with a multicopy plasmid because the level of overexpression in this case is already at the maximum tolerable by the cell. Immunoblotting confirmed that the increase in RNase D activity is due to an increase in rnd protein (not shown). These data indicate that RNase D expression is lowered by the presence of a UUG initation codon.

Analysis of the stem-loop structure led to unexpected findings. Although the hairpin resembled a transcription terminator, its removal or modification had relatively little effect on the level of rnd mRNA determined by dot-blots (Table 3) or primer extension (not shown). Thus, if this structure does act as a terminator, it is a very poor one. In contrast, the hairpin has profound effects on translation of rnd mRNA (Table 3). Deletion of half of the stem-loop structure or of the entire structure decreased RNase D expression about 75% with a multicopy plasmid and over 95% (after correcting for background activity with the plasmid alone) with the single copy plasmid. Likewise, disruption of the stem by two G→A point mutations had an identical effect, whereas changes in the loop had only a minor effect. In all cases examined rnd mRNA levels were normal or slightly elevated, and primer extension revealed that the normal transcription start site was used (not shown).

Table 3. Effect of mutations on RNase D expression

Mutation	RNase D specific activity (%) Multicopy	Single copy	rnd mRNA (%)
Wild type	100	100 (6.5-10)	100
UUG initiation codon to AUG	100	1100	-
Deletion (residues 126-139)	24	11	110-210
Deletion (residues 117-144)	25	12	100-200
Loop substitution (TTAAT to GTACC)	90	50	-
Stem substitution (G121 to A, G124 to A)	21	12	100-130
UUG to AUG and deletion (117- 144)	78	14	95-120
Multicopy plasmid (pUC18) alone	< 1	-	-
Single-copy plasmid (pOU61) alone	-	8-12	-

Multicopy or single copy plasmids with the indicated insert were transformed into an RNase-deficient strain. Log-phase cells were assayed for RNase D activity. The value of 6.5-10 in parenthesis is the wild type level of activity in the single-copy plasmid relative to the multicopy plasmid. All rnd mRNA levels have been normalized relative to bla mRNA.

The explanation for these unusual results is not yet understood, but it is clear that an intact stem-loop structure is necessary for proper translation of rnd mRNA. This requirement is not due to the presence of the UUG start codon because its conversion to AUG does not overcome the negative effect of deleting the hairpin (Table 3). From primer extension and dot-blot analysis it is also clear that the effect of the stem-loop is not on mRNA processing or stability, and that it is not a terminator for an upstream gene (not shown). Also, since normal levels of RNase D overexpression are obtained upon transformation of an RNase III$^-$ strain the hairpin is probably not a site for RNase III action. Finally, computer analysis has not revealed any new secondary structures that might block translation in mRNAs lacking the stem-loop structure. Thus, the most likely explanation for these results is that the hairpin acts as a positive effector of translaton.

It is now very clear that regions distinct from the Shine-Delgarno sequence and the initiation codon can influence the initiation of translation on prokaryotic mRNAs (Gold, 1988). The findings described here show that expression of the rnd gene is affected both by the initiation codon and by a stem-loop structure 30-50 nucleotides upstream of the initiation codon. However, the mechanism of these effects and the role of regulation of RNase D expression on the function of this nuclease in RNA metabolism require further study.

References

Amitsur M, Morad I and Kaufmann G (1989) In vitro reconstitution of anticodon nuclease from components encoded by phage T4 and Escherichia coli CTr5X. EMBO J. 8:2411-2415

Apirion D (1974) The fate of mRNA and rRNA in Escherichia coli. Brookhaven Symp. Biol. 26:286-306

Bardwell JCA, Regnier P, Chen SM, Nakamura Y, Grunberg-Monago M and Court DL (1989) Autoregulation of RNase III operon by mRNA processing. EMBO J. 8:3401-3407

Bechhofer DH and Zen KH (1989) Mechanism of erythromycin-induced ermC mRNA stability in Bacillus subtilis. J. Bacteriol. 171:5803-5811

Belasco JG and Higgins CF (1989) Mechanisms of mRNA decay in bacteria: A perspective. Gene 72:15-23

Beppu T and Arima K (1969) Induction by mercuric ion of extensive degradation of cellular ribonucleic acid in Escherichia coli. J. Bacteriol. 98:888-897

Bernstein P and Ross J (1989) Poly(A), poly(A) binding protein and the regulation of mRNA stability. TIBS 14:373-377

Cannistraro VJ and Kennell D (1989) Purification and characterization of RNase M and mRNA degradation in Escherichia coli. Eur. J. Biochem. 181:363-370

Chen CYA, Beatty JT, Cohen SN and Belasco JG (1988) An intercistronic stem-loop structure functions as an mRNA decay terminator necessary but insufficient for puf mRNA stability. Cell 52:609-619

Cudny H, Zaniewski R and Deutscher MP (1981) E. coli RNase D: catalytic properties and substrate specificity. J. Biol. Chem. 256:5633-5637

Deutscher MP (1984) Processing of tRNA in prokaryotes and eukaryotes. Crit. Rev. Biochem. 17:45-71

Deutscher MP (1985) E. coli RNases: Making sense of alphabet soup. Cell 40:731-732

Deutscher MP, Marlor CW and Zaniewski R (1985) RNase T is responsible for the end-turnover of tRNA in E. coli. Proc. Natl. Acad. Sci. U.S.A. 82:6427-6430

Deutscher MP, Marshall GT and Cudny H (1988) RNase PH: A new phosphate-dependent nuclease distinct from polynucleotide phosphorylase. Proc. Nat. Acad. Sci. USA 85:4710-4714

Deutscher MP (1988) The metabolic role of RNases. TIBS 13:136-139

Donovan WP and Kushner SR (1986) Polynucleotide phosphorylase and ribonuclease II are required for cell viability and mRNA turnover in Escherichia coli K-12. Proc. Natl. Acad. Sci. U.S.A. 83:120-124

Gilson E, Clement JM, Perrin D and Hofnung M (1987) Palindromic units: a case of highly repetitive DNA sequences in bacteria. Trends in Gen. 3:226-230

Gold L (1988) Posttranscriptional regulatory mechanisms in Escherichia coli. Ann. Rev. Biochem. 57:199-233

Hartley RW (1989) Barnase and barstar: two small proteins to fold and fit together. TIBS 14:450-454

Hayashi MN and Hayashi M (1985) Cloned DNA sequences that determine mRNA stability of bacteriophage ϕX174 in vivo are functional. Nucleic Acid Res. 13:5937-5948

Ito R and Ohnishi Y (1983) The roles of RNA polymerase and RNase I in stable RNA degradation in E. coli carrying the srnB$^+$ gene. Biochim. Biophys. Acta

739:27-34

King TC, Sirdeskmukh R and Schlessinger D (1986) Nucleolytic processing of RNA transcripts in procaryotes. Microbiol. Rev. 50:428-451

Lennette ET, Meyhack B and Apirion D (1972) A mutation affecting degradation of stable RNA in Escherichia coli. FEBS Lett. 21:286-288

Lundberg U, Melefors O and von Gabain A (To be published) Purification and characterization of a novel endoribonuclease controlling mRNA stability in E. coli. EMBO J.

Malter JS (1989) Identification of a AUUUA-specific messenger RNA binding protein. Science 246:664-666

Mullner EW, Neupert B and Kuhn LC (1989) A specific mRNA binding factor regulates the iron-dependent stability of cytoplasmic transferrin receptor mRNA. Cell 58:373-382

Newbury SF, Smith NH and Higgins CF (1987) Differential mRNA stability controls relative gene expression within a polycistronic operon. Cell 51:1131-1143

Nilsson G, Lundberg U and von Gabain A (1988) In vivo and in vitro identity of site specific cleavages in the 5' non-coding region of ompA and bla mRNA in Escherichia coli. EMBO J. 7:2269-2275.

Ohnishi Y and Schlessinger D (1972) Total breakdown of ribosomal and transfer RNA in a mutant of Escherichia coli. Nature New Biol. 238:228-231

Plunkett III G and Echols H (1989) Retroregulation of the bacteriophage lambda int gene: Limited secondary degradation of the RNase III-processed transcript. J. Bacteriol. 171:588-592

Portier C, Dondon L, Grunberg-Manago M and Regnier P (1987) The first step in the functional inactivation of the E. coli polynucleotide phosphorylase messenger is a ribonuclease III processing at the 5' end. EMBO J. 6:2165-2170

Shaw G and Kamer R (1986) A conserved AU sequence from the 3' untranslated region of GM-CSF mRNA mediates selective mRNA degradation. Cell 46:659-667

Schedl P, Roberts J and Primakoff P (1976) In vitro processing of E. coli tRNA precursors. Cell 8:581-594

Stevens A and Maupin MK (1987) A 5'→3' exoribonuclease of human placental nuclei: purification and substrate specificity. Nucleic Acids Res. 15:695-708

Takata R, Mukai T and Hori K (1987) RNA processing by RNase III is involved in the synthesis of Escherichia coli polynucleotide phosphorylase. Mol. Gen. Genet. 209:28-32

Tomlins RI and Ordal ZJ (1971) Precursor ribosomal RNA and ribosome accumulation in vivo during recovery of S. typhimurium from thermal injury. J. Bacteriol. 107:134-142

Uzan M, Favre R and Brody E (1988) A nuclease that cuts specifically in the ribosome binding site of some T4 mRNAs. Proc. Natl. Acad. Sci. U.S.A. 85:8895-8899

Zhang J and Deutscher MP (1988a) Cloning, characterization, and effects of overexpression of the Escherichia coli rnd gene encoding RNase D. J. Bacteriol. 170:522-527

Zhang J and Deutscher MP (1988b) Transfer RNA is a substrate for RNase D in vivo. J. Biol. Chem. 263:17909-17912

Zhang J and Deutscher MP (1988c) Escherichia coli RNase D: sequencing of the rnd structural gene and purification of the overexpressed protein. Nucleic Acids Res. 16:6265-6278

Zhang J and Deutscher MP (1989) Analysis of the upstream region of the E. coli rnd gene encoding RNase D. J. Biol. Chem. 264:18228-18233

DEGRADATION OF *puf* mRNA IN *Rhodobacter Capsulatus* AND ITS ROLE IN THE REGULATION OF GENE EXPRESSION

Gabriele Klug[1] and Stanley N. Cohen[2]
[1]Zentrum für Molekulare Biologie Heidelberg
Im Neuenheimer Feld 282
6900 Heidelberg
Federal Republic of Germany
and
[2]Department of Genetics
Stanford University School of Medicine
Stanford, California 94305-5120
U.S.A.

In procaryotes, genes for proteins that interact functionally are often organized into polycistronic operons that ensure their coordinated transcriptional activation or repression. Post-transcriptional regulation of the expression of individual genes within a polycistronic operon can result in the synthesis of different amounts of their gene products. The first evidence that such regulation can occur by means of segmental differences in mRNA stability was obtained for the *puf* operon of the photosynthetic bacterium, *Rhodobacter capsulatus* (Belasco et al., 1985), and these segmental differences were later shown to be important in regulating the stoichiometry of pigment protein complexes that comprise the photosynthetic apparatus (Klug and Cohen, 1988).

The demonstration (Belasco et al., 1985) that a single polycistronic mRNA species encodes the pigment protein complexes of the reaction center (RC) complex and of the light harvesting (LH) I complex of *R. capsulatus*, identified the *rxcA* operon, which more recently has been renamed *puf*. Studies of *puf* mRNA by Northern blotting and S1 protection analysis (Belasco et al., 1985) revealed that two size classes of mRNA are encoded by the distal half of the operon (Fig. 1): a 2.7 kb species having a half-life of five minutes was found to be homologous to the LHI genes *puf*B and *puf*A, to genes *puf*L and *puf*M which encode proteins of the RC complex, and to the open reading frame *puf*X, which codes for a gene product of unknown function. A second *puf* mRNA species, which encodes only the *puf*B and *puf*A proteins of the LH complex, is 0.5 kb in length and has a half-life greater than 20 minutes. The different stabilities of these *puf* mRNA species result in approximately a ninefold molar excess of LHI-encoding mRNA over RC-encoding mRNA.

NATO ASI Series, Vol. H 49
Post-Transcriptional Control of Gene Expression
Edited by J. E. G. McCarthy and M. F. Tuite

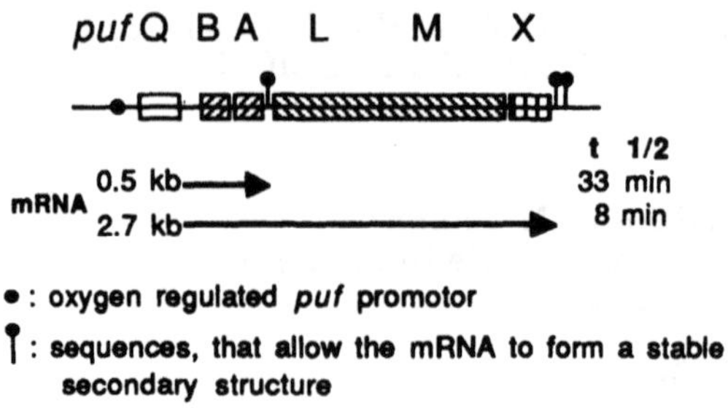

Figure 1: *puf* operon of *Rhodobacter capsulatus* and mRNA species found on Northern blots. The *puf* mRNA species were identified by Belasco et al. (1985). The half-life values are from Klug and Cohen (1990).

Pulse-chase experiments showed that the 0.5 kb *puf*BA mRNA is produced by degradation of the 2.7 kb *puf*BALMX mRNA segment (Belasco et al., 1985). An intercistronic hairpin loop structure located between *puf*A and *puf*L (Fig. 1) was predicted by computer analysis of the mRNA sequence (Belasco et al., 1985) and was shown to be responsible for the greater stability of the *puf*BA mRNA segment compared to the *puf*BALMX mRNA (Klug et al., 1987; Chen et al., 1988), possibly by protecting it from degradation by 3' to 5' exonucleases. Removal of the intercistronic hairpin loop structure resulted in continuous degradation of the 2.7 kb *puf* mRNA and consequent loss of the discrete 0.5 kb *puf* mRNA species on Northern blots (Figure 2; see also Klug et al., 1987; Chen et al.,

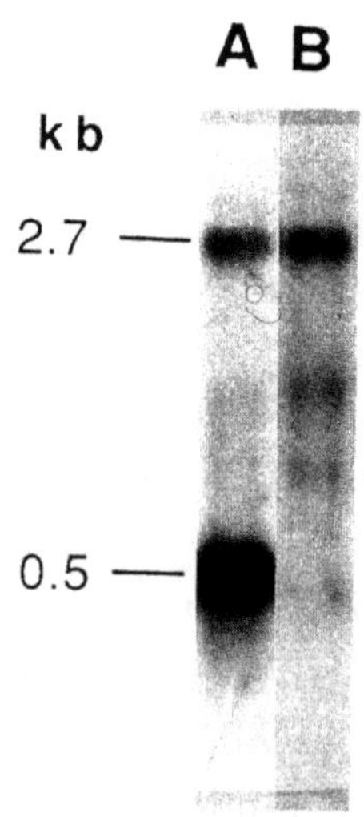

Figure 2: *puf* specific mRNA from strain UG304 (A, wild type *puf* sequence) and VG607 (B, intercistronic hairpin loop deleted from *puf* sequence).

1988). Later experiments have shown that the 2.7 kb *puf* mRNA species is actually the distal half of a transcript that starts far upstream and includes an additional protein-encoding reading frame, *pufQ* (Bauer et al., 1988; Adams et al., 1989). The RNA homologous to *pufQ* and the upstream noncoding region is highly unstable (Adams et al., 1989).

Biological consequences of the segmental differences in *puf* mRNA stability

Removal of the intercistronic hairpin loop structure results not in only a change in the molar ratio of LHI-specific to RC-specific mRNA sequences, but also in a changed composition of the photosynthetic membrane (Klug et al., 1987). *R. capsulatus* cells that have the wild type *puf* sequence present on the chromosome (NK3) or on a plasmid (UG304), but which lack the LHII complex, contain about eleven LHI complexes for each RC complex (Table 1). However, deletion of the intercistronic mRNA hairpin loop structure from *puf* transcripts reduced the LHI/RC ratio to about 40% of the control, consistent with the reduction in ratio of LHI-encoding to RC-encoding sequences observed as a consequence of decreased stability of the LHI-specific mRNA segment (cf. strains UC603 and UG304).

Table 1: Effect of the intercistronic mRNA structure on the stoichiometry of pigment protein complexes.

Strain	I nMol Bchl per nMol RC	II LHI complexes per RC complex
NK3 (wild type *puf* sequence on chromosome)	27.5 + 1.9	10.8
UG304 (wild type *puf* sequence on plasmid)	28.9 + 1.8	11.5
UG607 (*puf* sequence without hairpin loop on plasmid)	15.0 + 1.2	4.5

Values given in column II were calculated from the measurements in column I under the assumption that 2 Mol bacteriochlorophyll (Bchl) are bound to 1 Mol LHI and 4 Mol Bchl and two Mol of bacteriopheophytin are bound to 1 Mol of RC.

The LHI complexes surround the RC complex and transfer energy that originates from the absorption of light to the RC, where a charge separation occurs. In the absence of the LHII antenna complex, the amount of absorbed light energy depends mainly on the concentration of LHI antenna complex. Strain UG607, which has a decreased ratio of LHI to RC complexes showed a lower growth rate (doubling time 5 h 40 min) when grown photosynthetically than strain UG304 (doubling time 5 h 15 min), which has the wild type stoichiometry for LHI to RC complexes. Moreover, after a shift from growth at high oxygen tension to phototrophic growth in the absence of oxygen, strain UG304 carrying the wild type *puf* operon began rapid growth 2 to 3 hours before strain UG607 (Figure 3), which lacks the intercistronic hairpin structure in its *puf* operon.

The above data indicate that the observed segmental differences in the decay of *puf* mRNA are important in regulating the formation of the photosynthetic apparatus in *R. capsulatus*, and that these segmental differences are responsible for biologically relevant changes in cell function.

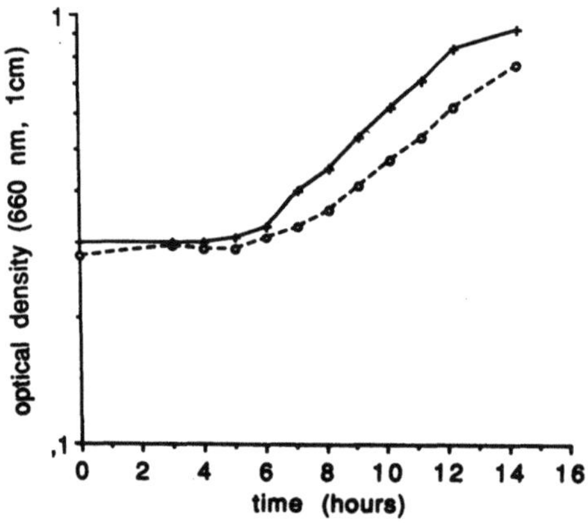

Figure 3: Cultures of strains UG304 (●—●—● wild type *puf* genes) and UG607 (●---●---● intercistronic hairpin loop removed from *puf*) were precultivated at high oxygen tension and shifted to phototrophic growth conditions at time 0. Strains UG304 and UG607 lack the LHII complex.

Endonucleolytic cleavage in the RC-coding region is rate limiting for decay of the *puf*LMX mRNA segment

Chen et al. (1988) have shown that insertion of the intercistronic hairpin loop region that confers high stability to the upstream *puf*BA mRNA segment at the 3' end of the *puf* operon does not increase stability of the 2.7 kb *puf*BALMX mRNA, which normally contains two 3' terminal regions of dyad symmetry that function as transcriptional terminators (see Fig. 1). Based on these data, these workers suggested that endonucleolytic cleavage 3' to the intercistronic hairpin loop structure, but 5' to the *puf* terminator structures, is the rate limiting step for degradation of the 2.7 kb *puf* mRNA segment.

Recent analysis of *puf* mRNA from constructs containing various *puf* operon deletions support the notion that endonucleolytic cleavage between the intercistronic hairpin loop structure and the dual transcription terminators downstream to *puf*X is the rate limiting step in decay of the distal half of the *puf* mRNA (Klug and Cohen, submitted for publication). Figure 4 illustrates that removal of a 1.4 kb segment from the RC encoding region leads to more than a doubling of the half-life of the remaining sequences in the RC-specific *puf* mRNA. However, separate deletion of individual segments within the 1.4 kb region fails to increase stability of the RC-encoding *puf* mRNA (Figure 4, pBSS51 and pBSS71), suggesting that multiple decay promoting loci are present within the 1.4 kb segment deleted in pΔRB6 and that removal of all of these sequences is required for prolongation of mRNA half-life.

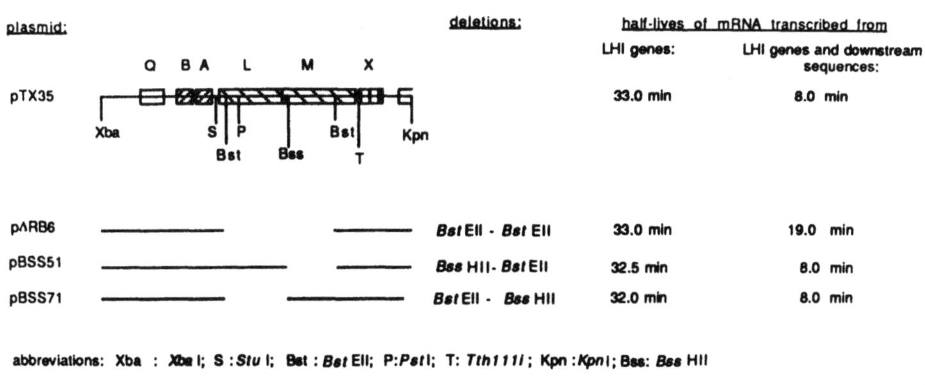

Figure 4: Half-life of *puf* mRNA transcribed from *puf* operon deletion mutants in a *puf*⁻ *R. capsulatus* strain.

Role of ribosomes in the degradation of *puf* mRNA

Figure 5 diagrams several plasmids in which segments of the *puf* genes are untranslated because of the absence of a translational start codon or the adventitious introduction of translational stop codons (Klug and Cohen, submitted for publication). When translation was blocked at the 5' end of *puf*L (pPS1, pRCL3, pΔRCL8), the RC-specific *puf* mRNA segment decayed so rapidly that it was not detected on Northern blots. However, when translation was allowed to proceed through at least 102 bp at the 5' end of *puf*L, the RC-specific mRNA segment decayed with wild type half-life, although a long stretch within the RC region still remained untranslated (Fig. 5, plasmid pAOL12). This result, which is similar to one obtained previously by insertion of translation stop codons in monocistronic transcripts (Nilsson et al., 1987) did not result from *rho*-dependent premature termination of transcription as a result of the translation stop, since no truncation of mRNA molecules was observed (Klug and Cohen,

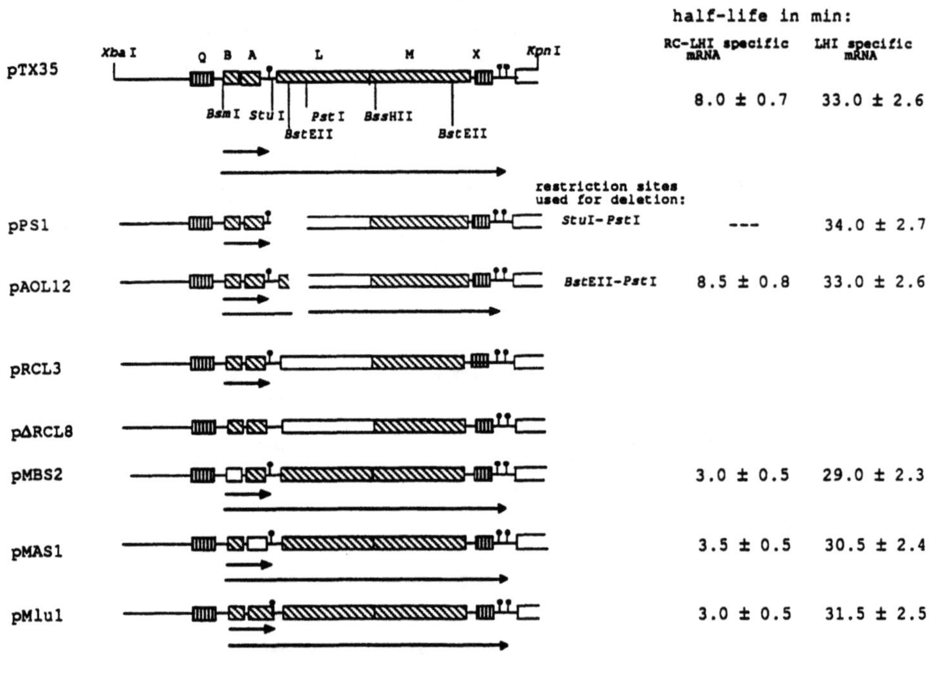

Figure 5: Plasmids containing *puf* sequences that were expressed in *puf R. capsulatus* strains in order to determine *puf* mRNA half-lives. The open boxes represent the segments of *puf* genes that are untranslated.

submitted for publication). When the intercistronic hairpin loop structure was deleted, the ribosomes present on the translated *puf*BA segment did not protect the segment against the rapid degradation that occurred downstream in the absence of translation of *puf*L. As seen in Figure 5, translational stops 5' to *puf*L affected the rate of decay of the RC-specific mRNA segments but had little if any effect on the stability of transcript segments upstream from the intercistronic hairpin. The different effects of translational stops on the degradation of mRNA segments upstream or downstream from the intercistronic hairpin loop suggest that the rate limiting step for the decay of the 0.5 kb *puf*BA mRNA segment is different from that responsible for rate limiting cleavage of the 2.7 kb *puf*BALMX mRNA segment.

Oxygen affects the decay of the *puf*LMX mRNA segment

Earlier work has shown that oxygen is a key factor in regulating the formation of the photosynthetic apparatus of *R. capsulatus*. Reduction of oxygen tension in previously aerobic cultures leads to the synthesis of pigments and pigment-binding proteins and the formation of photosynthetic membranes, even in the absence of light.

When the half-life of *puf* mRNA from wild type strain B10 was measured under high or low oxygen, different rates of decay for the 2.7 kb *puf* mRNA segment were observed (see also Figure 6; Klug and Cohen, 1990). The half-life of this segment was determined to be 3 to 3.5 min. in high oxygen cultures, but 7.5 to 9 min in cultures grown at low oxygen tension. In contrast, the kinetics of decay of the 0.5 kb *puf*BA mRNA segment were similar under high or low oxygen growth conditions. These findings are consistent with the notion that different mechanisms are rate limiting for decay of the 2.7 and 0.5 kb *puf* mRNA segments. Investigations of the stability of mRNA encoded by constructs containing deletions or insertions in the *puf* operon suggest the rate of endonucleolytic cleavage within the RC-coding region is affected by oxygen (Klug, submitted for publication).

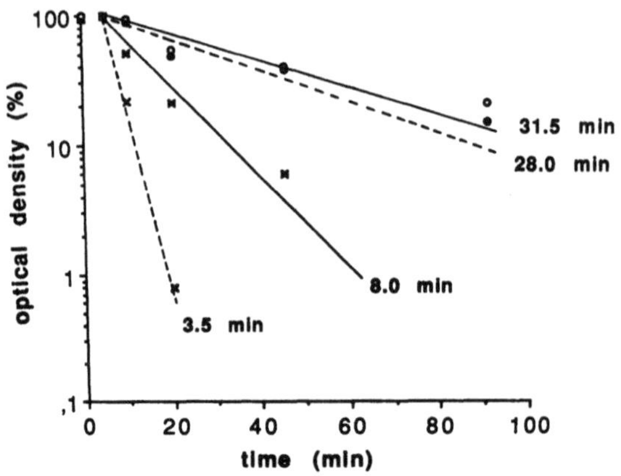

Figure 6: Decay of *puf* mRNA under high (20%, - - -) or low (1-2%, - - -) oxygen determined from Northrn blot experiments. Rifampicin was added at time 0. + 2.7 kb *puf* mRNA, o 0.5 kb *puf* mRNA.

References

Adams CW, Forrest ME, Cohen SN, Beatty JT (1989) Structural and functional analysis of transcriptional control of the *Rhodobacter capsulatus puf* operon. J Bacteriol 171:473-482

Bauer CE, Young DA, Marrs BL (1988) Analysis of the *Rhodobacter capsulatus puf* operon. Location of the oxygen-regulated promoter region and the identification of an additional *puf*-encoded gene. J Biol Chem 263:4820-4827.

Belasco JG, Beatty JT, Adams CW, von Gabain A, Cohen SN (1985) Differential expression of photosynthesis genes in *R. capsulata* results from segmental differences in stability within the polycistronic *rxc*A transcript. Cell 40:171-181

Chen CYA, Beatty JT, Cohen SN, Belasco JG (1988) An intercistronic stem-loop structure functions as an mRNA decay terminator necessary but insufficient for *puf* mRNA stability. Cell 52:609-619

Klug G, Adams CW, Belasco JG, Dörge B, Cohen SN (1987) Biological consequences of segmental alterations in mRNA stability: Effects of deletion of the intercistronic hairpin loop region of the *Rhodobacter capsulatus puf* operon. EMBO J 6:3515-3520

Klug G, Cohen SN (1988) Pleiotropic effects of localized *Rhodobacter capsulatus puf* operon deletions on production of light absorbing pigment-protein complexes. J Bacteriol 170:5814-5821.

Klug G, Cohen SN (1990) Rate-limiting endonucleolytic cleavage of the 2.7 kb *puf* mRNA of *R. capsulatus* is influenced by oxygen. *In:* Molecular Biology of Membrane-Bound Complexes in Photosynthetic Bacteria. G. Drews (ed.), Plenum Publishing Company Ltd. London.

Nilsson G, Belasco JG, Cohen SN and von Gabain A (1987) The effect of premature termination of translation on mRNA stability depends on the site of ribosomal release. Proc. Natl. Acad. Sci. USA 84:4890-4894.

Mutational Analysis of a RNase E Dependent Cleavage Site from a Bacteriophage T4 mRNA

C.EHRETSMANN, A.J.CARPOUSIS AND H.M.KRISCH
Department of Molecular Biology
30 Quai Ernest-Ansermet
CH-1211 Geneva 4
SWITZERLAND

SUMMARY

Recently E.coli endoribonuclease RNase E has been implicated in the functional and chemical decay of many bacteriophage T4 mRNAs. We have sought to identify the features of the phage RNA sequence that are necessary for RNase E dependent cleavage. A 117 bp synthetic sequence that contains the RNase E dependent processing site from the 5' leader of T4 gene 32 was assembled and inserted into pUC18. Correct RNase E dependent processing of a hybrid transcript containing this sequence was demonstrated in vivo. Mutations which reduce processing at this site have been obtained by in vitro mutagenesis.

INTRODUCTION

RNA degradation is an important determinant of gene expression. It is generally thought that mRNA decay is initiated by site specific endoribonucleases (Belasco & Higgins, 1988), but for most mRNAs the nuclease(s) that initiate decay are unknown. RNase III and RNase E have been implicated in the processing of a few specific mRNAs (Mudd

NATO ASI Series, Vol. H 49
Post-Transcriptional Control of Gene Expression
Edited by J. E. G. McCarthy and M. F. Tuite

et al. 1988, 1990; Portier et al. 1987). RNase E processes a precursor of 5S rRNA (Ghora & Apirion 1978) and RNA1, the inhibitor of ColE1 plasmid replication (Tomcsanyi & Apirion 1985). RNase E does not seem to have a general role in E.coli mRNA degradation (Apirion & Gitelman 1980). However the functional and chemical stability of many bacteriophage T4 mRNAs, for example the transcripts encoding gene 32, are greatly increased in an RNAse E mutant strain (Mudd et al. 1990).

The bacteriophage T4 gene 32 transcripts have three identified RNase E dependent cleavage sites (Mudd et al. 1988; Carpousis et al. 1989). Figure 1 presents a model in which RNase E dependent cleavages initiate the decay of the polycistronic gene 32 mRNA. RNAse E dependent processing at the -1340 and -71 sites (gene 32 initiation codon starts at +1) leads to rapid decay of the sequences upstream of these sites and to the accumulation of a processed gene 32 mRNA (Carpousis et al. 1989). These cleavages might initiate the decay of the RNA sequences 5′ of the sites by creating new 3′ ends which are susceptible to attack by 3′ to 5′ exonucleases. The relative stability of the downstream matured gene 32 mRNA must be a consequence of stability determinants located on the processed transcript. The 5′ leader sequence of the transcript is necessary for the unusual stability (Gorski et al. 1985). The protection provided by ribosomes loaded on this efficiently translated mRNA is probably also important. Finally the 3′ stem-loop structure of the gene 32 transcription terminator may act as a barrier to exonucleases (Newbury et al. 1987). The role of the RNase E dependent site within the gene 32 coding sequence at +831 is not clear, but this site might initiate degradation when translation is repressed by gene 32 protein (Krisch et al. 1974).

RNase E, itself, probably cleaves at these sites in the gene 32 polycistronic mRNA (Mudd et al 1990), but the in vitro processing of the transcript by the purified RNase E enzyme has not yet been demonstrated. Thus, the effect of RNase E on

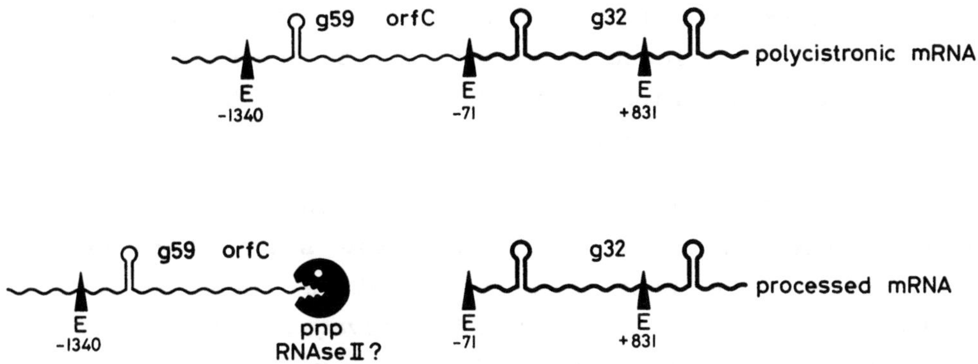

Fig.1. Model for the degradation of the bacteriophage T4 gene 32 polycistronic mRNA. RNase E dependent sites are indicated with a black wedge and their coordinates. Gene 59 and orfC are two of the upstream genes. Cleavage at the -71 site exposes a new 3′ end at which 3′ to 5′ exonucleases (pnp, polynucleotide phosphorylase; RNaseII) can degrade the upstream segment.

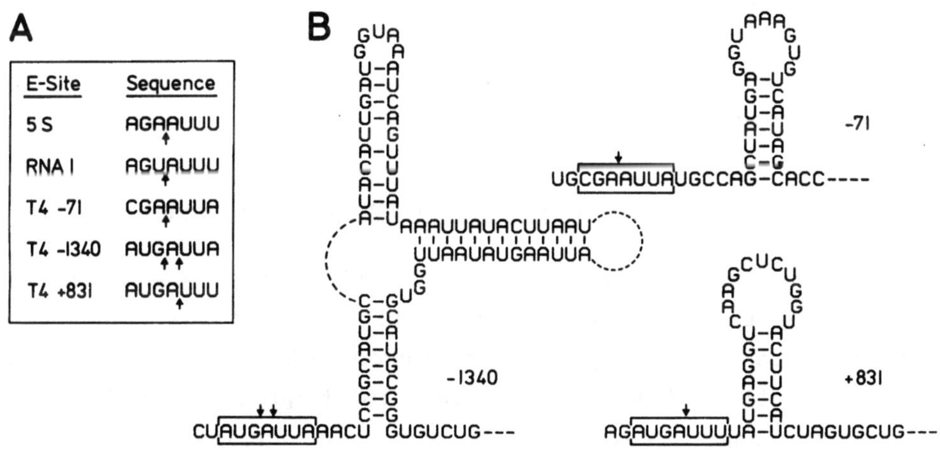

Fig.2 A and B. A comparison of RNase E dependent cleavage sites. A, Nucleotide sequences flanking the RNase E dependent cleavage sites of 5S rRNA, RNA 1 and T4 gene 32 polycistronic transcript. Arrows indicate the exact location of the cleavage. B, Potential secondary structures downstream of the three T4 RNase E dependent sites. A seven base region of sequence similarity around the cleavage site is boxed.

cleavage could be indirect. Nevertheless, the known sites in T4 mRNA have significant similarities with the RNase E site of 5S rRNA precursor which is cut in vitro by partially purified RNase E (Roy et al. 1983).

Figure 2A shows the sequence flanking all of the known RNase E dependent cleavage sites. These sites share certain similarities. The most conserved feature seems to be the Pu↓A↓UU sequence at the site of cleavage. For all of the sites, there is also the potential to form downstream double-stranded RNA close to the site of cleavage. Figure 2B illustrates the possible secondary structure of the sites in the gene 32 mRNA. These similarities suggest that RNase E recognition may require both a specific cleavage sequence and downstream secondary structure. To better define the sequence and structural requirements for RNase E dependent cleavage in vivo we have introduced mutations into the gene 32 processing site at -71.

RESULTS AND DISCUSSION

Construction of a vector carrying an RNase E dependent site

The DNA sequence from -97 to +20 of gene 32 was assembled from two pairs of overlapping oligonucleotides and then inserted into pUC18 (Yannisch-Perron et al. 1985). The sequence of this insert is shown in figure 3A and B. In addition to the cleavage site and the adjacent stem-loop (Fig.3A: inverted arrows), it contains the last 39 bps of the upstream T4 gene, orf C, as well as the gene 32 translation initiation region and the first twenty bps of this gene. Beyond +20, a HindIII site was added to the insert. To simplify the analysis of 5′ ends, the -45 promoter of gene 32 which is active in uninfected cells (Belin et al. 1987) was mutated by changing an A to a G at -52 (*). Furthermore, to facilitate manipulation of the two halves of the T4 insert, a single base pair change (*, T to C at -27) was made to create a unique HpaI site. The fragments were inserted in the pUC18 polylinker using the BamHI and HindIII sites.

In the resulting construct, pUCE10, transcription from the lac promoter yields an mRNA containing the 5' leader of lac mRNA followed by the T4 synthetic sequence and the lac alpha-peptide sequence (fig.3C). The translation initiating from lacZ is terminated by the orf C stop codon at -58, but a functional alpha-peptide sequence is initiated from the gene 32 initiation codon. E.coli was transformed with pUCE10 and transcription was induced from the lac promoter by the addition of IPTG. The RNA was extracted and the transcripts were then characterized by primer extension with an oligonucleotide complementary to +1 to +20 of gene 32.

To verify that the -71 site processing is RNase E dependent, we extracted RNA from wild type and RNase E ts mutant strains at both 30°C and 43°C. The results of primer extension analysis of these RNAs are shown in figure 4. A band corresponding to a 5' end at -71 (arrow) is detected both at 30°C and 43°C in the wild-type strain (rne^+; lanes 1,3). This -71 species is much reduced in the RNase E deficient strain (rne^-; lane 2) at the non-permissive temperature (43°C). Thus, the synthetic cleavage site is processed correctly and the cleavage clearly depends on RNase E activity. The processing does appear somewhat more efficient at 30°C than at 43°C. However, in contrast to the situation in bacteriophage T4 (Belin et al. 1987), the processing in the rne^+ strain at either temperature is not efficient. Perhaps the T4 sequence used in the construct does not include all the determinants necessary for very efficient cleavage.

In addition to the full length message and the processed -71 species, several other extension products appear in the gel. Some of these bands could arise from reverse transcriptase pausing artifacts or from non RNase E endonucleolytic cleavages. One of these bands, however, is

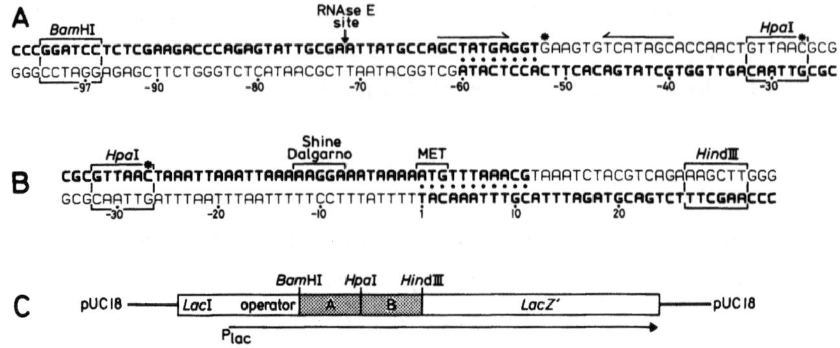

Fig. 3. A and B, Products of the mutually primed extension of two pairs of oligonucleotides. Sequences in bold type are the oligos and in light type are the product of elongation. Complementary regions of the oligos are shown by dots. C, T4 fragments A and B were inserted in pUC18 to give pUCE10.

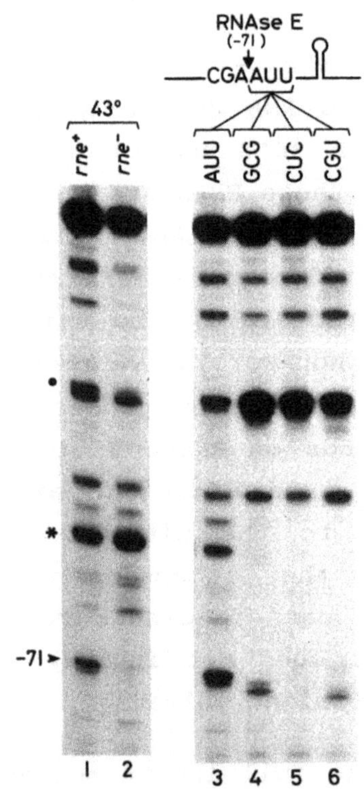

Fig.4. Analysis of the 5′ ends of the hybrid transcripts by primer extension and gel electrophoresis. Lanes 1, 2: pUCE10 in rne$^+$, rne$^-$ at 43°C. Lanes 3, 4, 5, 6: rne$^+$ at 30°C. The sequence changes in the mutants are indicated above each lane. Arrow, 5′ end at -71; *, see text; dot, 5′ end within the EcoRI site.

clearly affected in the RNase E mutant (fig.4; dot). It corresponds to a 5′ end located in the EcoRI site of the polylinker. Thus, this processing occurs at an AAUU sequence which also has a possible 5bps stem-loop structure just downstream. These features are compatible with the observed, but completely unexpected, RNase E dependent cleavage at this site. Other bands seem to be affected in the mutant strain but they have not been investigated further.

<u>Characterization of mutations in the synthetic -71 site</u>

To investigate the sequence requirements for cleavage, we introduced changes in the conserved AUU sequence. To make these changes, oligonucleotides containing altered sequence were used to construct a new BglII-HpaI fragment, which replaced the original BamHI-HpaI fragment in pUCE10. Mutant plasmids were identified by the loss of BamHI site, and the RNAs were sequenced to verify the changes in the AUU sequence (data not shown). Figure 4 (lanes 4, 5 and 6) shows the results that have been obtained in <u>rne</u>$^+$ stains at 30°C with three different mutants. The change of the AUU sequence to GCG (lane 4) completely abolishes cleavage at -71 (arrow). A weak band just below appears now, but it is not RNase E dependent (data not shown). The origin of this band is not clear but it could be due to reverse transcriptase pausing or a non RNase E endonuclease cut in the new sequence. The suppression of the processing at -71 is probably due to the sequence changes. The absence of processing could, however, be attributed to the sequestration of the mutated cleavage site in a possible weak secondary structure. However a mutation to CGC at the same site prevents the cleavage and is not involved in secondary structure (data not shown). It is striking that the absence of cleavage at -71 is associated with a large increase in RNase E dependent processing at the EcoRI site 50 bases upstream (fig.4: dot). A similar effect has also been observed in T4 where deleting the +831 site increased the processing at an upstream RNase E dependent site (A.J. Carpousis; personal communication). Among other possibilities, it may be that the processing at an RNase E site depends on cleavage at other adjacent RNase E sites. The loss, in all the mutant derivatives of pUCE10, of a 5′ end located 4 bases downstream of the BglII site (*) could be due to sequence changes from BamHI to BglII or/and to the absence of processing at -71.

Two other mutants in the AUU have been obtained: GCU and CUC (lanes 5 and 6). These double mutants also completely prevent the cleavage at -71. Together these three mutant sequences emphasize the importance of the AUU sequence for cleavage. Characterization of additional single base mutations in the cleavage sequence is now in progress. The requirement for secondary structure downstream of the cleavage site is also being investigated.

MATERIALS AND METHODS

Bacterial strains. *E.coli* RNase E^+ and RNase E^- isogenic strains N3433 (*rne*$^+$) and N3431 (*rne*$^-$ 3071 *ts*) (Goldblum & Apirion 1981) were grown in M9S containing 0.2% casamino acids as described by Mudd et al. (1988).

General methods. Restriction enzyme digests, ligation, transformation, plasmid preparation, ^{32}P end labelling of DNA primers with T4 polynucleotide kinase, gel electrophoresis, RNA extraction and 5′ end mapping by primer extension were performed essentially as described in Carpousis et al. (1989). Exponentially growing cells carrying the relevant plasmid were aerated at 30°C in M9S supplemented with ampicillin (20 µg/ml), and grown to a cell density of 2-3 x 10^8/ml. After centrifugation they were resuspended in the same medium at 4 x 10^8 cells/ml and preincubated for 10′ at 30°C. IPTG was added at 10^{-3} M and RNA was extracted after 5′. For RNA isolation at

43°C, the culture was transferred for 10' at 43°C prior to induction with IPTG.

Construction of the T4 gene 32 synthetic fragment. The two pairs of oligonucleotides were annealed at their complementary 3' ends and extended as described in Oliphant et al. (1987). Each of the two mutually primed extension reactions yielded a double stranded DNA fragment (A and B). Fragment A (BamHI and HpaI) and B (HpaI and HindIII) were then digested separately with the appropriate enzymes. The two segments were then mixed and ligated in presence of pUC18 digested with BamHI and HindIII. The sequence of the insert was verified by reverse transcriptase RNA sequencing with a primer complementary to the first 20 bps of gene 32. To construct the mutants, a BglII-HpaI fragment was synthesized with oligonucleotides mutated in the AUU box. These fragments were then used to replace the BamHI-HpaI region in pUCE10. Plasmids with mutations were identified by the loss of BamHI site, and their RNA sequenced to determine the sequence changes.

Acknowledgments. We thank Dick Epstein for discussions, advice on the manuscript, and for providing support and facilities. We also thank O. Jenni and F. Ebener for the skillful preparation of the figures. This work was supported by a grant from the Swiss National Science Foundation (no. 3.188.88). Additional support was provided by the Department of Public Instruction of the State of Geneva.

REFERENCES

Apirion D, Gitelman DR (1980) Decay of RNA in RNA processing mutants of E.Coli. Mol Gen Genet 177:339-343

Belasco JG, Higgins CF (1988) Mechanisms of mRNA decay in bacteria: A perspective. Gene 72:15-23

Belin D, Mudd EA, Prentki P, Yi-Yi Y, Krisch HM (1987) Sense and antisense transcription of bacteriophage T4 gene 32. J Mol Biol 194:231-243

Carpousis AJ, Mudd EA, Krisch HM (1989) Transcription and mRNA processing upstream of bacteriophage T4 gene 32. Mol Gen Genet 219:39-48

Ghora BK, Apirion D (1978) Structural analysis and in vitro processing to p5 rRNA of a 9S RNA molecule isolated from an rne mutant of E.Coli. Cell 15:1055-1066

Goldblum K, Apirion D (1981) Inactivation of the ribonucleic acid-processing enzyme ribonuclease E blocks cell division. J Bacteriol 146:128-132

Gorski K, Roch J-M, Prentki P, Krisch HM (1985) The stability of bacteriophage T4 gene 32 mRNA: a 5' leader sequence that can stabilize mRNA transcripts. Cell 43:461-469

Krisch HM, Bolle A, Epstein RH (1974) Regulation of the synthesis of bacteriophage T4 gene 32 protein. J Mol Biol 88:89-104

Mudd EA, Prentki P, Belin D, Krisch HM (1988) Processing of unstable bacteriophage T4 gene 32 mRNA into a stable species requires E.Coli ribonuclease E. EMBO J 7:3601-3607

Mudd EA, Carpousis AJ, Krisch HM (1990) E.Coli RNase E has a role in the decay of bacteriophage T4 mRNA. Genes & Dev, in press.

Newbury SF, Smith NH, Robinson EC, Hiles ID, Higgins CF (1987) Stabilization of translationally active mRNA by prokaryotic REP sequences. Cell 48:297-310

Oliphant RA, Struhl K (1987) The use of random-sequence oligonucleotides for determining consensus sequences. In: Wu R (ed) Methods in enzymology, vol 155. Academic Press, p 562

Portier C, Dondon L, Grunberg-Manago, Régnier P (1987) The first step in the functional inactivation of the E.Coli polynucleotide phosphorylase messenger is a ribonuclease III processing at the 5' end. EMBO J 6:2165-2170

Roy MK, Singh B, Ray BK, Apirion D (1983) Maturation of 5S rRNA: Ribonuclease E cleavages and their dependence on precursor sequences. Eur J Biochem 131:119-127

Tomcsanyi T, Apirion D (1985) Processing enzyme ribonuclease E specifically cleaves RNA 1. J Mol Biol 185:713-720

Yannisch-Perron C, Vieira J, Messing J (1985) Improved M13 phage cloning vector and host strains: nucleotides sequences of the M13mp18 and pUC19 vectors. Gene 33:103-119

The role of a novel site-specific endoribonuclease in the regulated decay of *E.coli* mRNA - a model for growth-stage dependent mRNA stability in bacteria

A. von Gabain, D. Georgellis, U. Lundberg, Ö. Melefors, L. Melin and O. Resnekov
Dept. of Bacteriology
Karolinska Institute
10401 Stockholm
Sweden

Abstract

The expression of many genes is adjusted in response to changes in bacterial growth rate or growth stage. One mechanism which affects gene expression is the control of mRNA stability. The *E.coli ompA* gene (encoding the outer membrane protein A) is an example where the stability of the mRNA changes in response to changes in the bacterial growth rate. In *B.subtilis* we have observed a reduction in the cellular concentration of the *sdh* (succinate dehydrogenase) and *odh* (2-oxoglutarate dehydrogenase) gene transcripts in cells approaching stationary phase which can be attributed to a specific decrease of mRNA stability. Furthermore, when we compared the cellular concentration and stability of the *aprE* (subtilisin) mRNA (preferentially expressed during stationary phase) with the *sdh* transcript (preferentially expressed during vegetative growth) we found that the growth-stage controlled appearance and disappearance of these two transcripts involves mechanism(s) which differentially alter their stability.

We have previously identified *in vivo* site-specific cleavages in the 5′non-coding region of the *ompA* transcript which seem to initiate its growth-regulated decay. We have also partially purified a novel *E.coli* endoribonuclease, RNase K. It catalyses site-specific cleavages in the 5′region of the *in*

NATO ASI Series, Vol. H 49
Post-Transcriptional Control of Gene Expression
Edited by J. E. G. McCarthy and M. F. Tuite

vitro ompA transcript which are in part identical to the cleavages found in the cellular *ompA* mRNA. The following results suggest that RNase K initiates degradation of the *ompA* mRNA: Using the HAK117 (*ams*^{ts} mutant) which has been previously characterized to have generally prolonged mRNA half-life, we noticed that the in *in vivo* RNase K cleavages are suppressed. RNase K *in vivo* cleavage products seem to have very short half-lives, which indicate they are decay intermediates rather than processing products. We also found that the difference in half-life between the *ompA* and *bla* (β-lactamase) gene transcripts is maintained in a mimiced *in vitro* decay with purified RNase K.

Introductory remark

Cellular transitions such as cell cycle, growth stage and differentiation rely on regulatory circuits that rapidly lower or shut off the expression of certain genes. Many genes which are expressed in a phase-specific manner, are found to recruit in addition to transcriptional, posttranscriptional mechanisms to shut down their expression. An efficient and perhaps reversible way to prevent expression of a gene is to block translation of the transcript (Oppenheim *et al.*, in this issue). Translational blockage seems to adjust the expression of the transferrin gene (Hentze *et al.* in this issue). Rapid decay of a transcript may give the same result. This can be achieved by destabilizing a transcript or by furnishing it with an intrinsically short half-life. Certain mRNA species have been reported to have an extremely short half-life and generally these transcripts seem to respond to cell growth or differentiation. Examples are the cellular oncogenes fos and myc (Belasco *et al.*, Laird-Offringa *et al.*, in this issue). Rapid changes of mRNA half-life have been found for the β-tubulin mRNA or the mRNA encoding the transferrin receptor (Harford *et al.*, Cleveland *et al.*, in this issue).

Results and Discussion

The stability of certain mRNA species follows cell growth in bacteria

The posttranscriptional control of gene expression in bacteria has not been studied as well as transcriptional control; probably because the average bacterial mRNA half-life seemed to make transcriptional shut off of gene expression an efficient mechanism. A few years ago, we discovered that mRNA stability seems to be a means to adjust the level of gene expression in regard to bacterial growth rate (Nilsson *et al.*, 1984). The half-life of the monocistronic *ompA* mRNA from *E.coli* decreases from 15 to 4 minutes when cells cultured in rich medium vs. poor medium are compared (Nilsson *et al.*, 1984). The changes of *ompA* mRNA stability is attributed to keep the level of OmpA protein in the outer membrane constant (Lugtenberg *et al.*, 1976, Lundberg *et al.*, 1988). Not all mRNA species are affected by such a mechanism; for example the monocistronic *E.coli bla* transcript has an invariant stability at different growth rates (Nilsson *et al.*, 1984).

In *B.subtilis* the cellular concentration of two Krebs cycle-specific transcripts (the *sdh* and *odh* mRNA species) drops four to sixfold after cells have entered stationary phase (Melin *et al.*, 1989, Melin *et al.*, submitt.a). The resulting reduction in the cellular concentration of the two mRNA species is in part accounted for by a drop of their respective half-lives (Melin *et al.*, submitt.a). We were interested to learn whether a transcript which preferentially appears in stationary phase, is also regulated post-transcriptionally. The *aprE* transcript encodes a protease which is preferentially expressed during early stationary phase. We found the *aprE* mRNA to be stable when it appears in stationary cells (Resnekov *et al.*, submitt.). Two hours further into stationary phase the cellular concentration of the transcript decreases, as does its half-life (Resnekov *et al.*, submitt.). It is worthwhile mentioning that the relatively short-lived *sdh* mRNA and the relatively long-lived *aprE* mRNA are both destabilized to about a fourth of their maximal half-life as their cellular concentration decreases (Melin *et al.*, 1989, Resnekov *et al.*, submitt.).

In order to investigate whether the cellular concentration of certain transcripts is adjusted in response to nutritional changes, we diluted *B.subtilis* stationary cells into fresh medium and we found that the cellular concentration of the *aprE* mRNA decreases extremely rapidly (Resnekov *et al.*, submitt.). Similarly, we discovered that the short half-life of the *E.coli ompA* mRNA at slow bacterial growth rate increases in less than four minutes fourfold when cells are diluted into fresh medium (Georgellis *et al.*, unpublished results). Thus, certain bacterial transcripts are stabilized or destabilized when nutritional conditions are altered.

Ribosomes and mRNA decay

A connection between mRNA stability and translating ribosomes is suggested by many results (Belasco and Higgins, 1988). It has been proposed that translating ribosomes either protect a transcript against nuclease attack or that the degradational machinery is actively mediated by translating ribosomes. Whether a reduction of translational efficiency is coupled to a drop in mRNA stability remains an open question for most transcripts (Belasco and Higgins, 1988). In the case of such a coupling, changes of translational efficiency would also cause alterations of the cellular concentrations of mRNA. The increase or decrease of expression is therefore a product of the two effects. When we tested the *ompA* and the *bla* mRNA at the two growth rates, we did not find any correlation between the differences in translational efficiency and the differences in half-life (Lundberg *et al.*, 1988).

We were also interested to learn whether premature translational abortion could make a transcript more vulnerable towards nucleolytic degradation. We found that a large section of the *bla* coding region (more than 70%) may be deprived of its translating ribosomes without affecting the stability of this transcript (Nilsson *et al.*, 1987). For the polycistronic *B.subtilis sdh* mRNA we made similar observations by analysing the first of the three cistrons, *sdhC* (Melin *et al.*, submitt.b). However, in both cases we found that complete deprivation of the cistrons of their translating ribosomes seems to destabilize the mRNA species (Nilsson *et al.*, 1987, Melin *et al.*, submitt.b). We interpret this to mean that the effect of ribosomes may be position-dependent. A possible explanation is that translating ribosomes prevent mRNA from folding in configurations that facilitate rapid degradation. It should be noted that in *B.subtilis* examples of mRNA species have been reported where the translation of short open reading frames in the 5′region preceding the major coding region are involved in the regulation of their stability (Bechhofer and Dubnau, 1987, Sandler and Weissblum *et al.*, 1988). It is also worthwhile mentioning that the *ompA* transcript has in its 5′region an AUG start upstream the *ompA* coding region which is preceded by a Shine Delgarno (Movva *et al.*, 1980) (Figure).

Chloramphenicol (cm), a drug which retards the elongation of translation, has been reported to stabilize mRNA in *E.coli* (Belasco and Higgins, 1988). Addition of cm to *E.coli* cells cultured in rich medium has been found to stabilize both the *bla* and the *ompA* gene transcripts by the same factor of two (Lundberg *et al.*, 1988). Thus, in the presence of cm the difference in stability remains between the two mRNA species. When cm is added to cells growing at a slow growth rate, the half-life of the *ompA* mRNA increases 10-fold, whereas the half-life of the *bla* mRNA only doubles (Lundberg *et al.*, 1988). One may speculate that the growth-rate-regulated degradation of the *ompA* mRNA at slow growth rate is translationally regulated. A similar conclusion has been drawn from results obtained for the decay of certain eukaryotic mRNA species (Belasco *et al.*, Cleveland *et al.*, Jakobson *et al.*, in this issue).

Our data suggest that ribosomes are involved in the regulated decay of mRNA. But, we also obtained data that indicates that differences in stability do not *per se* result from the degree of ribosomal protection of a transcript against nuclease attack.

The search for the step initiating mRNA decay

Like many investigators who tried to define whether an mRNA segment controls a transcript's stability (Yamamoto and Imamoto, 1975, Gorski *et al.*, 1985, Belasco and Higgins, 1988), we analysed the stability of fusion transcripts from mRNA species of different stability . Experiments with the two monocistronic *ompA* and *bla* gene transcripts revealed that exchanges of the 5'regions were most effective in influencing the stability of the hybrid transcripts (Belasco *et al.*, 1986). An *ompA-bla* fusion transcript, where the 5'region comes from the *ompA* gene was found to have a regulated stability, like the *ompA* wildtype messenger (Belasco ,personal communication). We obtained similar results when we analysed an *sdh-cat* hybrid-transcript in *B.subtilis* in response to growth stage (Melin *et al.*, submitt.b).

The 3' stem loop structures of the *bla* and *ompA* gene transcripts seem to function as a barrier against 3' to 5' exonucleases (Belasco *et al.*, 1986). The number of the 3' stem loop structures (between one and three) of the *bla* transcript seems to be irrelevant for the transcript's stability (von Gabain *et al.*, 1983). The REP structure has been reported to stabilize mRNA segments upstream of its insertion site (Newbury *et al.*, 1987a, Newbury *et al.*, 1987b). The addition of a REP sequence to the 3' end of the *bla* mRNA did not affect its stability (Higgins, *et al.*, in this issue).

While changes and exchanges of DNA segments helped us to define regions within an mRNA which determine stability, one can not conclude from such experiments that such segments are the target of the nucleolytic attack initiating decay. Furthermore treating a mRNA molecule as a linear entity does not take into account non-predictable changes of secondary and tertiary structure. In one of our *ompA-bla* hybrid-transcripts we noticed that a fraction of these mRNA molecules were more vulnerable towards degradation (Belasco *et al.*, 1986).

An approach to identify the segment(s) where decay initiates, is to analyse the decay of different segments of a transcript. We found some monocistronic and polycistronic mRNA species where all segments seem to decay at the same rate (von Gabain *et al.*, 1983, Byström *et al.*, 1989). Such a decay pattern indicates an 'all or nothing' decay mechanism, meaning that the initial step of decay is ensued by extremely rapid degradation of the RNA-molecule (von Gabain *et al.*, 1983). In a number of cases it was also found that polycistronic transcripts have segmental differences in stability (Barry *et al.*, 1980, Guarneros *et al.*, 1982, Belasco *et al.*, 1985, Newbury *et al.*, 1987a, Portier *et al.*, 1987, Båga *et al.*, 1988, Mudd *et al.*, 1988, Regnier *et al.*, in this issue). The *Rhodobacter capsulata puf* mRNA is an example where it is clear that its 3' L-M-X part is more labile than its 5' B-A part (Belasco *et al.*, 1985). Mechanisms and factors involved in this differential stability of the *puf* mRNA have been analysed (Chen *et al.*, 1988, Klug and Cohen, in this issue).

Interestingly, the 5'part of the *ompA* transcript seems to be slightly more stable than the 3' part in fast growing cells (when the *ompA* mRNA is most stable), while in slow growing cells (when the *ompA* mRNA is fourfold destabilized) this is not the case (Nilsson *et al.*, 1984). One may interpret this result to mean that events in the 5'region are initiating decay. We also obtained results with the polycistronic *sdh* transcript from *B.subtilis* that may be interpreted in a similar way (Melin *et al.*, submitt.b). It should however be noted that these segmental differences in stability are very subtle and that there are caveats related to the methods used to determine mRNA stability of mRNA segments.

We only identified an mRNA segment which initiates decay when we discovered cleavage products derived from the 5'region of the *ompA* transcript (Figure, Melefors and von Gabain, 1988). The discovery of these cleavage products was facilitated by the overproduction of an *ompA* mRNA variant (*tacompA*) using an inducible promoter (Melefors and von Gabain, 1988). For some of the cleavages we detected both the upstream and the downstream cleavage products, indicating the action of a site-specific endonuclease (Figure, Melefors and von Gabain, 1988, Lundberg *et al.*, in press). We identified a similar type of cleavages in the 5'region of the *bla* mRNA (Nilsson *et al.*, 1988).

One dilemma of identifying cleavages in mRNA is to show that they are related to mRNA decay and not the result of converting a precursor mRNA into a stable processing product (Blumer and Steege, 1984, Cannistraro *et al.*, 1986, Arraiano *et al.*, 1988, Mudd *et al.*, 1988, Båga *et al.*, 1988, Bardwell *et al.*, 1989, Steege *et al.*, in this issue). Three lines of evidence support the conclusion that at least some of the cleavages identified in the *ompA* mRNA initiate degradation (Figure).

- Firstly, the rate of decay of the *ompA* mRNA seems to be correlated to the ratio of cleavage products to the full length transcript when different bacterial growth rates are compared (Melefors and von Gabain, 1988).
- Secondly, in an *E.coli* mutant strain, HAK117 (*ams*^{ts}), which is defective in mRNA degradation at a non-permissive temperature, we could show that there is a connection between retarding the decay of the *ompA* mRNA and suppression of the cleavage products (Lundberg *et al.*, in press).
- Thirdly, after shifting HAK117 to a non-permissive temperature, the downstream cleavage products (which are 5' truncated *ompA* transcripts) disappear extremely rapidly Lundberg et., in press). In addition, we observed that a cleavage product differing in size from the *ompA* mRNA by only 8 5'-nucleotides is significantly less stable than the wild-type transcript (this cleavage product is derived from the *tacompA* hybrid transcript (Figure) (Lundberg *et al.*, in press). Thus, it seems that cleavages initiate rapid decay of the 'downstream' cleavage products.

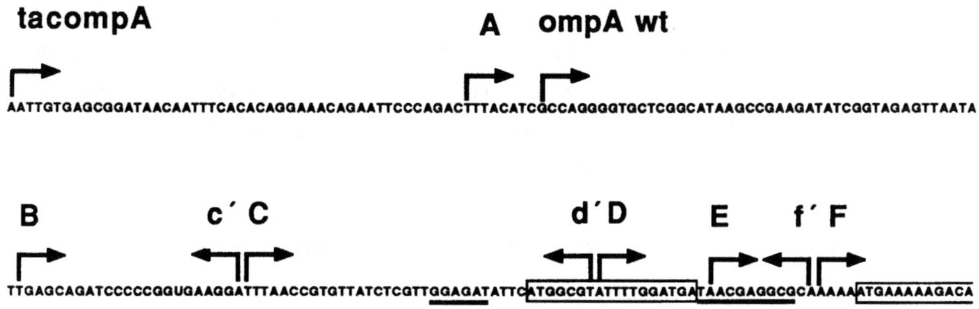

Figure: Sequence map of *ompA* and *tacompA* mRNA 5′ends. The start of the wild type *ompA* and *tacompA* transcripts and the 5′ends of the downstream cleavage products **A**, **B**, **C**, **D**, **E** and **F** are indicated with arrows pointing to the right. The 3′ends of the upstream cleavage products **c′**, **d′** and **f′** are indicated with arrows pointing to the left. The cleavage products affected in HAK117 (*ams*^{ts}) are **A**, **C** and **D**. The two potential Shine Delgarno sequences are underlined. The pro-OmpA and the potential upstream reading frame are boxed.

Characterization and partial purification of a site specific endoribonuclease catalyzing *ompA* and *bla* mRNA decay

By using *E.coli* extracts and *in vitro* synthesized *ompA* and *bla* transcripts we were able to show an endonucleolytic activity that gave similar cleavage products to those found *in vivo* (Nilsson *et al.*, 1988). This assay system was employed to purify this endonucleolytic activity (Lundberg *et al.*, in press). Analysing the purest fraction on a denaturing SDS gel followed by silver-staining, revealed five peptides; their sizes are 62, 60 25, 22 and 21 kDa (Lundberg *et al.*, in press). When fractionated over a Sephadex column the endonucleolytic activity is recovered at a position which corresponds to a globular protein of of 55 kDa (Lundberg *et al.*, in press). From these results we conclude that the activity is identical to either the 62 or 60 kDa peptide or to a complex of the three smaller peptides.

Characterization showed that the endonucleolytic activity is abolished by protease treatment, but not by micrococcas nuclease treatment. The activity which we named RNase K is dependent on Mg^{2+} ions, but is not affected by K^{+} ions. RNase K activity was found in extracts of *E.coli* that lack RNase III and RNase P (Nilsson *et al.*, 1988).

A comparison of *in vitro* RNase K upstream cleavage products with their *in vivo* counterparts on sequencing gels showed that they are identical to the *in vivo* upstream cleavage products, c and d (Figure). This was also found for the downstream cleavage products, C and D (Figure). RNase K therefore accounts at least for these two cleavages which initiate *ompA* mRNA degradation.

We also compared the *in vitro* efficiency of RNase K cleavage with different mRNA substrates. The *in vitro* difference in stability found between the *ompA* and *bla* transcripts reflects relatively well their difference in half-life in fast growing cells (Lundberg *et al.*, in press).

Phylogenetic comparison of the *ompA* 5′region from different bacterial species (e. g. *Serratia marcescens* vs. *E.coli*) shows a strong conservation of a secondary structure including a 5′ stem loop structure . We found that disturbing this structure seems to affect the specificity and rate of RNase K cleavage (Lundberg *et al.*, in press). A hypothesis which we would like to suggest, is that the stability of mRNA species controlled by RNase K cleavage depends on the higher order structure in the mRNA 5′region .

Conclusions and open questions

Our results fit best with a model that the regulated decay of the *ompA* mRNA is controlled by site-specific cleavages in the 5′region. We have shown that at least some of these cleavages are catalyzed by an endoribonuclease, RNase K. Our data also show that once the cleavage has occurred the decay of the cleavage products ensues extremely rapidly. The upstream cleavage products may then be degraded by 3′ to 5′ exonucleases (Mott *et al.*, 1985, Deutscher, in this issue). In order to explain the rapid disappearance of the downstream cleavage products one may postulate the existence of a 5′ to 3′ degrading ribonuclease which has yet to be identified in *E.coli* (Deutscher, in this issue). One may argue that cleavage in the 5′region simply works by destroying the transcripts′ capability to load on ribosomes, which therefore makes it vulnerable to all kind of nucleases. However, some of the cleavage sites where decay initiates map more than 100 nucleotides upstream of the ribosome binding site (Figure). It has also been shown that most of the region upstream of the *ompA* ribosome binding site can be deleted without significantly lowering the translational efficiency of the transcript (Emory and Belasco submitt.). We therefore prefer a a model that decay ensues by an active 5′ to 3′ mechanism.

Another way to imagine the role of the first cleavage is that it acts *in vivo* as a kind of ′knock out′ hit. Such a hypothesis would suggest that the mRNA together with ribosomes and some RNA-binding proteins is assembled in a kind of complex structure which instantly disintegrates after the first cleavage.

A question that we do not answer, is in what way the structure of the cleavage site mediates the specificity of the cleavages which in turn seems to determine half-life of the transcripts. We would particularly like to know what makes the cleavage sites of the *bla* transcript more vulnerable towards RNase K attack than the ones of the *ompA* mRNA. Another question is the relationship between the utilisation of the *ompA* cleavage sites and the bacterial growth rate. Finally, we would like to know the connection between the *ams* gene (Ono and Kuwano, 1979, Claverie-Martin *et al.*, 1989) and RNase K (Lundberg *et al.*, in press).

We have found for several instances in *B.subtilis* and *E.coli* that alterations of growth stage or media composition leads to changes of mRNA stability. In some cases, we found that these changes are extremely rapidly. The transferrin receptor mRNA provides a model how an RNA protein interaction controlled by i.e. iron may mediate a physiological change to a transcript

(Harford *et al.*, Hentze *et al.*, this issue). One may therefore speculate that there are multiple components that in response to different physiological changes bind to a transcript and affect its stability. Some mRNA binding proteins have already been identified and they may be candidates for regulated binding to mRNA (Platt *et al.*, Court *et al.*, in this issue). Ribosomes are such an example; their binding to mRNA may modulate structures (Gold *et al.*, in this issue) that in turn interfere with the nucleases initiating mRNA decay (Brody *et al.*, in this issue).

Acknowledgements

O.R is the recipient of an EMBO fellowship. This work was supported by grants from the Swedish Biotechnology Foundation (SBF), the Swedish Board of Technical Development (STU), the Swedish Cancer Society (RmC) and KABI AB (Stockholm).

References

Arraiano CM, Yancey SD and Kushner SR (1988) J Bacteriol. 170: 4625-4633

Båga M, Göransson M, Normark S and Uhlin BE (1988) Cell 52: 197-206

Bardwell JCA, Regnier P, Chen S-M, Nakamora Y, Grundberg-Manago M and Court DL (1989) EMBO J. 8:3401-3407

Barry G, Squires S and Squires CL (1980) Proc Natl Acad Sci USA 77:3331-3335

Bechofer DH and Dubnau D (1987) Proc Natl Acad Sci USA 84: 498-502

Belasco JG, Beatty JT, Adams CW, von Gabain A and Cohen SN (1985) Cell 40:171-181

Belasco JG, Nilsson G, von Gabain A and Cohen SN (1986) Cell 46:245-251

Belasco JG and Higgins CH (1989) Gene 72: 15-23

Blumer KJ and Steege DA (1984) Nucleic Acids Res. 12: 1847-1861

Byström A, von Gabain A and Björk G (1989) J. Mol. Biol. 209: 575-586

Cannistraro VJ, Subbarao MN and Kennell D (1986) J. Mol. Biol. 192:257-274

Chen C-YA, Beatty JT, Cohen SN and Belasco JG (1988) Cell 52: 609-619

Claverie-Martin F, Diaz-Torres MR, Yancey SD and Kushner SR (1989) J. Bacteriol. 171:5479-5486

Emory SA and Belasco JG (submitted)
von Gabain A, Belasco JG, Schottel JL, Chang ACY and Cohen SN (1983) Proc. Natl. Acad. Sci. USA 84:4890-4894
Gorski K, Roch J-M, Prentki P and Krisch HM (1985) Cell 43: 461-469
Guarneros G, Montanez C, Hernandez T and Court D (1982) Proc. Natl. Acad. Sci. USA 79:238-242
Lugtenberg B, Peters R, Bernheimer H and Berendsen W (1976) Molec. gen. Genet. 147:251-262
Lundberg U, Nilsson G and von Gabain A (1988) Gene 72:141-149
Lundberg U, von Gabain A and Melefors Ö (1990) EMBO J. in press.
Melefors Ö and von Gabain A (1988) Cell 52:893-901
Melin L, Rutberg L and von Gabain A (1989) J. Bacteriol. 171:2110-2115.
Melin L, Resnekov O, Hederstedt L, von Gabain A and Carlsson P submitted for publication (a)
Melin L, Dehlin E, Friden H, Rutberg L and von Gabain A submitted for publication (b)
Mott JE, Galloway JL and Platt T (1985) EMBO J. 4:1887-1891
Movva NR, Nakamura K and Inouye.M (1980) Proc. Natl. Acad. Sci. USA 77:3845-3849
Mudd EA, Prentki P, Belin D and Krisch HM (1988) EMBO J. 7:3601-3607
Newbury SF, Smith NH, Robinson EC, Hiles ID and Higgins CF (1987a) Cell 48:297-310
Newbury SF, Smith NH and Higgins CF (1987b) Cell 51:1131-1143
Nilsson G, Belasco JG, Cohen SN and von Gabain A (1984) Nature 312:75-77
Nilsson G, Belasco JG, Cohen SN and von Gabain A (1987) Proc. Nat. Acad. Sci. USA 84:4890-4994
Nilsson G, Lundberg U and von Gabain A (1988) EMBO J. 7:2269 -2275
Ono M and Kuwano M (1979) J. Mol. Biol. 129:343-357
Portier C, Dondon L, Grundberg-Manago M and Regnier P (1987) EMBO J. 6:2165-2170
Resnekov O, Rutberg L and von Gabain A Submitted for publication.
Sandler P and Weisblum B (1988) J. Mol. Biol. 203:905-915
Uzan M, Favre R and Brody E (1988) Proc. Natl. Acad. Sci. USA 85:8895-8899
Yamamoto T and Imamoto F (1975) J. Mol. Biol. 92:289-309

REGULATION OF mRNA STABILITY IN YEAST

Allan Jacobson, Agneta H. Brown, Janet L. Donahue, David Herrick[1], Roy Parker[2], and Stuart W. Peltz
Department of Molecular Genetics and Microbiology
University of Massachusetts Medical School
Worcester, Massachusetts 01655

Introduction

Differences in the decay rates of individual mRNAs can have profound effects on the overall levels of expression of specific genes. Although the potential importance of mRNA stability as a mechanism for regulating gene expression has been recognized (Ross, 1989), the structures and mechanisms involved in the determination of individual mRNA decay rates have yet to be elucidated. Several reports have suggested that the stability of individual mRNAs is related to basic mRNA properties such as size, poly(A) tail length, and translational efficiency (Bernstein and Ross, 1989; Losson and Lacroute, 1979; Santiago et al., 1986). Other experiments fail to reveal any correlation between these variables and the rate of mRNA turnover and suggest that differences in mRNA decay rates may be attributable to specific sequences within unstable mRNAs that promote their recognition by the cellular turnover machinery (Shapiro et al, 1988; Herrick et al, 1990).

We have begun an investigation of mRNA decay in the yeast *Saccharomyces cerevisiae*. Our initial objective was the identification of both stable and unstable mRNAs that were encoded by yeast genes that had already been well characterized. To accomplish this, we developed a simple and reliable assay in which mRNA decay was analyzed after thermal inactivation of RNA polymerase II in cells harboring the rpb1-1 ts allele (Nonet et al., 1987). Having identified stable and unstable yeast mRNAs we sought to exploit these mRNAs to determine whether: 1) non-specific determinants had a significant role in mRNA decay, 2) unstable mRNAs contained *cis*-acting "instability elements", 3) mutants altered in mRNA decay could be identified, and 4) mRNA structures important in the decay of mammalian mRNAs played a role in mRNA decay in yeast. The results of our experiments are summarized below.

---

[1]Current address: McArdle Laboratory for Cancer Research, University of Wisconsin-Madison, 450 North Randall Avenue, Madison, WI 53706; [2]Current Address: University Department of Molecular and Cellular Biology, University of Arizona, Tucson, Arizona 85721

NATO ASI Series, Vol. H 49
Post-Transcriptional Control of Gene Expression
Edited by J. E. G. McCarthy and M. F. Tuite

Measurement of mRNA decay rates in rpb1-1 mutants

Nonet et al. (1987) have constructed and characterized a conditionally lethal mutant (rpb1-1) with a temperature-sensitive lesion in the largest subunit of RNA polymerase II. In strains harboring the rpb1-1 allele, a shift to 36°C leads to the rapid and selective cessation of mRNA synthesis and to a reduction in the steady-state levels of pre-existing mRNAs (Nonet et al., 1987). The latter reduction is a reflection of ongoing mRNA turnover in the absence of new synthesis and serves as the basis for the procedure which we use routinely: cells growing at 24°C are abruptly shifted to 36°C by the addition of pre-warmed medium and culture aliquots are removed at different times after the temperature shift. RNA is isolated from each aliquot and the relative amounts of individual mRNAs are quantitated by RNA blotting methods. Figure 1 shows an example of the blotting assay and the resultant decay curves for the PGK1 mRNA ($t_{1/2}$=45 min) and the MATα1 mRNA ($t_{1/2}$=5 min).

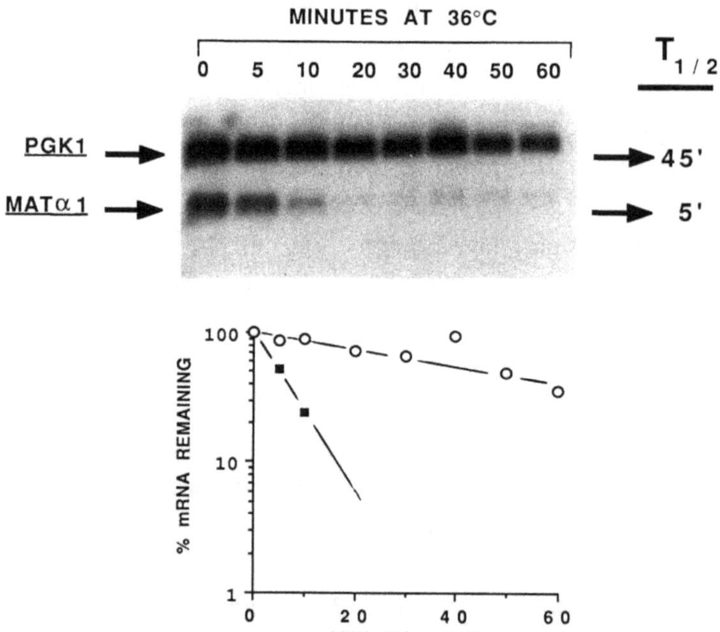

FIG. 1. Decay of stable and unstable mRNAs in yeast. Relative levels of the PGK1 and MATα1 mRNAs were measured at different times after a shift to 36°C in a temperature-sensitive RNA polymerase II mutant. Top: northern analysis of mRNA levels at different times after the temperature shift. Bottom: quantitation of the northern blot. mRNA levels are normalized to the level at t=0 and half-lives are obtained from the slope of the lines. Filled squares, MATα1, open circles, PGK1. From Parker and Jacobson (1990).

Using rpb1-1 mutants and the temperature shift procedure, half-lives were measured for over 20 different mRNAs and found to range from 2.5 to 45 min (Table 1). Within this range, we have arbitrarily defined three stability phenotypes: unstable (mRNAs with a $t_{1/2}<7$ min; e.g., STE2, STE3, HIS3, MFα1, and FUS1), moderately stable (mRNAs with a $t_{1/2}=10\text{-}20$ min; e.g., TCM1, PAB, and RP29), and stable (mRNAs with a $t_{1/2}>25$ min; e.g., ACT1, PGK1, and CYH2). As noted in Table 1, a significant number of the unstable mRNAs which we have identified are involved in mating-type regulation. This is not unexpected, since mating-type switching requires a rapid transition to a new set of mating-type-specific functions.

Table 1
Stability Phenotypes of Yeast mRNAs

Unstable ($t_{1/2}<7$ min)	**Moderately Stable** ($t_{1/2}=10\text{-}20$ min)	**Stable** ($t_{1/2}>25$ min)
HIS3	TCM1	ACT1
STE2*	RP29	PGK1
DED1	cDNA74	CYH2
cDNA90	PAB1	RP51a
STE3*	MFA1*	
MFα1*	LEU2	
MFA2*	GCN4	
MATα1*	CBP1	
FUS1*	HIS4	
URA3		
HTB1		

*Denotes mating type gene. From Herrick et al. (1990) and SWP, AHB, and AJ, unpublished exps.

mRNA decay rates measured in this manner have been compared to those obtained by two additional, independent procedures (approach to steady-state labeling and inhibition of transcription with thiolutin) and to those determined in the presence and absence of a heat-shock. We find that there are no significant differences in the mRNA decay rates measured in heat-shocked and non-heat-shocked cells and that, for most mRNAs, different procedures yield comparable relative decay rates (Herrick et al., 1990). Compared to other procedures used to measure mRNA decay rates, the temperature-shift procedure has several advantages, including: i) RNA polymerase II is selectively and rapidly inactivated; ii) the assay is relatively easy, i.e., there is no requirement for in vivo labeling or consideration of changes in pool sizes; iii) mRNAs of any abundance class can be readily detected and their decay rates measured; and iv) the use of northern blots allows the monitoring of mRNA integrity in parallel with the quantitation of mRNA decay rates, the detection of several mRNAs simultaneously, the reusability of blots for future experiments with different probes, and the quantitation of chimeric and parental mRNAs in the same experiment (Parker and Jacobson, 1990a).

Do non-specific determinants have a role in mRNA decay?

Possible determinants of mRNA stability can be subdivided into those which are specific and those which are non-specific. Specific determinants are mRNA sequences or structures which either interact directly with components of the cellular turnover machinery or target that machinery to other specific sites. Non-specific determinants are general mRNA features (such as the cap, poly(A) tail, or overall size) which could contribute to mRNA decay rates by promoting or hindering random interactions with non-specific nucleases. Having identified mRNAs which differ significantly in their respective decay rates, we compared these mRNAs with respect to possible differences in non-specific determinants. We find that: i) stable and unstable mRNAs do not differ significantly in their poly(A) metabolism; ii) removal of >90% of the poly(A) tail does not destabilize stable mRNAs; iii) there is no correlation between mRNA decay rate and mRNA size; iv) the degradation of both stable and unstable mRNAs is dependent on concommitant translational elongation; and v) the percentage of rare codons present in most unstable mRNAs is significantly higher than in stable mRNAs (Herrick et al, 1990). Observations (i)-(iv) lead us to conclude that non-specific determinants do not play a significant role in the decay of yeast mRNAs. The correlation alluded to in observation (v) may reveal an underlying mechanism for regulating mRNA decay rates (see below) or may simply reflect the fact that some highly expressed genes have independently evolved high rates of transcription and translation and slow rates of mRNA decay.

Identification of cis-acting determinants of mRNA instability

One explanation for the rapid decay of unstable mRNAs is the presence within these mRNAs of specific sequence elements which promote their rapid degradation. Our approach to delineating such sequences is to identify segments of unstable mRNAs which will promote rapid mRNA decay when transferred to mRNAs that are normally stable (Herrick, 1989; Parker and Jacobson, 1990 a,b). We have constructed numerous chimeric genes which encode portions of stable and unstable mRNAs, transferred the chimeric genes to yeast centromere plasmids, and used DNA-mediated transformation to introduce the resulting plasmids into cells harboring the rpb1-1 mutation. Decay rates of the parental and chimeric mRNAs were then measured by RNase protection or northern blotting assays using RNA extracted from temperature-shifted cells (see Figure 2). These experiments have demonstrated

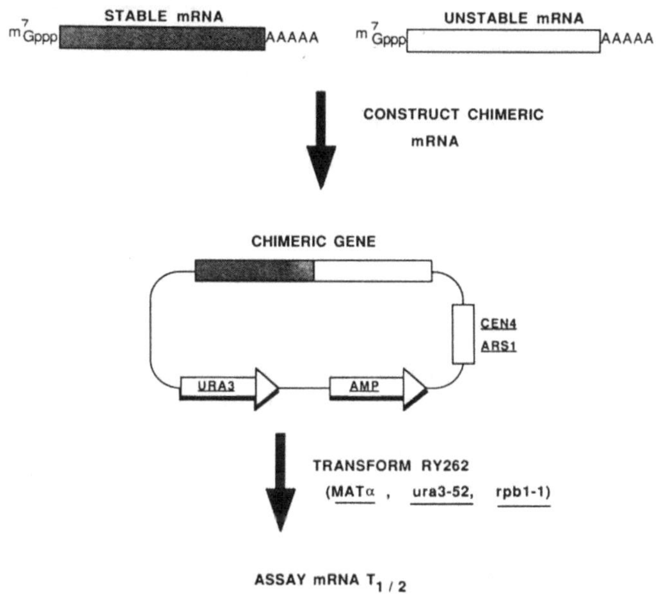

FIG. 2. Experimental strategy for the identification of *cis*-acting determinants of mRNA stability or instability. Sequences from stable mRNAs are shown as filled boxes and sequences from unstable mRNAs are shown as open boxes.

that: i) chimeric mRNAs consisting of fragments of stable mRNAs (ACT1 and PGK1) are stable (Parker and Jacobson, 1990b) and ii) fusion of fragments of unstable mRNAs to stable reporter mRNAs results in accelerated degradation of many of the hybrid mRNAs (Herrick, 1989; Parker and Jacobson, 1990 a,b). Deletion analyses demonstrate that the accelerated decay of hybrid mRNAs is not due to the loss of sequences within stable mRNAs which protect or stabilize those mRNAs, but is due to relatively discrete sequences derived from the unstable mRNAs. For the unstable HIS3, MATα1, and STE3 mRNAs such "instability elements" are found in coding regions; sequences from the

3'-UT regions of these mRNAs are not necessary or sufficient for rapid mRNA decay. By analyzing a series of deletions in which MATα1 coding sequences are progressively removed from a PGK1/MATα1 fusion gene, sequences required for the rapid decay of MATα1 mRNA have been localized to a 42 nucleotide segment between the endpoints of Δ52 and Δ94 (see Figure 3; Parker and Jacobson, 1990a). The sequences required for rapid decay of the STE3 and HIS3 mRNAs have, as yet, only been localized to much larger regions (400-500 nucleotides; see Figure 4; Herrick, 1989; Parker and Jacobson, 1990b). We have compared the sequence of the 42 nucleotide instability element of the MATα1 gene to that of other unstable yeast mRNAs and have yet to find any striking homologies in either the RNA or the protein sequence. This may reflect the possibility that the 42 nucleotide region is part of a larger element whose 3' boundary has yet to be defined. However, we do note that the 42 nucleotide region is rich in rare codons (8/14), where a rare codon is defined by its occurrence fewer than 13 times per 1000 yeast codons. Including one codon at the deletion boundary, this region also contains a stretch of six out of seven contiguous rare codons. This concentration of rare codons suggests that either a conserved sequence occurs within this region or that there may be a specific requirement for a translational pause at this site.

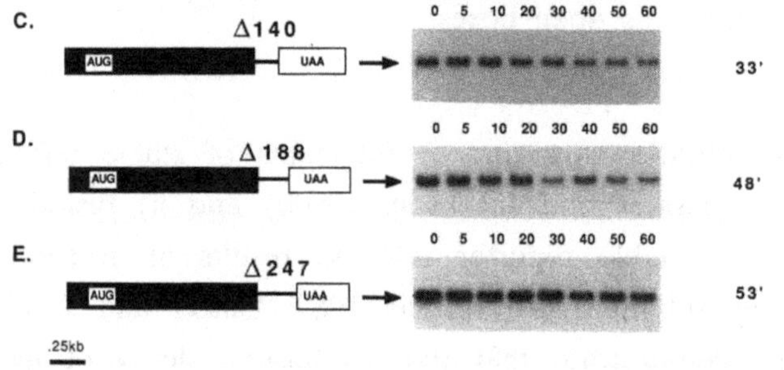

FIG. 3. Deletion analysis of MATα1 sequences contributing to rapid decay. Changes in mRNA decay rates resulting from deletion of MATα1 sequences from a PGK1/MATα1 fusion mRNA were quantitated as described in the legend to Fig.1. In the schematic, filled in boxes are PGK1 sequences, open boxes are MATα1 sequences, and deleted sequences are shown as single lines with the number of nucleotides deleted shown above each deletion. From Parker and Jacobson, 1990a.

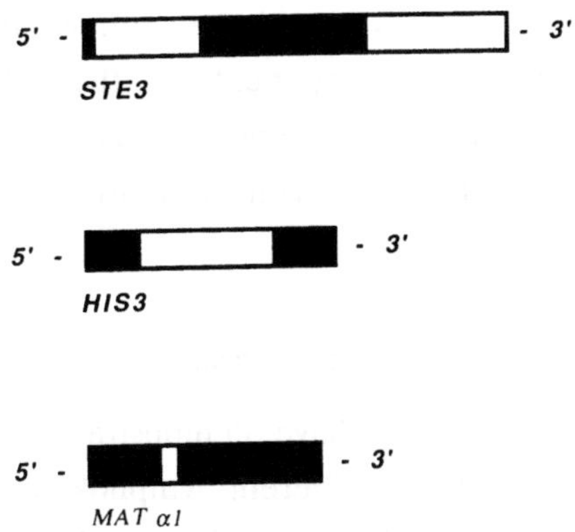

FIG. 4. Localization of instability elements in the STE3, HIS3, and MATα1 mRNAs.

Genetics of mRNA stability

To identify *trans*-acting factors involved in mRNA decay we have begun to characterize mutants that affect the accumulation of unstable mRNAs. Aebi et al. (1990) have described the isolation of a temperature-sensitive yeast mutant, ts352, which accumulates the moderately stable TCM1 and PAB1 mRNAs and the unstable STE2 mRNA after a shift from 23°C to 37°C. Ts352 has a defect in the gene coding for tRNA nucleotidyltransferase and, at the non-permissive temperature, this mutant accumulates tRNAs which are shorter than mature tRNAs (presumably because CCA termini are absent). Half-lives of stable, moderately stable, and unstable mRNAs measured in thiolutin-treated ts352 cells at 37°C are 3-5-fold slower than mRNA decay rates in ts^+ cells (JLD and AJ, unpublished exps.). Since cycloheximide treatment of cells has previously been shown to have comparable effects on mRNA decay rates in yeast (Herrick et al, 1990), our interpretation of this observation is that depletion of functional tRNAs in ts352 causes a reduction in translational elongation rates and that such an effect is formally equivalent to the effects of cycloheximide treatment (see below).

Having identified transferable, *cis*-acting elements capable of destabilizing mRNAs which contain them, we have initiated selections for mutants altered in the factors which must interact with such elements. The rationale for our selection is that destabilization of several different mRNAs in the same cell with the same element should create a situation in which the phenotype of two or more genes could be simultaneously "reverted" by a single mutation in a *trans*-acting component of the mRNA decay apparatus. This selection scheme assumes that element-specific degradative factors exist and that their loss will not affect cell viability. It remains to be seen whether these assumptions are valid.

Role of mammalian instability elements in yeast

Experiments in mammalian cells have demonstrated that AU-rich sequences within the 3'-UT regions of several different lymphokine and oncogene mRNAs can promote rapid mRNA decay (Shaw and Kamen, 1986; Peltz et al., 1990). To consider the possibility that sequence elements important for the regulation of mRNA decay rates in mammalian cells are also important in yeast, we have inserted the 3'-AU-rich element (ARE) from human GM-CSF mRNA into the 3'-UT regions of the yeast PGK1 mRNA and the *E. coli* lacZ mRNA (SWP, AHB, and AJ, unpublished exps.). Neither the decay rate of PGK1 mRNA nor the rate of synthesis of β-galactosidase in yeast were affected by these insertions, indicating fundamental differences between at least one aspect of mRNA turnover in yeast and mammals. Experiments in progress are designed to test whether stem-loop structures from the 3'-UT of the human transferrin receptor mRNA can affect yeast mRNA decay rates.

The translation connection

The demonstration that specific sequences within the MATα1, HIS3, and STE3 coding regions are required for rapid mRNA decay (see above) suggested that there may be a requirement for translation in the degradation of these mRNAs. To test whether translation is required for turnover of the MATα1 mRNA, we constructed two hybrid genes in which ribosome translocation through the MATα1 sequences was prevented by an amber stop codon at the junction between PGK1 and MATα1 sequences (Pα1.UAG mRNA) or at the junction of ACT1 and MATα1 sequences (Aα1.UAG mRNA; Parker and Jacobson, 1990a). In both cases the decay of the mRNAs containing the stop codons is significantly slower than the corresponding mRNAs in which translation continues through the MATα1 sequences. The Pα1.UAG mRNA ($t_{1/2}$=37 min) and the Aα1.UAG mRNA ($t_{1/2}$=17 min) respectively decay 5-6-fold and 2-3-fold slower than the same mRNAs lacking the UAG insertions (Pα1 mRNA, $t_{1/2}$=6 min; Aα1 mRNA, $t_{1/2}$=7.5 min; Parker and Jacobson, 1990a). Since an out of frame UAG normally exists within the 42 nucleotide

MATα1 instability element, we consider it unlikely that the inserted UAG simply provides a site for direct binding of some cellular factor. Rather, since the reduction in decay rates is attributable to the introduction of translational stop codons, we conclude that ribosome translocation through MATα1 sequences is required for facilitating rapid mRNA decay.

The requirement for translation in the degradation of the MATα1 mRNA is only one in a large set of independent observations which we and others have made, all of which point to an important role for translation in the mRNA decay process in yeast. These observations include: i) the reduction in the decay rates of all classes of mRNA in cells treated with cycloheximide (Herrick et al., 1990); ii) the stabilization of all classes of mRNA in the tRNA nucleotidyltransferase mutant, ts352 (Aebi et al., 1990; see above); iii) the destabilization of the URA3 (Losson and Lacroute, 1979), URA1 (Pelsy and Lacroute, 1984), and HIS4 (SWP, RP, and AJ, unpublished observations) mRNAs by premature nonsense codons within the 5' proximal portion of the respective coding regions; and iv) the unusually high concentration of rare codons within instability elements and in unstable mRNAs, in general (Herrick et al., 1990; Parker and Jacobson, 1990a).

The requirement of both a specific sequence and translation for mRNA decay suggests several non-exclusive mechanisms by which turnover could be affected: 1) the actual recognition event may occur, not at the RNA level, but as with mammalian tubulin mRNAs (Cleveland, 1988), as the nascent polypeptide emerges from the ribosome; 2) the ribosome itself may be critical to the degradation process, either by delivering a nuclease (Graves et al., 1987) or by triggering a catalytic event itself; or 3) the passage of a ribosome through this region may alter the secondary structure of the mRNA in such a way as to expose sequences containing nuclease recognition sites which would normally not be available (Graves et al., 1987). The concentration of rare codons in the sequences required for rapid decay, coupled with prevalence of rare codons in unstable yeast mRNAs and the known ability of rare codons to induce translational pausing (Wolin and Walter, 1988), suggest that a ribosome paused at a specific site may expose downstream nuclease recognition sites which could then be cleaved by either a soluble or a ribosome bound nuclease. We anticipate that specific tests of these hypotheses will be forthcoming in the near future.

Acknowledgments

This work was supported by a grant (GM27757) to A.J. from the National Institutes of Health and by postdoctoral fellowships to R.P. and S.W.P. from the National Institutes of Health and the American Cancer Society, respectively.

References

Aebi M, Kirschner G, Chen J-Y, Vijayraghavan U, Jacobson A, Martin NC, and Abelson J (1990) Isolation of a temperature-sensitive mutant with altered tRNA nucleotidyltransferase activities and cloning of the gene encoding tRNA nucleotidyltransferase in the yeast *Saccharromyces cerevisiae* (submitted for publication).

Bernstein P and Ross J (1989) Poly(A), poly(A) binding protein and the regulation of mRNA stability. TIBS 14:373-377.

Cleveland DW (1988) Autoregulated instability of tubulin mRNAs: a novel eukaryotic regulatory mechanism. TIBS 13:339-343.

Graves RA, Pandey NB, Chodchoy N, and Marzluff WF (1987) Translation is required for regulation of histone mRNA degradation. Cell 48:615-626.

Herrick D (1989) Structural determinants of mRNA turnover in yeast. Ph.D. Thesis, Univ. of Massachusetts Medical School, Worcester, Mass. USA.

Herrick D, Parker R, and Jacobson A (1990) Identification and comparison of stable and unstable mRNAs in *Saccharomyces cerevisiae*. Mol. Cell. Biol. 10 (in press).

Losson R and Lacroute F (1979) Interference of nonsense mutations with eukaryotic messenger RNA stability. Proc. Natl. Acad. Sci. USA 76:5134-5137.

Nonet M, Scafe C, Sexton J, and Young R (1987) Eucaryotic RNA polymerase conditional mutant that rapidly ceases mRNA synthesis. Mol. Cell. Biol. 7:1602-1611.

Parker R and Jacobson A (1990a) Translation and a forty-two nucleotide segment within the coding region of the mRNA encoded by the MATα1 gene are involved in promoting rapid mRNA decay in yeast. Proc. Natl. Acad. Sci. USA 87 (in press).

Parker R. and Jacobson A (1990b) Rapid decay of the mRNA encoded by the yeast STE3 gene is promoted by two separable elements (submitted for publication).

Pelsy F and Lacroute F (1984) Effect of ochre nonsense mutations on yeast URA1 stability. Current Genetics 8:277-282.

Peltz, SW, Brewer G, Bernstein P, Hart P, and Ross J (1990) Regulation of mRNA turnover in eukaryotic cells. Crit. Rev. Euk. Gene Exp. (in press).

Ross J (1989) The turnover of messenger RNA. Sci. Am. 260:48-55.

Santiago TC, Purvis IJ, Bettany AJE, and Brown AJP (1986) The relationship between mRNA stability and length in *Saccharomyces cerevisiae*. Nucleic Acids Res. 14:8347-8360.

Shapiro RA, Herrick D, Manrow RE, Blinder D, and Jacobson A (1988) Determinants of mRNA stability in *Dictyostelium discoideum* amoebae: differences in poly(A) tail length, ribosome loading, and mRNA size cannot account for the heterogeneity of mRNA decay rates. Mol. Cell. Biol. 8:1957-1969.

Shaw G and Kamen R (1986) A conserved AU sequence from the 3' untranslated region of GM-CSF mRNA mediates selective mRNA degradation. Cell 46:659-667.

Wolin SL and Walter P (1988) Ribosome pausing and stacking during translation of a eukaryotic mRNA. EMBO J. 7:3559-3569.

MUTATIONS INVOLVED IN mRNA STABILITY AND IN THE LENGTH OF THEIR POLY(A) TAILS IN THE YEAST *Saccharomyces cerevisiae*

L. Minvielle-Sebastia, A. Petitjean, B. Winsor*, N. Bonneaud and F. Lacroute
Centre de Génétique Moléculaire
Laboratoire propre du CNRS associé à L'Université P. et M. Curie
91190, GIF-sur-YVETTE
FRANCE

A selective approach has been used to isolate *S. cerevisiae* mutants specifically impaired in the processing of polyadenylated mRNAs.

The selection was based on increased cell sensitivity to cordycepin at 22° C (the 3'-deoxyadenosine, or cordycepin, is an antibiotic known to inhibit the production of poly(A) mRNA (Penman et al., 1970)), associated with a thermosensitive phenotype at 37° C. This selection allowed us to isolate two non-allelic mutants (Bloch et al., 1978), that have been recently renamed *rna14* and *rna15* for homonymy reasons. Three independent and recessive alleles for *rna14*, and two for *rna15*, are now available.

The *rna14* and *rna15* mutant phenotypes are essentially identical. Preliminary characterisation of the physiological function which is altered in these mutants revealed that the level of the poly(A)

---

* : L G M E , 1 1 rue Human , 67000 , STRASBOURG , FRANCE

NATO ASI Series, Vol. H 49
Post-Transcriptional Control of Gene Expression
Edited by J. E. G. McCarthy and M. F. Tuite

mRNAs, measured in pulse labeling experiments, decreases to about 15% of the wild-type level 10 minutes after a shift to the restrictive temperature (37° C) (Bloch et al., 1978) (Table 1). In the same experiment, no reduction in the non-polyadenylated RNA fraction was observed (Table1).

Table 1: Poly(A) RNAs synthesized at the permissive and the restrictive temperatures as a fraction of the total RNA

	22° C	37° C
W.T.	0.229	0.398
rna14	0.218	0.054
rna15	0.227	0.064

In addition, a pulse-chase experiment showed that the stability of poly(A) mRNAs is severely altered in both mutants compared to the wild-type, after a 10 minute shift to 37° C (Bloch et al., 1978) (Table 2).

Table 2: Stability at 37° C of poly(A) mRNAs synthesized at 22° C in the wild-type and in the mutant strains as a fraction of the total RNA

	22° C	37° C
W.T.	0.184	0.100
rna14	0.166	0.028
rna15	0.192	0.024

The data in Table 3 show that there is a general reduction in the amount of ADH1 mRNAs in the *rna15* mutant versus the wild-type. Although the difference is greater for the poly(A)$^+$ fraction than for the poly(A)$^-$ fraction, a partial effect on the rate of mRNA biosynthesis cannot be excluded.

Table 3: Amounts of ADH1 mRNA after a 2mn pulse at 37° C (in cpm/μ g)

	Poly(A)$^+$	Poly(A)$^-$
W.T.	4.08	3.19
rna15	0.43	1.49

In addition, total mRNAs decreases as the length of the pulse increases (Table 4)

Table 4: Ratios of ADH1 mRNA in mutant versus wild type as a function of the labeling time at 37° C

	1mn	2mn	4mn	8mn
rna15 / W.T.	0.41	0.27	0.21	0.15

More information was obtained from analysis of the poly(A) tail length of the mRNAs extracted from strains shifted to 37°C for increasing amounts of time. These RNAs were labeled *in vitro* at their

3'-end, digested with RNases that degrade all but the poly(A) tails, and analysed by polyacrylamide gel electrophoresis. After a 5 mn shift to 37°C, we observed that the poly(A) tail length significantly diminishes in the two mutants compared to the wild-type or to the same mutants at the permissive temperature. This very quick effect suggests that *RNA14* and *RNA15* proteins likely act directly on the regulation mechanism that adjusts the poly(A) tail length. After 60 mn at the non-permissive temperature, the longest poly(A) tails are in the range of about 20 residues. This corresponds to a dramatic shortening since, under normal conditions, the maximum length is about 80 nucleotides in *S. cerevisiae.* Our results completely agree with the majority of anterior results correlating mRNA stability and the poly(A) tail length.

Further characterization of the mutants has shown that their pools of triphosphate nucleotides are not modified, and that the *in vitro* RNA polymerase B(II) activity is slightly but significantly reduced, even with extracts made from cultures grown at the permissive temperature. Also, a slight impairement of actin mRNA splicing is seen at both permissive and restrictive temperatures. These last two effects are probably secondary consequences in the primary defect of our mutants.

In the previous study (Bloch et al., 1978), an impairment of the poly(A) polymerase(s) activity was excluded by *in vitro* measurements. Recent progress in the determination of this enzymatic activity allowed us to perform more sensitive assays which confirm that the poly(A) polymerase is not affected in our mutants

The cloning of *RNA14* and *RNA15* wild-type alleles has been done by complementing the temperature-sensitive phenotype with a pool of yeast genomic DNA inserted in the pFL1 multicopy shuttle vector. After subcloning the initial complementing inserts, the *RNA14* gene was localized in a 3.25 kb DNA fragment that, once sequenced, displays a 1911 bp open reading frame. The putative encoded protein (about 75.3 kD) has no homology with any protein sequence contained in the screened databanks, and does not show any obvious peculiarity. The codon bias measured according to Bennetzen et al. (Bennetzen and Hall, 1982) has a value of 10%, which generally corresponds to a moderately expressed gene; this is in good agreement with the strength of the hybridization signal of its 2.2 kb mRNA observed in Northern blots.

The *RNA15* gene is contained in a 2.35 kb DNA insert. Its sequence contains a 747 bp open reading frame. The putative protein (27.5 kD) does not resemble any protein found in the databanks or the *RNA14*-encoded protein. Nevertheless, an RNA-binding octapeptide consensus, close to those found in the poly(A)-binding protein (Grange et al., 1987; Sachs et al., 1986), is present in one copy number in the N-terminal end of the *RNA15* protein, and is followed by a stretch of glutamine and asparagine, which looks like the *opa* sequences found in many developmental genes (Wharton et al., 1985) (Table 6).

Table 5: Putative *RNA15*-encoded protein

1	MMFDPQTGRSKGYAFIEFRDLESSASAVRNLNGYQLGSRF	40
41	LKCGYSSNSDISGVSQQQQQQYNNINGNNNNNGNNNNNNSN	80
81	GPDFQNSGNANFLSQKFPELPSGIDVNINMTTPAMMISSE....	120

The sequences that are mentioned in the text are underlined.

The codon bias of the *RNA15* sequence is significantly lower (0%) than the *RNA14* codon bias, suggesting that *RNA15* is a low abundance protein.

Cloned into a single (or low) copy vector, *RNA14* or *RNA15* wild-type genes only complement the corresponding mutant. When subcloned in a 2μ based plasmid, *RNA14* wild-type gene suppresses the temperature-sensitivity of an *rna15* mutant, and conversely, the *RNA15* wild-type gene suppresses the *rna14* temperature-sensitive phenotype. As the proteins encoded by these two genes, which have similar mutant phenotypes, do not show any obvious isology, and have very different lengths and presumed abundancies, the most likely hypothesis is that they both participate in the control of messenger RNA poly(A) tail length, directly interacting with each other. Our mutants would be impaired in this interaction, lowering their reciprocal affinities, and this could be circumvented either by the wild-type protein corresponding to the mutant, or by an increased dosage of the other interacting protein.

Knowing that either *RNA14* or *RNA15* is able to reciprocally suppress both mutations when cloned on a high copy number plasmid, we have searched for proteins directly interacting with this complex. These proteins should be capable of correcting our two original mutations when overexpressed on a multicopy vector. At present, two such suppressors have been found for the *rna14* mutant, but none has yet been found for the *rna15* mutant. One of the two suppressors of the *rna14* mutation has been sequenced and could encode a protein having similarities with a ribosomal protein.

It is interesting to note that neither of our mutants are complemented, even partially, by the overexpression of the poly(A)-binding protein (PAB) wild-type gene (Sachs et al., 1986), suggesting that there is no direct interaction between the PAB protein and the *RNA14* and *RNA15* proteins. This is not surprising because, as shown by Sachs and coworkers (Sachs and Davis, 1989), the phenotype of some PAB conditional mutants is opposite to that of *rna14* and *rna15*: they observe an <u>increase</u> of the poly(A) tail length under restrictive conditions.

Our present hypothesis is that the *RNA15* protein may directly bind the mRNA poly(A) tail, in association with the *RNA14* protein which could itself be in contact with other proteins; such a complex could be directly involved in the regulation of the poly(A) tail length and also be coupled to the translation apparatus. Since we are presently purifying antibodies raised against our proteins, it should be possible to localize them in the cell, in order to test this hypothesis.

As shown by Sachs and coworkers (Sachs and Davis, 1989), the poly(A)-binding protein is required for both poly(A) tail shortening and protein synthesis initiation. Supressors of mutants in this protein allow translation to initiate but do not change the poly(A) tract length; this is probably because the lengthening of the poly(A) tail is a secondary effect of the lack of protein synthesis initiation.

Regine Losson measured the stability of different wild-type and mutant messenger RNAs in normal conditions and when protein synthesis is inhibited either by a temperature-sensitive mutation, or by cycloheximide (Losson et al., 1988). She found that naturally unstable mRNAs were clearly stabilized in the cells blocked in protein synthesis. Non sense mutations located near the N-terminal end, and involving a destabilization of the transcript, were also stabilized in these experiments. Interestingly, she recently showed that the mRNAs of a thermosensitive poly(A)-binding mutant are more stable after a shift to the restrictive temperature (unpublished results).

Taken together, these data suggest that the 5'- and the 3'-end of mRNAs could be in close contact in polysomes actively engaged in protein biosynthesis. As a model, the functional unit of cytoplasmic free polysomes could be circular, allowing a direct re-initiation of protein biosynthesis by the last small ribosomal subunit of the translation line. A mechanism, tightly coupled to translation via this small ribosomal subunit, would regulate messenger RNA stability, by random cuts in the poly(A) tail. This mechanism would have been selected during evolution to allow an optimum half-life for each type of mRNA. According to the structure of the 3'- and the 5'-ends, the probability and the position of the cuts would be modulated. The translation coupling would avoid inactivation of travelling or stocked mRNAs before enough encoded proteins could be synthesized.

Such a circular structure would explain many of the results presented above, namely (i) the effect of translation inhibitors on mRNA stability (Hartwell and McLaughlin, 1969); (ii) the lengthening of the poly(A) tail associated with the inhibition of translation initiation (Sachs and Davis, 1989), and the stabilization of mRNAs in some poly(A)-binding protein mutants (Losson, unpublished results); (iii) the involvement of the 5'- and the 3'-end of the messenger RNAs in their stability (Eick et al., 1985; Furuichi et al., 1977; Huez et al., 1978; Shaw and Kamen, 1986).

Acknowledgments

We thank Linda Sperling for her critical reading of the manuscript. L.M.S. is supported by a grant from Le Ministère de la Recherche et de la Technologie. A.P. is supported by a post-doctoral E.M.B.O. long-term fellowship.

References

Bennetzen JL, Hall BD (1982) Codon selection in yeast. J Biol Chem 257:3026-3031

Bloch JC, Perrin F, Lacroute F (1978) Yeast temperature-sensitive mutants impaired in processing of poly(A)-containing RNAs. Mol Gen Genet 165:123-127

Eick D, Piechaczyk M, Henglein B, Blanchard J, Traub B, Kofler E, Wiest S, Lenoir GM, Bornkamm GW (1985) Aberrant c-*myc* RNAs of Burkitt's lymphoma cells have longer half-lives. EMBO J 4:3717-3725

Furuichi Y, LaFiandra A, Shatkin AJ (1977) 5'-Terminal structure and mRNA stability. Nature 266:235-239

Grange T, Martins de Sa C, Oddos J, Pictet R (1987) Human mRNA polyadenylate binding protein: evolutionary conservation of a nucleic acid binding motif. Nuc Acids Res 15:4771-4787

Hartwell LH, McLaughlin CS (1969) A mutant of yeast apparently defective in the initiation of protein synthesis. Proc Natl Acad Sci USA 62:468-474

Huez G, Marbaix G, Gallwitz D, Weinberg E, Devos R, Hubert E, Cleuter Y (1978) Functional stabilisation of HeLa cell histone messenger RNAs injected into *Xenopus* oocytes by 3'-OH polyadenylation. Nature 271:572-573

Losson R, Windsor B, Lacroute F (1988) mRNA stability in *S. cerevisiae.* Yeast 4:S2

Penman S, Rosbach M, Penman M (1970) Messenger and heterogeneous nuclear RNA in HeLa cells. Proc Natl Acad Sci 67:1878-1885

Sachs AB, Bond MW, Kornberg RD (1986) A single gene from yeast for both nuclear and cytoplasmic polyadenylate-binding proteins: domain structure and expression. Cell 45:827-835

Sachs AB, Davis RW (1989) The poly(A) binding protein is required for poly(A) shortening and 60S ribosomal subunit-dependent translation initiation. Cell 58:857-867

Shaw G, Kamen R (1986) A conserved AU sequence from the 3' untranslated region of GM-CSF mRNA mediates selective mRNA degradation. Cell 46:659-667

Wharton KA, Yedvobnick B, Finnerty VG, Artavanis-Tsakonas S (1985) *opa*: a novel family of transcribed repeats shared by the *notch* locus and other developmentally regulated loci in D. melanogaster. Cell 40:55-62

ERRATUM:

An error in the sequence of *RNA15* gene was found after writing of this manuscript. The actual length of the open reading frame of the gene is 891 bp, encoding a 32,8 kDa protein. This modification do not change the features of the protein (the RNA-binding consensus domain and the opa sequence) but lengthened the protein at its N-terminus. The sequences data will be soon available on the EMBL, GenBank and MIPS databases with the accession numbers X55943 and X55944 for *RNA14* and *RNA15* genes, respectively.

RAPID DEGRADATION OF THE c-FOS PROTO-ONCOGENE TRANSCRIPT

Joel G. Belasco, Ann-Bin Shyu, and Michael E. Greenberg
Department of Microbiology and Molecular Genetics
Harvard Medical School
Boston, MA 02115 USA

SUMMARY: The c-*fos* proto-oncogene transcript is targeted for rapid decay in mammalian fibroblasts by two mRNA degradation pathways. Each of these pathways recognizes a distinct, functionally independent element within the c-*fos* message. One determinant of c-*fos* mRNA instability is an AU-rich element within the c-*fos* 3' untranslated region. The other is located within the protein-coding region of the message. The existence of two pathways for rapid degradation of c-*fos* mRNA may be important for ensuring tight regulation of c-*fos* gene expression.

Although the instability of messenger RNA has long been recognized as an important aspect of gene regulation in all organisms, comparatively little is yet understood about the mechanisms and structural determinants of mRNA degradation in either eukaryotic or bacterial cells. One strategy for elucidating the process of mRNA decay is to investigate the degradation of especially labile or stable messages in an effort to understand the basis for their unusual lifetimes in vivo. One of the least stable mammalian messages known is the c-*fos* proto-oncogene transcript, whose extremely short cytoplasmic lifetime may be crucial to the ability of cells to regulate the c-*fos* gene tightly and prevent it from triggering uncontrolled cell growth.

The c-*fos* gene was originally identified as the cellular homologue of the v-*fos* oncogene of the transforming retrovirus FBJ osteosarcoma virus (Curran et al., 1983). Recently, the c-*fos* gene has been shown to encode a transcription factor (Chiu et al., 1988; Rauscher et al., 1988;

NATO ASI Series, Vol. H 49
Post-Transcriptional Control of Gene Expression
Edited by J. E. G. McCarthy and M. F. Tuite

Sassone-Corsi et al., 1988; Halazonetis et al., 1988; Kouzarides and Ziff, 1988). The Fos protein is now known to associate with the product of the *c-jun* proto-oncogene to form a heterodimer that activates transcription by binding to specific sites in a number of mammalian promoters.

Transient expression of the *c-fos* gene is one of the first cellular responses to stimulation with growth factors, phorbol esters, neurotransmitters, and membrane-depolarizing agents (Greenberg and Ziff, 1984; Cochran et al., 1984; Muller et al., 1984; Kruijer et al., 1984; Curran and Morgan, 1985; Greenberg et al., 1985, 1986). In fact, the *c-fos* gene has come to be regarded as a paradigm for a large class of "immediate early" genes that are turned on moments after growth-factor stimulation. Within minutes after stimulation of NIH3T3 fibroblasts with serum or purified growth factors, the cytoplasmic concentration of *c-fos* mRNA rises dramatically from an initially undetectable level, reaching a maximum level within 30 min after induction. Then, just as suddenly, the concentration of the *c-fos* message begins to fall, so that by 2 hr after growth-factor stimulation, no *c-fos* mRNA remains in the cell. The transient nature of *c-fos* gene expression is a consequence of multiple phenomena. First of all, *c-fos* transcription is rapidly induced by growth-factor stimulation (Greenberg and Ziff, 1984). Transcription of the *c-fos* gene is transient, and it ceases within about 30 min after it begins (Greenberg and Ziff, 1984). Secondly, the rapid disappearance of *c-fos* mRNA from the cytoplasm shortly after its arrival is due to the extreme instability of this message, whose half-life of 15 min makes it one of the most labile mammalian mRNAs known (Greenberg and Ziff, 1984; Shyu et al., 1989). (A more typical mammalian transcript would survive in the cytoplasm for hours.) The Fos protein is also very unstable and itself decays with a half-life of only 2 hr (Vosatka et al., 1989). Thus, expression of the *c-fos* gene is very tightly regulated to produce a brief burst of Fos protein that helps to trigger programs for cell growth and differentiation.

We have sought to determine the basis for the extreme instability of *c-fos* mRNA in vivo. To achieve this goal, we have employed a transient-induction assay (Treisman, 1985; Shyu et al., 1989) to examine the decay of mutant *c-fos* messages in growth-factor stimulated fibroblasts. Mouse NIH3T3 cells are transiently co-transfected with the human *c-fos* gene or a variant thereof and with an α-globin gene that is under the control of an SV40 enhancer. (The α-globin gene provides an internal standard that allows correction for variations in transfection efficiency and sample handling.) The cells are made quiescent by serum-starvation, and then stimulated with serum or with a purified growth factor to induce transient transcription from the c-*fos* promoter. Total cytoplasmic RNA is isolated at time intervals after stimulation and analyzed by RNase-protection, using probes complementary to the transfected gene transcripts. In NIH3T3 cells transiently transfected with the wild-type human c-*fos* gene, the transfected human and endogenous mouse c-*fos* mRNAs are both transiently induced by serum stimulation, and both decay rapidly with the same 15 min half-life, as determined from a semi-logarithmic plot of RNase-protection data from samples collected after cessation of c-*fos* transcription (Shyu et al., 1989).

This assay offers distinct advantages over other approaches to measuring mRNA stability. Foremost among these is that, by monitoring mRNA decay in growth-factor-stimulated cells and in the absence of transcription inhibitors such as actinomycin D, c-*fos* mRNA degradation can be examined under physiological conditions that are relevant to normal c-*fos* gene expression.

Using this assay, we set out to identify the structural elements in c-*fos* mRNA responsible for its marked instability. An attractive candidate was a 75-nt AU-rich element (ARE) in the c-*fos* 3′ untranslated region (UTR), an element that resembled a 3′ UTR element previously implicated in the rapid degradation of granulocyte-monocyte colony stimulating factor (GM-CSF) mRNA (Shaw and Kamen, 1986). On the basis of this similarity, it was expected that deletion of this element would stabilize the c-*fos* message. Therefore, we were surprised to find that either deleting this element precisely or deleting almost the entire c-*fos* 3′ UTR had almost no effect on the instability of c-*fos* mRNA in serum-stimulated NIH3T3 cells. All of these mutant messages decayed rapidly with a 20-min half-life resembling that observed for wild-type c-*fos* mRNA (Shyu et al., 1989). Similarly, replacement of the entire c-*fos* 3′ UTR with that of the stable rabbit β-globin message also failed to increase the lifetime of the c-*fos* message (Kabnick and Housman, 1988; Shyu et al., 1989). Thus, no element in the c-*fos* 3′ UTR is necessary for the lability of c-*fos* mRNA.

Nevertheless, the c-*fos* 3′ UTR can destabilize messages that carry it. This destabilizing activity was demonstrated by showing that replacement of the 3′ UTR of an otherwise stable message with that of c-*fos* results in a marked reduction in mRNA half-life (Kabnick and Housman, 1988; Shyu et al., 1989). The stable message selected for this purpose was rabbit β-globin mRNA. Precise fusion of the c-*fos* promoter to the β-globin transcriptional unit allowed wild-type β-globin mRNA to be synthesized transiently in response to growth-factor stimulation. This gene fusion was named BBB to indicate that it encodes a message whose 5′ UTR, coding region, and 3′ UTR all are derived from the β-globin gene (in this three letter nomenclature for messages transcribed from the c-*fos* promoter, the first letter refers to the origin of the 5′ UTR [B if from β-globin and F if from c-*fos*], the second letter indicates the source of the coding region, and the third letter refers to the source of the 3′ UTR). When NIH3T3 cells transiently transfected with the BBB gene were stimulated with serum, the cytoplasmic concentration of β-globin mRNA, like that of c-*fos* mRNA, rose rapidly at first. However, in contrast to the labile c-*fos* message, the level of β-globin message was found to plateau when BBB transcription ceased about 30 min after growth-factor stimulation, and the concentration of BBB mRNA remained high for at least one full day thereafter due to its very long half-life (>24 hr) (Kabnick and Housman, 1988; Shyu et al., 1989).

Replacement of the BBB 3′ UTR with that of c-*fos*, to generate a gene fusion called BBF, had a dramatic destabilizing effect, reducing the half-life of β-globin mRNA from >24 hr down to just 28 min (Shyu et al., 1989). This effect was due primarily to the presence within the c-*fos* 3′ UTR of the 75-nt c-*fos* AU-rich element, as demonstrated by the marked stabilizing effect of deleting the c-*fos* ARE from BBF mRNA (Shyu et al., 1989). Moreover, simple insertion of the c-*fos* ARE into the BBB 3′ UTR reduced the β-globin mRNA half-life to just 37 min (Shyu et al., 1989); thus, introduction of this short AU-rich element had nearly as large a destabilizing effect as substitution of the entire c-*fos* 3′ UTR.

Therefore, although the c-*fos* ARE is sufficient to mediate rapid degradation of messages that carry it, it is entirely dispensable for the instability of the c-*fos* message. The finding that this potent mRNA destabilizing element is unnecessary but sufficient for rapid degradation of c-*fos* mRNA suggested that the c-*fos* message contains at least two independent determinants of instability, each of which alone is sufficient to mediate rapid degradation of the message. One is the AU-rich element within the c-*fos* 3′ UTR. The other must be located upstream of the 3′ UTR.

To determine whether this second destabilizer was located in the c-*fos* 5′ UTR or in the protein-coding region, the decay of two additional hybrid transcripts was examined. In one, FBB, the β-globin 5′ UTR was replaced with the corresponding c-*fos* segment; in the other, BFB, the β-globin protein-coding region was replaced with the c-*fos* coding region. The extremely slow decay rate of FBB mRNA (half-life >10 hr) showed that the c-*fos* 5′ UTR cannot confer instability on β-globin mRNA (Shyu et al., 1989). In contrast, the short 17 min half-life of BFB mRNA indicates that the c-*fos* coding region contains a potent destabilizing element that is even more effective than the c-*fos* AU-rich element at mediating rapid mRNA degradation (Shyu et al., 1989).

Thus, the c-*fos* message apparently contains two instability determinants that can function independently, one in the coding region and the other in the 3′ UTR. Each of these destabilizing elements seems to be the target of a distinct mRNA degradation pathway (Shyu et al., 1989). The evidence for this is two-fold. First of all, the c-*fos* coding region contains neither a long AU-rich segment nor the AUUUA sequence motif found in all known AU-rich destabilizing elements. In addition, treatment of serum-stimulated cells with either of two transcription inhibitors, actinomycin D or 5,6-dichloro-1-β-D-ribofuranosyl-benzimidazole (DRB), gradually and selectively blocks degradation mediated by the c-*fos* AU-rich element but has little effect on the decay pathway that recognizes the destabilizing element in the c-*fos* coding region (Shyu et al., 1989).

It is worth considering why a single message should be the target of two different degradation pathways. One possible explanation is that each of these degradation pathways may be subject to regulation in some cell types or under certain growth conditions. The ability of cells to turn on and off a pathway for mRNA degradation would allow gene expression to be modulated at the level of mRNA stability for messages that are targeted by a single degradation pathway. However, it may be essential that certain gene transcripts be short-lived under all growth conditions. Thus, the existence of two degradation pathways for c-*fos* mRNA may serve as a fail-safe mechanism to ensure rapid decay of this message under growth conditions or in cell types where one or the other of these pathways may be inactive. Such a fail-safe mechanism might be especially important in the case of the c-*fos* gene, whose tightly regulated expression may be essential for maintaining the correct pattern of cellular gene regulation during cell growth and differentiation. Indeed, Meijlink et al. (1985) have shown that deregulated expression of the c-*fos* gene can convert it from a benign proto-oncogene into a full-fledged oncogene capable of transforming cells in culture.

Recently, we have begun to identify more precisely the features of the c-*fos* AU-rich element and the c-*fos* coding region that are important for their ability to direct rapid message decay. We have also examined the mechanism by which each of the instability determinants in the c-*fos* message causes mRNA degradation. These findings will be reported elsewhere (Shyu et al., in preparation).

In summary, our data indicate that the c-*fos* transcript is targeted for rapid decay by two distinct mRNA degradation pathways. Each of these pathways recognizes a different structural element within the c-*fos* transcript. One determinant of c-*fos* mRNA instability is the AU-rich element within the c-*fos* 3′ UTR. The other is located within the protein-coding region of the message. The removal of either of these elements from the c-*fos* message has little effect on the lifetime of c-*fos* mRNA because each can function independently to cause rapid mRNA degradation. Further clarification of the mechanisms and structural determinants of c-*fos* mRNA decay should be invaluable both for understanding how this crucial proto-oncogene is regulated and for elucidating two important pathways for mRNA degradation in mammalian cells.

References

Chiu R, Boyle WJ, Meek J, Smeal T, Hunter T, Karin M (1988). The c-Fos protein interacts with c-Jun/AP-1 to stimulate transcription of AP-1 responsive genes. Cell 54: 541-552.

Cochran BH, Zullo J, Verma IM, Stiles CD (1984). Expression of the c-*fos* gene and a c-*fos* related gene is stimulated by platelet derived growth factor. Science 226: 1080-1082.

Curran T, MacConnell WP, van Straaten F, Verma IM (1983). Structure of the FBJ murine osteosarcoma virus genome: molecular cloning of its associated helper virus and the cellular homolog of the v-*fos* gene from mouse and human cells. Mol. Cell. Biol. 3: 914-921.

Curran T, Morgan JI (1985). Superinduction of the *fos* gene by nerve growth factor in the presence of peripherally active benzodiazepines. Science 229: 1265-1268.

Greenberg ME, Ziff EB (1984). Stimulation of 3T3 cells induces transcription of the c-*fos* proto-oncogene. Nature 311: 433-438.

Greenberg ME, Greene LA, Ziff EB (1985). Nerve growth factor and epidermal growth factor induce transient changes in proto-oncogene transcription in PC12 cells. J. Biol. Chem. 260: 14101-14110.

Greenberg ME, Ziff EB, Greene LA (1986). Stimulation of neuronal acetylcholine receptors induces rapid gene transcription. Science 234: 80-83.

Halazonetis TD, Georgopoulos K, Greenberg ME, Leder P (1988). c-Jun dimerizes with itself and with c-Fos, forming complexes of different DNA binding affinities. Cell 55: 917-924.

Kabnick KS, Housman DE (1988). Determinants that contribute to cytoplasmic stability of human c-*fos* and β-globin mRNAs are located at several sites in each mRNA. Mol. Cell. Biol. 8: 3244-3250.

Kouzarides T, Ziff E (1988). The role of the leucine zipper in the fos-jun interaction. Nature 336: 646-651.

Kruijer W, Cooper JA, Hunter T, Verma IM (1984). Platelet derived growth factor induces rapid but transient expression of the c-*fos* gene and protein. Nature 312: 711-716.

Meijlink F, Curran T, Miller AD, Verma IM (1985). Removal of a 67-base-pair sequence in the noncoding region of protooncogene *fos* converts it to a transforming gene. Proc. Natl. Acad. Sci. USA 82: 4987-4991.

Muller R, Bravo R, Burckhardt J, Curran T (1984). Induction of c-*fos* gene and protein by growth factors precedes activation of c-*myc*. Nature 312: 716-720.

Rauscher FJ, Cohen DR, Curran T, Bos TJ, Vogt PK, Bohmann D, Tjian R, Franza BR (1988). Fos-associated protein p39 is the product of the *jun* proto-oncogene. Science 240: 1010-1016.

Sassone-Corsi P, Lamph WW, Kamps M, Verma IM (1988). *fos*-associated cellular p39 is related to nuclear transcription factor AP-1. Cell 54: 553-560.

Shaw G, Kamen R (1986). A conserved AU sequence from the 3' untranslated region of GM-CSF mRNA mediates selective mRNA degradation. Cell 46: 659-667.

Shyu A-B, Greenberg ME, Belasco JG (1989). The c-*fos* transcript is targeted for rapid decay by two distinct mRNA degradation pathways. Genes Dev. 3: 60-72.

Treisman R (1985). Transient accumulation of c-*fos* RNA following serum stimulation requires a conserved 5' element and c-*fos* 3' sequence. Cell 42: 889-902.

Vosatka RJ, Hermanowski-Vosatka A, Metz R, Ziff EB (1989). Dynamic interactions of c-*fos* protein in serum-stimulated 3T3 cells. J. Cell. Physiol. 138: 493-502.

POST-TRANSCRIPTIONAL CONTROL OF GENE EXPRESSION IN CHLOROPLASTS

David B. Stern, Chen Hsu-Ching, Cynthia C. Adams and Karen L. Kindle[1]
Boyce Thompson Institute for Plant Research
Cornell University
Tower Rd.
Ithaca, NY 14853
USA

Introduction

In higher plants, photosynthetically active chloroplasts differentiate from progenitor organelles that lack chlorophyll and thylakoid membrane structures. Chloroplast development requires light which initiates the coordinate expression of nuclear and chloroplast genes. The mechanisms by which these genes are regulated have been the subject of investigations in many laboratories (for reviews see Mullet, 1988; Tobin and Silverthorne, 1985).

In plastids, the accumulation of mRNAs is regulated by both transcriptional and post-transcriptional mechanisms (for reviews see Gruissem, 1989a; Gruissem *et al.*, 1988; Mullet, 1988). Generally, transcriptional control appears to be at the level of overall genome transcription and promoter strength (Deng *et al.*, 1987); however several cases have been described in which individual genes may be transcriptionally controlled (Deng *et al.*, 1987; Haley and Bogorad, 1990). Regulation of overall transcription could be accomplished by changes in the level of RNA polymerase and/or in the availability or conformation of template. *In vitro* studies in higher plants (Lam and Chua, 1987) and *in vivo* studies with *Chlamydomonas* (Thompson and Mosig, 1987) suggest that template supercoiling may play an important role in regulating transcription rates. Recently, several reports have implicated selective methylation as a regulator of chloroplast gene transcription during chromoplast development in tomato and in cultured sycamore cells (Kobayashi *et al.*, 1990; Ngernprasirtsiri *et al.*, 1988).

General transcriptional control mechanisms, however, cannot explain the differential accumulation of mRNAs during plastid development. For example, whereas the spinach *rbc*L mRNA accumulates only 2-fold relative to plastid rRNA during chloroplast development, *psb*A mRNA increases at least ten-fold[2]. Run-on transcription experiments indicate that the relative transcription rates of these genes are constant at all developmental stages, suggesting that mRNA accumulation is controlled at the level of transcript stability. In this article, we

[1] Cornell-NSF Plant Science Center, Biotechnology Building, Cornell University, Ithaca, NY 14853

[2]*rbc*L encodes the large subunit of ribulose-1,5-bisphosphate carboxylase; *psb*A encodes the 32 kd reaction center protein of photosystem II

NATO ASI Series, Vol. H 49
Post-Transcriptional Control of Gene Expression
Edited by J. E. G. McCarthy and M. F. Tuite

summarize what is known about mRNA stability determinants in chloroplasts, and present a possible mechanism for the control of mRNA stability *via* these determinants.

Results

The 3' inverted repeat is a common feature of plastid mRNAs

Sequence analysis and S1 nuclease protection experiments have demonstrated that the 3' termini of plastid mRNAs contain inverted repeat (IR) sequences capable of forming stem/loop structures (e.g. Sijben-Muller *et al.*, 1986; Zurawski *et al.*, 1981). By analogy to bacterial systems, it was suggested that these structures might serve as transcription terminators, mRNA processing signals, and/or RNA stability determinants. Using *in vitro* transcription/RNA processing systems, it has now been demonstrated that these structures are both mRNA processing signals and RNA stability determinants, but are unlikely to play an important role as terminators of transcription (Chen and Orozco, 1988; Stern and Gruissem, 1987).

3′ Inverted repeats stabilize plastid RNAs *in vitro*

We have used a spinach *in vitro* system to analyze different IR sequences. T7 transcripts with deletion and point mutations in the 3' IR sequence have been tested for stability in protein extracts, to determine whether thermodynamic stability or other structural features of the IR contribute to the relative stabilities of different plastid mRNAs. It was found that IRs lacking the potential to form a stable stem/loop structure are rapidly degraded *in vitro* (Stern *et al.*, 1989). However, the relative stabilities of 3' IR-RNAs *in vitro* do not correspond to their theoretical free energies of formation (see Figure 1). Thus it can be concluded that the potential to form a stem/loop structure is necessary to stabilize mRNA segments *in vitro*, but thermodynamic considerations alone cannot explain the differential stabilities of the RNAs examined.

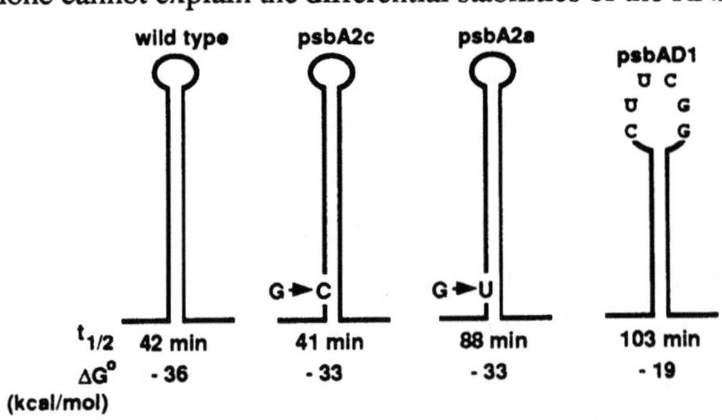

Figure 1. Structures of wild type and mutant spinach *psb*A 3' IR-RNAs. Mutations were introduced into the spinach *psb*A 3' IR (Stern and Gruissem, 1987), and T7 transcripts corresponding to these 3' IR-RNAs were tested for stability in the chloroplast protein extract according to Stern and Gruissem (1987). The calculated ΔG° and $t_{1/2}$ are shown.

Mutagenic analysis of the spinach *psb*A 3' IR suggests that nucleotide sequences in both the stem and the loop influence RNA stability both *in vitro* and in electroporated chloroplasts. For example, a single G→U (but not G→C) change near the base of the stem increased the half-life *in vitro* from 42 to 88 minutes (Figure 1; Adams and Stern, unpublished data). Furthermore, changing the loop sequence so as to include a CUUCGG motif that imparts extraordinary stability to prokaryotic mRNAs (Tuerk *et al.*, 1988) increased half-life from 42 to 103 minutes. When wild-type and mutant RNAs were electroporated into chloroplasts, this hierarchy of stabilities was maintained (Adams and Stern, unpublished data). These results suggest that higher order structure of the stem/loop influences mRNA stability, perhaps by altering nuclease sensitivity (see below).

Removal of the *atp*B 3' IR reduces mRNA accumulation *in vivo*

To demonstrate that a 3' IR is required for plastid mRNA stability *in vivo*, we have used the photosynthetic alga *Chlamydomonas reinhardtii*. *Chlamydomonas* chloroplasts can be transformed by bombarding cells with DNA-coated tungsten pellets, and the DNA is integrated into the chloroplast genome by homologous recombination (Boynton *et al.*, 1988).

Using this transformation technology, we have introduced deletions into the 3' untranslated region of the *atp*B gene, which encodes a subunit of the chloroplast ATPase (Woessner *et al.*, 1986). We find that destabilization or removal of the potential stem/loop structure in this region is correlated with a substantial reduction in the amount of detectable mRNA (Figure 2) and protein product, although no differences are detected in relative transcription rates *in vivo* (Table 1).

Table 1
AtpB expression in *Chlamydomonas* transformants

	strain	growth on HS	*atpB* mRNA	*atpB* protein
	w.t.	+++	100%	100%
	Δ21	+++	100%	100%
	Δ27	+	<10%	~25%
	Δ31	-	n.d.	0%

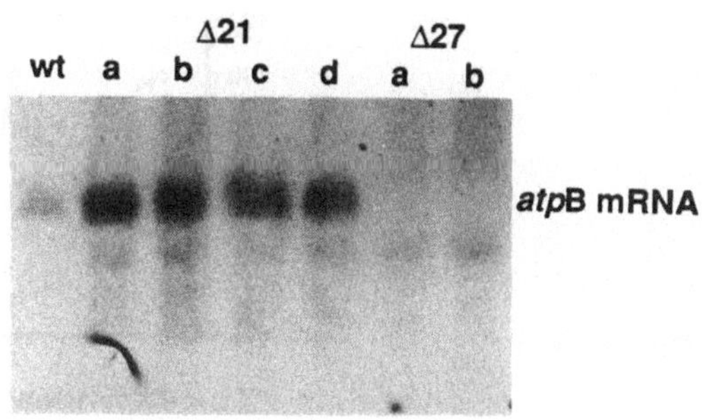

Table 1 and Figure 2. *atp*B mRNA levels in *Chlamydomonas* transformants with altered 3' untranslated regions. Total RNA isolated from transformants with the *atp*B gene structures indicated in Table 1 was probed with a gene-specific probe. In the Table, the thick bar shows the *atp*B protein coding region, and a possible stem/loop in the 3' UTR is indicated. Growth on HS indicates growth of transformants on selective medium that requires photosynthesis and thus expression of *atp*B. *atp*B mRNA levels were quantified from filter hybridizations, and *atp*B protein from Western blots probed with an anti-*atp*B antibody (a kind gift of Dr. R.E. McCarty) using an equal number of cells from each strain.

Additionally, the *atp*B mRNA of transformants lacking IRs is heterogeneous in size as determined by filter hybridizations (Kindle and Stern, unpublished data). This suggests that in *Chla-*

mydomonas, the *atp*B 3' IR functions both in mRNA 3' end formation and in transcript stabilization. Introduction of the spinach chloroplast *pet*D IR into these deleted transformants restores both transcript homogeneity and abundance (Kindle and Stern, unpublished data).

Plastid 3' IRs are protein-associated *in vitro* and *in vivo*

RNA structure alone cannot explain the regulation of mRNA stability in chloroplasts, since chloroplast mRNA accumulation changes in response to light, even though relative transcription rates are unaltered. We have hypothesized that RNA:protein interactions contribute to the regulation of RNA stability. Using gel shift and UV-crosslinking assays, we have demonstrated that several proteins interact with plastid 3' IR-RNAs *in vitro* (Stern and Gruissem, 1989; Stern *et al.*, 1989). A plastid protein that binds the mustard chloroplast *trn*K precursor mRNA *in vitro* has also been reported (Nickelsen and Link, 1989). If spinach chloroplast mRNAs are also protein-associated *in vivo*, these mRNAs should be present as ribonucleoprotein particles. Figure 3 shows a gradient analysis of native spinach chloroplast nucleic acids. In both cesium sulfate and glycerol gradients, the native mRNA bands at a different position from deproteinized RNA, consistent with the presence of mRNPs. Specifically, native RNA is less dense than control RNA in cesium sulfate gradients (data not shown), and sediments more rapidly than control RNA in glycerol gradients.

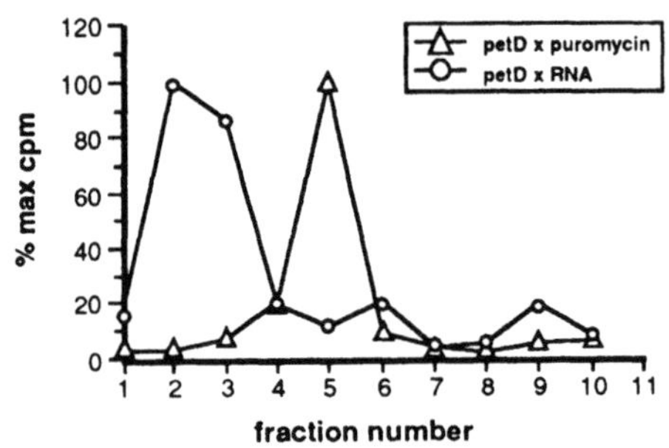

Figure 3. Total spinach chloroplast nucleic acids (open circles) or a puromycin-treated preparation of spinach chloroplast polysomes (open triangles) were sedimented in 15-40% glycerol gradients. Gradient fractions were analyzed for mRNA corresponding to *pet*D, a chloroplast gene encoding subunit IV of the cytochrome b_6/f complex. The lowest fraction number is from the top of the gradient.

Although the functions of these RNA-binding proteins are not yet well-defined, current data have implicated several proteins in mRNA 3' end formation. This is suggested by the observation that the candidate proteins bind precursor 3' IR-RNAs *in vitro*, but not the 3' IRs of mature RNAs (Stern *et al.*, 1989). At least one polypeptide associates with the IR and flanking sequences *in vitro* (Stern *et al.*, 1989; Chen and Stern, unpublished data) and may play a role in modulating IR stability. Interestingly, a comparison of 3' IR-RNA-binding proteins for three genes demonstrated that some bind in a gene-specific manner (Stern *et al.*, 1989) and could therefore have gene-specific regulatory functions.

If one or more of these proteins is involved in the control of mRNA decay, it might be expected to be developmentally regulated in the spinach system. This possibility is currently under investigation. One way in which such a protein might function would be to affect RNA conformation and therefore nuclease sensitivity. To study this, we are currently purifying several RNA-binding proteins, and are analyzing their effect on nuclease susceptibility as described below.

Nuclease activities in chloroplasts

Previously identified ribonucleases in chloroplasts include several tRNA-processing enzymes (Gruissem, 1989b) and also a nuclease activity that processes mRNA precursors (Stern and Gruissem, 1987; Stern and Gruissem, 1989). Our experiments show that if a precursor 3' IR-RNA is made that contains two tandem IRs, only the 3' IR is processed (Figure 4). This is consistent with an exonuclease, but not an endonuclease activity.

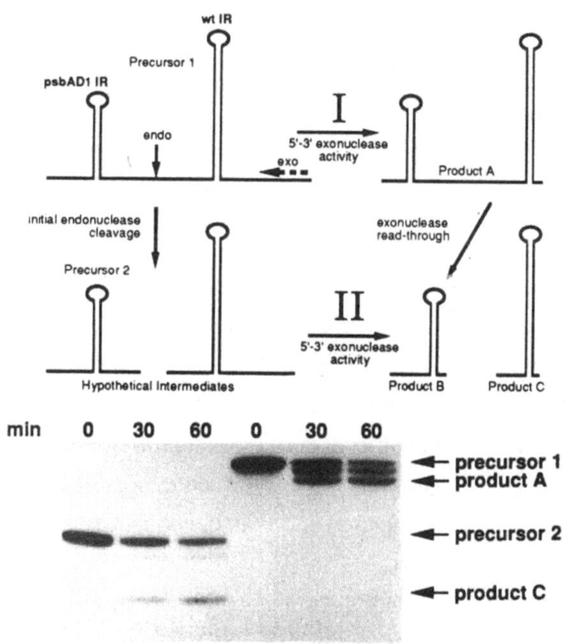

Figure 4. Schematic (upper) and gel (lower) analysis of *in vitro* 3' IR-RNA processing products from tandem IR constructs. Precursors transcribed with T7 RNA polymerase contained either the psbAD1 IR (Precursor 2) or the psbAD1 IR in tandem with the wild type IR (Precursor 1). Two possible processing pathways for the tandem IR-RNA are indicated; in (I), the 3' IR is processed exonucleolytically yielding predominantly product A; a small amount of digestion through the wt IR generates product B. In (II), there is an endonuclease cleavage between the two IRs; subsequent exonucleolytic digestion yields equimolar amounts of products B and C. Gel analysis (right) shows a predominance of product A, consistent with pathway I; low amounts of product B are seen upon prolonged exposures (data not shown). The band migrating between Precursor 1 and Product A is an exonuclease pause site.

In bacterial systems, exonucleases such as RNAse II and polynucleotide phosphorylase are general mRNA decay enzymes (Donovan and Kushner, 1986). Since the plastid genetic system has many prokaryotic features, it is possible that the chloroplast exonuclease(s) does not have a regulatory role. Therefore, we have also searched for endonuclease activities that may be important in mRNA decay; *E. coli* ribonucleases such as RNAse III have such functions (Brawerman, 1987; Cannistraro and Kennell, 1989; Nilsson *et al.*, 1988). Our preliminary data from *in vitro* assays show evidence for single-strand endoribonuclease activities. This activity is not inhibited by polyuridylic acid, which completely inhibits the exonuclease activity associated with mRNA processing (Stern and Gruissem, 1987). Figure 5 shows the major endonuclease cleavage sites for *pet*D and *psb*A 3' IR-RNAs, which will be described in detail elsewhere (Adams and Stern, unpublished data; Chen and Stern, unpublished data). A potential role for endonucleolytic cleavages by these activities would be a rate-limiting step that initiates mRNA decay.

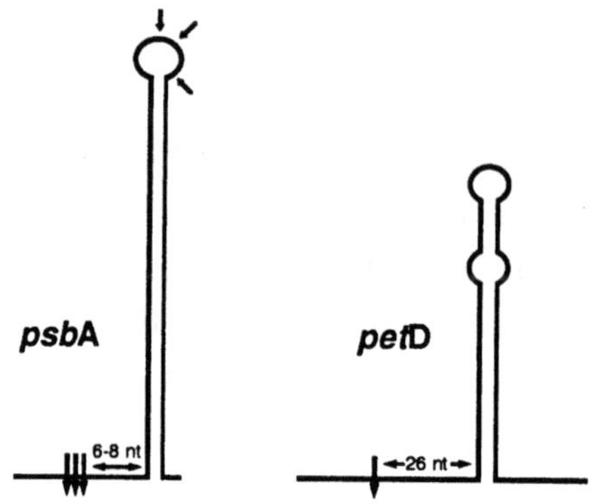

Figure 5. Endonuclease cleavages observed *in vitro* using spinach chloroplast protein extract and 3' IR-RNAs corresponding to *psb*A (left) and *pet*D (right). 5' end-labeled RNAs were incubated under standard processing conditions (Stern and Gruissem, 1987), except that the RNA:protein ratio was 1 fmol RNA:21 µg protein and 3 µg poly-U were included to inhibit exonuclease activity.

Discussion

Control of RNA stability in chloroplasts

Through *in vivo* and *in vitro* investigations of RNA decay in chloroplast systems, we have begun to understand some of the important components that determine RNA stability. However, we still do not understand what are the rate-limiting step(s) that initiate transcript degradation, and how degradation is controlled at an mRNA-specific level during plastid differentiation. At least three components appear to participate in mRNA metabolism: 1) structural features of mRNA, in particular the 3' IR; 2) mRNA-binding proteins, and 3) endo- and exonuclease activities.

Both *in vivo* and *in vitro* results obtained with *Chlamydomonas* and spinach, respectively, suggest that the 3' IR is necessary for mRNA stability; however whether it is sufficient has not been established. Furthermore, several cases exist in higher plant chloroplasts where mRNAs that lack 3' IRs accumulate to significant levels (Neuhaus *et al.*, 1989; Ruf and Kossel, 1988). Thus, an IR does not appear to be required in all cases. Chloroplast transformation will allow the definition of sequence requirements for transcript stability.

Ascertaining the functions of chloroplast mRNA-binding proteins may provide critical insights into mRNA stability. For example, it is well known that protein binding can influence RNA conformation and consequently translation and/or RNA stability. One such instance is the

IRE-binding protein of animal cells, which negatively affects ferritin mRNA translation and positively affects the stability of transferrin receptor mRNA (Casey *et al.*, 1988). Analogous functions may be found for chloroplast mRNA-binding proteins. In addition to analyzing mRNA-binding proteins *in vitro*, it will be important to demonstrate that such RNA:protein associations occur *in vivo*. Although mRNPs have been relatively well-studied in animal cells, there is virtually no information available for chloroplasts or bacteria at the present time.

Ribonucleases found in chloroplasts most likely play an important role in the control of mRNA accumulation. Our results reported here and elsewhere (Stern and Gruissem, 1987; Stern and Gruissem, 1989; Adams and Stern, unpublished data; Chen and Stern, unpublished data) indicate that exonuclease activity is present that may be analogous to the mRNA-degrading enzyme RNAse II of *E. coli*. However, RNAse II is thought to be a general degradative nuclease, and the gene-specific patterns of transcript stability observed during chloroplast development suggest that more complex mechanisms are used, probably involving interactions between nucleases and RNA-binding proteins. These binding proteins could prevent exonucleases from degrading mRNA molecules from their 3' ends, or the proteins could alter RNA secondary and/or tertiary structure to modulate endonuclease susceptibility. Endonuclease cleavages appear to be rate-limiting in degrading *E. coli* mRNAs such as *omp*A and *rnc* (Nilsson *et al.*, 1988; Bardwell *et al.*, 1989) and similar mechanisms may exist in chloroplasts.

Developmental regulation of plastid mRNA stability

One particularly interesting feature of the chloroplast system is that mRNA stability is developmentally regulated, and such regulation can be transcript specific (Deng and Gruissem, 1987; Mullet and Klein, 1987). Although it can be argued that regulation of individual transcripts at the level of stability is inefficient, it is already known in *Chlamydomonas* that specific nuclear genes are required for the translation (Rochaix *et al.*, 1989), processing (Choquet *et al.*, 1988) or stability (Kuchka *et al.*, 1989) of particular chloroplast messages.

In order to study developmental regulation of chloroplast mRNA stability, one would like to compare *in vitro* RNA decay, specific RNA-binding proteins and nucleases in plastid extracts from different developmental stages. This is difficult in dicotyledonous plants such as spinach, since amyloplasts from dark-grown plants are extremely fragile and plastid extracts are contaminated with high amounts of non-specific nuclease activity. Antibodies directed against specific mRNA-binding proteins and improvements in nuclease assays will help to overcome these problems.

Comparison of the chloroplast with bacterial and other eukaryotic systems

It is generally accepted that the chloroplast is a eubacterial endosymbiont (Gray and Doolittle, 1982). It is therefore not surprising to find similarities in the pathways of RNA decay between bacteria and chloroplasts. These features appear to include the frequent presence of an IR at the mRNA 3' end and the finding of exonuclease activities that process 3' IR-RNA precursors. In *E. coli*, it has been proposed that mRNA decay is initiated by endonuclease cleavages that follow traversing ribosomes (King *et al.*, 1986). Such a decay mechanism is plausible in chloroplasts, but unlike *E. coli*, non-polysomal mRNAs can accumulate to high levels in chloroplasts (Klein *et al.*, 1988). These messages would have to be protected from endonuclease attack in a ribosome-independent manner.

The control of stability for cytosolic eukaryotic mRNAs is likely to differ significantly from control in chloroplasts, since RNA transport and polyadenylation are not found in chloroplasts. However, both chloroplast and cytosolic mRNAs can respond to environmental or hormonal signals at the level of stability (reviewed in Brawerman, 1987; Gruissem, 1989a). Several examples of RNA 3' IR-binding proteins exist for cytosolic RNAs, and these instances may serve as models on which to base experiments in the plastid system. The two best characterized examples are the IRE-binding protein that regulates ferritin and transferrin receptor mRNAs (Casey *et al.*, 1988), and a protein that binds non-polyadenylated histone mRNA and is involved in its processing and stability (Mowry and Steitz, 1987).

Future prospects

It remains to be seen what the functions of plastid mRNA-binding proteins are, and what interplay exists between translation and RNA stability. With the ability to transform both chloroplast and nucleus (Boynton *et al.*, 1988; Kindle, 1990) in *Chlamydomonas*, genetic approaches to these questions will be possible. Similarly, the utility of spinach for biochemical approaches will facilitate the purification of regulatory molecules and the reconstitution of *in vitro* decay systems.

References

Bardwell JCA, Regnier P, Chen S, Nakamura Y, Grunberg-Manago M, Court DL (1989) Autoregulation of RNase III operon by mRNA processing. EMBO J. 8:3401-3407

Boynton JE, Gillham NW, Harris EH, Hosler JP, Johnson AM, Jones AR, Randolph-Anderson BL, Robertson D, Klein TM, Shark KB, Sanford JC (1988) Chloroplast transformation in *Chlamydomonas* with high velocity microprojectiles. Science 240:1534-1538

Brawerman G (1987) Determinants of messenger RNA stability. Cell 48:5-6

Cannistraro VJ, Kennell D (1989) Purification and characterization of ribonuclease M and mRNA degradation in *Escherichia coli*. Eur. J. Biochem. 181:363-370

Casey JL, Hentze MW, Koeller DM, Caughman SW, Rouault TA, Klausner RD, Harford JB (1988) Iron-responsive elements: Regulatory RNA sequences that control mRNA levels and translation. Science 240:924-928

Chen L, Orozco EM Jr. (1988) Recognition of prokaryotic transcription terminators by spinach chloroplast RNA polymerase. Nucl. Acids Res. 16:8411-8431

Choquet Y, Goldschmidt-Clermont M, Girard-Bascou J, Kuck U, Bennoun P, Rochaix J-D (1988) Mutant phenotypes support a *trans*-splicing mechanism for the expression of the tripartite *psa*A gene in the *C. reinhardtii* chloroplast. Cell 52:903-913

Deng X, Stern DB, Tonkyn JC, Gruissem W (1987) Plastid run-on transcription: Application to determine the transcriptional regulation of spinach plastid genes. J. Biol. Chem. 262:9641-9648

Deng XW, Gruissem W (1987) Control of plastid gene expression during development: the limited role of transcriptional regulation. Cell 49:379-387

Donovan WP, Kushner SR (1986) Polynucleotide phosphorylase and ribonuclease II are required for cell viability and mRNA turnover in *Escherichia coli* K-12. Proc. Natl. Acad. Sci. USA 83:120-124

Gruissem W (1989a) Chloroplast gene expression: How plants turn their plastids on. Cell 56:161-170

Gruissem W (1989b) Chloroplast RNA: Transcription and processing. In: A. Marcus (ed) The Biochemistry of Plants: A Comprehensive Treatise, vol. 15. Academic Press, New York, p 151-191

Gruissem W, Barkan A, Deng X, Stern DB (1988) Transcriptional and post-transcriptional control of plastid mRNA levels in higher plants. Trends Genet. 4:258-263

Haley J, Bogorad L (1990) Alternative promoters are used for genes within maize chloroplast polycistronic transcription units. The Plant Cell 2:323-333

Kindle KL (1990) High-frequency nuclear transformation of *Chlamydomonas reinhardtii*. Proc. Natl. Acad. Sci. USA 87:1228-1232

King TC, Sirdeskmukh R, Schlessinger D (1986) Nucleolytic processing of ribonucleic acid transcripts in procaryotes. Microbiol. Rev. 50:428-451

Klein RR, Mason H, Mullett JE (1988) Light-regulated translation of chloroplast proteins. I. Transcripts of *psa*A-*psa*B, *psb*A, and *rbc*L are associated with polysomes in dark-grown and illuminated barley seedlings. J. Cell. Biol. 106:289-301

Kobayashi H, Ngernprasirtsiri J, Akazawa T (1990) Transcriptional regulation and DNA methylation in plastids during transitional conversion of chloroplasts to chromoplasts. EMBO J. 9:307-313

Kuchka MR, Goldschmidt-Clermont M, van Dillewijn J, Rochaix J-D (1989) Mutation at the *Chlamydomonas* nuclear *NAC*2 locus specifically affects stability of the chloroplast *psb*D transcript encoding polypeptide D2 of PSII. Cell 58:869-876

Lam E, Chua NH (1987) Chloroplast DNA gyrase and *in vitro* regulation of transcription by template topology and novobiocin. Plant Mol. Biol. 8:415-424

Mowry KL, Steitz JA (1987) Both conserved signals on mammalian histone pre-mRNAs associate with small nuclear ribonucleoproteins duing 3' end formation in vitro. Molec. Cell. Biol. 7:1663-1672

Mullet JE (1988) Chloroplast development and gene expression. Annu. Rev. Plant Physiol. Plant Molec. Biol. 39:475-502

Mullet JE, Klein RR (1987) Transcription and RNA stability are important determinants of higher plant chloroplast RNA levels. EMBO J. 6:1571-1579

Neuhaus H, Scholz A, Link G (1989) Structure and expression of a split chloroplast gene from mustard (*Sinapis alba*): ribosomal protein gene *rps*16 reveals unusual transcriptional features and complex RNA maturation. Curr. Genet. 15:63-70

Ngernprasirtsiri J, Kobayashi H, Akazawa T (1988) DNA methylation as a mechanism of transcriptional regulation in nonphotosynthetic plastids in plant cells. Proc. Natl. Acad. Sci. USA 85:4750-4754

Nickelsen J, Link G (1989) Interaction of a 3' RNA region of the mustard *trn*K gene with chloroplast proteins. Nucl. Acids Res. 17:9637-9648

Nilsson G, Lundberg U, von Gabain A (1988) *In vivo* and *in vitro* identity of site specific cleavages in the 5' non-coding region of *omp*A and *bla* mRNA in *Escherichia coli*. EMBO J. 7:2269-2275

Rochaix J-D, Kuchka M, Mayfield S, Schirmer-Rahire M, Girard-Bascou J, Bennoun P (1989) Nuclear and chloroplast mutations affect the synthesis or stability of the chloroplast *psb*C gene product in *Chlamydomonas reinhardtii*. EMBO J. 8:1013-1021

Ruf M, Kossel H (1988) Structure and expression of the gene coding for the α-subunit of DNA-dependent RNA polymerase from the chloroplast genome of *Zea mays*. Nucl. Acids Res. 16:5741-5754

Sijben-Muller G, Hallick RB, Alt J, Westhoff P, Herrmann RG (1986) Spinach plastid genes coding for initiation factor IF-1, ribosomal protein S11 and RNA polymerase α-subunit. Nucl. Acids Res. 14:1029-1045

Stern DB, Gruissem W (1987) Control of plastid gene expression: 3' inverted repeats act as mRNA processing and stabilizing elements, but do not terminate transcription. Cell 51:1145-1157
Stern DB, Gruissem W (1989) Chloroplast mRNA 3' end maturation is biochemically distinct from prokaryotic mRNA processing. Plant Molec. Biol. 13:615-625
Stern DB, Jones H, Gruissem W (1989) Function of plastid mRNA 3' inverted repeats: RNA stabilization and gene-specific protein binding. J. Biol. Chem 264:18742-18750
Thompson RJ, Mosig G (1987) Stimulation of a *Chlamydomonas* chloroplast promoter by novobiocin in situ and in *E. coli* implies regulation by torsional stress in the chloroplast DNA. Cell 48:281-287
Tobin EM, Silverthorne J (1985) Light regulation of gene expression in higher plants. Ann. Rev. Plant Physiol. 36:569-593
Tuerk C, Gauss P, Thermes C, Groebe DR, Gayle M, Guild N, Stormo G, d'Aubenton-Carafa Y, Uhlenbeck OC, Tinoco IJ, Brody EN, Gold L (1988) CUUCGG hairpins: Extraordinarily stable RNA secondary structures associated with various biochemical processes. Proc. Natl. Acad. Sci. USA 85:1364-1368
Woessner JP, Gillham NW, Boynton JE (1986) The sequence of the chloroplast *atp*B gene and its flanking regions in *Chlamydomonas reinhardtii*. Gene 44:17-28
Zurawski G, Perrot B, Bottomley W, Whitfeld PR (1981) The structure of the gene for the large subunit of ribulose-1,5-bisphosphate carboxylase from spinach chloroplast DNA. Nucl. Acids Res. 9:3251-3270

Differential mRNA Stability: A Regulatory Strategy for Hsp70 Synthesis

Robert B. Petersen and Susan Lindquist
The Howard Hughes Medical Institute
The Department of Molecular Genetics and Cell Biology
The University of Chicago
5841 S. Maryland Avenue
Chicago, Illinois 60637 USA

Introduction

Many organisms face the challenge of a changing environment. Individual cells in a multicellular organism face similar challenges. To cope with this, cells and organisms have developed the ability to mount transient responses in which subsets of genes appropriate to current conditions are activated for an appropriate time. One particularly well studied example of a transient response is provided by the heat-shock response (for general reviews see Lindquist, 1986; Lindquist and Craig, 1988). This induction of a small number of highly conserved proteins (known as the heat shock proteins or hsps) occurs when cultured cells or whole organisms are exposed to temperatures above their normal growth range. When the cells or organisms are returned to normal temperatures normal patterns of synthesis are restored.

The hsps, and close relatives of the hsps which are produced at normal temperatures, participate in an extraordinary variety of cellular processes, including DNA synthesis, protein secretion, the response to steroid hormones, and the "chaperoning" of newly synthesized proteins. The variety and complexity of these interactions are only beginning to be unravelled. At any rate , the proteins are vital for growth at moderately high temperatures and for survival at extreme temperatures (Lindquist and Craig, 1988; Sanchez and Lindquist, in press).

In keeping with the vital roles of these proteins, a variety of regulatory mechanisms have evolved to regulate their synthesis. In all organisms studied, heat shock genes are preferentially transcribed at high temperatures. Regulatory mechanisms also act at the levels of RNA processing, translation, and mRNA degradation. These mechanisms are less well understood but have an equal, if not greater, effect on gene expression. The regulation of message degradation plays a particularly important role in the regulation of hsp70 synthesis in Drosophila. This is the subject of this article.

NATO ASI Series, Vol. H 49
Post-Transcriptional Control of Gene Expression
Edited by J. E. G. McCarthy and M. F. Tuite

Regulation of Recovery from Heat Shock

The pattern of hsp70 synthesis during the heat shock response

The hsp70 genes of Drosophila are repressed during normal growth and development. After a temperature shift from 25°C to 37°C the hsp70 genes are activated and the transcription of normal cellular genes is repressed. At the same time, preexisting polysomes disappear in the cytoplasm although the preexisting messages are stably maintained in the cell (McKenzie, 1976; Storti et al., 1980; Findly and Pederson, 1981; Petersen and Mitchell, 1981; Lindquist, 1981; DiDomenico et al., 1982b). As the newly synthesized hsp transcripts accumulate in the cytoplasm polysomes are reformed and the hsps are virtually the only detectable translation products in the cell (Lindquist, 1980). The preferential translation of the hsp transcripts is dependent on sequences in the 5' untranslated region (UTR) of the messages (Klemenz et al., 1985; McGarry and Lindquist, 1985; Hultmark et al., 1986). Preferential synthesis of hsps continues as long as the cells are maintained at the elevated temperature. When the cells are returned to their normal growth temperature, hsp synthesis is repressed and normal cellular protein synthesis is resumed (DiDomenico et al., 1982a,b).

The return to normal patterns of synthesis

The restoration of normal protein synthesis during recovery differs in several ways from the sudden induction of hsp synthesis during heat shock. First, it is a much more gradual process and has two distinct phases. Immediately after the return to 25°C, cells continue to produce hsps exclusively. The length of time required to repress hsp70 synthesis and restore normal protein synthesis is proportional to the severity of the prior heat treatment. For example, when cells are returned to 25°C after a 30 minute heat treatment at 35°C, repression of hsp70 synthesis begins within an hour. However, after a 30 minute heat treatment at 39°C, repression of hsp70 synthesis begins after 7 hours. Thus, the return of normal protein synthesis after heat shock may take from one to many hours.

During heat shock 25°C messages are inactivated as a group and they are reactivated as a group during recovery. Since these messages are heterogeneous, the similarity in their behavior is striking. In sharp contrast, hsp mRNAs do not behave as a group and are repressed very asynchronously. Nevertheless, there is a distinct and reproducible pattern to their repression (DiDomenico et al., 1982b, Lindquist and DiDomenico, 1985). Hsp70 is always the first protein to be repressed and hsp82 is always the last. In some cases, the repression of these two messages is completely asynchronous. The repression of hsp70 may be nearly complete at a time when expression of hsp82 is still continuing to increase.

Another major difference in the behavior of 25°C messages and hsp messages is the mechanism of their repression. During heat shock, translationally inactive 25°C mRNAs are quantitatively retained in the cell. During recovery, hsp mRNAs are degraded as they are repressed. In fact, translational repression of hsp mRNAs is so tightly coupled to their degradation, the two processes are indistinguishable (DiDomenico, et al . 1982b; Petersen and Lindquist, 1988).

The regulation of hsp70 message stability

The degradation of hsp70 messages is very highly regulated. Cells that are maintained at high temperatures produce hsps, and virtually nothing but hsps, until they begin to die (Lindquist, S., unpublished). If actinomycin D is added one hour after the cells are shifted to the elevated temperature the same pattern of protein synthesis is observed. Thus, the continued synthesis of hsps at high temperatures is not dependent on continued transcription, but on the continued translation of stabilization of stable hsp mRNAs.

As discussed above, when cells are returned to normal temperatures hsp synthesis is repressed. Again, the same phenomenon is observed if actinomycin D is added shortly after the heat shock (Fig. 1). From this we can conclude that stability of hsp mRNAs is regulated and not simply a function of temperature. Thus, with a more severe heat shock, presumably, more hsp is required and hsp mRNAs are stabilized for a longer time.

The hsp70 message is the most abundant of the heat shock messages and shows the greatest degree of regulation. We have, therefore, concentrated our attention on this message. There are two different schemes by which the degradation of hsp70 mRNA might be governed. One possibility is that it is an inherently unstable message which is stabilized by heat shock. During recovery, then, the message would be returned to its natural, unstable state. Alternatively, it might be an inherently stable message. A special mechanism would then be activated during recovery for degrading the message. We tested these possibilities by placing modified hsp70 coding sequences under the control of the metallothionein promoter. In this way we could induce synthesis of an hsp mRNA at normal temperatures, with very low concentrations of copper. Under these conditions the message was extremely short lived, with a half-life of less than 15 minutes. If the cells were heat shocked after induction, however, the message was stabilized and had a half-life of greater than 4 hours. Therefore, the mechanism responsible for degradation of hsp70 mRNA present at normal temperatures is inactivated during heat shock (Petersen and Lindquist, 1988). Presumably it is the

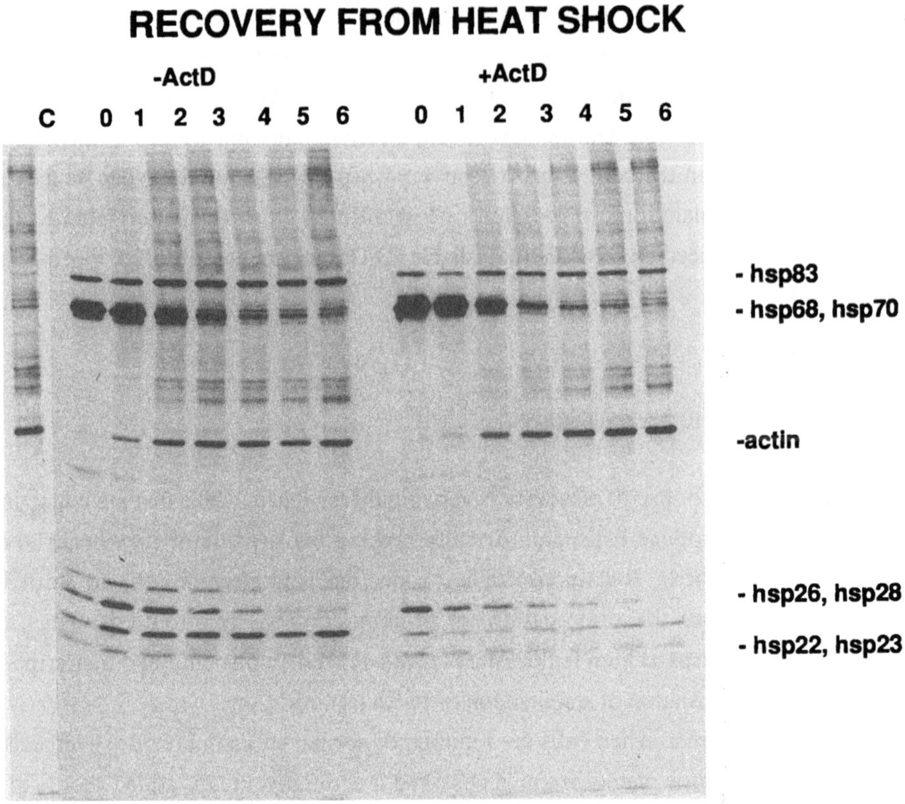

Fig.1 Cells were heat-shocked at 37°C and then returned to 25°C for recovery. Actinomycin D was added to the cells on the right-hand side of the figure one hour after the shift down to 25°C. Cells were labelled with ^{3}H-leucine at hourly intervals beginning immediately after heat shock (lane 0). Lane C is a control sample. The position of the major heat shock proteins and actin are indicated.

reactivation of this mechanism during recovery which is responsible for regulated degradation.

The sequences which target this message for regulated degradation have been partially identified. The first indication of what sequences might be important was provided by an X-ray induced deletion. This deletion replaced the 3' half of the hsp70 gene with an unknown sequence. The message produced by this gene was more stable during recovery than the wild-type message (Simcox et al. 1985). More recently, a variety of mutations saturating the hsp70 gene were created. Mutations were placed in the message leader, protein coding sequences, and 3' UTR, and examined for their effects on mRNA degradation. Neither the 5' end mutations nor the protein coding mutations affected the stability of the message in cis. (Some of the protein coding mutations affected the recovery process as a whole because they produced aberrant proteins, but none specifically affected the mutated message. See below; Rossi, 1988; Rossi, J., J.S. Solomon, K. Golic, T. McGarry, R. B. Petersen and S. Lindquist, manuscript in preparation).

Mutations which replaced the 3' UTR of the hsp70 message with sequences from an actin message altered the stability of the resulting message (McGarry, T., unpublished). This suggested that these sequences were important for regulation of hsp70 mRNA during recovery . To determine if the 3' UTR was sufficient to target a message for regulated degradation, two hsp70-alcohol dehydrogenase gene fusions were examined (Petersen and Lindquist, 1989). The first contained the hsp70 promoter and message leader region fused to adh coding and 3' UTR. This gene produced a message which was translated well at high temperatures. During recovery, even from a mild heat shock, it was very stable and continued to be translated at a high rate while endogenous hsp70 messages were translationally repressed and degraded. The second adh construct was identical to the first except that the hsp70 3' UTR was used to replace the adh 3' UTR of the original construct. This gene produced a message that was degraded with the same kinetics as the hsp70 message during recovery from a mild heat shock. Moreover, the degradation of the message was delayed, in the same way that degradation of the hsp70 message was delayed, after a more severe heat shock (Petersen and Lindquist, 1989).

Another difference in the regulation of these genes was observed when constitutive levels of synthesis were compared. Both genes had a detectable constitutive level of adh expression. However, the gene which carried the hsp70 3'UTR had a much lower level of adh expression. Thus the 3' UTR of the hsp70 message plays a critical role in regulating its degradation, both in recovery from heat shock and in preventing the inappropriate expression of hsp70 in the absence of heat shock at normal temperatures. Thus, the hsp70 promoter is active at a low level at normal temperatures. However, transcripts which carry the hsp70 3' UTR are then immediately targeted for degradation. We believe that this degradation plays an important role in lowering the constitutive expression of hsp70 in cells growing at their normal temperature. In Drosophila cells, we have found that expression of hsp70 in cells growing at 25°C has a deleterious effect on growth, underscoring the biological importance of this regulatory mechanism.

The 3' UTR of the hsp70 message is AU rich and contains sequence elements that resemble those implicated in the turnover of unstable messages in mammalian cells, such as c-myc, c-fos, and various lymphokine mRNAs. Specifically, an AUUUA-like motif is present 54 nucleotides downstream of the translation stop codon and 104 nucleotides upstream of the polyadenylation site. Although the nucleotides specifically responsible for destabilizing the hsp70 message have not been identified, it seems more than a coincidence that both the c-myc and c-fos messages are stabilized by heat shock in mammalian cells (Sadis et al., 1988). Moreover, Theodorakis and Morimoto (1987) have demonstrated that the constitutively synthesized hsp70 message of human cells is much more stable after heat shock than before heat shock.

We propose that 1) the degradation mechanism employed to reduce constitutive expression of hsp70 mRNAs is highly conserved and is the same mechanism that is employed to degrade other unstable messages, 2) the inactivation of this mechanism by heat shock, together with the transcriptional activation of the heat shock genes and the selective translation of hsp mRNAs, facilitates the extraordinarily rapid accumulation of heat-shock proteins, 3) it is the reactivation of this degradation mechanism, together with the transcriptional repression of heat shock genes (Ashburner, 1970; DiDomenico et al, 1982a), that mediates the repression of hsp synthesis during recovery (Fig. 2).

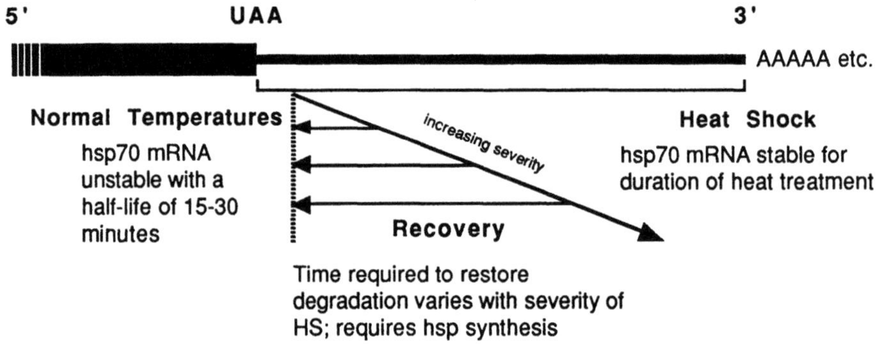

Fig 2. Regulation of hsp70 synthesis by selective RNA degradation. Hsp70 mRNA is unstable at normal temperatures. With heat shock hsp70 mRNA is stabilized for the duration of the heat treatment. After cells are returned to normal temperatures, the stability of hsp70 mRNA is directly related to the severity of the preceding heat shock. This process is controlled by sequences in the hsp70 3' UTR.

The role of hsps in the recovery process

The repression of hsp mRNAs and the reactivation of 25°C messages during recovery is a remarkably precise and highly reproducible process (DiDomenico et al., 1982a). Several lines of evidence suggest that hsps themselves play an important role in it. First, when cells are given a mild heat treatment, which induces the synthesis of hsps, and are then given a second, severe heat treatment, they recover 25°C protein synthesis much more rapidly than do cells that are exposed to the severe heat treatment alone (Petersen and Mitchell, 1981). Similarly, when cells are gradually raised to high temperatures they recover from that exposure much more rapidly than cells that are suddenly raised to the same temperature.

Certainly, this regimen of heat treatment more closely mimics conditions organisms encounter in their natural environment and presumably reflects the protective effects of hsp synthesis.

Second, for a given heat treatment, a specific quantity of heat-shock protein is always made before 25°C protein synthesis is restored (DiDomenico et al., 1982a; Lindquist and DiDomenico, 1985). A variety of treatments which interfere with the synthesis of hsps, also interfere with the recovery process. For example, in one series of experiments, cells were treated with actinomycin D at various times during heat shock, to limit the quantity of hsp message they produced. When these cells were returned to 25°C for recovery, those which produced hsps at a slower rate (because they had a reduced concentration of hsp message) took proportionately longer to return to normal patterns of synthesis (DiDomenico et al., 1982a). In another series of experiments, cells were treated with cycloheximide, to block hsp synthesis, immediately prior to a mild heat shock. When cells were returned to 25°C, hsp70 messages were stable for at least 10 hours. When the addition of cycloheximide was delayed until after the heat shock, allowing the cells to produce a substantial quantity of hsps, the message was rapidly degraded (DiDomenico et al., 1982a).

Although it is not yet clear which hsps are required for recovery, the evidence suggests that hsp70, at least, plays an important role. First, hsp70 is the only protein whose accumulation is quantitatively related to the recovery of 25°C protein synthesis (DiDomenico et al., 1982a). That is, for a given heat shock a specific quantity of hsp70 is always produced before recovery begins. The synthesis of other hsps is considerably more variable. Second, when the synthesis of hsp70 is selectively reduced, by transforming cells with genes constructed to produce antisense hsp70 RNA, the rate at which they return to normal patterns of synthesis is reduced in proportion (Lindquist et al., 1988). Reducing the rate of hsp26 synthesis with antisense RNA has no such effect (McGarry and Lindquist, 1986). Third, cells transformed with certain hsp70 protein coding mutations show a marked delay the recovery of normal protein synthesis after heat shock (Rossi, J., 1988; Rossi, J., J.S. Solomon, K. Golic, T. McGarry, R.B. Petersen and S. Lindquist, manuscript in preparation).

Superficially there are two ways to view the involvement of hsps in restoring the normal pattern of protein synthesis: First, the block in the translation of 25°C messages and the block in the degradation of hsp messages might simply be regulatory mechanisms that are employed to maximize the synthesis of hsps. They would be released only when a sufficient quantity of hsp has been produced. Alternatively, high temperatures might damage the translation and message-degradation machinery of the cell. Hsps, then would be required to repair the damage. The difference between these two views, however, is largely artificial. Heat damages a wide variety of cellular processes. The heat-shock genes have evolved within this context to bypass the damage. Not surprisingly, then, they effectively take advantage of the inability of the 25°C genes and mRNAs to function to maximize their own expression .

Since the general role of the heat-shock proteins is to repair normal cellular processes damaged by heat, certain of their functions are, by their very nature, autoregulatory. That is, they remove those conditions which they had evolved to exploit for their own synthesis. The challenge ahead lies in defining the nature of these regulatory mechanisms and their interactions with hsps in precise molecular terms.

Concluding Remarks

Transcriptional regulation plays an important role in the heat-shock response. In some organisms, most notably Drosophila, the preferential translation of hsp mRNAs at high temperatures is also a well established phenomenon. Most recently, mechanisms acting at the levels of RNA processing and selective message degradation have been shown to exert a profound effect on gene expression both during heat shock and during recovery from heat shock. These mechanisms have been most carefully studied in Drosophila, but they may be nearly universal in their application. The concerted action of all of these mechanisms is responsible for the remarkably rapid and intense induction of hsps at high temperatures. That so many different mechanisms are coordinated to accomplish this induction is in keeping with the crucial roles these proteins play in providing protection from the toxic effects of heat. However, these mechanisms are probably not restricted to the heat shock response. They are almost certainly related to mechanisms employed in other transient responses.

Acknowledgements

The authors would like to thank the following individuals for sharing unpublished data: K. Golic, T. McGarry, J. Rossi, Y. Sanchez, and J. Solomon. Portions of this article originally appeared in: Yost HJ, Petersen RB, Lindquist S (to be published) Posttranscriptional regulation of heat shock protein synthesis in Drosophila. In: Stress Proteins in Biology and Medicine. Cold Spring Harbor Press, Cold Spring Harbor, NY.

REFERENCES

Ashburner M (1970) The genetic analysis of puffing in polytene chromosomes of Drosophila. Proc Roy Soc London B 176: 319- 327

DiDomenico BJ, Bugaisky GE, Lindquist S (1982a) The heat shock response is self-regulated at both the transcriptional and posttranscriptional levels. Cell 31: 593- 603

DiDomenico BJ,Bugaisky GE, Lindquist S (1982b) Heat shock and recovery are mediated by different translational mechanisms. Proc Natl Acad Sci USA 79: 6181-6185
Findly RC, Pederson T (1981) Regulated transcription of the genes for actin and heat-shock proteins in cultured Drosophila cells. J Cell Biol 88: 323-328
Hultmark D, Klemenz R, Gehring W (1986) Translational and transcriptional control elements in the untranslated leader of the heat-shock gene hsp22. Cell 44: 429- 438
Klemenz R, Hultmark D, Gehring W (1985) Selective translation of heat shock mRNA in Drosophila melanogaster depends on sequence information in the leader. EMBO J 4: 2053-2060
Lindquist S (1980) Varying patterns of protein synthesis in Drosophila during heat shock: implications for regulation. Dev Biol 77: 463-469
Lindquist S (1981) Regulation of protein synthesis during heat shock. Nature (London) 294, 311-314
Lindquist S and DiDomenico B (1985) Coordinate and Noncoordinate Gene Expression During Heat Shock: A Model for Regulation. New York: Academic Press
Lindquist S (1986) The Heat-shock Response. Ann Rev Biochem 55: 1151-1191
Lindquist S, McGarry T, Golic K (1988) Use of antisense RNA in studies of the heat shock response. In: Melton DA (ed) Current Communications in Molecular Biology: Antisense RNA and DNA, Cold Spring Harbor Laboratory, Cold Spring Harbor, New York, p 71
Lindquist S, Craig EA (1988) The heat-shock proteins. Ann Rev Genet 22: 631- 677
McGarry T, Lindquist S (1986) Inhibition of heat shock protein synthesis by heat-inducible antisense RNA. Proc Natl Acad Sci USA 83: 399- 403
McGarry TJ, Lindquist SL (1985) The preferential translation of Drosophila hsp70 mRNA requires sequences in the untranslated leader. Cell 42: 903-911
McKenzie SL (1976) Protein and RNA synthesis induced by heat in Drosophila. Ph.D. thesis Harvard University
Petersen N, Mitchell HK (1981) Recovery of protein synthesis following heat shock; pre-heat treatment affects mRNA translation. Proc Natl Acad Sci USA 78: 1708-1711
Petersen R, Lindquist S (1988) The Drosophila hsp70 message is rapidly degraded at normal temperatures and stabilized by heat shock. Gene 72: 161- 168
Petersen RB, Lindquist S (1989) Regulation of hsp70 synthesis by messenger RNA degradation. Cell Reg 1: 135- 149
Rossi J (1988). Investigations Into the Function of Heat Shock Proteins. Ph.D. Thesis, University of Chicago
Sadis S, Hickey E, Weber LA (1988) Effect of heat shock on RNA metabolism in HeLa cells. J Cell Physiol 135: 377- 386
Sanchez Y, Lindquist S (to be published) Hsp104 is required for induced thermotolerance. Science
Simcox AA, Cheney CM, Hoffman EP, Shearn A (1985) A deletion of the 3' end of the Drosophila melanogaster HSP70 gene increases stability of mutant RNA during recovery from heat shock. Mol Cell Biol 5: 3397- 3402
Storti RV, Scott MP, Rich A, Pardue ML (1980) Translational control of protein synthesis in response to heat shock in D. melanogaster cells. Cell 22: 825-834
Theodorakis NG, Morimoto RI (1987) Posttranscriptional regulation of hsp70 expression in human cells: effects of heat shock, inhibition of protein synthesis, and adenovirus infection on translation and mRNA stability. Mol Cell Biol 7: 4357- 4368

POST-TRANSCRIPTIONAL CONTROL OF IS*10* TRANSPOSASE EXPRESSION: ANTISENSE RNA BINDING AND OTHER CONFORMATIONAL CHANGES AFFECTING MESSENGER RNA STABILITY AND TRANSLATION

C. Ma, J.E. Gonzalez, C.C. Case, T. Sonnabend, J. Rayner and R.W. Simons
Department of Microbiology and Molecular Genetics and the Molecular Biology Institute
University of California at Los Angeles
Los Angeles, CA 90024

INTRODUCTION

Insertion sequence IS*10* (Fig. 1) encodes a transposase function (*tnp*) that is rate-limiting for transposition (Morisato *et al.*, 1983). Transposase expression is controlled at both the transcriptional and post-transcriptional levels. The *tnp* promoter is among the weakest of well-characterized promoters (Simons *et al.*, 1983; Case *et al.*, 1988) and its activity is inhibited by DNA-adenine methylation (Roberts *et al.*, 1985) and integration host factor binding (J. Krull and R.W.S., in preparation). Here, we are concerned with three different post-transcriptional mechanisms that operate to control *tnp* translation (Fig. 1). All three manifest their effects by altering RNA secondary structure at or near the *tnp* ribosome binding site. First, a small antisense RNA (Fig. 1B) binds to the 5' end of short (nascent) the *tnp* mRNAs (Fig. 1A) to block ribosome binding (Fig 1C). Second, RNA secondary structure in the full-length *tnp* mRNA sequesters the 5' end (Fig. 1D), resulting in two effects: insensitivity to antisense RNA binding and decreased *tnp* translation. Third, *tnp* mRNAs initiated at external promoters (Fig. 1E) are not translated efficiently because RNA secondary structure sequesters the *tnp* ribosome binding site. The salient features of each control mechanism are described below.

ANTISENSE RNA CONTROL OF TRANSPOSASE EXPRESSION

Antisense RNAs have been shown or proposed to control expression of a number of prokaryotic genes by mechanisms that include premature transcription termination, facilitated message decay and translation inhibition (Simons and Kleckner, 1988; Simons, 1990). In the IS*10* case, antisense control directly blocks ribosome binding at the *tnp* translation initiation site. IS*10* has two promoters near its outside end (Fig. 1): pIN, which specifies the *tnp* mRNA (also termed RNA-IN), and the opposing and somewhat

NATO ASI Series, Vol. H 49
Post-Transcriptional Control of Gene Expression
Edited by J. E. G. McCarthy and M. F. Tuite

stronger pOUT promoter, which specifies the antisense RNA (also termed RNA-OUT). RNA-OUT has a simple stem-loop secondary structure (Kittle *et al.*, 1989; Case *et al.*, 1988). The secondary structure of RNA-IN is more complicated and plays an important role in *tnp* expression in at least two ways (below). The 5' ends of RNA-OUT and RNA-IN are complementary to one another across a region that includes the ribosome binding site for the *tnp* gene. Genetic studies suggested that RNA-OUT controls *tnp* expression in *trans* at a post-transcriptional level (Simons and Kleckner, 1983; Case *et al.*, 1990). More recently, we examined the consequences of RNA-OUT/RNA-IN pairing in detail, *in vivo* and *in vitro* (Case *et al*, 1990; Ma and Simons, 1990).

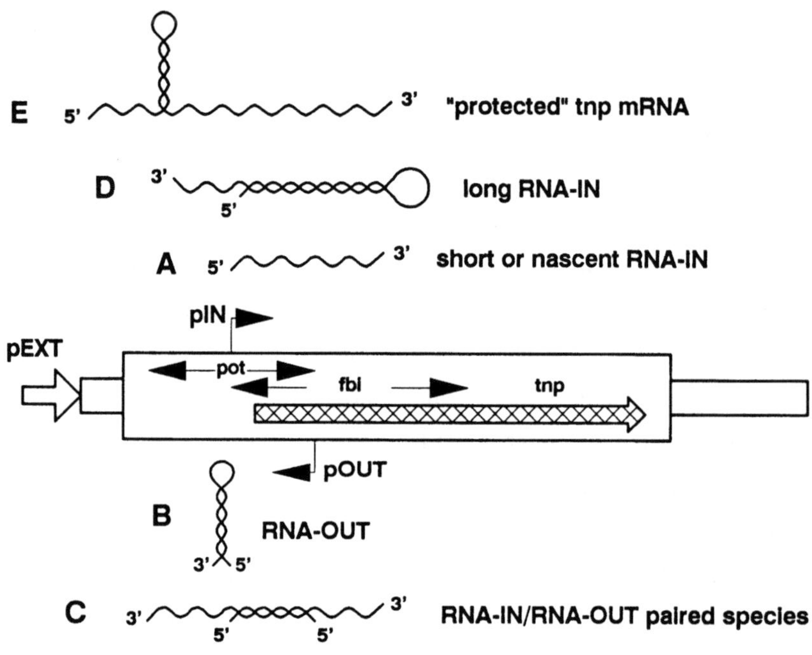

Figure 1. Genetic structure of IS*10*. The IS*10* transposase (*tnp*) gene (cross-hatched arrow), pIN and pOUT promoters (bent arrows) and two important symmetries (*pot* and *fbi*) are shown. pEXT depicts an external promoter that might transcribe the *tnp* gene. Five RNA species are also depicted (see text): **A**, short RNA-IN transcripts, which are relatively unstructured; **B**, the RNA-OUT antisense RNA, which has a stem-loop structure; **C**, the paired species that forms between RNA-OUT and short RNA-IN transcripts; **D**, full-length RNA-IN which folds into the *fbi* structure; **E**, hybrid *tnp* mRNAs (from pEXT) whose translation is prevented ("protected") by the *pot* structure.

The transposase mRNA is destabilized during antisense control. When both RNA-IN and RNA-OUT are expressed *in vivo* under conditions where antisense control is observed, the RNA-OUT/RNA-IN paired species is efficiently and specifically cleaved by ribonuclease III (RNaseIII), which leads to destabilization of both RNA-OUT and RNA-IN (Case *et al.*, 1990). Similar cleavage is seen *in vitro* with purified RNaseIII and RNA-OUT/RNA-IN paired species. RNaseIII is the principal double-strand RNA specific endoribonuclease in *Escherichia coli* (King *et al.*, 1986) and is required for antisense RNA control in two systems: inhibition of bacteriophage λ *c*II gene expression by the λ OOP RNA (Krinke and Wulff, 1987) and inhibition of plasmid R1 *repA* expression by the *copA* antisense RNA (Blomberg *et al.*, 1990). Nevertheless, RNaseIII cleavage is <u>not</u> required in the IS*10* case; in *E. coli* host cells deficient for RNaseIII, where cleavage and destabilization are not observed, antisense control remains essentially normal (Case *et al.*, 1990).

RNA-OUT/RNA-IN pairing blocks ribosome binding at the transposase translation initiation site. Larry Gold and his colleagues have developed a simple assay for the detection of the ternary complexes that form between a 30S ribosomal subunit, the initiator tRNA and an mRNA (Hartz *et al.*, 1988). In this procedure, termed "toe-printing", reverse transcriptase extension of a downstream primer pauses when the 3' margin (or "toe") of the 30S subunit is encountered. This gives rise to a characteristic extension product corresponding to a position 15 nucleotides downstream of the start codon (AUG+15). In the toe-print assay, ribosome binding at the *tnp* translation initiation site generates an AUG+15 (Fig. 2), which was authenticated by mutations that alter *tnp* translation *in vivo*: the *97G* mutation, which improves the Shine/Dalgarno sequence, increases *tnp* translation *in vivo* and the toe-print signal *in vitro* ≈5fold; the *CJ109* mutation, which changes the AUG initiation codon to ACG, decreases *tnp* translation and ribosome binding >10 fold. RNA-OUT/RNA-IN binding is also detected by primer extension. Fortunately, these signals, termed the "OUT-print", are distinct from the toe-print; they correspond to a cluster of sites near the 5' end of bound RNA-OUT (Fig. 2).

With these two assays we have been able to show that the OUT-print signal increases at the expense of the toe-print signal as the level of RNA-OUT/RNA-IN pairing increases, showing that 30S ternary complexes are unable to form at the *tnp* translation initiation site when RNA-IN is bound by RNA-OUT. Pairing also inhibits 70S complex (mRNA+30S+50S+tRNA+IFs) formation. Finally, mutations that prevent pairing or alter its specificity have corresponding effects in this *in vitro* assay. We propose inhibition of ribosome binding also occurs *in vivo*, and that it is sufficient to account for IS*10* antisense control.

```
                            97G          CJ109              AUG
      RNA-IN                 G             C                +15
                             |             |                 |
         5'-GCGAAAAAUCAAUAAUCAGACAACAAGAUGUGCGAACUCGAUAUUUUACACGAC--
                          (AGGAGG)              |ΔΔ          ΔΔΔ| toe-print
                                          |Δ  ΔΔΔΔ ΔΔΔ | OUT-Print
--UAGUUUUACCAAAGCGCUUUUUAGUUAUUAGUCUGUUGUUCUACACGCUU-5'

                                                       RNA-OUT
```

Figure 2. The toe-print and OUT-print signals on RNA-IN. The 5' terminal ≈55nt of RNA-IN and RNA-OUT are shown. The *tnp* Shine/Dalgarno sequence (consensus in brackets) and initiation codon are underlined. The primary toe-print and OUT-print signals are indicated with Δs. The 97G and CJ109 point mutations increase and decrease, respectively, *tnp* translation and the toe-print signal.

RNA-IN SECONDARY STRUCTURE AFFECTS ANTISENSE RNA CONTROL

When RNA-IN transcripts are longer than about the first 315 nt, they assume a secondary structure that probably plays a significant role in two aspects of IS*10* gene expression. This effect first became apparent during *in vitro* analysis of RNA-IN/RNA-OUT pairing, where it was observed that short RNA-IN transcripts (<315 nt) pair efficiently to RNA-OUT whereas longer species (>315 nt) did not (Kittle, 1988; J.E.G. and R.W.S., unpublished; Table I). Inspection of these long transcripts revealed a sequence, termed *fbi* (*f*old *b*ack *i*nteraction), which is complementary to the 5' end of RNA-IN. Computer-assisted analysis (Zucker *et al.*, 1981) predicted that long RNA-IN transcripts would form a secondary structure in which the *fbi* sequence is paired to the 5' end (Fig. 3). Such sequestration of the RNA-IN 5' end would make it unavailable for pairing by RNA-OUT; other genetic and physical studies show that the 5' end of RNA-IN must be free in order for RNA-OUT/RNA-IN pairing to occur (Kittle *et al.*, 1989).

The *fbi* structure also prevents RNA-OUT/RNA-IN pairing *in vivo*. We examined the possibility that *fbi* prevents RNA-OUT pairing *in vivo* in the following way. Expression of a multicopy *tnp'-lacZ'* protein fusion is normally under antisense control (Case *et al.*, 1990). If the level of RNA-IN present in the cell is increased in *trans* (from a second, compatible plasmid), without increasing the level of RNA-OUT, antisense control of the fusion is "titrated" and fusion expression increases ≈10fold (Simons and Kleckner, 1983). However, such titration occurs only if the RNA-IN expressed in *trans* is less than 315 nt in length (Δ*fbi*); longer transcripts (*fbi*+) do not titrate efficiently (Table I). Furthermore, an 8 bp insertion in the *fbi* sequence (*fbi-ins*1), which is predicted to disrupt secondary structure, enables long transcripts to titrate efficiently *in vivo* and to pair efficiently *in vitro* (Table I). Finally, this assay facilitated the isolation of point mutations that increase titration by long RNA-IN transcripts. Two such mutations, *fbi*1 and *fbi*2 (Fig. 3) map to the 5' end of RNA-IN and are predicted to disrupt secondary structure such that the 5' end is no longer sequestered. Importantly, both *fbi*1 and *fbi*2 increase pairing between RNA-OUT and long RNA-IN transcripts *in vitro* (Table I). Thus, titration *in vivo* correlates very well with pairing *in vitro*.

Table I. The effect of *fbi* mutations on RNA-OUT/RNA-IN pairing *in vitro* and *in vivo*.

RNA-IN genotype[1]	Relative rate of RNA-OUT/RNA-IN pairing *in vitro*[2]	Relative level of titration *in vivo*[3]
Δ*fbi*	1.0	10.0
fbi+	0.03	1.0
*fbi-ins*1	1.0	11.0
*fbi*1	0.17	3.5
*fbi*2	0.14	3.0

[1] Δ*fbi*, the first 259 nt of RNA-IN; *fbi*+, the first 529 nt; *fbi-ins*1, the first 529 nt with an 8 nt insertion at 314/315; *fbi*1 and *fbi*2, the first 529 nt with the indicated point mutations (Fig. 3).

[2] Apparent second order rate constants for pairing between RNA-IN (with indicated genotype) and RNA-OUT were determined as described (Ma and Simons, 1990). Values are relative rates, where 1 = 2.7×10^5 mol^{-1} sec^{-1}.

[3] Titration by multicopy plasmids capable of expressing the indicated RNA-IN species (see text). Values are relative levels of fusion expression where 1 = 1.7 units of β-galactosidase activity.

The secondary structure of *fbi*+ and *fbi*- transcripts. We have partially determined the secondary structures of *fbi*+ and *fbi-ins*1 RNA-IN transcripts *in vitro* by dimethyl sulfate (DMS) modification and cobra venom nuclease studies (together with free energy considerations and genetic data). The secondary structure of the *fbi*+ transcript (Fig. 3)

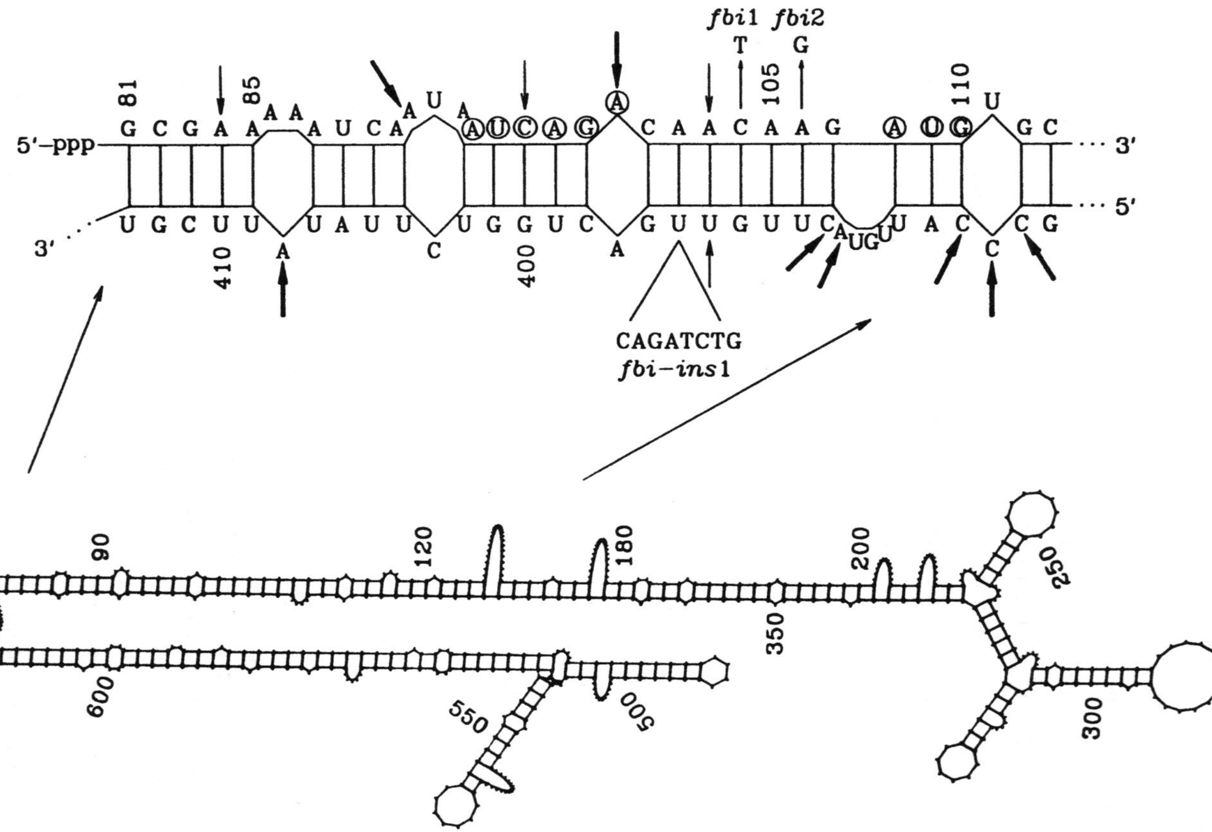

Figure 3. The *fbi*$^+$ secondary structure. The lower portion shows the computer predicted *fbi*$^+$ secondary structure, partially confirmed by DMS modification and cobra venom nuclease studies. The upper portion shows detailed structure between the 5' end of RNA-IN (top strand) and the *fbi* sequence (bottom strand). The dark and light arrows indicate strong and weak DMS modifications. The *fbi*1, *fbi*2 and *fbi-ins*1 mutations are also indicated. Nucleotides corresponding to the *tnp* Shine/Dalgarno and start codon are circled. Numbering is according to IS*10* coordinates.

shows that the *fbi* sequence (bp 380-413 bp of IS*10* or nt 300-333 of RNA-IN) is stably paired to the 5' end of RNA-IN. The *tnp* Shine/Dalgarno and AUG are also sequestered within this structure. In contrast, the 5' end of the *fbi-ins*1 transcript has a predominantly single-stranded RNA character (not shown). Thus, the actual secondary structure of RNA-IN affects RNA-OUT/RNA-IN pairing such that long transcripts are insensitive to antisense control. We assume that *fbi*1 and *fbi*2 change secondary structure in similar ways.

The *fbi* structure reduces translation of RNA-IN transcripts. In addition to its effect on antisense RNA control, the *fbi* site appears to decrease *tnp* translation. In the absence of antisense control (abolished by point mutation), expression of a *tnp'-lacZ'* protein fusion containing an intact *fbi* site is ≈4 fold lower than that of an isogenic *fbi-ins*2 protein fusion (*fbi-ins*2 = *fbi-ins*1 + 4 nt). In contrast, no large effect is seen with the corresponding transcriptional (operon) fusions. Furthermore, analysis of fbi^+ and fbi^- RNA-IN transcripts expressed *in vivo* reveals no differences in message degradation that might explain decreased translation. Together, these results suggest that the *fbi* site directly inhibits translation of the *tnp* gene by sequestering the *tnp* ribosome binding site. Thus, the *fbi* interaction would decrease the functional half-life of the RNA-IN message by restricting efficient translation initiation to the few seconds it takes to synthesize the downstream *fbi* sequence. It follows that antisense RNA control need act only on nascent RNA-IN transcripts. These models are currently being examined by *in vitro* ribosome binding studies and by the introduction of point mutations that increase or decrease intrinsic *tnp* translation rates.

It is worth noting that restriction of RNA-IN translation to nascent transcripts might increase transposase concentration in the vicinity of the element from which it is synthesized. Such an effect might partially explain the preferential *cis*-action of the IS*10* transposase (Morisato *et al.*, 1983).

PROTECTION FROM EXTERNAL TRANSCRIPTION

When an IS*10* element inserts downstream of an active host promoter such that transcription might proceed across the *tnp* gene, activation of transposase expression (and transposition) is expected. However, such is not the case, even with strong, externally situated promoters (Davis *et al.*, 1985). This phenomenon results from a combination of three different effects: (i) a decrease in the activity of the transposase binding site at the IS*10* terminus, (ii) 4-fold termination of external transcription somewhere within the outer 70 bp of IS*10*, and (iii) inefficient translation of any externally generated transcripts that do proceed across the *tnp* gene. Here, we are concerned with the latter of these three effects.

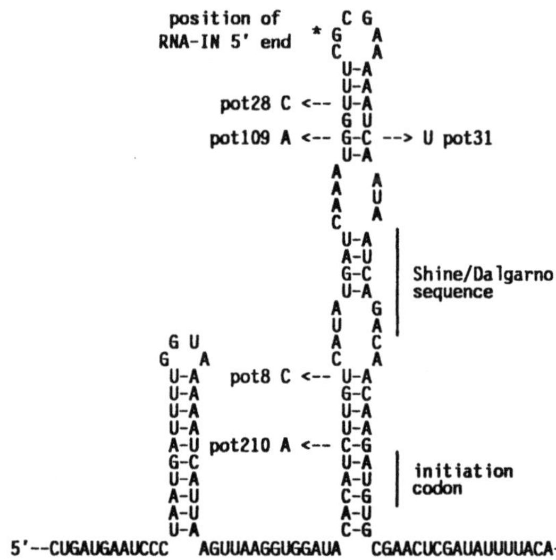

Figure 4. The *pot* secondary structure. Shown is the predicted secondary structure of *tnp* mRNAs generated from an externally positioned promoter (preliminary DMS modification studies support this interpretation). The position of the 5' terminus of RNA-IN (specified by pIN) is indicated (*) for comparison. Note that the Shine/Dalgarno sequence and initiation codon of the *tnp* gene are sequestered. The nucleotide changes in several *pot*⁻ mutants are also shown.

Secondary structure in externally generated *tnp* mRNAs blocks ribosome binding

Computer-assisted analysis suggests that externally generated transcripts will assume a secondary structure in which the *tnp* translation initiation site is sequestered, thereby directly blocking ribosome binding (Fig. 4). On the other hand, RNA-IN transcripts (from pIN) do not assume this structure. Indeed, nascent RNA-IN transcripts are predicted to have their ribosome binding sites free of secondary structure (above). When we examined ribosome binding we found that externally generated transcripts fail to bind 30S ribosomal subunits in the toe-print assay, whereas native RNA-IN transcripts toe-print normally (C.M. and R.W.S., unpublished). We term this inhibitory secondary structure *pot* (*p*rotection from *o*utside *t*ranscription).

Table II. The effect of *pot* mutations on translation of externally generated *tnp* mRNAs.

IS*10* genotype	Genetic fusion[1]	Relative level of fusion expression[2]
wild type	p*lac*UV5-*tnp'*-*lacZ'* (protein)	1
*pot*8	"	42
*pot*31	"	20
*pot*109	"	21
*pot*31/*pot*109	"	17
wild type	p*lac*UV5-*tnp'*-*lacZ*$^+$ (operon)	1
*pot*8	"	1.4
*pot*31	"	1.2
*pot*109	"	1.2
*pot*31/*pot*109	"	2

[1] In the protein fusions, the first 24 *tnp* codons are fused in frame to the *lacZ* coding sequence. In the operon fusion, the same IS*10* sequences are fused to an intact *lacZ* gene that retains its normal translation initiation signals. In both fusion types, transcription is specified by the an externally positioned *lac*UV5 promoter (as for pEXT in Fig. 1). Thus, for each isogenic case, the ratio of protein and operon fusion expression reflects differences in *tnp* translation.

[2] Values are relative levels of fusion expression where 1 = 7.8 and 6160 units of β-galactosidase activity from, respectively, the wild-type protein and operon fusions.

Mutations in the *pot* structure reveal that its rate of formation is important. Point mutations that decrease the *pot* effect are predicted to destabilize the *pot* structure (Fig. 4). They increase expression of *tnp'-lacZ'* protein fusions transcribed from external promoters but have little effect on isogenic operon fusions (Table II), strongly suggesting that they increase *tnp* translation. One interpretation is that *pot* mutations destabilize *pot* structure, making the *tnp* translation initiation site more accessible for ribosome binding. However, two other observations argue against this simple view. First, when the complementary *pot*109 and *pot*31 mutations (Fig. 4) are combined inhibition is not restored (Table II) even though the *pot* structure is probably restored. Second, *pot* mutations do not improve ribosome toe-printing on externally generated transcripts (C.M. and R.W.S., unpublished). These results suggest that *pot* mutations either slow the rate of inhibitory structure formation *in vivo*, with little effect on the thermodynamic stability of the structure once it does form, or they perturb the binding of some important function (protein?) present *in vivo* but absent in the toe-print assay. These possibilities are under investigation.

ACKNOWLEDGEMENTS

We thank our colleagues for helpful discussions and E. Simons for expert technical assistance. Authors were supported by grants or fellowships from NIH (C.M.), UCLA (J.E.G.) or the American Cancer Society (C.C.C., J.R., R.W.S.). This research was supported by grants from NIH and the American Cancer Society.

REFERENCES

Blomberg P, Wagner EGH and Nordstrom K (1990) Control of replication of plasmid R1: The duplex between the antisense RNA, CopA, and its target, CopT, is processed specifically *in vivo* and *in vitro* by RNase III. EMBO J (In press)

Case CC, Roels SM, Gonzalez JE, Simons EL and Simons RW (1988) Analysis of the promoters and transcripts involved in IS*10* antisense RNA control. Gene 72:219-236

Case CC, Simons EL, and Simons RW (1990) The IS*10* transposase mRNA is destabilized during antisense RNA control. EMBO J 9:1259-1266

Davis MA, Simons RW, and Kleckner N (1985) IS*10* protects itself at two levels from fortuitous activation by external promoters. Cell 43:379-387

Hartz D, McPheeters DS, Traut R and Gold L (1988) Extension inhibition analysis of translation initiation complexes. Meth Enzymol 164:419-425

King TC, Sirdeskmukh R and Schlessinger D (1986) Nucleolytic processing of ribonucleic acid transcripts in procaryotes. Microbiol Rev 50:428-451

Kittle JD (1988) PhD thesis. Harvard University, Cambridge, MA

Kittle JD, Simons RW, and Kleckner N (1989) IS*10* antisense pairing initiates by an interaction between the 5' end of the target RNA and a loop in the antisense RNA. J Mol Biol 210:561-572

Krinke L and Wulff DL (1987) OOP RNA, produced from multicopy plasmids, inhibits λ *c*II gene expression through an RNase III-dependent mechanism. Genes Dev 1:1005-1013

Ma C and Simons RW (1990) The IS*10* antisense RNA blocks ribosome binding at the transposase translation initiation site. EMBO J 9:1267-1274

Morisato P, Way JC, Kim HJ and Kleckner N (1983) Tn*10* transposase acts preferentially on nearby transposon ends *in vivo*. Cell 32:799-807

Roberts D, Hoopes BC, McClure WR and Kleckner N (1985) IS*10* transposition is regulated by DNA adenine methylation. Cell 43:117-130

Simons RW (1990) Natural antisense RNA control in bacteria, phage and plasmids. In: Applications of antisense nucleic acids. van der Krol AR and Mol JNM (eds) Marcel Dekker, New York (In press)

Simons RW, Hoopes BC, McClure WR and Kleckner N (1983) Three promoters near the termini of IS*10*: pIN, pOUT and pIII. Cell 34:673-682

Simons RW and Kleckner N (1983) Translational control of IS*10* transposition. Cell 34:683-691

Simons RW and Kleckner N (1988) Biological regulation by antisense RNA in prokaryotes. Annu Rev Genet 22:567-600

Zucker M and Stiegler P (1981) Optimal computer folding of large RNA sequences using thermodynamics and auxiliary information. Nucl Acids Res 9:133-148

The Antisense Approach and Early *Xenopus* Development

A Colman, C Baker and J Shuttleworth
School of Biochemistry
University of Birmingham
Birmigham B15 2TT
United Kingdom

Abstract

Antisense RNAs or oligodeoxynucleotides (oligos) have been used to suppress expression from specific genes during the development of various eukaryotic organisms, from the slime mold up to the mouse. We have been using oligos to ablate selected mRNAs in *Xenopus* oocytes. Using the endogenous histone H4 mRNA as a target we find that the degree of ablation obtained depends on the choice of oligo, and that the relative effectiveness of the different oligos can be mimicked in an in vitro reaction. Substitution of phosphorothioate for phosphodiester linkages in one oligo greatly increases its efficacy at low concentrations. Finally we describe experiments where the injection into oocytes of an oligo complementary to a new cdc2-related kinase mRNA whose cDNA we have recently cloned, accelerates the rate of progesterone-induced maturation.

Introduction

Use of both genetics and experimental manipulation of embryos is essential for a full understanding of the embryology of any metazoan. However the relative importance of these two approaches varies according to features of each organism and the way in which it develops. For example, the small genome size and short life cycle of invertebrates such as *Drosophila* and *Caenorhabditis,* makes them more amenable to a genetic approach than a slowly growing organism with a large genome, such as the frog *Xenopus*. Conversely, the large size of *Xenopus* embryos makes them ideal for surgical techniques. Such techniques have demonstrated that interactions between the cells of a developing vertebrate are important in establishing the basic body plan. However only in a few instances (eg retinoic acid in bird limb formation, Eichele, 1989) have the inducers (the molecules which convey the instructions) been identified. The unequivocal identification of inducers is one of many areas in vertebrate embryology where the absence of a conventional genetic approach is a handicap. This has led to much interest in alternative,"reverse" genetic strategies. For example,the overexpression of a specific gene can be effected by its introduction, either as DNA (Geiblehaus et al.1988) or RNA (McMahon and Moon, 1989), into the embryo. The resultant, altered phenotype may

NATO ASI Series, Vol. H 49
Post-Transcriptional Control of Gene Expression
Edited by J. E. G. McCarthy and M. F. Tuite

shed some light on the normal role of the encoded protein. More directly, the advent of stem cell culture together with sophisticated vectors and selection techniques, have made the disruption of specific genes possible in mammals (Rossant and Joyner, 1989). A third strategy that has been successfully employed is the antisense approach. Essentially this entails the introduction into the nucleus or cytoplasm of a cell, of nucleotide sequences which are complementary to the transcripts of selected target genes. The antisense sequences can be RNA, produced in embryos from introduced DNA templates or directly injected; in these cases it is believed that the hybridisation of the antisense strand interferes sterically with the processing of transcripts or their translation. Alternatively, oligodeoxynucleotides can be injected into embryonic cells where they are believed to mediate a RNAse H-like cleavage of the hybridised RNA strand (Dash et al. 1987; Shuttleworth and Colman, 1988).

The antisense approach is technically simpler than gene disruption and offers the advantage over overexpression that a null mutant phenotype may in principle be obtained. The first demonstration of its potential in early embryos was made by Melton (1985) using the injection of *Xenopus* oocytes with sense and antisense RNAs. This was soon followed by the production of <u>Kruppel</u> (Rozenberg et al., 1985) and <u>Wingless</u> (Cabrerra et al., 1987) phenocopies in *Drosophila* embryos through the injection of the respective antisense RNAs. One might expect *Xenopus* to be a particularly favourable testing ground for this approach in vertebrates, since amongst the stockpile of mRNAs synthesised from the maternal genome, are species of potential developmental importance which are never expressed in somatic cells. Surprisingly, the use of antisense RNA has proved disappointing in *Xenopus*. It was initially thought that the presence of a RNA duplex unwinding activity in the cytoplasm of unfertilised eggs and embryos was responsible for the lack of success (Bass and Weintraub, 1987; Rebagliati and Melton, 1987). However this activity corrupts the translational reading frame of hybridised mRNA coding regions through the covalent modification of adenine to inosine (Bass and Weintraub, 1988). Since this conversion would effectively render a mRNA untranslatable, one might expect the strategy to work. The problem seems to be that the injected antisense RNA fails to hybridise with its endogenous target mRNA.

An alternative approach, using antisense DNA, was first pioneered in *Xenopus* by Kawasaki (1985) who showed that translation from injected RNAs could be inhibited by injection of short complementary oligodeoxynucleotides (oligos) . Subsequently several other groups have shown that endogenous mRNAs are also accessible to injected oligos (Dash et al. 1987; Shuttleworth and Colman, 1988). It was found that considerable differences existed between the abilities of oligos of different sequence to mediate the ablation of the same target RNA. A particular problem arose when attempts were made to grow oligo-treated embryos since the doses of oligo required to remove all detectable mRNA were found to compromise subsequent development. In this paper we describe our attempts to remedy the above situations. Specifically, we wished to see if the search for the ideal oligo could be simplified by referral to models of RNA secondary structure or, failing this,by the development of rapid, in vitro test systems. A second objective was to obtain a modified oligo which could cause efficient cleavage at a much lower ratio of oligo to target RNA. We report some progress towards both objectives: an in vitro ablation protocol has been devised which should allow the relative in vivo efficacies of different oligos to be predicted; and we also find that oligos containing totally substituted phosphorothioate backbones are effective at lower concentrations probably because of increased stability in vivo. Finally we show that in some circumstances even partial loss of an RNA can have striking biological consequences. We describe a modulation of the meiotic cell cycle in oocytes brought about the injection of an oligo directed against a cdc 2-related kinase which we have recently cloned from a *Xenopus* cDNA library.

Methods

General

Isolation of Xenopus oocytes, microinjection procedures, oligo synthesis, RNA synthesis, extraction, and Northern analysis have been described elsewhere (Shuttleworth and Colman,1988).

In vitro cleavage reactions

Buffers (x1);**A1**: 100mM KCl, 20mM Tris.HCl pH 7.4; **A2**: 100mM KCl, 20mM Tris.HCl pH 7.4, 3mM $MgCl_2$, 2mM DTT, 0.1mg/ml BSA; **B**: 100mM KCl, 20mM Tris.HCl pH 7.4, 1.5mM $MgCl_2$,1mM DTT, 0.05 mg/ml BSA.

For "annealing" reactions 5µg of *Xenopus* total RNA (extracted as above) was dissolved in water. 2µg of oligo and then 2x A1 buffer were added to produce a final volume of 10µl in 1x A1 buffer. This was incubated at 60°C for 20 minutes, then briefly centrifuged to collect condensation from the sides of the tube. 10µl 1x A2 buffer was added and digestion started by addition of *E.coli* RNAseH (BRL) to 10u/ml. The enzyme was stored in 50% v/v glycerol, 1x buffer B at 20x working strength. Digestion was at 21°C for 1 hour. The reaction was then stopped by phenol extraction and the RNA recovered by ethanol precipitation.

For non-annealing reactions, 5µg RNA and 2µg oligo were mixed in a final volume of 20µl Buffer B. RNAseH was usually added to this immediately and digestion was as for annealed reactions. We have shown that this simplified protocol produces the same results as following the annealing protocol with the 20 minute incubation at 21°C instead of 60°C.

Oligos

Sequences of oligos are: H4-1,5'ATGCGCTTGACTCCCCCTCT3'; H4-2, 5'CACCACATCCATGGCGGTAA3'; H4-3, 5'CGTAGAGAGTGCGGCCCTGG3'; 15-1, 5'TTCTTGACTGGCTAAAGTC3'. The 25nt and 30nt oligos all had the same 5' end as the oligo H4-1and extended 20nt, 25nt or 30nt through the below sequence; 5'ATGCGCTTGACTCCCCCTCTCCGGGCCAGG3'.

Results and Discussion

1. Endogenous histone H4 mRNA is cleaved in oocytes by an antisense oligo

RNA molecules both in vivo and in vitro contain regions of secondary structure and various computer algorithms have been designed in order to predict these regions from the primary sequence. Within the cell most RNA molecules associate with proteins to form ribonucleoprotein complexes (RNPs). We might anticipate that both secondary structure and protein binding would reduce the effectiveness of some of the oligos directed against a specific target; indeed recent studies using oligo hybridisation to small nuclear RNPs have confirmed that only small regions of the RNA component of the complex are accessible to the oligos (Cotten et al., 1989). In addition, the effectiveness of a particular oligo will be a function of its size (which influences hybrid stability) and nucleotide composition (which affects hybrid stability, resistance to cell nucleases, RNAse H activity, and oligo secondary structure).

In order to address some of the above factors we have initiated a series of experiments designed to test the accessibility in vivo of the oocyte histone H4 mRNA to oligos of different size and sequence (Shuttleworth and Colman, 1988; Shuttleworth et al., 1988). We chose this RNA because it is relatively abundant (>100pg/oocyte) and also shows the interesting property of translational control. Core histone synthesis (ie. including H4) increases 20-fold as the oocyte matures into the unfertilised egg while general protein synthesis increases 2-fold. Using 25ng/oocyte of a 20-mer oligo, H4-1 (this amount represents a 1500x molar excess of oligo over mRNA), we obtained cleavage of over 95% of the mRNA within 60 min. Neither continued incubation nor additional oligo injections resulted in further mRNA cleavage and we speculate that the residual ~5% represents a H4 mRNA pool which is totally inaccessible to H4-1 oligo possibly due to protein binding (see below).

The ablation had all the hallmarks of a RNAse H-mediated reaction with the production of 5' and 3' fragments of predicted size, the 5' fragment being the more stable. We next tested a nested set of oligos based on the H4-1 sequence,where the sizes of the oligos increased by increments of two bases, from a 6-mer to a 20-mer. Cleavage in vivo was first detected with a 10-mer and reached a maximum with a 12-mer.

Irrespective of the size of oligo, maximal H4 mRNA cleavage was only seen after the injection of ~1500 fold molar excess of oligo over target mRNA. Similar experiments with oligos against the localised maternal mRNA, Vg1, showed that ratios in excess of 30,000 (100ng oligo/cell) were required for maximal cleavage (in this case, of all detectable Vg1 mRNA). This is partly due due to the sensitivity of phosphodiester oligos to the endogenous nucleases in oocytes; half lifes of the order of 10min have been reported. We found that these high levels of oligo often disrupted early development and therefore compromised our experiments. Although we noted artefactual effects with several oligos including control (sense) sequences, Kloc et al (1989) have succeeded in raising embryos from oocytes which had been totally depleted of mRNA after the injection of 20 ng of complementary oligo. Since we have occasionally obtained normal development after such doses we attribute this variability in oligo toxicity to the batch of oocytes used. Clearly it would be desirable to circumvent potential variables of this nature, and for this reason we have repeated some of the above experiments using modified oligos.

2 Phophorothioate-substituted oligos are more effective than phosphodiester oligos

Of the several chemically modified oligo structures that have been used in oligo-mediated RNA cleavage experiments in the last few years, phosphorothioate oligos (S-oligos) seem to be the most promising alternative to the phophodiester forms (Marcus -Sekura et al., 1987)), offering the same advantage of specificity and the formation of RNAse H-sensitive hybrids, without the disadvantage of nuclease sensitivity. In some pilot experiments, using a three hour incubation period, we found that a fully substituted, phosphorothioate H4-1 oligo (H4-1S) was less effective than its phosphodiester counterpart in cleaving H4 mRNA. However the presence of both 5' and 3' fragments at the end of the incubation, in oocytes injected with H4-1S, indicated that a slower reaction was probably occuring under these circumstances. The prolonged activity of H4-1S is confirmed by the experiment shown in Fig 1. In this experiment the comparison between H4-1 and H4-1S was performed at sub-maximal oligo doses (5ng/oocyte) and after 3 and 24 hours of incubation, so that any advantage from prolonged cleavage activity could be detected. It is clear that the net cleavage at 24h was considerably greater using H4-1S. In addition, when 25-mer and 30-mer S-oligos which included the H4-1S sequence were used , even greater cleavage was seen. We conclude that the phosphorothioate linkage is beneficial to oligo action. However whilst these amounts of oligo (5ng/oocyte) had no effect on general protein synthesis, 50ng doses were quite toxic to the cells. Since S-oligos form less stable hybrids with RNA than unmodified oligos (Stein et al.,1988), we attribute the effect of increasing size on the S-oligo action to the greater hybrid stability obtained. We do not yet know whether these S-oligos will be toxic to developing embryos.

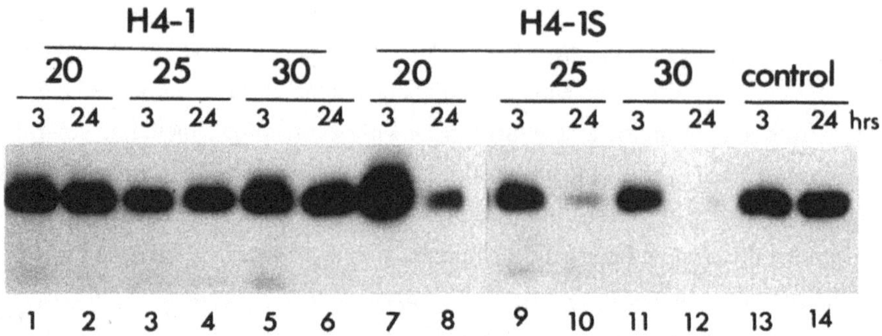

Fig 1.Phophorothioate- versus phosphodiester- oligo mediated cleavage. Oocytes were injected with 5ng of phosphodiester- (H4-1) or phosphorothioate-(H4-1S) containing 20-,25- or 30-mer oligos.After 3 or 24h incubation, RNA was extracted, and analysed by Northern blot using an antisense H4 ^{32}P RNA probe.

3. Relative in vivo efficacies of different oligos are preserved in vitro

We have previously noted (Shuttleworth et al, 1988) that two different anti-H4 oligos cleave mRNA with different efficiencies. This observation is confirmed in Fig 2a, and extended by the inclusion of a third oligo, H4-3. Interestingly this pattern of cleavage efficiency, with H4-1>H4-3>H4-2, is preserved in a completely **in vitro** analysis (Fig 2b). Prior annealing in vitro results in all H4 mRNA being cleaved by each oligo (Fig 2b). Since calculations (according to the equation of Meinkoth and Wahl, 1984) on the hypothetical stability of these oligos indicates that the oligos should form hybrids with dissociation temperatures (Td) of 64oC, H4-1; 62oC, H4-2; and 68oC,H4-3. Ranking these oligos in order of increasing Td would not lead us to predict their performance correctly. We believe that the feature limiting the effectiveness of oligos in vitro (and probably in vivo), is the secondary structure of the target RNA. Unfortunately, as we have stated before, computer predictions from the known H4 mRNA sequence give an unreliable guide to the behavior of our oligos. If we assume that single-stranded regions are more accessible, then treatment of the H4 sequence using the algorithms of Zuker and Steigler (1981) would give a ranking of H4-3>H4-1>H4-2. Furthermore using the same algorithms on a synthetic H4 RNA only 88% homologous (at the nucleotide level) to native H4 mRNA, a strikingly different structure was predicted (data not shown) Yet when this synthetic RNA was challenged with H4-1 and H4-2, both in vivo and in vitro, H4-1 was much better in both situations (Fig 2c).

The observation that the deproteinised RNA used in the in vitro assay and the same RNAs in vivo show a similar sensitivity to cleavage by different oligos, indicates that the target regions in vivo are naked or that a protein binding does not prevent oligo access. In either case the results suggest that the in vitro assay may provide a simple means of testing and discriminating between the abilities of different oligos to mediate cleavage in vivo.

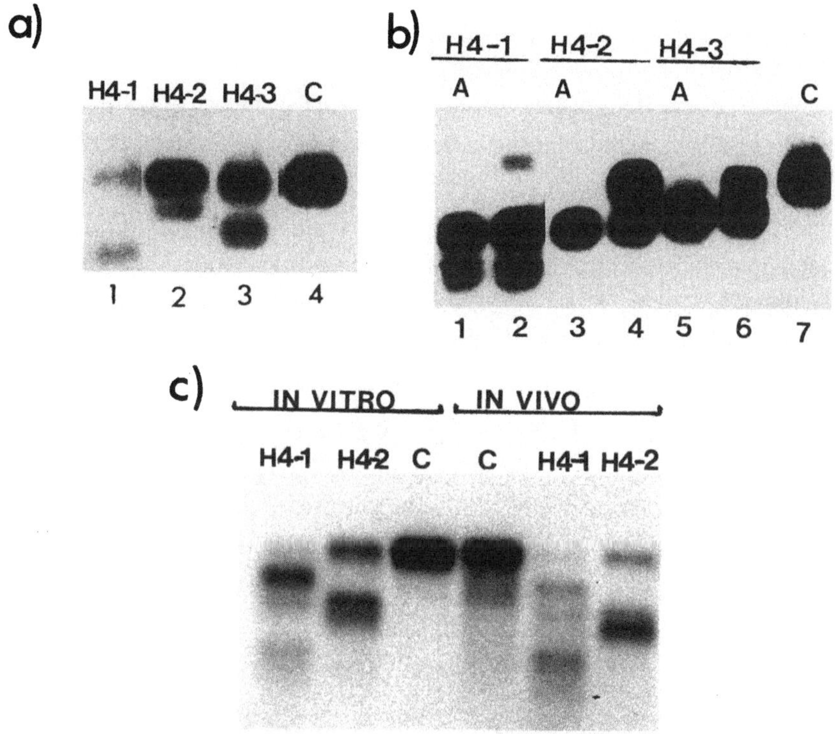

Fig 2 Cleavage by different oligos in vivo and in vitro
a) oocytes were injected with 50ng of oligos H4-1, H4-2 or H4-3 and cultured for 3h before analysis as in Fig 1. **b)** Total oocyte RNA was digested in vitro (see methods) with oligos H4-1, H4-2 or H4-3 before or after annealing (A) and analysed as above. **c)** 150pg ^{32}P synthetic H4 RNA was injected into oocytes or used to supplement the IN VITRO reaction as in b) without annealing. The oocytes were injected 3h later with 50 ng oligos H4-1 or H4-2 (IN VIVO). After a further 3h total RNA was prepared from the oocytes, and the RNA samples from both treatments run on denaturing gels which were then autoradigraphed. Abbrevation: C, control.

4. Oligos and the oocyte cell cycle

Stage VI oocytes are arrested in prophase of meiosis 1. After hormonal stimulation, meiosis resumes and proceeds until metaphase of meiosis 2 is reached whereupon a further arrest occurs. This process is called maturation. A major regulator of the process is maturation promoting factor (MPF). Recently MPF has been shown to contain a heterdimer consisting of cyclin B and $p34^{cdc2}$, the *Xenopus* homologue of the yeast cell cycle control protein cdc2 (see Gautier et al., 1990). $p34^{cdc2}$ is a serine/threonine kinase and in its inactive state contains phophorylated serine, threonine and tyrosine residues. Activation is associated with dephosphorylation of tyrosine. Although many details of maturation are poorly understood, it is clear that a cascade of phosphorylation and dephosphorylation reactions occur. Oligo-mediated ablation studies have recently provided some insight into some of the steps involved. Sagata et al (1988) found that oligo-mediated depletion of the mRNA encoding the $p39^{c\text{-}mos}$ proto-oncogene product resulted in inhibition of maturation. Sagata et al. (1989) also found found that injection of this RNA caused maturation even in the absence of hormonal stimulation. These results were intepreted as demonstrating the involvement of $p39^{c\text{-}mos}$ protein in the activation of MPF.

Another protein implicated in maturation is the translation product of oocyte D7 mRNA. Smith et al (1988) showed that oligo-depletion of this RNA delayed hormonally-induced maturation. Recently we have cloned the cDNA, $p40^{MO15}$, corresponding to an oocyte mRNA encoding a putative cdc2-related protein kinase (Shuttleworth et al., in preparation). When oocytes were injected with a complementary oligo (oligo15-1), about 33% of the MO15 mRNA was cleaved (Fig 3,inset). A control oligo (H4-1) had no effect. When the kinetics of maturation were monitored using the assay of germinal vesicle (oocyte nucleus) breakdown (GVBD), it was found that MO15 RNA depletion caused maturation to occur earlier (Fig 3). We conclude that $p40^{MO15}$ is involved in negatively regulating meiosis in *Xenopus* oocytes.

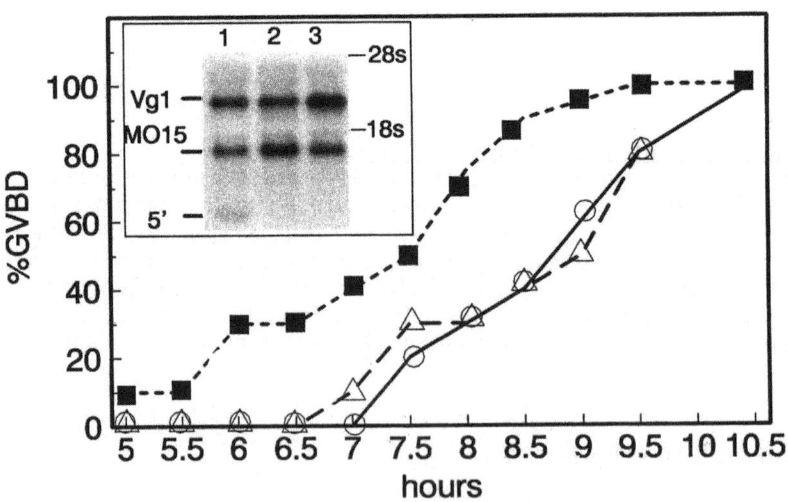

Fig 3 Effect of MO15 RNA depletion on oocyte maturation
Oocytes were injected with 50ng 15-1 oligo (-■- and track 1,inset), H4-1 oligo (-△- and track 2,inset) or had no injection (-○- and track 3, inset) and cultured overnight. RNA was then

extracted from oocytes in each sample, and analysed for the presence of Vg1 and M015 RNA by Northern blot using the appropriate ^{32}P antisense probes (inset); the oligo-generated M015 5'fragment is shown as are the positions in the gel of 18 and 28S ribosomal RNAs. 10µg/ml progesterone were added to the culture media surrounding the remaining oocytes at 0h, and maturation assessed by the appearance of a white spot on the oocyte surface,which was due to germinal vesicle breakdown (GVDB).

5 Concluding remarks

Antisense RNAs or oligonucleotides can be usefully employed to understand developmental phenomena. The use of either reagent still has a degree of empiricism and with each of them it is difficult to remove or functionally inactivate all of the target mRNA. However a complete loss of function may not always be necessary to achieve an effect, and a number of genes [eg *Drosophila* notch (Wright, 1970)] show phenotypic effects due to only partial loss. Thus the inducible synthesis of an antisense RNA against the ribosomal protein rpA1 gene in *Drosophila* severely disrupted oogenesis despite the fact that no hybrid could be detected (Qian et al, 1988). It was already known that modest reductions in the production of other ribosomal proteins was highly deleterious to oogenesis. In addition, chemical modifications may improve the efficacy of oligos in general, and the emergence of cleavage-competent, catalytic antisense RNAs (ribozymes, Haseloff and Guerlach,1988) may herald a new era in antisense applications.

References

Bass B and Weintraub H (1987) A developmentally regulated activity that unwinds RNA duplexes. Cell 48 607-613.

Bass B and Weintraub H (1988) An unwinding activity that covalently modifies its double-stranded RNA substrate. Cell 55 1089-1098.

Cabrera C V, Alonso M C, Johnston P, Phillips R G, Lawrence, P A (1987) Phenocopies induced with antisense RNA identify the wingless gene. Cell 50. 659-663

Cazanave, C., Stein, C.A., Loreau, N., Thuong, N.T., Neckers, L.M.,

Cotten, M Schaffner G and Birnstiel M (1989) Ribozyme, antisense DNA inhibition of U7 small nuclear ribonucleoprotein-mediated histone pre-mRNA processing in vitro. Mol Cell Biol 9 4479-4487

Dash P, Lotan, Knapp, M. Kandel, E and Goelet P (1987) Selective elimination of mRNAs in vivo: complementary oligonucleotides promote RNA degradation by an RNAse H-like activity Proc Natl Acad Sci USA 84 7896-7900

Eichele G (1989 Retinoids and vertebrate limb pattern formation Trends in Genetics 5 246-250

Gautier,J., Minshull,J., Kohka,M., Glotzer,M., Hunt,T. & Maller,J. (1990) Cyclin is a component of maturation promoting factor from *Xenopus*. Cell **60**, 487-494

Gieblehaus D, Eib D and Moon R (1988) Antisense RNA inhibits expression of membrane skeleton protein 4.1 during embryonic development of *Xenopus* Cell 53 601-615

Haseloff J and Guerlach W (1988) Simple RNA enzymes with new and highly specific endoribonuclease activity Nature 334 585-591

Kawasaki,E.S. (1985) Quantitative hybridization-arrest of mRNA in *Xenopus* oocytes using single-stranded complementarty DNA or oligonucleotide probes. Nucleic Acids Res. **13**, 4991-5004

Kloc M Miller M Carraco A Eastman E and Etkin L (1989) The maternal store of the xlgv 7 mRNA in full grown oocytes is not required for normal development in **Xenopus.** Development 107 899-907
Marcus-Sekura, C.J., Woerner, A.M., Shinozuka, K., Zon, G. and Quinnan Jr, G.V. (1987) Nucleic Acids Res 15, 5749-5761
McMahon A and Moon R (1989) Ectopic expressionof the proto-oncogene *int-1* in *Xenopus* embryos leads to duplication of the embryonic axis Cell 58 1075-1084
Meinkoth, J. and Wahl, G. (1984) Anal. Biochem. 138, 267-284
Melton D (1985) Injected antisense RNAs specifically block messenger translation in vivo. Proc Natl Acad Sci USA 82 144-148
Qian S, Hongo S and Jacobs-Lorena M (1988) Antisense ribosomal protein gene expression specifically disrupts oogenesis in *Drosophila melanogaster* Proc Natl Acad Sci USA 85 9601-9605
Rebegliati M and Melton D (1987) Antisense RNA injections in fertilised frog eggs reveal an RNA duplex unwinding activity Cell 48 599-605
Rossant J and Joyner A (1989) Towards a molecular genetic analysis of mammalian development. Trends in Genetics 5 277-282
Rozenberg U B, Preiss A, Seifert E, Jackle H and Knipple D C (1985) Production of phenocopies by Kruppel antisense RNA injection into *Xenopus* embryos Nature 313 703-706
Sagata,N., Daar,I., Oskarson,M., Showalter,S.D. & Vande Woude,G.F. (1989) The product of the c-mos proto-oncogene as a candidate initiator for oocyte maturation. Science **245**, 643-645
Sagata,N., Oskarson,M., Copeland,T., Brumbaugh,J. & Vande Woude,G.F. (1988) Function of c-mos proto-oncogene product in meiotic maturation in *Xenopus* oocytes. Nature **335**, 519-525
Shuttleworth,J. & Colman, A. (1988) Antisense oligonucleotide-directed cleavage of mRNA in *Xenopus* oocytes and eggs. EMBO J. **7**, 427-434
Shuttleworth,J., Matthews,G., Dale,L., Baker,.C. & Colman,A. (1988) Antisense oligodeoxynucleotide-directed cleavage of maternal mRNA in *Xenopus* oocytes and embryos. Gene **72**, 267-275
Smith,R.S., Dworkin,M.B. & Dworkin,E. (1988) Destruction of a translationally controlled mRNA in *Xenopus* oocytes delays progesterone induced maturation. Genes and Development **2**, 1296-1306
Stein, C.A., Subasinghe, C., Shinozuka, K. and Cohen, J.S. (1988) Nucleic Acids Res 16, 3209-3221
Subasinghe, C., Helene, C., Cohen, J.S. and Tolume, J.-J. (1989) Nucleic Acids Res. 17, 4255-4273
Wright T (1970) The genetics of embryogenesis in *Drosophila* Adv Genet 15 261-395
Zuker,M. & Stiegler,P. (1981) Optimal computer folding of large RNA sequences using thermodynamics and auxillary information. Nucleic Acids Res. **9**, 133-148

THE ROLE OF BOXA IN TRANSCRIPTION OF RIBOSOMAL RNA OPERONS OF *ESCHERICHA COLI*: CHANGES IN THE PROCESSIVITY OF RNA POLYMERASE

R. J. Gaudino and E. A. Morgan
Department of Experimental Biology
Roswell Park Memorial Institute
666 Elm Street
Buffalo, N.Y. 14263
USA

Termination of transcription by the RNA polymerase of *E. coli* is generally assumed to be quite limited in the absence of deliberate transcription termination signals. However, the inherent termination rate of RNA polymerase in the absence of deliberate termination signals (what we will refer to as the processivity of RNA polymerase) has not been carefully quantitated. Therefore, it is not known if the inherent processivity of RNA polymerase is an important limitation to transcription, or whether processivity can be increased by post-initiation modifications of RNA polymerase in some operons.

While termination in the absence of deliberate termination signals is generally probably low, it is not always the properties of the RNA polymerase alone that are responsible for low termination. For example, it has long been apparent that termination is sometimes kept low by the coupling of transcription to translation, as termination signals (generally Rho-dependent) are unmasked in some operons when translation of mRNA is prevented. This latter phenomena (known as polarity) can be regarded as a post-initiation modulation of the ability of RNA polymerase to terminate transcription. It remains to be determined if this modulation is caused by a direct modification of RNA polymerase itself, or simply results from the interference of ribosomes with the complex interaction of Rho, RNA, and RNA polymerase. Despite these uncertainties, the existance of the phenomena of polarity suggests that the cell may engage in interesting and complex mechanisms which modulate processivity for biologically important reasons.

In this contribution we quantitively characterize the processivity of RNA polymerase transcribing several types of DNA, and show that a sequence present in rRNA operons alters the processivity of RNA polymerase by a post-initiation modification of RNA polymerase or the transcript. Our studies suggest a richness of means by which the processivity of RNA polymerase is modified.

The first proposal that rRNA operons might have a mechanism to reduce premature termination was based solely on the theoretical argument that transcription of all rRNA species should be equimolar, and that rRNA operons were likely to have an undiscovered mechanism to achieve equimolar synthesis by preventing premature termination because mRNA operons had such a mechanism (the coupling of transcription and translation)

NATO ASI Series, Vol. H 49
Post-Transcriptional Control of Gene Expression
Edited by J. E. G. McCarthy and M. F. Tuite

which rRNA operons obviously did not have. This proposal was supported by several early experiments in which sequences causing termination in mRNA-encoding operons were shown to be relatively ineffective in causing termination in intact rRNA operons (reviewed in Morgan, 1986). Subsequently, operon fusions employing termination signals of various types were used to show that transcribed region of rRNA operons which preceeded the first structural sequences (the leader region) was responsible for this decrease in termination. Further studies revealed the presence of sequences in the leader region which bore a resemblance to sequences present in the nut regions of lambdoid phages (reviewed in Morgan, 1986). The nut regions were known to be involved in antitermination. The complete extent of the regions with similarity to lamdoid phages remains controversial, but one region, the boxA sequence, is undeniably present, functionally important, and probably functionally related to the boxA sequence present in lambdoid phages. The argument that this sequence is an important sequence hinges on the fact that the sequence is present twice in rRNA operons in remarkably similar locations, is identical at 13 of 14 positions in the two locations, is absolutely conserved in all sequenced rRNA operons, and is the only precursor-specific sequence other than base-paired regions involved in formation of the RNaseIII processing stems which is highly conserved in the seven rRNA operons of E. coli. Additional evidence for a functional role for boxA is presented below. Some of the arguments about conservation of sequences in and near boxA have been presented in more detail elsewhere (Berg et. al., 1989; reviewed in Morgan, 1986).

In Fig. 1, the location of boxA sequences in rRNA operons is shown. As can be seen from this figure, the two boxA sequences abut RNaseIII processing stems preceeding the 16S and 23S rRNA genes. The sequences comprising the two RNase III recognition stems have no primary sequence similarity to each other.

The direct abutment of the boxA sequences to the base of the stems involved in RNase III processing hints at a possible correlation between the roles of boxA and the roles of RNase III stems, but this has not yet been proven. As we clearly show below, the boxA sequences do not require the RNase III stems to function in at least some assay systems, and abnormalities of rRNA synthesis do not seem to result if the spatial proximity of these elements is disturbed. Thus, many otherwise interesting models of how these sequences might function together seem unattractive at present.

To determine the role of boxA in the synthesis of rRNA, we developed a novel system to measure the function and expression of rRNA. In this system, a multicopy plasmid containing rrnC was constructed. A spectinomycin resistance mutation was introduced into the 16S rRNA gene and an erythromycin resistance mutation was introduced into the 23S rRNA gene of this plasmid (Sigmund et al., 1988). rRNA function could thenn be easily assessed by determining the resistance of cells to spectinomycin and erythromycin, which provides a hypersensitive assay responsive to changes in the percentage of rRNA in the cell of the resistant type (total rRNA per cell changes little or not at all due to autogenous feedback regulation of transcription). The amount of rRNA which accumulates can also be very accurately measured by a primer extension method capable of quantitating two rRNA molecules which differ by a single base (Sigmund et al., 1988).

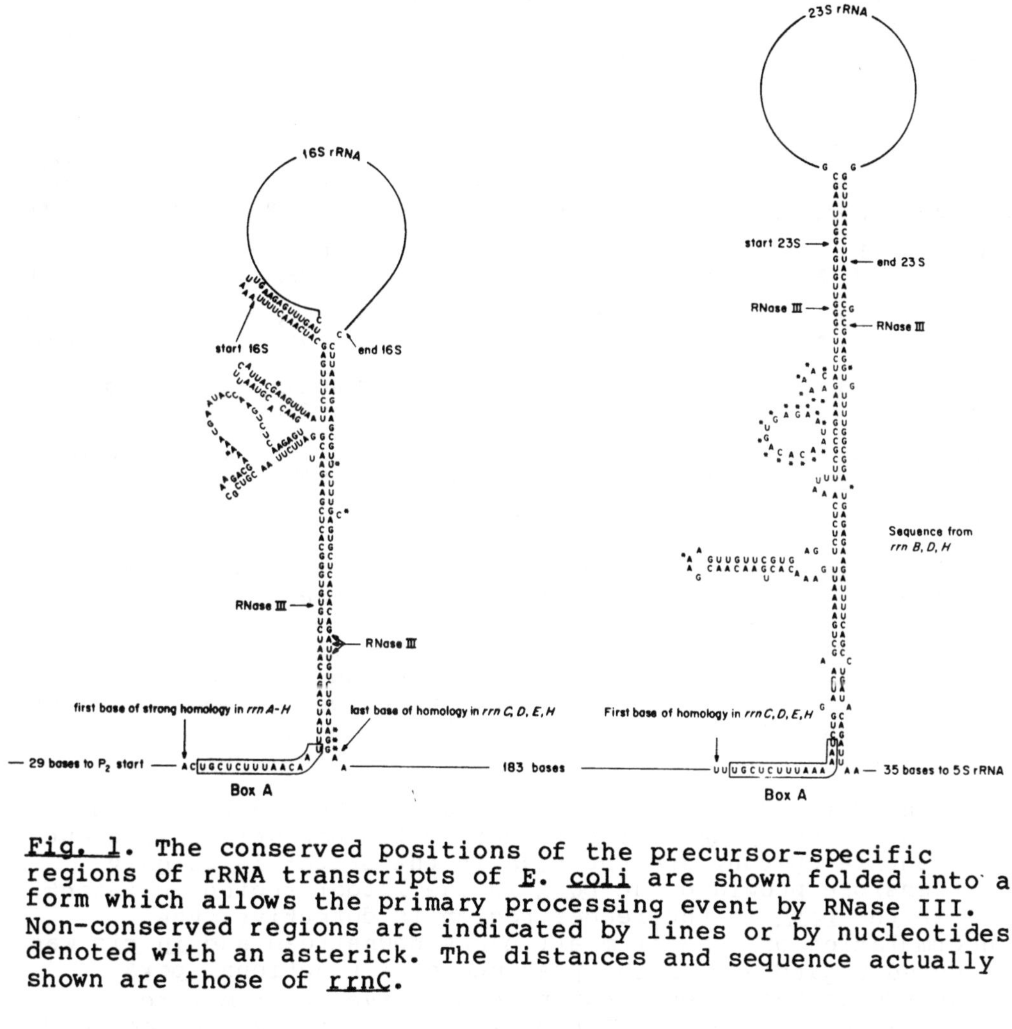

__Fig. 1__. The conserved positions of the precursor-specific regions of rRNA transcripts of __E. coli__ are shown folded into a form which allows the primary processing event by RNase III. Non-conserved regions are indicated by lines or by nucleotides denoted with an asterick. The distances and sequence actually shown are those of __rrnC__.

Mutations were then introduced into two of the most conserved nucleotides in __boxA__s of this plasmid by converting the second and third bases of __boxA__ from GC to TA by oligonucleotide-directed mutagenesis. This limited mutational alteration was chosen in preference to larger alterations to minimize the possibility of unintentional alterations in RNA processing or ribosome assembly. Mutations were introduced into the __boxA__ in the leader region, into the __boxA__ between the 16S and 23S rRNA genes (the spacer region), and into both __boxA__s. As shown below in experiments employing operon fusions, it is likely that these alterations of boxA sequence completely abolish boxA function. Quantitative measurements of the effects of these mutations on rRNA accumulation indicated that the mutations had a measureable effect on the rRNA accumulation of downstream genes (Table 1). The quantitative effects of boxA mutations on rRNA accumulation were also faithfully reflected in the changes in spectinomycin and erythromycin resistance of the cells containing these mutant plasmids (data not shown).

	rRNA ACCUMULATION: PERCENT FROM PLASMID						
BoxA Mutations	16S (ddA)	16S (ddG)	16S (average)	23S (ddA)	RELATIVE ACCUMULATION	23S/16S	RNA "LOST" PER 1000 bp TRANSCRIBED
None	83.2 +/- 0.6%	80.2 +/- 0.8%	81.7 +/- 0.7%	66.2 +/- 1.4%	100% of 16S 100% of 23S	0.81	0% in 16S 0% in 23S
Leader and Spacer	64.2 +/- 2.1%	56.9 +/- 1.7%	60.5 +/- 1.6%	27.6 +/- 1.1%	74% of 16S 42% of 23S	0.46	17% in 16S 12% in 23S
Leader only	65.5 +/- 1.2%	62.6 +/- 0.8%	64.1 +/- 0.8%	45.8 +/- 0.6%	78% of 16S 69% of 23S	0.72	14% in 16S 3% in 23S
Spacer only	85.5 +/- 2.2%	79.6 +/- 2.3%	82.6 +/- 1.8%	64.9 +/- 2.6%	101% of 16S 98% of 23S	0.79	0% in 16S 1% in 23S

Table 1. The effect of the boxA mutations on rRNA accumulation is shown. The values of accumulation use the wild-type rRNA from the chromosome as an internal standard. Two values for 16S rRNA are given because the value for 16S rRNA has been independently measured by primer extension terminated with either of two dideoxynucleotides. The values given are +/- the SEM. The RNA is described as "lost" because this data does not formally prove loss through premature termination.

The data obtained by this method proves the following points: 1) the boxA in the leader region can work throughout the entire operon, and the boxA in the spacer is largely dispensible provided the boxA in the leader region is intact. 2) if the boxA in the leader region is defective, the accumulation of rRNA from all genes downstream and *in cis* was reduced regardless of the presence of a functional boxA sequence in the spacer region, suggestive of premature termination in the 16S rRNA gene. 3) if the boxA in the leader region is defective, the boxA in the spacer region functions to prevent any loss of 23S rRNA in excess of what is probably lost by premature termination in the 16S rRNA gene. 3) the fact that the spacer boxA is not needed for efficient 23S rRNA synthesis in the presence of an intact leader boxA indicates that the proximity of RNase III stems to boxA does not limit the function of boxA to sequences between the base-paired halves of the stem which abuts a boxA. 4) In the absence of boxA there is an increase in termination which can be estimated at approximately 15% per thousand base pairs throughout the 16S and 23S rRNA genes. 5) given that all rRNA molecules which are defective or in molar excess due to premature termination are probably rapidly degraded (Siehnel and Morgan, 1985), the two boxA sequences will more than double the number of ribosomes made per transcription initiation event.

As is the case for small differences in mortgage interest rates over long periods of time, small increases in processivity can have very meaningful effects in large operons where equimolar synthesis of all portions of the transcript is important.

The primer extension measurements and antibiotic resistance levels described above formally measure rRNA accumulation rather than transcription termination. Therefore, cells containing these plasmids were pulse-labeled with tritiated uridine for 10 seconds (incorporation into RNA occurs only during the last 5 seconds), followed by rapid and efficient RNA extraction by pipetting the culture into a boiling mixture of SDS, EDTA, and phenol. The resulting RNA was mixed with gel-purified, full-length P32-labeled rrnC transcript made in vitro using a bacteriophage RNA polymerase, and the relative synthesis of portions of the rRNA transcript determined from the H3/P32 ratio of RNA hybridizing to DNA probes from rrnC. The positions of the probes are shown in Fig. 2. and the hybridization results in Fig. 3.

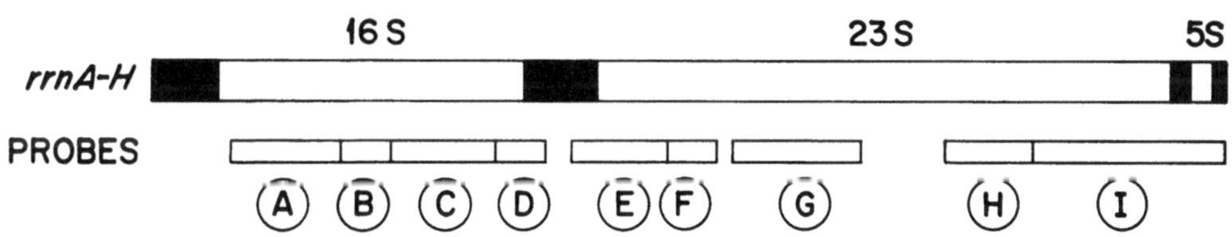

Fig. 2. DNA probes used for hybridization to pulse-labeled RNA mixed with a full-length P32-labeled RNA internal standard.

The hybridization results shown in Fig. 3 are affected by a non-reducible and very reproducible uneveness-of-curve which apparently is inherant in the hybridization of pulse-labeled, forcibly prematurely terminated transcripts prepared in this manner (steady-state labeled RNA yields a flat curve indicating virtually equimolar amounts for all rRNA regions). As a result, conclusions made exclusively from this data must remain somewhat tentative, as the difference in the curves may not solely reflect the amount of labeled RNA complementary to each probe, and thus may not be exclusively indicative of the amount of premature termination within the rRNA operon. Nevertheless, these results strongly suggest that the greatest amount of termination occurs in rRNA operons with both leader and spacer boxA mutations, followed by operons with leader boxA mutations. This hybridization data indicates that boxA mutations cause about 20% termination per thousand base pairs transcribed. Altogether, measures of termination obtained by hybridization are very similar to the values for termination suggested by primer extension measurements of accumulation, lending confidence to the conclusion that the results are correct.

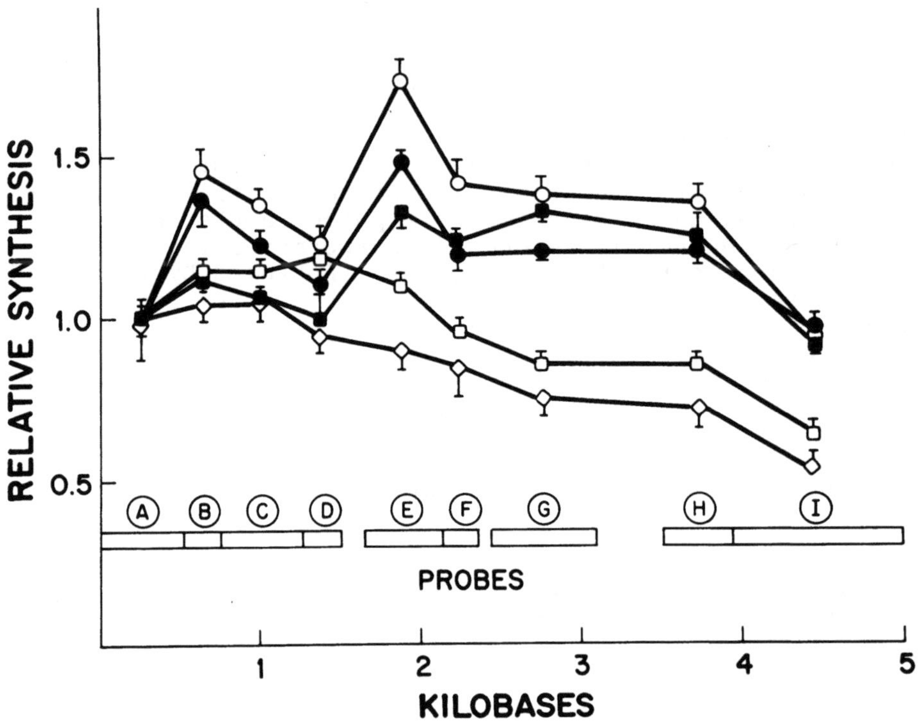

Fig 3. Normalized relative synthesis of the regions of rRNA operons in cells containing plasmids with rrnC. The error bars indicate the SEM and reflect the accumulative error of all aspects of sample preparation and hybridization. Open circles - no boxA mutations and no antibiotic resistance mutations in 16S or 23S rRNA genes. Filled circles - antibiotic resistance mutations, but no boxA mutations. Filled squares - antibiotic resistance mutations and a spacer boxA mutation. Open squares - antibiotic resistance mutations and a leader boxA mutation. Open diamonds - antibiotic resistance mutations, leader and spacer boxA mutations.

We have attempted to determine if the termination in rRNA operons as measured above is Rho-dependent by redoing these experiments using mutant plasmids transformed into a rho15(ts) strain. Unfortunately, we have recently discovered that only strains partially or fully reverted to rho wild-type live after being transformed with plasmids containing an rRNA operon (the vector we used does replicate in Rho- strains). Although the results are therefore of a preliminary nature, the fact that the boxA mutations still had very strong effects in all partially Rho- strains tested may indicate that the termination which boxA counteracts is not strongly Rho-dependent.

We also attempted to define the important regions involved in boxA function and further characterize the role of boxA in affecting termination or processivity by constructing and anlysing operon fusions in which a promoter was followed by either an intact or mutant boxA (the mutations used were the same mutation used in the intact rRNA operon) and then a mRNA coding region. The promoters used were either the Ptac promoter, the rrnC P1P2 promoter or the rrnC promoter with the rrnC leader attached without alteration of the wild-type sequence between the promoter and leader (P1P2-leader). The sequences containing boxA were an rrnC leader cassette spanning from the transcription start of the rrnC P2 promoter to within the 16S rRNA gene (Xleader), or artificial boxA cassettes. The cassettes are described in Fig. 4. The mRNA coding region used was a trp-lac fusion operon described in Fig. 5.

rrnC **leader**

WILD-TYPE	.........GGCACTGCTCTTTAACAATTTATCAGACA.........
MUTANT	.........GGCACTtaTCTTTAACAATTTATCAGACA.........

synthetic ***boxA*****s**

24 GC	gggACTGCTCTTTAACAATTTATCAGAag
24 AT	gggACTtaTCTTTAACAATTTATCAGAag
8 GC	gggtaTGCTCTTTcctggatctccgaaag
8 AT	gggtaTtaTCTTTcctggatctccgaaag

*Sma*I *Hind*III

Fig.4. Sequence of the DNAs containing BoxA used in this study. Sequences homologous to the rrnC leader are capitalized, while sequences which differ are in small letters. The boxA regions are underlined. Thus, the 24 GC synthetic boxA sequence has 24 positions analogous to the rrnC leader and includes all of boxA. The 24 AT synthetic boxA is identical except for a two-base change in boxA.

Pulse labeled mRNA was prepared from cells containing the various plasmid constructions and hybridized to probes using a full length, gel purified, P32-labeled reference RNA in a procedure very similar to that used to analyse rRNA synthesis as described above. RNA levels from strains completely isogenic except for mutations in boxA were compared to insure that only the properties of boxA affected the results obtained. Representative data is shown in Fig. 6, and all the data is summarized in Table 2.

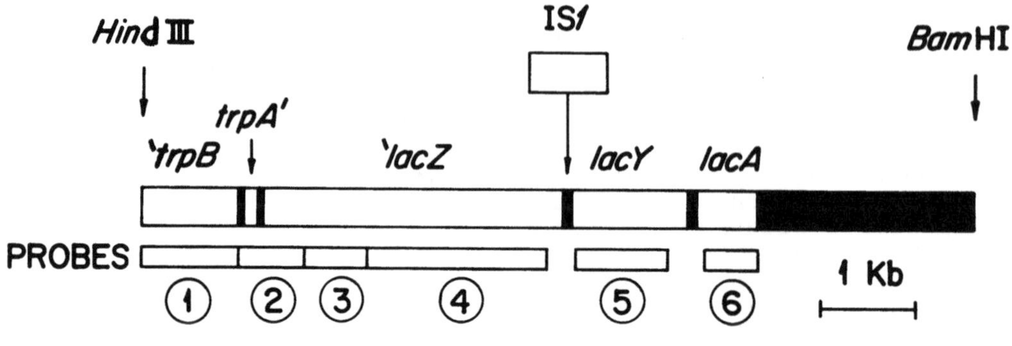

Fig. 5. The trp-lac fusion operon used to assess boxA function, and the probes used in hybridization measurement of the relative molar yields of different regions of RNA. In all of the constructs made, trpB and IS1 are probably rarely translated, the trpA'-'lacZ fusion gene is poorly translated due to translational coupling between trpA and trpB, and lacY and lacA are fully translated (Holben and Morgan, 1985).

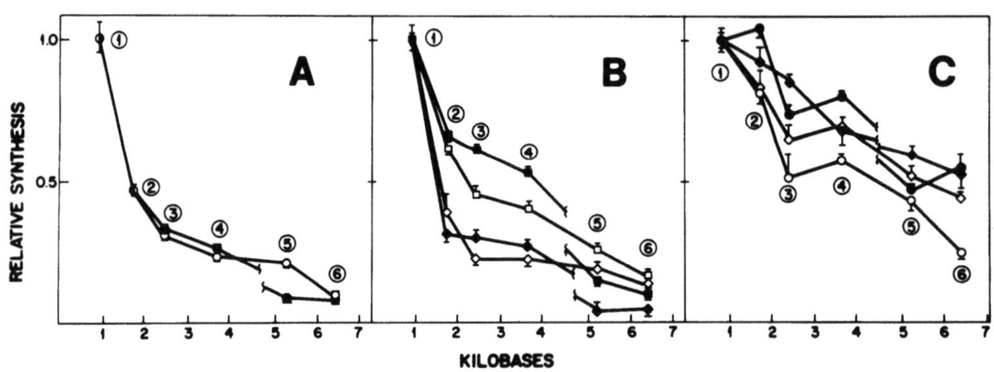

Fig. 6. Normalized relative RNA synthesis from the trp-lac fusion operon. The numbers in circles indicate the probes described in Fig. 5. In all panels, the tac promoter is directing synthesis, open symbols are from operon fusions without and IS1 insertion, and filled symbols are from operon fusions with an IS1 insertion. In panel A, there is no leader or artificial boxA sequence between the promoter and trp-lac fusion operon. In panel B, squares indicate that an intact rrnC leader cassette (Xleader) has been placed between the promoter and the fusion operon, and circles indicate that the Xleader with a boxA mutation is at the same position. In panel C, the symbols denote the same constructions, but with

synthesis occurring in a rho15(ts) strain at 37 C. In all panels, the error bars are the SEMs which indicate the errors arising from of all aspects of sample preparation and hybridization.

	PERCENT TERMINATION IN RHO+ PER KB				PERCENT TERMINATION IN RHO- PER KB			
REGION	N	BoxA w.t.	BoxA mut.	diff-erence	N	BoxA w.t.	BoxA mut.	diff-erence
1-2 (0.9 Kb)	10	42	68	26	2	7	14	7
2-4 (1.9 Kb)	10	11	7	-4	2	13	10	-3
4-5 (2.4 Kb)	5	10	5	-5	1	6	8	2
IS1 (0.8 Kb)	5	-6	14	20	n.d.	n.d.	n.d.	n.d.
5-6 (1.2 Kb)	10	6	3	-3	n.d.	n.d.	n.d.	n.d.

Table 2. Summary of the percent termination per kilobase transcribed of the trp-lac fusion operons in the regions between the midpoints of the indicated probes. N is the number of samples. Here, the data from all promoters analysed has been averaged together because the promoters behaved indistinguishably. Similarly, the data from the intact rrnC leaders and 24 nucleotide synthetic boxAs has been averaged together, since the reduction in termination caused by both was indistinguishable.

Several salient points emerged from this analysis. 1) premature termination is high only in non-translated regions of the operon fusion (IS1 and trpB), consistent with proposals that transcriptional-translational coupling reduces termination efficiently. 2) the intact rrnC leader and the 24 GC version of boxA containing little more than the 14 bp boxA consensus sequence work equally well in reducing termination in the non-translated regions, whereas the 8 GC synthetic boxA containing only 8 bp of boxA does not work at all. The functional boxA sequences may also reduce termination in the translated regions, but the rate of termination is too low, and the effect of boxA too weak, to determine this from the available data. 3) the GC to TA mutation used througout this study completely abolishes the function of boxA in this assay. 4) boxA is very poor at reducing termination (much worse than translation appears to be), but the actual reduction of termination per kilobase transcribed is similar to that which boxA causes in rRNA operons. 5) it is not clear if the termination boxA reduces is Rho-dependent or not. What is clear is that boxA is quite poor at reducing termination that is demonstrably Rho-dependent. 6) excess termination due to the absence of translation is largely Rho-dependent, as previously reported by many others. 7) translation, *boxA*, and removal of Rho by mutation fail to eliminate all termination, even in combination.

Several questions remain, and their answers will help to determine how important processivity is to the function of RNA polymerase. 1) If *boxA* is so poor at reducing Rho-dependent termination and does not affect Rho-independent termination or termination unaffected by translation, exactly what type of termination *does* it reduce efficiently to enable it to eliminate all measurable termination in *rrn* operons? In this work, as well as all previous reports in the literature where termination efficiency can be assessed (Morgan, 1980; Brewster and Morgan, 1981; Siehnel and Morgan, 1983; Holben and Morgan, 1984; Holben et al., 1985) or guessed from the data presented, *boxA* has been very poor at reducing both Rho-dependent and Rho-independent termination caused by legitimate terminators. 2) Why do intact *rrn* operons have essentially no premature termination, whereas the *trp-lac* fusion operon has uniformly distributed, detectable termination despite the reduction of termination by *boxA*, translation, and *rho* mutations? 3) How have *rrn* operons weeded out all types of termination except termination sensitive to *boxA*? 4) Would other operons be likely to have a mechanism to increase processivity? 5) Does having mechanisms increasing processivity in a few operons mean that low processivity is an advantage in most? 4). Do large operons with a requirement for low processivity preferably use antitermination mechanisms or high processivity RNA polymerases as part of their transcriptional or regulatory strategies because the low processivity of RNA polymerase is otherwise a significant problem? 5) Do the classical antitermination mechanisms affect processivity in the absence of classical terminators?

Berg KL, Squires C, Squires CL (1989) Ribosomal RNA antitermination: Function of leader and spacer region BoxB-BoxA sequences and their conservation in diverse micro-organisms. J Mol Biol 209:345-358

Brewster JM, Morgan EA (1981) Tn*9* and IS*1* inserts in a ribosomal RNA operon of *Escherichia coli* are incompletely polar. J Bacteriol 148:897-903

Holben WE, Morgan EA (1984) Antitermination of transcription from an *Escherichia coli* ribosomal RNA promoter. Proc Natl Acad Sci USA 81:6789-6793

Holben WE, Prasad SM, Morgan EA (1985) Antitermination by both the promoter and leader regions of an *Escherichia coli* ribosomal RNA operon. Proc Natl Acad Sci USA 82:5073-5077

Morgan EA (1980) Insertions of Tn*10* into an *E. coli* ribosomal RNA operon are incompletely polar. Cell 21:257-265

Morgan EA (1986) Antitermination mechanisms in ribosomal RNA operons of Escherichia coli. J Bacteriol 168:1-5

Siehnel RJ, Morgan EA (1983) Efficient read-through of Tn9 and IS1 by RNA polymerase molecules that initiate at ribosomal RNA promoters. J Bacteriol 153:672-684

Sigmund CD, Ettayebi M, Borden A, Morgan EA (1988) Antibiotic resistance mutations in ribosomal RNA genes of *Escherichia coli*. Meth Enzymol 164:673-690

Siehnel RJ, Morgan, EA (1985) Unbalanced ribosomal RNA gene dosage and its effects on ribosomal RNA and ribosomal protein synthesis. J Bacteriol 163:476-486

REGULATION OF λ N-GENE EXPRESSION

Luis Kameyama, Leonor Fernandez, [1]Gabriel Guarneros, Donald L. Court
Molecular Control and Genetics Section
Laboratory of Chromosome Biology
ABL-Basic Research Program
Bldg. 539 NCI/FCRF
Frederick, MD 21701
USA

INTRODUCTION

N-antitermination

The N gene product of phage λ is a positive regulatory factor for the transcription of other λ genes during phage development. It acts by modifying the RNA polymerase transcription complex. In the absence of N function, RNA polymerase initiates at the early λ promoters, p_L and p_R, but terminates transcription shortly after initiation at terminators, t_{L1} and t_{R1} respectively. The action of N prevents polymerase from terminating at t_{L1} and t_{R1} as well as at other more distant terminators in the p_L and p_R operon (see review by Friedman, 1988). In each of these operons, between the promoter and the first terminator, there is a target sequence which is also required for the N-antitermination process (Friedman et al., 1973; Rosenberg et al, 1978). These targets, nutL and nutR, have been defined genetically by mutations to block antitermination in the respective operons (Salstrom and Szybalski, 1978; Olson, et al., 1984). These mutations have defined two regions of nut, boxA and boxB (Friedman & Gottesman, 1983). The boxA site is conserved in sequence among other phages and *E. coli* whereas the box B site is unique for each phage type and is thought to be the N recognition site (Lazinski, et al., 1989) (See Figure 1).

Host functions, Nus, are also involved in the N-antitermination process as defined by the 5 Nus mutants (Nus A, B, C, D and E) that are blocked for N action *in vitro* (Friedman, 1971; Georgopoulos, 1971). One or more of these Nus factors is thought to interact with the common boxA sequence (Friedman & Gottesman, 1983). Nus C is defined by a mutation in the beta-subunit of RNA polymerase itself. Nus A protein binds to RNA polymerase during elongation (Greenblatt & Li, 1981). The Nus D mutant is altered in transcription

---

[1]Department of Genetics; CINVESTAV del IPN, Mexico 14 DF, Mexico

NATO ASI Series, Vol. H 49
Post-Transcriptional Control of Gene Expression
Edited by J. E. G. McCarthy and M. F. Tuite

termination factor Rho (Das et al., 1983), and the Nus E mutant is altered in the 30 S ribosomal protein S10 (Friedman et al., 1981). Nus B has not been specifically assigned a function but is essential for E. coli as cold sensitive lethal mutations exist (Georgopoulos et al., 1980).

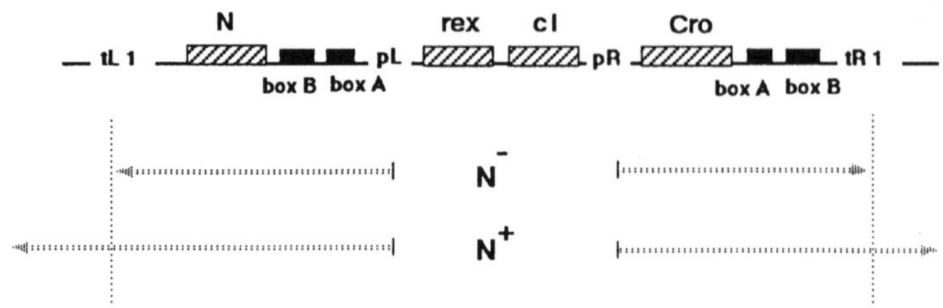

FIG 1. λ early transcription. λ genes N, rex, cI, and cro are indicated by cross-hatched rectangles. The boxA and boxB sites of nutL and nutR are shown in the p_L and p_R operons. Arrows in divergent directions represent transcription from p_L and p_R promotors by RNA polymerase in the absence or presence of λ N function.

Both genetic and biochemical results show that N and the Nus factors recognize the nut sequence after it has been transcribed into RNA (Olson et al., 1982; Warren and Das, 1984; Zuber et al., 1987; Barik et al., 1987). It has been shown that RNA polymerase in a crude in vitro system form a large complex with N and Nus factors for antitermination. (Horwitz et al., 1987; Barik et al., 1987). However, in a purified system under appropriate conditions, only N and Nus A are necessary to cause RNA polymerase to antiterminate (Whalen et al., 1981). It has also been shown that boxA is not always essential (Zuber et al., 1987). Thus, depending upon conditions, there may be different requirements for antitermination.

The P_L-N-operon

N is the first gene in the p_L - operon, and the N AUG initiation signal is located 223 nucleotides from the start of transcription (Franklin & Bennett, 1979; Franklin; 1985). Thus N has a long leader RNA which has several interesting landmarks (Figure 2). The nutL target is located in the leader, it is composed of a boxA sequence (CGCUCUUAA) and a boxB stem-loop structure beginning 34 and 50 nucleotides from the promoter start, respectively. Also in the leader, between nutL and N, the RNA forms a structure sensitive to RNaseIII processing. Two specific sites of RNaseIII cleavage between nucleotides 88 and 89 and nucleotides 196 and 197 of the leader were determined both in vivo (Lozeron et al., 1977) and in vitro

(Steege et al., 1987). These two sites are located near each other in the base paired RNA stem structure of Figure 2.

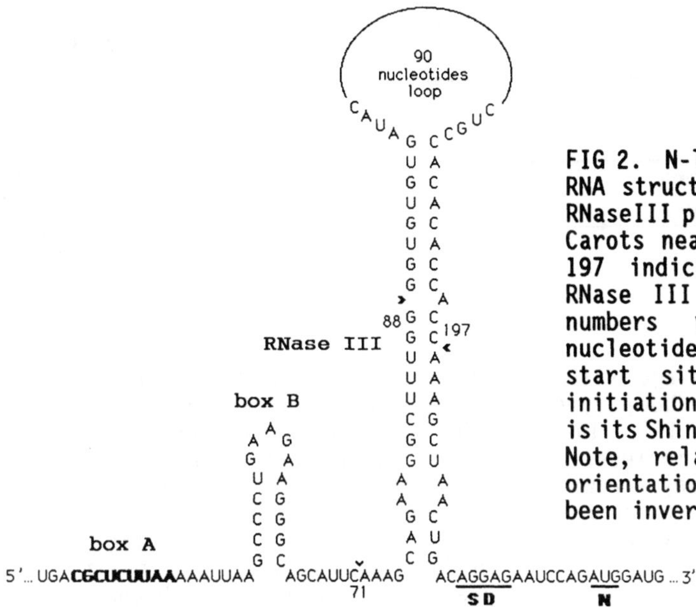

FIG 2. N-leader region. Potential RNA structures nutL (boxB) and an RNaseIII processing site are shown. Carots near positions 71, 88, and 197 indicate positions at which RNase III cleaves the RNA. The numbers represent distance in nucleotides from the p_L-promoter start site. The N gene AUG initiation codon is underlined as is its Shine-Dalgarno (SD) sequence. Note, relative to Figure 1, the orientation of the p_L transcript has been inverted in Figure 2.

Processing of the N-leader region has no obvious effect on the stability of the N-mRNA (Anevski and Lozeron, 1981). Several predictions have been made as to whether RNaseIII processing of the leader transcript might affect N gene expression (Steege et al., 1987; Anevski and Lozeron, 1981). We have examined this question and also looked at the potential effect of N on its own expression.

RESULTS

RNaseIII processing of the N-leader RNA

The N leader region was cloned into the HincII-BamHI region of the pGEM4 vector (Promega Co.) so that it might be transcribed in vitro with T7 RNA polymerase. Labeled transcripts were made that begin 62 nucleotides upstream of the p_L start and end -390 nucleotides away within the N gene (See Figure 3). These were processed using a purified RNaseIII (Chen et al., 1990; L. Fernandez and D. Court, unpublished results). Previous results of in vitro processing indicated two cut sites. According to our assay method, three RNaseIII cut sites were found. We detect those beyond nucleotides 88 and 196 relative to the p_L start, plus another at nucleotide 71. This site at 71 had been suggested from in vivo results, but the 71 nucleotide transcript was later postulated to be an exonuclease decay product from the longer 88

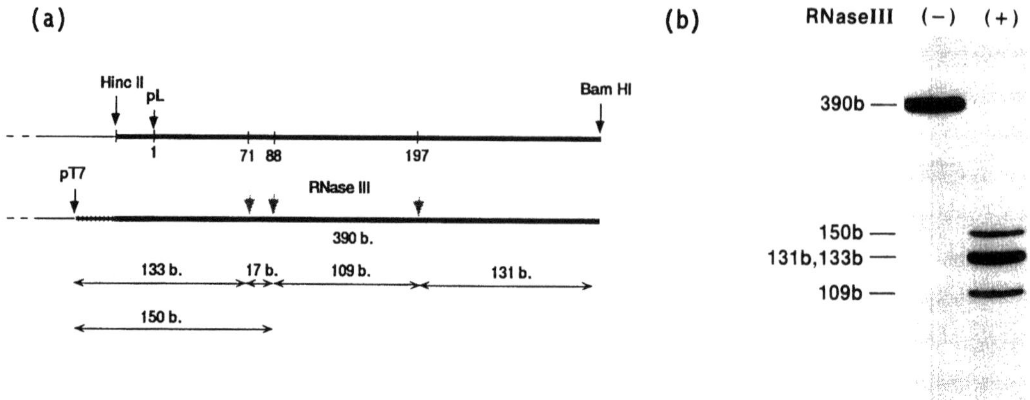

FIG 3. RNase III cleavage sites in the N leader region. (a) Schematic representation of the T7 RNA polymerase transcript of the N leader region, and it's processing by RNase III. The black bar from HincII to BamH1 corresponds to λ DNA (360bp) which includes part of the pL promotor, the N leader and about one third of N. Grey arrows indicate RNaseIII cut sites, and 133b, 17b, 109b, 131b, 150b lines correspond to the processed RNA fragments. Transcription in vitro was made according to Promega Co., instructions. RNA was labelled with 50 μCi of α-^{32}P-CTP in 25μl.(Amershan). (b) The in vitro transcription reaction was made in the presence (+) and in the absence (-) of RNaseIII. T7 RNA polymerase transcribed a 390b run-off RNA. In the presence of RNaseIII, four major RNA fragments were generated: 150b, 133b, 131, and 109b. The 133b and 131 transcripts migrate as one band but with a double intensity.

nucleotide transcript (Steege et al., 1987). We now confirm that it is generated in vitro by RNaseIII processing. RNaseIII cuts the precursor RNA completely at positions 88 and 196 but only partially at 71. The minor 17 nucleotide long transcript cut from between 71 and 89 can be seen on 20% acrylamide gels (data not shown). The 109 nucleotide transcript is cut from between positions 88 and 197. We believe that processing at 71 requires formation of a transient structure during the transcription. If RNase III is added much later most processing at this site is lost.

RNaseIII activates N gene expression

To examine N regulation, lacZ fusions have been made to the N gene. Both operon and protein fusions have been made at the same restriction site in N (See Figure 4), so as to be able to monitor and distinguish effects on transcription or translation of the N gene.

The p_L promoter of the N-lacZ fusion plasmids was under the control of the λ repressor which was supplied by a prophage copy of the λ cI gene. A mutant heat sensitive repressor was used so that at the 32°, p_L is repressed, but at 42° p_L is active.

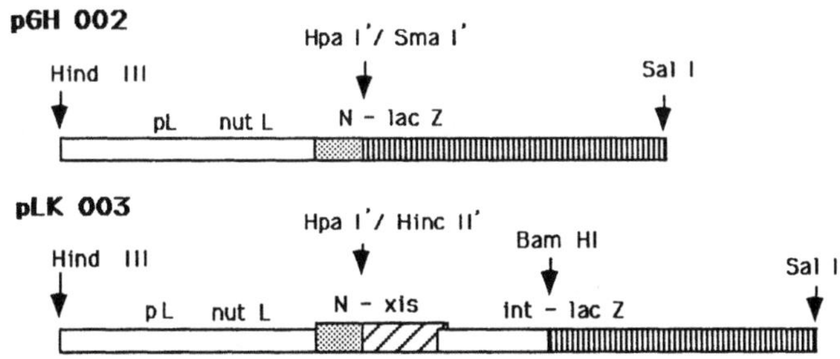

FIG 4. N-lacZ protein and operon fusions. Plasmid pGH002 contains a N-lacZ protein fusion that joins the 33rd codon of N in frame to lacZ deleted for its first eight codons. Plasmid pLK003 contains an N to lacZ operon fusion in which lacZ is fused in frame to the int gene. Upstream of int, N is joined to the xis gene using the same HpaI site in N used to make the N-lacZ fusion in pGH002. The HindIII to SalI fragment shown here replaced the small HindIII to SalI fragment of pBR322.

The effect of RNaseIII processing of the N-leader was examined by using a host cell that was either wild type (rnc^+) or mutant (rnc105) for the RNase III gene. Results showed that RNaseIII had no effect on lacZ expression in the operon fusion but increased the expression in the protein fusion to N (Table 1). Thus RNaseIII stimulates N translation (L. Kameyama, L. Fernandez, D. Court and G. Guarneros, manuscript in preparation).

	RNase III	
	−	+
pL N-lacZ	22	280
pL N-xis int-lacZ	245	280

TABLE 1. RNaseIII effect on expression of β-galactosidase in protein and operon fusions. Strain N5656 (SA500 his^-, ilv^-, galE490 pglΔ8, lacZ xA21::tn 10 (λ [int-ral]Δ N^+ cI857 ΔHI) and its isogenic derivative N5656 glyA::tn5 rnc105 were grown at 32° in L broth to an OD_{650} = 0.2. At that time, λ repressor was inactivated by shifting the culture to 42°. Measurements of β-galactosidase (Miller, 1972) were made after induction for 30 minutes.

N is autoregulated

To examine the effect of N on its own regulation, the prophage that supplied the temperature sensitive repressor was made either N^+ or N^-. At 32°, cI-repressor prevents expression of the prophage N gene and the plasmid N-lacZ constructs, their expression from p_L is induced at 42°. These studies were carried out in rnc105 hosts to avoid possible complications of RNaseIII. N like RNaseIII affects the lacZ expression only in the protein fusion, however, N inhibits expression (L. Kameyama, G. Guarneros and D. Court, in prep).

Sites Involved in RNaseIII and N Action

Using a unique BsmI endonuclease restriction site between nutL and the RNaseIII sensitive sites, independent deletions were made that either entered the nutL or RNase III region (Figure 5). The effect of each deletion was examined in the protein fusion system.

Deletion of the RNaseIII processing region, ie, the 5′ side of the RNA stem structure, increased N-lacZ expression. This higher level was independent of whether the cell was rnc^+ or rnc105. This deletion mutant however, was still regulated negatively by N.

Deletion of the nutL site, ie, the 3′ side of the boxB stem structure, also increased N-lacZ expression. In this case however, the higher level of expression was insensitive to N regulation but could be stimulated by RNaseIII (L. Kameyama and D. Court, unpublished results).

Plasmid	nut L	RNase III process region	CONTROL RNase III	CONTROL N
pGH 002	+	+	Yes	Yes
pLK 246	Δ	+	Yes	No
pLK 251	+	Δ	No	Yes

FIG 5. Schematic representation of nutL and the left side of the RNaseIII processed stem. Numbers 1, 2, and 3 indicate sites for EcoNI, BsmI, and BspMII restriction endonucleases, respectively. The cut sites at 1, 2, and 3 are 56, 64, and 129 nucleotides from the 5′ start of the p_L transcript. The nutL deletion was made by removing DNA between 1 and 2; the RNaseIII site deletion was made by removing DNA between 2 and 3. N-lacZ protein fusion plasmids with these deletion mutants were tested in cells for their ability to respond to the presence of RNaseIII or N protein. In the table after each plasmid, + indicates the site is present and a delta indicates it is deleted. If RNaseIII is able to stimulate, or N is able to inhibit lacZ expression, it is indicated by yes in the appropriate CONTROL column.

CONCLUSIONS

In summary, the primary sites controlling N gene expression are distinct for RNaseIII and N and reside at positions expected for RNaseIII and N to interact with the RNA.

RNaseIII processing removes the base paired RNA structure upstream of the AUG initiation signal (See Figure 2). The deletion of the left side of this stem disrupt this secondary structure. This suggests that the stem structure by being close to the initiation site inhibits N translation. Its removal by either processing or deletion eliminates the inhibition.

It is more difficult to make a simple explanation of how the nutL site and N down regulate N gene translation. Models are being examined that analyze whether N is acting through the transcription antitermination complex to affect translation or whether N can directly interact with ribosomes through the RNA site at nutL. The latter model is supported by results (Das and Wolska, 1984) that show N can be found associated with the 30S ribosome and that this association is dependant upon the function of the S10 protein.

Acknowledgments We thank Carolyn Redmond for typing the manuscript. These studies were supported in part, by a grant from CONACyT (Consejo Nacional de Ciencia y Tecnologia) of Mexico, and in part by the National Cancer Institute, DHHS, under contract No. N01-CO-74101 with ABL. The contents of this publication do not necessarily reflect the views or policies of the Department of Health and Human Services, nor does mention of trade names, commercial products, or organizations imply endorsement by the U.S. Government.

References

Anevski PJ, Lozeron HA (1981) Multiple pathways of RNA processing and decay for the major leftward N-independent RNA transcript coliphage lambda. Virology 113:39-53

Barik S, Ghosh B, Whalen W, Lazinski D, Das A (1987) An antitermination protein engages the elongating transcription apparatus at a promoter-proximal recognition site. Cell 50:885-899

Chen Su-Min, Takiff HE, Barber AM, Dubois GC, Bardwell JCA, Court DL (1990) Expression and Characterization of RNase III and Era Proteins. J Bio Chem 265:2888-2895

Das A, Gottesman ME, Wardwell J, Trisler P, Gottesman S (1983) A mutation in the Escherichia coli rho gene that inhibits the N protein activity of phage λ. Proc Natl Acad Sci USA 80:5530-5534

Das A, Wolska K, (1984) Transcription antitermination in vitro by lambda N gene product; requirement for a phage nut site and products of host nusA, nusB, nusE genes. Cell 38:165-173

Franklin NC (1985) Conservation of genome form but not sequence in the transcription antitermination determinants of bacteriophage λ, Φ21 and P22. J Mol Biol 181:75-84

Franklin NC, Bennett GN (1979) The N protein of bacteriophage lambda, defined by its DNA sequence, is highly basic. Gene 8:107-119

Friedman DI (1988) Regulation of Phage gene expression by termination and antitermination of transcription. In: Calendar R (ed) The Bacteriophages. Plenum Publishing Corp, Vol 2, pp 263-319

Friedman DI, Jolly CT, Mural RJ (1973) Interference with the expression of the N gene product of phage λ in a mutant of Escherichia coli. Virology 51:216-226

Friedman DI, Schauer AT, Baumann MR, Baron LS, Adhya SL (1981) Evidence that ribosomal protein S10 participates in the control of transcription termination. Proc Natl Acad Sci USA 78:1115-1118

Friedman DI, Gottesman M (1983) Lytic mode of lambda development. In: Hendrix RW, Roberts JW, Stahl FW, Weisberg RA (eds) Lambda II. Cold Spring Harbor Laboratory, Cold Spring Harbor, New York, pp 21-51

Georgopoulos CP, Swindle J, Keppel F, Ballivet M, Bisig R, Eisen H (1980) Studies on the E. Coli groNB (nusB) gene which affects bacteriophage λ N gene function. Mol Gen Genet 179:55-61

Greenblatt J, Li J (1981) Interaction of the sigma factor and the nusA gene protein of E. Coli with RNA polymerase in the initiation-termination cycle of transcription. Cell 24:421-428

Horwitz RJ, Li J, Greenblatt J (1987) An elongation control particle containing the N gene transcriptional antitermination protein of bacteriophage lambda. Cell 51:631-641

Lazinski D, Grzadzielska E, Das A, (1989) Sequence-specific recognition of RNA hairpins by bacteriophage antiterminators requires a conserved arginine - rich motif. Cell 59:207-218

Lozeron HA, Anevski PJ, Apirion D (1977) Antitermination and absence of processing of the leftward transcript of coliphage lambda in the RNase III difficient host. J Mol Biol 109:359-365

Miller JH (1972) Experiments in molecular genetics, Cold Spring Harbor Laboratory, Cold Spring Harbor, New York pp 352-355

Olson ER, Flamm EL, Friedman DI (1982) Analysis of nutR: A region of phage lambda required for antitermination of transcription. Cell 31:61-70

Olson ER, Tomich CC, Friedman DI (1984) The nusA recognition site: Alteration in its sequence or position relative to upstream translation interferes with the action of the N antitermination function of phage λ. J Mol Biol 180:1053-1063

Rosenberg M, Court D, Shimatake H, Brady C, Wulff DL (1978) The relationship between function and DNA sequence in an intercistronic regulatory region of phage λ. Nature 272:414-423

Salstrom JS, Szybalski W (1978) Coliphage λnutL-: A unique class of mutants defective in the site of gene N product utilization for antitermination of leftward transcription. J Mol Biol 124:195-221

Steege DA, Cone KC, Queen C, Rosenberg M (1987) Bacteriophage λ N gene leader RNA. J Biol Chem 262:17651-17658

Warren F, Das A (1984) Formation of termination-resistant transcription complex at phage lambda nut locus: effects of altered translation and a ribosomal mutation. Proc Natl Acad Sci USA 81:3612-3616

Whalen W, Ghosh B, Das A (1988) NusA protein is necessary and sufficient in vitro for phage λ N gene product to suppress a Rho-independent terminator placed downstream of nutL. Proc Natl Acad Sci USA 85:2494-2498

Zuber M, Patterson TA, Court DL (1987) Analysis of nutR: A site required for transcription antitermination in phage lambda. Proc Natl Acad Sci USA 84:4514-4518

MESSENGER RNA 3' END FORMATION IN E. COLI AND S. CEREVISIAE

T. Platt, C. A. Brennan, J. S. Butler*, D. A. Campbell, P. P. Sadhale+, P. Spear, E. J. Steinmetz, S.-Y. Wu, and F. M. Zalatan+
Departments of Biochemistry, *Microbiology & Immunology, and +Biology
University of Rochester
Rochester, NY 14642.

Transcription termination and RNA processing can both participate in generating the 3' ends of mRNA molecules in living cells. Our own work is currently focused on rho-dependent transcription termination in E. coli, and on mRNA 3' end processing in S. cerevisiae. This article will summarize and integrate our current results with some general concepts in the field.

I. Rho-dependent Termination of Transcription in E. coli

In contrast to termination of transcription by RNA polymerase at the familiar hairpin-loop "intrinsic" termination sites, it is generally accepted that termination mediated by rho factor is complex. The mechanism must involve interactions between rho protein and its RNA target, directional action from a distance to disrupt the nascent RNA-DNA duplex, and concurrent pausing and dissociation of RNA polymerase from the template DNA. Curiously, no consensus sequences or structures have yet emerged, and whether additional factors, such as nusA protein, are also necessary is not yet known. Our model system is the rho site trp t' at the end of the trp operon, spanning ca. 200 nt, of which the 5' half contains sufficient information for rho specificity, while the 3' mRNA endpoints fall into the 3' half.

A. Removal of cytosines does not eliminate termination. The elements responsible for determining rho-dependent termination have not been identified, but rho does appear to require an RNA transcript relatively lacking in classical secondary structure, and with a moderate abundance of cytosine residues. To evaluate the participation of cytosines in termination, we used bisulfite to mutagenize the promoter-proximal region of trp t' (ca. 100 nt), inserted in an m13

NATO ASI Series, Vol. H 49
Post-Transcriptional Control of Gene Expression
Edited by J. E. G. McCarthy and M. F. Tuite

vector between the trp promoter and a lacZ reporter gene. This yielded more than 30 variant terminators with a spectrum of C-to-U changes at the 28 possible positions in this region: in general, as more cytosines are converted to uracil, the extent of readthrough transcription increases (Zalatan & Platt, in prep.). With the simple premise that each C contributes independently, we observe that about one-third are unimportant and a third are of uncertain significance; the remaining third are important, based on their consistent absence in the inefficiently terminating mutants, but display no interpretable pattern.

Surprisingly, the derivative in which all 28 C residues were changed to U still retains about 30% of its original termination efficiency, relative to an appropriate control. What remains unknown in this, as in the other bisulfite mutants, is whether the termination endpoints are altered, in addition to changes in termination efficiency. When taken together with the fact that 5-iodocytosine substituted RNA fails to function in the helicase assay (Brennan et al., 1987), these results support the traditional view that at least some cytosines are important, but also imply that more than one type of interaction is responsible for generating the affinity and specificity of rho factor for its RNA target, in agreement with Faus and Richardson (1989). Determination of the specific trp t' recognition elements has been hampered by the difficulty of identifying any rho "footprint" -- in our hands, we cannot discern any significant protection of RNA by rho from (i) T1 or pancreatic RNase, (ii) hydroxyradical cleavage, (iii) methylation, or (iv) hybridization of short oligonucleotides. Possibly the rapid exchange of binding to two different sites in the rho hexamer is responsible for this, reflecting alternating occupancy of ATP sites, as only 3 sites per hexamer are active at any one time (for discussion see Dombroski & Platt, 1989).

B. RNA binding and NTP hydrolysis by rho factor. The complementary problem is how rho recognizes its target RNA, and several lines of evidence indicate that an N-terminal "headpiece" domain of ≤130 amino acids is primarily involved in RNA binding. First, this rho headpiece has considerable similarity to the RNA binding consensus regions identified in a number of other proteins, and when rho is bound to polyC, the only region resistant to mild tryptic digestion is the N-terminal F3 fragment of 123-130 residues (Brennan et al., 1990; Dolan et al., 1990). Second, renaturation of a 151 amino acid fragment regains all the RNA binding ability and specificity of the intact rho molecule (see Dombroski &

Platt, 1989). Third, UV crosslinking of trp t' RNA to rho is exclusively to the headpiece, and appears localized to a proteolytic fragment spanning the Gly-Phe-Gly-Phe consensus sequence (Brennan & Platt, in prep.). To test the importance of this consensus, we converted both Phe residues into Leu, but found no difference in rho function in vitro. We conclude that if the Phe residues are crucial, Leu can effectively substitute; conversion and evaluation of Phe to Ala changes are underway.

RNA binding is an absolute prerequisite for NTP hydrolysis in vitro, hence there must be conformational changes in the ATP-binding domain that result from RNA interactions with the headpiece. In turn, ATP (or NTP) hydrolysis must be coupled to the directional action of rho at a distance. Partial uncoupling of this reaction (relative to ATP) can be observed in vitro with the 3 other nucleotides, and the inefficient use of energy (for ATP, ca. 4000 molecules are hydrolyzed per RNA molecule released in the helicase assay) suggests that only fractional amounts are productively converted to helicase action (Brennan et al., 1990). These differences correlate with changes in rho conformation as probed by mild trypsin digestion, and suggest possible models for energy transduction and the relationships between the RNA-binding headpiece of rho and the separate ATP-binding domain (see Dombroski & Platt, 1989). In any event, the nucleotide-dependent partial uncoupling of hydrolysis from helicase action implies that NTP hydrolysis, while necessary, is not sufficient for effective helicase function.

Site-directed rho mutations in the ATP-binding region that affect hydrolysis and helicase activity in vitro (but not RNA binding) display parallel losses of transcription termination activity in vivo, measured at 3 different termination sites (Dombroski & Platt, 1990). This confirms the importance of rho factor's NTPase activity in vivo and the premise that determinations of rho function in vitro accurately reflect its behavior in intracellular termination events.

C. Models for rho can be tested with the helicase assay. Our development of the helicase assay demonstrated that rho has an intrinsic ability to disrupt an RNA-DNA (or RNA-RNA) duplex, dependent on hydrolysis of NTPs and a rho site (trp t') located 5' to the duplex region (Brennan et al., 1987). The usefulness of this assay, independent of transcription and RNA polymerase, was diminished by the stoichiometric inefficiency and high ATP concentrations needed in our original conditions. Changes to lower salt and Mg^{++} concentrations, plus the realization that optimal [Mg^{++}] and [ATP] were interdependent, yielded a much more efficient assay (Brennan et al., 1990), though for unknown reasons, we cannot yet obtain release stoichiometry better than 1:1 (rho:duplex). Under these improved conditions, several mechanistic models for rho action can be tested.

First, the ability of rho to act with trp t' at increasing distance from the 3' duplex was tested by inserting rho-neutral spacer sequences between trp t' and the duplex. Even at the longest distances tested (ca. 450 nt), the rate of RNA release was barely altered, and the stoichiometry of release was unchanged. This rules out models invoking cooperative action between adjacent rho hexamers, but raises the possibility that rho might act at a distance by looping out on its RNA tether to find the duplex on which it then could act. A serendipitous discovery provided the test for this hypothesis: at 0.4 mM Mg^{++}, rho disrupts the RNA-DNA duplex even more readily than at 4 mM, but fails to disrupt an RNA-RNA duplex region. A spacer RNA (200 nt) interposed into the normal substrate, with a 50 nt fragment of complementary RNA annealed to it, blocks rho helicase function at 0.4 but not at 4.0 mM Mg^{++}. In the control reaction lacking the 50 nt RNA, disruption of the 3' RNA-DNA duplex is efficient in both cases, as expected (Steinmetz, Brennan & Platt, submitted). This indicates that rho must maintain semi-continuous contact with single-stranded RNA, as it would if it translocated 5'-3' from its initial binding site to the 3' RNA-DNA duplex, and rules out simple looping models. These results also suggest that rho is not simply a juggernaut, moving inexorably along the RNA through any duplex structure it encounters, because a complementary RNA fragment cannot be displaced at low Mg^{++}.

Overall, a tracking model with 5'-3' translocation fueled by NTP hydrolysis remains the most likely one. Whether tracking is continuous, or involves interrupted contact with specific spaced sequences, has not been determined. One remaining question is whether rho remains anchored to some portion of its initial binding site. A weak footprint signal at a fixed point in λtR1 RNA suggests that the rest of the RNA may be tracked through the "secondary site", forming a loop of increasing size as rho translocates towards the 3' region (Faus & Richardson, 1990). Protection, however, may be due to a short translocation time relative to the time spent at the initial site; hence it is not easy to distinguish between anchored and unanchored tracking. The specific details may turn out to depend on the particular termination site being examined.

II. Messenger RNA 3' End Formation and Termination in Yeast

For some time after the report by Zaret and Sherman (1982) of a 38 bp deletion (cyc1-512) affecting the 3' end formation of S. cerevisiae CYC1 mRNA, the mechanism(s) responsible for generating mature 3' ends of yeast polymerase II transcripts remained uncertain. On the one hand, all other eukaryotes examined utilized endonucleolytic cleavage and subsequent polyadenylation of a longer pre-mRNA to produce mature transcripts (except for the non-polyadenylated cell-cycle specific histone mRNAs). On the other hand, the ubiquitous and essential AAUAAA signal of higher eukaryotes has been convincingly shown not to be a part of the signal in yeast, and sequence (or secondary structure) comparisons of the 3' regions of yeast genes revealed no other compelling universal consensus. Since longer pre-mRNAs had not been detected in yeast, and the yeast histone mRNAs are also polyadenylated, circumstantial evidence superficially favored mechanisms involving transcription termination to generate a 3' end, followed by promiscuous polyA addition. We were drawn to this problem by our historical interest in termination, and the prospect of examining mRNA 3' end formation by harnessing the power of yeast genetics to its biochemical accessibility.

A. A genetic screen for defects in 3' end formation. We devised a genetic approach to obtain mutations without any prejudice as to mechanism. An integrated plasmid in which the yeast actin gene (with its intron) had been fused to the HIS4 gene conferred the ability to grow on histidinol to a host strain deleted for HIS4. When an 83 bp fragment from the 3' region of CYC1 was interposed within the actin intron, it diminished HIS4C (histidinal dehydrogenase) expression and the strain became Hol$^-$, unable to grow on histidinol, presumably due to premature 3' end formation within the insert (Ruohola et al., 1988). We hoped that selecting for increased HIS4C expression (Hol$^+$) would capture some mutations altering mRNA 3' end formation at the interposed CYC1 segment.

Our preliminary examination of Hol$^+$ mutations that were recessive and trans-acting revealed three complementation groups (Ruohola et al., 1988). We have now identified three more complementation groups, and confirmed two of the three that were previously identified (we were unable to work with isolates from the third). Of the five genes (now designated REF, for RNA end formation), three are genetically linked to the monitoring locus at HIS4 on chromosome III, but are separated by 5 cM (REF3), 23 cM (REF5), and 29 cM (REF1) from the integrated indicator plasmid; the two other genes (REF2, REF4) are unlinked to chromosome III (Campbell, Baker & Platt, in prep.). Northern blot analyses indicate that in some of these mutants there is a marked reduction of the truncated (by CYC1 insert) mRNA transcript, and attempts to identify and characterize the REF genes are underway.

B. A helicase search revealed 3' processing events in vitro. The success of the helicase assay for E. coli rho factor prompted us to search for a similar activity in yeast extracts, reasoning that displacement of the final 3' region of nascent RNA from the DNA template was a universal problem. The search itself was temporarily abandoned (RNase H activity in the extracts destroyed the helicase substrate), but not before obtaining indications that the input RNA was also being cleaved near the normal polyadenylation site. We went on to show that a capped SP6 transcript corresponding to the 3' region of the S. cerevisiae CYC1 gene was accurately cleaved and polyadenylated at its normal poly(A) site in whole cell extracts (Butler & Platt, 1988). Cleavage was energy dependent, but independent of subsequent poly(A) addition. Failure to cleave a pre-mRNA carrying cyc1-512 (a 38 bp deletion that eliminates correct 3' end formation in vivo) or a similarly sized E. coli trp RNA, indicated that this processing event is specific and confirmed the importance of signals (as yet unidentified) within this 38 bp region.

The possibility that cleavage might be peculiar to CYC1 was ruled out by examining other yeast pre-mRNAs, including those for GAL1, GAL7, GAL10, HIS4, CBP1, PRT2, and histone H2B2. We have now shown that synthetic RNAs corresponding to the 3' regions of each of these genes can be endonucleolytically cleaved and polyadenylated in whole cell yeast extracts, although variation in processing efficiency is seen among the different substrates (Butler et al., 1990; Sadhale, Butler & Platt, in prep.) These assay results in vitro prove that 3' end formation does not depend on transcription termination (though the reverse may be true, and indirect effects have not been ruled out). Moreover, since the same extract can process a variety of unrelated pre-mRNAs, it is likely that cleavage and polyadenylation are the final 3' maturation steps for all pol II transcripts in S. cerevisiae. Moreover, they appear to be closely coupled, because our extracts fail to polyadenylate a "pre-cleaved" RNA, in contrast to results obtained in mammalian extracts.

Further characterization of the processing activity reveals that, as in higher eukaryotic systems, substrate RNA is very rapidly sequestered in high molecular weight complexes that can be readily detected on native acrylamide gels. Preliminary purification attempts indicate that the polyadenylation activity is labile, but the cleavage activity can be followed through several steps. UV crosslinking reveals a small number of proteins radiolabeled by input RNAs, with some specificity evident after fractionation of the activity on a DEAE column. We have also begun examining extracts from various mutants for defects in activity. Our only success thus far has been to retrieve one mutant from the Hartwell temperature-sensitive collection with a ts defect for polyadenylation in vitro, which cosegregates with the ts phenotype through several outcrosses (Butler, unpub.). Examination of extracts from a number of the Hol$^+$ REF mutants, and from the mutants prp20 and rna1-1 (both display lethal temperature sensitive defects in RNA metabolism), has thus far failed to reveal significant differences in 3' end processing in vitro, compared to control strains.

C. Signal elements for yeast 3' end formation remain elusive. The AAUAAA polyadenylation signal of higher eukaryotes is non-functional in S. cerevisiae. Since our extracts also fail to cleave a mammalian α-globin SP6 pre-mRNA, even the full complement of mammalian signals is not sufficient to direct mRNA 3' end formation in yeast, at least in vitro. Despite a commonality of function in the 3' regions of yeast genes, the lack of sequence similarities, even among just the genes tested, suggests that signals used in the processing of these substrates may differ from one another or be comprised of multiple elements. One hopeful clue stems from experiments examining processing susceptibility of CYC1 RNAs with base analog replacements. Cleavage is eliminated when 5-BrUTP replaces UTP during synthesis of the pre-mRNA substrate, while processing of substrates made with ITP (instead of GTP) or 5-iodoCTP (instead of CTP) proceeds normally (Wu & Platt, unpub.).

Another approach to elucidating comparative signals is offered by the structural genes in galactose metabolism, since they respond differentially to the ts mutation rna1-1: under nonpermissive conditions the mRNAs of GAL1 and GAL10, but not of GAL7, are longer than normal at their 3' ends. Pre-mRNAs corresponding to these three genes are processed in vitro, but the efficiency of processing is reduced compared to other transcripts we have examined (CYC1, HIS4 and H2B2), and the signals for GAL7 premessage processing seem to be confined to a much shorter 3' region than those involved in processing of GAL1 and GAL10 mRNAs (Sadhale, Butler & Platt, in prep.). It is puzzling that antisense transcripts for both GAL7 and GAL10 are also processed in vitro, although no corresponding mRNAs have been reported in vivo.

III. Ramifications and Ruminations

Although the specific details of mRNA 3' end formation in yeast and E. coli are probably quite different, there are likely to be common mechanistic themes governing these macromolecular interactions. A deeper understanding of RNA-protein recognition and the interplay of transcription with RNA processing should also be applicable to more complex higher eukaryotic organisms.

A. How is RNA-protein recognition achieved? Interactions between nucleic acids and proteins are important molecular events in living cells. Questions about how RNA is recognized and bound by proteins have been neglected compared to those about DNA-protein interactions, despite the key role of RNA sequence and structure in many crucial aspects of cellular function. Most work thus far has focused on protein recognition of rRNA, tRNA, viral RNA, and hairpin loops in mRNA, all of which are highly structured. The major challenge, in both our systems, is that signals critical for function lie within short (ca. 100 nt) regions of RNA that are unstructured and lack obvious homologies. We infer that some significant element(s) other than linear homology and classical base-paired structure must play an important role in function, and that higher order structure may be involved. As model systems these RNA molecules may illuminate a new set of unusual elements, if we can formulate the right questions. Such elements could include combinations of small but precisely spaced signals, such as the 12 nt C-spacing proposed for rho recognition (Platt, 1986), or the tripartite sequence in CYC1 (Zaret & Sherman, 1982), though neither of these specific examples remains tenable. Components of RNA recognition might also include pseudoknots, tertiary interactions, or structures of higher order.

To examine protein recognition of RNA, the small RNA-binding domain of the bacterial rho factor may provide a useful model system. UV crosslinking of the trp t' RNA target occurs exclusively within this headpiece (Brennan & Platt, unpubl.), but in neither the RNA nor the protein are any specific contacts, structures, or regions involved in recognition yet known. The yeast system, with UV-crosslinks of CYC1 RNA to several proteins in large macromolecular complexes, is also promising though less well advanced.

B. What is the relationship between processing and termination?

At the 3' end of the E. coli trp operon, termination of transcription at either of the two tandem sites (at the rho-independent trp t hairpin or the rho-dependent trp t' site) is the major event in mRNA maturation (see Platt, 1986). At trp t', subsequent exonucleolytic processing (probably by RNase II) generates a slightly shorter but stable mRNA whose 3' end coincides with the trp t hairpin. There is no evidence, however, that defects in either termination or processing at trp t' have any effect on trp expression.

By contrast, processing events appear to dominate in yeast, with specificity inherent in the RNA transcript implying independence from transcriptional events. The fate of RNA polymerase II at the ends of yeast genes is unknown, but evidence suggests that signals for transcription termination are also closely associated with the region altered by the cyc1-512 deletion (Zaret & Sherman, 1982; Osborne & Guarente, 1989), and that pol II does not proceed far beyond the poly(A) site in vivo (Russo & Sherman, 1989). It is still an open question whether termination events are specified primarily by interactions with the mRNA (as in prokaryotes), or have components dependent on DNA recognition by the transcriptional apparatus. Preliminary experiments with whole cell and nuclear extracts have failed to detect any gel retardation of DNA fragments spanning the CYC1 3' region, nor do these fragments display any aberrant mobility that would indicate bends or other unusual structure (Platt, unpub.). In both E. coli and S. cerevisiae, the questions of functional signal redundancy (reminiscent of mammalian enhancer elements), the participation of higher order RNA structure, and the principles of protein-RNA recognition are among the dominant challenges for future work. The years ahead promise to reveal exciting answers to some of these questions.

References

Brennan CA, Dombroski AJ, Platt T (1987) Transcription Termination Factor Rho Is an RNA-DNA Helicase. Cell 48:945-952

Brennan CA, Steinmetz EJ, Spear P, Platt T (1990) Specificity and Efficiency of Rho-factor Helicase Activity Depends on Magnesium Concentration and Energy Coupling to NTP Hydrolysis. J Biol Chem 265:5440-5447

Butler JS, Sadhale P, Platt T (1990) RNA Processing in vitro Produces Mature 3' Ends of a Variety of Saccharomyces cerevisiae mRNAs. Mol Cell Biol 10:in press (June)

Butler JS, Platt T (1988) RNA Processing Generates the Mature 3' End of Yeast CYC1 Messenger RNA in Vitro. Science 242:1270-1274

Dolan JW, Marshall NF, Richardson JP (1990) Transcription Termination Factor Rho Has Three Distinct Structural Domains. J Biol Chem 265:5747-5754

Dombroski AJ, Platt T (1989) Structure and Function of Rho Factor and Its Role in Transcription Termination. In: Adolph K (ed) Molecular Biology of Chromosome Function. Springer-Verlag, New York, p 224

Dombroski AJ, Platt T (1990) Mutations in the ATP-Binding Domain of E. coli Rho Factor Affect Transcription Termination in vivo. J. Bact. 172:in press (May)

Faus I, Richardson JP (1989) Thermodynamic and Enzymological Characterization of the Interaction between Transcription Termination Factor Rho and λ cro mRNA. Biochem 28:3510-3517

Faus I, Richardson JP (1990) Structural and Functional Properties of the Segments of Lambda cro mRNA that Interact with Transcription Termination Factor Rho. J Mol Biol:in press

Osborne BI, Guarente LP (1989) Mutational analysis of a yeast transcriptional terminator. Proc Nat Acad Sci USA 86:4097-4101

Platt T (1986) Transcription Termination and the Regulation of Gene Expression. Annu Rev Biochem 55:339-372

Ruohola H, Baker SM, Parker R, Platt T (1988) Orientation-dependent function of a short CYC1 DNA fragment in directing mRNA 3' end formation in yeast. Proc Nat Acad Sci USA 85:5041-5045.

Russo P, Sherman F (1989) Transcription terminates near the poly(A) site in the CYC1 gene of the yeast Saccharomyces cerevisiae. Proc Nat Acad Sci USA 86:8348-8352

Zaret K, Sherman F (1982) DNA sequence required for efficient transcription termination in yeast. Cell 28:563-573

CULTURE CONDITIONS AFFECT DIFFERENTLY THE TRANSLATION OF INDIVIDUAL *ESCHERICHIA COLI* mRNAs

N. Jacques, M. Chevrier-Miller, J. Guillerez and M. Dreyfus
Laboratoire de Génétique Moléculaire
Ecole Normale Supérieure
46 rue d'Ulm
75005 Paris
France

Introduction

Individual bacterial mRNAs are translated at very different rates (Ray and Pearson, 1975; McCarthy, et al., 1985). Extensive work during the last decade has revealed that these unequal performances largely stem from sequence differences in the region contacted by the ribosome during initiation (Ribosome Binding Site or RBS: see Steitz, 1969). Thus, the nature of the initiation codon, the length of the Shine-Dalgarno sequence (SD), and the spacing between them, all contribute to the translational efficiency (Gold, 1988). Additional RBS sequence elements, less characterized to date, are probably also important (McCarthy et al., 1985; Petersen et al., 1988). Finally, the extent of secondary structure around RBSs profoundly affects their efficiencies (Gold, 1988; de Smit and van Duin, 1990). However, the role of these different elements has not been investigated as a function of growth conditions. The concentration of ribosomes, factors, precursors, etc, vary according to the metabolic state of the cell (Bremer and Dennis, 1987), and in vitro such changes affect differently the translational yields from individual mRNAs (Gualerzi et al., 1988). Should this also occur in vivo, then the translational yields from individual mRNAs would be expected to change according to growth conditions.

Quite a few mRNAs have been reported to be differently translated depending upon the metabolic state of the cell, in the absence of a specific regulatory loop. The leader sequence of the ampC mRNA, which encodes an E. coli ß-lactamase, harbours a putative ribosome binding site which varies in occupancy according to growth conditions, thereby modulating the ampC expression via an attenuation-like mechanism (Jaurin et al., 1981). The concentration of the central metabolism enzyme 6-phosphogluconate dehydrogenase, encoded by the gnd gene, rises under rapid growth, due to the increased translatability of its mRNA

NATO ASI Series, Vol. H 49
Post-Transcriptional Control of Gene Expression
Edited by J. E. G. McCarthy and M. F. Tuite

(cf. Jones et al, 1990). However, the underlying mechanism has not yet been worked out in detail, nor is it known whether the ampC and gnd cases are unique or represent first examples of an as yet unrecognized regulatory mechanism.

We have recently described the construction of a series of E.coli strains which are completely isogenic except for the chromosomal region corresponding to the lacZ RBS. In its place, we inserted small DNA fragments (60-80 nt) encompassing the RBSs from various other genes, together with some flanking sequence (cf. Fig. 2A). The lac promoter as well as most of the transcribed sequence are invariant throughout this collection of strains, and therefore the level of ß-galactosidase synthetized by each of them should primarily reflect the efficiencies of the corresponding RBSs. As expected, these levels varied from one construct to the next (Dreyfus, 1988). More surprisingly, however, they also varied relatively to one another depending upon which medium is used for growth - i.e. rich medium or minimal medium with a poor carbon source. In particular, under slow growth, those strains showing the lowest activities were invariably favored over others.

While suggestive of an effect of metabolism upon translation efficiencies, this preliminary experiment did not reveal which feature(s) of the mRNAs was responsible for their variable translatability. Moreover, two possible sources of complications were apparent. First, the mRNAs and hybrid proteins under comparison are not identical in sequence, and therefore may have different intrinsic stabilities. Second, in E. Coli, translation is strongly coupled to transcription and mRNA degradation: the less efficient the lacZ translational start, the higher the probability that transcription stops before reaching the end of the gene (transcriptional polarity; Adhya and Gottesman, 1978). Moreover, mRNAs are generally more prone to degradation when poorly translated (Kennell, 1986). Therefore the observed changes in ß-galactosidase levels might reflect metabolic effects on transcription termination and/or mRNA degradation, rather than translation.

The lamB and galE RBSs

To determine which properties of the mRNAs are responsible for their variable translatability, we started from a given RBS and introduced into it point mutations that alter the SD element, the initiation codon, or the secondary structure. This approach also minimized the sequence differences between the mRNAs under comparison.

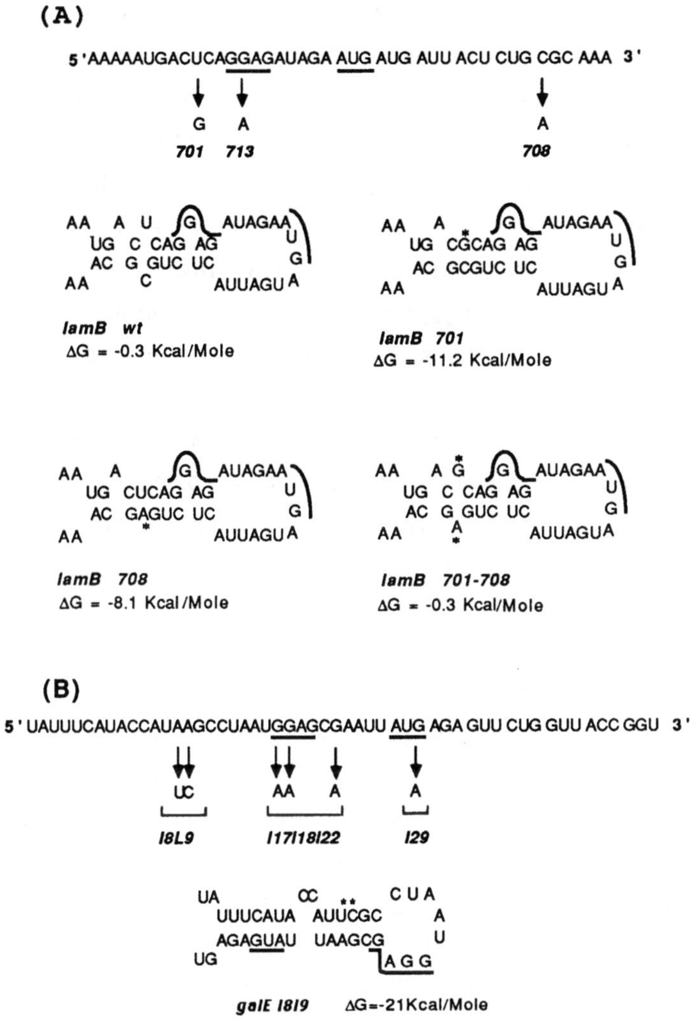

Figure 1. A- Sequence of fragments carrying the lamB (A) and galE (B) RBSs, which were cloned in-phase with lacZ (cf. Fig. 2). Arrows indicate the position of the mutations studied herein, named after Hall et al. (1982) and Busby and Dreyfus (1983). The Shine-Dalgarno sequence and initiation codon are underlined.. The energies of the putative secondary structures were computed according to Kaneisha and Goad (Nucl. Acids Res. 10:265-278 (1982)).

Several nearby mutations affecting the translation of lamB, the gene coding for the phage λ receptor protein, have been isolated. One of them (named 713) alters the SD homology, while the other two (701 and 708) result in a similar secondary structure which sequestrates the SD (Fig. 1). Ultimate proof of the reality of these structures came from the construction of the double mutant (701-

708), in which the wild-type phenotype was restored (Hall et al., 1982). Many mutations depressing translation have also been isolated in the galE gene, which encodes UDP-galactose 4-epimerase. In particular, Busby and Dreyfus (1983) isolated a triple and a single mutant, denoted /17/18/22 and /29, which altered the SD element and the initiation codon, respectively. Finally, Bingham et al (1988) described a double mutation (/8/9) which presumably results in a stable hairpin sequestrating the initiation region (Fig.1). Small DNA fragments, carrying the wild-type or mutated lamB and galE RBSs, were prepared and substituted for the genuine lacZ RBS into the chromosomal lac operon. In all cases, the effects of the lamB and galE RBS mutations on the expression of the lac hybrid gene were qualitatively similar to those originally reported. However, we noted that the double mutant 701-708 produced hardly more ß-galactosidase than the 708 mutant, and markedly less than the wild-type. Conceivably, in the lac context, the double mutant is inhibited by an uncharacterized secondary structure; alternatively, the 701 or 708 mutations per se may reduce the initiation rate.

Elimination of polarity

In order to cope with the polarity problem, we put the lamB-lacZ or galE-lacZ hybrid genes under the control of a promoter recognized by the polarity-insensitive T7 RNA polymerase (Studier and Moffat, 1986; Lindahl et al., 1989). A general outline of these constructs, which are described in detail elsewhere (Chevrier-Miller et al., submitted), is depicted on Fig. 2B. It has previously been reported that E. Coli will not sustain the simultaneous presence of a multicopy T7-transcribed gene, and of an active T7 RNA polymerase gene (Studier and Moffat, 1986). In contrast, we found that, when a single copy of the T7-transcribed lacZ gene is present on the chromosome, balanced growth is easily achieved. To check that the T7 transcription is truly insensitive to polarity, we used the fact that the nearby malQ gene, which encodes the easy-to-assay enzyme amylomaltase, was expressed from the T7 promoter due to some readthrough across the T7 terminator (Fig. 2B): most strikingly, this expression was independent of the presence or absence of an RBS in front of lacZ. This conclusion was extended by Northern blot analysis: no difference in mRNA steady state or degradation pattern was seen, irrespective of the level of translation of the mRNA. Hence, in this system, translation is truly uncoupled from either transcription or mRNA degradation. In contrast, when the same hybrid genes, inserted in the natural lac

(A)

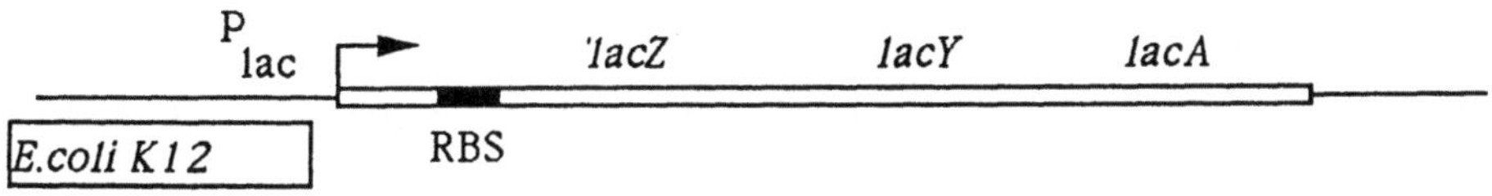

RBS	*lacZ*	*lacA*
galE wt	2600 u	11.5 u
galE 1819	10 u	0.5 u

(B)

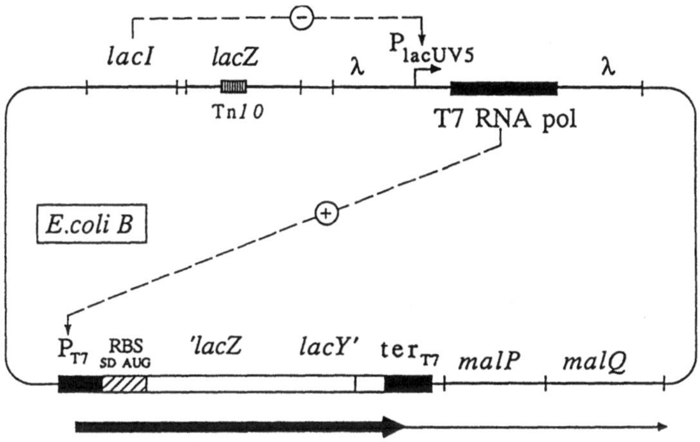

RBS	*lacZ*	*malQ*
galE wt	13000 u	400 u
galE 1819	29 u	490 u

Figure 2. A- (up) Small DNA fragments (50-80 nt.) encompassing the RBSs from various E.coli genes, have been inserted in place of the genuine lacZ RBS, on the E.coli chromosome (see Dreyfus, 1988). (Bottom) ß-galactosidase (lacZ) and transacetylase (lacA) activities from strains carrying either the wild type galE RBS, or the down mutant *1819* (Fig.1). Cells were grown in rich medium. Note the strong polar effect of the *1819* mutation on lacA expression. B- (up) The galE-lacZ and lamB-lacZ hybrid genes are fused to the T7 gene 10 promoter and terminator (Rosenberg et al (1987) Gene 56:125-135) and inserted onto the malA region of the chromosome by homologous recombination (Raibaud et al., 1984). The recipient strain is a derivative of BL21 (λDE3) (Studier and Moffat, 1986), an E.coli B strain which harbours an IPTG inductible T7 RNA polymerase gene. (Bottom) ß-galactosidase (lacZ) and amylomaltase (malQ) activities from strains carrying the same two RBS as in A. Cells were grown in minimal glycerol medium. Note the lack of polarity in this case.

chromosomal location, were transcribed by the E. Coli RNA polymerase, down mutations in the galE or lamB RBS caused extensive polarity, as judged from the expression of the promoter-distal gene lacA. This holded particularly true for the very inefficient galE RBS mutants (cf. Fig. 2A).

The translational efficiencies of the galE and lamB RBS are affected differently by changes in growth conditions

The ß-galactosidase levels from the lamB and galE series of mutants, grown either in fully supplemented glucose medium, or in minimal acetate medium, is shown on Fig. 3. It is seen that, compared to wild-type, the mutant RBSs which are structure-inhibited (701, 708, and l8l9) are more active in acetate medium, while those lacking an efficient SD element or initiation codon (713, l17l18l22, and l29) are less so (Fig. 3, left). However, this did not hold when the E. coli RNA polymerase was used to transcribe the galE series: in this case, all three mutants rised in acetate medium, although the secondary structure mutant was still favored over the other two (Fig. 3, middle). We believe that this odd behaviour stems from a partial relief of polarity in acetate medium. Indeed, polarity effects are known to be less severe when catabolic repression is relieved (Ulmann et al., 1979).

The above effects were not specific for two particular growth media. Rather, the ratios of ß-gal activities between mutants differed in all media examined. However, in the lamB series, the ratios were nearly levelled off in all minimal media. For the galE series, the bulk of the medium effect was obtained when comparing the fully supplemented glycerol medium (μ=1.5 h^{-1}) to the same medium lacking Leu, Ile, Val and Pro (gly/AA-4 medium) (μ=1 h^{-1}). It seems, therefore, that the physiological parameter which is relevant here is not the growth rate per se, but rather the supply of aminoacids. This conclusion was consistent with the result of starvation experiments (Fig. 3, right). Compared to unstarved cultures, cultures starved for isoleucine in gly/AA-4 medium showed the same changes in ß-galactosidase expression as acetate-grown cultures.

Possible interpretations

In aminoacid-starved Rel^+ cells, the frequency of translation initiation on individual mRNAs is reduced, and the elongation cycle is lengthened (see Cashel and Rudd, 1987, for review). Both features are exploited by the cells to reduce the level of protein synthesis, in order to cope with the short aminoacid supply.

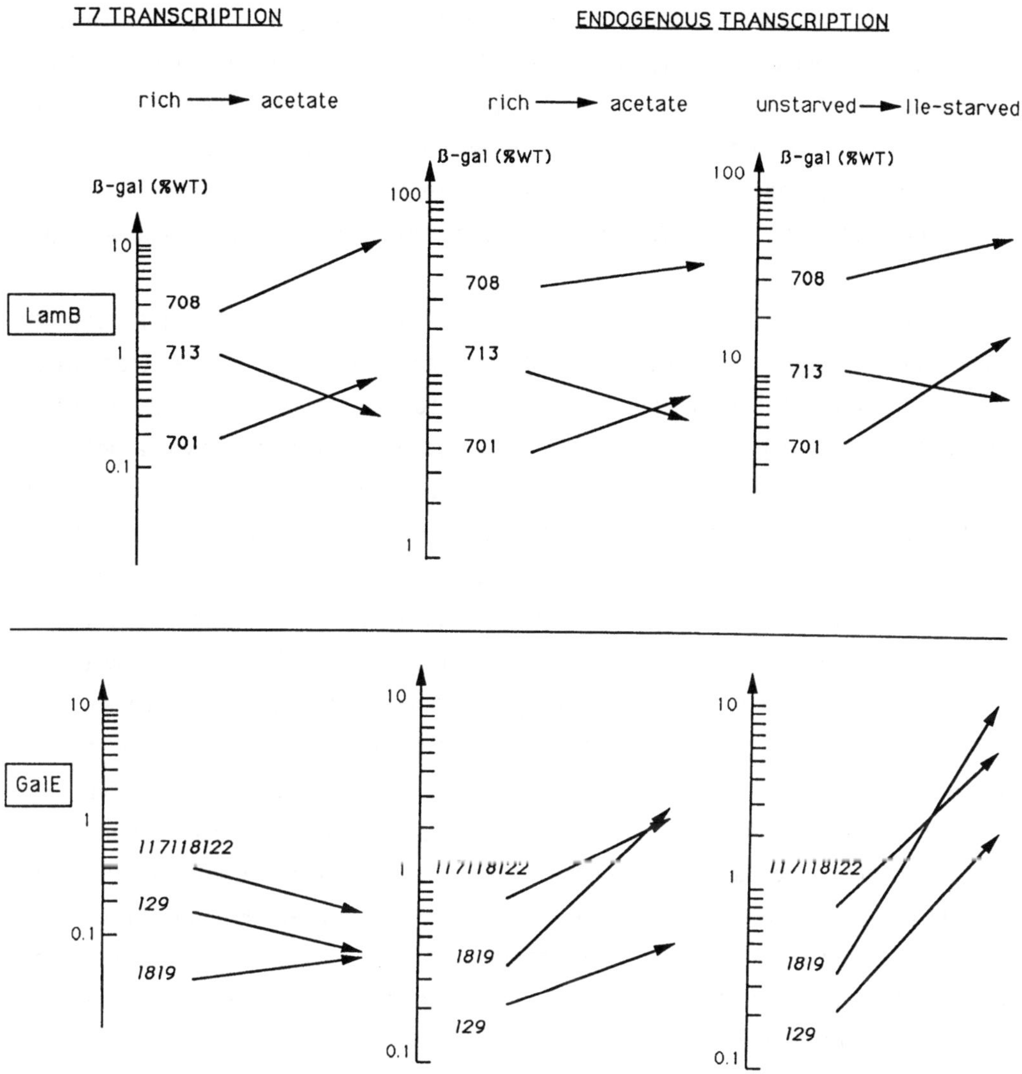

Figure 3. ß-galactosidase activities from hybrid genes carrying the mutated galE and lamB RBSs, under different growth conditions (100% corresponds to the wild-type activity in each case). All cultures were grown in MOPS medium (Neidhardt et al. (1974) J. Bact. 119:736-747). "rich" is a fully supplemented medium with 0.2% glucose as the carbon source, and "acetate" is a minimal medium with 0.4% sodium acetate. "unstarved" cells were grown in rich medium; "starved" cells were grown in (gly/AA-4) medium (see text) and subsequently treated with valine for 10 hours before harvest. All cultures were continuously induced with IPTG, except "starved" cultures, which were provided with IPTG 5mn after valine addition.

Similarly, during balanced growth in minimal medium, the elongation rate is reduced with respect to fully-supplemented medium (Bremer and Dennis, 1987). Moreover, in this case also, an uncharacterized mechanism seems to operate which adjusts the overall translation initiation to a level compatible with the resources of the medium: thus, this level is insensitive to artificial manipulations of the concentration of mRNAs and free ribosomes inside the cell (Nomura et al., 1987).

It cannot be excluded that the changes in translation efficiencies reported herein are a mere consequence of the general metabolic control over translation initiation. Conceivably, a mechanism able to govern the overall initiation frequency, could also alter the specificity of ribosomes towards individual mRNAs. Nevertheless, we presently favor an alternative interpretation in which variations of the elongation rate play the major role. In Fig. 4C, we illustrate a possible mechanism whereby a slowed ribosome would favor structured starts over less structured ones, or over starts which bind ribosomes slowly, such as the SD^- mutants. According to this interpretation, it should be possible to mimic the effect of changing growth conditions, simply by manipulating the elongation rate. This, indeed, proved to be correct for the lamB series of mutants. In these experiments, we supplemented the cultures with fusidic acid, an antibiotic which inhibits the elongation factor EF-G. Alternatively, we introduced into our strains a mutated fus gene which encodes a thermosensitive EF-G protein, and allowed cells to grow at the semipermissive temperature. Using both approaches, we observed that, when the elongation rate was decreased in fully supplemented medium, the secondary structure mutants were favoured over the SD^- one, as during growth in acetate medium (Fig. 4).

Conclusions

In this work, we have shown that the translational yields from individual mRNAs can be differently affected by the metabolic state of the cell. The observed effects can be quite large - well over an order of magnitude in some cases (N.J. and M.D., unpublished). While it seems likely that this phenomenon is general, we ignore whether it serves special regulatory purposes in some cases, or whether it is simply a passive effect which is buffered by specific regulatory loops.

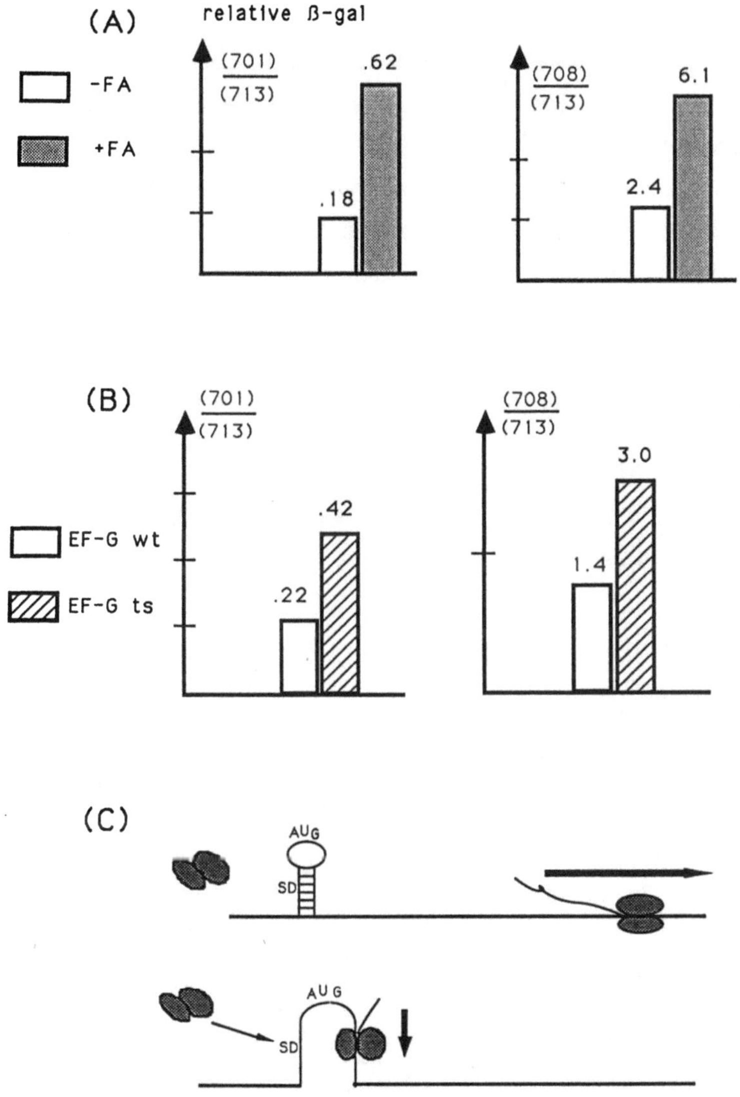

Figure 4. The activities of the mutated lamB RBSs are affected differently by a slowed translation elongation. (A) BL21 derivatives carrying the RBS 701, 708, or 713 (T7 transcription: cf. Fig. 2B) were grown in "rich" medium (cf. legend of Fig. 3) containing or not 330 μg/ml fusidic acid (FA) (μ= 0.33 h^{-1} and 2.2 h^{-1}, respectively), and then assayed for ß-galactosidase. (B) A fus allele encoding a thermosensitive elongation factor G (Tocchini-Valentini and Mattoccia (1968) Proc. Natl. Acad. Sci. USA 61:146-151) has been introduced into the E.coli K12 series harbouring the mutated lamB RBSs (Fig. 2A). Cultures were grown at 32°C, which resulted in a 3-fold lower elongation rate in cells carrying a thermosensitive EF-G, compared to wild-type. (C). A simple model which might explain how a slow elongation comparatively favors translation initiation from structured RBSs.

We have also shown that in vivo, the initiation and elongation of translation are coupled in some cases, and we propose that this coupling is at least in part responsible for the metabolic variations in translational yields. Such coupling is paradoxical since we have observed it with mRNAs endowed with inefficient RBS, for which translation initiation, rather than elongation, should be rate-limiting. We propose that elongation can melt the structures which inhibit initiation: the longer the melting of the structure by forefront ribosomes, the higher the probability of additional initiations. The validity of this model, which is reminiscent of the translational attenuation mechanism which regulates the expression of the cat and ermC genes in Bacilli (cf. Lovett (1990) for references), remains to be tested.

Acknowledgments

We are much indebeted to our colleagues O. Raibaud, M. Springer, and M. Uzan for their help and useful advices.

References

Adhya S and Gottesman M (1978) Control of transcription termination, Ann. Rev. Biochem. 47:967-996.

Bingham A., Fulford F., Murray P., Dreyfus M., and Busby S. (1988) Translation initiation in the Escherichia Coli galE gene, in Genetics of Translation, Tuite H.F., Picard M., and Bolotin-Fukuhara M. Eds, Springer-Verlag, Berlin, pp 307-316.

Bremer H and Dennis PP (1987) Modulation of the chemical composition and other parameters of the cell by growth rate, in Escherichia Coli and Salmonella Typhimurium: Cellular and molecular biology, FC Neidhardt et al. eds., ASM, Washington DC, pp 1527-1542.

Busby S and Dreyfus M (1983) Segment-specific mutagenesis of the regulatory region of the galactose operon, Gene 21:123-133.

Cashel M and Rudd KE (1987) The stringent response, in Escherichia Coli and Salmonella Typhimurium: Cellular and molecular biology, F.C. Neidhardt et al. eds, ASM, Washington DC, pp 1410-1438

Dreyfus M (1988) What constitute the signal for the initiation of protein synthesis on E. Coli mRNAs? J. Mol. Biol. 204:79-94

Gold L (1988) Posttranscriptionnal regulation mechanisms in E. coli, Ann. Rev. Biochem. 57:199-233.

Gualerzi CO, Calogero RA, Canonaco RA, Brombach M and Pon CL (1988) Selection of mRNA by ribosomes during procaryotic translational initiation, in Genetics of Translation, Tuite H.F., Picard M., and Bolotin-Fukuhara M. Eds, Springer-Verlag, Berlin, pp 317-330.

Hall MN, Gabay J, Débarbouillé M and Schwartz M (1982) A role for mRNA secondary structure in the control of translation initiation, Nature 295:616-618.

Jaurin B, Grundström T, Edlund T, and Normark S (1981) The E. coli ß-lactamase attenuator mediates growth rate-dependent regulation, Nature 290:221-225

Jones WR, Barcak GJ, and Wolf RE (1990) Altered growth-rate dependent regulation of 6-Phosphogluconate deshydrogenase (...), J. Bact 172:1197-1205

Kennell DE (1986) The instability of mRNAs in bacteria, in Maximizing gene expression,W. Reznikoff and L. Gold Eds., Butterworth, USA, pp 101-142.

Lindahl L, Archer RH, McCormick JR, Freedman LP and Zengel JM (1989) Translational coupling of the two proximal genes in the S10 ribosomal protein operon, J. Bact. 171:2639-2645.

Lovett PS (1990) Translational attenuation as the regulator of inducible cat genes, J. Bact 172:1-6.

McCarthy JEG, Schairer HU and Sebald W (1985) Translational initiation frequency of atp genes from E. coli: identification of an intercistronic sequence that enhances translation, EMBO J., 4:519-526.

Nomura M, Bedwell DM, Yamagishi M, Cole JR and Kolb JM (1987) RNA polymerase and regulation of RNA synthesis in E. coli: RNA polymerase concentration, stringent control, and ribosome feedback regulation, in RNA polymerase and the regulation of transcription W. Reznikoff et al. eds. Elsevier pp137-149..

Petersen GB, Stockwell PA, and Hill DF (1988) mRNA recognition in E. coli: a possible second site of interaction with 16S ribosomal RNA, EMBO J., 7:3957-3962.

Raibaud O, Mock M and Schwartz M (1984) A technique for integrating any DNA fragment into the chromosome of E.coli, Gene 29:231-241.

Ray PN and Pearson ML (1975) Functional inactivation of bacteriophage λ morphogenetic gene mRNA, Nature 253:647-650.

de Smit MH and van Duin J (1990) Control of procaryotic translational initiation by mRNA secondary structure, Prog. Nucl. Acid Res. and Mol. Biol. 38:in press.

Steitz JA (1969) Polypeptide chain initiation: nucleotide sequences of the three ribosome binding sites in Bacteriophage R17 RNA, Nature 224:957-964.

Studier FW and Moffat BA (1986) Use of bacteriophage T7 RNA polymerase to direct selective high-level expression of cloned genes, J. Mol. Biol. 189:113-130

Ulmann A, Joseph E, and Danchin A (1979) Cyclic AMP as a modulator of polarity in polycistronic transcriptional units, Proc. Natl. Acad. Sci. USA 76:3194-3197.

POST-TRANSCRIPTIONAL CONTROL IN E.coli: THE TRANSLATION AND DEGRADATION OF mRNA

John E.G. McCarthy, Birgit Gerstel, Karsten Hellmuth, Volker Lang, Brian Surin and Peter Ziemke
Gesellschaft für Biotechnologische Forschung
Mascheroder Weg 1
D-3300 Braunschweig
FRG

Introduction

Both the translation and the degradation of mRNA are subject to strict control. The efficiency of translation in E.coli can vary from gene to gene by up to perhaps a factor 1000. Transcript stability, on the other hand, varies by at least a factor 50. Thus these two variables alone can provide for an enormous range of gene expression rates. What structural attributes of the mRNA are responsible for these variations in functional properties, and how can these properties be manipulated? This short review attempts to outline answers to these questions drawing from current knowledge in the field. Given the breadth of the subject matter, we have concentrated mainly on general principles of the control of translation and mRNA stability, with particular reference to the influence of the polycistronic environment on structure/function relationships. The distinction betwen control and regulation observed here has been explained previously (McCarthy,1990).

Translational control

There has been much debate about the steps in translation where rate control is exercised. In particular, the argument that variations in codon usage between genes reflect a strategy to control the efficiency of translation has received considerable support. Yet the rate of completion of polypeptide chains must be determined by the rate of loading/release of ribosomes. Only where the rate of elongation is so slow that a queue of ribosomes develops that stretches back physically to inhibit the binding of 30S subunits at the start of a gene would elongation normally be expected to limit translational efficiency (assuming that long-range interactions between different regions of the mRNA are usually insignificant; see later). But how are we to imagine the control of translational initiation? A functionally defined translational initiation region (TIR) is undoubtedly a useful concept in this respect (McCarthy and Gualerzi, 1990; Fig.1).

NATO ASI Series, Vol. H 49
Post-Transcriptional Control of Gene Expression
Edited by J. E. G. McCarthy and M. F. Tuite

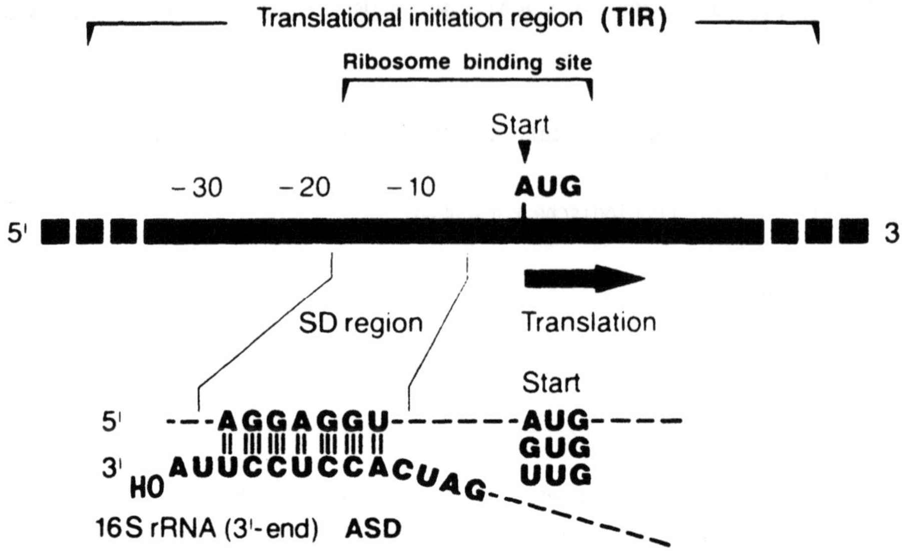

Fig.1 The prokaryotic (E.coli) ribosome-binding site (RBS) and the concept of the translational initiation region (TIR). The Shine-Dalgarno (SD) region is complementary to bases proximal to the 3′-end of the 16S rRNA (the ASD region). The distance between the SD region and the start codon in recognised TIRs is 3-13 bases. The minimal definition of the 'classical' RBS comprises the SD region and the start codon plus the bases in between. At the other extreme the RBS might be defined as the expanse of mRNA protected from ribonuclease digestion by bound ribosomes. The 5′ part of the TIR may comprise an untranslated leader or alternatively may include the end of an upstream gene (on a polycistronic message), and thus a translational start codon. The primary sequence of the TIR contains not only recognition elements but also encodes 'blue-prints' for possible modes of higher order structure. If the TIR (as opposed to the RBS) is to be defined as that region determining both the site and the efficiency of initiation, the definition will include bases beyond the 5′ and 3′ limits of the RBS.

It has become clear that bases both upstream and downstream of the 'classically' defined ribosomal binding site (RBS) influence translational initiation efficiency (Fig.1). Moreover, apart from the recognised and proven recognition elements, viz. the start codon and Shine-Dalgarno (SD) region, several alternative recognition sequences have been proposed (Table 1). Most of these sequences are suggested to interact with the 16S rRNA, one of them with the $tRNA^{fMet}$ a-loop, and one with the ribosomal S1 subunit. None of them can substitute for the SD region, but they may help the 30S subunit locate the initiation site and/or through their binding energy increase the affinity of the 30S protein for the TIR (see later). The sites on the 16S rRNA summarised in Table 1 are discussed further by Firpo and Dahlberg (this volume), while Gold and Hartz discuss the S1 interaction (this volume).

It should be stressed that in general the significance of the alternative recognition sequences listed in Table 1 remains uncertain. None of the RNA-RNA interactions have as yet been confirmed. Moreover, a positive effect of such sequences on the translational efficiency of heterologous genes need not necessarily be attributable to a direct interaction with specific structural components of the ribosome. The introduction of these generally A/T-rich sequences is likely to decrease the stability of secondary or tertiary structure involving the TIR, and it has become clear that higher order structure plays a dominant role in controlling initiation efficiency. Indeed, the most radical 'minimal' model of rate control through the TIR leaves little room for significant participation of extra primary sequence interactions (McCarthy and Gualerzi, 1990). Granted the theoretical maximum initiation rate allowed by a given specified combination of a defined Shine-Dalgarno region and a defined start codon, this model states that the actual rate is dictated by the degree to which access to the TIR is hindered by secondary (higher order) structure(s). It is therefore clear that an unequivocal assessment of the roles of the putative signals will only be possible in the light of experimental data that allow differentiation between direct sequence interactions and the more indirect effects of the presence of these signals on the higher order structure of the TIR. Attempts to analyse the function of the atpE sequence in this way have indicated that a direct RNA:RNA interaction with the 16S rRNA is likely to be of lesser importance in relation to the stability of secondary structure in the TIR (Schauder and McCarthy, 1989). The existence of an S1:mRNA interaction, on the other hand, has been confirmed, although its significance for initiation efficiency remains incompletely defined (see Gold and Hartz, this volume; Boni et al., 1990; and compare McCarthy, 1987).

The definition of the TIR given in Fig.1 assumes that the control of translational initiation, especially in terms of higher order structure, is a localised phenomenon. Recent examinations of the functioning of isolated TIRs are consistent with this assumption (see e.g. Dreyfus, 1988; Lang et al., 1989). Moreover, both the tight coupling between transcription and translation in prokaryotes, as well as the typical ribosomal loading density on the mRNA, will usually mean that longer range interactions will be of lesser significance. This principle might break down where mRNA is poorly translated, or uncoupled from transcription (Petersen, 1989; Jacques and Dreyfus, 1990). The TIR concept presented here also has important practical consequences for strategies of the optimisation of foreign gene expression (McCarthy, 1990b).

TABLE 1 CONFIRMED/PROPOSED RECOGNITION ELEMENTS IN THE *E.coli* TIR

SEQUENCE ELEMENT	POSITION RELATIVE TO START CODON	(PROPOSED) FUNCTION	EVIDENCE/COMMENTS
Confirmed recognition sequences			
Start codons (AUG, GUG, UUG)	+1 to +3	Initiation via interaction with fMet-$tRNA^{fMet}$	Usually only possible choices (but see AUU in *infC*) [1]
Shine-Dalgarno region: 3 or more of UAAGGAGGUGA	Usually within the region -15 to -2	Interaction with 3′ end of 16S rRNA (bases 1532-1542, especially 1535-1539)	Clearest evidence: 'specialised' ribosome systems [2,3]
Proposed (additional) recognition elements			
Preference for specific bases, especially As	Spacer region, 5′ of SD, 3′ of start	See 'enhancing' sequences (below)	Sequence analyses/expression with manipulated TIRs [4-9]
Complementarity to bases adjac. to ASD: any 3 or more of UGAUCC	Within region -58 to -4	Extra recognition element compl. to 16S rRNA (bases 1529-1534) 'initiation promoting site'	Sequence analyses/assumed initiation efficiencies [10]
At least 3 of: UCAAACUCUUCAAUUU	Within region +4 to +25	Extra recognition element compl. to 16S rRNA (bases 1-18)	Sequence analyses [11]
'Omega' sequence - core is: ACAAUUAC	Region not clearly defined	Extra recognition element active in various organisms	Data obtained using sequence from tobacco mosaic virus [12]
Extra *infC* interaction sites: AAGGUAUUAA and GGCGG	-5 to +5 and +7 to +11, resp.	Interaction with 16S rRNA (bases 463-472 and 1399-1403)	Importance of AUU start for IF3 feedback confirmed [1,13]
Extended complementarity to $tRNA^{fMet}$: GUUAUGAGC	-3 to -1 and +4 to +6	More stable interaction with $tRNA^{fMet}$ a-loop	Sequence analysis [14]

Other 'enhancing' sequences (e.g. in TIRs of <u>atpE</u>, bacteriophage T7 gene 10)	Within region ca. -50 to +50	At least to some extent due to highly accessible SD/start region (open TIR);see also [20]	Data obtained using various natural and manipulated TIRs [15,16]
S1 binding site: polyU	In vicinity of SD	Ribosomal subunit S1 binding	Analyses of interactions with mRNA [17-19]

References:

[1] Gold, 1988
[2] Hui and de Boer, 1987
[3] Jacob et al., 1987
[4] Dreyfus, 1988
[5] Stormo, 1986
[6] Gold et al., 1981
[7] Scherer et al., 1980
[8] Kozak, 1983
[9] Hui et al., 1984
[10] Thanaraj and Pandit, 1989
[11] Petersen et al., 1988
[12] Gallie and Kado, 1989
[13] Gold et al., 1984
[14] Ganoza et al., 1985
[15] McCarthy et al., 1985
[16] Olins and Rangwala, 1989
[17] Boni et al., 1990
[18] McCarthy, 1987
[19] Subramanian, 1983
[20] Firpo and Dahlberg, this volume

The molecular basis of translational control

How is control exercised at the molecular level? The 'minimal' model mentioned above assumes that the accessibility of bases that can interact with the 30S part of the ribosome controls initiation efficiency. We can therefore imagine a type of 'thermodynamic' control in which the rate of initiation on a specific mRNA is proportional to the fraction of the molecules that are in an open conformation that allows binding of the 30S subunits. This leads to the prediction of a logarithmic relationship (within certain limits) between the rate of initiation and the free energy of intramolecular base-pairing responsible for occlusion of the start site. Obviously, any base-pairing that can interfere with the correct positioning of the 30S subunit on the TIR, irrespective of the direct involvement of the SD region or of the start codon, is relevant here. The relevance of the standard thermodynamic relationship:

$$\Delta G_{folding} = -RT\ln K_{folding}$$

to translation was pointed out initially by Stormo (1986). The MS2 coat protein TIR has proved to be an ideal system for testing the predicted relationship between rate and $\Delta G_{folding}$ (de Smit and van Duin, this volume). The results of this latter study underline the dominant role of secondary (higher order) structure in the TIR suggested by many previous analyses of predicted structure/function in vivo (see references in de Smit and van Duin, this volume, and in Schauder and McCarthy, 1989).

Under this type of 'thermodynamic control', variations in translational efficiency should be reflected in differences in apparent affinities between 30S subunits and TIRs, although not necessarily in a linear fashion. In fact, studies of the apparent affinities of 30S ribosomal subunits for stretches of the atp polycistronic mRNA bearing different atp TIRs have revealed a correlation with initiation efficiency (Lang et al., 1989). An efficiently translated region of the mRNA (including the atpE TIR) bound 30S ribosomal subunits more tightly than did TIRs belonging to less efficiently translated genes. The TIR of the poorly translated atpG gene, on the other hand, which is able to form much more stable secondary structure, was very weakly bound by 30S ribosomal subunits. These data must, of course, be understood in the context of the competition between different TIRs at individual concentrations (as in the cell) well below saturation level for binding to 30S ribosomal subunits. Moreover, further experimental tests of the principle of 'thermodynamic control' need to be performed with other systems before we start to generalise too much about the significance of such data. Finally, the discussed mechanism of control does not rule out the possibility that kinetic constants on the translational initiation pathway can be influenced by mRNA structure. This type of 'kinetic' control could theoretically operate in parallel to the 'thermodynamic' control described.

Translational coupling

Most genes in E.coli lie on a polycistronic message. Thus individual TIRs often do not act as isolated functional units, but rather are influenced by the translation of neighbouring genes. Indeed, translational coupling (in cis) is probably widespread. In the ribosomal subunit operons coupling mediates the 'communication' of translational repression exercised by ribosomal subunits on single genes to further genes in the same operon, but there are many cases of coupling in operons that are not regulated in this way (McCarthy and Gualerzi, 1990; Stormo, 1986). Coupling is likely to occur wherever genes on a polycistronic message can structurally interact with each other at the mRNA level, and may thus represent an inevitable consequence of the selective pressure to conserve DNA (and RNA) in bacteria (McCarthy, 1990a). There are at least two ways in which coupling can occur, being mediated either by reinitiation or by means of a secondary structure coupling mechanism. The latter type of mechanism allows more flexibility in the stoichiometry of coupling, so that the downstream gene in a coupled pair may even be much more efficiently translated than its upstream partner (Stormo,1986; Gerstel and McCarthy, 1989). The possibility that specific primary recognition signals may play a role in translational coupling has not been ruled out, but neither is there any hard evidence for their existence. In summary, the influence of translational coupling on TIR function has in general received too little attention, and the mechanism of this phenomenon deserves more intensive investigation.

Manipulating mRNA stability

A picture of the pathway of mRNA degradation in E.coli has been slowly developing over the last twenty years (see mRNA stability section in this volume and the references in McCarthy, 1990a and b). Although a general outline is apparent, few details are known, and the story is complicated by the fact that the course of the pathway can vary from gene to gene. We have been interested in the role of differential mRNA stability in determining expression rates of genes on polycistronic messages, and in particular of the genes in the atp operon (Fig.2). atpI and atpB are relatively poorly expressed and also quickly degraded at the mRNA level (McCarthy et al., 1988). The remaining genes show widely varying rates of translation while displaying a considerably higher level of stability. Two questions have featured in recent work: first, can atpI and/or atpB be stabilised by improving the efficiency of their translation or by manipulating their 3′ flanking sequences?; and second, how is the stability of the message downstream of atpB affected by the introduction of elements known to exert a stabilising or destabilising effect in other environments? The second line of investigation was intended to yield information about rate control in the more stable part of the operon.

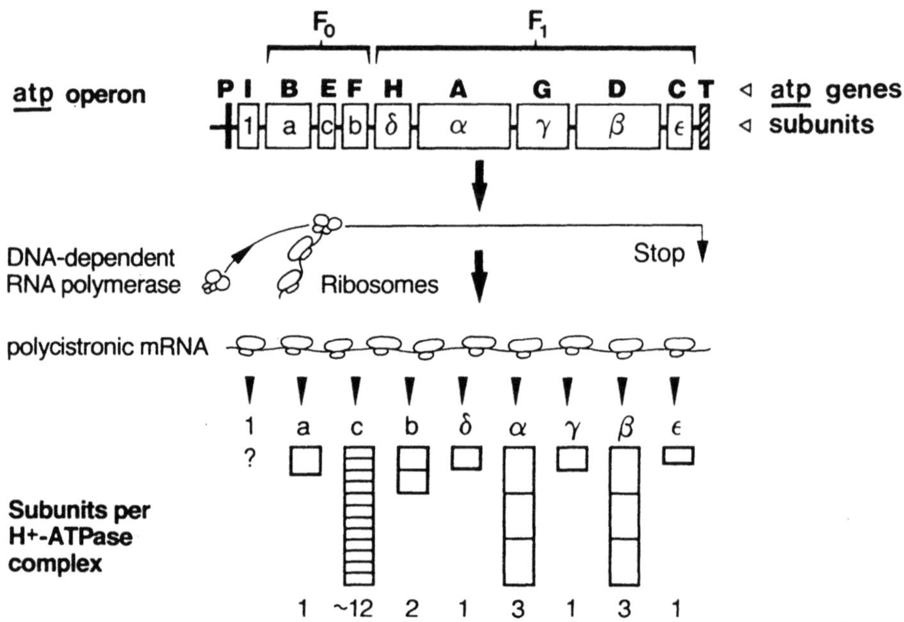

Fig.2 Schematic representation of atp gene expression. The major route is via synthesis of a 7 kb polycistronic mRNA bearing all nine cistrons, whereby atpE through to atpC show similar functional half-lives. The atpIB region is less stable. In general, however, control at the translational level is primarily responsible for determining the overall rates of synthesis of the respective H^+-ATPase subunits. Capital letters indicate the positions of the major promoter (P), the major ρ-independent terminator (T), and of the atp genes, respectively. The identities of the subunits encoded by the atp genes are given within the boxes. The atp genes are organised into groups encoding the membrane-integrated (F_0) subunits, and the F_1 sector subunits, respectively, of the H^+-ATPase. atpI encodes an additional protein of as yet unknown function.

In order to be able to study both the chemical and the functional half-lives of the atp mRNA more easily, experiments have been performed with transcripts specified by the plasmid-borne operon. Under these conditions, all of the proteins specified by the atp operon except the atpI product can be visualised by fluorography after loading a total extract from cells labelled with radioactive amino acids on an SDS-polyacrylamide gel. The atpI product can be seen, however, if the sequence upstream of the normally poorly translated atpI gene is replaced by the atpBE non-coding intercistronic region. The increase in translational efficiency

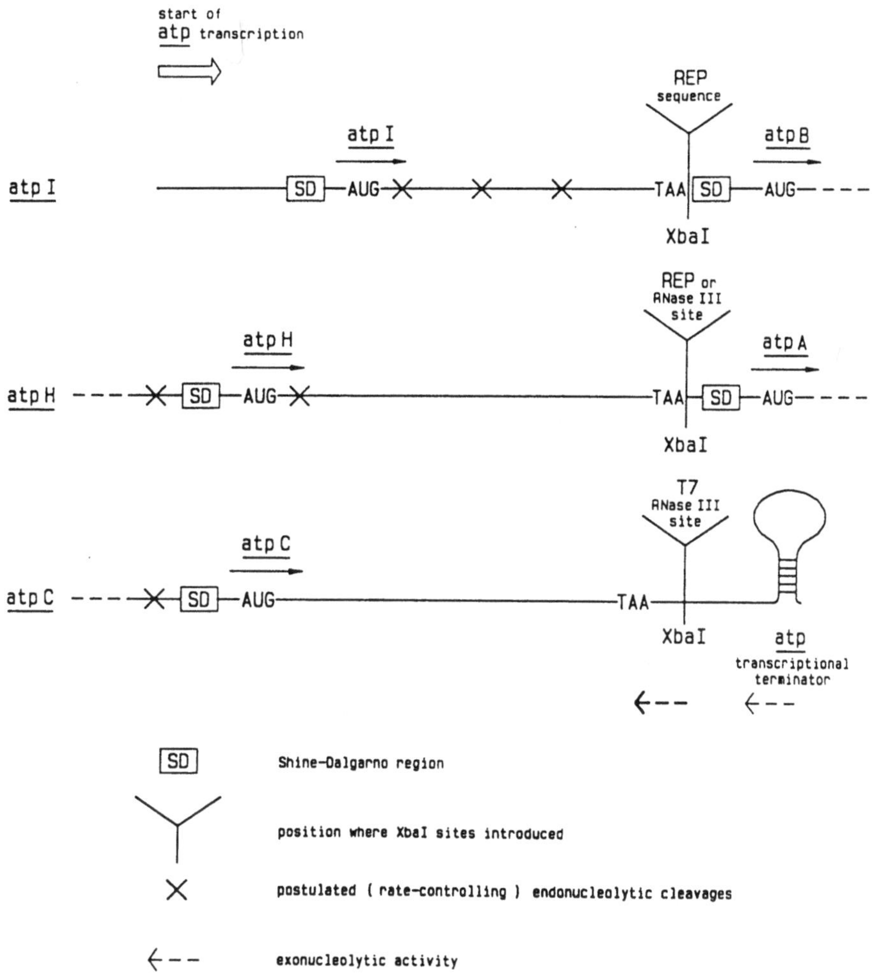

Fig.3 The natural and manipulated decay of atp genes at the mRNA level. The diagram indicates hypothetical models explaining the degradation behaviour of three atp genes as described in the text. The drawings are not to scale. XbaI sites were introduced at the positions indicated and used for insertion of the sequences described. atpI is rapidly attacked (apparently in the coding region) by endonucleases and its functional decay rate is therefore relatively insensitive to the influence of structures 3′ of the stop codon (and evidently also 5′ of the start codon). Further experiments have revealed that sequences 3′ of the other genes are normally not involved in their functional inactivation. For example, the introduction of a REP sequence (Newbury et al., 1987) immediately 3′ of atpH has no influence on its stability. Similarly, when an RNase III site that is cleaved to leave a structure that stabilises 3′ ends (Panayotatos and Truong, 1985) is inserted at the same site, there is also no change in atpH stability, although now the atp polycistronic mRNA is cleaved at this point. Finally, atpC could be destabilised by trimming back the operon DNA to delete the ρ-independent terminator and inserting an XbaI site 5′ of the usual terminator position. Thus where normally progressive 3′-5′ digestion from the end of the message is relatively insignificant, the atpC mRNA becomes destabilised. However, insertion of a 'stabilising' RNase III site at this point (see above) reverses at least some of this effect.

that is responsible for the higher rate of synthesis of the atpI product does not, however, result in a readily measurable increase in the half-life of the atpI mRNA, which remains at less than 60 seconds. This behaviour in the polycistronic environment confirms earlier findings with individually cloned atp genes (McCarthy and Bokelmann, 1988); in other words, changes in translational efficiency need not necessarily influence mRNA stability. Von Gabain et al. also arrive at a similar conclusion (this volume).

The possibility that manipulation of intercistronic regions could change the pathway and rate of mRNA stability was tested (Fig.3). The atpI mRNA could not be stabilised by inserting a REP sequence just 3′ of its stop codon (compare Klug and Cohen, this volume). Primer extension studies (Lang et al., 1989; Schaefer et al., 1989) have previously indicated the positions of potential endonuclease cleavage sites within atpI. Both observations are consistent with the occurrence of rate-controlling cleavages within the N-terminal region of the gene's coding sequence. The results of further experiments in the same vein (Fig.3) also indicated that degradation does not normally initiate from the 3′-ends of the other atp genes, although it is possible to destabilise them through manipulation downstream of their respective stop codons. Indeed, the data are consistent with a pathway of decay initiated at the 5′-end of, or within, each gene. However, the exact sites of the initial changes remain to be determined.

Final remarks

Understanding of the control of initiation and of TIR function in E.coli has advanced considerably over recent years, although the modulation of this function through coupling effects deserves more attention. It seems as though surprisingly simple models can allow an explanation of the majority of the structure/-function data relating to mRNA translation. The control of mRNA stability, on the other hand, remains more of a mystery. However, advances in the genetics and biochemistry of the degradation process promise to yield important insights into the mechanism of degradation. Given the typical location of E.coli genes in polycistronic organisational units, the influence of this microenvironment should not be forgotten. For example, the intercistronic areas may need to fulfil functions that depend in different ways on structural features of the mRNA. Whereas a stem-loop structure can protect the 3′-end of an upstream gene from degradation, it can simultaneously strongly reduce the translational efficiency of a downstream gene and/or render the downstream gene dependent on translational coupling. Such interrelationships are likely to influence the molecular evolution of both coding and non-coding regions in polycistronic systems.

References

Boni IV, Isaera DM, Musychenko ML, Tzareva NV (1990) Ribosome-messenger recognition: mRNA target sites for ribosomal protein S1. Nucleic Acids Res in press

Dreyfus M (1988) What constitutes the signal for the initiation of protein synthesis on Escherichia coli mRNAs? J Mol Biol 204: 79-94

Gallie DR, Kado CI (1989) A translational enhancer derived from tobacco mosaic virus is functionally equivalent to a Shine-Dalgarno sequence. Proc Natl Acad Sci USA 86: 129-132

Ganoza M, Marliere P, Kofoid E, Louis BG (1985) Initiator tRNA may recognise more than the initiation codon in mRNA: a model for translational initiation. Proc Natl Acad Sci USA 82: 4587-4591

Gerstel B, McCarthy JEG (1989) Independent and coupled translational initiation of atp genes in Escherichia coli: experiments using chromosomal and plasmid-borne lacZ fusions. Molec Microbiol 3: 851-859

Gold L (1988) Posttranscriptional regulatory mechanisms in Escherichia coli. Ann Rev Biochem 57: 199-233

Gold L, Pribnow D, Schneider T, Shinedling S, Singer BS, Stormo G (1981) Translational initiation in prokaryotes. Ann Rev Microbiol 35: 365-403

Gold L, Stormo G, Saunders R (1984) Escherichia coli translational initiation factor IF3: a unique case of translational regulation. Proc Natl Acad Sci USA 81: 7061-7065

Hui A, de Boer H (1987) Specialised ribosome system: preferential translation of a single mRNA species by a subpopulation of mutated ribosomes in Escherichia coli. Proc Natl Acad Sci USA 84: 4762-4766

Hui A, Hayflick J, Dinkelspiel K, de Boer HA (1984) Mutagenesis of the three bases preceding the start codon of the ß-galactosidase mRNA and its effect on translation in Escherichia coli. EMBO J 3: 623-629

Jacob WF, Santer M, Dahlberg AE (1987) A single base change in the Shine-Dalgarno region of 16S rRNA of Escherichia coli affects translation of many proteins. Proc Natl Acad Sci USA 84: 4757-4761

Jacques N, Dreyfus M (1990) Translation initiation in Escherichia coli: old and new questions. Molec Microbiol in press

Kozak M (1983) Comparison of initiation of protein synthesis in procaryotes, eucaryotes and organelles. Microbiol Rev 47: 1-45

Lang V, Gualerzi C, McCarthy JEG (1989) Ribosomal affinity and translational initiation in Escharichia coli. J Mol Biol 210: 659-663

McCarthy JEG (1987) Expression of the unc genes in Escherichia coli. J Bioenerg Biomemb 20: 19-39

McCarthy JEG (1990a) Post-transcriptional control in the polycistronic operon environment: studies of the atp operon of Escherichia coli. Molec Microbiol in press

McCarthy JEG (1990b) Optimising post-transcriptional steps of gene expression in Escherichia coli. In: Greenaway PJ (ed) Advances in Gene Technology, Vol II, JAI Press, London

McCarthy JEG, Bokelmann C (1988) Determinants of translational initiation efficiency in the atp operon of Escherichia coli. Molec Microbiol 2: 456-465

McCarthy JEG, Gualerzi C (1990) Translational control of prokaryotic gene expression. Trends in Genetics 6: 78-85

McCarthy JEG, Schairer HU, Sebald W (1985) Translational initiation frequency of atp genes from Escherichia coli: identification of an intercistronic sequence that enhances translation. EMBO J 4: 519-526

McCarthy JEG, Schauder B, Ziemke P (1988) Post-transcriptional control in Escherichia coli: translation and degradation of the atp operon mRNA. Gene 72: 131-139

Newbury SF, Smith NH, Higgins CF (1987) Differential mRNA stability controls relative gene expression within a polycistronic operon. Cell 51: 1131-1143

Olins PO, Rangwala SH (1989) A novel sequence element derived from bacteriophage T7 mRNA acts as an enhancer of translation of the lacZ gene in Escherichia coli. J Biol Chem 264: 16973-16976

Panayotatos N, Truong K (1985) Cleavage within an RNase III site can control mRNA stability and protein synthesis in vivo. Nucleic Acids Res 13: 2227-2240

Petersen C (1989) Long-range translational coupling in the rplJL-rpoBC operon of Escherichia coli. J Mol Biol 206: 323-332

Petersen GB, Stockwell PA, Hill DF (1988) Messenger RNA recognition in Escherichia coli: a possible second site of interaction with 16S ribosomal RNA. EMBO J 7: 3957-3962

Schaefer EM, Hartz D, Gold L, Simoni RD (1989) Ribosome-binding sites and RNA-processing sites in the transcript of the Escherichia coli unc operon. J Bacteriol 171: 3901-3908

Schauder B, McCarthy JEG (1989) The role of bases upstream of the Shine-Dalgarno region and in the coding sequence in the control of gene expression in Escherichia coli: translation and stability of mRNAs in vivo. Gene 78: 59-72

Scherer GFE, Walkinshaw MD, Arnott S, Morre DJ (1980) The ribosome binding sites recognised by E.coli ribosomes have regions with signal character in both the leader and protein coding segments. Nucleic Acids Res 8: 3895-3907

Stormo GD (1986) Translation initiation. In: Maximizing Gene Expression, Reznikoff W, Gold L (eds), Butterworths, Boston, p195

Subramanian AR (1983) Structure and functions of ribosomal protein S1. Progr Nucleic Acid Res Mol Biol 28: 101-142

Thanaraj TA, Pandit MW (1989) An additional ribosome-binding site on mRNA of highly expressed genes and a bifunctional site on the colicin fragment of 16S rRNA from Escherichia coli: important determinants of the efficiency of translation-initiation. Nucleic Acids Res 17: 2973-2985

CONTROL OF TRANSLATIONAL INITIATION BY mRNA SECONDARY STRUCTURE: A QUANTITATIVE ANALYSIS

Maarten H. de Smit and Jan van Duin
Department of Biochemistry
Gorlaeus Laboratories
University of Leiden
P.O. Box 9502
2300 RA Leiden
The Netherlands

Introduction

There is good evidence that mRNA secondary structure is one of the main factors determining the efficiency of translational initiation in prokaryotes (reviewed by Stormo, 1986; Gold, 1988; de Smit & van Duin, 1990). For example, Hall *et al.* (1982) found that mutations stabilizing a potential hairpin structure in the ribosome binding site of the *lamB* gene inhibited its expression and this inhibition could be relieved by second-site destabilizing mutations. They further suggested that the level of expression in the different mutants was related to the relative stability of the helix. Similarly, others showed the expression of heterologous genes in *E. coli* to be related to the stability of defined secondary structures involving the ribosome binding sites (Buell *et al.*, 1985; Tessier *et al.*, 1984; Spanjaard *et al.*, 1989).

In this paper, we present a quantitative analysis of the relationship between the stability of a local secondary structure and the efficiency of translational initiation. As a model system, we have chosen the coat gene of RNA bacteriophage MS2. Firstly, analysis with structure-sensitive enzymes and chemical reagents, as well as phylogenetic sequence comparison, have provided evidence that its ribosome binding site adopts a defined hairpin structure (Fig. 1) (Skripkin *et al.*, 1990). The existence of this structure enabled us to vary its stability in a predictable way through site-directed mutagenesis. Secondly, because antiserium against the coat protein is available, the expression could be detected and quantified by Western blotting. Thirdly, the high efficiency of this ribosome binding site, combined with the use of the strong inducible P_L-promoter from phage λ, allows detection of the protein output over a range of four order of magnitude.

Our results reveal a strictly linear relationship between the efficiency of translational initiation and the fraction of RNA molecules in which the ribosome binding site is unfolded. Incorporation of this finding in a simple thermodynamic model of translational initation shows that expression is completely dependent on spontaneous unfolding of the ribosome binding site. We conclude that ribosomes will only bind to single-stranded RNA.

NATO ASI Series, Vol. H 49
Post-Transcriptional Control of Gene Expression
Edited by J. E. G. McCarthy and M. F. Tuite

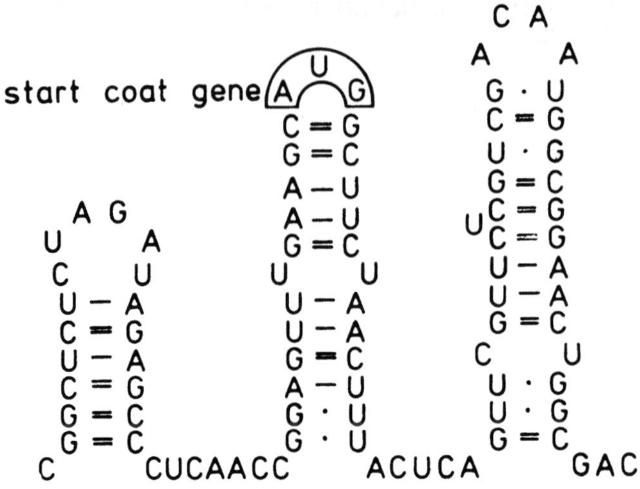

Fig. 1. Secondary structure of the translational initiation region of the coat gene of bacteriophage MS2.

Results

The secondary structure of the initiation region of the MS2 coat protein gene is shown in Fig. 2, together with the mutations introduced. All mutations leave the Shine-Dalgarno (SD) region and the amino acid sequence of the coat protein intact. In Fig. 3 a representative Western blot of five mutants is shown, illustrating how relative expression levels were determined.

For several mutants we have verified that the observed differences in expression arise at the translational level, by cloning a reporter gene immediately downstream of the MS2 information. For this purpose we used a cDNA copy of the coat protein gene of tobacco rattle virus (Angenent *et al.*, 1989). All analyzed mutants produced this protein at the same level as judged by Western blotting (not shown), under full control of the P_L-promoter.

Only helices stronger than the wild type reduce expression. In the first set of mutants base pair III was varied (Fig. 2). The twelve mutants obtained (nos. 2 to 13, Table 1) show that weakening of the helix has no detectable effect on coat protein production. In contrast, replacement of the A·U pair by C·G or G·C reduces expression to 20% and

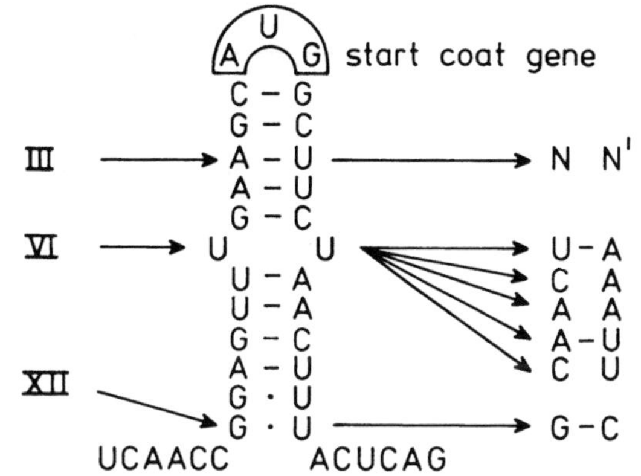

Fig. 2. Mutations introduced in the coat initiator hairpin. Mutated base pairs are denoted by Roman numerals (cf. Table 1).

4%, respectively (nos. 5 and 11). This reduction cannot be due to the nucleotide substitutions *per se*, since the same mutations are present individually in the destabilized mutants.

Multiple stabilizing mutations close the ribosome binding site completely. To obtain the maximal stabilization allowed by the amino acid sequence, we simultaneously changed base pair III to G·C, pair VI to U·A and pair XII to G·C. In this mutant (no. 14), coat protein production had dropped below the limits of detection, *i.e.* to less than 0.003% of the wild type. Clearly, the highly efficient ribosome binding site of the coat gene can be shut off completely by a secondary structure of sufficient stability.

Compensatory mutations restore expression. To confirm that translation in the above-described mutant no. 14 is prevented by the proposed secondary structure, we have introduced a number of second-site destabilizing mutations. First, base pair III was opened up by changing G·C into A·C (no. 17). This mutation indeed raised expression from below detection to 3% of the wild type. Replacement of the U·A pair at position VI by an A·A or C·A mismatch resulted in an expression level of 0.05 to 0.1% (nos. 15 and 16). When both base pairs were disrupted simultaneously, wild type expression was restored (nos. 18 and 19), again demonstrating that expression is related to the secondary structure, rather than to the nucleotide sequence.

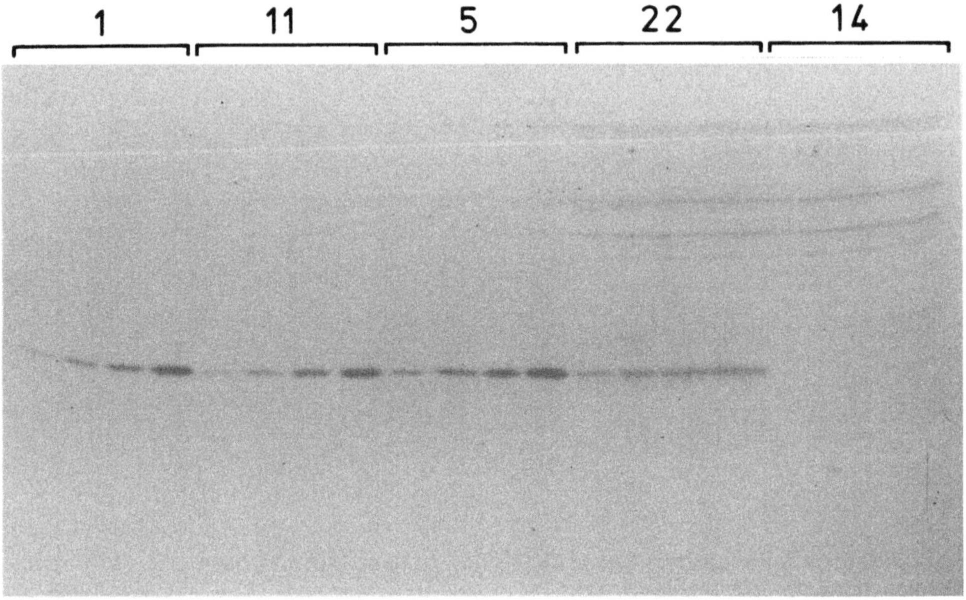

Fig. 3. Representative Western blot, showing the coat protein in serial dilutions of extracts of five mutants. Ratios between the sample sizes of the different mutants were 1:8:32:1024:1024. Within each series, every lane contains twice as much extract as the previous one. Mutant numbers correspond to Table 1. More accurate estimates of relative protein production were obtained by comparing mutants with similar expression levels.

Changes in expression by single substitutions at base pair VI or XII. When the terminal G·U pair (XII) was altered to G·C, expression dropped to 6% (no. 20), confirming the participation of this base pair in the helical structure. The importance of the U·U mismatch in the middle of the helix (base pair VI) was examined by replacing it by C·U, A·U and U·A, respectively. As expected, the change to C·U had no effect at all (no. 21). In the A·U and U·A mutants, on the other hand, expression was reduced to a mere 0.2 to 0.3% (nos. 22 and 23). It is known that mismatches drastically affect helix stability and the dramatic changes in expression reflect this fact. Evidently, the mismatch plays an essential role in allowing efficient translation.

Table 1. Mutations introduced in the MS2 coat initiator hairpin

No.	Base pair[1] III	VI	XII	ΔG_f^o [2]	expression (% of w.t.)[3]
1	(A-U)	(U U)	(G·U)	-2.4	100
2	G G	-	-	+1.0	100
3	G A	-	-	+1.0	100
4	G·U	-	-	-1.8	100
5	G-C	-	-	-4.4	4
6	A A	-	-	+1.0	100
7	A C	-	-	+1.0	100
8	U·G	-	-	-2.0	100
9	U U	-	-	+1.0	100
10	U C	-	-	+1.0	100
11	C-G	-	-	-3.9	20
12	C U	-	-	+1.0	100
13	C C	-	-	+1.0	100
14	G-C	U-A	G-C	-10.7	<0.003
15	C-C	A A	G-C	-7.2	0.1
16	G-C	C A	G-C	-7.2	0.05
17	A C	U-A	G-C	-5.3	3
18	A C	A A	G-C	-1.8	80
19	A C	C A	G-C	-1.8	100
20	-	-	G-C	-5.2	6
21	-	C U	-	-2.4	100
22	-	A-U	-	-6.0	0.2
23	-	U-A	-	-5.9	0.3

[1]Numbering as in Fig. 2. Base pairs in the wild type structure (no. 1) are shown in parentheses. Dashes indicate the presence of the wild type base pair in the mutants.

[2]ΔG_f^o of the hairpin, calculated using the energy parameters of Freier *et al.* (18) (1 cal = 4.184 J).

[3]Relative expression, determined by comparing serial dilutions of induced cultures on Western blots, as shown in Fig. 3.

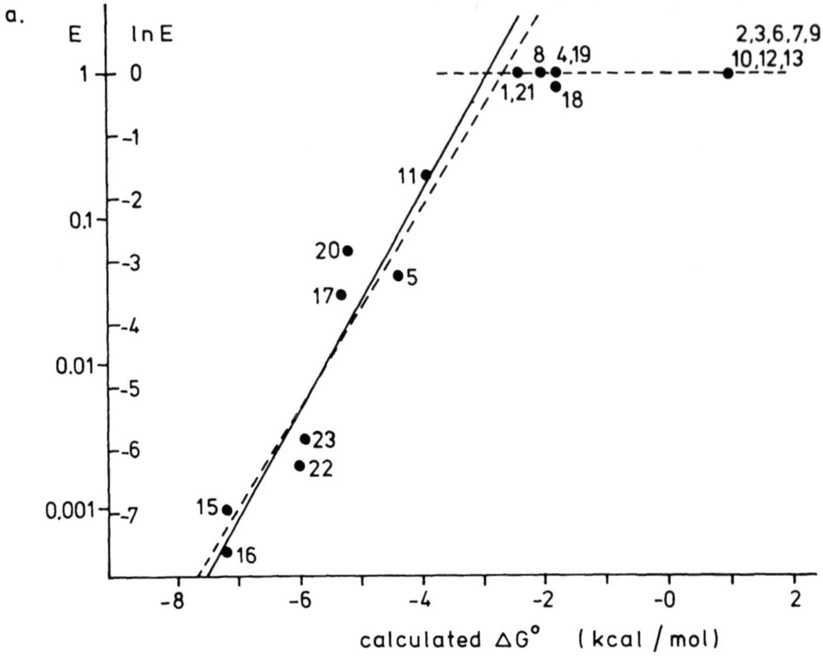
a.
E
ln E
1
0.1
0.01
0.001
0
-1
-2
-3
-4
-5
-6
-7
2,3,6,7,9
10,12,13
8
4,19
1,21
18
11
20
5
17
23
22
15
16
-8
-6
-4
-2
-0
2
calculated ΔG° (kcal / mol)

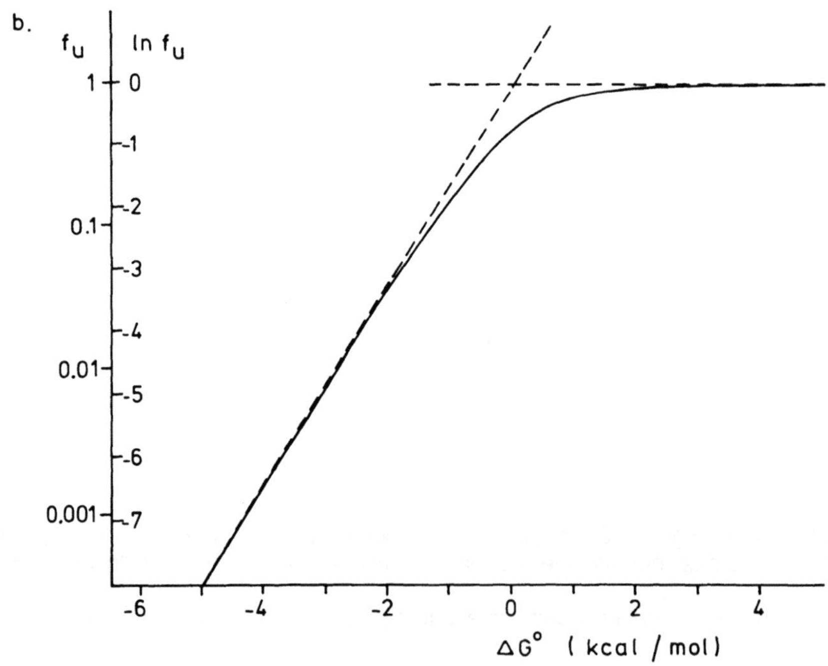
b.
f_u
ln f_u
1
0.1
0.01
0.001
0
-1
-2
-3
-4
-5
-6
-7
-6
-4
-2
0
2
4
ΔG° (kcal / mol)

Discussion

Expression in stabilized mutants is proportional to the fraction of unfolded mRNA molecules. We have shown that stabilization of the secondary structure of the coat gene ribosome binding site reduces its translational efficiency to a level that seems determined by the strength of the structure. To approach this apparent relationship in a more quantitative manner, we have calculated the stability (ΔG_f^o or free energy of formation) of the coat gene initiator hairpin in each of our mutants, using the free energy parameters of Freier *et al.* (1986). As shown in Table 1, a good correlation between the ΔG_f^o of the hairpin and the coat protein production is indeed observed.

Since the ΔG_f^o is a measure for the fraction of molecules in the unfolded state, the observed correlation suggests that the translational efficiency is directly determined by the availability of unfolded ribosome binding sites. This idea, proposed earlier by several authors (*e.g.* Looman *et al.*, 1987; Draper, 1987; Gold, 1988), can now easily be tested, using the equation

$$\Delta G_f^o = - RT \ln K_f , \qquad (1)$$

where R = gas constant, T = temperature (315 K) and K_f = equilibrium constant of helix formation. At low ΔG_f^o values, where most molecules are in the hairpin conformation, the fraction of unfolded molecules, f_u , approximately equals $1/K_f$, yielding the equation.

$$\ln f_u \approx \Delta G_f^o / RT \qquad (2)$$

If, as we suspect, the expression is linearly related to f_u , then plotting the natural logarithm of the expression (ln E) as a function of the ΔG_f^o should yield a straight linewith slope $1/RT$. Fig. 4a shows that our results are in good agreement with this prediction. The regression line through the lower eight points (solid, slope = 1.7) virtually coincides with the dashed line, showing the theoretical slope of $1/RT$ = 1.6.

It is worth noticing, that in this region an increase in stability of only -2.8 kcal/mol is sufficient to reduce the expression by two orders of magnitude. This roughly corresponds to the replacement in a helix of a G·U pair by a G·C pair.

Fig. 4. (a) Relationship between the stability of the coat initiator hairpin and coat protein production. Expression is plotted on a natural-logarithmic scale. The solid line is based on linear regression of the lower eight points. Dashed lines indicate the theoretical slope of $1/RT$ and the plateau (see text). Mutant numbers correspond to Table 1. (b) Relationship between the stability of an RNA helix and the fraction of RNA molecules in which it is unfolded. This fraction is plotted on a natural-logarithmic scale. Dashed lines indicate the asymptotes.

Expression in destabilized mutants is not limited by a cellular component. Having confirmed that expression in the stabilized clones is proportional to the fraction of unfolded RNA molecules, we are left to explain why destabilization beyond the wild type does not result in increased expression, *i.e.* why Fig. 4a displays a plateau. Given the high efficiency of the promoter and the ribosome binding site, we considered the possibility that some cytoplasmic factor such as initiator-tRNA or initation factors might become limiting. To test this, we have compared the expression of the wild type to that of several destabilized mutants under conditions where only small amounts of the messenger RNA are synthesized (by repression of the P_L-promoter at 28°C). We found that while wild type expression dropped to around 1% of the induced (42°C) value, the level was just as low in the destabilized mutants (not shown). We must conclude, therefore, that expression in the wild type and the destabilized mutants is not limited by the availability of some cellular component. Apparently, the plateau corresponds to full accessibility of the ribosome binding site. This is in fact predicted by theory, as with increasing ΔG_f^o values, the fraction of unfolded molecules approaches 1 in an asymptotic fashion (see below).

It is rather intriguing that the stability of the wild type hairpin is just on the verge of becoming inhibitive. Apparently, the evolution of this helix has resulted in the highest stability compatible with maximal expression. Earlier reports have already shown, that a certain degree of base-pairing in a ribosome binding site, as revealed by protection against nuclease attack, does not necessarily impair the translational efficiency (Rosa, 1981; Schmidt *et al.*, 1987).

The shape of the expression curve is determined by the unfolding of the secondary structure. If, as we have proposed in the previous section, the plateau of Fig. 4a also reflects the accessibility of the ribosome binding site, the direct relationship between expression and fraction of unfolded molecules should extend over the whole ΔG_f^o range. To test this idea, the approximation used above is not sufficient and we must introduce a more precise mathematical approach.

Basic thermodynamics reveals the following relationship between the equilibrium constant (K_f) and the fraction of the unfolded molecules (f_u):

$$f_u = 1/(K_f + 1). \tag{3}$$

Combination of equations (1) and (3) gives:

$$\ln f_u = -\ln(e^{-\Delta G_f^o/RT} + 1). \tag{4}$$

In Fig. 4b $\ln f_u$ is plotted as a function of ΔG_f^o. This yields a curve with two asymptotes,

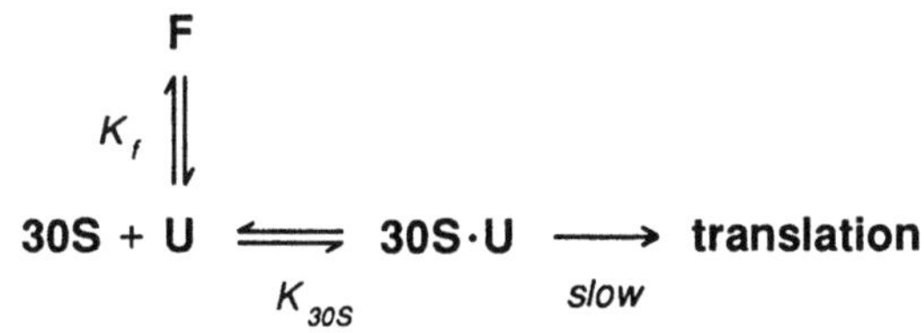

Fig. 5. The two equilibria that determine the fraction of mRNA bound to 30S subunits. K_f is the equilibrium constant of the equilibrium between the unfolded (U) and the folded (F) form of the ribosome biding site. K_{30S} is the association constant of 30S with the unfolded mRNA.

one identical with the straight line described in the first section (for $\Delta G_f^o \rightarrow -\infty$) and the other corresponding to the plateau described in the second section (for $\Delta G_f^o \rightarrow +\infty$). The finding that the theoretical curve of Fig. 4b is identical in shape to the experimental curve of Fig. 4a confirms that the relationship between the expression and the fraction of unfolded molecules extends over the whole range of ΔG_f^o values.

A simple thermodynamic model of translational initiation. Although the expression curve has the same shape as the theoretical one (Fig. 4), it is clearly shifted to the left. In part, this may be due to a systematic error in the calculated helix stabilities. However, up to now we have not taken into account that ribosomes bind to the unfolded mRNA molecules and thereby shift the equilibrium.

To a first approximation, the system may be regarded as composed of two equilibria; one between the folded (F) and the unfolded (U) form of the ribosome binding site, the other representing the reversible association of a free 30S subunit with the unfolded messenger (Fig. 5) (Stormo, 1986; Draper, 1987). This is legitimate, since the step following this association appears relatively slow (Gualerzi *et al.*, 1977; Ellis & Conway, 1984). The equilibria are characterized by K_f , the aforementioned equilibrium constant of helix formation, and K_{30S} , the association constant of the 30S subunits with the ribosome binding site.

Since expression is directly proportional to the concentration of the 30S·mRNA complex (30S·U in Fig. 5) (Draper, 1987; Ellis & Conway, 1984) the relative expression (E) equals the fraction of mRNA present in this complex:

$$E = [30S{\cdot}U] \, / \, ([U] + [F] + [30S{\cdot}U]) \tag{5}$$

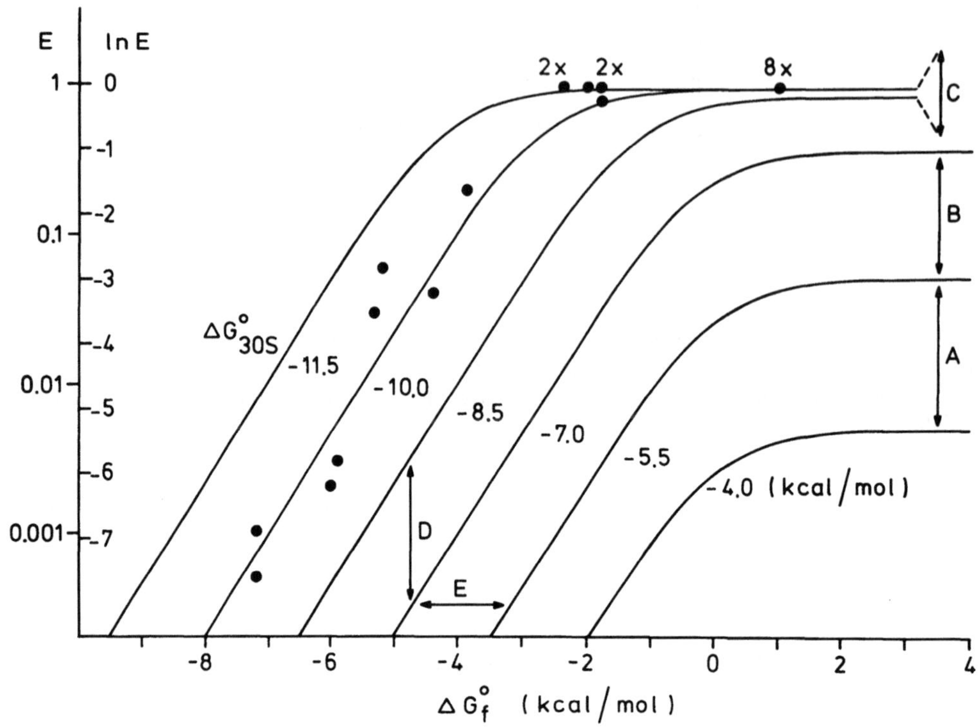

Fig. 6. Relationship between the stability of the secondary structure in a ribosome binding site and the relative expression predicted by the scheme of Fig. 5, for various affinities (see text). Experimental data are shown as in Fig. 4a. Arrows A to E are explained in the text.

In this equation, the concentrations can be substituted, using the definitions of the equilibrium constants and this results in:

$$E = K_{30S} \cdot [30S] \,/\, (1 + K_f + K_{30S} \cdot [30S]) \tag{6}$$

To compare this to the experimental curve in Fig. 4a, E has to be plotted on a natural logarithmic scale as a function of ΔG^o_f, which is related to K_f through equation (1). In a similar way, K_{30S} is related to ΔG^o_{30S}, which is a measure for the affinity between the 30S subunit and the ribosome binding site. [30S] can be estimated from literature data. At a growth rate of 2.9, the intracellular concentration of translating ribosomes is about 44 μM (Gouy & Grantham, 1980). 77% of the total amount of ribosomes are actively translating mRNA while 15% are present as free, vacant subunits (Forchhammer & Lindahl, 1971). Thus, the concentration of free 30S subunits is about 8.5 μM in our cells. The true value might be somewhat lower since not all free 30S subunits may carry the initiation factors.

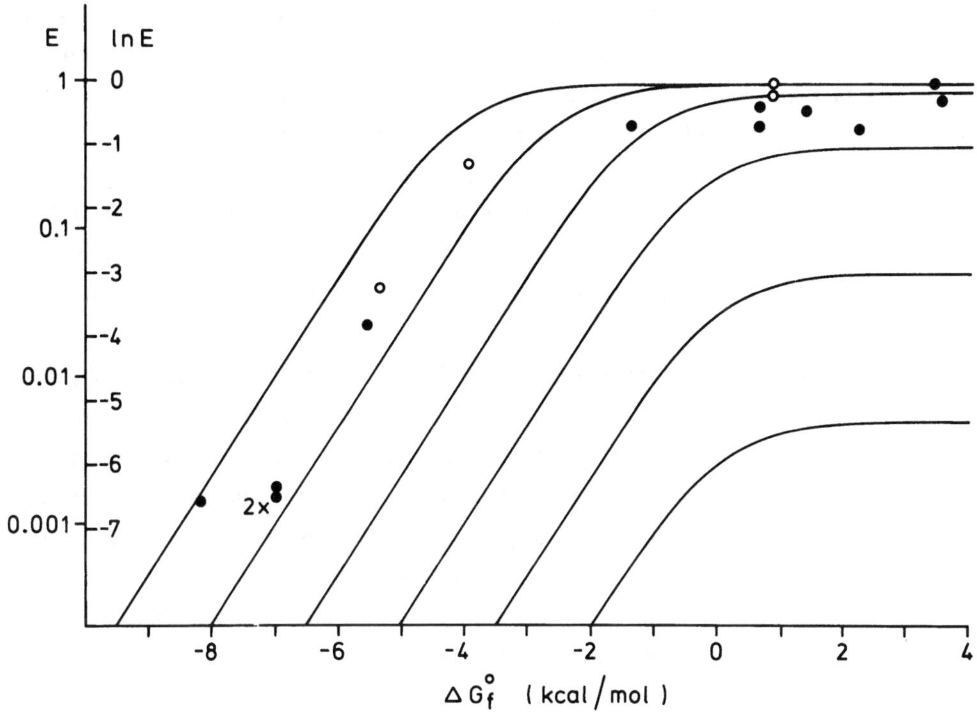

Fig. 7. Same as Fig. 6, but showing experimental results from a synthetic somatomedin C gene (filled circles; Buell *et al.*, 1985) and from the *lamB* gene (open circles; Hall *et al.*, 1982). Highest expression levels were set at 100%.

In Fig. 6, the resulting curves are represented for various values of ΔG^o_{30S}. Each curve is identical in shape to Fig. 4b, but its position depends on ΔG^o_{30S}. The higher the affinity of the ribosome for its binding site, the further the curve shifts to the left.

Our experimental data, entered in Fig. 6, fit a ΔG^o_{30S} value of about -10 kcal/mol for the coat gene ribosome binding site, corresponding to a K_{30S} of about 10^7 M^{-1}. This agrees well with *in vitro* data on association constants of other ribosome binding sites, which vary between 10^7 M^{-1} and 3.10^7 M^{-1} (Ellis & Conway, 1984; Draper, 1987; Calogero *et al.*, 1988).

General applicability of the thermodynamic approach and the parameters. To examine whether this thermodynamic approach can be applied to other ribosome binding sites as well, we analyzed two cases from the literature where differences in expression could be unambiguously attributed to changes in the stability of a defined secondary structure. The first set of data (filled circles in Fig. 7) concerns the expression in *E. coli* of a synthetic gene, coding for the human growth factor somatomedin C (Buell *et al.*, 1986). These data appear to fit the model well, and suggest roughly the same affinity constant as for the coat gene start. The second set (open circles) consists of three mutants in the *lamB* gene start region (Hall *et al.*, 1982) and also shows a good fit, with about the same affinity. The fact that our data points show the expected slope of $1/RT$ and an affinity value close to that obtained by independent *in vitro* measurements indicates that the free energy parameters of Freier *et al.* (1986) yield ΔG^o_f values that are close to the true stabilities *in vivo*. When we plotted our data using either the rules of Tinoco *et al.* (1971) or the rules of Salser (1977) and Cech *et al.* (1983) we found a much higher scatter which made it virtually impossible to draw a line through the points thus obtained (not shown).

In a preliminary report of these results, we have suggested that the stability of the coat initiator hairpin may be influenced by its RNA context (de Smit & van Duin, 1990). This idea was based on our finding that the expression curve (Fig. 4a) shifted to the right when we deleted most of the upstream MS2 sequences. Further analysis revealed, however, that the plateau value had also gone down and this, together with other recent findings, has now convinced us that the upstream sequences in MS2 RNA do not lower the stability of the initiator helix (ΔG^o_f) but increase the affinity of ribosomes (ΔG^o_{30S}) for this initation region by an as yet unknown mechanism.

Implications for the recognition of translational initiation sites by ribosomes. Our findings clearly show that ribosomes do not recognize mRNA secondary structures and they are contradictory to earlier ideas that the location of an AUG codon in the loop of a hairpin or exposure of just the SD-region could facilitate initiation (Iserentant & Fiers, 1980; Selker & Yanofsky, 1979; Queen & Rosenberg, 1981). As the effects of base-pair changes are independent of their position within the ribosome binding site, initiation apparently requires spontaneous unfolding of the whole helix. The simplest interpretation of these findings is, that ribosomes can only bind to single-stranded RNA. We find no evidence for ribosome binding to a partially unfolded ribosome binding site, nor for an active unfolding of the secondary structure by initiating 30S subunits. These processes, if playing a significant role in initiation, would raise the expression and therefore cause an apparent shift of the expression curve (Fig. 6) to the left (Draper, 1987). The 'melting activity' of initiating ribosomes is nothing more than the trapping of the mRNA in the unfolded state, an activity which ultimately results in a shift of the equilibrium to the unfolded side.

Ribosome binding and the efficiency of translational initiation. Closer examination of the curves of Fig. 6 reveals a number of interesting features. Firstly, it refutes the wide-spread misconception that the expression from an unstructured ribosome binding site must be proportional to its association constant with 30S subunits (Stormo, 1986). Equation (6) shows that the uninhibited expression level (plateau) equals

K_{30S}·[30S] / (1 + K_{30S}·[30S]). When K_{30S}·[30S] = 100 the mRNA is virtually saturated with initiating ribosomes and a further increase no longer results in the production of more protein (Fig. 6, arrow C). Thus, it is not at all surprising that two ribosome binding sites differing more than tenfold in K_{30S} , may show only a twofold difference in expression (Calogero *et al.*, 1988). This also implies that small variations in the affinity will only be reflected in the expression level if a ribosome binding site is weak (arrows A and B) or involved in secondary structure (arrow D).

Although an increased affinity does not necessarily raise the plateau value, it does help to overcome the negative effects of base-pairing in the ribosome binding site. The model predicts that a tenfold increase in K_{30S} results in a 'gain' of 1.4 kcal/mol (Fig. 6, arrow E). In this light, the alledged helix unwinding property of ribosomal protein S1 (reviewed by Subramanian, 1983) may simply be the consequence of its large positive effect on the 30S binding constant. This effect may also be the basis for the claim that increasing the strength of the SD-interaction has a much more pronounced effect if the ribosome binding site is involved in secondary structure (Munson *et al.*, 1984). Unfortunately, the quantitative role of the SD-sequence in 30S binding has yet to be determined.

Finally, it should be noted that the absolute value of the maximal expression level (E = 1) of any mRNA is determined by the rate at which ternary complexes enter the elongation cycle (assuming that initiation remains the rate limiting step in product formation) (Gualerzi *et al.*, 1988). Variations in this rate between different mRNAs may result from differential proofreading by IF3 when alternative initiation codons (GUG, UUG, AUU) are used (Gualerzi *et al.*, 1988; Berkhout *et al.*, 1986), or from unequal kinetics in the binding of the second aminoacyl tRNA. Whether these variations are significant remains to be seen.

Nucleotide sequence versus secondary structure in the recognition of initiation sites. As described in the results section, we have performed mutagenesis at positions -6, -3, +6, +9 and +15 relative to the AUG codon. The fact that all resulting changes in expression can be explained in terms of the stability of the secondary structure implies that there are no effects of the nucleotide changes *per se* at these positions. The same conclusion was reached by Precup & Parker (1987) for positions +12 and +13. Nevertheless, statistical analyses have demonstrated that the nucleotide choice throughout natural ribosome binding sites is strongly biased (Scherer *et al.*, 1980; Stormo *et al.*, 1982). Furthermore, shotgun cloning of *E. coli* DNA revealed that efficient expression (> 3% of wild type *lacZ*) could only be obtained from true ribosome binding sites, in spite of the presence of large numbers of 'false starts' (AUG or GUG, preceded by a reasonable SD-sequence) in the DNA used (Dreyfus, 1988). It was concluded that the used fragments of 60 to 80 nucleotides contained all information necessary to discriminate between true and false sites, and that this information must therefore lie in the nucleotide sequence outside the initiation codon and the SD-region.

How can these seemingly conflicting data be reconciled? We presume that evolutionary selection against initiation at false sites usually results in a low affinity (low K_{30S}) combined with sequestration by secondary structure (high K_f) (Ganoza *et al.* 1987). At the same time the nucleotide bias in true sites is probably related to the need to preclude stable secondary structures (Looman *et al.*, 1987) and interfering Shine-Dalgarno-like sequences.

We may learn from evolution that the secondary structure is probably the most crucial factor to consider in designing efficient ribosome binding sites. This point was poignantly illustrated by the work of Buell *et al.* (1985) and Spanjaard *et al.* (1989), who initially optimized their ribosome binding sites at the nucleotide- and codon usage levels, but experienced that expression was ultimately restricted by secondary structure. Only through a change in the stability of the structure can a one-nucleotide substitution outside the initiation codon and the SD-sequence lead to a 500-fold reduction in expression as in our mutant no. 22 (Table 1).

Conclusion

Our results demonstrate that translational initiation fully depends on spontaneous unfolding of the ribosome binding site. The model that we have presented for the first steps in this process is obviously simplified. Nevertheless, it fits the available data and it allows a number of predictions that can be tested in future research. Our work shows that it is possible to investigate the thermodynamics of these processes *in vivo*, i.e. under the precise conditions that play a role in molecular biology.

Acknowledgements

We thank Dr. W. Fiers and coworkers for the MS2 cDNA and expression vectors, Dr. T. Kunkel for strain BW313 and Gerco Angenent for TRV cDNA. Molly Hughes is acknowledged for her assistance and support in the initial stages of this work.

References

Angenent GC, Posthumus E, Bol JF (1989) Biological activity of transcripts synthesized in vitro from full-length and mutated DNA copies of tobacco rattle virus RNA2. Virology 173:68-76

Berkhout B, van der Laken, CJ, van Knippenberg, PH (1986) Formylmethionyl-tRNA binding to 30S ribosomes programmed with homopolynucleotides and the effect of translational initiation factor 3. Biochim Biophys Acta 866:144-153

Buell G, Schultz M-F, Selzer G, Chollet A, Movva NR, Semon D, Escanez S, Kawashima E (1985) Optimizing the expression in *E. coli* of a synthetic gene encoding somatomedin-C (IGF-I). Nucleic Acids Res 13:1923-1938

Calogero RA, Pon CL, Canonaco MA, Gualerzi CO (1988) Selection of the mRNA translation initiation region by Escherichia coli ribosomes. Proc Natl Acad Sci USA 85: 6427-6431

Cech TR, Tanner NK, Tinoco I, Weir BR, Zuker M, Perlman PS (1983) Secondary structure of the Tetrahymena ribosomal RNA intervening sequence: Structural homology with fungal mitochondrial intervening sequences. Proc Natl Acad Sci USA 80:3903-3907

de Smit MH, van Duin J (1990) Control of prokaryotic translational initiation by mRNA secondary structure. Progr. Nucleic Acid Res Mol Biol 38:1-35

Draper DE (1987) Translational regulation of ribosomal proteins in Escherichia coli. In: Ilan J (ed) Translational regulation of gene expression. Plenum New York (pp. 1-23)

Dreyfus M (1988) What constitutes the signal for the initiation of protein synthesis on Escherichia coli mRNAs? J Mol Biol 204:79-94

Ellis S, Conway TW (1984) Initial velocity kinetic analysis of 30S initiation complex formation in an in vitro translation system derived from Escherichia coli. J Biol Chem 259:7607-7614

Forchhammer J, Lindahl J (1971) Growth rate of polypeptide chains as a function of the cell growth rate in a mutant of Escherichia coli 15. J Mol Biol 55:563-568

Freier SM, Kierzek R, Jaeger JA, Sugimoto N, Caruthers MH, Neilson T, Turner DH (1986) Improved free-energy parameters for predictions of RNA duplex stability. Proc Natl Acad Sci USA 83:9373-9377

Ganoza MC, Kofoid EC, Marlière P, Louis BG (1987) Potential secondary structure at translation-initiation sites. Nucleic Acids Res. 15:345-360

Gold L (1988) Posttranscriptional regulatory mechanisms in Escherichia coli. Ann Rev Biochem 57:199-233

Gouy M, Grantham R (1980) Polypeptide elongation and tRNA cycling in Escherichia coli: A dynamic approach. FEBS Lett 115:151-155

Gualerzi C, Risuleo G, Pon CL (1977) Initial rate kinetic analysis of the mechanism of initiation complex formation and the role of initiation factor IF-3. Biochemistry 16:1684-1689

Gualerzi CO, Calogero RA, Canonaco MA, Brombach M, Pon CL (1988) Selection of mRNA by ribosomes during prokaryotic translational initiation. In: Tuite MF, Picard M, Bolotin-Fukuhara M (eds) Genetics of translation. Springer Heidelberg (pp. 317-330)

Hall MN, Gabay J, Débarbouillé M, Schwartz M (1982) A role for mRNA secondary structure in the control of translation initiation. Nature 295:616-618

Iserentant D, Fiers W (1980) Secondary structure of mRNA and efficiency of translation initiation. Gene 9:1-12

Looman AC, Bodlaender J, Comstock LJ, Eaton D, Jhurani P, de Boer HA, van Knippenberg PH (1987) Influence of the codon following the AUG initiation codon on the expression of a modified lacZ gene in Escherichia coli. EMBO J 6:2489-2492

Munson LM, Stormo GD, Niece RL, Reznikoff WS (1984) lacZ Translation initiation mutations. J Mol Biol 177:663-683

Precup J, Parker J (1987) Missense misreading of asparagine codons as a function of codon identity and context. J Biol Chem 262:11351-11355

Queen C, Rosenberg M (1981) Differential translation efficiency explains discoordinate expression of the galactose operon. Cell 25:241-249

Rosa MD (1981) Structure analysis of three T7 late mRNA ribosome binding sites. J Mol Biol 147:55-71

Salser W (1977) Globin mRNA sequences: Analysis of base pairing and evolutionary implications. Cold Spring Harbor Symp Quant Biol 42:985-1002

Scherer GFE, Walkinshaw MD, Arnott S, Morré DJ (1980) The ribosome binding sites recognized by E. coli ribosomes have regions with signal character in both the leader and protein coding segments. Nucleic Acids Res 8:3895-3907

Schmidt BF, Berkhout B, Overbeek GP, van Strien A, van Duin J (1987) Determination of the RNA secondary structure that regulates lysis gene expression in bacteriophage MS2. J Mol Biol 195:505-516

Selker E, Yanofsky C (1979) Nucleotide sequence of the trpC-trpB intercistronic region from Salmonella typhimurium. J Mol Biol 130:135-143

Skripkin EA, Adhin MR, de Smit MH, van Duin J (1990) Secondary structure of the central region of bacteriophage MS2 RNA. J Mol Biol 211:447-463

Spanjaard RA, van Dijk MCM, Turion AJ, van Duin J (1989) Expression of the rat interferon-α1 gene in Escherichia coli controlled by the secondary structure of the translation-initiation region. Gene 80:345-351

Stormo GD (1986) Translation initiation. In: Reznikoff W, Gold L (eds) Maximizing gene expression. Butterworths Boston (pp. 195-224)

Stormo GD, Schneider TD, Gold LM (1982) Characterization of translational initiation sites in E. coli. Nucleic Acids Res 10:2971-2996

Subramanian A-R (1983) Structure and functions of ribosomal protein S1. Progr Nucleic Acid Res Mol Biol 28:101-142

Tessier L-H, Sondermeyer P, Faure T, Dreyer D, Benavente A, Villeval D, Courtney M, Lecocq J-P (1984) The influence of mRNA primary and secondary structure on human IFN-γ gene expression in E. coli. Nucleic Acids Res 12:7663-7675

Tinoco I, Borer PN, Dengler B, Levine MD, Uhlenbeck OC, Crothers DM, Gralla J (1973) Improved estimation of secondary structure in ribonucleic acids. Nature New Biol 246:40-41

THE ROLE OF RIBOSOMAL RNA IN THE CONTROL OF GENE EXPRESSION

Matthew A. Firpo
Albert E. Dahlberg
Section of Biochemistry
Brown University
Providence, Rhode Island 02912
U.S.A.

It is now accepted that ribosomal RNA (rRNA) has an important functional role during the translation of messenger RNA (mRNA). There is evidence for base paired interactions between mRNA and rRNA during initiation, elongation and termination. At initiation the relative level of translation of different mRNAs is determined, in part, by the complementarity between the Shine-Dalgarno sequence of the mRNA and the Anti-Shine-Dalgarno sequence near the 3' end of 16S rRNA. Several groups have proposed that other regions of 16S rRNA may be involved in similar base paired interactions with mRNA prior to initiation to selectively enhance translation of particular mRNAs. Here we summarize the current data on translational enhancers and identify the proposed sites in 16S rRNA to which they might base pair. The data are presented in a model of the 30S subunit showing a potential pathway for mRNA as it passes over putative rRNA-mRNA contact points during the mRNA selection process, as well as during translation.

Many factors influence the selection of specific mRNAs for translation. The Shine-Dalgarno sequence and an initiation codon (AUG or GUG), as well as a suitable spacing between the two are necessary for translation initiation (Shine, Dalgarno, 1974; Steitz, Jakes, 1975). Additional sequences around the initiation codon must also be involved, as has been noted by several groups as long as ten years ago (Sherer, et al., 1980; Gold, et al., 1981; Gren, 1984; Schneider, et al., 1986; Gold, et al., 1984; Kastelein, et al., 1983). A statistical analysis by Schneider, et al. (1986), showed that translational initiation domains were non-random, both upstream and downstream of the AUG start codon (from -20 to +15). More recently, there have been a number of reports about translational enhancers, specific sequences located both 5' and 3' to the Shine-Dalgarno and AUG start codon which promote selection of specific mRNAs by the E. coli ribosome. These sequences "search" for complementary sequences in 16S rRNA to which they can base pair transiently when mRNA makes its initial contact with the

NATO ASI Series, Vol. H 49
Post-Transcriptional Control of Gene Expression
Edited by J. E. G. McCarthy and M. F. Tuite

TABLE 1

POSSIBLE STAGES OF TRANSIT BY mRNA ON THE 30S RIBOSOMAL SUBUNIT PRIOR TO INITIATION

	STAGE	PUTATIVE BASE PAIRING INTERACTIONS: mRNA	PUTATIVE BASE PAIRING INTERACTIONS: 16S rRNA
1.	INITIAL CONTACT BETWEEN mRNA AND 30S RIBOSOMAL SUBUNIT.	OLINS' UPSTREAM EPSILON SEQUENCE (UUAACUUUA) IN T7 GENE 10; MCCARTHY'S ENHANCER IN atp E GENE (UUAACU); GOLD'S GENE 32 ENHANCER (UUAAAUUAA). (EPSILON ALSO FUNCTIONS JUST DOWNSTREAM OF AUG START CODON.)	460 REGION
2.	mRNA ENTERS CLEFT OF 30S SUBUNIT.	GALLIE'S UPSTREAM OMEGA SEQUENCE (ACAAUUAC).	5' REGION
3.	mRNA 5' END EXITS PLATFORM.	PETERSEN'S DOWNSTREAM ENHANCER SEQUENCES (3 OR 4 NUCLEOTIDES).	5' REGION
		THANARAJ AND PANDIT'S UPSTREAM TP SITE (UGAUCC).	ANTI-TP (A-TP) REGION
4.	mRNA IN RIBOSOME BINDING SITE TO BEGIN TRANSLATION.	SHINE-DALGARNO SEQUENCE (GGAGG).	ANTI-SHINE-DALGARNO (A-SD) REGION
		INITIATION CODON (AUG OR GUG).	fMET-tRNA$_i$ IN P SITE

ribosome. The transient on/off base pairing is thought to bring about a relative increase in the concentration of the particular mRNA in the region of the 30S subunit, thus enhancing the potential for this mRNA to be translated.

In early reports by McCarthy, et al. (1985, 1986), a translational enhancer region was identified in the E coli. atpE gene. This region was AU rich and contained sequences which are similar to the Epsilon sequence reported more recently by Olins and coworkers (1988, 1989). The Epsilon sequence, located just 5' to the Shine-Dalgarno region in phage T7 gene 10, produces a dramatic enhancement of expression of this gene product and also functions successfully when placed in front of foreign genes. The similarity between this 9 nucleotide sequence (UUAACUUUA) and sequences in McCarthy's enhancer (UUAAUUUAC and UUAACU) are striking, as is the seven of nine nucleotide identity in the gene 32 sequence between -14 and -22 (UUAAAUUAA) as noted by Gold (personal communication). The fact that the sequence itself is important was shown by Olins and coworkers since randomization caused considerable decrease in effectiveness as an enhancer. The pyrimidine-rich composition of these enhancers may be of some importance, however, since ribosomal protein S1 is known to have an affinity for such polypyrimidine tracts (Dahlberg, Dahlberg, 1985; Szer, et al., 1975). Protein S1 is located on the solvent side of the 30S subunit but in proximity to the putative enhancer binding sequences in 16S rRNA (Brimacombe, et al., 1988; Stern, et al., 1988). As the focus of this chapter is on rRNA-mRNA interactions, however, we will not review what is certainly an important functional role for ribosomal proteins and initiation factors during initiation. Suffice it to say that proteins S1 and S21 and initiation factor 3 have the capacity to unwind RNA secondary structure. This may help to properly orient the mRNA with respect to the protein and the 30S subunit as it searches for rRNA binding sites and, ultimately, the initiation region to begin translation.

Olins' group was the first to propose a specific site in rRNA for a base-paired interaction with an enhancer sequence (Olins, Rangwala, 1989). They noted that the Epsilon sequence complements bases 458 to 466 in 16S rRNA of E. coli, the 460 region shown in Figure 1. While most of the rRNA sequence is contained in a base paired stem structure, they argue that the strength of the mRNA-rRNA hybrid exceeds that of the intramolecular rRNA structure, and note that the 460 loop region is on an exposed surface of the ribosomal subunit (Moazed, et al., 1986). Quite striking was their finding that the Epsilon sequence

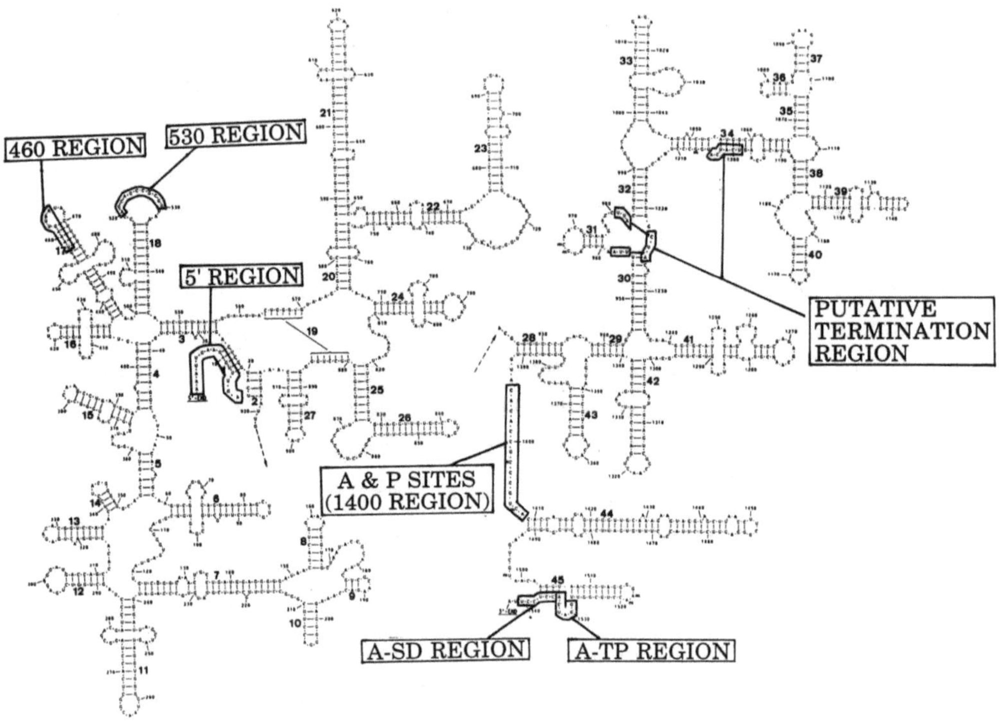

Figure 1. Secondary structure model of E. coli 16S rRNA by Maly and Brimacombe (1983). Potential regions of contact with mRNA are indicated. A-SD and A-TP represent Anti-Shine-Dalgarno and Anti-Translational Promoter, respectively.

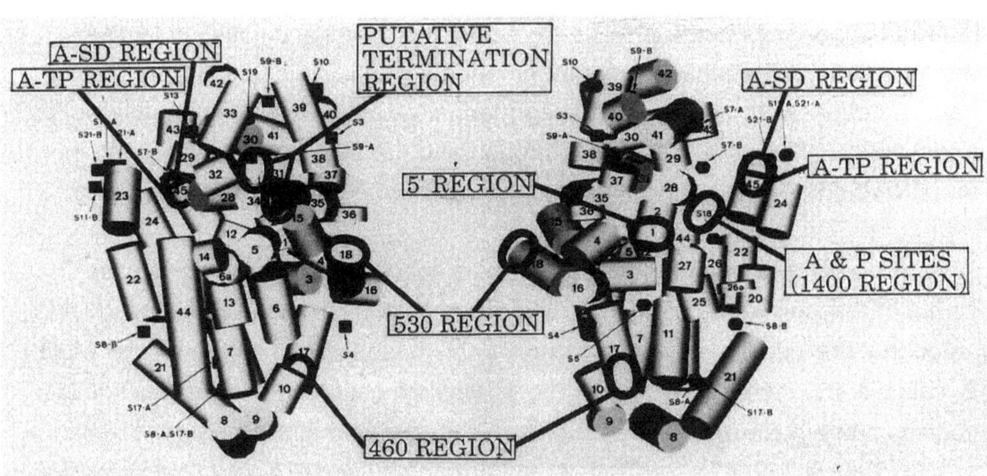

Figure 2. Arrangement of 16S rRNA in the 30S ribosomal subunit according to Brimacombe, et al. (1988).

stimulated translation even when placed immediately downstream as well as upstream of the AUG initiation codon.

Gallie, et al. (1987, 1989), have identified a 68 nucleotide sequence in the noncoding 5' leader region of tobacco mosaic virus RNA which they call Omega. This region enhances translation of both eucaryotic and procaryotic translation systems *in vitro* and *in vivo*. If Omega functions by a base pairing interaction with rRNA, similar to that proposed for Epsilon, then it must recognize a sequence that is present in both eucaryotic and procaryotic rRNAs. The Omega sequence does not complement the 460 region of 16S rRNA although, like Epsilon, it is AU rich. It is functionally equivalent to the Shine-Dalgarno sequence in cases where that sequence is lacking in mRNA and it can enhance translation 40- to 120-fold, although the level of translation is still 10 times less than with a normal ribosome binding site. In the presence of a normal ribosome binding site, the Omega sequence enhances translation only 1- to 4-fold. The position of the Omega sequence in the mRNA can range from immediately upstream of the start site to as far as 110 bases upstream, and tandem copies of the sequence more than double the enhancement of translation in E. coli (Sleat, et al., 1988). Gallie and coworkers noted that the sequence ACAAUUAC is repeated three times in the Omega region and would appear to be important. It may be significant that CAAUU, within this sequence, complements AAUUG, located at the 5' end of 16S RNA. This region of the 5' end is strongly conserved in bacterial species and Petersen, et. al (1988), have proposed that it is a potential base pairing site for translational enhancer sequences <u>downstream</u> of the start codon in mRNA.

Petersen and coworkers have done a rather thorough study of potential base pairing between mRNA and 16S rRNA using published sequence data (Petersen, et al., 1988). They note that short (3 or 4 nucleotides) sequences early in the coding region of the mRNA (nucleotides +4 to +21) complement three or more nucleotides in the first 16 nucleotides at the 5' end of 16S rRNA. They propose that base pairing may occur between the 5' end of 16S rRNA and these short sequences in the coding region. Because the distance between these sequences and the start codon is variable it is most probable that the putative base pairing occurs prior to the point where the mRNA recognizes the ribosome binding site (AUG and the Anti-Shine-Dalgarno sequence). A similar proposal involving base pairing of the 5' region of 16S rRNA with the A protein and coat protein mRNAs of RNA coliphages R17 and Qβ was put forth by van Knippenberg (1975) a number of years ago.

More recently, Thanaraj and Pandit (1989), have proposed the existence of yet another computer-determined translational enhancer in mRNA. Unlike Petersen, who looked for enhancers in the coding region, Thanaraj and Pandit searched in the upstream, nontranslated region of mRNAs, between -55 and -1, 5' to the AUG codon. They restricted their search to sequences which would complement sequences near the 3' terminus of 16S rRNA (the 49 terminal nucleotides in the colicin E3 fragment). They identified a sequence UGAUCC, or a portion thereof, in highly expressed genes, but not present in genes less frequently expressed. This sequence, called the TP site for Translation-initiation Promoting site, complements the sequence GGAUCA (nucleotides 1529-1534) adjacent to the Shine-Dalgarno region at the 3' end of 16S rRNA. In many cases the TP site is located outside the region normally protected by the ribosome (more than 20 bases from the start site). However, the occurrence of this site is quite restricted to the more highly expressed genes, those with greater than 9,000 molecules/genome/cell.

It is apparent that translational enhancers have been identified by two quite different approaches. The first involved actual sequences in known mRNAs which were shown to have translational enhancer activity. The activity could be destroyed by mutagenesis or transferred to other mRNAs. The second approach involved computer searches for sequence homology between rRNA and regions near the start codons of mRNAs. The thinking in all cases is that any base pairing that occurs between these two RNAs is of a transient nature and it results in the mRNA remaining in closer proximity to the ribosome, thus enhancing its competition for available ribosomes and increasing the likelihood of its being translated by the ribosome. This on/off base pairing does not restrict a mRNA to only one translational enhancer. Indeed, as the mRNA moves across and through the ribosome there may be several points of contact at different positions on the ribosome which might enhance the opportunity for that particular mRNA to find its correct ribosome binding site. Certainly the evidence is gathering that mRNA has multiple contact sites with rRNA during translation (Noller, 1984; Dahlberg, 1989), and we know from the early studies (Steitz, Jakes, 1975; Kang, Cantor, 1985) that the ribosome protects at least 20 nucleotides upstream of the AUG start site as well as downstream.

Having reviewed the information about the different translational enhancers in E. coli, we will now try to bring together what is known about the locations of the putative binding sites in the rRNA. These data will be incorporated into a model which will show a possible pathway the mRNA might take from the time it makes an initial contact with the ribosome and weaves its way into and through the cleft, until it finds the correct contact points to begin initiation of translation. We start by showing Brimacombe's secondary structure model of 16S rRNA (Figure 1) (Maly and Brimacombe, 1983). Note that the stem structures are numbered and putative contact points between rRNA and translational enhancers in mRNA have been identified (460 region, Anti-Shine-Dalgarno region, 5' region, and Anti-TP region). Additional sites are also identified which are thought to be contact points with mRNA during translation. These include the decoding region with A and P sites in the 1400 region (Prince, et al., 1982), and the 530 loop region which is one of three sites (along with the 1400 and Anti-Shine-Dalgarno regions) thought by Trifonov (1987) to be involved in proofreading mRNA by transient base pairing to keep it in proper reading frame during translation. Finally the putative termination region (nucleotides 1199-1204, 956-958, 981-983, and 1223-1225) is thought to be involved in proofreading mRNA for stop codons prior to their entry into the A site. These sequences complement the stop codons with which they are thought to base pair, as the initial step in the termination of translation (Murgola, et al., 1988). Figure 2 shows the placement of the 16S rRNA in the 30S ribosomal subunit according to Brimacombe, et al. (1988), again with the putative mRNA contact points identified. It is apparent that all of these regions are fairly well exposed on the surface of the subunit and presumably available for base pairing. A more stylized model of the 30S subunit, according to Lake (1985), is shown in Figure 3. Finally, the placement of mRNA on the 30S subunit is depicted in Figure 4 and shows a sequence of steps (1 through 4) which the mRNA might take as it makes contact with and moves through the ribosomal subunit. These steps are described in Table 1 and represent an attempt to order the arrangement of translational enhancers (upstream and/or downstream) with the topological arrangement of contact points in the 30S subunit. The "fit" appears to be quite good as upstream enhancers, such as the Epsilon sequence, make the initial contact with the 460 region of the ribosomal subunit, which sticks out like a chin into a pool of mRNAs. Contact involving the upstream Omega sequence occurs subsequent to transport of the mRNA towards the cleft, past the 5' region. This is followed by Petersen's downstream sequences making contact with the same 5' region and the TP site contacting the Anti-TP region further up in the platform of the subunit. Obviously the distribution of translational enhancers varies among mRNAs, and anywhere from none to all of these contacts might occur for a particular mRNA. The model

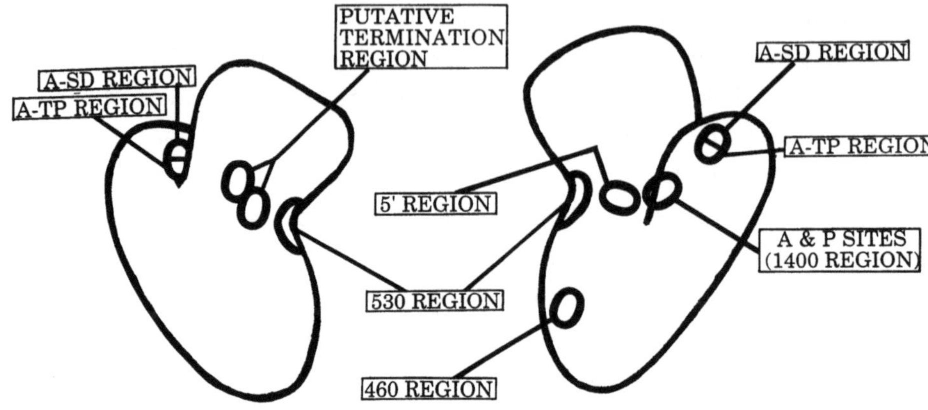

Figure 3. Model of the 30S ribosomal subunit according to Lake (1985) showing potential regions of contact with mRNA.

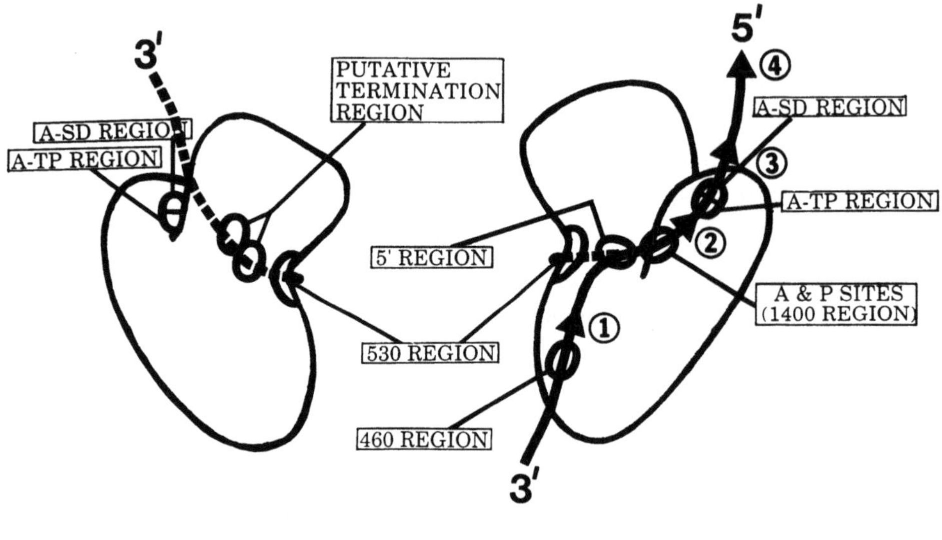

Figure 4. Hypothetical model showing potential contact points and movement of mRNA (solid line) through the ribosome <u>prior</u> to initiation of translation. 1. through 4. represent progressive stages of mRNA movement. See Table 1 for description. The dashed line shows the proposed reorientation of the 3' (downstream) portion of mRNA <u>during</u> translation.

in Figure 4 is not intended to imply that the mRNA moves like a snake across the ribosome. It may be more like a series of transient on/off contacts (hopping across the ribosome), occasionally enhanced by rRNA-mRNA base pairing and facilitated, as well, by ribosomal proteins.

Once the mRNA has reached the proper ribosome binding site with the AUG start codon in the P site and the upstream Shine-Dalgarno sequence paired to the Anti-Shine-Dalgarno region, the 50S subunit binds to this initiation complex to form a 70S ribosome. At this point the 50S subunit would displace the 3' end of mRNA upward from the 460 region at the base of the body of the 30S subunit if it had not already moved, relocating in the neck region to make contact with the 530 and putative termination regions, as shown by the dashed line in Figure 4. The mRNA is now placed in a position which can account for all of the rRNA/mRNA contacts thought to occur during translation: The 1400 decoding region's A and P sites; the upstream Anti-Shine-Dalgarno region; the downstream 530 region; and the putative termination region. Evidence in support of a functional interaction between mRNA and rRNA during elongation comes from data by Weiss, et al. (1987, 1988), in which they studied the mechanism of programmed ribosome frameshifting with the RF2 mRNA. There is no evidence for involvement of the Anti-TP region or the 5' region during translation, and base pairing contact with the 530 and 1400 regions remains speculative (Trifonov, 1987). However, there is evidence for a functional interaction between mRNA and rRNA on the solvent side of the 30S subunit involving base pairing between stop codons and complementary sequences in the putative termination region (Goringer, et al., submitted).[1]

In the end, however, we must acknowledge that there is no direct evidence for base pairing between enhancer sequences in mRNA and complementary sequences in rRNA except for the Shine-Dalgarno/Anti-Shine-Dalgarno

---

[1] The orientation of mRNA on the 30S subunit in Figure 4 is very similar to that proposed by Olson, et al. (1988), using data from structural (antibody probing) rather than functional interactions. A comprehensive review of this literature by a number of excellent investigators using a variety of techniques, is unfortunately, beyond the scope of this article. The reader is referred to references cited by Olson, et al. (1988).

interaction (Hui, deBoer, 1987; Jacob, et al., 1987). Studies similar to these, directed at the putative enhancer binding sites in rRNA, are needed. The loss of functional activity as a result of site-directed mutagenesis of enhancer sequences, and the subsequent restoration of activity by mutagenesis of the rRNA to restore complementary base pairing, would provide strong support for the idea of translational enhancers. Such experiments are currently in progress in several laboratories.

Bibliography

Brimacombe R, Atmadja J, Stiege W, Schuler D (1988) J. Mol. Biol., 199, 115-136.

Dahlberg AE, Dahlberg JE (1975) Proc. Natl. Acad. Sci. USA, 72, 2940-2944.

Dahlberg AE (1989) Cell, 57, 525-529.

Gallie DR, Sleat DE, Watts JW, Turner PC, Wilson TMA (1987) Nucleic Acids Res., 15, 3257-3273.

Gallie DR, Kado CI (1989) Proc. Natl. Acad. Sci. USA, 16, 129-132.

Gold LD, Pribnow D, Schneider T, Shinedling S, Singer BS, Stormo G (1981) Ann. Rev. Microbiol., 35, 365-403.

Gold L, Stormo G, Saunders R (1984) Proc. Natl. Acad. Sci. USA, 81, 7061-7065.

Goringer HU, Murgola EJ, Hijazi K, Dahlberg AE (1990) submitted.

Gren EJ (1984) Biochimie, 66, 1-29.

Hui A, deBoer HA (1987) Proc. Natl. Acad. Sci. USA, 84, 4762-4766.

Jacob WF, Santer M, Dahlberg AE (1987) Proc. Natl. Acad. Sci. USA, 84, 4757-4761.

Kang C, Cantor CR (1985) J. Mol. Biol., 181, 241-251.

Kastelein RA, Berkhaut B, Overbeek GP, Van Duin J (1983) Gene, 23, 245-254.

Lake J (1985) Ann. Rev. Biochem., 54, 507-530.

Maly P and Brimacombe R (1983) Nucleic Acids Res., 11, 7263-7286.

McCarthy JEG, Schairer HU, Sebald W (1985) EMBO J., 4, 519-526.

McCarthy JEG, Sebald W, Gross G, Lammers R (1986) Gene (Amst.), 41, 201-206.

Moazed D, Stern S, Noller HF (1986) J. Mol. Biol., 187, 399-416.

Murgola EJ, Hijazi KA, Goringer HU, Dahlberg AE (1988) Proc. Natl. Acad. Sci. USA, 85, 4162-4165.

Noller HF (1984) Ann. Rev. Biochem., 53, 119-162.

Olins PO, Devine CS, Rangwala SH, Kavka KS (1988) Gene (Amst.), 73, 227-235.

Olins PO, Rangwala SH (1989) J. Biol. Chem., 264, 16973-16976.

Olson HM, Lasater LS, Cann PA, Glitz DG (1988) J. Biol. Chem., 263, 15196-15204.

Petersen GB, Stockwell PA, Hill DF (1988) EMBO J., 7, 3957-3962.

Prince JB, Taylor BH, Thurlow DL, Ofengand J, Zimmermann RA (1982) Proc. Natl. Acad. Sci. USA, 79, 5450-5454.

Scherer GE, Walkinshaw MD, Arnott S, Morre DJ (1980) Nucleic Acids Res., 8, 3895-3907.

Schneider TD, Stormo GD, Gold L, Ehrenfeucht A (1986) J. Mol. Biol., 188, 415-431.

Shine J, Dalgarno L (1974) Proc. Natl. Acad. Sci. USA, 71, 1342-1346.

Sleat DE, Hull R, Turner PC, Wilson TMA (1988) Eur. J. Biochem., 175, 75-86.

Steitz JA, Jakes K (1975) Proc. Natl. Acad. Sci. USA, 72, 4734-4738.

Stern S, Weiser B, Noller HF (1988) J. Mol. Biol., 204, 447-481.

Szer W, Hermoso JM, Leffler S (1975) Proc. Natl. Acad. Sci USA, 72, 2325-2329.

Thanaraj TA, Pandit MW (1989) Nucleic Acids Res., 17, 2973-2985.

Trifonov EN (1987) J. Mol. Biol., 194, 643-652.

van Knippenberg PH (1975), Nucleic Acids Res., 2, 79-85.

Weiss RB, Dunn DM, Atkins JF, Gesteland RF (1987) Cold Spring Harbor Symp. Quant. Biol., 52, 687-693.

Weiss RB, Dunn DM, Dahlberg AE, Atkins JF, Gesteland RF (1988) EMBO J., 7, 1503-1507.

THE PHAGE f1 GENE VII START SITE AND ITS MUTANTS REVEAL THAT TRANSLATIONAL COUPLING CAN CONFER FUNCTION TO INHERENTLY INACTIVE INITIATION SITES

D. A. Steege and M. Ivey-Hoyle[1]
Department of Biochemistry
Duke University Medical Center
Durham, N.C. 27710 U.S.A.

The abundant mRNA species observed in *Escherichia coli* infected with the filamentous phages f1, fd or M13 provide a number of interesting cases of post-transcriptional and translational controls. The mRNAs all derive from the region of the genome containing the tightly linked genes II, X, V, VII, IX and VIII (Fig. 1; reviewed in Model and Russel, 1988). This series of overlapping RNAs have common 3' ends generated by termination at the rho-independent site (T) just beyond gene VIII, and unique 5' ends produced in two ways. Three of the RNAs, which for phage f1 are denoted species A, B and H, are primary transcripts from strong phage promoters. The others, RNAs C, D, E, F and G, are the 3' products of post-transcriptional cleavage (Cashman and Webster, 1979; Cashman *et al.*, 1980). S1 nuclease protection and primer extension methods have mapped the 5' ends of the processed RNAs in uridine-rich regions in the f1 sequence: those for RNAs C, D, E and F are within the gene II coding region, and the 5' endpoint of RNA G is within gene VII (Blumer and Steege, 1984, 1989). The nuclease activity or activities that mediate these cleavages are encoded by the host (Blumer and Steege, 1984), and are distinct from the endonucleases RNase III, RNase P, and RNase E (Cashman and Webster, 1979; Cashman *et al.*, 1980; Smits *et al.*, 1980). Of the RNA species, the larger have half-lives of less than two minutes, while the others are relatively long-lived: RNA E has a half-life of about 2.5 minutes, RNA F, 6 minutes, RNA G, 8 minutes and RNA H, 10 minutes (Cashman *et al.*, 1980).

Despite the fact that the abundance of these transcripts and their relative

[1]Present address: Department of Molecular Genetics, Smith Kline and French Laboratories, King of Prussia, PA. 19406. U.S.A.

NATO ASI Series, Vol. H 49
Post-Transcriptional Control of Gene Expression
Edited by J. E. G. McCarthy and M. F. Tuite

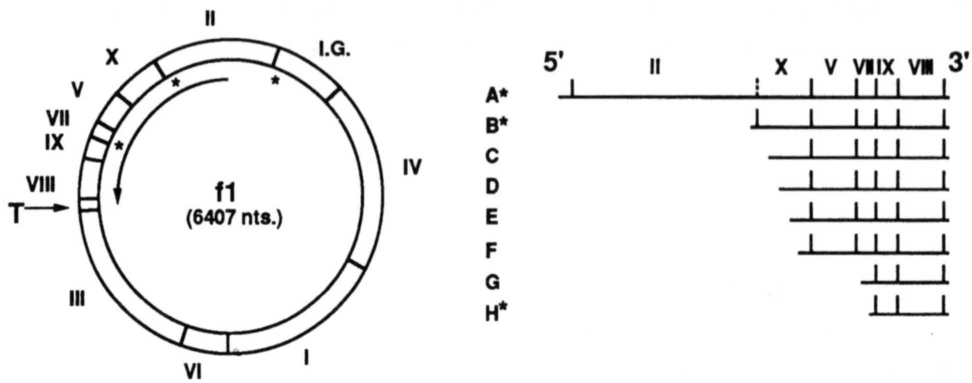

Fig. 1. Genetic map of the filamentous phage and the major mRNAs observed in f1-infected cells. The RNA diagram draws from work of many investigators (reviewed in Model and Russel, 1988). An arrow indicates the direction of transcription; asterisks denote primary transcripts.

stabilities maintain message for genes V, VII, IX and VIII at high levels in the phage-infected cell, very different amounts of the protein products are actually observed. Whereas the single-stranded DNA binding protein (gene V) and the major coat protein (gene VIII) are present in excess of 10^5 copies, probably only about 1000 copies each of the gene VII and gene IX products are made per generation. These products are small hydrophobic proteins of 33 and 32 amino acids, respectively. In the virion these proteins are found in about five copies each on the end where assembly begins, and they still bear the formyl group on the initiating methionine (Grant *et al.*, 1981; Simons *et al.*, 1981; Lopez and Webster, 1983).

From early studies of phage-specific translation, the indication was that expression of genes VII and IX is limited at this step of gene expression. The isolated *in vivo* RNAs, when translated *in vitro*, yielded substantial amounts of gene V and gene VIII proteins, but no detectable gene VII or gene IX protein (LaFarina and Model, 1978; Cashman and Webster, 1979). Ribosome binding experiments using RNA synthesized *in vitro* from replicative form f1 DNA detected only two of the four initiation sites, gene V and gene VIII (Pieczenik *et al.*, 1974; Ravetch *et al*, 1977). A series of ribosome binding experiments carried out with the purified *in vivo* mRNA species D, E, F, G and H (Blumer *et al.*, 1987) confirmed these results, and in addition revealed the likely basis for inefficient translation of gene IX. The gene V and gene VIII initiation sites were recognized efficiently in each mRNA in which they were present. Binding to the gene IX site could be detected, but at a level about tenfold lower. However, binding increased at least tenfold if RNA G was trimmed with RNase to remove

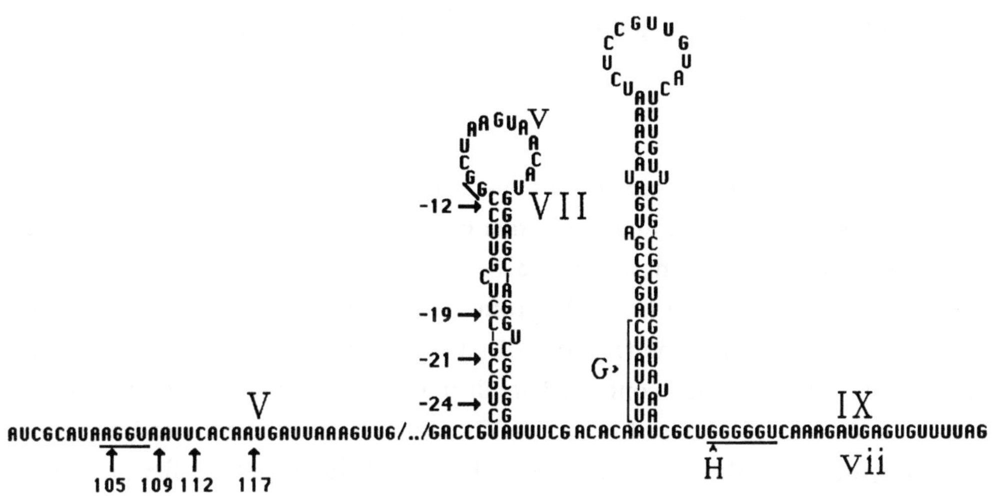

Fig. 2. Partial sequence of the gene V-VII-IX region (Hill and Petersen, 1982), displaying the most stable predicted RNA structures (Zuker and Stiegler, 1981). Sequences are keyed as follows: start and stop codons for phage coding regions, Roman numerals in upper and lower case letters, respectively; Shine-Dalgarno sequences, underlines; 5' endpoints for RNAs G and H, arrowheads. Deletions are numbered as follows: gene V initiation site, basepairs (bp) removed from the 5' end of the f1 insert, gene VII site, nucleotides remaining in each deletion derivative upstream of the AUG.

specifically those nucleotides which comprise the ascending arm of the stable hairpin structure predicted to form (and seen as T1 RNase-resistant material) just upstream of the gene IX site (Fig. 2). This result suggested that while the IX site is inherently one with substantial initiation activity, it is masked in structure in each of the phage mRNAs. Structural masking would also explain earlier genetic data raising the possibility that gene IX expression depends on ribosomal movement to unfold RNA secondary structure: an amber mutation at codon 3 of M13 gene VII results in no expression of downstream gene IX, but one at codon 12 does (Simons *et al.*, 1982). Further support for RNA structure's role in limiting expression of gene IX is the homologous stem-loop found at the same position upstream of the poorly expressed gene IX in the DNA sequence of the distantly related filamentous phage IKe (Peeters *et al.*, 1985). The series of ribosome binding experiments with the purified *in vivo* f1 mRNAs (Blumer *et al.*, 1987), while yielding insight on translational regulation of gene IX, provided no information about how the gene VII initiation site functions. The VII site showed no detectable interactions with ribosomes, even when it was the sole initiation site present on a small RNA fragment.

INITIATION ACTIVITY FROM THE VII SITE DEPENDS ON UPSTREAM TRANSLATION

Since translational activity from gene VII could not be detected in *in vitro* translation or ribosome binding experiments, *lac* fusions have been employed to determine its requirements. Initially constructs were made in which transcription was driven by the *lacIQ* promoter and translation of β-galactosidase depended on initiation at the f1 gene VII site or, as a control, the V site (Blumer *et al.*, 1987). The plasmids contained small fragments bearing the V site (143 bp, pMI5), the VII site (105 bp, pMI7), or the VII site with entire gene V coding region upstream (402 bp, pMI5-7). Present by itself, the VII site in pMI7 failed to initiate synthesis of β-galactosidase at levels above background, whereas the V site gave ~8000 units of activity. However, with all of gene V present upstream of the VII site in pMI5-7, the VII site did function, but at an apparent efficiency tenfold lower than the V site. This result indicated that f1 sequences present in pMI5-7 were required for VII site activity. In view of early genetic evidence that gene V amber mutations were polar on both gene VII and gene IX (Lyons and Zinder, 1972; Simons *et al.*, 1982), the likely explanation for the upstream sequence requirement was that gene VII is translationally coupled to gene V.

By altering the sequences and translational patterns upstream of the VII site, it was subsequently established that genes V and VII are indeed a translationally coupled gene pair and that activity from the VII site is absolutely dependent on the process of translation having proceeded to a stop codon immediately upstream (Ivey-Hoyle and Steege, 1989). The region of the f1 insert in pMI5-7 required for VII site activity proved from deletion analysis to be the translation initiation site for gene V. As deletions removed its Shine-Dalgarno sequence, activity from the downstream VII site dropped sharply, and activity was abolished when deletions extended to the gene V initiator AUG (Fig. 2). The presence of an amber mutation at codon 22 in gene V abolished expression of the VII site-*lacZ* fusion, and the *serU132* amber suppressor coordinately restored gene V protein and β-galactosidase activity to 25-30% of their wild-type levels. Ruling out the possibility that gene V protein was required as a *trans*-acting factor to activate the VII site, a short region from the very beginning of the *lacZ* coding region served as a substitute source of translation upstream of the VII site. These constructs also showed that very little of the gene V sequence upstream from the stop codon (18 nucleotides) need be present to obtain coupled translation from the VII site at its normal *in vivo* level.

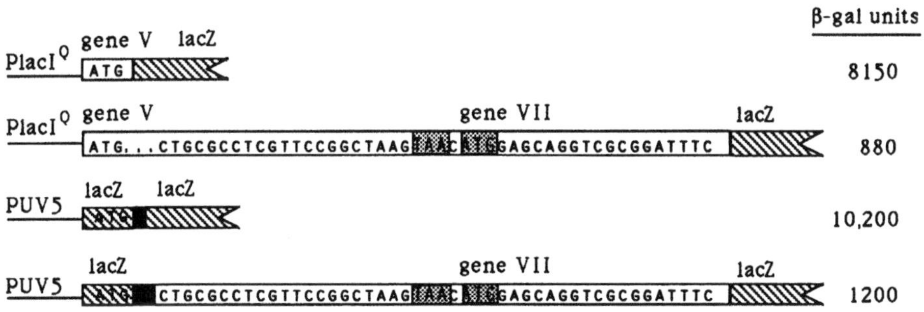

Fig. 3. Translational coupling at the V-VII junction at the wild-type spacing of one nucleotide is inefficient. In each pair of constructs, the lower shows the level of coupled translation (β-gal) from the VII site, and the upper provides a measure of translation of the upstream cistron, gene V or *lacZ*. The data are from Ivey-Hoyle and Steege (1989).

ATYPICAL CHARACTERISTICS OF COUPLING AT THE V-VII JUNCTION

Although gene VII's dependence on upstream translation is a property common to all distal genes in translationally coupled gene pairs, the VII site's coupling properties otherwise were found to be quite atypical (Ivey-Hoyle and Steege, 1989). First, coupling efficiency is low, considering the spacing of only 1 nucleotide between the gene V UAA stop codon and the gene VII AUG. Both when gene V is in its natural configuration upstream and when the beginning of the *lacZ* gene is used as a substitute source of upstream translation, initiation from the VII site is about tenfold lower than upstream initiation efficiency (Fig. 3). This suggests that most ribosomes terminating one nucleotide away and therefore already in contact with the VII site do not recognize it as an initiator region sequence. The second striking characteristic of coupling at the V-VII intercistronic junction is that VII site activity remains proportional to upstream translation levels over such a broad range (Fig. 4). Since coupled translation from the VII site is so much less efficient than initiation at upstream gene V, one might have expected to see what occurs at other coupled initiation sites, namely full activity from the VII site once a modest number of ribosomes are traversing gene V. By contrast, Lindahl *et al.* (1989) find for the ribosomal protein gene pair S10-L3 that translation of the distal gene L3 is nearly maximal even when S10 translation is only 20-50% of its wild-type level. Third, VII site activity is far more sensitive to the spacing between termination and initiation codons than the prototypic distal gene of a coupled gene pair. When the normal 1-nucleotide

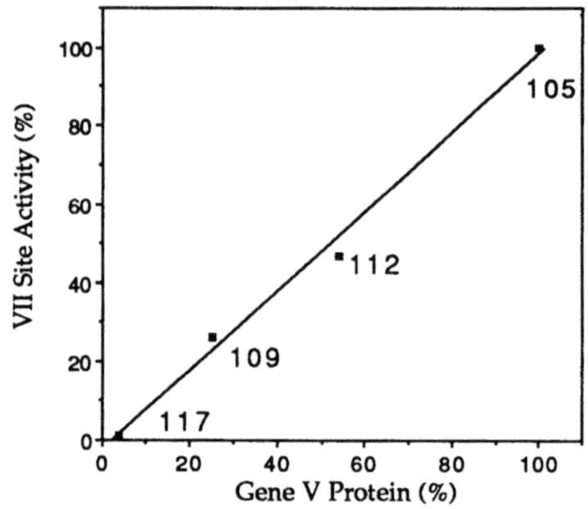

Fig. 4. VII site activity is proportional to the amount of gene V protein synthesized. Mutants from the pMI5-7 deletion series are as in Fig. 2. Gene V protein levels were quantified using anti-gene V protein IgG (Ivey-Hoyle and Steege, 1989). Activities are expressed relative to the strain bearing mutant 105.

spacing is increased to only 5 nucleotides, VII site activity drops by 80%. Finally, a strategy which reveals substantial activity from other coupled and structurally occluded initiation sites does not for the VII site. If the 5' junction between f1 sequence and the *lac* fusion vector is such that the -1 reading frame of the gene V sequence is used, upstream translation proceeds to a UGA codon spaced 33 nucleotides before the gene VII AUG. As a result, the position of terminating ribosomes should prevent the GC-rich lower segment of a structure predicted for the V-VII intercistronic region (Fig. 2) from nucleating, but leave the VII site accessible to new ribosomes from the pool. In this situation, however, VII site activity at only 6% of its full level is observed.

WHY DOES THE VII SITE SHOW NO INDEPENDENT INITIATION ACTIVITY?

Two possible explanations for the VII site's lack of function in both *in vitro* and *in vivo* assays of independent initiation are: 1) the site has inherent activity but is masked in structure, or 2) the site lacks the elements necessary for ribosomal recognition. Early on, the secondary structure predicted by computer folding for the V-VII intercistronic junction appeared to provide support for the first possibility. In the hairpin shown in Fig. 2, the gene VII AUG is in the loop, but the remainder of the VII site sequence forms the stem. However, an experiment designed to ask if this structure is what prevents ribosome binding gave a negative result (Ivey-Hoyle and Steege, 1989). Deletions were made to sequentially remove the nucleotides of the ascending arm of the stem, and a set of eight variants in which 12-24 nucleotides remained upstream of the gene VII AUG were characterized. For none was VII site activity above background observed, even though the predicted structure disappears from the RNA folding pattern once the lower 4-5 nucleotides of the ascending arm are gone, and in the rest of the deletion series the VII site is far more accessible. If the VII site indeed had independent activity, one would have expected to see it in one of the sequence contexts provided by the deletion series.

Adding to the suggestion from the above experiment that the VII site is instead a defective initiator region is its primary sequence, which differs considerably from a prototypic translation initiation site. It has a poor match to all consensus sequences derived for prokaryotic initiation sites, and its GC-content is threefold higher than an initiation site should be. Its potential for base-pairing with the 3' end of 16 S rRNA is minimal. At an appropriate spacing of 9 nucleotides from the AUG, there is only the dinucleotide GG. There is also a UAAG, but probably at too close a spacing for pairing with the complementary nucleotides at the extreme 3' end of 16 S rRNA.

In view of the VII site's behavior in translational coupling and its atypical sequence, the most likely basis for its inactivity then was that it lacks the information required for recognition by the 30 S ribosomal subunit. To address this possibility, designed and random changes have been made in the VII site sequence to obtain derivatives with independent activity (Ivey-Hoyle and Steege, unpublished). The potential complexity of such an analysis has been reduced by using as a target for mutagenesis a truncated VII site sequence which still contains 19 nucleotides upstream of the AUG, but is missing the first 7 nucleotides of the ascending arm sequence in the structure of Fig. 2. Designed changes giving the site appropriately spaced Shine-Dalgarno sequences GGAG or GGAGG make activity detectable, but at quite low levels. An additional A at +4 further enhances activity threefold. Random changes conferring low levels of activity are found throughout the site: in general, they either lengthen the 2-base Shine-Dalgarno complementarity present in the wild-type site or change G and C residues to A or T. A number of mutations together, combined with a strong Shine-Dalgarno sequence, are required to generate a modestly efficient initiation site. Thus, the effects of point mutations add to the evidence that the VII site's defect is its primary sequence. Rather than pinpointing a particular region or element that interferes with activity or is responsible for the lack of activity, the data demonstrate that at least 15 positions in the site are suboptimal.

CONCLUSIONS AND PERSPECTIVES

Taken together, available evidence is most consistent with the conclusion that the VII site is inherently inactive as an initiation site. The site comes into use only when ribosomes approach from along the upstream mRNA and terminate within its sequence. Translational coupling thus plays a novel role at the V-VII intercistronic junction. Rather than exposing the initiation site or enhancing its natural activity, it actually confers activity to the region containing the gene VII initiator AUG. In essentially all respects, VII site's properties are identical to translational restart sites activated by nonsense mutations within genes (Napoli *et al.*, 1981; Cone and Steege, 1985; Matteson and Steege, submitted). With no evidence that the VII site is used by free ribosomes, this is a prime candidate for an instance of translational coupling in which translation of the distal cistron occurs through obligate reuse of components coming in from the upstream cistron. If this is the case, note that the mechanism does not guarantee coupling at a 1:1 ratio, as has been suggested (Yates and Nomura, 1981). From this and the results of others (Das and Yanofsky, 1984; Spanjaard and vanDuin, 1989; Matteson and Steege, submitted), it is clear that the sequence of the distal initiation site is an important determinant of coupling efficiency. In light of evidence that the 3' end of 16 S rRNA scans mRNA during elongation (Weiss *et al.*, 1988), the expectation is that coupling efficiency will prove to be a function of how well the 30 S particle makes contact with the mRNA sequence as the ribosome is undergoing termination in the upstream cistron.

If the properties of the f1 gene VII initiation site are considered in the more general context of how the major f1 mRNAs are translated, the principal elements of the regulatory strategies become apparent. Coupling at the V-VII junction determines the fraction of translational activity occurring on gene V that propagates into gene VII. Ribosomes moving through gene VII then unfold an RNA structure that limits ribosome access to the gene IX site. It is not yet clear whether the observed initiation efficiencies for genes VII and IX suffice to account for the final product yields observed in the phage-infected cell, or whether further downregulation occurs during elongation through these very small coding regions (Fig. 2). Another intriguing issue is whether these translational mechanisms have evolved to ensure low-level production of the two proteins, which as membrane proteins might well be detrimental to the cell. In this regard, it will be important to look at translation of the equivalents of genes V and VII encoded by the filamentous phage IKe. IKe closely resembles f1 in phage structure, life cycle and genome organization. However, the phages differ with respect to host specificities, and their sequences are only 55% identical (Peeters *et al.*, 1985). At the V-VII intercistronic region, both f1 and IKe sequences are predicted to assume similar secondary structures which involve sequences at the gene VII start. However, the presumptive VII site of IKe resembles a prototypic initiation site to a greater extent than the VII site of f1.

Finally, since the VII site demonstrates that translational coupling can confer activity to an inherently defective sequence, gene starts within polycistronic clusters are not constrained to have the features or functional properties of independent initiation sites. Although many of them do, start sites which look quite dissimilar to standard initiator region sequences are appearing at an increasing rate as more and more *E. coli* and phage genes are sequenced.

ACKNOWLEDGMENT

Research in the author's laboratory (D.A.S.) is supported by National Institutes of Health Grant GM33349.

REFERENCES

Blumer KJ, Ivey MR, Steege DA (1987) Translational control of f1 gene expression by differential activities of the gene V, VII, IX, and VIII initiation sites. J Mol Biol 197:439-451

Blumer KJ, Steege DA (1984) mRNA processing in *Escherichia coli*: an activity encoded by the host processes bacteriophage f1 mRNAs. Nucl Acids Res 12:1847-1861

Blumer KJ, Steege DA (1989) Recognition and cleavage signals for mRNA processing lie within local domains of the phage f1 RNA precursors. J Biol Chem 264:20770-20777

Cashman JS, Webster RE (1979) Bacteriophage f1 infection of *Escherichia coli*: Identification and possible processing of f1-specific *in vivo* mRNAs. Proc Nat Acad Sci USA 76:1169-1173

Cashman JS, Webster RE, Steege DA (1980) Transcription of bacteriophage f1: The major *in vivo* RNAs. J Biol Chem 255:2554-2562

Cone KC, Steege DA (1985) Functional analysis of *lac* repressor restart sites in

translational initiation and reinitiation. J Mol Biol 186:733-742
Das A, Yanofsky C (1984) A ribosome binding site sequence is necessary for efficient expression of the distal gene of a translationally coupled gene pair. Nucl Acids Res 12:4757-4768
Grant RA, Lin TC, Konigsberg W, Webster RE (1981) Structure of the filamentous bacteriophage f1. Location of the A, C and D minor coat proteins. J Biol Chem 256:539-546
Hill DF, Petersen GB (1982) Nucleotide sequence of bacteriophage f1 DNA. J Virol 44:32-46
Ivey-Hoyle M, Steege DA (1989) Translation of phage f1 gene VII occurs from an inherently defective initiation site made functional by coupling. J Mol Biol 208, 233-244
LaFarina M, Model P (1978) Transcription in bacteriophage f1-infected *Escherichia coli*. I. Translation of RNA *in vitro*. J Mol Biol 86:368-375
Lindahl L, Archer RH, McCormick JR, Freedman LP, Zengel JM (1989) Translational coupling of the two proximal genes in the S10 ribosomal protein operon of *Escherichia coli*. J Bacteriol 171:2639-2645
Lopez J, Webster RE (1983) Morphogenesis of filamentous bacteriophage f1: Orientation of extrusion and production of polyphage. Virology 127:177-193
Lyons LB, Zinder ND (1972) The genetic map of the filamentous bacteriophage f1. Virology 49:45-60
Matteson RJ, Steege DA (1990) Translational reinitiation in the *lac* repressor mRNA: bridging by out-of-frame translation and RNA secondary structure, primary sequence effects. Nucl Acids Res, submitted.
Model P, Russel R (1988) Filamentous Bacteriophage. In: Calendar R (ed) The Bacteriophages, vol 2. Plenum Press, New York, 375-456
Napoli C, Gold L, Singer BS (1981) Translational reinitiation in the rIIB cistron of bacteriophage T4. J Mol Biol 149:433-449
Peeters BPH, Peters RM, Schoenmakers JGG, Konings RNH (1985) Nucleotide sequence and genetic organization of the genome of the N-specific filamentous bacteriophage IKe. Comparison with the genome of the F-specific filamentous phages M13, fd, and f1. J Mol Biol 181:27-39
Pieczenik G, Model P, Robertson HD (1974) Sequence and symmetry in ribosome binding sites of bacteriophage f1 RNA. J Mol Biol 90:191-214
Ravetch J, Horiuchi K, Model P (1977) Mapping of bacteriophage f1 ribosome binding sites to their cognate genes. Virology 81:341-351
Simons GFM, Konings RNH, Schoenmakers JGG (1981) Gene VI, gene VII, and gene IX of phage M13 code for minor capsid proteins of the virion. Proc Nat Acad Sci USA 78:4194-4198
Simons GFM, Veeneman GH, Konings RNH, VanBoom JH, Schoenmakers JGG (1982) Oligonucleotide-directed mutagenesis of gene IX of bacteriophage M13. Nucl Acids Res 10:821-832
Smits MA, Schoenmakers JGG, Konings RNH (1980) Expression of bacteriophage M13 DNA *in vivo*. Isolation, identification and characterization of phage-specific messenger RNA species. Eur J Biochem 112:309-321
Spanjaard RA, vanDuin J (1989) Translational reinitiation in the presence and absence of a Shine-Dalgarno sequence. Nucl Acids Res 17:5501-5507
Weiss RB, Dunn DM, Dahlberg AE, Atkins JF, Gesteland RF (1988) Reading frame switch caused by base-pair formation between the 3' end of 16 S rRNA and the mRNA during elongation of protein synthesis. EMBO J 7:1503-1507
Yates JL, Nomura M (1981) Feedback regulation of ribosomal protein synthesis in *E. coli*: Localization of the mRNA target sites for repressor action of ribosomal protein L1. Cell 24:243-249
Zuker M, Stiegler P (1981) Optimal computer folding of large RNA sequences using thermodynamics and auxiliary information. Nucl Acids Res 9:133-148

Measurement of translation rates *in vivo* at individual codons and implication of these rate differences for gene expression

Michael A. Sørensen, Kaj Frank Jensen*) and Steen Pedersen

Institute of Microbiology and
Institute of Biological Chemistry*),
University of Copenhagen,
Øster Farimagsgade 2A, DK-1353 and
Sølvgade 83, DK-1307 *),
Copenhagen K, Denmark.

Introduction

One of the problems in discussing codon usage is that our language is not equipped with the proper terms to cope with such diversity. A codon which is common in one organism may be a rare codon in another organism and even within the same species, terms like "rare codon" do often not include the same set of codons, when used by different authors. We will restrict ourselves to discuss results only from *Escherichia coli* in order to avoid this and other complications as much as possible.

We will focus our attention mainly on the translation rate *in vivo* and come to a model for the function of the codon usage in the determination of the protein expression level under different growth conditions. It is clear, however, that the choice of synonymous codons may influence a variety of cellular processes other than translation. Thus, signals can be generated, while the same amino acid is being coded for, to be used in the determination of: mRNA halflife, mRNA secondary structures mediating transcription pauses and perhaps even replication pauses and, through coupling to the leading ribosome, transcription termination. In addition, codon usage may possibly influence frameshifting [28], protein folding, transport and cotranslational modification [19; 22] and translation errors [17]. Other aspects of codon usage have been reviewed recently [1].

NATO ASI Series, Vol. H 49
Post-Transcriptional Control of Gene Expression
Edited by J. E. G. McCarthy and M. F. Tuite

Codon usage

When the first genes [6; 7] were sequenced it was observed that not all synonymous codons were used with the same frequency. The number of sequenced codons was small, and only when the sequences of the first ribosomal protein genes were obtained [21] the clear bias in codon usage became evident and has later been confirmed by numerous studies and reviewed extensively e.g. [10; 24].

In the early discussions the ribosomal type codons were supposed to be translated more efficiently, presumably meaning faster or more accurate than the infrequent codons and the thought was that the biased, ribosomal type codon usage caused the high expression. This hypothesis was never tested experimentally although it must be evident that the expression level can only be determined by the initiation rate and not by differences only affecting the elongation rate: all ribosomes which has initiated translation will generate a protein, no matter how long time they spend in the process because, in general, the events giving rise to unfinished proteins, e.g. frameshifting and drop-off, probably is not a major event in the average translation.

In the analysis of the function of codon usage, it is interesting to note, that it is the strong bias towards the use of the frequent codons in the ribosomal protein genes which seems to be selected for, and not the absence of these codons in the repressor type proteins [24].

The translation rate is sequence specific

Using incorporation of very short pulses of radioactivity into various nascent peptides [23; 29] or a direct measurement of the rate of translation of mRNAs with different codon usage [18] it was found that either part of a mRNA or different whole mRNAs might be translated with different rates. Patches of slow translation generate descrete nascent peptides and by determining the amount of the nascent fragments and their lengths, a correlation between the translation rate and the codon use was found [29]. However, from the migration distance alone it was difficult to determine the precise length of the nascent peptide and this introduced a degree of uncertainty in this work. Nevertheless, these authors suggested a model where the tRNA concentration exclusively determined the rate of the translation because the cognate tRNA had to compete with the non-cognate tRNAs for the A site of the ribosome [29]. This model was supported by the demonstration that the duration of a translation pause on the codon AGG was sensitive to the concentration of $tRNA^{ARG}$ [16].

The method used by Pedersen [18] is a pulse-chase experiment which determines the time it takes the ribosome to traverse from the first methionine residue, which will appear in the full length peptide, to the end of the protein and it was possible to show that some mRNAs e.g. the *lacI* mRNA, were translated aprox. 30% slower than mRNAs coding for ribosomal protein and translation factors. The most likely explanation was that the <u>kind</u> of codons, abundant in the *lacI* gene, but not in the ribosomal protein mRNAs caused the difference though other possibilities existed. Recently we tested, whether or not, mRNA structure was involved in determining the translation rate and found (in the case investigated) that codon usage was the major, and perhaps only, determinant of differences in translation rate [26]. To obtain this result our method to measure the translation rate had to be refined. This involved labelling of mixtures of cultures and fitting the data to several sets of theoretical curves, each calculated from an average translation rate and the position of methionine residues in the protein, that was measured. The result was that differences in translation time as short as 1 sec can now be reliably determined. We were able to show that the average common codon, defined by an insert of 24 codons from the *rpsA* gene was translated at a rate of 12 amino acids/sec which is 6-fold faster than the translation time for the average infrequent codon, being defined by a set of frame-shifted codons from the same insert [26]. These directly measured differences in rate between frequent and infrequent codons are of such magnitude that they can explain the previously observed differences in translation rate between the individual mRNAs [18].

The tRNA concentration is not the sole determinant of translation rate

Our data (unpublished) suggests that translation rate differences can be observed even among mRNAs having the ribosomal, high biased codon usage. This lead us to suspect that significant rate differences must occur within the group of codons pairs read by the same tRNA, a model which is now also supported by *in vitro* experiments [27]. We [25; 3] have now tried to verify this hypothesis.

Coupling between the first ribosome and the RNA polymerase have been shown to be very important for the regulation of some genes for biosynthesis of nucleoside triphosphate precursors, e.g. the *pyrE* gene. The translation rate in the leader peptide have been suggested to be of importance for this coupling via an attenuator and the system have been developed to give a measure for the relative translation rate of individual codons [2]. A similar method, based on the competition between the rate of elongation at a particular codon and the rate of termination in a sequence analogous to the regulatory frameshifting sequence of the release factor 2 gene, have been developed to give a measure of the relative translation rates [4].

Our pulse-chase method [25] is a direct measurement and the only to give the absolute *in vivo* translation rates. So far we have only precise determinations of the translation rate on the two glutamic acid codons, GAA and GAG, and on two codons CGA (arg) and CCG (pro) which were present in the inserts. The data for these translation rates will be published elsewhere [25], but briefly GAA was found to be fast (approx. 22 aa/sec) whereas GAG,CCG and CGA were all slow (approx. 7, 6 and 3.5 aa/sec respectively). The tRNA concentrations for these codons have been determined [11] and our results indicate that translation rate and tRNA concentration do not seem to be correlated. In the case of the two glutamic acid codons they both are read by the same tRNAs of which $tRNA^{GLU}_2$ is the predominant species. The same qualitative result was obtained by Bonekamp [3] for the translation rate at several codons read by the same tRNA.

A comparison with the results of Curran and Yarus [4] shows that we find reasonable correspondance in the case of CCG (slower than the average codon), but we find the arginine codon CGA to be slow (3.5 aa/sec) while it was classified amongst the fastest codons by Curran and Yarus [4]. A possible explanation [4] is that their system measures the rate of entry of a ternary complex to the ribosome, and that either the translocation step for this codon is very slow or the ternary complex on the CGA codon is predominantly rejected by the proofreading mechanism.

Effect on expression level:

To produce an effect on the expression level, the codon usage as mentioned above must influence the rate of translation initiation. This could be the case if the slow translation gave a delay in the time it takes the ribosome to clear the ribosome binding site, thereby inhibiting the initiation by the succeding ribosome for a considerable period, [14; 18]. This effect will be more dramatic, the closer the translation pause is located to the ribosome binding site: Consider for instance a patch of 10 slow codons situated 120 codons downstream from the start of the *lacZ* gene and let these be translated in 3 seconds, whereas the rest of the gene is translated with a rate of 12.5 codons per sec [26]. The average initiation frequency on the *lacZ* mRNA is estimated to be 1 per 4 seconds [13] and the next ribosomes will not form a queue because the preceding ribosome have translated through the stretch of slow codons. Even if the slow translation of the ten codons takes 5 seconds, the pause will not have any effect on the expression level. A queue will start to back up towards the initiation site but it will have to be ten ribosomes wide before it start to mask the ribosome binding site (assuming the ribosome diameter to be about 12 codons, [9; 31]). A ribosome queue has been demonstrated *in vivo* and shown not *per se* to reduce the translation yield [30; 25]. If the translation pause was really a stop, this buildup takes 50 seconds, but if ribosomes leak through the streach of slow codons with a rate of one per 5 seconds, the queue will only be about two ribosomes long and therefore not interfere with the translation initiation rate, because ten ribosomes will have left the queue in these 50 sec. This picture is generally true even considering statistical variations in the initiation rate or in the elongation rate on individual mRNAs. In a mRNAs average lifetime, believed for the *lacZ* mRNA to allow 30 initiations [13], we will therefore see no effect on the expression level of a translation pause located this far downstream from the start. The length of the patch with the slow codons will have no effect because ribosomes will still traverse a ribosome diameter with the same rate. Going through such calculations with different locations of the pause etc shows that the closer the pause is located to the initiation site, the larger the effect on the initiation rate will be. Also, the parameters mRNA halflife and the rate of initiation per mRNA are very important in determining the length of the ribosome queue and therefore the ability to interfere with the initiation rate.

From our measurements we can now start to estimate how the variation in the clearing time, defined as the time the ribosome takes to translate a ribosome diameter, affects the translation yield from a mRNA: The fastest codon we have come across has a translation rate of approx. 22 aminoacids/sec (GAA) while the slowest is 3.5 codons/sec (CGA)[25]. Unfortunately, our knowledge of the individual translation rates are not yet detailed enough to calculate the clearing time for any individual ribosome binding site. Therefore, in the following we will estimate the consequences of having slow, average or fast codons in the beginning of the structural gene, but evidently we suspect to find the fast codons in the ribosomal genes and, conversely, the slowest codons in the genes with unbiased codon usage. The time required for the initiation complex between the 30S ribosomal subunit and mRNA to find the 50S subunit is not known, but we estimate it to be in the range of 0.2 to 1 sec. The fastest possible clearing time will therefore be expected to be 0.5 sec, the average around 1.5 to 2.5 sec and the slowest quite possibly approximately 4.5 seconds. In the following we will investigate this effect of a codon specific variation in the clearing time in this range.

The figure shows the effect of variations in the clearing time on the rate of initiation. These curves were calculated by

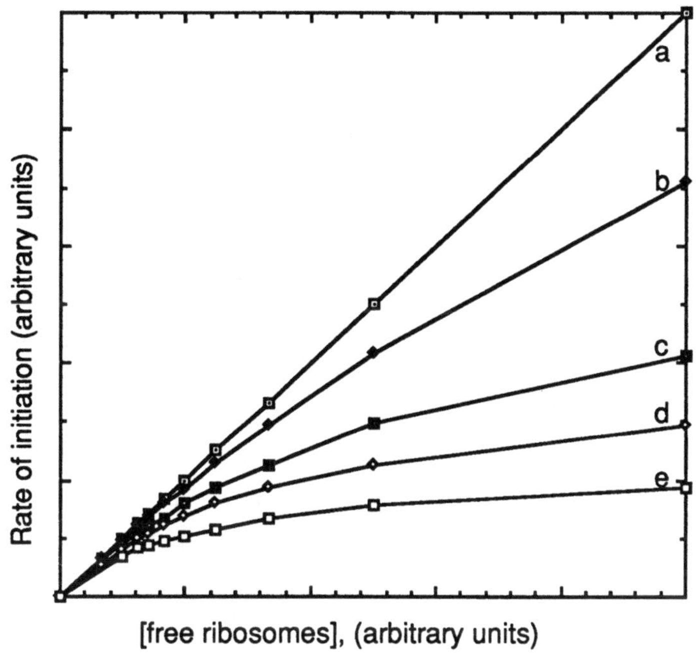

simulating random initiation with an average initiation frequency at different free ribosome concentrations which are given in arbitrary units, but which are indicated by the initiation frequency on a ribosome binding site which has the property of being capable of infinitely fast to let ribosomes translate the first ribosome diameter (curve a). We then calculate the effect of variations in the clearing time, i.e. the time a ribosome binding site will be blocked by the preceding ribosome. The curves for clearing times of 0.5, 1.5, 2.5 and 4.5 sec are shown (curves b,c,d,e respectively). Not surprisingly we see, that the higher the concentration of free ribosomes, the more drastic is the effect of increasing the clearing time on the initiation rate and that these variations in effect changes the V_{max} and K_m for the initiation reaction.

From these considerations we suggest that the codon specific variation in the time it takes the ribosome to traverse the first couple of ribosome diameters, almost exclusively will determine the effect of codon choice on the expression level This seem not to have been generally noticed and e.g. the results of Looman [15] can be explained as an interference with the clearing time instead of an effect on the ribosome binding reaction.

Growth rate regulation

Recently we described a model for the growth rate regulation of the *E.coli* cell [12]. A consequence of this model is that the concentration of free ribosomes in the cell or at least the initiation frequency per ribosome will increase with increasing growth rate. From the analysis represented in the figure we will therefore propose that codon usage, even for a constant SD sequence can be used to specify

i) a low, but ribosome concentration independent initiation frequency as exemplified by initiation site e in figure 1

ii) at least part of the growth rate regulation of ribosomal protein operons, because only translation initiation sites with low clearing times can use the higher ribosome concentration present in rich media to increase the rate of initiation.

Previously, Gausing [8] stated that the translation yield from the ribosomal protein mRNAs was independent of the growth rate. Her data, however are best interpreted by assuming that the translation efficiency for the ribosomal protein mRNAs compared to the average translation efficiency show a slight increase (about 15% between acetate and rich medium) . Because the ribosomal type proteins constitutes 60% of total proteins in the fast growing cells [20] they contribute themselves heavily to the average. Therefore it is possible, in our opinion, that the translation efficiency of ribosomal protein mRNAs is perhaps 30-50% higher in the rich medium compared to the minimal medium. This could have many reasons for instance that the extent of the ribosomal feedback control mechanism varies with the growth rate, but it might also be because the ribosomal type protein mRNAs have low clearing times and therefore are the only mRNAs that are able to use the increased ribosome concentration for initiation.

Previously, Ehrenberg and Kurland [5] proposed, that a more efficient utilisation of the translation apparatous at the high growth rates was the selective pressure for the bias in codon usage in the ribosomal protein genes. This we believe still to be true, but in addition we propose the same mechanism to be used specifically to regulate the synthesis of ribosomal proteins at the high growth rate in addition to the other control mechanisms regulating these genes.

We therefore believe that all effects of codon usage can be explained by kinetic considerations caused by differences in their translation rate. The original hypothesis that the bias in codon usage was related to the high expression level is therefore right and explained by the clever use of codons with different rates in the beginning of the structural genes.

Acknowledgement

This work was supported by the Danish Center of Microbiology.

References

1. Anderson, S. G. E. and C. G. Kurland. 1990. Codon Preferences in freeliving microorganisms. Microbiol. Rev. in press.
2. Bonekamp, F., H. D. Andersen, T. Christensen and K. F. Jensen. 1985. Codon-defined ribosomal pausing in *Escherichia coli* detected by using the *pyrF* attenuator to probe the coupling between transcription and translation. Nucleic Acids Res. **13**: 4113-4123.
3. Bonekamp, F., H. Dalbøge, T. Christensen and K. F. Jensen. 1989. Translation rates of individual codons are not correlated with tRNA abundance or with frequencies of utilisation in *Escherichia coli*. J. Bacteriol. **171**: 5812-5816.
4. Curran, J. and M. Yarus. 1989. Rates of amino acid selection at 29 sense codons *in vivo*. J. Mol. Biol. **209**: 65-77.
5. Ehrenberg, M. and C. Kurland. 1984. Costs of accuracy determined by a maximal growth constraint. Q. Rev Biophys. **17**: 45-82.
6. Farabaugh, P. J. 1978. Sequence of the *lacI* gene. Nature. **274**: 765-767.
7. Fiers, W., R. Contreras, F. Duerinck, G. Haegeman, D. Iserentant and J. Merregaert. 1976. Complete nucleotide sequence of bacteriophage MS2 RNA: primary and secondary structure of the replicase gene. Nature. **260**: 500-507.
8. Gausing, K. 1977. Regulation of Ribosome Production in *Escherichia coli*: Synthesis and Stability of Ribosomal RNA and Ribosomal Protein Messenger RNA at Different Growth Rates. J. Mol. Biol. **115**: 335-354.
9. Gold, L., D. Pribnow, T. Schneider, S. Shinedling, Swebilius, B. Singer and G. Stormo. 1981. Annu. Rev. Microbiol. **35**: 365-403.
10. Grosjean, H. and W. Fiers. 1982. Preferential codon usage in procaryotic genes: the optimal codon-anticodon interaction energy and the selective codon usage in efficientley expressed genes. Gene. **18**: 199-209.
11. Ikemura, T. 1981. Correlation between the abundance of Escherichia coli transfer RNAs and the occurence of the respective codons in its protein genes. J. Mol. Biol. **146**: 1-21.
12. Jensen, K. F. and S. Pedersen. 1990. Metabolic growth rate control in *Escherichia coli* may be a consequence of subsaturation of the macromolecular biosynthetic apparatous with substrates and catalytic components. Microbiol. Rev., **2** in press:
13. Kennell, D. and H. Riezman. 1977. Transcription and translation initiation frequencies of the *E. coli lac* operon.

J. Mol. Biol. **114**: 1-21.
14. Liljenström, H. and G. von Heijne. 1987. Translation rate modification by preferential codon usage: Intragenic position effects. J .theoret. Biol. **124**: 43-55.
15. Looman, A. C., J. Bodlaender, L. J. Comstock, D. Eaton, P. Jhurani, H. A. deBoer and P. H. vanKnippenberg. 1987. Influence of the codon following the AUG initiation codon on the expression of a modified *lacZ* gene in *Escherichia coli*. EMBO J. **6**: 2489-2492.
16. Misra, R. and P. Reeves. 1985. Intermediates in the synthesis of TolC protein include an incomplete peptide stalled at a rare Arg codon. Eur. J. Biochem. **152**: 151-155.
17. Parker, J. 1989. Errors and alternatives in reading the universal genetic code. Microbiol. Rev. **53**: 273-298.
18. Pedersen, S. 1984. *Escherichia coli* ribosomes translate *in vivo* with variable rate. EMBO J. **3**: 2895-2898.
19. Pedersen, S. 1984. "In *Escherichia coli* individual genes are translated with different rates *in vivo*." in Gene expression, A. Benzon symposium 19. pp 101-111. eds. Clark, B. F. C. and H. U. Petersen. Munksgaard. Copenhagen.
20. Pedersen, S., P. L. Bloch, S. Reeh and F. C. Neidhardt. 1978. Patterns of protein synthesis in E.coli: a catalog of the amount of 140 individual proteins at different growth rates. Cell. **14**: 179-190.
21. Post, L. E., G. D. Strycharz, M. Nomura, H. Lewis and P. P. Dennis. 1979. Nucleotide sequence of the ribosomal protein gene cluster adjacent to the gene for RNA polymerase subunit ß in *Escherichia coli*. Proc. Natl. Acad. Sci. USA. **76**: 1697-1701.
22. Purvis, I. J., A. J. E. Bettany, T. C. Santiago, J. R. Coggins, K. Duncan, R. Eason and A. J. P. Brown. 1987. The efficiency of folding of some proteins is increased by controlled rates of translation *in vivo*: A hypothesis. J. Mol. Biol. **193**: 413-417.
23. Randall, L. L., L.-G. Josefson and S. J. S. Hardy. 1980. Novel intermediates in the synthesis of maltose-binding protein in *E.coli*. Eur. J. Biochem. **107**: 375-379.
24. Sharp, P. M. and W. Li. 1986. Codon Usage in regulatory genes in *Escherichia coli* does not reflect selection for 'rare' codons. Nucl. Acids Res. **14**: 7737-7749.
25. Sørensen, M. and S. Pedersen. 1990. *In vivo* translation rates of individual codons in *E. coli*: three-fold difference at the GAA and GAG glutamic acid codons read by the same tRNA. Manuscript in preparation.
26. Sørensen, M. A., C. G. Kurland and S. Pedersen. 1989. Codon usage determines the translation rate in Escherichia coli. J. Mol. Biol. **207**: 365-377.
27. Thomas, L. K., D. B. Dix and R. C. Thompson. 1988. Codon choice and gene expression: Synonymous codons differ in their ability to direct aminoacetylated-transfer RNA binding to ribosomes *in vitro*. Proc. Natl. Acad. Sci. USA. **85**: 4242-4246.
28. Trifonov, E. N. 1987. Translation Framing Code and Frame-monitoring Mechanism as Suggested by the Analysis of mRNA and 16S rRNA Nucleotide Sequences. J. Mol. Biol. **194**: 643-652.
29. Varenne, S., J. Buc, R. Lloubes and C. Lazdunski. 1984. Translation is a non-uniform process: effect of tRNA availability on the rate of elongation of nascent polypeptide chains. J. Mol. Biol. **180**: 549-576.
30. Vind, J. 1990. Thesis, University of Copenhagen.
31. von Heine, G., L. Nielson and C. Blomberg. 1978. Models for mRNA translation: Theory versus experiment. Eur. J. Biochem. **92**: 397-402.

REGULATION OF GENE EXPRESSION BY MINOR CODONS IN *Escherichia coli*; MINOR CODON MODULATOR HYPOTHESIS

Masayori Inouye and Giafen Chen
Department of Biochemistry
University of Medicine and Dentistry of New Jersey
Robert Wood Johnson Medical School
675 Hoes Lane
Piscataway, New Jersey 08854 USA

Non-random usage of synonymous codons in various organisms has been well documented (Ikemura 1981a and 1981b; Maruyama et al., 1986). In *Escherichia coli*, codon usage in abundant proteins such as ribosomal proteins and major outer membrane proteins is extremely biased, and such non-random usage was speculated to be due to preferential usage of major isoaccepting species of tRNA (Post et al., 1979; Nomura et al., 1980). From the analysis of codon usage in various *E. coli* proteins and relative quantities of tRNAs, a strong positive correlation between tRNA content and the occurrence of respective codons has been demonstrated, and it was proposed that *E. coli* genes encoding abundant protein species selectively use the "optimal codon" or major codons as determined by the abundance of isoaccepting tRNA (Ikemura, 1981a).

Because of these observations it is generally believed that protein production from a gene containing minor codons or non-optimal codons is less efficient than that from a gene containing no minor codons. However, contrary to this notion, some researchers have failed to demonstrate the negative effect of minor codons on gene expression (Holm, 1986; Sharp and Li, 1986). Further, it was recently demonstrated that the difference in overall translation time between the genes with and without minor codons was small, although there is a difference in

NATO ASI Series, Vol. H 49
Post-Transcriptional Control of Gene Expression
Edited by J. E. G. McCarthy and M. F. Tuite

translation rate between common codons and minor codons (Sorensen, et al., 1989; Kurland, 1987). Therefore, it was concluded that the minor codon usage does not contribute to significant effects on gene expression at the level of translation. On the other hand, in Saccharomyces cerevisiae it has been demonstrated that replacing an increasing number of major codons with synonymous minor codons at the 5'-end of the coding sequence of the gene for phosphoglycerate kinase caused a dramatic decrease of the gene expression (Hoekema et al., 1987). In this work decreasing efficiency of mRNA translation was attributed to mRNA destabilization.

What is the actual effect of minor codons on gene expression? Are there any general roles of minor codons in gene expression? In order to answer these questions, we first focused our attention on AGA/AGG codons for arginine (Chen and Inouye, 1990). Among codons used in E. coli AGA and AGG are the least used codons, and in most of the cases arginine is coded by CGU (49%), CGC (38%) CGG (6%) or CGA (4%). The usages of AGA and AGG are only 1.7% and 0.8%, respectively. It is also known that AGA and AGG codons are recognized by a very rare tRNA, and interestingly the gene for this tRNA has been identified to be dnaY (Garcia et al., 1986). When we analyzed 678 polypeptides available in GenBank, 452 proteins lack both AGA and AGG codons (Chen and Inouye, 1990). Among the remaining 226 proteins, 132 have either a single AGA or AGG codon, and to our surprise it became very clear that in these proteins either the AGA or AGG codon is preferentially used within the first 25 codons. Although it is less significant, the same tendency was observed for the proteins containing more than one AGA/AGG codon (Chen and Inouye, 1990). It is important to note that approximately 40% of the genes containing a single AGA or AGG codon have the minor codon within the first 25 codons. This preferential usage of minor codons within the first 25 codons was observed not only for AGA/AGG codons but also for other minor codons such as UCA plus AGU for serine, AUA for isoleucine, ACA for threonine, CCC for proline and GGG for glycine (Chen and Inouye, 1990).

Minor Codon Molulator Hypothesis/AGA-AGG Modulator Hypothesis

The analysis described above raises the intriguing possibility that gene expression in E. coli may be modulated by AGA/AGG and other minor codons by their placement near the initiation codon. It is reasonable to assume that the availability of charged tRNAs for minor codons becomes more limited than those for major codons as cell growth slows down. In particular, the dnaY gene, the tRNA gene for AGA and AGG (Garcia et al., 1986) may be under the stringent control. As a result the rate of translation at the minor codons becomes slower due to pausing or stalling of a ribosome at a minor codon. As pointed out by Pedersen (1990), the closer such a translation pause is located to the initiation codon, the more severe the effect of the pause becomes on the inhibition of translational initiation by the succeeding ribosome. If the pause site is far from the translational initiation site, a queue of ribosomes maybe formed from the pause site towards the initiation codon. However, the queue may not reach the translational initiation site. Assuming a paused ribosome passes the minor codon at a specific rate, the length of the queue is determined by the translation rate at the minor codon, the translation rate of the preceding coding region, and the translational initiation rate (the ribosome-binding-site clearing time; Pedersen, 1990). These factors are dependent upon the growth conditions, the codon usage in the mRNA and the nature of the ribosome-binding site. When two ribosomes are lined up at a minor codon, the queue can inhibit the entry of a succeeding ribosome only if the minor codon is located within 24 codons of the initiation codon; 18 codons are equivalent to the length covered by one and a half ribosomes, assuming that a ribosome covers 12 codons on a mRNA (Gold et al., 1981). If the minor codon causing the pause is further away from the initiation codon, the queue can be longer without blocking translational initiation.

The queuing concept described above (Pedersen, 1990) is essential to understand the proposed minor codon modulator hypothesis. The minor codons cause ribosomal pause on a mRNA if the availability of their tRNAs becomes limited under certain growth conditions. The closer the pause occurs to the initiation site, the more effective the inhibitory effect of the paused ribosome is on translational initiation. Not only AGA/AGG but in addition most of the other minor codons are preferentially located within the first 25 codons of the initiation codon in _E_. _coli_ mRNAs. This strongly suggests that _E_. _coli_ mRNAs have evolved to most effectively attenuate translation when cell growth becomes limited; a paused ribosome binding to mRNA near the translational initiation site prevents the entry of another ribosome, thus working as a repressor at the level of translation. Since the paused ribosome also blocks polysome formation, the mRNA is likely to be more vulnerable to nuclease attack. Thus the minor codon modulator hypothesis also predicts that the mRNAs containing a minor codon near the initiation codon become unstable under limited growth conditions.

Experimental Approach

On the basis of the hypothesis described above, we examined the effects of AGG codons on gene expression by inserting between one to five

AGG codons after the tenth codon from the initiation codon of the lacZ gene (Chen and Inouye, 1990). We found that the production of β-galactosidase decreased as increasing numbers of AGG codons were inserted. With five AGG codons, β-galactosidase production completely ceased after a mid-log phase of cell growth. After 22 hr upon lacZ induction, the overall production of β-galactosidase was only 11% of the production in control cells without insertion of any arginine codons. In contrast, when five CGU codons (the major arginine codon) were inserted instead of AGG, the production of β-galactosidase continued even after stationary phase and the overall production was 66% of the control.

It is important to note that there are two distinct phases for the β-galactosidase production from the lacZ gene containing five AGG codons during cell growth. During the initial early growth phase (phase I), the minor codons exhibit no effect, whereas in phase II, after mid-log phase, β-galactosidase production completely stops. The dramatic inhibitory effect of the AGG codons in phase II is likely due to the limited availability of charged tRNAs for AGG. However, on the basis of the AGA/AGG modulator hypothesis, the inhibitory effect was observed possibly because the minor codons were positioned very close to the initiation codon in the lacZ gene. If this is the case, one should be able to alleviate the inhibitory effect of the five AGG codons on the lacZ expression by increasing the distance between the initiation codon and the site of the AGG codons. Indeed, as the distance was increased by inserting sequences of various lengths between the initiation codon and the site of the AGG codons, β-galactosidase production increased almost linearly up to eight-fold. When the distance became equivalent to 76 amino acid residues, the yield of β-galactosidase production approached the level obtained with the control gene lacking any arginine codon insertions (Chen and Inouye, 1990). In other words, the further away the minor codons were located from the initiation codon, the more the inhibitory effect of the minor codons was suppressed as predicted from the AGA/AGG modulator hypothesis.

The proposed minor codon (or AGA/AGG) modulator hypothesis suggests a novel mechanism for global regulation of gene expression and cellular functions. During nutritional deprivation, the AGA/AGG codons or other minor codons function as a modulator to inhibit the production of key proteins in E. coli involved in DNA replication, protein synthesis and cell division. Under such growth-rate limiting conditions, the distance between the initiation codon and the AGA/AGG codons may play an essential role in the synthesis of these key proteins, thereby regulating global cellular activities. It is interesting to point out that among a group of proteins containing a single AGA or AGG codon within the first 25 codons, included are various essential gene products such as (a) a protein required for DNA replication (single-strand DNA binding protein), (b) proteins associated with protein synthesis (glycyl-tRNA synthetase α-subunit, phenylalanine tRNA synthetase β-subunit, and ribosomal S10 protein), (c) proteins associated with gene regulation (adenylate cyclase, LexA protein, and IHF, leu operon leader peptide), and (d) other proteins (FtsA protein, EnvZ protein, SecY protein, protein component of ribonuclease P, and succinate dehydrogenase) (see Table 1). It should be noted that most of the proteins listed in the Table contain an AGA or AGG codon located within the tenth codon. This indicates that the ribosome which had just undergone translational initiation has to pause within ten codons from the initiation site when charged tRNA for

AGA and AGG becomes limited. Since this paused ribosome still covers the initiation codon or is located very closely to the initiation codon, the second ribosome is unable to initiate translation. In this way, the ribosome bound to the site proximal to the initiation codon now functions as a translational repressor. In this case, the rate of translational initiation is equal to the rate of translation at the minor codon. Therefore, the production of the protein is directly controlled by the availability of charged tRNA for AGA and AGG.

Acknowledgement

The authors are grateful to Dr. S. Pedersen for valuable discussions and suggestions, and for letting his manuscript in this book available before publication.

References

Aiba H, Mori K, Tanaka, M, Ooi T, Roy A, and Danchin A (1984) The complete nucleotide sequence of the adenylate cyclase gene of Escherichia coli Nucl Acids Res 12:9427-9440

Cerretti DP, Dean D, Davis GR, Bedwell DM, and Nomura M (1983) The spc ribosomal protein operon of Escherichia coli: Sequence and cotranscription of the ribosomal protein genes and a protein export gene. Nucl Acids Res 11:2599-2616

Chen G, and Inouye M (1990) Suppression of the Negative Effect of Minor Arginine Codons on Gene Expression; Preferential Usage of Minor Codons within the First 25 Codons of the Escherichia coli Genes. Nucleic Acids Res, 18:1465-1473

Flamm E, and Weisberg RA (1985) Primary structure of the hip gene of Escherichia coli and of its product, the beta-subunit of integration host factor

Garcia GM, Mar PK, Mullin DA, Walker JR and Prather NE (1986) The E. coli dnaY gene encodes an arginine transfer RNA. Cell 45:453-459

Gold L, Pribnow D, Scheider T, Shinedling S, Singer BS and Stormo G (1981) Translational initiation in prokaryotes. Ann Rev Microbiol 35:365-403.

Hansen FG, Hansen EB, and Atlung T (1985) Physical mapping and nucleotide sequence of the rnpA gene that encodes the protein component of ribonucleases P in Escherichia coli. Gene 38:85-93

Hoekema A, Kastelein RA, Vasser M and de Boer HA (1987) Codon replacement in the pGK1 gene of Saccharomyces cerevisiae: Experimental approach to study the role of biased codon usage in gene expression. Mol Cell Biol 7:2914-2924

Holm L (1986) Codon usage and gene expression. Nucl Acids Res 14:3075-3087

Horii T, Ogawa T, and Ogawa H (1981) Nucleotide sequence of the lexA gene of E. coli. Cell 23:689-697

Hull EP, Spencer ME, Wood D, and Guest JR (1984) Nucleotide sequence of the promoter region of the citrate synthase gene (gltA) of Escherichia coli FEBS Lett 156:366-370 (revised)

Ikemura T (1981a) Correlation between the abundance of Escherichia coli transfer RNAs and the occurrence of the respective codons in its protein genes. J Mol Biol 146:1-21

Ikemura T (1981b) Correlation between the abundance of Escherichia coli transfer RNAs and the occurrence of the respective codons in its protein genes: A proposal for a synonymous codon choice that is optimal for the E. coli translational system. J Mol Biol 151:389-409

Keng T, Webster TA, Sauer RT, and Schimmel P (1982) Gene for Escherichia coli glycyl-tRNA synthetase has tandem subunit coding regions in the same reading frame. J Biol Chem 257:12503-12508

Kurland CG (1987) Strategies for efficiency and accuracy in gene expression. Trends Biochem 12:126-128

Markham BE, Little JW, and Mount DW (1981) Nucleotide sequence of the lexA gene of Escherichia coli K-12. Nucl Acids Res 9:4149-4161

Maruyama T, Gojobori T, Aota SI and Ikemura T (1986) Codon usage tabulated from the GenBank genetic sequence data. Nucl Acids Res 14:r151-4197

Mizuno T, Wurtzel ET, and Inouye M (1982) Osmoregulation of gene expression: II. DNA sequence of the envZ gene of the ompB operon of Escherichia coli and characterization of its gene product. J Biol Chem 257:13692-13698

Normura M, Yates JL, Dean D and Post LE (1980) Feedback regulation of ribosomal protein gene expression in Escherichia coli: Structural homology of ribosomal RNA and ribosomal protein mRNA. Proc Natl Acad Sci USA 77:7084-7088

Olins PO, and Nomura M (1981) Regulation of the S10 ribosomal protein operon in E. coli: Nucleotide sequence at the start of the operon. Cell 26:205-211

Pedersen, S (1990) in this volume.

Post LE, Stycharz GD, Nomura M, Lewis H, and Dennis PP (1979) Nucleotide sequence of the ribosomal protein gene cluster adjacent tot he gene for RNA polymerase subunit b in Escherichia coli. Proc Natl Acad Sci USA 76:1697-1701

Robinson AC, Kenan DJ, Hatfull GF, Sullivan NF, Spiegelberg R, and Donachie WD (1984) DNA sequence and transcriptional organization of exential cell division genes ftsQ and ftsA of E. coli evidence for overlapping transcriptional units. J Bacteriol 160:546-555

Sancar A, Williams KR, Chase JW and Rupp WD (1981) Sequences of the ssb gene and protein. Proc Natl Acad Sci USA 78:4274-4278

Sharp PM and Li WH (1986) Codon usage in regulatory gene in Escherichia coli does not reflect selection for "rare" codons. Nucl Acids Res 14:7737-7749

Sorensen MA, Kurland CG and Pedersen S (1989) Codon usage determines translation rate in Escherichia coli. J Mol Biol 207:365-377

Wessler SR, and Calvo JM (1981) Control of leu operon expression in Escherichia coli by a transcription attenuation mechanism. J Mol Biol 149:579-597

Wurtzel ET, Chou MY, and Inouye M (1982) Osmoregulation of gene expression: I. DNA sequence of the ompR gene of the ompB operon of

Escherichia coli and characterization of its gene product. J Biol Chem 257:13685-13691

Yi QM, Rockenbach S, Ward JE Jr, and Lutkenhaus J (1985) Structure and expression of the cell division genes ftsQ, ftsA and ftsZ. J Mol Biol 184:399-412

TABLE I: GENES CONTAINING ONLY ONE AGA or AGG CODON EXISTING WITHIN THE FIRST 25 CODONS[a]

	(Protein)	(Position)	Total (arginine)	(Ref)[b]
DNA Replication:	Single-strand DNA binding protein	4th(AGA)	10	1
Protein Synthesis:	Gly-tRNA synthetase α-subunit	7th(AGG)	16	2
	Phe-tRNA synthetase β-subunit	2nd(AGG)	24	3
	Ribosomal protein S10	5th(AGA)	12	4
Gene Regulation:	IHF integration host factor	9th(AGA)	8	5
	Adenylate cyclase	11th(AGA)	25	6
	LexA protein	7th(AGG)	15	7
	leu operon leader peptide	21st(AGA)	4	8
Other:	FtsA protein	7th(AGA)	21(22)	9(10)
	EnvZ protein	2nd(AGG)	47	11
	SecY protein	22nd(AGA)	21	12
	Ribonuclease P (protein component)	8th(AGG)	23	13
	Succinate dehydrogenase:			
	large subunit	6th(AGA)	40	14
	small subunit	2nd(AGA)	17	14

[a]: Source: From the analysis of data available from Genbank Release 59, March: 1989.

[b]: 1(Sancar et al., 1981), 2(Keng et al., 1982), 3(Mechulam et al., 1985), 4(Olins and Nomura, 1981), 5(Flamm and Weisberg), 6(Aiba et al., 1984), 7(Horii et al., 1981; Markham et al., 1981), 8(Wessler and Calvo, 1981), 9(Robinson et al., 1984), 10(Yi et al., 1985), 11(Wurtzel et al., 1982; Mizuno et al., 1982), 12(Cerretti, et al., 1983), 13(Hansen et al., 1985), 14(Hull et al., 1984).

A SHORT REVIEW OF SCANNING

Marilyn Kozak
Department of Biochemistry
University of Medicine and Dentistry of New Jersey
Robert Wood Johnson Medical School
675 Hoes Lane
Piscataway, New Jersey 08854 USA

The scanning model for initiation of translation in higher eukaryotes postulates that the 40S ribosomal subunit (carrying Met-tRNA$_i^{met}$ and an imperfectly defined set of initiation factors) enters at the 5' end of the mRNA and then migrates, stopping at the first AUG codon in a favorable context for initiation. Here I review some of the key supporting observations, some complications, implications, and some outstanding questions about scanning. An earlier review of these subjects, with more extensive documentation, was published in the Journal of Cell Biology (Kozak, 1989a).

1. *End-dependent initiation of translation*

The earliest direct evidence for end-dependence was the demonstration that efficient translation of many viral mRNAs in vitro requires a methylated cap structure (Shatkin, 1976). Later in vivo experiments confirmed the nearly absolute dependence of translation on the m7G cap (Horikami et al., 1984; Fuerst & Moss, 1989; Malone et al., 1989).[1]

With natural or synthetic mRNAs that have an unstructured leader sequence the cap requirement is obviated, but the translation of such mRNAs still requires an end. This was shown by testing the ability of various templates to bind to wheat germ and reticulocyte ribosomes after circularization of the templates by RNA ligase (Kozak, 1979; Konarska et al., 1981). The use of templates of two different sizes would seem to negate the criticism that the circles failed to bind to eukaryotic ribosomes because they were too small rather than because they lacked a free end.[2] Thus, the inability of circular

---

[1] *The cap is known to stabilize mRNAs as well as to enhance their translation. The two effects are distinguishable by the fact that an unmethylated cap is sufficient to confer stability while stimulation of translation requires methylation.*

[2] *If the UpG(pA)$_n$ circles (40-80 nt) were at the cutoff point—able to bind to prokaryotic but too small to bind to eukaryotic ribosomes—then the (A,U,G)$_n$ circles which were mostly bigger than 80 nt should easily have bound to eukaryotic ribosomes. Indeed some of the (A,U,G)$_n$ circles which failed to bind to eukaryotic ribosomes were big enough to form disomes with E. coli ribosomes (Kozak, 1979).*

NATO ASI Series, Vol. H 49
Post-Transcriptional Control of Gene Expression
Edited by J. E. G. McCarthy and M. F. Tuite

templates to bind to eukaryotic ribosomes remains a key observation in support of an end-dependent (scanning or perhaps threading) mechanism.

Recent reports that certain viral (Dolph et al., 1988; Carrington & Freed, 1990) and cellular (Sarnow, 1989) mRNAs can be translated in vivo without the usual requirement for cap binding proteins should not be taken as evidence for direct internal initiation by ribosomes. Remember, the circle experiments show that, even when the m7G cap is not needed, the mRNA still has to have an end. Thus, the ability of some mRNAs to be translated without cap binding protein(s) is merely an indication of an efficient (because it is unstructured) leader sequence. That ribosomes load *from the 5' end* onto such unstructured sequences is suggested from reconstitution experiments in which the facilitating effect of the heat shock leader, for example, requires that it be at the 5' end (McGarry & Lindquist, 1985). In short, although the usual requirement for a methylated cap constitutes evidence for end-dependent initiation, absence of cap-dependence is not a priori evidence for internal initiation.

2. *Migration of 40S ribosomal subunits*

From experiments carried out in crude wheat germ extracts[3] we can say three things about the scanning step:

(a) 40S ribosomal subunits can migrate. This was deduced from the formation of abnormal, polysome-like complexes when reovirus and other mRNAs were incubated with wheat germ ribosomes in the presence of the oligopeptide antibiotic edeine (Kozak & Shatkin, 1978). The complexes were shown to consist of multiple 40S ribosomal subunits bound *throughout the length* of each (monocistronic) mRNA under circumstances which suggested that 40S subunits could *enter* only at the 5'-end of the mRNA. Thus, we deduced that the 40S ribosome/initiation factor complexes migrate away from the 5'-end after binding. We postulated that edeine allowed the abnormal complexes to form because the drug impaired recognition of the AUG codon; but we could not rigorously disprove the possibility that edeine induced the migration. Thus, it was crucial to establish the next point.

---

[3] *Although some of these experiments have been repeated in reticulocyte lysates, the wheat germ system is preferable for two reasons: (a) The experiments require loading of multiple 40S ribosomal subunits onto each mRNA. In this regard wheat germ S23 extracts work better than reticulocyte lysates because the translational capacity is considerably higher in the wheat germ system. (b) The experiments require that the functional mRNA be intact. Although the extent of mRNA degradation might be no less in the wheat germ than in the reticulocyte system, in wheat germ extracts the resulting mRNA fragments are inert. In contrast, the reticulocyte system often translates mRNA fragments derived by cleavage.*

(b) 40S ribosomal subunits do migrate from the cap to the AUG codon, in experimental transcripts. Using two types of experiments which were carried out without drugs,[4] we showed that 40S subunits could be *trapped upstream from the AUG codon*. The first approach was to deplete the extract of ATP and show that 40S subunits failed to reach the AUG codon, as shown by protection experiments with appropriately labeled templates (Kozak, 1980). The second approach was to introduce a very stable hairpin structure between the cap and the first AUG codon (Kozak, 1989b). The result was that a 40S ribosomal subunit could still bind to ^{32}P-labeled "hp7" mRNA, but the 40S subunit stalled on the 5' side of the hairpin, exactly as the scanning model predicts.
(c) 40S ribosomal subunits must migrate to reach the AUG initiator codon in most natural mRNAs from higher eukaryotes. Our finding that the 40S ribosome/factor complex can protect up to 65 nucleotides (nt) near the 5' end of reovirus mRNAs (Kozak & Shatkin, 1976) has provoked the suggestion that mRNA leader lengths are "matched" to the rather large size of the initiation complex. If that were true, there would be no need for scanning: when the 40S ribosome/factor complex binds at the 5' end (the argument goes), the ribosome would be at the AUG initiator codon. But the actual numbers do not fit that simplistic idea.

The average length of vertebrate (cellular) mRNA leader sequences is ∿90 nt (from Table 3, Kozak, 1987a), which is much bigger than the ∿50 nt maximum size[5] of 40S ribosome-protected fragments. Moreover, if the 40S ribosome/factor complex were to bind directly to an AUG codon positioned at the "optimal"[6] distance from the cap, the efficiency of initiation should fall precipitously when the leader length departs from that magic number. In fact, ribosomes initiate without difficulty on leader sequences that are (for example) only 25 nt long; and mRNAs with leader sequences of >200 nt may, under some circumstances, actually be translated better than matched mRNAs with leader sequences of average (70 to 90 nt) length (Kozak, 1988a).

3. *Linear migration as opposed to looping*

Recent in vitro experiments with mRNAs that contain a very stable hairpin (e.g. hp7, which has a ΔG of -61 kcal/mol and is positioned 72 nt from the cap)

[4] *The only antibiotic used was sparsomycin, which inhibits elongation.*

[5] *The AUG codon usually lies 12 to 15 nt from the 3' edge of the ribosome-protected fragment. Thus, the ∿65 nt fragment protected by the 40S ribosome/factor complex extends maximally ∿50 nt upstream from the AUG codon.*

[6] *The word is in quotes because there is, in fact, no optimal leader length for mRNAs in higher eukaryotes.*

strongly support the hypothesis that 40 subunits advance *linearly* until they reach the AUG codon. The principal finding, as mentioned above, is that a 40S ribosomal subunit binds very efficiently to the 5' end of hp7 mRNA and then stalls on the 5' side of the hairpin (Kozak, 1989b). Apparently the 40S ribosome/factor complex cannot hop across the base of a hairpin.

A second, more extensive line of evidence concerns the strong tendency of eukaryotic ribosomes to initiate at the first AUG codon. (a) Among natural mRNAs, the functional initiator codon is usually the AUG triplet closest to the 5' end. This is true at least 90% of the time (Kozak, 1987a) and there are rules (see below) to predict the deviations from the first-AUG rule. (b) When upstream AUG codons are inserted experimentally into leader sequences, the site of initiation nearly always shifts to the adventitious upstream site (reviewed by Kozak, 1989a). That body of evidence is impressive because it was compiled using many different constructs in many different laboratories, but a caveat accompanies those experiments: the sequence around the silent downstream AUG codon always differs from the sequence around the functional upstream site. Thus, a skeptic could argue that the upstream AUG codon, which just happens to be first, occurs in a better context for initiation. (c) To rigorously test the importance of position, we produced a transcript in which one particular sequence—the sequence that initiates rat preproinsulin—was repeated three or four times near the 5' end of a chimeric transcript (Figure 1, left). The ability of mammalian ribosomes to initiate at each AUG triplet in the tandem array was monitored in vivo by searching for insulin-related polypeptides with N-terminal amino acid extensions. The out-

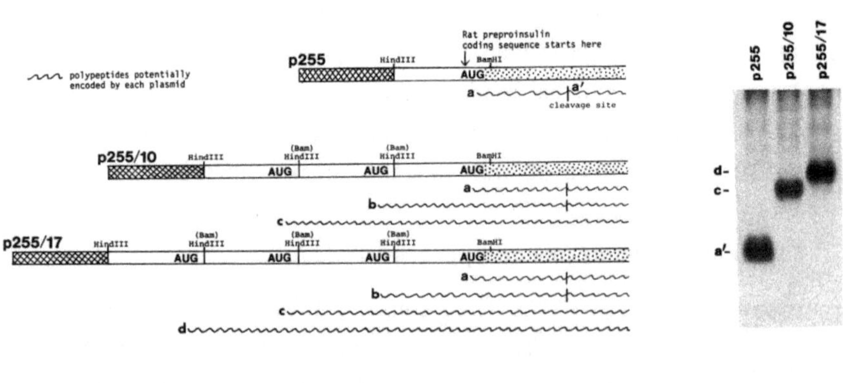

Figure 1. Potential (left) and actual (right) patterns of translation of insulin-related polypeptides from mRNAs with tandemly reiterated copies of the initiation site for rat preproinsulin. The experiment is a repeat of one published previously (Kozak, 1983). The band marked a' in the polyacrylamide gel (right panel) is wild-type proinsulin.

come was that ribosomes initiated exclusively at the first AUG codon in each tandem array. (Figure 1, right).

Although limitations of space prevent me from including much relevant data from lower eukaryotes, one experiment carried out in yeast cannot be overlooked. Cigan et al. (1988) have shown that, when the anticodon sequence in one of the $tRNA_i^{met}$ genes is changed from 3'-UAC-5' to 3'-UCC-5', the mutant form of $tRNA_i^{met}$ directs ribosomes to initiate at (the first!) AGG instead of the usual AUG codon. This constitutes direct proof that the initiator codon is recognized primarily by base pairing with the anticodon in Met-$tRNA_i^{met}$, and the experiment is compelling proof of scanning.

The main conclusion from all these experiments is that selection of the initiator codon is systematic—too systematic to be explained by "looping" or bending of the mRNA after the 40S ribosomal subunit binds at the 5' end. In the latter case, one might expect to see a gradient in which AUG codons nearer the 5' end are somewhat preferred over those farther downstream. But what one actually sees is more than a gradient: what one sees is *exclusive* initiation at the first AUG codon in ∿90% of eukaryotic mRNAs, and there are recognizable features (detailed in the next section) that explain the deviations in the remaining 10%.

4. *Exceptions to the first-AUG rule*

Considerable study has centered on the subset of eukaryotic mRNAs that violates the first-AUG rule. The outcome, in short, is that a scanning process still seems to operate[7] but, for one reason or another, some ribosomes gain access to the second AUG codon or, rarely, an AUG codon even farther downstream. As outlined in Figure 2, scanning is "leaky"—i.e. the first AUG is not recognized efficiently—when the flanking sequence is suboptimal,[8] or when the first AUG triplet lies too near the cap, or when the first AUG codon

---

[7]For a handful of viral mRNAs there are claims that ribosomes initiate directly at internal sites rather than scanning down to the AUG codon. Several early claims along such lines have been abandoned as more data emerged (Kozak, 1989a). More recent claims of direct internal initiation continue to be defended—and could be correct; but the published experiments have serious loopholes (Kozak, 1989a). A sequence or structure that allows direct internal binding has not been defined for even one mRNA, and there is no hint of a common structure among all the mRNAs for which direct internal initiation is facilely claimed.

[8]The extent to which ribosomes bypass the first AUG codon depends on both the primary sequence (Kozak, 1986a, 1987b) and the occurrence of secondary structure shortly downstream (Kozak, 1989c). The effect of downstream secondary structure is to improve recognition of the preceding AUG codon when the flanking primary sequence is suboptimal.

occurs very close to the second AUG. Of these three conditions, the first—suboptimal context around the first AUG—is the most common. A large group of viral (Kozak, 1986b) and a smaller number of cellular mRNAs make use of this device to produce two proteins from one transcript.

The ability of eukaryotic ribosomes to reinitiate translation is another route to reaching an AUG codon that is not first-in-line. We know little about the reinitiation mechanism, which occurs constitutively in some mammalian cells (Kozak, 1987c). In yeast, reinitiation is regulable, and deduction of the cis and trans-acting requirements is an interesting, ongoing endeavor (Miller & Hinnebusch, 1989; Williams et al., 1989).

It is hard to estimate the number of mRNAs from higher eukaryotes in which reinitiation is expected to occur, because reports of cDNA sequences with upstream AUG codons are often in error (Kozak, 1987a, 1989a). A simple rule-of-thumb is that the more upstream AUG triplets one sees in a cDNA se-

---

	Favored by:
Efficient recognition of 1st AUG	*Strong context: GCC$^{A}_{G}$CC<u>AUGG</u> Downstream secondary structure
Less efficient recognition of 1st AUG allows some initiation from 2nd AUG	*Weak context around 1st AUG — *See bifunctional viral mRNAs - Cell 47:481* 1st AUG too near cap (≤10 nt) 2nd AUG close to 1st (≤ 4 nt)
Efficient recognition of 1st and 2nd AUG codons, the latter by REINITIATION #1 #2 ---AUG--UAA___*n*___AUG-----	¶Terminator codon between 1st and 2nd AUGs Increasing the distance *n* *(MCB 7:3438)* Strong context around 2nd AUG *(Cell 44:283)* **Special features at termination site *(Hinnebusch, Genes & Dev. 3:1217)* **Modulation of/by eIF2 *(Hinnebusch, Donahue PNAS 86:7515)*
5'-minicistron	SV40 agnoprotein (61 amino acids) - *Khoury, Nature 291:346* Yeast CPA1 gene (25 amino acids) - *Pierard, Cell 49:805*
5'-full cistron	Synthetic transcript/tobacco *(Angenon, MCB 9:5676)* - but INEFFICIENT! CaMV *(Hohn, Cell 59:1135)* - but TRANS-ACTING viral protein needed! EBV, EBNA2 *(Kieff, JV 61:945)* - but second mRNA not ruled out.

Figure 2. Structural features in eukaryotic mRNAs that favor exclusive initiation at the first AUG codon (group 1), or initiation at the second AUG codon at the expense of the first (group 2), or efficient expression from the first and second AUG codons by reinitiation (group 3). As noted at the bottom, reinitiation has been shown clearly to occur following a short "minicistron," but reinitiation appears to be difficult following the translation of a full-length cistron. For discussion of possible reasons and for more information about reinitiation, see Kozak, 1987c. (NOTES: *not evident in S. cerevisiae; ¶absolute requirement; **demonstrated so far only in yeast) quence, the more likely it is that the 5' sequence will turn out to contain an intron (Brown & Feix, 1990; Hayflick et al., 1989; Tedder et al., 1988, 1989) or a cloning error (Coffinó, 1988) or some other artifact. There are annoying reports in which a cDNA with many upstream AUG codons is claimed to represent the functional mRNA; but, for the accompanying expression studies, the authors chose to truncate the long, AUG-burdened leader sequence! (See Fuller et al., 1988; Burtis & Baker, 1989; Zerial et al., 1989.)

5. *Some implications of scanning*

One consequence of scanning is that eukaryotic mRNAs are typically monocistronic, although the aforementioned circumventions of the first-AUG rule (leaky scanning, reinitiation, etc) do enable a small number of mRNAs to produce two proteins (reviewed by Kozak, 1989a). In other cases where the first-AUG rule applies strictly, one gene sometimes produces two overlapping mRNAs, one of which initiates from a far-upstream promoter and includes an "extra" upstream, in-frame AUG codon. Subsequent initiation of translation from the first AUG codon in each transcript generates two versions of the encoded protein, the longer of which has an N-terminal amino acid extension that affects the intracellular distribution and/or function of the protein (Kozak, 1988b).

The fact that 40S ribosomal subunits enter at the 5' end rather than at the AUG codon shifts the principal site of regulation to the upstream sequences. Thus, changing the context around the AUG codon would not be expected to, and does not, affect competition among eukaryotic mRNAs (Kozak, 1989c). However, the extent to which the 5' end of the mRNA is exposed should be, and is, a key determinant of translational efficiency (Kozak, 1989b). Leader length is also important for translational efficiency: the longer the leader sequence, the more efficiently the mRNA is translated under conditions of competition (Kozak, 1988a). A possible explanation which we are currently testing is that extra 40S subunits might accumulate on long, unstructured leader sequences.

One function of cap binding proteins (CBP) might be to stabilize the association after a 40S ribosomal subunit contacts the 5' end of mRNA. Stabilization by CBP might prevent the 40S subunit from detaching, which could be especially important when the leader sequence of the mRNA is structured and therefore the 40S subunit cannot advance immediately. Apart from the CBP, it is hard to imagine how the binding of other proteins to mRNA sequences could enhance translation. Rather, it seems likely that regulatory proteins will act as repressors of translation in eukaryotes, as they do in prokaryotes (Kozak, 1988b). One anticipated difference, however, is that eukaryotic repressor proteins should be more effective when they bind near the 5' end than when they bind near the AUG codon.

6. *Outstanding questions*

Although the overall direction of scanning is 5' to 3', the possibility that the 40S subunit flutters (moving back and forth over a given stretch of the leader sequence) has not been ruled out. The processivity of scanning also has not been evaluated. We don't know whether 40S subunits advance one nucleotide at a time, or three nucleotides at a time, or in some other measure. If 40S subunits advance triplet by triplet, however, each entering ribosome must pick at random any one of the three possible frames. Were scanning uniquely phased, a one- or two-nucleotide insertion between the cap and the AUG codon should drastically impair translation; in fact, such mutations are usually innocuous. An interesting possibility is that 40S subunits might be nudged into the correct phase by the GCC or ACC motif that immediately precedes the AUG codon in vertebrate mRNAs. That would explain the ability of $^{A}_{G}$CC to enhance initiation when the purine occurs in position -3 or (less effectively) position -6, but not when the $^{A}_{G}$CC motif is shifted out-of-phase with respect to the AUG codon (Kozak, 1986a, 1987b). An alternative possibility is that the ACC or GCC motif might facilitate recognition of the AUG triplet simply by slowing scanning. How the $^{A}_{G}$CC motif is recognized (e.g. by interaction with rRNA or with proteins) is a major outstanding question.

The amount of ATP required, and why it is required, remain unknown. The demonstration that migration of 40S ribosomal subunits requires ATP (Kozak, 1980) might mean that ATP powers the movement of 40S subunits. Alternatively, some other ATP-dependent step (e.g. unwinding of mRNA structure) might be a precondition for the 40S subunit to advance.

Finally, we know little about how initiation factors fit into the scheme. The most reliable information has come from Tom Donahue's clever application of genetics in the yeast system. The biochemical analyses carried out with more-or-less pure mammalian initiation factors, however, are difficult to interpret. The mRNA/factor binding studies that have been done are limited in that one (or, at best, a few) factors have been tested in isolation. The interactions with mRNA might be very different in the presence of ribosomes and the full set of initiation factors, just as interaction of the signal sequence with the 54 kd subunit of SRP has been shown to occur only when SRP is associated with ribosomes (Bernstein et al., 1989). eIF-4A together with -4B might have helicase activity, as recent experiments suggest; but whether that activity facilitates the *entry* of a 40S subunit at the 5' end or the subsequent *migration* (or both, or neither) is pure speculation.

References

Bernstein HD, Rapoport TA, Walter P (1989) Cytosolic protein translocation factors: is SRP still unique? Cell 58:1017-1019

Brown JWS, Feix G (1990) A functional splice site in the 5' untranslated region of a zein gene. Nuc Acids Res 18:111-117

Burtis KC, Baker BS (1989) Drosophila doublesex gene controls somatic sexual differentiation by producing alternatively spliced mRNAs encoding related sex-specific polypeptides. Cell 56:997-1010

Carrington JC, Freed DD (1990) Cap-independent enhancement of translation by a plant potyvirus 5' nontranslated region. J Virol 64:1590-1597

Cigan AM, Feng L, Donahue TF (1988) $tRNA_i^{met}$ functions in directing the scanning ribosome to the start site of translation. Science 242:93-97

Coffino P (1988) Probable cloning artefacts previously interpreted as unusual leader sequences of rodent ODC mRNAs. Gene 69:365-368

Dolph PJ, Racaniello V, Villamarin A, Palladino F, Schneider RJ (1988) The adenovirus tripartite leader may eliminate the requirement for cap-binding protein complex during translation initiation. J Virol 62:2059-2066

Fuerst TR, Moss B (1989) Structure and stability of mRNA synthesized by vaccinia virus-encoded bacteriophage T7 RNA polymerase in mammalian cells: importance of the 5' untranslated leader. J Mol Biol 206:333-348

Fuller F, Porter JG, Arfsten AE, Miller J, Schilling JW, Scarborough RM, Lewicki JA, Schenk DB (1988) Atrial natriuretic peptide clearance receptor. J Biol Chem 263:9395-9401

Hayflick JS, Adelman JP, Seeburg PH (1989) The complete nucleotide sequence of the human gonadotropin-releasing hormone gene. Nuc Acids Res 17:6403

Horikami SM, DeFerra F, Moyer SA (1984) Characterization of the infections of permissive and nonpermissive cells by host range mutants of VSV defective in RNA methylation. Virol 138:1-15

Konarska M, Filipowicz W, Domdey H, Gross HJ (1981) Binding of ribosomes to linear and circular forms of the 5'-terminal leader fragment of tobacco mosaic virus RNA. Eur J Biochem 114:221-227

Kozak M (1979) Inability of circular mRNA to attach to eukaryotic ribosomes. Nature 280:82-85

Kozak M (1980) Role of ATP in binding and migration of 40S ribosomal subunits. Cell 22:459-467

Kozak M (1983) Translation of insulin-related polypeptides from messenger RNAs with tandemly reiterated copies of the ribosome binding site. Cell 34:971-978

Kozak M (1986a) Point mutations define a sequence flanking the AUG initiator codon that modulates translation by eukaryotic ribosomes. Cell 44:283-292

Kozak M (1986b) Bifunctional messenger RNAs in eukaryotes. Cell 47:481-483

Kozak M (1987a) An analysis of 5'-noncoding sequences from 699 vertebrate messenger RNAs. Nuc Acids Res 15:8125-8148

Kozak M (1987b) At least six nucleotides preceding the AUG initiator codon enhance translation in mammalian cells. J Mol Biol 196:947-950

Kozak M (1987c) Effects of intercistronic length on the efficiency of reinitiation by eucaryotic ribosomes. Mol Cell Biol 7:3438-3445

Kozak M (1988a) Leader length and secondary structure modulate mRNA function under conditions of stress. Mol Cell Biol 8:2737-2744

Kozak M (1988b) A profusion of controls. J Cell Biol 107:1-7

Kozak M (1989a) The scanning model for translation: an update. J Cell Biol 108:229-241

Kozak M (1989b) Circumstances and mechanisms of inhibition of translation by secondary structure in eucaryotic mRNAs. Mol Cell Biol 9:5134-5142

Kozak M (1989c) Context effects and inefficient initiation at non-AUG codons in eucaryotic cell-free translation systems. Mol Cell Biol 9:5073-5080

Kozak M, Shatkin AJ (1976) Characterization of ribosome-protected fragments from reovirus messenger RNAs. J Biol Chem 251:4259-4266

Kozak M, Shatkin AJ (1978) Migration of 40S ribosomal subunits on messenger RNA in the presence of edeine. J Biol Chem 253:6568-6577

Malone RW, Felgner PL, Verma IM (1989) Cationic liposome-mediated RNA transfection. Proc Natl Acad Sci USA 86:6077-6081

McGarry TJ, Lindquist S (1985) The preferential translation of Drosophila hsp70 mRNA requires sequences in the untranslated leader. Cell 42:903-911

Miller PF, Hinnebusch AG (1989) Sequences that surround the stop codons of upstream open reading frames in GCN4 mRNA determine their distinct functions in translational control. Genes and Devel 3:1217-1225

Sarnow P (1989) Translation of glucose-regulated protein 78/immunoglobulin heavy-chain binding protein mRNA is increased in poliovirus-infected cells at a time when cap-dependent translation of cellular mRNAs is inhibited. Proc Natl Acad Sci USA 86:5795-5799

Shatkin AJ (1976) Capping of eucaryotic mRNAs. Cell 9:645-653

Tedder TF, Streuli M, Schlossman SF, Saito H (1988) Isolation and structure of a cDNA encoding the B1 (CD20) cell-surface antigen of human B lymphocytes. Proc Natl Acad Sci USA 85:208-212

Tedder TF, Klejman G, Schlossman SF, Saito H (1989) Structure of the gene encoding the human B lymphocyte differentiation antigen CD20 (B1). J Immunol 142:2560-2568

Williams NP, Hinnebüsch AG, Donahue TF (1989) Mutations in the structural genes for eukaryotic initiation factors 2α and 2β of Saccharomyces cerevisiae disrupt translational control of GCN4 mRNA. Proc Natl Acad Sci USA 86:7515-7519

Zerial M, Toschi L, Ryseck R-P, Schuermann M, Muller R, Bravo R (1989) The product of a novel growth factor activated gene, fosB, interacts with JUN proteins enhancing their DNA binding activity. EMBO J 8:805-813

YEAST mRNA STRUCTURE AND TRANSLATIONAL EFFICIENCY

H.A. Raué, J.J. van den Heuvel and R.J. Planta
Biochemisch Laboratorium
Vrije Universiteit
de Boelelaan 1083
1081 HV Amsterdam
The Netherlands

INTRODUCTION

The ultimate level of a protein in a living cell is determined by a complex set of regulatory processes that exert their influence at all levels of gene expression, from transcription through RNA processing to translation and even the modification and turn-over of the final protein product. Knowledge of these manifold regulatory circuits and their mutual interaction, therefore, is of prime importance to understand how the numerous cellular proteins are produced in the correct relative amounts.

Analysis of the regulation of gene expression in eukaryotic cells started to grow past the age of childhood about fifteen years ago with the advent of recombinant DNA technology. Initially, attention was focussed primarily on control of transcription. It soon became clear, however, that significant differences exist also in the efficiency with which different mRNAs are recruited as templates for translation (Lodish, 1976). Moreover, there is a steadily growing number of genes whose expression in response to changing physiological conditions is found to be regulated in large part at the level of translation. Consequently, interest in the factors controlling translational efficiency of mRNA has soared and our knowledge of these factors is increasing rapidly.

THE SCANNING MODEL

In contrast to the situation in prokaryotes, the large majority of eukaryotic mRNAs is monocistronic (Kozak, 1987). Translation initiation by eukaryotic ribosomes in most cases starts at the AUG codon closest to the 5'-end of the mRNA (Kozak, 1987; Cigan & Donahue, 1987). According to the latest version of the scanning model (Kozak, 1989a), first the 40S ribosomal subunit, carrying Met-$tRNA_i^{met}$, GTP and several initiation factors (43S initiation complex) binds at or near the 5'-terminal m^7GpppN cap structure, a process also requiring the participation of a number of cap-binding proteins (reviewed by Proud, 1986; Rhoads, 1988). The 43S complex then migrates along the 5'-untranslated (leader) region, helped by the ATP-dependent unwinding activity of eIF-4A and eIF4F, until it encounters the first AUG codon. At that point, in principle, it initiates translation by pausing (Wolin & Walter, 1988) and associating with a 60S subunit as well as the elongation factors. However, the efficiency with which the AUG codon is recognized depends upon its immediate context, both up- and downstream. A suboptimal context causes a portion of the scanning 40S subunits to bypass the AUG codon and initiate at one located farther downstream. This "leaky scanning" was invoked to explain the presence of so-called "upstream AUG codons" in the leader sequences of a number of eukaryotic mRNAs, the incidental production of two proteins having different N-terminal sequences from the same (usually viral) mRNA and the translation of downstream open reading frames in the minority class of polycistronic eukaryotic mRNAs. However, contrary to the original belief, recent evidence strongly suggests that eukaryotic ribosomes are also capable of initiating

NATO ASI Series, Vol. H 49
Post-Transcriptional Control of Gene Expression
Edited by J. E. G. McCarthy and M. F. Tuite

translation at specific sites located internally on an mRNA (reviewed by Sonenberg & Pelletier, 1989; Herman, 1989). Most compelling in this respect are the experiments on the naturally uncapped poliovirus and encephalomyocarditis viral mRNAs, which seem to use this mechanism to appropriate the host ribosomes for their own purposes. Internal initiation has also been reported for the capped Sendai virus P/C mRNA (Curran and Kolakofsky, 1989). The precise sequence requirements for internal binding of 40S subunits are not yet clear. Internal translation initiation in eukaryotic mRNA still differs from that in prokaryotes since the site recognized by the 40S subunit and the AUG codon can be a considerable distance apart. Thus, again the AUG codon is probably reached by scanning. Whether internal translation initiation also occurs on cellular eukaryotic mRNAs is still being debated.

CONTROL AT VARIOUS LEVELS OF TRANSLATION

Termination Control of mRNA translation can apparently take place in all three phases of this process. Regulation at the level of termination, although at first sight improbable, does occur (reviewed by Valle & Morch, 1988). It involves naturally occurring suppressor tRNAs, which have been detected in both plant and animal cells. These tRNAs probably play a role in regulation of viral gene expression by partially suppressing a termination codon in an otherwise open reading frame that, in this way, encodes two proteins of different length but identical N-terminal sequence. Frame-shifting as a means to skip a termination codon and produce different proteins from partially overlapping coding sequences might also be considered regulation of termination. This mechanism is again frequently encountered in viral mRNAs, in particular those of retroviruses, but also in the Ty-mRNA of yeast (Kingsman & Kingsman, 1988).

Elongation Translational efficiency may also be influenced at the level of elongation. One item of debate in this respect is the role of codon bias. Various organisms display an unmistakable preference in their use of alternative codons for the same amino acid, a preference that may also vary with the level of expression of the gene in question. Codon bias in yeast for instance, is manifestly correlated with expression rate, "minor codons" that correspond to low-abundance tRNA species being more frequently encountered in genes expressed at low levels than in highly expressed genes (Bennetzen & Hall, 1982; Sharp *et al.*, 1986). However, it is not at all clear whether this is a causal relationship, that is to say to what extent the presence of such "minor codons" indeed curtails translation. On the one hand, extensive replacement of the natural codons of the yeast *PGK* gene by minor counterparts did reduce translational efficiency by about a factor three (Hoekema *et al.*, 1987) and a synthetic interferon-γ gene containing only preferred codons was expressed in yeast cells at a considerably higher rate than the wild-type gene (Jones *et al.*, 1985). On the other hand, expression of several other heterologous genes in yeast apparently is not affected by poor codon bias (Kingsman *et al.*, 1985; Ernst, 1988) or only when the gene is present in a large number of copies (Purvis *et al.*, 1987a). Experiments to test the effect of codon bias on translational efficiency are fraught with difficulties anyway, since they entail alterations in the mRNA that may also influence mRNA stability and secondary structure. The presence of secondary structural elements within the coding region can negatively affect translational efficiency (Baim *et al.*, 1985; Shatkin & Liebhaber, 1986; Kozak, 1986a; see also below).

Initiation Several structural features of an mRNA are known to influence the rate of translation initiation. We have already mentioned the effect of context on the efficiency with which the scanning 43S initiation complex recognizes an AUG codon as the start site for translation. In vertebrate mRNAs the sequence affecting AUG recognition appears to cover position -6 to +4 (the A of the AUG codon being +1) and may even extend to -9 (see Kozak, 1989a for references). Translational initiation is affected most severely by the nature of the nucleotide at position -3. A purine (preferably an A) at this position causes about a fivefold higher initiation rate than a pyrimidine (Kozak, 1986b). Changes in the nature of the residues at other positions have much lesser effects, the G at +4 being the next most important nucleotide. All in all, there is about a twentyfold difference in translational efficiency between a vertebrate containing an AUG in the most favorable (ACCAUGG) and one containing this codon in the most unfavorable (UUUAUGU) context. The distinct preference for an A residue at position -3 extends to plant (Lütcke *et al.*, 1987)), fungal and yeast (Cigan and Donahue, 1987) mRNAs, although the remainder of the consensus context is different in all four cases. In yeast, replacing the purine at -3 by a pyrimidine has a much less dramatic effect than in animal cells, reducing translational efficiency by at most a factor 2-3 (Baim and Sherman, 1988; Cigan *et al.*, 1988a; Van Den Heuvel *et al.*, 1989). In a systematic study on context effects of the yeast *HIS4* AUG start codon, Cigan *et al.* (1988a) found a U at position -3 to have the strongest negative influence (-40%), followed by G (-25%). Replacement of the preferred A at +3 by C left translation initiation unaffected. The nature of the nucleotide at position +4 did not significantly influence translational initiation at the *HIS4* start codon, despite the fact that yeast mRNAs display a clear preference for a U at this position. More important appears to be the nature of the second codon as a whole. UCU reduced translation by some 30% compared to UCG or to the GUU codon that is normally present in *HIS4* mRNA (Cigan *et al.*, 1988a). This result is somewhat surprising in view of the fact that yeast mRNAs display a clear preference for UCU as the second codon.

Eukaryotic cells, in contrast to prokaryotes, virtually exclusively use AUG as the translational start codon. In higher eukaryotes *in vivo* initiation at a different codon has been observed for only three naturally occurring mRNAs: translation of adeno-associated virus capsid protein is initiated at an ACG codon (Becerra et al., 1985) as is that of Sendai virus C' protein (Curran and Kolakofsky, 1988), while the c–*myc*–1 mRNA probably uses CUG (Hahn et al., 1988). *In vitro* as well as *in vivo* initiation at codons other than the above-mentioned is also possible in higher eukaryotic cells, but only at low efficiency (Peabody, 1989; Kozak, 1989b). In yeast, there are so far no naturally occurring examples of initiation at codons other than AUG. Inefficient initiation *in vivo* at other codons has been shown to take place, however. UUG was even reported to support translation initiation on a *CYC1-GALK* fusion mRNA at 6.9% of the rate found for an AUG codon (Zitomer *et al.*, 1984). Using *CYC1-βGAL* fusions, however, Clements et al. (1988) did not detect efficiencies higher than 0.5% (for GUG). Donahue and coworkers (Cigan *et al.*, 1988b; Donahue *et al.*, 1988) have shown that both the initiator $tRNA_i^{met}$ and the β subunit of initiation factor eIF–2 play a central role in selection of the start codon. A change in the anticodon of the initiator tRNA from 3'-UAC-5' to 3'-UCC-5' allowed translational initiation at an AGG - and no other - codon. On the other hand, mutations in the Zn(II) finger motif present in eIF–2β allow the scanning 43S complex, of which both the initiator tRNA and eIF-2 are part, to initiate at a UUG instead of an AUG codon. Similarly, a mutation in the yeast eIF-2α subunit permits initiation at UUG as does a mutation in the as yet unidentified *SUI-1* gene, which may encode a protein involved in modification of eIF-2 (Donahue et al., 1988). Apparently, complementarity between the AUG and the anticodon of the initiator tRNA is an important criterium in start codon selection. The stringency with which this criterium is applied, however, appears to depend on the structure of the eIF-2 protein. The precise role of context nucleotides in AUG selection remains to be clarified as does the possible involvement of additional *trans*-acting factors.

Directly linked to the question of AUG start codon selection is the effect of so-called "upstream AUG codons" on translational efficiency. Although still a minority, there is a significant number (~10%) of eukaryotic - both viral and cellular - mRNAs in which the coding region does not start at the 5'-proximal but at a more distal AUG codon (Kozak, 1987,1989a; Cigan and Donahue, 1987). In addition there are a number of viral mRNAs that contain overlapping coding regions starting at different AUG codons. Generally, upstream AUG codons have a negative effect on initiation at their downstream counterparts because, according to the scanning model, they sequester some of the scanning subunits, thus reducing the number that can initiate at the downstream start codon. In accordance with this idea, an upstream AUG codon in an unfavorable context has a less severe negative effect on translational efficiency than the same codon in a favorable context (Kozak, 1984,1986b,1989b). This strategy of limiting the access to the authentic AUG start codon by the presence of (an) upstream AUG codon(s) appears to have been used to control the expression of a number of critical eukaryotic genes, including several proto-oncogenes, in a constitutive fashion (see Kozak, 1989a for references).

Apart from simply bypassing the upstream AUG, ribosomes can also start at a downstream AUG codon by reinitiation, provided translation from the upstream AUG is terminated by a stop codon (Peabody & Berg, 1986; Kozak 1989a). The efficiency of reinitiation increases with increasing distance between the upstream open reading frame and the downstream AUG (Kozak, 1989a). However, reinitiation is possible even when the stop codon is located at a short distance downstream of the next AUG (Peabody & Berg, 1986).

In yeast about 5% of the known mRNAs carry one or more upstream AUG codons (Cigan & Donahue, 1987). The best studied are the mRNAs encoding GCN4, the transcription factor involved in general amino acid control, and CPA1, an enzyme from the arginine biosynthetic pathway. *GCN4* mRNA contains four short (2-3 codons) upstream ORFs, while *CPA1* mRNA possesses a single, 25 codons long, ORF in its leader region. In both cases, the upstream ORF or ORFs are crucial for translational regulation of the mRNA (see Hinnebush, 1988a,b for reviews). In the case of *GCN4* mRNA, the two 5'-proximal and two 5'-distal ORFs have distinctly different functions. The latter serve to block access to the legitimate GCN4 start codon, whereas the former somehow counteract this negative effect, provided the *trans*-acting factor GCD1 is absent or its action blocked by *trans*-acting factors GCN2 and GCN3 as occurs during amino acid starvation. The current model envisages that under these conditions translation of the 5'-proximal ORFs increases the ability of the ribosomes to reinitiate at downstream AUG codons, something that normally yeast ribosomes are virtually incapable of (Sherman and Stewart, 1982).

In contrast to the situation for *GCN4* mRNA, the peptide encoded by the upstream ORF of the *CPA1* mRNA may be directly involved in translational regulation of this mRNA. It has been suggested, though not yet proven, that this peptide in conjunction with the product of the *CPAR* gene and arginine freezes ribosomes that have translated the upstream ORF on the mRNA, thus blocking access to the AUG codon starting the *CPA1* coding region (Werner *et al.*, 1987). In the absence of arginine this would not happen, leaving the way to the start codon open. It is not quite clear how, in the latter case, ribosomes reach the *CPA1* coding region. The upstream ORF does start with an AUG codon in an unfavorable context (U at -3) allowing it to be bypassed (Cigan *et al.*, 1988a). However, as discussed above, context effects in yeast are relatively small so the negative influence of the upstream ORF on production of the CPA1 protein may still be considerable even in the absence of arginine. Possibly, therefore, reinitiation occurs also on the *CPA1* mRNA. The role of upstream AUG codons in some other yeast mRNAs, *e.g.* those encoding the regulatory proteins PHO4 and PPR1, may be to limit translation in a constitutive fashion.

Numerous examples document the negative influence of secondary structure in the leader sequence on translational initiation (see Kozak 1989a,c for references). Secondary structure can be due to short-range base pairing (hairpins), long-range interactions between the leader and sequences elsewhere in the coding or trailer regions of the mRNA (Spena *et al.*, 1985; Kozak, 1989c; Van Den Heuvel *et al.*, 1990; see also below) or even to base-pairing with anti-sense RNA (Melton, 1985). Depending on its position a secondary structural element may either hinder access of the 43S initiation complex to the cap (Lee *et al,* 1983, Lawson *et al.*, 1986, Kozak, 1989c) or it may interfere with the migration of the scanning complex towards the AUG codon (Baim & Sherman, 1988; Cigan *et al.*, 1988a, Bettany *et al.*, 1989; Kozak, 1989c). The extent to which a hairpin structure limits scanning varies with its thermodynamic stability (Kozak, 1986a,1989c; Baim & Sherman, 1988; Cigan *et al.*, 1988a; Guan & Weiner, 1989). Moreover, vertebrate 43S initiation complexes are less sensitive to such structures than their counterparts from yeast. In vertebrate mRNAs the calculated ΔG of the hairpin has to exceed -30 kcal/mole before an effect on translation is observed, while a complete block is achieved only when ΔG is around -50 kcal/mole or more (Kozak, 1986; Guan & Weiner, 1989). In yeast, on the other hand, a hairpin having a ΔG of –21 kcal/mole already reduced translational efficiency by 95% (Cigan *et al.*, 1988) and hairpins of even lower stability were found to have a severe effect (Baim & Sherman, 1988). As mentioned above, secondary structure in the coding region also reduces translational efficiency by slowing down elongation. However, 80S ribosomes can disrupt more stable structures than 40S initiating subunits, since introduction of a hairpin having a ΔG of -61 kcal/mole downstream of the AUG codon of prepro-insulin mRNA still allowed a detectable amount of *in vitro* translation by either reticulocyte or wheat germ ribosomes, whereas the presence of the same hairpin upstream of the AUG codon blocked translation completely (Kozak, 1989c). Yeast 80S ribosomes, however, are less potent in resolving secondary structure (Baim *et al.*, 1985; Shakin & Liebhaber, 1986).

Although hairpin structures generally decrease translational efficiency, there are some recent indications that they can have a stimulating effect as well, when occurring immediately downstream of an AUG codon in a suboptimal context (Kozak, 1989b) or surrounding such an AUG codon, with the AUG in the loop (Gough *et al.*, 1985).

The 5'-cap structure is crucial for efficient initiation of translation in eukaryotes since it serves as the primary element recognized by initiation factor eIF-4E which starts formation of the complex to which the initiating 40S subunit binds (reviewed by Rhoads, 1988). Virtually all eukaryotic mRNAs are capped, the only exceptions being some viral mRNAs such as those from polio- and encephalomyocarditis virus which, as discussed above, appear to circumvent the need for a cap by employing internal initiation. It is not known whether differences in cap structure, *i.e.* the nature of the second nucleotide and the degree of methylation, found in eukaryotic mRNAs act as discriminatory factors in translation.

As has become clear from the foregoing discussion, the structure of the leader is of particular importance in determining the translational efficiency of a eukaryotic mRNA. Two aspects of leader structure that remain to be inspected are leader length and its nucleotide composition.

EFFECT OF LEADER LENGTH

In both vertebrates and yeast the lengths of the large majority of the mRNA leaders fall between 25 and 95 nucleotides (n), only a few being considerably longer (up to 1000 n) or shorter (down to no leader at all). Long leaders do not appear to be detrimental to translation *per se* (Johansen *et al.*, 1985; Kozak, 1987; Cigan *et al.*, 1988a) and, provided they are unstructured, may even confer a selective advantage on the messenger under conditions of mRNA competition (Kozak, 1988). On the other hand, *in vitro* studies on mammalian mRNAs have provided some indication that optimal translation requires a certain minimum leader length (see Van Den Heuvel, 1989 for references).

We have carried out a systematic analysis of the effect of leader length on translational efficiency *in vivo* using the yeast phosphoglycerate kinase (*PGK*) gene (Van Den Heuvel *et al.*, 1989). This gene was

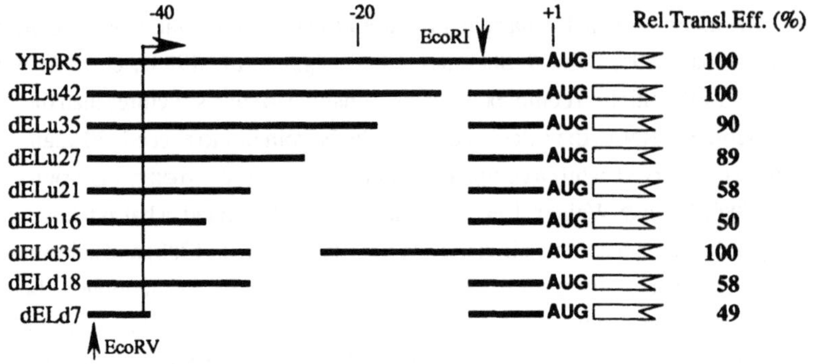

Fig.1. Effect of leader length on translational efficiency of *PGK* mRNA. The various deletions introduced in the leader of the *PGK* mRNA are schematically indicated. Deletions were created either by BAL31 digestion starting from the EcoRI site in YEpR5 followed by religation using an EcoRI linker (dELu series) or by insertion of deoxyoligonucleotides between the EcoRI site and an EcoRV site directly upstream of the transcription start site (dELd series). The latter is indicated by the bent arrow. For details see Van Den Heuvel *et al.*, 1989. The numbers in the names of the various mutants indicate the number of nucleotides left in the leader. The relative translational efficiency of the various mutant mRNAs is shown to the right. The value for the YEpR5 mRNA was arbitrarily set at 100%.

chosen because it is one of the few yeast genes not having multiple transcription start sites that enormously complicate this type of study. Starting with a slightly modified version of the gene containing an EcoRI site directly upstream of the ATG codon (Hoekema *et al.*, 1987), we introduced deletions of various sizes extending upstream from this site (dELu; Fig.1). In order to be able to exclude possible effects due to the removal of specific sequence elements a second - smaller - series of mutants was created that contain deletions extending downstream of a position 3' to the transcription start site (dELd; Fig.1). In a final mutant the transcription start site was moved as close to the EcoRI site as possible, resulting in a leader only 7 n long. In all cases the context of the AUG codon was left intact for obvious reasons. Reverse transcription analysis of the various mRNAs confirmed the expected structure of the mutant leaders in all cases.

The effect of the leader deletions on translational efficiency was quantified by assaying the level of mutant mRNA by Northern hybridization and reverse transcription and of PGK protein by gel electrophoresis and scanning of the stained gel, using yeast transformants carrying about twelve copies of the mutant gene in question. The results, summarized in Fig.1, show that reducing leader length from 45 to 27 n does not have any effect. A further shortening to 21 n, however, reduced translational efficiency by about a factor 2. This effect is due solely to the diminished length of the leader and not to removal of a positive *cis*-acting element because removal of the same region of the leader did not have any effect on translation of the dELd35 mutant mRNA. Translational efficiency did not drop any further when the length of the leader was pared down to either 16 or 7 n.

We have envisaged two possible explanations for the observed reduction in translation. One supposes the 43S initiating complex to "overshoot" AUG codons lying too close to the cap because they are in a sterically unfavorable position with respect to the anticodon of the Met-tRNA$_i^{met}$ when the complex associates with the 5'-end of the mRNA. The other puts the blame on the assembly of the 80S ribosomes at the AUG start codon. Such assembly entails pausing of the ribosome (Wolin & Walter, 1988) which, if occurring too close to the 5'-end of the messenger, could temporarily block access to the cap. The fact that 80S ribosomes, depending upon their origin, occupy 25-40 n of the mRNA (Wolin & Walter, 1988) is an argument in favor of the latter interpretation. Of course it also requires the assumption that the rate at which 40S subunits bind to the cap is closely similar to the rate at which 80S ribosomes are assembled at the AUG codon.

EFFECT OF NUCLEOTIDE COMPOSITION OF THE LEADER

Apart from the AUG context rule discussed above, there does not appear to exist any need for specific sequence elements in the leader of eukaryotic mRNA in order to ensure a maximum rate of translation initiation (Cigan *et al.*, 1988a; Baim & Sherman, 1988; Van Den Heuvel *et al.*, 1989). Nevertheless, yeast mRNA leaders show a distinct bias in nucleotide composition, being rich in A, relatively poor in C and U and extremely poor in G. Furthermore, there are significant differences in this bias between mRNAs derived from genes expressed at high and low rates respectively, suggesting that the nucleotide composition of the leader may constitutively influence gene expression in yeast (Cigan & Donahue, 1987).

In order to analyze the effect of nucleotide composition of the leader on translational efficiency in a systematic manner, we replaced most of the leader of the *PGK* mRNA by various defined sequences through insertion of synthetic deoxyoligonucleotides into the EcoRI site of the dELu16 mutant shown in Fig.1. These oligonucleotides introduced either 18 n long homopolymer stretches or 20 n long heteropolymer sequences of the type $[X_nY_m]_4$ (in which n+m= 5) and thus restored leader length to a value above the critical one of ~30 n. Again, the context of the transcription start point and AUG start codon was left undisturbed (Van Den Heuvel *et al.*, 1990).

The effect of the various insertions on translational efficiency was assayed as described in the previous section and the results are summarized in Fig.2. Insertion of a polyA tract (dINpA) into the dELu16 leader restored translational efficiency to the level of the original mRNA (YEpR5) in agreement with the effect of leader length discussed above and the A-rich nature of naturally occurring yeast mRNAs (Cigan & Donahue, 1987). The same effect was caused by insertion of a polyC tract (dINpC), showing that, despite the relative dearth of C residues in yeast mRNA leaders, a high abundance of C does not diminish translation. In agreement with these results, inserting sequences that consist of mixtures of A and C in any proportion also resulted in wild-type values for translational efficiency (not shown).

In contrast, insertion of a polyU tract into the dELu16 leader (dINpU) did not restore translational efficiency to the level of the control, despite the increased size of the leader region. This phenomenon can most likely be explained by the occurrence of base pairing between the polyU tract in the leader and the polyA tail of the mRNA. Similar long-range interactions have been invoked to explain the reduced translation of zein mRNA *in vitro* (Spena *et al.*, 1985) and of mutant prepro-insulin mRNA *in vivo*, albeit in the latter case only under stress conditions (Kozak, 1989c). We analyzed the possible effect of such long-range interactions on translation in yeast further by constructing a mutant *PGK* mRNA that carries both a polyC tract in the leader and an equally long polyG tract in its trailer region (pC-YEpR5-Δ76-pG). While neither insertion by itself affected translation, their combination reduced translational efficiency by a factor five (Fig.2D) lending further support to the notion that both ends of a eukaryotic mRNA are sufficiently close to allow base pairing to take place between leader and trailer, including the polyA tail.

Interruption of the polyU tract in the leader with a G at every fifth position appears to disturb base pairing sufficiently to allow unhampered passage of the scanning 43S initiation complex, since the $[U_4G]_4$ insertion into the dELu16 leader did restore translational efficiency to the level of the control (Fig.2B). Increasing the proportion of G residues diminishes translation again to an ultimate value of only a few percent when a polyG tract is inserted. This effect, previously observed also for plant (Galili *et al.*, 1986) and mammalian (Rao *et al.*, 1988) as well as the yeast *CYC1* (Sherman *et al.*, 1986) mRNA, is probably due to the ability of G-rich sequences to form a self-complementary structure with a high thermodynamic stability (Zimmerman *et al.*, 1975) that can not be disrupted by the scanning 40S subunits. However, as shown in Fig.2B a leader containing as much as 40% G residues interspersed with U $\{[U_3G_2]_4\}$ still allows translation to proceed at about 80% of the value shown by one containing hardly any G residues at all. Even a G content as high as 60% $\{[U_2G_3]_4\}$ leaves a significant level of translation. These results were confirmed with the $[A_nG_m]_4$ series of leader insertions (data not shown). These results make it unlikely that the difference in average G content existing betweenleaders of highly and lowly expressed yeast genes (20% *v.* 2% respectively) can be the cause of their different levels of expression. Of course a number of consecutive G residues in an otherwise G-poor leader can cause a drastic reduction in translation (Sherman *et al.*, 1986) but such stretches do not occur as a general event in mRNA leaders of yeast genes expressed at low rates (Cigan & Donahue, 1987).

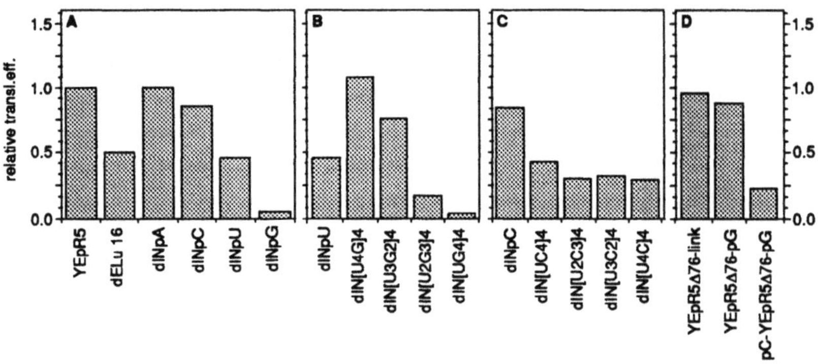

Fig. 2. Effect of nucleotide composition of the leader on translational efficiency of *PGK* mRNA. Nucleotide composition of the *PGK* mRNA leader was systematically altered by introduction of deoxyoligonucleotides into the EcoRI site of the dELdu16 mutant (Fig.1). The nature of the insertion is indicated by the name of the mutant. The relative translational efficiency of the various mutant mRNAs is indicated. Panel A: homopolymer insertions. Panel B: Effect of increasing G content. Panel C: insertions containing mixtures of U and C. Panel D: Effect of complementary sequences in leader and trailer. YEpR5-Δ76-link: a mutant in which the 76 n long ClaI-SspI fragment from the trailer has been replaced by an EcoRI linker; YEpR5-Δ76-pG contains an 18 n long polyG tract inserted into the EcoRI site created by the linker; pC-YEpR5-Δ76-pG contains both an 18 n long polyC tract in the leader (identical to dINpC) and the 18 n long polyG tract in the trailer. For details see Van Den Heuvel *et al.*, 1990.

Rather surprisingly, insertion of sequences containing any mixture of U and C residues into the dELu16 leader did not increase translational efficiency (Fig.2C). Either such sequences posses some intrinsic (conformational) property that negates the effect of the increase in leader length, or this type of leader is also able to base pair to sequences elsewhere in the (coding or 3'-noncoding regions) *PGK* mRNA. Computer analysis using the FOLD program (Zuker & Stiegler, 1981) indeed showed the possibility for such base pairing to be present. Analysis of the effect of these leaders on translation when fused to another coding region should allow us to discriminate between these two possibilities. Such experiments are currently in progress in our laboratory.

EFFECT OF TRAILER STRUCTURE

Structural changes in the trailer region have been shown to affect translational efficiency of both the *CYC*1 (Zaret & Sherman, 1984) and the *PYK* (Purvis *et al*, 1987b) mRNA in yeast, in the first case positively, in the second negatively. In neither case, however, has the basis for the observed effect been explained. In the course of our experiments on the *PGK* mRNA we have introduced several different structural changes into its trailer sequence, varying from a deletion of about half this region to insertion of homopolymer stretches at various points as well as insertion of a very stable hairpin structure. In none of these cases did we observe a significant effect on translational efficiency of the mutant mRNA. Only the introduction of a sequence complementary to an insertion in the leader, as discussed in the previous section, drastically diminished translation. Thus, we conclude that the trailer region of the *PGK* messenger as such is not likely to play a role in determining its rate of translation. Whereas the effect of trailer structure may vary with the particular mRNA under consideration, the polyA tail does appear to be generally important for translational initiation. Adding polyA to an *in vitro* translation system inhibits translation of polyA$^+$, but not polA$^-$ mRNA provided the mRNA is added free of protein (Jacobsen & Favreau, 1983; Grossi de Sa *et al.*, 1988). The inhibition could be reversed by the addition of polyA-binding protein (PAB) suggesting that it is due to sequestering of PAB by the excess polyA. Sachs & Davis (1989) have recently shown that depletion of PAB in yeast cells results in inhibition of translation initiation *in vivo*. Seven unlinked suppressor mutations were isolated that allow translation to proceed even in the absence of PAB. All seven are cold-sensitive mutations affecting the amount of 60S ribosomal subunits. One was traced to the gene encoding ribosomal protein L46. These results indicate that efficient translation initiation requires the binding of PAB to the polyA tail, a requirement mediated through the 60S ribosomal subunit

ACKNOWLEDGEMENTS

We would like to thank R.J.M. Bergkamp, R. Kieft, R. Giltay, K. van Vliet, R. Hartong, and Y. van Delft for excellent technical assistance, P. Vreken and A. de Maat for providing the set of YEpR5-Δ76 deletion and insertion mutants and Dr J. Maat for fruitful discussions. This work was supported in part by the Programmacommissie Industriële Biotechnologie (Commission for Industrial Biotechnology) with financial aid from the Ministry of Economic Affairs.

REFERENCES

Baim SB, Pietras DF, Eustice DC, Sherman F (1985) A mutation allowing an mRNA secondary structure diminishes translation of *Saccharomyces cerevisiae* iso-1-cytochrome c. Mol Cell Biol 5: 1839-1846

Baim SB, Sherman F (1988) mRNA structures influencing translation in the yeast *Saccharomyces cerevisiae*. Mol Cell Biol 8: 1591-1601

Becerra, SP, Rose JA, Hardy, M, Baroudy BM, Anderson C (1985) Direct mapping of adeno-associated virus capsid proteins B and C: a possible ACG initiation codon. Proc Natl Acad Sci USA 82: 7919-7923

Bennetzen JL, Hall BD (1982) Codon selection in yeast. J Biol Chem 257: 3026-3031

Bettany AJE, Moore PA, Cafferkey R, Bell LD, Goodey AR, Carter BLA, Brown AJP (1989) 5'-Secondary structure formation, in contrast to a short string of non-preferred codons, inhibits the translation of the pyruvate kinase mRNA in yeast. Yeast 5: 187-198

Cigan AM, Donahue TF (1987) Sequence and structural features associated with translational initiator regions in yeast - A review. Gene 59: 1-18

Cigan AM, Pabich EK, Donahue TF (1988a) Mutational analysis of the *HIS4* translational initiator region in *Saccharomyces cerevisiae*. Mol Cell Biol 8: 2964-2975

Cigan AM, Feng L, Donahue TF (1988b) $tRNA_i^{met}$ functions in directing the scanning ribosome to the start site of translation. Science 242: 93-97

Clements JM, Laz TM, Sherman F (1988) Efficiency of translation by non-AUG codons in *Saccharomyces cerevisiae*. Mol Cell Biol 8: 4533-4536

Curran J, Kolakofsky D (1988) Ribosomal initiation from an ACG codon in the Sendai virus P/C mRNA. EMBO J 7: 245-251

Curran J, Kolakofsky D (1989) Scanning independent ribosomal initiation of the Sendai virus Y proteins *in vitro* and *in vivo*. EMBO J 8: 521-526

Donahue TF, Cigan AM, Pabich EK, Valavicius BC (1988) Mutations at the Zn(II) finger motif in the yeast eIF-2ß gene alter ribosomal start-site selection during the scanning process. Cell 54: 621-632

Ernst JF (1988) Codon usage and gene expression. TIBTECH 6: 196-199

Galili G, Kawata EE, Cuellar RE, Smith LD, Larkins BA (1986) Synthetic oligonucleotide tails inhibit *in vitro* and *in vivo* translation of SP6 transcripts of maize zein cDNA clones. Nucl Acids Res 14: 1511-1524

Gough NM, Metcalf D, Gough J, Grail D, Dunn AR (1985) Structure and expression of the mRNA for murine granulocyte-macrophage colony stimulating factor. EMBO J 4: 645-653

Grossi de Sa MF, Standart N, Martins de Sa C, Akhayat O, Huesca M, Scherrer K (1988) The poly(A)-binding protein facilitates *in vitro* translation of poly(A)-rich mRNA. Eur J Biochem 176: 521-526

Guan KL, Weiner H (1989) Influence of the 5'-end region of aldehyde dehydrogenase messenger RNA on translational efficiency - potential secondary structure inhibition of translation *in vitro*. J Biol Chem 264: 17764-17769

Hahn SR, King MW, Bentley DL, Anderson CW, Eisenman RN (1988) A non-AUG translational initiation in *c-myc* exon 1 generates an N-terminally distinct protein whose synthesis is disrupted by Burkitt's lymphomas. Cell 52: 185-195

Herman AC (1989) Alternatives for the initiation of translation. Trends Biochem Sci 14: 219-222

Hinnebush AG (1988a) Mechanisms of gene regulation in the general control of amino acid biosynthesis in *Saccharomyces cerevisiae*. Microbiol Rev 52: 248-273

Hinnebush AG (1988b) Novel mechanisms of translational control in *Saccharomyces cerevisiae*. Trends Genet 4: 169-174

Hoekema A, Kastelein RA, Vasser M, De Boer HA (1987) Codon replacement in the *PGK1* gene of *Saccharomyces cerevisiae*: experimental approach to study the role of biased codon usage in gene expression. Mol Cell Biol 7: 2914-2924

Jacobsen A, Favreau M (1983) Possible involvement of poly(A) in protein synthesis. Nucl Acids Res 11: 6353-6368

Johansen H, Schümperli D, Rosenberg M (1984) Affecting gene expression by altering the length and sequence of the 5' leader. Proc Natl Acad Sci USA 81: 7698-7702

Jones M, Koski R, Egan K, Bitter GA (1985) The effects of codon bias on IFN-gamma expression in yeast *Saccharomyces cerevisiae*. J Cell Biochem Suppl 0: 217

Kingsman AJ, Kingsman SM (1988) Ty: a retro-element moving forward. Cell 53: 333-335

Kingsman SM, Kingsman AJ, Dobson MJ, Mellor J, Roberts NA (1985) Heterologous gene expression in *Saccharomyces cerevisiae*. Biotechnol Gen Eng Rev 3: 377-416

Kozak M (1984) Selection of initiation sites by eucaryotic ribosomes: effect of inserting AUG triplets upstream from the coding sequence for preproinsulin. Nucl Acids Res 12: 3873-3893

Kozak M (1986a) Influences of mRNA secondary structure on initiation by eukaryotic ribosomes. Proc Natl Acad Sci USA 83: 2850-2854

Kozak M (1986b) Point mutations define a sequence flanking the AUG initiator codon that modulates translation by eukaryotic ribosomes. Cell 44: 283-292

Kozak M (1987) An analysis of the 5'-noncoding sequences from 699 vertebrate messenger RNAs. Nucl Acids Res 15: 8125-8148

Kozak M (1988) Leader length and secondary structure modulate mRNA function under conditions of stress. Mol Cell Biol 8: 2737-2744

Kozak M (1989a) The scanning model for translation: an update. J Cell Biol 108: 229-241

Kozak M (1989b) Context effects and inefficient initiation at non-AUG codons in eucaryotic cell-free translation systems. Mol Cell Biol 9: 5073-5080

Kozak M (1989c) Circumstances and mechanisms of inhibition of translation by secondary structure in eucaryotic mRNAs. Mol Cell Biol 9: 5134-5142

Lawson TG, Ray BK, Dodds JT, Grifo JA, Abramson RD, Merrick WC, Betsch DF, Weith HL, Thach RE (1986) Influence of 5' proximal secondary structure on the translational efficiency of eukaryotic mRNAs and on their interaction with initiation factors. J Biol Chem 261: 13979-13989

Lee KAW, Guertin D, Sonenberg N (1983) mRNA secondary structure as a determinant in cap recognition and initiation complex formation. J Biol Chem 258: 707-710

Lodish H (1976) Translational control of protein synthesis. Annu Rev Biochem 45: 39-72

Melton DA (1985) Injected anti-sense RNAs specifically block messenger RNA translation *in vivo.* Proc Natl Acad Sci USA 82: 144-148

Peabody DS (1989) Translation initiation at non-AUG triplets in mammalian cells. J Biol Chem 264: 5031-5035

Peabody DS, Berg P (1986) Termination-reinitiation occurs in the translation of mammalian cell mRNAs. Mol Cell Biol 6: 2695-2703

Proud CG (1986) Guanine nucleotides, protein phosphorylation and the control of translation. Trends Biochem Sci 11: 73-77

Purvis IJ, Loughlin L, Bettany AJE, Brown AP (1987a) Translation and stability of an *Escherichia coli* ß-galactosidase mRNA expressed under the control of pyruvate kinase sequences in *Saccharomyces cerevisiae.* Nucl Acids Res 15: 7963-7974

Purvis J, Bettany AJE, Loughlin L, Brown AJP (1987b) The effects of alterations within the 3' untranslated region of the pyruvate kinase messenger RNA upon its stability and translation in *Saccharomyces cerevisiae.* Nucl Acids Res 15: 7951-7962

Rao CD, Pech M, Robbins KC, Aaronson SA (1988) The 5' untranslated sequence of the c-*sis*/platelet-derived growth factor 2 transcript is a potent translational inhibitor. Mol Cell Biol 8: 284-292

Rhoads RE (1988) Cap recognition and the entry of mRNA into the protein synthesis initiation cycle. Trends Biochem Sci 13: 52-56

Sachs AB, Davis RW (1989) The poly(A) binding protein is required for poly(A) shortening and 60S ribosomal subunit-dependent translation initiation. Cell 58: 857-867

Sharp PM, Tuohy TMF, Mosurski KR (1986) Codon usage in yeast: cluster analysis clearly differentiates highly and lowly expressed genes. Nucl Acids Res 14: 5125-5143

Shakin SH, Liebhaber SA (1986) Destabilization of messenger RNA/complementary DNA duplexes by the elongating 80S ribosome. J Biol Chem 261: 16018-16025

Sherman F, Baim SB, Hampsey DM, Goodhue CT, Friedman LR, Stiles JI (1986) Properties of protein translation determined with altered forms of the yeast *CYC1* gene. In: Matthews MB (ed) Translational control. CSH Laboratories, Cold Spring Harbor NY, p 42

Sherman F, Stewart JW (1982) Mutations altering initiation of translation of yeast iso-1-cytochrome *c*; Contrast between the eukaryotic and prokaryotic initiation process. In: Strathern JN, Jones EW, Broach JR (eds) Molecular Biology of the Yeast *Saccharomyces*: Metabolism and Gene Expression. CSH Laboratories, Cold Spring Harbor NY, p 301

Sonenberg N, Pelletier, J (1989) Poliovirus translation: A paradigm for a novel initiation mechanism. BioEssays 11: 128-132

Spena A, Krause E, Dobberstein B (1985) Translation efficiency of zein mRNA is reduced by hybrid formation between the 5'- and 3'-untranslated region. EMBO J 4: 2153-2158

Valle RPC, Morch M-D (1988) Stop making sense or: Regulation at the level of termination in eukaryotic protein synthesis. FEBS Lett 235: 1-15

Van Den Heuvel JJ, Bergkamp RJM, Planta RJ, Raué HA (1989) Effect of deletions in the 5'-noncoding region on the translational efficiency of phosphoglycerate kinase mRNA in yeast. Gene 79: 83-95

Van Den Heuvel JJ, Planta RJ, Raué HA (1990) Effect of leader primary structure on the translational efficiency of phosphoglycerate kinase mRNA in yeast. Yeast : in press

Werner M, Feller, A, Messenguy, F, Piérard, A (1987) The leader peptide of yeast gene *CPA1* is essential for the translational repression of its expression. Cell 49: 805-813

Wolin SL, Walter P (1988) Ribosome pausing and stacking during translation of a eukaryotic mRNA. EMBO J 11: 3559-3569

Zaret KS, Sherman F (1984) Mutationally altered 3' ends of yeast *CYC1* mRNA affect transcript stability and translational efficiency. J Mol Biol 176: 107-135

Zimmerman SB, Cohen GH, Davies DR (1975) X-ray fiber diffraction and model building study of polyguanylic acid and polyinosinic acid. J Mol Biol 92: 181-192

Zitomer RS, Walthall DA, Rymond BC, Hollenberg CP (1984) *Saccharomyces cerevisiae* ribosomes recognize non-AUG initiation codons. Mol Cell Biol 4: 1191-1197

Zuker M, Stiegler P (1981) Optimal computer folding of large RNA sequences using thermodynamics and auxiliary information. Nucl Acids Res 9: 133-148

HUMAN FETAL $^{G}\gamma$- AND $^{A}\gamma$-GLOBIN mRNA AND ADULT β-GLOBIN mRNA EXHIBIT DISTINCT TRANSLATION EFFICIENCY AND AFFINITY FOR EUKARYOTIC INITIATION FACTOR 2

Susan Marsh, Yona Banai, Mira Na'amad and Raymond Kaempfer
Department of Molecular Virology
The Hebrew University-Hadassah Medical School
91010 Jerusalem, Israel

Introduction

In the human fetus, beginning at 8 weeks of gestation, embryonic globins are gradually replaced by the α-globin chain and two β-like chains, $^{G}\gamma$-globin and $^{A}\gamma$-globin, that differ only in presence of Gly or Ala at position 136 (Schroeder *et al.*, 1968). $\alpha_2\gamma_2$ is the predominant hemoglobin expressed during fetal life. Starting shortly before birth, the γ-globin chains are replaced gradually by the adult β- and δ-globin chains. Six months after birth, about 98% hemoglobin is $\alpha_2\beta_2$. Besides the dramatic γ to β switch, a second switch, $^{G}\gamma$-globin to $^{A}\gamma$-globin, is observed postnatally, when the $^{G}\gamma/^{A}\gamma$-globin ratio declines progressively (Huisman *et al.*, 1974; Comi *et al.*, 1980). Synthesis of $^{G}\gamma$-, $^{A}\gamma$-, and β-globin can take place in a single erythroid colony (Terasawa *et al.*, 1980; Comi *et al.*, 1980; Peschle *et al.*, 1980), raising the question, how switching is regulated.

Among the hemoglobinopathies in which γ-chain synthesis is increased, β-thalassemia and hereditary persistence of fetal hemoglobin (HPFH) are most prominent (Weatherall and Clegg, 1979; Maniatis *et al.*, 1980). In β-thalassemia, γ-chain synthesis is observed, inversely proportional in extent to the expression of the β-globin gene (e.g., Di Segni *et al.*, 1978). A variety of deletions overlapping parts of γ and/or β genes also lead to high levels of $\alpha_2\gamma_2$ (Maniatis *et al.*, 1980). Greek HPFH is caused by a point mutation at position -117, 5' to the $^{A}\gamma$-globin gene (Gelinas *et al.*, 1985; Collins *et al.*, 1985), while $^{G}\gamma\beta^{+}$ HPFH is associated with a point mutation at position -202, 5' to the $^{G}\gamma$-globin gene (Collins *et al.*, 1984), implying transcriptional control in the γ to β switch. While regulation of transcription is clearly involved in hemoglobin switching, a precise explanation is still lacking. Moreover, the possible contribution of translational control has so far not been investigated.

Eukaryotic initiation factor 2 (eIF-2) is essential for the binding of mRNA to ribosomal subunits, a key step in translational control of gene expression. eIF-2 forms a ternary complex with Met-tRNA$_f$ and GTP that joins the 40S ribosomal subunit, yielding an obligatory intermediate for binding of mRNA. eIF-2 can also

NATO ASI Series, Vol. H 49
Post-Transcriptional Control of Gene Expression
Edited by J. E. G. McCarthy and M. F. Tuite

undergo a direct and specific interaction with mRNA (Kaempfer *et al.*, 1978a,b; Kaempfer *et al.*, 1979; Kaempfer *et al.*, 1981; Perez-Bercoff and Kaempfer, 1982). All mRNA species tested possess a high-affinity binding site for eIF-2 (Kaempfer *et al.*, 1978a,b), while RNA species not serving as mRNA, such as negative-strand viral RNA (Kaempfer *et al.*, 1978a), tRNA, or rRNA (Barrieux and Rosenfeld, 1977), do not contain such a site. In satellite tobacco necrosis virus RNA (Kaempfer *et al.*, 1981) and Mengovirus RNA (Perez-Bercoff and Kaempfer, 1982), eIF-2 protects a nucleotide sequence that is virtually identical with the ribosome binding site sequence. These findings suggested that eIF-2 may interact directly with mRNA during initiation of translation, guiding the ribosomal subunit to its binding site in mRNA. In support of this view, studies of translational competition have revealed a direct correlation between the ability of an mRNA species to compete in translation and its affinity for eIF-2 (Di Segni *et al.*, 1979; Rosen, *et al.*, 1982). α-Globin mRNA competes more weakly than β-globin mRNA in translation and this competition is relieved when eIF-2 is added in excess; α-globin mRNA binds more weakly than β-globin mRNA to eIF-2 (Di Segni *et al.*, 1979). On a molar basis, Mengovirus RNA competes 35-fold more effectively in translation than globin mRNA, in a manner that is relieved by excess eIF-2, and this RNA exhibits a 30-fold higher affinity than globin mRNA for eIF-2 (Rosen *et al.*, 1982). Moreover, eIF-2, but no other initiation factor, promotes the selection of the 5'-proximal translation initiation site by ribosomes (Dasso *et al.*, 1990). In yeast, mutants permitting an altered initiation codon to be recognized map into the α- and β-subunits of eIF-2 (Donahue *et al.*, 1988; Cigan *et al.*, 1989).

Here, we show that the human mRNA species encoding α-, $^G\gamma$-, $^A\gamma$-, and β-globin differ in their ability to compete at initiation of translation, effectiveness increasing in the order $\alpha < {}^G\gamma < {}^A\gamma < \beta$. Translational competition between these mRNA species is readily relieved by an excess of eIF-2. β- and γ-globin mRNA compete distinctly in their binding to eIF-2, β-globin mRNA exhibiting significantly higher affinity. The difference in translation efficiency of $^G\gamma$- and $^A\gamma$-globin mRNA is determined by regions in mRNA distant from the AUG initiation codon context. These findings show that in addition to transcriptional control, differential translation efficiency of mRNA species and affinity for eIF-2 may play an important role in regulating fetal and adult human globin synthesis.

Results

α-, β- and γ-Globin mRNA Exhibit Distinct Abilities to Compete in Translation

To examine the translation efficiencies of authentic human fetal and adult globin mRNA species, poly $(A)^+$ mRNA was isolated from cord blood. Such mRNA preparations contain the globin mRNA species that encode α-globin and β-globin, as well as the two fetal chains, $^A\gamma$-globin and $^G\gamma$-globin. These four globin chains

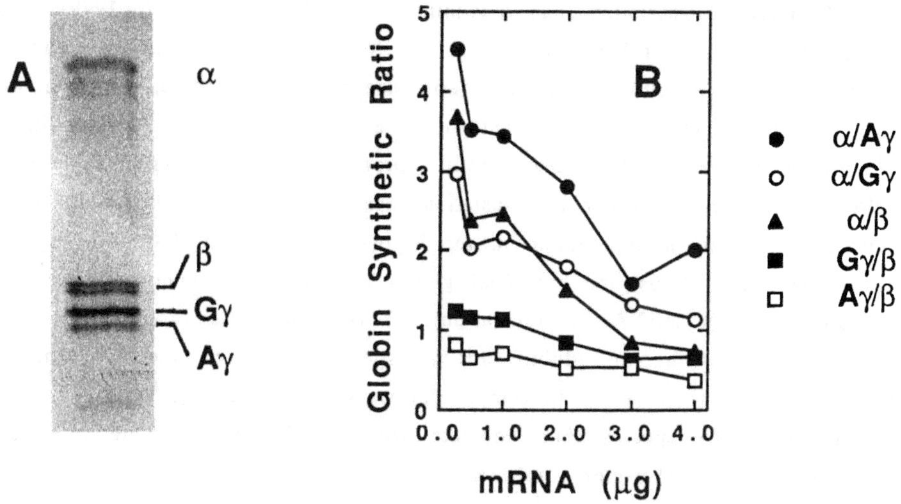

Fig. 1. *Translation of α-, $^{G}\gamma$-, $^{A}\gamma$-, and β-globin mRNA species under increasing competition pressure.* (A) Isoelectric focusing analysis (Peschle *et al.*, 1980). Autoradiogram shows ^{35}S-labeled products of translation in reticulocyte lysate (Rosen *et al.*, 1982), directed by total cord blood mRNA. Position of individual globin chains is indicated. (B) Upon translation of cord blood mRNA in the indicated amounts per reaction mixture of 16 μl, globin synthetic ratios were obtained by microdensitometry of autoradiograms as shown in (A).

migrate with essentially identical electrophoretic mobility in SDS-polyacrylamide gels, but they may be separated by isoelectric focusing. Figure 1A depicts the pattern of ^{35}S-labeled polypeptides, translated from total cord blood mRNA in a micrococcal nuclease-treated reticulocyte lysate. $^{A}\gamma$-globin and $^{G}\gamma$-globin are well resolved from each other and from α- and β-globin. An identical pattern of resolution is observed when authentic globins from cord blood lysate are focused and the gel is stained (not shown).

To study the ability of individual globin mRNA species to compete in translation, total cord blood mRNA was subjected to competition pressure analysis. Otherwise identical cell-free translation reaction mixtures, containing micrococcal nuclease-treated rabbit reticulocyte lysate, were programmed with increasing concentrations of cord blood mRNA, containing a constant proportion of the different globin mRNA species. As the concentration of added mRNA increases, translation of more strongly competing mRNA templates is favored progressively at the expense of the translation of more weakly competing ones (Lodish, 1976). Hence, the ratio of products encoded by a more weakly competing template and a more strongly competing one will decline with increasing mRNA concentration. Using this strategy, even small differences in competing ability between two mRNA species can be resolved.

This approach was used previously to show that rabbit β-globin mRNA competes more effectively in translation than α-globin mRNA (Di Segni *et al.*, 1979). Figure 1B depicts globin synthetic ratios as a function of increasing competition pressure. The α/β globin synthetic ratio declines as expected, by over 4-fold across the range of mRNA concentrations examined. The $\alpha/^{A}\gamma$-globin and $\alpha/^{G}\gamma$-globin synthetic ratios also decline progressively, showing that α-globin mRNA competes more weakly in translation than $^{A}\gamma$-globin and $^{G}\gamma$-globin mRNA. A clear, though less steep, decline in the $^{G}\gamma/\beta$-globin and $^{A}\gamma/\beta$-globin synthetic ratios is observed. Thus, the two fetal globin mRNA species compete more weakly than adult β-globin mRNA in translation. This result was reproducible. From Figure 1B, the 4 human globin mRNA species analyzed clearly exhibit an ascending order in their ability to compete in translation: $\alpha < [^{G}\gamma, ^{A}\gamma] < \beta$. The two fetal globin mRNA species thus compete with efficiencies that are intermediate between those of α- and β-globin mRNA.

$^{G}\gamma$–, $^{A}\gamma$– and β–Globin mRNA Compete in Ascending Order in Translation

To compare the translation efficiencies of $^{G}\gamma$-globin and $^{A}\gamma$-globin mRNA and that of β-globin mRNA more directly, we used hybridization selection by specific DNA sequences to obtain enriched preparations of γ- and β-globin mRNA. Figure 2A shows isoelectric focusing analysis of the translation products directed by mRNA that was hybrid-selected with $^{G}\gamma$- and β-globin DNA. The high degree of nucleotide sequence homology between $^{G}\gamma$-globin and $^{A}\gamma$-globin mRNA, 93% (Efstratiadis *et al.*, 1980; Slightom *et al.*, 1980), prevented the resolution of these two species, but they could be enriched significantly over adult globin mRNA. A reconstituted mixture of these two mRNA preparations was then translated at increasing concentrations. Total protein synthesis reached plateau at approximately 0.8 μg of the mRNA mixture, and was not stimulated in the presence of an excess of the initiation factor, eIF-2, at any mRNA concentration tested (not shown). Competition in translation between rabbit α-globin and β-globin mRNA *in vitro* is relieved by excess eIF-2, in conditions where this factor does not affect total protein synthesis (Di Segni *et al.*, 1979). In Figure 2B, the stimulatory effect of added eIF-2 on the globin synthetic ratio is plotted as a function of mRNA concentration. Addition of eIF-2 has a strong stimulatory effect on both $^{G}\gamma/\beta$-globin and $^{A}\gamma/\beta$-globin synthetic ratios, up to 0.8 μg of mRNA per reaction mixture. This stimulatory effect is abrogated at higher mRNA concentrations. Apparently, stimulation by eIF-2 becomes more pronounced as competition pressure increases, up to the point where even added eIF-2 no longer can relieve competition. This result shows that $^{G}\gamma$-globin and $^{A}\gamma$-globin mRNA compete more weakly than β-globin mRNA for eIF-2 in translation. The $^{G}\gamma/^{A}\gamma$-globin synthetic ratio is increasingly stimulated by eIF-2, showing that $^{G}\gamma$-globin mRNA competes more weakly than $^{A}\gamma$-globin mRNA in translation and that this competition is

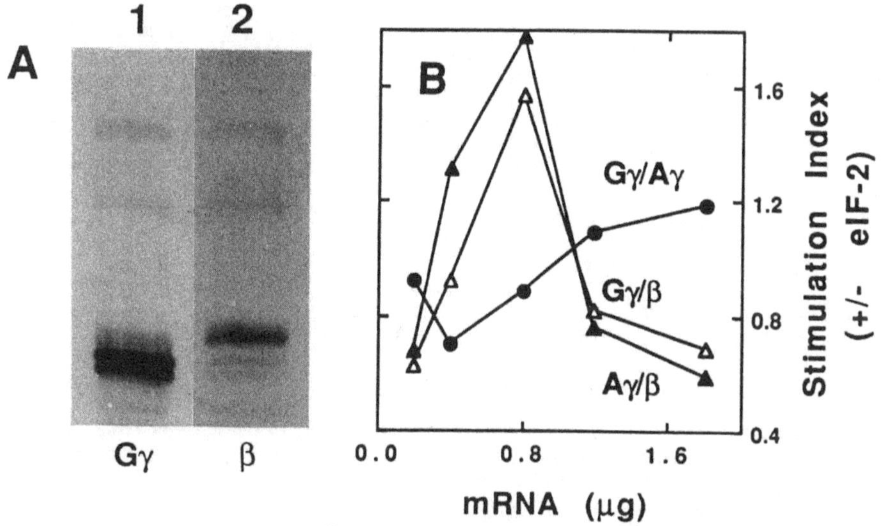

Fig. 2. *Competition between Gγ-globin, Aγ-globin and β-globin mRNA in translation and relief by eIF-2.* (A) Translation products directed by mRNA, hybrid-selected with genomic Gγ-globin DNA (Lawn *et al.*, 1980) and β-globin cDNA (pJW102), in lanes 1 and 2, respectively, separated by isoelectric focusing. (B) A 1:1 mixture of these two mRNA preparations was translated in the indicated amounts per 30 μl translation mixture, in the presence of 5 μg of purified eIF-2 (Rosen and Kaempfer, 1979; Rosen *et al.*, 1982) or buffer, and incorporation of [^{35}S]L-methionine was measured. (C) Upon isoelectric focusing analysis and microdensitometry of the autoradiogram, globin synthetic ratios were determined. The index of stimulation (+/-eIF-2) is shown.

relieved by eIF-2 (Figure 2B). The more gradual and extended response of the Gγ/Aγ ratio to eIF-2 suggests that the difference in competing ability between the two fetal mRNA species is smaller than that between each of these mRNAs and β-globin mRNA.

Even greater resolution of the translation properties of Gγ-globin and Aγ-globin mRNA is obtained in Figure 3. Here, total translation did not yet reach plateau values in response to progressive concentrations of a mixture of hybrid-selected γ- and β-globin mRNA; translation was not stimulated by addition of eIF-2 (Figure 3A). Absolute amounts of Gγ-globin, Aγ-globin, and β-globin translation products were quantitated upon translation in the absence or presence of added eIF-2, as shown in Figs. 3B and C, respectively. Addition of excess eIF-2 did not significantly affect total γ-globin synthesis. However, it had a pronounced and opposite effect on synthesis of Gγ-globin and Aγ-globin: the translation yield of Gγ-globin increased at the expense of that of Aγ-globin. In the absence of added eIF-2, the Gγ/Aγ-globin synthetic ratio declined with increasing competition pressure (Figure 3D), showing directly that Gγ-globin mRNA competes more weakly in translation than Aγ-globin mRNA. In the presence of excess eIF-2, this decline in Gγ/Aγ-globin synthetic ratio was prevented. As seen in Figure 3D, the extent of stimulation of the Gγ/Aγ-globin synthetic ratio by eIF-2 increases almost

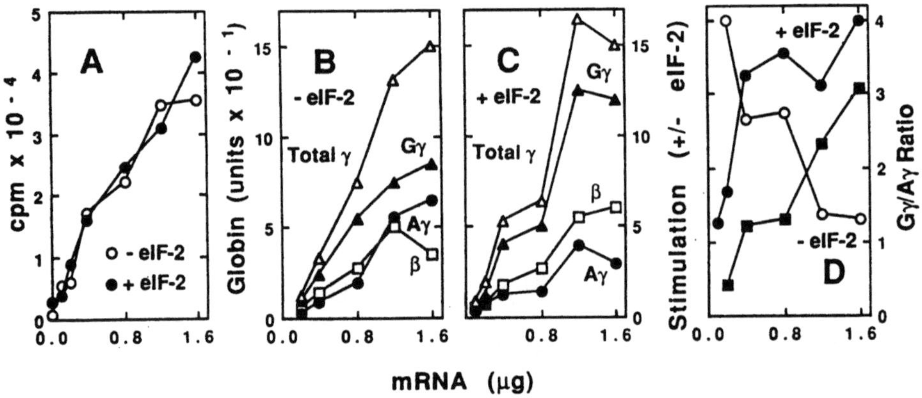

Fig. 3. *Competition between* $^{G}\gamma$*-globin and* $^{A}\gamma$*-globin mRNA in translation and relief by eIF-2.* (A) A mixture of hybrid-selected γ- and β-globin mRNA preparations was translated in the indicated amounts per 30 µl translation mixture, in the presence of 4 µg of purified eIF-2 or buffer, and incorporation of [^{35}S]L-methionine was measured. After translation without (B) or with added eIF-2 (C), β-, $^{A}\gamma$-, $^{G}\gamma$- and the sum of $^{A}\gamma$- and $^{G}\gamma$-globin (total γ) chain synthesis was determined by isoelectric focusing and microdensitometry of the autoradiogram. (D) Graph depicts $^{G}\gamma/^{A}\gamma$-globin synthetic ratio upon translation without (o) or with added eIF-2 (•), as well as index of stimulation (+/- eIF-2) for this ratio (■).

8-fold with increasing competition pressure. The two fetal mRNA species thus compete distinctly in translation, with $^{G}\gamma$-globin mRNA the weaker template. This competition is relieved by eIF-2.

γ- and β-Globin mRNA Exhibit Distinct Affinities for eIF-2

$^{G}\gamma$-globin, $^{A}\gamma$-globin and β-globin mRNA species show an ascending order in their ability to compete in translation, and compete in a manner that can be relieved by excess eIF-2. These findings raise the question, whether differences in competing ability are reflected in different affinities of the individual mRNA species for eIF-2. Rabbit α- and β-globin mRNA bind directly to eIF-2 (Di Segni *et al.*, 1979). ^{32}P-labeled, complete human β-globin mRNA SP6 transcript, carrying 30 A residues at its 3' end, is effectively retained on nitrocellulose filters in the presence of increasing amounts of eIF-2. At saturating levels of eIF-2, over 80% of input radioactivity can be retained (data not shown).

The affinities of total cord blood mRNA, β-globin mRNA, and γ-globin mRNA for eIF-2 are compared in Figure 4. In Figure 4A, constant amounts of eIF-2 and a ^{32}P-labeled mRNA probe are incubated with increasing amounts of unlabeled, competing RNA species. Competition of these mRNAs for eIF-2 results in the displacement of bound labeled RNA. Hybrid-selected, authentic β-globin mRNA is seen to compete far more effectively for eIF-2 than total cord blood mRNA: half-maximal competition by β-globin mRNA is attained at about 10-fold lower RNA concentration.

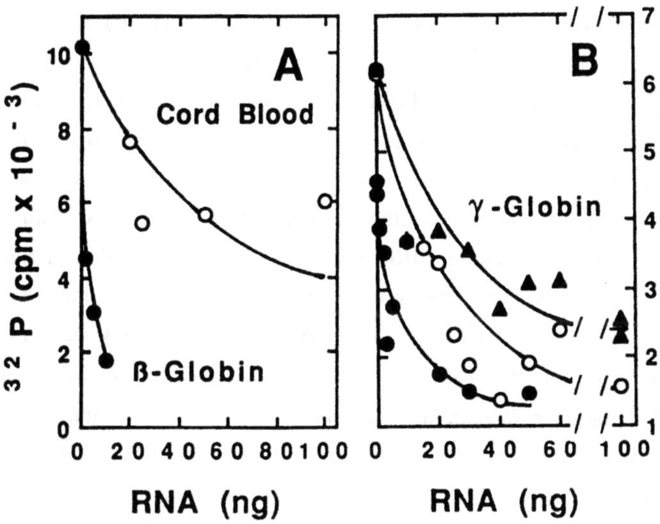

Fig. 4. *Competition between total cord blood mRNA, β-globin mRNA, and γ-globin mRNA in binding to eIF-2.* (A) Constant amounts of eIF-2 and of a 5' end-labeled, single-stranded mRNA probe from *Penicillium chrysogenum* viral RNA (25,120 cpm; Gonsky *et al.*, 1990) were incubated with the indicated amounts of unlabeled, total cord blood mRNA or hybrid-selected, authentic β-globin mRNA. (B) Constant amounts of eIF-2 and of ^{32}P-labeled β-globin mRNA SP6 transcript (17,300 cpm) were incubated with the indicated amounts of unlabeled, hybrid-selected, authentic γ-globin mRNA (▲), total cord blood mRNA (o), or unlabeled β-globin mRNA transcript (●). Binding was assayed on nitrocellulose filters (Kaempfer *et al.*, 1978a).

In Figure 4B, the labeled probe was β-globin mRNA (SP6 transcript). Unlabeled β-globin mRNA transcript also competes about 10-fold more effectively than total cord blood mRNA. Hence, authentic β-globin mRNA and β-globin mRNA transcript bind to eIF-2 with comparable affinity. By contrast, hybrid-selected, authentic total γ-globin mRNA competes more weakly than cord blood mRNA and at least 20-fold more weakly than β-globin mRNA for eIF-2 (Figure 4B).

Discussion

These results demonstrate that differential translation efficiency of mRNA species and competition for eIF-2 contribute to regulation of fetal and adult human globin synthesis. The mRNA species encoding α-, $^{G}\gamma$-, $^{A}\gamma$-, and β-globin possess distinct abilities to compete at initiation of translation, effectiveness increasing in the order $\alpha < {}^{G}\gamma < {}^{A}\gamma < \beta$. Translational competition between these mRNA species is readily relieved by an excess of eIF-2. γ- and β-globin mRNA species compete distinctly in their binding to this initiation factor, γ-globin mRNA exhibiting a significantly lower affinity than β-globin mRNA.

Considering the conservation of the protein domains between the corresponding globin chains and the strong conservation of gene organization in the globin gene family (Efstratiadis *et al.*, 1980), it is indeed noteworthy that mRNA species encoded by individual globin genes differ so greatly in their competing ability and in their affinity for eIF-2. However, human β-globin mRNA shows significant nucleotide sequence divergence from the γ-globin mRNA species in the 5' UTR (49%), coding region (92%), as well as the 3' UTR (68%), and the average divergence of the α- and γ-globin mRNA sequences is 49% (Efstratiadis *et al.*, 1980). These differences can account for the distinctly different ability to compete in translation observed for α-, β- and γ-globin mRNA species.

By contrast, the clear difference in translation efficiency observed for $^{G}\gamma$- and $^{A}\gamma$-globin mRNA is unexpected, since their nucleotide sequences differ in only 8 out of 581 positions (Slightom *et al.*, 1980). In the 53-nt long 5' UTR, only a single nucleotide difference occurs, at position 25, which is G in $^{G}\gamma$-globin mRNA and A in $^{A}\gamma$-globin mRNA. Another single nucleotide difference occurs in the 438-nt long coding region, at codon 136 (GGA vs. GCA). The remaining 6 differences are found in the 90-nt long 3' UTR, four of them in a block just after the termination codon and one between the AAUAAA sequence and the start of the poly (A) tail (Poon *et al.*, 1978; Forget *et al.*, 1979; Slightom *et al.*, 1980). The change at codon 136 gives rise to the distinct isoelectric properties of $^{G}\gamma$- and $^{A}\gamma$-globin (Figure 1), but it has not been shown rigorously if differences occurring in the other positions are gene-specific or represent genetic polymorphisms. Two groups obtained identical sequences for the coding region of $^{G}\gamma$-globin mRNA (Forget *et al.*, 1979; Slightom *et al.*, 1980) and the 3' UTR of $^{A}\gamma$-globin mRNA (Poon *et al.*, 1978; Slightom *et al.*, 1980), but observed differences in the 3' UTR of $^{G}\gamma$-globin mRNA (Forget *et al.*, 1979; Slightom *et al.*, 1980). There is no doubt, however, that the sequence context (Kozak, 1986) of the AUG initiation codon at positions 54-57 is identical for the two fetal globin mRNA species over a range of more than 400 nt, from positions 26 through 462. The distinct difference in translation efficiency and ability to compete for eIF-2 in translation exhibited by $^{G}\gamma$-globin and $^{A}\gamma$-globin mRNA must, therefore, be determined by sequence difference(s) located far from the AUG initiation codon.

The observed differences in translation efficiency between $^{G}\gamma$-, $^{A}\gamma$-, and β-globin mRNA, as well as in affinity for eIF-2 between γ- and β-globin mRNA, have been confirmed with different mRNA preparations, each prepared from pooled samples of human cord blood. Hence, these differences cannot be due to genetic polymorphisms, but must reflect intrinsic properties of these globin mRNA species.

The translation behavior of $^{G}\gamma$-, $^{A}\gamma$-, and β-globin mRNA, observed in conditions of competition pressure *in vitro*, fits with the developmental changes in globin synthetic ratio observed *in vivo*, both in terms of the γ to β switch and in terms of the $^{G}\gamma$ to $^{A}\gamma$ switch, as reviewed in the Introduction.

Translation of the more strongly competing β-globin mRNA template increases *in vivo* at the expense of translation of the more weakly competing γ-globin mRNA templates. Synthesis of $^{G}\gamma$-globin declines before that of $^{A}\gamma$-globin, leading to a developmental change in $^{G}\gamma/^{A}\gamma$-globin ratio. Our findings suggest that in addition to transcriptional control, differential translation efficiency of mRNA species and affinity for eIF-2 play an important role in regulating fetal and adult human globin synthesis.

In this study, the translation yield of individual globin mRNA species was determined in translation reactions containing increasing concentrations of an mRNA mixture of constant composition. The increasing translation competition pressure thus generated allowed us to resolve even small differences in competing ability between individual mRNA species. This approach was used previously to show that β-globin mRNA competes more effectively than α-globin mRNA in translation (Di Segni *et al.*, 1979) and that liver mRNAs encoding hemopexin, ferritin and albumin compete in ascending order (Kaempfer and Konijn, 1983). The difference in competing ability between $^{G}\gamma$-globin and $^{A}\gamma$-globin mRNA, resolved here by analysis of hybrid-selected mRNA, is smaller than that between α- and β-globin mRNA, and smaller yet than that between the α and γ, or γ and β species (Figure 1).

Results of Figure 4 show that the affinities of β- and total γ-globin mRNA for eIF-2 differ considerably. Since translation efficiency is generated by repeated cycles of initiation, while binding of mRNA to eIF-2 is a one-step process, it is difficult to compare the extent of the difference in these two properties directly. Yet, the fact that the globin mRNA species compete in translation in a manner that can be relieved by eIF-2, and the overall correlation observed between translation efficiency and affinity for eIF-2, emphasize the role of eIF-2 in determining translation efficiency of these mRNA species.

At physiological salt concentrations, ternary complex formation between eIF-2, GTP and Met-tRNA$_f$ is essentially irreversible, while mRNA/eIF-2 complexes are in rapid equilibrium (Kaempfer *et al.*, 1978b; Rosen *et al.*, 1981). Hence, free eIF-2 molecules will tend to form stable ternary complexes with GTP and Met-tRNA$_f$ that bind to the 40S ribosomal subunit, an obligatory step before binding of mRNA can occur. It is at this stage that competition between mRNA species for eIF-2 may take place (Di Segni *et al.*, 1979; Rosen *et al.*, 1982). eIF-2 possesses a high-affinity binding site for ATP (Gonsky *et al.*, 1990). Binding of ATP to eIF-2 causes Met-tRNA$_f$ to dissociate from eIF-2 and enhances the binding of mRNA to this factor (Gonsky *et al.*, 1990). During initiation of translation, binding of ATP may thus facilitate the conversion of eIF-2 from a Met-tRNA$_f$-binding protein into an mRNA-binding protein (Gonsky *et al.*, 1990).

The concept that eIF-2 guides the ribosome to the translation initiation site in mRNA is supported by several, independent lines of evidence (see Introduction). The distinct ability of $^G\gamma$- and $^A\gamma$-globin mRNA to compete for eIF-2 in translation, demonstrated here, shows that regions located far from the AUG initiation codon context contribute to the structure in mRNA recognized by eIF-2.

Acknowledgements

We thank Drs. B.G. Forget, T. Maniatis and R. Spritz for plasmids and the Departments of Obstetrics at Shaarei Zedek and Hadassah University Hospitals for cord blood. Supported by grant 88-00267 from the US-Israel Binational Science Foundation.

References

Barrieux A, Rosenfeld MG (1977) Characterization of GTP-dependent Met-tRNA$_f$ binding protein. J Biol Chem 252: 392-398

Cigan AM, Pabich EK, Feng L, Donahue TF (1989) Yeast translation initiation suppressor sui2 encodes the α subunit of eukaryotic initiation factor 2 and shares sequence identity with the human α subunit. Proc Natl Acad Sci. USA 86:2784-2788

Collins FS, Metherall JE, Yamakawa M, Pan J, Weissman SM, Forget BG (1985) A point mutation in the $^A\gamma$-globin gene promoter in Greek hereditary persistence of fetal haemoglobin. Nature 313:325-326

Collins FS, Stoeckert CJ Jr, Serjeant G.R, Forget BG, Weissman SM (1984) $^G\gamma\beta^+$ Hereditary persistence of fetal hemoglobin: Cosmid cloning and identification of a specific mutation 5' to the $^G\gamma$ gene. Proc Natl Acad Sci USA 81:4894-4898

Comi P, Giglioni B, Ottolenghi S, Gianni A, Polli E, Barba P, Covelli A, Migliaccio G, Condorelli M, Peschle C (1980) Globin chain synthesis in single erythroid bursts from cord blood: studies on $\gamma \rightarrow \beta$ and $^G\gamma \rightarrow {}^A\gamma$ switches. Proc Natl Acad Sci USA 77:362-365

Dasso MC, Milburn SC, Hershey JWB, Jackson RJ (1990) Selection of the 5'-proximal translation initiation site is influenced by mRNA and eIF-2 concentrations. Eur J Biochem 187:361-371

Di Segni G, Kerem H, Cividalli G, Rachmilewitz EA, Kaempfer R (1978) Absence of functional β-globin messenger RNA in Kurdish Jews with β°-thalassemia. Israel J Med Sci 14:1116-1123

Di Segni G, Rosen H, Kaempfer R (1979) Competition between α- and β-globin messenger ribonucleic acids for eukaryotic initiation factor 2. Biochemistry 18: 2847-2854.

Donahue TF, Cigan AM, Pabich EK., Valavicius BC (1988) Mutations at a Zn(II) finger motif in the yeast eIF-2β gene alter ribosomal start-site selection during the scanning process. Cell 54:621-632

Efstratiadis A, Posakony JW, Maniatis T, Lawn RM, O'Connell C, Spritz RA, DeRiel JK, Forget BG, Weissman SM, Slightom JL, Blechl AE, Smithies O, Baralle FE, Shoulders CC, Proudfoot NJ (1980) The structure and evolution of the human β-globin gene family. Cell 21:653-668

Forget BG, Cavallesco C, DeRiel JK, Spritz RA, Choudary PV, Wilson JT, Wilson LB, Reddy VB, Weissman SM (1979) Structure of the human globin genes. In: Axel R, Maniatis T, Fox CF (eds) Eucaryotic Gene Regulation, 14th ICN-UCLA Symposium on Molecular and Cellular Biology. Academic Press, New York, pp. 367-381

Gelinas R, Endlich B, Pfeiffer C, Yagi M, Stamatoyannopoulos G (1985) G to A substitution in the distal CCAAT box of the $^{A}\gamma$-globin gene in Greek hereditary persistence of fetal haemoglobin. Nature 313:323-325

Gonsky R, Lebendiker MA, Harary R, Banai Y, Kaempfer R (1990) Binding of ATP to eukaryotic initiation factor 2: Differential modulation of mRNA-binding actitivity and GTP-dependent binding of methionyl-tRNA$_f^{Met}$. J Biol Chem 265: in press.

Huisman THJ, Schroeder WA, Efremov GD, Duma H, Meadenovski B, Hyman CB, Rachmilewitz EA, Bouver N, Miller A, Brodie A, Shelton JR, Appel G (1974) The present status of the heterogenity of fetal hemoglobin in β-thalassemia: An attempt to unify some observations in thalassemia and related conditions. Ann NY Acad Sci 232:107-124

Kaempfer R, Konijn M (1983) Translational competition by mRNA species encoding albumin, ferritin, haemopexin and globin. Eur J Biochem 131:545-550

Kaempfer R, Hollender R, Abrams WR, Israeli R (1978a) Specific-binding of messenger RNA and methionyl-tRNA$_f^{Met}$ by the same initiation factor for eukaryotic protein synthesis. Proc Natl Acad Sci USA 75:209-213

Kaempfer R, Rosen H, Israeli R (1978b) Translational control: Recognition of the methylated 5' end and an internal sequence in eukaryotic mRNA by the initiation factor that binds methionyl-tRNA$_f^{Met}$. Proc Natl Acad Sci USA 75:650-654

Kaempfer R, Hollender R, Soreq H, Nudel U (1979) Recognition of messenger RNA in eukaryotic protein synthesis: Equilibrium studies of the interaction between messenger RNA and the initiation factor that binds methionyl-tRNA$_f$. Eur J Biochem 94:591-600.

Kaempfer R, van Emmelo J, Fiers W (1981) Specific binding of eukaryotic initiation factor 2 to satellite tobacco necrosis virus RNA at a 5'-terminal sequence comprising the ribosome binding site. Proc Natl Acad Sci USA 78:1542-1546

Kozak M (1986) Point mutations define a sequence flanking the AUG initiator codon that modulates translation by eukaryotic ribosomes. Cell 44:283-292.

Lawn RM, Efstratiadis A, O'Connell C, Maniatis T (1980) The nucleotide sequence of the human β-globin gene. Cell 21: 647-651.

Lodish, H.F. (1976) Translational control of protein synthesis. Annu Rev Biochem 45:39-72.

Maniatis T, Fritsch EF, Lauer G, Lawn RM (1980) The molecular genetics of human hemoglobins. Ann Rev Genet 14:145-178

Perez-Bercoff R, Kaempfer R (1982) Genomic RNA of Mengovirus: Recognition of common features by ribosomes and eukaryotic initiation factor 2. J Virol 41:30-41

Peschle C, Migliaccio G, Covelli A, Lettieri F, Migliaccio R, Condorelli M, Comi P, Pozzoli ML, Giglioni B, Ottolenghi S, Cappellini MD, Polli E, Gianni AM

(1980) Hemoglobin synthesis in individual bursts from normal adult blood: All bursts and subcolonies synthesize $^{G}\gamma$- and $^{A}\gamma$-globin chains. Blood 56:218-226

Poon R, Kan YW, Boyer HW (1978) Sequence of the 3' noncoding and adjacent coding regions of human γ-globin mRNA. Nucleic Acids Res 5: 4625-4630

Rosen H, Kaempfer R (1979) Mutually exclusive binding of messenger RNA and initiator methionyl transfer RNA to eukaryotic initiator factor 2. Biochem Biophys Res Commun. 91:449-455

Rosen H, Knoller S, Kaempfer R (1981) Messenger RNA specificity in the inhibition of eukaryotic translation by double-stranded RNA. Biochemistry 20: 3011-3020

Rosen H, Di Segni G, Kaempfer R (1982) Translational control by messenger RNA competition for eukaryotic initiation factor 2. J Biol Chem 257: 946-952

Schroeder WA, Huisman T, Shelton J, Shelton JB, Kleihauer E, Dozy A, Robberson B (1968) Evidence for multiple structural genes for the γ chain of human fetal hemoglobin. Proc Natl Acad Sci USA 60: 537-544.

Slightom JL, Blechl AE, Smithies O (1980) Human fetal $^{G}\gamma$- and $^{A}\gamma$-globin genes: Complete nucleotide sequences suggest that DNA can be exchanged between these duplicated genes. Cell 21:627-638.

Terasawa T, Ogawa M, Porter PN, Karam JD (1980) $^{G}\gamma$ and $^{A}\gamma$ globin-chain biosynthesis by adult and umbilical cord blood erythropoietic bursts and reticulocytes. Blood 56: 93-97.

Weatherall DJ, Clegg JB (1979) Recent developments in the molecular genetics of human hemoglobin. Cell, 16:467-479.

EFFECTS OF THE 5'-LEADER SEQUENCE OF TOBACCO MOSAIC VIRUS RNA, OR DERIVATIVES THEREOF, ON FOREIGN mRNA AND NATIVE VIRAL GENE EXPRESSION

T. Michael A. Wilson, Keith Saunders[1], Mandy J. Dowson-Day[2], David E. Sleat[3], Hans Trachsel[4] and Karl W. Mundry[5]
Center for Agricultural Molecular Biology,
Rutgers University,
New Brunswick,
NJ 08903-0231
U.S.A.

INTRODUCTION

The primary function of a viral genome is to replicate to produce progeny virions. To achieve this, the exquisitely compact genetic information must perform several functions very efficiently and it is not uncommon for one sequence to fulfil several unrelated tasks. One such pleiotropic sequence, the 5'-untranslated leader of tobacco mosaic virus (TMV) RNA, is the main subject of this article.

Located at the structurally concave end of the ribonucleoprotein particle and creating the structural core for 1.45 turns of the right-handed single-start viral helix, the so-called "omega" (Ω) sequence of TMV RNA must perform at least three functions: (a) interact weakly with the viral coat protein to initiate nucleocapsid disassembly, (b) recruit tobacco 40S ribosomal subunits to initiate viral gene expression, and (c) create at least part of the 3'-terminal promoter on the complementary (minus-strand) RNA for RNA-dependent RNA polymerase activity to begin synthesis of progeny plus-strand, viral-sense RNA molecules. Only the first two of these functions are relevant to the subject of this Workshop and have been shown to be interconnected, mechanistically [Wilson, 1984; 1988].

Twenty-five years ago, it was first reported [Mundry, 1965] that a unique, 70-residue-long, RNaseT1-resistant oligonucleotide (called Ω) existed within the 6395 nucleotide (nt) genome [Goelet *et al.* 1982] of the common (U1 or *vulgare*) strain of TMV. Successively more precise mapping techniques eventually located this G-less tract to the 5'-extremity of the viral genome [Mandeles, 1968; Mundry and Priess, 1971; Garfin and Mandeles, 1975; Richards *et al.* 1977; 1978]. Thus, long before the advent of cDNA clones of TMV RNA, the native, capped 5'-leader sequence of type strain TMV RNA was determined to be (bold = Ω):

m^{7}GpppG**UAUUUUUACAACAAUUACCAACAACAACAAACAACAAAC**
AACAUUACAAUUACUAUUUACAAUUACA<u>**AUG**</u>....

---

[1]Department of Virus Research, John Innes Institute, AFRC IPSR, Colney Lane, Norwich NR4 7UH, U.K., [2] Nitrogen Fixation Laboratory, AFRC IPSR, Sussex University, Brighton BN1 9RQ, U.K., [3] Department of Plant Pathology, Cornell University, Ithaca NY 14853 U.S.A., [4] Institut fur Biochemie und Molekularbiologie der Universitat Bern, Buhlstrasse 28, CH-3012 Bern, Switzerland, [5] Biologisches Institut, Universitat Stuttgart, Pfaffenwaldring 57, D-7000 Stuttgart 80, FRG.

NATO ASI Series, Vol. H 49
Post-Transcriptional Control of Gene Expression
Edited by J. E. G. McCarthy and M. F. Tuite

From the earliest RNA-fingerprinting data [Mundry and Priess, 1971; Garfin and Mandeles, 1975] discrepancies or variants were noted in this sequence, and more recently [Goelet *et al.* 1982] an apparent polymorphism at the 5'-end of U1 TMV RNA was shown to be due to contamination by the tomato strain [Meshi *et al.* 1983; Beck *et al.* 1986]. Sequence data also exist for the leaders of tomato strain (L or SPS), U2 and Dahlemense TMV [Kukla *et al.* 1979; Ohno *et al.* 1984]. Between the first transcribed G and the start codon for a large non-structural protein (126kDa; RNA replicase and/or capping enzyme [Dunigan and Zaitlin, 1990]) all these leaders lack G residues and differ in length by multiples of 3 nucleotides. Both these features have significant implications for the structure of native virus (discussed below).

Traditionally, TMV RNA was (and is) one of the most popular and easily obtained mRNA "standards" against which to compare the expression of other protein coding RNAs. During the development of *in vitro* translation systems TMV RNA was used routinely. Rarely does one find a message which directs higher incorporation of ^{35}S-, ^{14}C-, or ^{3}H-amino acids into polypeptide(s). In part, this trivially reflects the size of the primary translation product (126 kDa), but as we now understand, it also depends upon having a highly efficient, competitive mRNA leader.

Given the gene expression strategy of TMV, which involves production of two (or more) subgenomic mRNAs, each with its own, potentially regulatable leader sequence transcribed from internal promoters on full-length minus-sense RNA, it seems illogical for the virus to provide for such efficient expression of the first open-reading frame (ORF). This ORF encodes a large non-structural (hence presumably "catalytic") polypeptide, now shown to have biochemical properties compatible with the viral RNA-guanylyltransferase (capping) enzyme [Dunigan and Zaitlin, 1990] and probably also 7-methyltransferase activity, by analogy with Sindbis virus [Mi *et al.* 1989; Scheidel *et al.* 1989]. Although the latter has not yet been demonstrated, experiments are in progress. Nevertheless, during TMV infection extremely large amounts of the 126kDa polypeptide are made, no doubt as a consequence of the high efficiency of the 68-nt leader, or Ω. The 126kDa product accumulates in large subcellular structures (1 per cell) called viroplasms or X-bodies [Saito *et al.* 1987; Hills *et al.* 1987] which approximate in size to the cell nucleus, are visible in the light microsope [Iwanowski, 1903], and attach to the nuclear membrane [Wijdeveld *et al.* 1989]. Teleological arguments concerning the relative efficacies of different viral mRNAs (and hence features of their 5'-regulatory sequences) are therefore unfounded.

Stoichiometrically, each progeny plus-strand RNA of 6395 nts requires 2131 or 2132 coat protein (CP) subunits (17.6kDa) for complete encapsidation. The levels of CP expressed in systemically virus-infected plants are extremely high. Total CP accumulates up to 10% of the soluble leaf protein. In part, this may be due to transcriptional controls operating at the most 3'-proximal ORF [Lehto and Dawson, 1990; Lehto *et al.* 1990] however the short leader sequence on the CP subgenomic mRNA: **m^{7}GpppGUUUUAAAU<u>AUG</u>** may also enhance CP expression. Although this sequence has not been tested extensively, substantial translational enhancement of foreign reporter gene transcripts has been achieved with *cis*-active subgenomic mRNA leaders from other plant viruses [Jobling and Gehrke, 1987; Gallie *et al.* 1987b; Jobling *et al.* 1988]. However, in our experience, the genomic RNA leader from TMV (Ω or its derivative Ω', lacking the 3'-AUG) remains the most efficient and universally applicable enhancer of mRNA translation.

In competitive RNA translation experiments both the uncapped RNA of satellite tobacco necrosis virus (STNV) and the 5'-capped genomic RNA of potato virus X (PVX) will outcompete TMV RNA (Ω) [Herson *et al.* 1979; J.G. Atabekov, Univ. of Moscow, personal communication] in wheat germ extract (WG) or rabbit reticulocyte lysate (RRL), respectively. The 84-nt leader of PVX RNA lacks G in the first 42 residues and has the initiation codon context CAA**AUG**G (in TMV RNA the 126kDa context is ACA**AUG**G). To date, we have neither constructed nor tested the STNV or PVX leaders in chimaeric mRNAs.

STRUCTURAL STUDIES ON NATIVE TMV LEADER

Between 1976-78, it was shown that the 5'-end of TMV particles was intrinsically less stable to agents such as SDS, mild alkali, urea or DMSO, and that progressive virus disassembly proceeds in a polar fashion [reviewed in Wilson and Shaw, 1985]. More recently, native TMV particles or pH 8-pretreated TMV particles, not significantly uncoated, have been shown to disassemble cotranslationally both *in vivo* and *in vitro* in a wide variety of plant- or animal-based systems [Wilson, 1984; 1988; Shaw *et al.* 1986] and even in extracts of *Escherichia coli* [Wilson, 1986].

To comply with the rules of eukaryotic mRNA translation [Kozak, 1989], exposure of the 5' leader sequence at least to the (first) AUG codon for the 126kDa ORF would seem an essential pre-requisite for 40S subunit scanning. We have obtained direct evidence that all of Ω, and about 130-nts downstream are exposed in pH 8-"destabilized" virions.

One source of information was the binding of 10-15 residue-long oligodeoxy-ribonucleotides to treated virions in the presence of RNaseH. Some oligos targetted to sequences beyond the AUG of the 126kDa ORF completely inhibited both naked TMV RNA (control) and pH 8-treated virus expression [Mundry *et al.* manuscript submitted]. However other oligos failed to affect either template, even at molar ratios of 1000:1 - possibly for reasons described below.

For over 10 years it has been known that TMV RNA (and several other viral RNAs) can accommodate two 80S ribosome complexes on their extended 5' leader sequence when peptidyltransferase is inhibited by sparsomycin or anisomycin [Filipowicz and Haenni, 1979; Konarska *et al.* 1981; Tyc *et al.* 1984]. By incubating pH 7.5-treated TMV *particles* in WG or RRL in the presence of sparsomycin, then viewing and scoring 200 of the resulting complexes in the electron microscope, we detected 10% with one 80S bound and 10% as disomes (Fig. 1). While the origin of the former is unclear (RNase cleavage, shearing during EM grid preparation, or limited initiation events), the latter could only arise by virtue of the complete Ω sequence being exposed. Recent independent studies [Roenhorst *et al.* 1989] using uninhibited WG translation reactions programmed with pH 8.1-dialysed TMV particles reported complexes with up to 20 ribosomes per virion (they estimated one 80S per 140-nts unencapsidated), and that only 5% of virus particles participated in cotranslational disassembly. Thus the proportion of virions participating seems to vary, but falls in the 5-20% range estimated previously [Wilson, 1984].

Using uniformly ^{32}P-labelled TMV of the U1 or U2 strain, treated with 1% (w/v) SDS at 37°C and sampled between 15 secs and 15 mins, exposure of Ω (70-nts and 55-nts, respectively) and its release by RNaseT1 was seen to be an almost immediate and synchronous event (Fig. 2A). Furthermore, after the rapid (immediate) exposure of Ω (and about another 100-nts) there was a lag phase of 2 mins before further RNase-sensitive cpm were released (Fig. 2B).

Recently, refined X-ray fibre diffraction data to 2.9Å resolution have confirmed that positioning a G-base in the 5'-proximal nucleotide binding site on each TMV CP

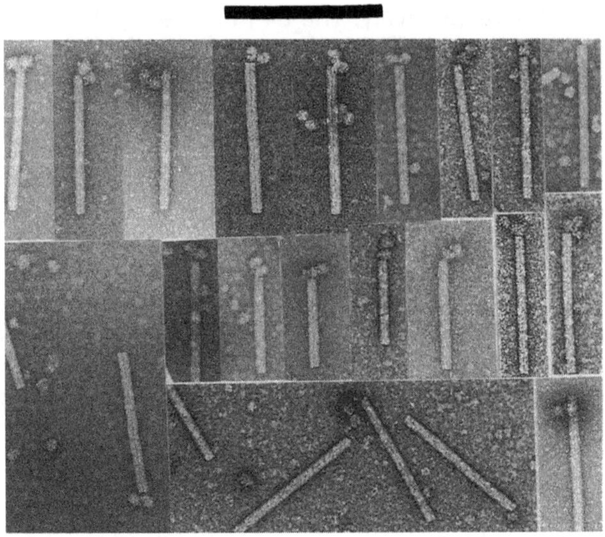

Fig.1. Selected initiation complexes arrested with sparsomycin from WG and RRL programmed with pH 7.5-treated TMV (*vulgare* strain) particles. Bar = 300nm.

subunit provides for additional stability through two "bonus" H-bonds between O-6 and Arg122, and between N-2 and Asp115 [Namba *et al.* 1989]. Hence, lacking G, Ω interacts less favourably with viral CP and can easily be fully exposed. Modelling the phased encapsidation of genomic RNA in trinucleotides from the known origin-of-assembly sequence where G occurs at every third residue [Turner and Butler, 1988] reveals that, in U1 strain TMV, the first transcribed G (i.e. that preceding Ω) resides in the 5'-position of the last CP subunit to be added to the growing virion in this direction. Hence the m^7Gppp motif, predicted to be unable to interact correctly with CP in the highly conserved trinucleotide binding site, would remain exposed. ^{32}pCp-Labelling studies and anti-cap antibody binding studies have been used to confirm this prediction [Wu *et al.* 1984], and thereby reduce virus infectivity by over 90%. A second effect of this likely-preferred RNA packaging phase is that the 126kDa ORF is read in a different frame to being encapsidated. Doubtless the conservation of 3n nucleotides between the first G and the AUG in four strains of TMV also has some structural significance in this respect.

There are some indirect indications that the upstream 80S-binding site in Ω, AUU residues 15-17 in TMV RNA (14-16 in Ω; [Tyc *et al.* 1984]), may serve as a cryptic start codon for a polypeptide related to the legitimate 126kDa species (N-f-Met-Ala) but with an NH_2-terminal extension of 18 amino acids (starting N-f-Met-Thr) [Hunter *et al.* 1976; Pelham, 1978]. However, immunological methods and direct Edman microsequencing have failed to show that, in the absence of sparsomycin, protein synthesis initiates at this site. Given the ease with which all of Ω becomes uncoated in virions (Figs.1,2A,2B), there seems little need to postulate use of a more 5'-proximal (cryptic) start codon to initiate cotranslational virus disassembly.

Although there are G-residues in the 5'-leader of a legume strain of TMV (Cc; Sunn hemp mosaic virus) which prevent appearance of an Ω-like fragment upon RNaseT1 digestion [Kukla *et al.* 1979] and uncoating of this strain proceeds 3'-5' in SDS [Beachy *et al.* 1976], treatment of Cc TMV particles in pH 7.2 buffer must also expose sufficient 5'-leader to recruit 40S ribosomal subunits for cotranslational disassembly [Wilson and Watkins, 1985].

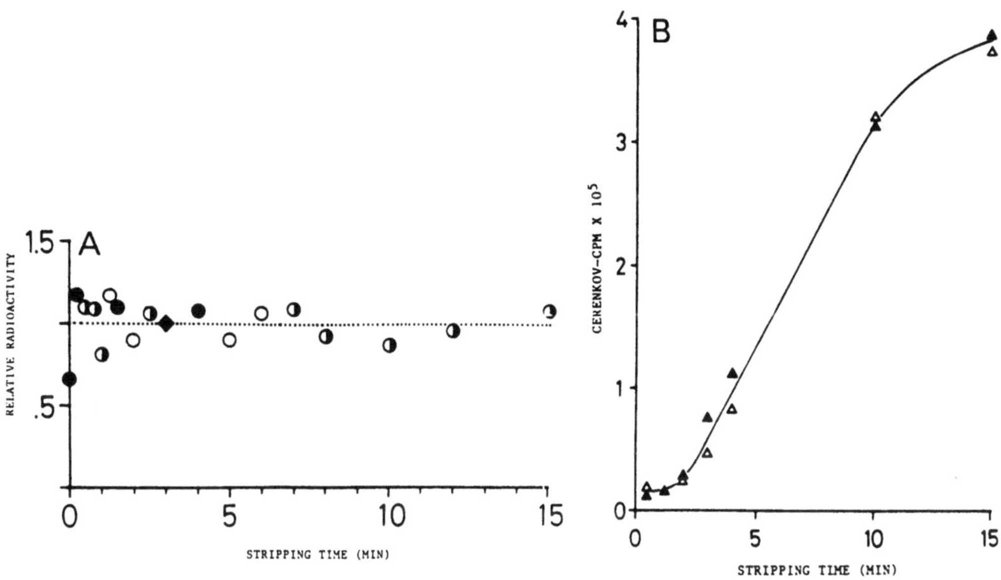

Fig. **2A**. Recovery of Ω as a function of time of exposure of ^{32}P-TMV U1 to 1% SDS. Data from 5 replicate experiments are included as averages from 5 (♦), 3 (●), 2 (◑) or single (o) estimates. The absolute amount of radioactive Ω recovered remained constant at about 2 x 10^3 cpm. Values are expressed relative to the cpm recovered at 3 mins (average of 5 replicate values). **2B**. Release of RNaseT1 (▲) or RNaseA (Δ) sensitive material from uniformly ^{32}P-TMV U1 by 1% SDS as a function of time of exposure. Assuming the plateau of 4 x 10^5 cpm corresponds to completely uncoated TMV RNA, then the meta-stable species up to 2 mins have exposed 160-nts.

When the first 250, 500 or 1000 nts of TMV (U1) RNA were analysed for predicted minimum free energy structures [Zuker and Stiegler, 1981], we were impressed by the consistently low folding index of Ω (Fig. 3.). Substituting several computer generated, "random" 68-nt leader sequences composed only of C,U and A residues produced highly folded conformations. Conversely, creating a C-to-G transversion of native Ω creates an exactly superimposable image of Fig. 3. It will be interesting to compare such sequences with a recently improved algorithm [Jaeger *et al.* 1989], however these consistently unstructured computer-analyses of Ω [Sleat *et al.* 1988] suggest that lack of G residues is of secondary importance to the base sequence itself (see below). In contrast, much of the downstream sequence is extensively folded which may account for the inefficient binding to (and RNaseH inactivation of) TMV RNA by some short synthetic oligodeoxyribonucleotides. In virions, there are no RNA:RNA interactions since each successive turn of the helix is separated by a layer of CP subunits. Thus the first ribosome to translocate has no template secondary structure to overcome. Instead, there must be some mechanism to dislodge CP subunits progressively. We still have no molecular information on this process.

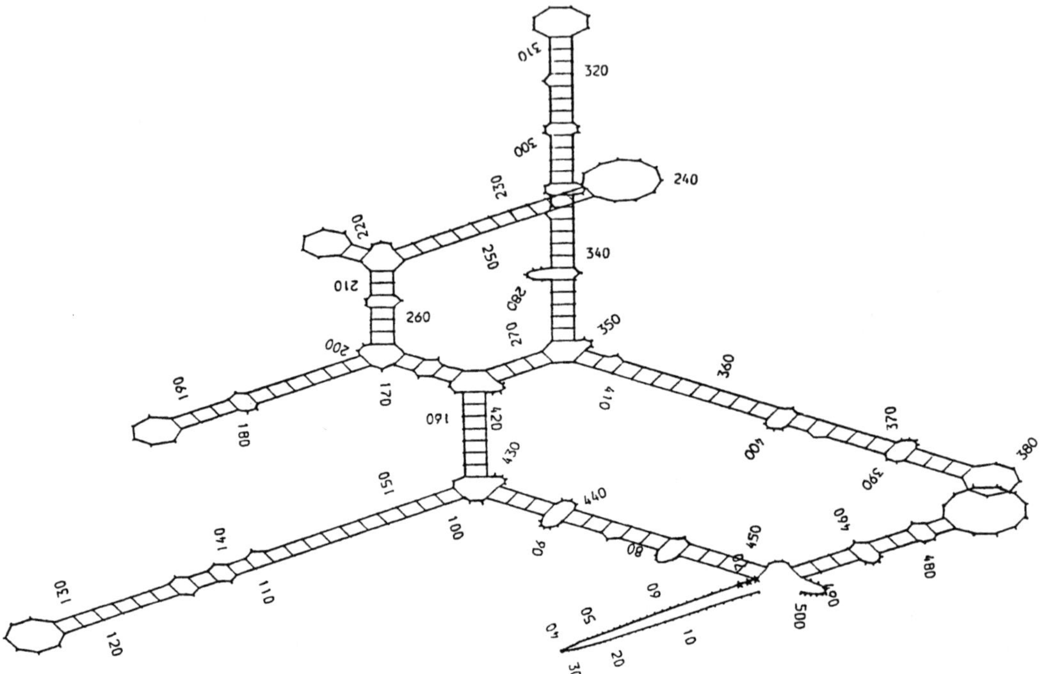

Fig.3. Computer-predicted minimum energy structure for nts 1-500 of native TMV RNA (U1 strain). Energy = -124.2 Kcal/mol. Asterisks mark the AUG codon for the 126kDa ORF.

TRANSLATIONAL ENHANCER EFFECT OF NATIVE Ω OR DERIVATIVES

Since 1987, numerous reports have appeared in which a conveniently tailored derivative of true Ω (called Ω') or deletions/base substitutions thereof, all with HindIII and SalI linker sequences, have been cloned next to various reporter gene cassettes to examine the effect on expression of the resulting chimaeric mRNA [Gallie *et al.* 1987a,b; Sleat *et al.* 1987;1988; Gallie *et al.* 1988a,b; 1989; Gallie and Kado, 1989; Jobling *et al.* 1988]. In every case some translational enhancement was noted compared with "natural" transcripts lacking the Ω'-derived sequences. However, in all these examples, and in analogous experiments with the 36-nt leader of alfalfa mosaic virus RNA4 [Jobling and Gehrke, 1987; Jobling *et al.* 1988], caution should be applied in interpreting "fold-stimulations" without defining and optimizing the translation conditions [Sleat *et al.* 1988; Dasso and Jackson, 1989] and using the complete native reporter gene leader as a baseline. Nevertheless, some useful valid comparisons can be made by varying other parameters.

Recently, mRNAs encoding chloramphenicol acetyltransferase (CAT) with or without the Ω' leader insertion [Gallie *et al.* 1987a] were translated in yeast extracts (*Saccharomyces cerevisiae*) dependent on exogenous eIF-4E or eIF-4A when preincubated at elevated temperatures or grown on glucose. From the results shown in Fig. 4 and Table1, it is clear that the presence of Ω' conferred eIF-4E-, but not eIF-4A-independent translation of CAT mRNA [Altmann *et al.* 1990].

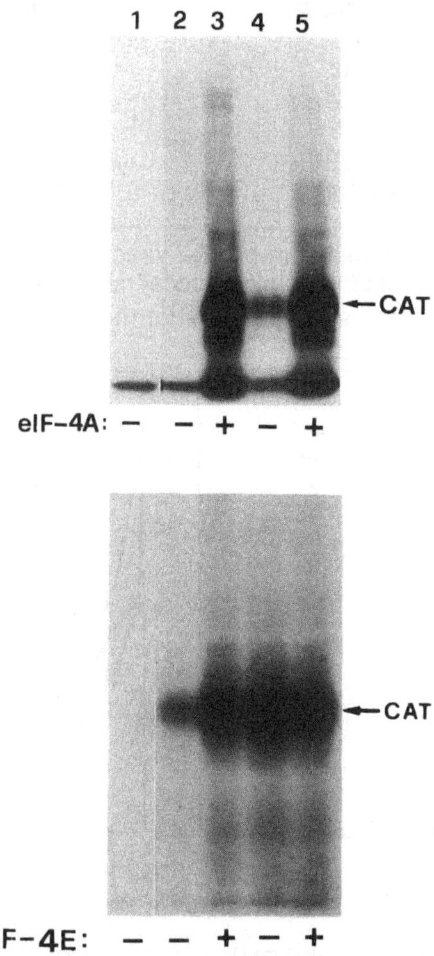

Fig.4. Translation of 5'-capped CAT mRNA (50ng, lanes 2 & 3) or 5'-capped-Ω'-CAT mRNA (50ng, lanes 4 & 5) in eIF-4A (top) or eIF-4E (bottom) dependent yeast extracts supplemented with (top panel) 1 µg eIF-4A or (lower panel) 50ng eIF-4E where indicated by (+) signs below. Control incubations (-mRNA) are shown in lanes 1. Arrows indicate ^{35}S-Met-labelled CAT.

Earlier observations by ourselves and others [Richards *et al.* 1978; K.E. Richards and G.P. Lomonossoff, personal communications] that recovery of Ω released by RNaseT1 was usually low and rarely better than 50% of the theoretical yield concur with computer predictions of its unfolded, RNase-hypersensitive configuration.

Thus we found that Ω' did not confer enhanced stability to chimaeric mRNAs encoding neomycin phosphotransferase (NPTII) or CAT (Fig. 5 modified from [Sleat *et al.* 1988]).

Table 1: mRNA translation in eIF-4E-dependent extracts of *Saccharomyces cerevisiae*

Template	- eIF-4E	+eIF-4E	Fold
No mRNA	4644	6088	1.31
Total yeast mRNA	13646	92780	6.80
m7G-CAT	11902	40472	3.40
m7G-Ω'-CAT	46834	44700	0.95

Data presented as TCA-insoluble cpm per 3μl yeast extract after 45 min at 23°C. Yeast extract was pre-digested with micrococcal nuclease and preincubated for 10 min at 37°C, the non-permissive temperature for the *ts* eIF-4E mutation.

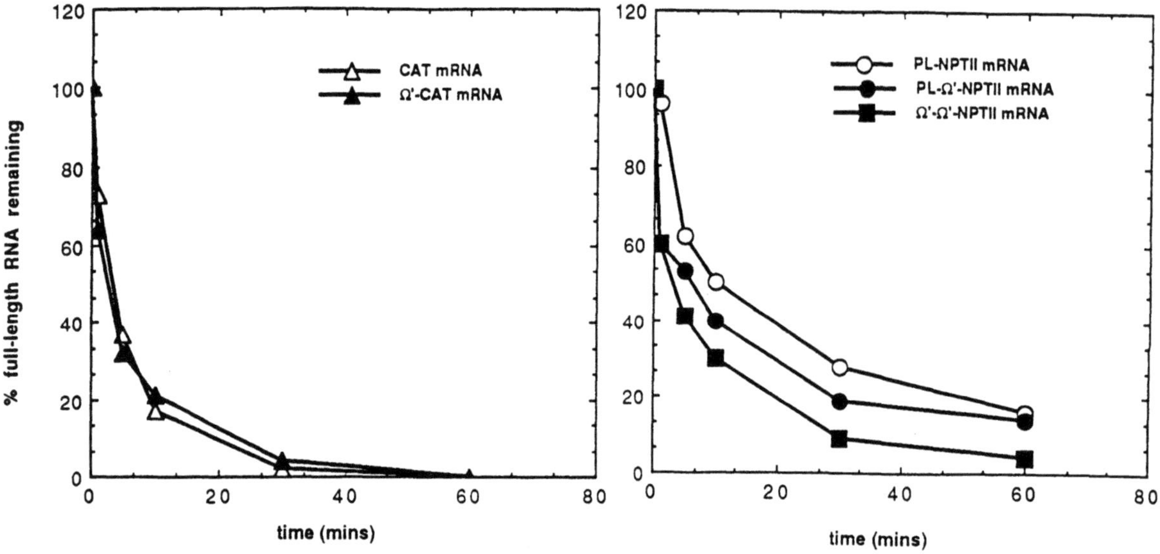

Fig. 5. Effect of Ω' on CAT or NPTII mRNA half-life in WG extract. The amount of each gel-purified full-length SP6 transcript remaining after various times of incubation is expressed as a percentage of that recovered at t=0. Symbols are self-explanatory (PL = pSPT19 42bp polylinker region).

In a naive attempt to understand the mechanism of action of Ω', we created several gross deletions and base substitutions focussed on the sparsomycin-dependent AUU-disome formation site and the central C,A-rich region [Gallie *et al.* 1988a; Δ3 was one residue less than reported, i.e. ...UUACCΔCAUUA...]. As can be seen from Table 2, no clear pattern of enhancement emerged to identify a particular sub-sequence within Ω' as the dominant, universal enhancer motif.

Table 2: Effect of deletions in Ω' on CAT or GUS mRNA expression in animal, plant or bacterial systems. Values expressed relative to Ω'.

Leader on mRNA	CAT in oocytes	Native GUS in tobacco protoplasts	Kozak* GUS in tobacco protoplasts	Kozak GUS in E. coli (trp promoter)
-Ω'	0.10	0.05	0.02	0.12
Δ1	0.49	1.00	0.92	0.96
Δ2	0.29	0.91	1.08	8.63
Δ3	0.53	0.58	0.52	8.91
Δ4	0.42	0.98	0.84	0.95
Δ5	0.54	0.98	1.00	1.31
A_{14} to C_{14}	0.45	1.25	1.12	6.22
(C,A) to (U)	0.99	0.02	0.13	0.57
+Ω'	**1.00**	**1.00**	**1.00**	**1.00**

* Mutagenized to create ...ACCAUGG... context [Jefferson, 1987] from native GUS ...CUUAUGU...

Extensive experiments had implicated a simple lack of secondary structure in Ω or Ω' as the primary cause of its enhancing capacity, at least in eukaryotic expression systems [Sleat *et al.* 1988]. To test this further, and to continue to seek for a sub-sequence in Ω which may be important, we designed and constructed artificial leaders (i) composed only of C,U and A residues, but extensively folded on computer analysis; or (ii) containing highly reiterated copies of short repetitive sequence motifs found in native Ω. Structural data concerning these can be found in Tables 3 & 4.

Table 3: Information on T7-promoter plasmid GUS mRNA leader sequences.

Description	Short Form	Plasmid Name (+GUS gene)	Frequency # in native Ω	Length of 5'-ss RNA§ / total leader
Highly Repetitive				
$(CAA)_n$	HR1	pJII1031	31/66*	61/68
$(CA)_n$	HR2	pJII1041	27/67	63/68
$(ACAAUUAC)_n$	HR3	pJII1061	3/61	24/51
$(AUC)_n$	HR4	pJII1051	1/66	55/68
Partially Repetitive				
Folded $(C,U,A)_n$	PR1	pJII1021	-	1/68
"Native" Leaders				
TMV RNA (Ω)	NL1	pJII1001	1/1	68/68
C/G transversion of (Ω)	NL2	pJII1091	-	68/68
AlMV RNA4	NL3	pJII1081	-	34/36

\# Frequency includes all circular permutations of the sequence shown.
* Theoretical maximum of 67 dimers, 66 trimers or 61 octamers in a 68-nt leader sequence.
§ Length of unfolded 5'-tail as a fraction of total leader length.

The effects of these leaders on *in vitro* translation of GUS mRNA were tested extensively [Saunders and Wilson, manuscript submitted]. In each case, the precise leader-GUS ORF fusion was through an NcoI restriction site and the first transcribed base was always G (or m^7GpppG). The actual leader sequences are shown in Table 4, together with those of some relevant earlier constructs [Gallie *et al.* 1987a; Sleat *et al.* 1987]. Typical translation results are shown in Table 5, expressed relative to transcripts from pJII140 for simplicity.

Table 4: T7-promoter plasmid leader sequences (NcoI site underlined)

NL1	GUAUUUUUACAACAAUUACCAACAACAACAAACAACAAACAACAUU ACAAUUACUAUUUACAAUUACC**AUG**G
NL2	GUAUUUUUAGAAGAAUUAGGAAGAAGAAGAAAGAAGAAAGAAGAUU AGAAUUAGUAUUUAGAAUUACC**AUG**G
NL3	GUUUUUAUUUUUAAUUUUCUUUCAAAUACUUCCACC**AUG**G
HR1	GCAACAACAACAACAACAACAACAACAACAACAACAACAACAACAAC AACAACAACAACAACAACACC**AUG**G
HR2	GCACACACACACACACACACACACACACACACACACACACACACAC ACACACACACACACACACACACCC**AUG**G
HR3	GACAAUUACACAAUUACACAAUUACACAAUUACACAAUUACACAAU UACCC**AUG**G
HR4	GAUCAUCAUCAUCAUCAUCAUCAUCAUCAUCAUCAUCAUCAUCAU CAUCAUCAUCAUCAUCAUCAUCC**AUG**G
PR1	GCACUCCAUCUUUCCUCCACCUCCCCCCACAUUUUCUACAAUCAC CACUUAAUAUCUUUCUACCCACC**AUG**G

pJII139 (SP6-HindIII-SalI-GUS)
GAAUACAAGCUUGGGCUGCAGGUCGACC**AUG**G

pJII140 (SP6-HindIII-Ω'-SalI-GUS)
GAAUACAAGCUUUAUUUUUACAACAAUUACCAACAACAACAAAC
AACAAACAACAUUACAAUUACUAUUUACAAUUACAGUCGACC**AUG**G

pJII102 (SP6-HindIII-Ω'-SalI-CAT)*
GAAUACAAGCUUUAUUUUUACAACAAUUACCAACAACAACAAAC
AACAAACAACAUUACAAUUACUAUUUACAAUUACAGUCGACGAG
AUUUUCAGGAGCUAAGGAAGCUAAA**AUG**G

* Used as internal reference mRNA for *in vitro* translation dose-response experiments with the various T7-leader-GUS constructs of Table 3 (data not included).

Table 5: Stimulation* of uncapped GUS mRNA expression *in vitro* by various leaders compared to Ω' (= 1.00)

	MDL	WG	E. coli
HR1	1.29	1.61	2.00
HR2	0.50	0.43	0.08
HR3	0.57	0.94	0.18
HR4	0.73	0.77	0.19
PR1	0.38	0.28	0.13
pJII139 (-Ω')	0.36	0.36	0.07
Absolute cpm/μl (x 10^{-4})			
pJII140 (Ω'-GUS mRNA)	3.7	1.1	11.7

* Expressed as: TCA-insoluble cpm incorporated by leader-GUS mRNA/ cpm with pJII140 (Ω') mRNA.

From these experiments we conclude that, in the absence of any other beneficial feature, a leader of low secondary structure (here composed of repeat motifs from native Ω) enhances GUS mRNA expression in eukaryotic cell-free systems. Work with intact cell systems is underway. More significantly, HR3, a 6-fold reiteration of the disome-forming sequence reputed to be important [Gallie *et al.* 1988a; Gallie and Kado, 1989] is less active than $(CAA)_n$ in both 80S and 70S ribosome systems. Only the $(CAA)_n$ leader worked well in *E. coli* probably because of its relatively high purine content [Shine and Dalgarno, 1974]. Thus the central C,A-rich portion of Ω may account for its efficacy in 70S ribosome systems [Wilson, 1986; Gallie *et al.* 1987a,b; Sleat *et al.* 1988; Gallie and Kado, 1989] rather than speculative homologies to T7 gene10 ribosome-binding motifs (-UUAACUUUA-; [Olins *et al.* 1988]).

Producing workable quantities of mRNAs with the native TMV (Ω) or AlMV RNA4 leaders from a T7 promoter proved difficult, as might have been predicted from the inefficiency of incorporation of U-residues in the first 8 nucleotides transcribed (see NLs1-3 in Table 4; [Martin *et al.* 1988]).

Recently, using a CaMV 35S promoter plasmid vector based on pBI120 [Bevan, 1985] and pCaP35J [Yamaya *et al.* 1988] which provided a StuI site at the +1 position and synthetic oligonucleotides to produce (by blunt-end ligation into the StuI site) a native Ω leader but with a 3'-NcoI site for precise insertion of a reporter gene (GUS), we have obtained some encouraging data regarding the effect of "native" Ω on mRNA expression *in vivo*. By mutagenesis, one of us (M.J.D-D.) also created pTZ18-based vectors with +4 or +7 nucleotides of the natural CabbS 35S RNA (manuscript in preparation). The leader structures of these plasmids are given in Fig.6., and their relative levels of GUS gene expression are summarized in Table 6. Earlier studies [D.E. Sleat, Ph.D. thesis, University of Liverpool 1987] in protoplasts electroporated with supercoiled or linearized pUC-based CaMV 35S promoter-CAT or Ω'-CAT plasmids also revealed about a 2-fold stimulation due to the TMV leader. Very recently, maize transformation using an anthocyanin regulatory gene (R) as reporter was successful utilizing the Ω'-sequence from pJII101 in the gene construction [Ludwig *et al.* 1990].

```
-7      +1                                     +40
GGAGAGGACACGCTGAAATCACCAGTCTCTCTCTACAAATCTATCTC        w.t. CabbS
       +1
GGAGAGAACACGGGGGATCTCGACCATG                           pBI221.2
       +1
GGAGAGGCCTGACCTGCAG                                    pCaP35J
       Stu I

   +1
    GTATTTTTACAACAATTACCAACAACAACAAACAACAAACAACATTACAATTACTATT
    TACAATTACCATGG                                            Ω*
             NcoI
   +1
    ACACΩ*                                                 +4 Ω*
    ACACGCTΩ*                                              +7 Ω*
```

Fig.6. CaMV 35S promoter transcriptional fusions in binary vectors relevant to data shown in Table 6. Ω* is identical to NL1 of Table 4, being native Ω with an A-to-C transversion to create the NcoI cloning site.

Table 6: Transient expression of GUS activity in tobacco mesophyll protoplasts electroporated with various linearized, CaMV 35S promoter-leader-GUS pTZ18-derived plasmid constructs.

	Experiment #		
	1	2	3
pBI120 (rbcS-GUS)	≤0.07	-	-
pBI221.2 (~ pRAJ275)	1.00	1.00	1.00
pMJD36 Ω*-GUS	2.88	1.44	3.49
pMJD50 +4-Ω*-GUS	4.67	4.83	4.56
pMJD70 +7-Ω*-GUS	5.28	3.51	4.50

All values are expressed relative to pBI221.2 [Jefferson, 1987].
Both pBI vectors produce mRNAs with +5 nucleotides of the CaMV 35S RNA.
Ω* = the untranslated leader sequence of TMV RNA (U1 strain) including the native 5'-G residue, but converting the final A to C, to create an NcoI cloning site [...CCAUGG...] in place of the native TMV sequence [...CAAUGG...].

The utility of *cis*-active leader sequences from efficiently expressed viral genes is becoming widely accepted. Ω and its derivatives, as well as the AlMV RNA4 leader sequence, and the "ribosome landing pad" from encephalomyocarditis virus [Parks *et al.* 1986] have been tested by numerous basic research laboratories and commercial companies for all manner of applications. It is to be hoped that this trend continues and that the feedback continues to be positive.

ACKNOWLEDGEMENTS

We thank Kitty Plaskitt for the electron micrograph shown in Fig.1. Work on Ω, Ω' and its derivatives was supported in part by Diatech Ltd. London. Financial support was also provided by the Deutsche Forschungsgemeinschaft to K.-W.M., the Nuffield Foundation (Grant SCI/168/2097/G) to T.M.A.W., and by the Swiss National Foundation (Grant 31-25565.88) to H.T. Ω and its derivatives are the subject of a patent application (British patent 86 13481).

REFERENCES

Altmann M, Blum S, Wilson TMA, Trachsel H (to be published) The 5'-leader sequence of TMV RNA mediates initiation factor 4E-independent, but not initiation factor 4A-independent translation in yeast extracts. Gene

Beachy RN, Zaitlin M, Bruening G, Israel HW (1976) A genetic map for the cowpea strain of TMV. Virology 73:498-507

Beck DL, Knorr DA, Dawson WO (1986) Sequence variations among cDNA clones of TMV (Abstract). J Cellular Biochem 10D:281

Bevan MW, Mason SE, Goelet P (1985) Expression of tobacco mosaic virus coat

protein by a cauliflower mosaic virus promoter in plants. EMBO J 4:1921-1926
Dasso MC, Jackson RJ (1989) On the fidelity of mRNA translation in the nuclease-treated rabbit reticulocyte lysate system. Nucleic Acids Res 17:3129-3144
Dunigan DD, Zaitlin M (to be published) Capping of tobacco mosaic virus RNA: analysis of viral-coded guanylyltransferase-like activity. J Biol Chem
Filipowicz W, Haenni A-L (1979) Binding of ribosomes to 5'-terminal leader sequences of eukaryotic messenger RNAs. Proc Natl Acad Sci USA 76:3111-3115
Garfin DE, Mandeles S (1975) Sequences of oligonucleotides prepared from tobacco mosaic virus ribonucleic acid. Virology 64:388-399
Gallie DR, Sleat DE, Watts JW, Turner PC, Wilson TMA (1987a) The 5'-leader sequence of tobacco mosaic virus RNA enhances expression of foreign gene transcripts *in vitro* and *in vivo*. Nucleic Acids Res 15:3257-3273
Gallie DR, Sleat DE, Watts JW, Turner PC, Wilson TMA (1987b) A comparison of eukaryotic viral 5'-leader sequences as enhancers of mRNA expression *in vivo*. Nucleic Acids Res 15:8693-8711
Gallie DR, Sleat DE, Watts JW, Turner PC, Wilson TMA (1988a) Mutational analysis of the tobacco mosaic virus 5'-leader for altered ability to enhance translation. Nucleic Acids Res 16:883-893
Gallie DR, Walbot V, Hershey JWB (1988b) The ribosomal fraction mediates the translational enhancement associated with the 5'-leader of tobacco mosaic virus. Nucleic Acids Res 16:8675-8694
Gallie DR, Kado, CI (1989) A translational enhancer derived from tobacco mosaic virus is functionally equivalent to a Shine-Dalgarno sequence. Proc Natl Acad Sci USA 86:129-132
Gallie DR, Lucas WJ, Walbot V (1989) Visualizing mRNA expression in plant protoplasts: factors influencing efficient mRNA uptake and translation. The Plant Cell 1:301-311
Goelet P, Lomonossoff GP, Butler PJG, Akam ME, Gait MJ, Karn J (1982) Nucleotide sequence of tobacco mosaic virus RNA. Proc Natl Acad Sci USA 79:5818-5822
Herson D, Schmidt A, Seal S, Marcus A, van Vloten-Doting L (1979) Competitive mRNA translation in an *in vitro* system from wheat germ. J Biol Chem 254:8245-8249
Hills GJ, Plaskitt KA, Young ND, Dunigan DD, Watts JW, Wilson TMA, Zaitlin M (1987) Immunogold localization of the intracellular sites of structural and nonstructural tobacco mosaic virus proteins. Virology 161:488-496
Hunter TR, Hunt T, Knowland J, Zimmern D (1976) Messenger RNA for the coat protein of tobacco mosaic virus. Nature (Lond) 260:759-764
Iwanowski D (1903) Uber die Mosaikkrankheit der Tabakspflanzen. Pflanzenkr. 13:1-41
Jaeger JA, Turner DH, Zuker M (1989) Improved predictions of secondary structures for RNA. Proc Natl Acad Sci USA 86:7706-7710
Jefferson RA (1987) Assaying chimeric genes in plants: the GUS gene fusion system. Plant Mol Biol Reporter 5:387-405
Jobling SA, Gehrke L (1987) Enhanced translation of chimaeric messenger RNAs containing a plant viral untranslated leader sequence. Nature (Lond) 325:622-625
Jobling SA, Cuthbert CM, Rogers SG, Fraley RT, Gehrke L (1988) *In vitro* transcriptional and translational efficiency of chimeric SP6 messenger RNAs devoid of 5'-vector nucleotides. Nucleic Acids Res 16:4483-4498
Konarska M, Filipowicz W, Domdey H, Gross HJ (1981) Binding of ribosomes to

linear and circular forms of the 5'-terminal leader fragment of tobacco-mosaic-virus RNA. Eur J Biochem 114:221-227

Kozak M (1989) The scanning model for translation: an update. J Cell Biol 108:229-241

Kukla BA, Guilley HA, Jonard GX, Richards KE, Mundry K-W (1979) Characterization of long guanosine-free RNA sequences from the Dahlemense and U2 strains of tobacco mosaic virus. Eur J Biochem 98:61-66

Lehto K, Dawson WO (1990) Changing the start codon context of the 30K gene of tobacco mosaic virus from "weak" to "strong" does not increase expression. Virology 174:169-176

Lehto K, Grantham GL, Dawson WO (1990) Insertion of sequences containing the coat protein subgenomic RNA promoter and leader in front of the tobacco mosaic virus 30K ORF delays its expression and causes defective cell-to-cell movement. Virology 174: 145-157

Ludwig SR, Bowen B, Beach L, Wessler SR (1990) A regulatory gene as a novel visible marker for maize transformation. Science 247:449-450

Mandeles S (1968) Location of unique sequences in tobacco mosaic virus ribonucleic acid. J Biol Chem 243:3671-3674

Meshi T, Ishikawa M, Takamatsu N, Ohno T, Okada Y (1983) The 5'-terminal sequence of TMV RNA. Question on he polymorphism found in *vulgare* strain. FEBS Lett 162:282-285

Mi S, Durbin R, Huang HV, Rice CM, Stollar V (1989) Association of the Sindbis virus RNA methyltransferase activity with the nonstructural protein nsP1. Virology 170:385-391

Mundry K-W (1965) A model of the coat protein cistron of tobacco mosaic virus and its biochemical investigation: the model, the experimental approach, and the isolation of a long oligonucleotide from TMV-RNA. Z Vererbungsl 97:281-296

Mundry K-W, Priess H (1971) Structural elements of viral ribonucleic acid and their variation II ^{32}P-oligonucleotide maps of large G-lacking segments of RNA of tobacco mosaic virus wild strains. Virology 46:86-97

Mundry K-W, Watkins PAC, Ashfield T, Fernandez A-G, Plaskitt KA, Eisele-Walter S, Wilson TMA (to be published) Complete uncoating of the 5'-leader sequence of tobacco mosaic virus RNA occurs rapidly and is required to initiate cotranslational virus disassembly *in vitro*. J Gen Virol

Namba K, Pattanayek R, Stubbs G (1989) Visualization of protein-nucleic acid interactions in a virus: refined structure of intact tobacco mosaic virus at 2.9Å resolution by X-ray fiber diffraction. J Mol Biol 208:307-325

Ohno T, Aoyagi M, Yamanashi Y, Saito H, Ikawa S, Meshi T, Okada Y (1984) Nucleotide sequence of the tobacco mosaic virus (tomato strain) genome and comparison with the common strain genome. J Biochem 96:1915-1923

Olins PO, Devine CS, Rangwala, SH, Kavka KS (1988) The T7 phage gene 10 leader RNA, a ribosome-binding site that dramatically enhances the expression of foreign genes in *Escherichia coli*. Gene 73:227-235

Parks GD, Duke GM, Palmenberg AC (1986) Encephalomyocarditis virus 3C protease: efficient cell-free expression from clones which link viral 5' noncoding sequences to the P3 region. J Virol 60:376-384

Pelham HRB (1978) Leaky UAG termination codon in tobacco mosaic virus RNA. Nature (Lond) 272:469-471

Richards K, Guilley H, Jonard G, Keith G (1977) Leader sequence of 71 nucleotides devoid of G in tobacco mosaic virus RNA. Nature (Lond) 267:548-550

Richards K, Guilley H, Jonard G, Hirth L (1978) Nucleotide sequence at the 5'

extremity of tobacco-mosaic-virus RNA 1. The noncoding region (nucleotides 1-68). Eur J Biochem 84:513-519
Roenhorst JW, Verduin BJM, Goldbach RW (1989) Virus-ribosome complexes from cell-free translation systems supplemented with cowpea chlorotic mottle virus particles. Virology 168:138-146
Saito T, Hosakawa D, Meshi T, Okada Y (1987) Immunocytochemical localization of the 130k and 180k (putative replicase components) of tobacco mosaic virus. Virology 160:477-481
Saunders K, Wilson TMA (to be published) Functional analysis of synthetic mRNA leaders derived from the 5'-noncoding sequence of TMV RNA. Nucleic Acids Res
Scheidel LM, Durbin RK, Stollar V (1989) SV_{LM21}, a Sindbis virus mutant resistant to methionine deprivation, encodes an altered methyltransferase. Virology 173:408-414
Shaw JG, Plaskitt KA, Wilson TMA (1986) Evidence that tobacco mosaic virus particles disassemble cotranslationally *in vivo*. Virology 148:326-336
Shine J, Dalgarno L (1974) The 3'-terminal sequence of *Escherichia coli* 16S ribosomal RNA: complementarity to nonsense triplets and ribosome binding sites. Proc Natl Acad Sci USA 71:1342-1346
Sleat DE, Gallie DR, Jefferson RA, Bevan MW, Turner PC, Wilson TMA (1987) Characterisation of the 5'-leader sequence of tobacco mosaic virus RNA as a general enhancer of translation *in vitro*. Gene 60:217-225
Sleat DE, Hull R, Turner PC, Wilson TMA (1988) Studies on the mechanism of translational enhancement by the 5'-leader sequence of tobacco mosaic virus RNA. Eur J Biochem 175:75-86
Turner DR, Joyce LE, Butler PJG (1988) The tobacco mosaic virus assembly origin RNA. Functional characteristics defined by directed mutagenesis. J Mol Biol 203:5312-547
Tyc K, Konarska M, Gross HJ, Filipowicz W (1984) Multiple ribosome binding to the 5'-terminal leader sequence of tobacco mosaic virus RNA. Assembly of an 80S ribosome-mRNA complex at the AUU codon. Eur J Biochem 140:503-511
Wijdeveld MMG, Goldbach RW, Verduin BJM, van Loon LC (1989) Association of viral 126-kDa protein in X-bodies with nuclei in mosaic-diseased tobacco leaves. Arch Virol 104:225-239
Wilson TMA (1984) Cotranslational disassembly of tobacco mosaic virus *in vitro*. Virology 137:255-265
Wilson TMA (1986) Expression of the large 5'-proximal cistron of tobacco mosaic virus by 70S ribosomes during cotranslational disassembly in a prokaryotic cell-free system. Virology 152:277-279
Wilson TMA (1988) Structural interactions between plant RNA viruses and cells. In: Miflin BJ (ed) Oxford Surveys of Plant Molecular and Cell Biology. Oxford University Press, Oxford, 5:89-144
Wilson TMA, Shaw JG (1985) Does TMV uncoat cotranslationally *in vivo*? Trends in Biochem Sci 10:57-60
Wilson TMA, Watkins PAC (1985) Cotranslational disassembly of a cowpea strain (Cc) of TMV: evidence that viral RNA-protein interactions at the assembly origin block ribosome translocation *in vitro*. Virology 145:346-349
Wu A-Z, Dai R-M, Shen X-R, Sun Y-K (1984) The location and function of the 5'-cap structure of TMV-RNA in the virion. Acta Biochim et Biophys Sinica 16:501-507
Yamaya J, Yoshioka M, Meshi T, Okada Y, Ohno T (1988) Expression of tobacco mosaic virus RNA in transgenic plants. Mol Gen Genet 211:520-525
Zuker M, Stiegler P (1981) Optimal computer folding of large RNA sequences using thermodynamics and auxiliary information. Nucleic Acids Res 9:133-148

TRANSLATIONAL CONTROL OF THE *c*III GENE OF BACTERIOPHAGE LAMBDA

Amos B. Oppenheim, Shoshy Altuvia, Daniel Kornitzer, and Dinah Teff
Department of Molecular Genetics
The Hebrew University-Hadassah Medical School
POB 1172
Jerusalem 91010 Israel

INTRODUCTION

In the life cycle of bacteriophage λ an interplay between a large number of genes takes place. Among them is a set of phage genes which participate in the decision between the lytic and the lysogenic pathway. The level of expression of these genes is regulated at the transcription initiation level by regulatory functions encoded by the *c*I and *cro* genes. The function of these proteins and their target sites was recently elegantly summarized (Ptashne, 1986). Additional regulatory functions that are essential for the lysogenic pathway are encoded by the *c*II and *c*III genes. The former acts as a positive regulator activating several phage promoters that allow rapid synthesis of *c*I repressor and λ integrase. The latter, a 54 amino acid long polypeptide, acts by stabilizing the *c*II protein, thereby enhancing *c*II activity (Altuvia and Oppenheim, 1986; Hoyt et al, 1982; Rattray et al, 1984).

We have shown that *c*III translation is subject to unique requirements (Altuvia and Oppenheim, 1986; Altuvia et al, 1987; Kornitzer et al, 1989; Altuvia et al, 1989). We found that the expression of the *c*III gene requires sequences located both upstream and downstream of the Shine and Dalgarno sequence and the AUG initiating codon. We proposed a model in which the equilibrium between two alternative mRNA secondary structures, A and B, determines the rate of translation initiation.

$$A \rightleftharpoons B$$

According to this hypothesis, in structure B the mRNA is efficiently translated, whereas in the alternative structure, A, it is unavailable for ribosome binding. In this chapter we describe the results leading to this hypothesis.

NATO ASI Series, Vol. H 49
Post-Transcriptional Control of Gene Expression
Edited by J. E. G. McCarthy and M. F. Tuite

RESULTS

a. Alternative mRNA structures: Genetic evidence

We have isolated and studied a set of mutants affecting *c*III gene activity (Altuvia and Oppenheim, 1986; Kornitzer et al, 1989) . All the available *c*III mutations can be divided into three groups on the basis of their location on a 100-nucleotide-long sequence around the *c*III translation initiation site. Group one includes mutations in the RNaseIII recognition region located upstream of *c*III. Group two includes mutations in the translation initiation signal. These mutations (mutations 2 and 17) affect the Shine-Dalgarno sequence and the AUG initiation codon, respectively.

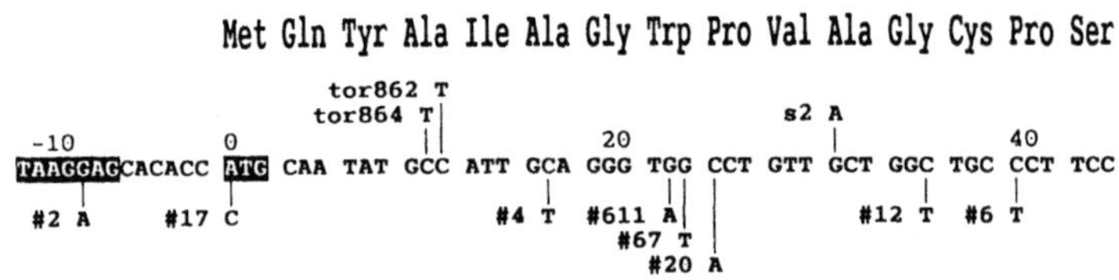

Figure 1. Nucleotide and amino acid sequence of the 5' end of the *c*III gene. The overexpressing mutants are indicated above the sequence, and the low-level expression mutants are indicated under the sequence.

Group three includes mutations within the coding region. These mutations can be divided further into three categories. The first consists of mutations that dramatically reduce *c*III translation. The second category includes mutations which result in elevated expression of *c*III. Finally, amber mutation 611, identical to the *c*III *r*2 mutation, results in the termination of *c*III translation. All the mutations within the *c*III coding region cluster in a sequence of 40 nucleotides, which code for the first 14 amino acids of the CIII protein. Surprisingly, no mutation affecting the activity of the CIII protein without affecting its synthesis was isolated. We have recently introduced an arginine to proline mutation at amino acid 21, generating a mutant protein which is efficiently synthesized, but inactive.

The experimental data indicate that the mutations affect *c*III expression at the level of translation. Two alternative mRNA structures can be drawn from the sequence described in Fig. 1 (Fig. 2). All the mutations located within the coding region which lead to a reduction of *c*III translation lower the stability of structure B. Some of these mutations enhance the stability of structure A. On the other hand, all the mutations causing elevated expression of *c*III interfere with the formation of structure A. Thus, it seems that one mRNA structure, B, correlates with high-level expression of *c*III, whereas structure A correlates with low levels of *c*III expression. The mutational data lead us to propose that the *c*III mRNA exists in equilibrium between two alternative structures. In one of these structures, B, the mRNA can serve as an efficient template in the translation initiation process, whereas the alternative structure, A, is unavailable for ribosome binding. In the overexpressing mutants, structure B predominates, whereas in most of the low-level expression mutants, A is favored.

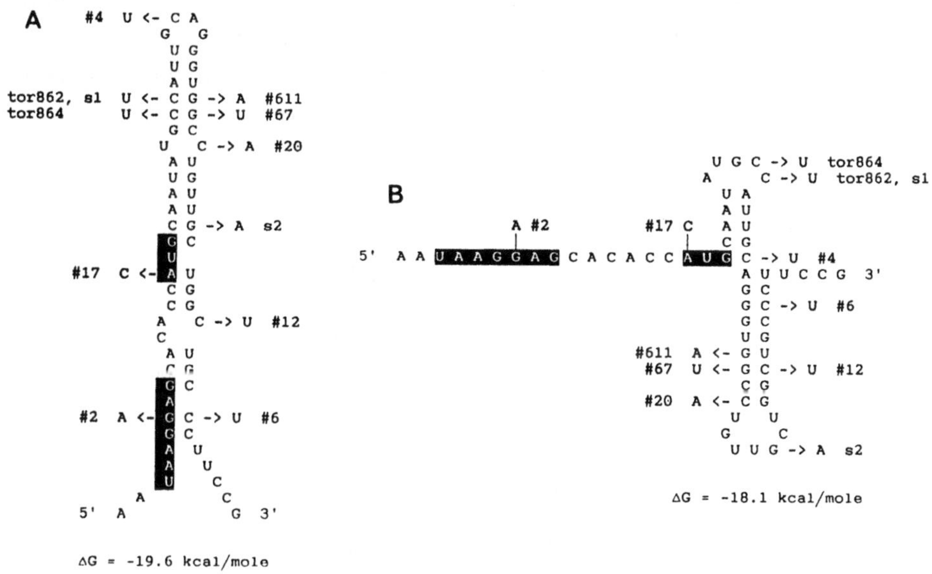

Figure 2. The A and B structures of *c*III mRNA. The figure includes the Shine-Dalgarno sequence and initiation codon (shaded), and the location of the various mutations. The predicted stability is indicated under each structure.

b. Alternative mRNA structures: Biochemical evidence

To probe the mRNA secondary structure, we cloned the appropriate DNA fragments under the phage T7 promoter. The *in vitro*-synthesized RNA was subjected to partial cleavage by the single-strand-specific nucleases T_1 and S_1, the double-strand-specific cobra venom nuclease, or the chemical modification by dimethyl sulfate which reacts with unpaired nucleotides. The treated RNA samples were anealed to a radioactively-labeled primer, and reverse-transcribed. The reaction products were then separated by gel electrophoresis and visualized by autoradiography. The detailed findings were recently published (Altuvia et al., 1989), and a summary of the results is given in Fig. 3. Our results clearly show that the mRNA of *tor*862 is present exclusively in structure B, the mRNA of mutant 12 is found in structure A, while the wild type mRNA is present in both alternative structures. For convenience in this discussion, we divided the sequence into three domains (I, II, and III) according to the three stem and loop structures, starting from the 5' end.

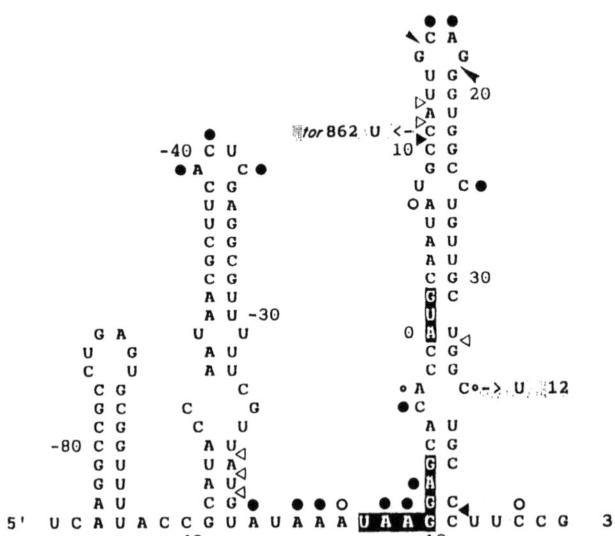

(a)

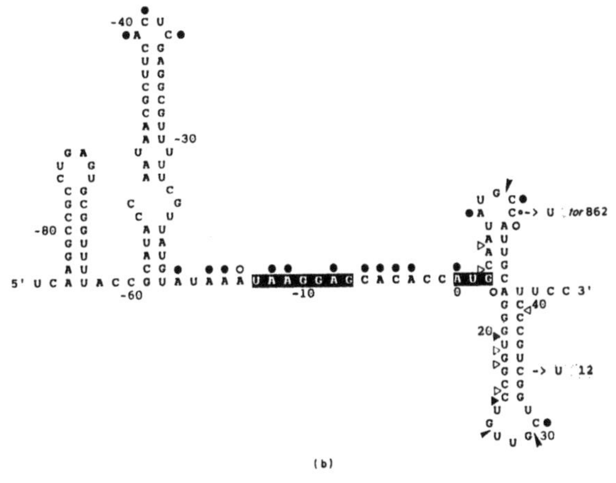

Figure 3. Summary of structural mapping of wild-type, mutant 12 and mutant *tor*862 mRNA at neutral pH. (a) The digestion/modification sites of mutant 12 mRNA are indicated on structure A; (b) the sites of *tor*862 mRNA are indicated on structure B. The digestion/modification sites obtained with wild-type mRNA correspond to the sites displayed for both 12 and *tor*862. The arrowheads indicate strong (large arrowhead) and weak (small arrowhead) T1 and S1 digestion sites. The triangles indicate strong (filled triangles) and weak (open triangles) cobra venom digestion sites. The circles indicate strong (filled circles) and weak (open circles) dimethyl sulfate modification sites. The small open circles indicate dimethyl sulfate modification sites displayed only by the wild-type mRNA.

c. 30S ribosomes can recognize and bind to mRNA in structure B

To analyze 30S ribosomal subunit binding to the various mRNA species we used a technique, recently developed by L. Gold, termed extension inhibition ("toeprint"; Winter et al., 1987; Hartz et al., 1989). In this method, the binding of a 30S ribosomal subunit particle to the mRNA blocks the elongation of a DNA primer by reverse transcriptase. Analysis of the reaction products on a sequencing gel allows determination of the precise location of the 3' end of the RNA region covered by the ribosome and quantification of the rate of ribosome binding to this site.

Kinetic analysis of such a ribosome binding experiment is shown in Fig. 4 (see Altuvia et al., 1989). The results show that the 30S ribosomal subunits are unable to bind to #12 RNA, but bind rapidly to *tor*862 RNA. The binding to the wild type RNA shows a continuous increase over a period of 60 minutes. This result may indicate that the ribosome, by trapping the bound RNA, shifts the A - B equilibrium of the wild type RNA towards structure B.

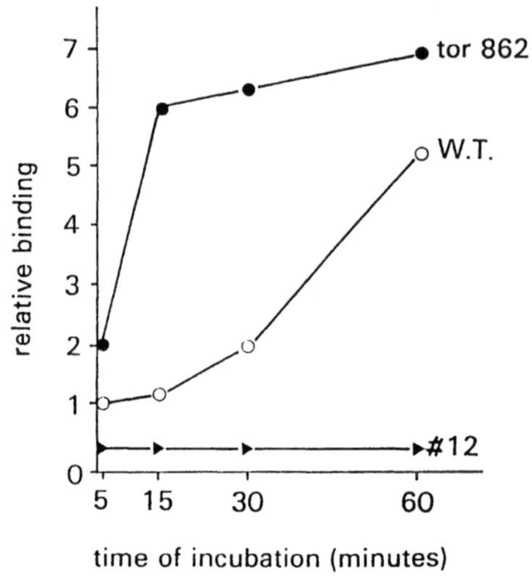

Figure. 4. Analysis of 30S ribosomal subunit binding to *c*III mRNA by primer extension inhibition. Kinetics of 30S ribosomal subunit binding to wild-type (open circles), mutant *tor*862 (filled circles), and mutant 12 (triangles) mRNA.

d. The participation of RNaseIII in *c*III expression

We found that phages grown on an RNaseIII-deficient host (*rnc*-105) greatly favor the lytic pathway. Phage mutants (λ*tor*; see Fig. 1) that can efficiently lysogenize the mutant host map in the *c*III gene (Altuvia and Oppenheim, 1986). We further found that the translation of *c*III-lacZ protein fusions is specifically reduced by about 10-fold when expressed in the RNaseIII deficient host. In the same background, the level of the corresponding operon fusion was reduced by only 2-fold (Altuvia et al.,, 1987). (We recently found that the translation of the *c*III protein was also reduced by 10-fold in the absence of RNaseIII.) An RNaseIII cleavage site was identified in domain II upstream of the Shine and Dalgarno sequence. We proposed that RNaseIII stimulates *c*III translation by binding to the upstream region, thereby exposing the *c*III ribosome binding site. Our recent findings (unpublished) indicate that the mechanism of activation of *c*III expression by RNaseIII is more complex. We recently found that RNaseIII can also cleave domain III of the RNA at low efficiency. Furthermore, the deletion of part of the stem of domain II did not affect the *rnc* dependence of the *c*III-lacZ gene fusion expression. We postulate that either the effect of RNaseIII on *c*III translation is indirect or that binding of RNaseIII to domain III directly activates translation initiation.

e. Evolutionary conservation of mRNA structures in the *c*III region of lambdoid phages

Bacteriophages belonging to the lambdoid family follow similar genetic organization and programming of gene functions. It has been suggested that these viral genomes are made of functionally distinct modules (Campbell and Botstein, 1983). Comparison of the DNA sequences of three phages, λ, HK022, and P22 (see Daniels et al.,1983; Oberto et al., 1989; Semerjian et al., 1989, respectively), in the region around the *c*III gene, shows the conservation of "micro-modules", or domains. Domain I (see Fig. 3) is almost completely conserved between λ and P22. The corresponding sequence in HK022 is not homologous to the two former; however, it is predicted to form a similar stem and loop structure. The function of this RNA structural domain is not clear. Domain II, which was shown in λ to serve as an RNaseIII processing site, is almost completely conserved between λ and HK022. The corresponding sequence in P22 is not homologous to the two other phages; however, RNA secondary structure prediction yields a similar stem-bulge-stem structure. All three phages contain a 6 to 7 bp region of complementarity to the 3' end of the 16S rRNA (Shine - Dalgarno sequence) followed by a 6 bp spacing to the initiation codon. In all three phages the sequence coding for the amino-terminal portion of *c*III (domain III) can form alternative mRNA secondary structures, although the primary sequences are different. One can therefore conclude that the three structural domains described for phage λ are conserved in the other lambdoid phages.

SUMMARY

Genetic and biochemical experiments allowed us to establish a structure-function relationship for the *c*III mRNA. Our experiments demonstrate that the *c*III mRNA is present in two conformations at equilibrium. Only one conformation permits translation initiation. We suggest that the regulation of this equilibrium provides a mechanism for the control of *c*III gene expression. The enzyme RNaseIII may participate, directly or indirectly, in governing the A to B equilibrium.

ACKNOWLEDGMENTS

We thank Dr. H. Giladi for many useful discussions, S. Koby for technical help, and K. Simon Baruch for helping to bring this manuscript to its final form. This work was supported by a grant from the National Council for Research and Development, Israel and the Gesellschaft fur Biotechnologische Forschung mbH, Braunschweig, and by grant GM38694 from the National Institutes of Health. This research was performed in the Irene and Davide Sala Laboratory for Molecular Genetics.

REFERENCES

Altuvia S, Kornitzer D, Teff D, Oppenheim AB (1989) Alternative mRNA Structures of the *c*III Gene of Bacteriophage λ Determine the Rate of Its Translation Initiation. J Mol Biol 210:265-280

Altuvia S, Locker-Giladi H, Koby S, Ben-Nun O, Oppenheim AB (1987) RNaseIII stimulates the translation of the *c*III gene of bacteriophage λ. J Bacteriol 84:6511-6515

Altuvia S, Oppenheim AB (1986) Translational regulatory signals within the coding region of the bacteriophage λ *c*III gene. J Bacteriol 167:415-419

Campbell A, Botstein D (1983) Evolution of the Lambdoid Phages. In: Hendrix RW, Roberts JW, Stahl FW, Weisberg RA (eds) Lambda II. Cold Spring Harbor Laboratory, Cold Spring Harbor, New York pp 365-380

Daniels D, Schroeder J, Szybalski W, Sanger F, Coulson A, Hong G, Hill D, Petersen G, Blattner F (1983) Complete annotated lambda sequence. In: Hendrix RW, Roberts JW, Stahl FW, Weisberg RA (eds) Lambda II. Cold Spring Harbor Laboratory, Cold Spring Harbor, New York pp 519-676

Hartz D, McPheeters DS, Traut R, Gold L (1989) Extension Inhibition Analysis of Translation Initiation Complexes. Methods Enzymol 163, in press

Hoyt MA, Knight DM, Das A, Miller I, Echols H (1982) Control of phage λ development by stability and synthesis of *c*II protein: role of the viral *c*III and host *hfl*A, *him*A, and *him*D genes. Cell 31:565-573

Kornitzer D, Teff D, Altuvia S, Oppenheim AB (1989) Genetic analysis of bacteriophage λ *c*III gene: mRNA structural requirements for translation initiation. J Bacteriol 171:2563-2572

Oberto J, Weisberg RA, Gottesman ME (1989) Structure and Function of the *nun* Gene and the Immunity Region of the Lambdoid Phage HK022. J Mol Biol 207:675-693

Ptashne M (1986) A Genetic Switch, Cell and Blackwell, Cambridge Massachusetts

Rattray A, Altuvia S, Mahajna G, Oppenheim AB, Gottesman M (1984) Control of bacteriophage lambda *c*II activity by bacteriophage and host functions. J Bacteriol 159:267-288

Semerjian A, Malloy DC, Poteete AR (1989) Genetics Structure of the Bacteriophage P22 pL Operon. J Mol Biol 207:1-13

Winter RB, Morrissey L, Gauss P, Gold L, Hsu T, Karam J (1987) Bacteriophage T4 regA protein binds to mRNAs and prevents translation initiation. Proc Natl Acad Sci USA 84:7822-7826

THE *mom* OPERON OF BACTERIOPHAGE MU IS REGULATED BY A COMBINATION OF TRANSCRIPTIONAL AND TRANSLATIONAL CONTROLS

F. G. Wulczyn, F. Schmidt, A. Seiler† and R. Kahmann
Institut für Genbiologische Forschung Berlin GmbH
Ihnestraße 63
D-1000 Berlin 33
FRG

I. Introduction

The *mom* operon of bacteriophage Mu has proven to be an attractive model system for the study of prokaryotic gene regulation. Of particular interest is the integration of transcriptional and post-transcriptional controls in the regulation of *mom* gene expression. *mom* regulation has been recently reviewed (Kahmann and Hattman, 1987), in this chapter we will focus on progress made in the last few years. Transcriptional regulation will be briefly discussed, followed by a more comprehensive treatment of post-transcriptional control. New data on the mRNA sites involved in post-transcriptional regulation will be presented, and the overexpression and purification of the regulatory Com protein will be described.

The *mom* operon is the rightmost transcriptional unit on the Mu chromosome (Fig. 1). The operon contains two, overlapping genes: the *mom* gene, which encodes a site-specific DNA modification function, and the *com* gene, which encodes an activator for *mom* gene translation. The two genes are co-transribed from a single promoter, with transcription initiating at position 971 (Fig. 1, Bölker et al., 1989).

The *mom* gene encodes a unique DNA modification function that is responsible for the insensitivity of Mu to a number of host restriction systems (Toussaint 1976). *mom* expression can thus be monitored in an in vivo assay by comparing phage plating efficiency on restricting and non-restricting strains. The DNA modification has been shown to be the conversion of adenine to α-*N*-(9-ß-D-2´-deoxyribofuranosylpurin-6-yl) glycinamide (Swinton et al. 1983). The recognition sequence for Mom is 5´...(C/G)-A-(C/G)-N-Py...3´ (Kahmann 1984), in fully modified DNA approximately 15% of the adenines are modified (Hattman 1979).

A critical observation for the understanding of *mom* regulation was the finding that constitutive *mom* expression is lethal to the host, and that premature expression during phage development leads to a reduction in phage burst size (Kahmann et al., 1985 and M. Bölker, personal communication). In order to cope with the lethal effects of *mom*, we believe that Mu has

†Max Planck Institut für molekülare Genetik, Ihnestr. 73, D-1000 Berlin 33 FRG

NATO ASI Series, Vol. H 49
Post-Transcriptional Control of Gene Expression
Edited by J. E. G. McCarthy and M. F. Tuite

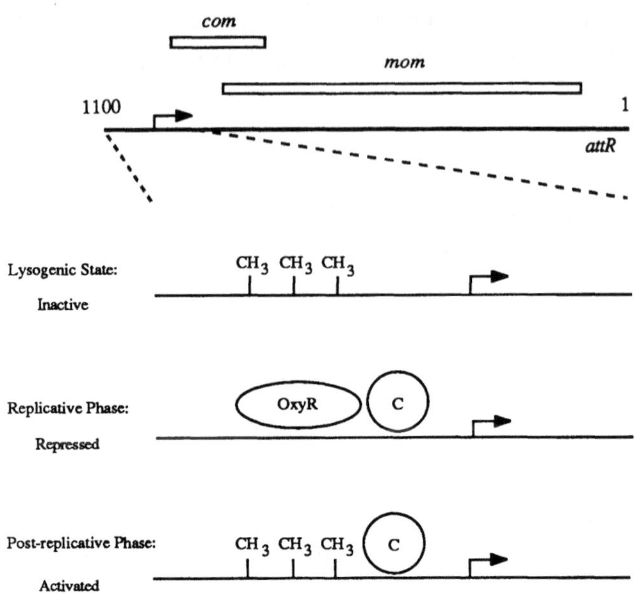

Figure 1. A model for transcriptional control of *mom*. A schematic diagram of the *mom* operon is shown at the top. The heavy bar represents the right end of the Mu chromosome, the arrow marks the initiation point of the *mom* operon transcript. The positions of the *com* and *mom* genes are given by open rectangles. The overlap of the two genes is 47 base pairs. An extended view of the transcriptional control region from position 1100 to 900 is shown below, including transcription start site, C binding site and Region I. The methylation state of the three GATC sites in Region I during three stages of Mu development is indicated. The phage C protein is represented by a circle, it is first expressed during the replicative phase of Mu growth. The host OxyR protein (ellipse) is present throughout the three stages, we propose that OxyR can occupy its binding site only in the active phase of phage replication (see text) due to the sensitivity of OxyR binding to methylation. Methylation of Region I releases repression by OxyR and allows activation by C.

evolved an intricate regulatory scheme designed to limit *mom* expression to a brief window in time after phage DNA replication and before phage DNA packaging. Much of this temporal regulation occurs at the level of transcription initiation, which will be outlined in the next section.

Temporal Control of the *mom* Promoter

mom belongs to the class of Mu late genes and the *mom* promoter is completely dependent on the phage C protein, the activator of Mu late transcription. The C protein contains a putative helix-turn-helix motif, and has been shown to bind the *mom* promoter from position -35 to -53 (Bölker et al., 1989). The mechanism of activation is believed to be analogous to activation of

the lambda P_{RM} promoter by the cI protein (Hochschild et al., 1983, Bushman et al., 1989). Transcriptional activation by C is, in turn, negatively regulated by the *E. coli* OxyR protein. OxyR has been shown to bind to a site in the upstream regulatory region of the *mom* promoter known as Region I, immediately upstream of the C binding site (Bölker and Kahmann, 1989). The mechanism of repression is currently under investigation. One question that is raised by the interplay between OxyR and C is: how is repression by OxyR released to allow *mom* transcription? The answer to this question lies in the sensitivity of OxyR binding to DNA methylation in Region I. Region I contains three GATC recognition sites for the *E. coli* Dam methyltransferase. It has long been known that *mom* is not expressed in *dam* strains (Toussaint, 1977), but that deletion or mutation of the GATC sites alleviates the requirement for Dam (Kahmann, 1983; Seiler et al., 1986). This implies that OxyR binds Region I preferentially when it is unmethylated, a prediction that has been confirmed in vitro by testing methylated and unmethylated *mom* promoter fragments for OxyR binding (Bölker and Kahmann, 1989).
Based on these results, we propose the model for the role of methylation in the regulation of *mom* expression diagrammed in Fig. 1. In the lysogenic state, and in the early phase of phage development, the *mom* promoter is inactive due to its dependence on the C protein. During the replicative phase of Mu development, C is expressed and begins to activate the phage late promoters. We believe that during this phase OxyR has access to its site in Region I and blocks the action of C. This would occur if the GATC sites in Region I were transiently undermethylated as the pace of Dam-methylation lags behind that of DNA replication. The effect would be to delay *mom* transcription until Dam succeeds in methylating Region I, displacing OxyR, and thus tie the onset of transcription to a precise timepoint in phage development.

II. Post-transcriptional Control

The *com* gene is a positive regulator of *mom* gene expression

Although the proposed model for transcriptional control might appear to satisfy the need to restrict *mom* expression temporally, transcriptional control is in fact supplemented by control at the post-transcriptional level. The first gene in the *mom* operon, the *com* gene, has been shown to be a positive regulator of *mom* translation. The role of the *com* gene in *mom* regulation was initially revealed by cloning experiments. Constructs containing the complete operon fused to constitutive vector promoters could not be obtained, due to the lethal effect of constitutive Mom-modification. Clones in which the *com* gene was truncated, however, were viable, suggestive of a stimulatory effect of *com* on *mom* expression (Kahmann et al., 1985). This was confirmed by testing *mom* expression under the control of the inducible P_L promoter. Whereas a clone carrying the intact operon overexpressed the Mom protein, a clone deleted for the 5′ end of *com* produced significantly lower amounts of Mom. These results demonstrated that an intact *com* gene is required for efficient *mom* expression, and that the effect is promoter independent (Kahmann et al., 1985).

Stimulation by *com* occurs in trans

The possibility that *mom* translation is passively coupled to translation of the upstream *com* gene was ruled out by fusing the 3′ end of *com* in frame to the 5′ end of the highly expressed MS-2 polymerase gene. Despite efficient expression of the MS-2-Com fusion protein, production of Mom could not be detected by SDS-PAGE (Kahmann et al., 1985, see Fig. 2). The Mom protein was expressed from the fusion plasmid, however, when Com was supplied in trans from a second plasmid (Fig. 2). This result is in agreement with complementation experiments using Com^+ and Com^- phage to restore *mom* expression from cloned operons carrying mutations in *com* (Kahmann et al., 1985). *com* must, therefore, code for a regulatory protein acting in trans to stimulate *mom* expression post-transcriptionally.

Com is not a transcriptional antiterminator

Two studies used search plasmids for transcriptional termination signals to identify possible Com-dependent termination sites in the *mom* operon (Wulczyn and Kahmann, 1987; Hattman et al., 1987). Premature termination in the absence of Com affected at most 33% of transcription through the operon, as compared to the presence of Com (Wulczyn and Kahmann, 1987). This low level is insufficient to account for the dependence of *mom* expression on Com, and is more likely to reflect an increase in rho-dependent termination in the absence of Com. Indeed, Hattman et al. (1987) reported that Com acted to overcome a strong rho-dependent termination site located in vector sequences between *mom* sequences and a reporter gene. The authors concluded that a role for Com as a translational activator was the most likely explanation for their results.

Stimulation by Com is accompanied by increased accumulation of *mom* mRNA

In addition to the possibility of translational activation, Com could conceivably enhance *mom* expression by mediating either a mRNA processing event or mRNA stability. The effect of Com on *mom* mRNA accumulation was determined by comparing RNA from Mu phage carrying either a wildtype *mom* operon or one with a nine base pair deletion in *com* (Wulczyn and Kahmann, 1987). This deletion in *com* abolishes Com function (Kahmann et al., 1985). Compared to Com^- phage, in cells carrying Com^+ phage 20-fold higher steady-state levels of full length *mom* operon mRNA were observed in Northern blots. In Com^- cells the reduction in the full-length transcript was not compensated by the appearance of alternative transcripts. This is in agreememt with the results of S_1 mapping experiments (Hattman and Ives, 1984) and

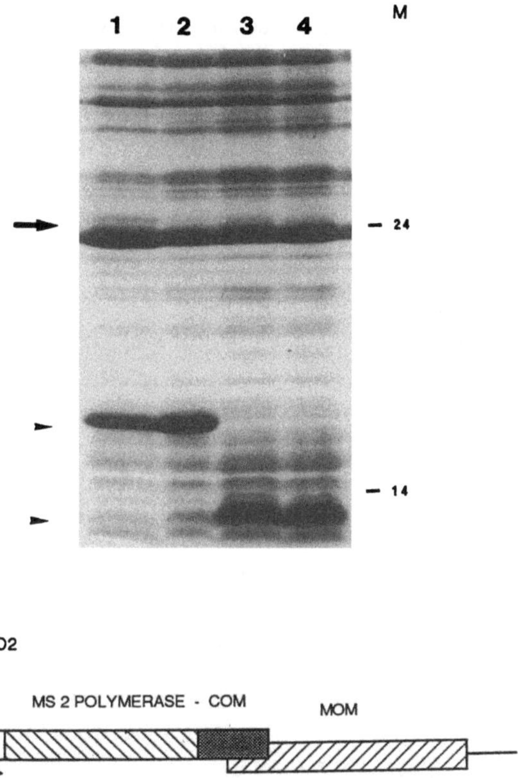

Figure 2. Stimulation of *mom* expression by Com. Total cell extracts prepared 2 hours after heat induction were separated by SDS-PAGE, positions of marker protein migration are given on the right in kilodaltons. Extracts were from cells containing either the MS2-Com fusion plasmid pLc24-HD2 or the fusion vector without insert, pLc24 (Kahmann et al., 1985), together with either pACYC184 or the Com expression plasmid pCom4. Lane 1: pLc24-HD2 and pCom4; 2: pLc24-HD2 and pACYC184; 3: pLc24 and pCom4; 4: pLc24 and pACYC184.The arrow points to the position of the Mom protein visible in Lane 1. The arrowheads point to the position of the MS2-Com fusion protein (Lanes 1 and 2) or the MS2 polymerase fragment encoded by pLc24 (Lanes 3 and 4). The MS2-Com fusion gene and *mom* gene carried on pLc24-HD2 are shown schematically below the gel.

primer extension analysis (Wulczyn et al., 1989) that indicate that Com is not involved in processing the *mom* mRNA to an active form. S_1 mapping identified only one 5´ end, initiated at the *mom* promoter. Primer extension revealed that, although the levels of mRNA are different, in both Com^+ and Com^- cells the major transcript extends to the *mom* promoter, and no significant difference in the pattern of minor 5´ ends was observed.

mom mRNA accumulation is linked to translation

The Northern blot experiments described above could not distinguish between two possible mechanisms by which Com could enhance *mom* mRNA accumulation. Com might directly inhibit a rate limiting step in *mom* mRNA decay, for example by protecting the mRNA from an endonucleolytic cleavage. Alternatively, translational activation by Com could stabilize the mRNA if a rate limiting step in decay were slowed by translating ribosomes. The role of translation in prokaryotic mRNA degradation is controversial (reviewed in Belasco and Higgins, 1988), certain mRNA´s have been reported to be stable in the absence of translation on the major part of the message (Nilsson et al., 1987; Gorski et al., 1985; and articles in this volume). On the other hand, enhanced mRNA degradation has been observed upon reduction of translation by translational repressor proteins (Brot et al., 1980; Cole and Nomura, 1986). Our approach to test if translation is involved in *mom* mRNA degradation was to insert a stop codon in *mom* and determine the effect on mRNA accumulation.

This experiment was performed with the help of a *lacZ* assay for *mom* expression. First, ß-galactosidase expression from *mom-lacZ* translational fusion plasmids, transcribed from the lacUV5 promoter, was shown to be dependent on in intact *com* gene. *mom-lacZ* expression from constructs lacking *com* was stimulated when *com* was provided in trans from a second plasmid (Wulczyn et al., 1989). Using this assay, the effect of an artificial stop codon in *mom* on *mom-lacZ* mRNA accumulation was tested for a Com^{+} and a Com^{-} construct. The distance of the stop codon from the *mom* initiation codon and *com* termination codon was 64 and 17 base pairs, respectively. A Northern blot of mRNA from these plasmids revealed that accumulation of full-length *mom-lacZ* mRNA was dependent on Com <u>and</u> on translation of the *mom* gene (Wulczyn et al., 1989). This demonstrated that stabilization of the *mom* mRNA by Com is dependent on translation, and that translation of the *mom* mRNA can slow a rate limiting step in *mom* mRNA degradation. It is not known if this rate limiting step involves 3´ to 5´ exonuclease digestion or endonucleolytic cleavage(s) at a site downstream of the stop codon insertion. Com binding to a site in the *mom* mRNA has, interestingly, been inferred to protect the upstream RNA from 3´ to 5´ exonucleases (see Wulczyn et al., 1989 for a discussion).

Support for the interpretation that Com is a positive regulator of *mom* translation was obtained by fusing *com* in frame to *lacZ*. Com in trans had no effect on the expression of a *com-lacZ* fusion gene, although the construct contained all of the sequences required for stimulation of *mom* expression (Wulczyn et al., 1989). Stimulation by Com, therefore, is specific for the *mom* reading frame, a result incompatible with a direct interaction of Com with a processing nuclease, and strongly suggestive of a role in translation.

Sequence requirements for Com-dependent *mom* expression

Use of the *mom-lacZ* assay for stimulation by Com allowed the sequence requirements for stimulation to be defined. A 53 nucleotide sequence centered at the *mom* initiation codon was found to be necessary and sufficient for conveying Com-dependent *mom-lacZ* expression

A

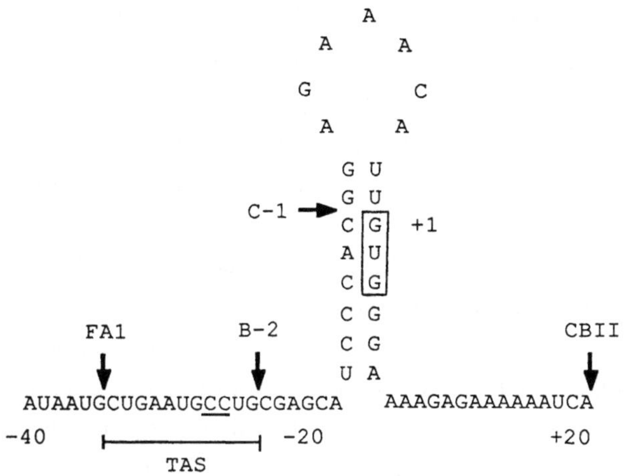

B

	-30 -20 -10 1 +10 +20	ß-gal units Com⁻	Com⁺
pComD1	GAAUGCCUGCGAGCAUCCCACGGAGAAACAUUGUGGGAAAAGAGAAAAAAATCACG	571	1684
pComDL1	AUGCCUGCGAGCAUCCCACGGAGAAACAUUGUGGGAAAAGAGAAAAAAATCACG	770	780
pComS	GAUUGCCUGCGAGCAUCCCACGGAGAAACAUUGUGGGAAAAGAGAAAAAAATCACG	135	207
pComD1Δ	GAAUGCCU*CGAGCAUCCCACGGAGAAACAUUGUGGGAAAAGAGAAAAAAATCACG	110	931
pComD2	GAAUGCCUGCGAGCAUCCCACGGAGAAACAUUGUGGGAAAAGAGAAAAA	380	1060
pComD3	GAAUGCCUGCGAGCAUCCCACGGAGAAACAUUGUGGGAAAAGAGAA	310	850
pComD4	GAAUGCCUGCGAGCAUCCCACGGAGAAACAUUGUGGGAAAAGA	210	375
pComXL	GAAUGCCUGCGAGCAGAACACGGAGAAACAUUGUGGGAAAAGAGAAAAAAATCACG	4082	4021
pComXR	GAAUGCCUGCGAGCAUCCCACGGAGAAACAUUGUGTTCAAAGAGAAAAAAATCACG	4675	4553
pComZL	GAAUGCCUGCGAGCAAGGCACGGAGAAACAUUGUGGGAAAAGAGAAAAAAATCACG	2307	2672
pComZR	GAAUGCCUGCGAGCAUCCCACGGAGAAACAUUGUGTCCAAAGAGAAAAAAATCACG	7532	6668

Figure 3. The *mom* TIR. A) The sequence of the *mom* TIR is shown, with the structure of a potential stem and loop structure (see text). The *mom* GUG initiation codon is boxed. Arrows indicate deletion endpoints of the *mom-lacZ* fusion plasmids pTRFA1, pTRB-2, pTRC-1 and pTRCBII (Wulczyn et al., 1989). The two cytosines deleted in pTRMAΔ2 are underlined (Wulczyn et al., 1989). B) The sequence of mutants in the *mom* TIR is shown. An internal deletion (pComD1Δ) is symbolized by *, base substitutions are underlined. Mutants were generated by cloning synthetic oligonucleotides into the plasmid pTSVK. Complementary oligonucleotides were synthesized, the sense strand with an EcoRI overlap on the 5´ end, the antisense strand with a BamHI overlap at the 3´ end. After hybridization, the oligonucleotides were ligated into the EcoRI and BamHI sites of pTSVK resulting in the fusion of the *mom* GUG to *lacZ*. pTSVK is a derivative of the *lacZ* fusion vector pTSV11 (Wyckoff et al., 1986), in which the *tet* gene of pTSV11 is replaced by *neo*. ß-galactosidase assays from cells carrying pACYC184 (Com⁻) or the Com expression plasmid pCom4 (Com⁺) were as previously described (Wulczyn et al., 1989).

(Fig. 3). The 5´ end at position -32 is defined by the clone pTRFA1, the 3´ end at position +20 relative to the *mom* GUG by the clone pTRCBII (The numbering scheme used differs from that used in the original publication, Wulczyn et al., 1989). This region has the potential to form a stem loop structure that has been implicated in the control of translational initiation (Fig. 3A). The *mom* GUG initiation codon and Shine-Dalgarno sequence are sequestered in this structure, so that formation of this stem-loop would be expected to inhibit ribosome binding. The positions of various deletions and mutations introduced into the *mom* TIR in relation to the stem-loop structure is also shown in Figure 3A.

Upstream of the stem-loop, deletions into the 5′ end to position -22 (pTRB-2) or -30 (pComDL1) led to a loss of stimulation by Com. Similarly, deletion of nucleotides -26 and -27 (pTRMAΔ2), or mutation of the A at position -30 to U (pComS), destroyed or reduced stimulation, respectively. Taken together, these results show that sequences between -32 and –22 are required for Com to stimulate *mom* translation. We call this sequence the *t*ranslational *a*ctivation *s*equence, or TAS. The mutations place the 5′ end of TAS to between -32 and -30. The mutations do not define the 3′ end of TAS, it is not clear if sequences in the stem-loop or at the 3′ end of the TIR interact with Com. An interesting mutation in this regard is pComD1Δ. pComD1Δ is deleted for the G at position -24, yet can be stimulated by Com very efficiently (Fig. 3).

The role of secondary structure in regulating *mom* translation was revealed by deleting upstream sequences to position -12 (pTRC-1). *mom-lacZ* expression from pTRC-1 was very efficient and independent of Com. This is consistent with a role for the stem-loop in repressing initiation, since the deletion in pTRC-1 would destroy the ability of this structure to form. This conclusion is supported by the mutants pComXL, pComXR, pComZL and pComZR (Fig. 3B). In each of these mutants base pairing in the potential stem-loop is reduced, and in each case translation of *mom* is derepressed. Combining the mutations of pComXR with pComXL, or pComZR with pComZL, however, does not restore repression (data not shown). This is presumably due to competing secondary structures that can be formed by the mutant RNA′s, leaving the *mom* RBS in an open conformation (data not shown). It will be necessary to test additional complementary exchange mutants to confirm the role of secondary structure in regulating *mom* translation. Recently, ribonuclease mapping of secondary structure has been used to show that the predicted stem-loop does form in vitro (F. G. Wulczyn and R. Kahmann, in preparation).

Deletions entering into the 3′ end of the *mom* TIR from position +20 revealed the importance of an A rich stretch for maximal initiation from the *mom* RBS. Progressive deletions (plasmids pComD1 to pComD4, Fig 3B) led to a gradual loss of ß-galactosidase expression without curtailing stimulation by Com. Computer predictions of *mom* mRNA secondary structure suggest that this sequence is single-stranded, which may explain its ability to enhance ribosome binding.

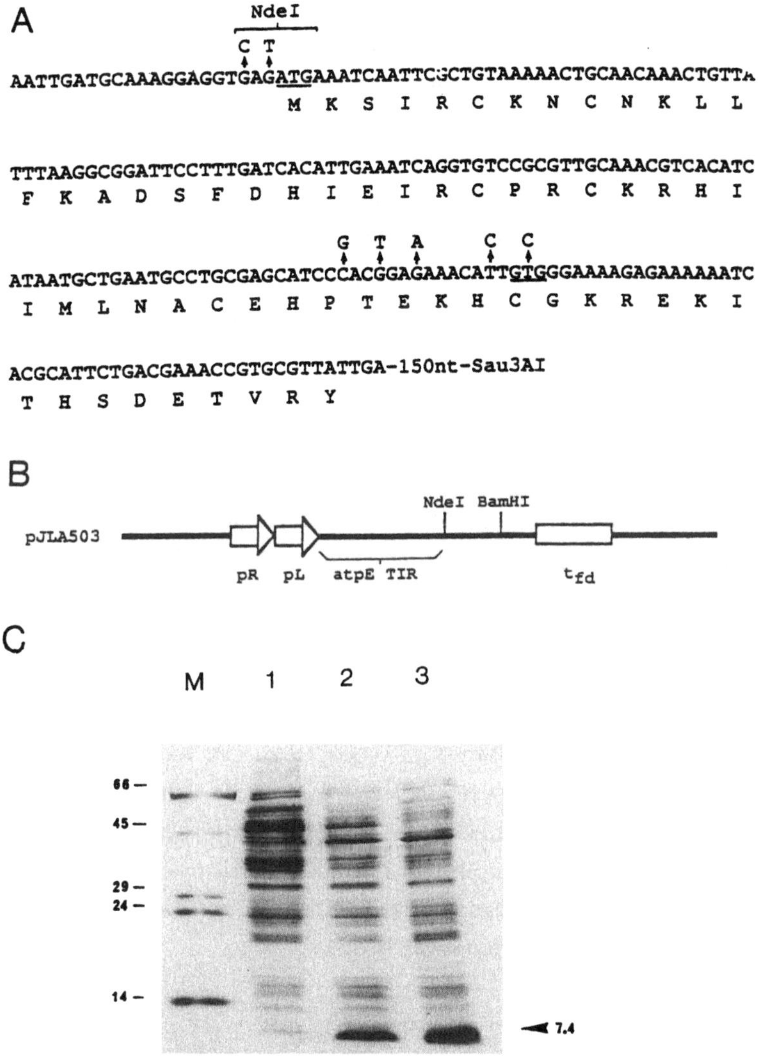

Figure 4. Overexpression of the Com protein. A) The nucleotide and amino acid sequences of the *com* gene are shown. The insertion of a NdeI site at the 5′ end of *com* was done by oligonucleotide mutagenisis using the method described by Klippel et al., (1988), as was the introduction of a number of silent mutations in *com* designed to eliminate initiation at the *mom* TIR and improve codon usage in *com*. Base changes are indicated by arrows. After introduction of the NdeI site, the *com* gene was excised as a NdeI-Sau3A fragment and inserted into pJLA503 (Schauder et al., 1987). B) Schematic diagram showing relevant features of pJLA503. C) SDS-PAGE of cell extracts from 1) XL–1(pJLA503); 2) XL–1(pComix); 3) C600*lon*(pComix). The position of Com is given by an arrowhead. Cells were grown at 28°C to OD 0.6 (650 nm), then shifted to 42°C for three hours and harvested. SDS-PAGE was on a 5 cm, 10-20% gradient polyacrylamide gel prepared according to Fling and Gregerson (1986).

Overexpression and purification of Com

Despite the excellent match to the consensus Shine-Dalgarno sequence preceding the *com* start codon (see Fig. 4A), initial attempts to overexpress Com using a T_7 expression vector failed. We reasoned that the problem might be inefficient translation from the *com* TIR, particularly in light of the low level of expression of *com-lacZ* fusion genes (Wulczyn et al., 1989). We therefore fused an intact *com* gene to the highly efficient *atpE* TIR on the vector pJLA503 (Fig. 4B), yielding plasmid pComix (Fig. 4). Using pComix, high level overexpression of Com was obtained both in DH5α and C600 *lon* strains (Fig. 4C). Extracts from C600 *lon* cells were used for the purification of Com.

Although we have found that crude cell extracts can be used in coupled in vitro transcription/translation extracts and RNA binding studies using the gel retention assay (Wulczyn and Kahmann, in preparation), it was desirable to have samples of Com free of contaminating ribonucleases. We developed a two step purification scheme that yielded Com in excess of 90% purity and essentially free of ribonuclease activity (data not shown). Details of the procedure are given in the legend to Figure 5. Briefly, cells carrying pComix were harvested after induction and disrupted in a french pressure cell. The lysate was cleared by centrifugation and the supernatant was heat treated. After renewed centrifugation, the supernatant was applied to a Mono Q cation exchange column for FPLC. Com was identified as a 7000 dalton protein eluting with a peak at 650 mM NaCl by SDS-PAGE (Fig. 5A). To confirm that this protein visible is Com, Mono Q fractions were screened by western blotting with a monoclonal antibody specific for Com. This antibody was raised against the MS2-Com fusion protein described above and isolated as a positive clone that did not react with MS2 protein alone (A. Seiler and F. G. Wulczyn, in preparation). The antibody detected Com in the flowthrough fractions and in a peak centered at fraction 20 (Fig 5B).

Summary and Perspectives

Initiation of translation at the *mom* TIR is inefficient in the absence of Com. This inefficiency is due to a stem and loop in the TIR, since mutations in the TIR that reduce secondary structure increase the rate of initiation and make initiation Com independent. At least part of the sequence that is required by Com to stimulate *mom* translation (TAS) lies upstream of the proposed stem and loop. One possible mechanism for activation is that the binding of Com to TAS destabilizes the stem and loop. Com would thus stimulate translation by presenting the *mom* TIR in an open conformation. The translational repressor phage R17 coat protein has been shown to do exactly the opposite, it stabilizes an inhibitory secondary structures (Gralla et al., 1974). Another model would propose that Com stimulates ribosome binding by direct interaction with the ribosome. Interaction between Com bound at the TAS and the initiation machinery might

A

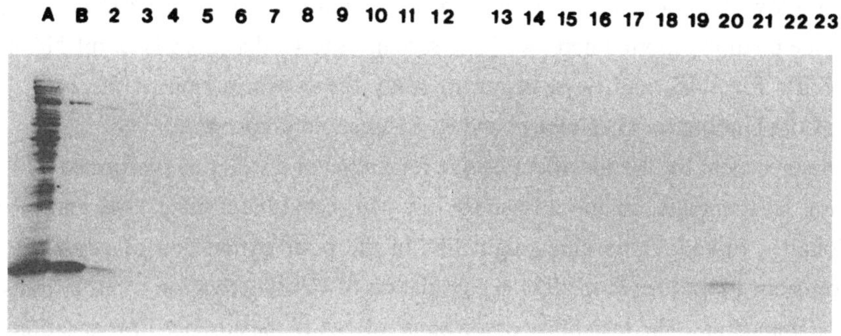

B

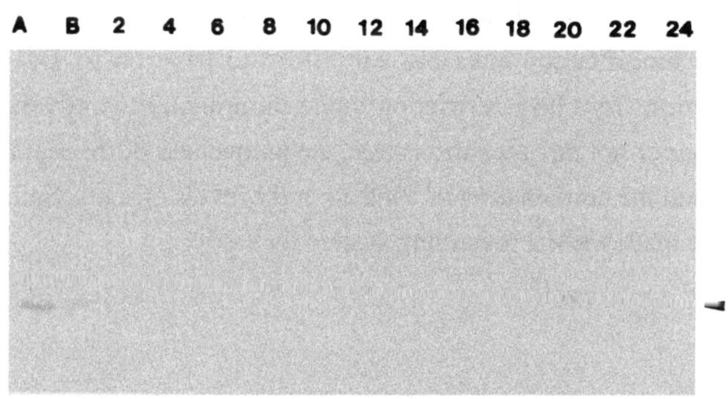

Figure 5. Purification of Com. 100ml cultures of C600*lon*(pComix) were grown to OD 0.6 (650nm) at 28°C, then shifted to 42°C for 3 hours. Cells were harvested and the cell paste was taken up in 5ml Buffer 1 (5% glycerol, 50 mM Tris hydrochloride, pH 7.5, 1 mM EDTA, 100 mM DTT). Phenylmethylsulphonyl flouride was added to a concentration of 0.2 mg/ml and the cells were disrupted in a french pressure cell (15,000 psi) (Lane A). The lysate was centrifuged 1 hour at 55,000 rpm in a Beckmann 100.3 rotor. 1 ml aliquots of supernatant were heated in a 100°C water bath for 5 minutes and the precipitate removed by a 5 min. spin in an Eppendorf centrifuge (Lane B). The supernatant was applied to a Mono Q HR5/5 column equilibrated with Buffer A (25 mM Tris hydrochloride, pH 7.5, 1 mM EDTA, 20 mM DTT) containing 100 mM NaCl. The column was washed with 10 ml Buffer A containing 100 mM NaCl, followed by elution with a 13 ml linear gradient of 100 mM NaCl to 1 M NaCl in Buffer A. 1 ml fractions were collected and in A) an aliquot analysed on a 5cm, 10-20% gradient polyacrylamide gel prepared according to Fling and Gregerson (1986) (Lanes 1-22). B) A Western blot of even numbered Mono Q fractions (see A) probed with a monoclonal antibody specific for Com is shown. Lanes are marked as in A. Gel conditions were as in A, conditions for western blotting will be described in detail elsewhere (A. Seiler and F.G.Wulczyn, in preparation). The arrow marks the position of Com. The faint band migrating above Com is not always detected in Westerns and has not been observed by silver staining. It may represent an oligomeric form of Com that was incompletely denatured prior to electrophoresis.

stimulate ternary initiation complex formation despite unfavorable secondary structure. Naturally, the two mechanisms are not mutually exclusive. The availibility of purified Com protein will allow the question of mechanism to be addressed biochemically. Using Com purified by the method reported here, we have recently shown that Com is an mRNA binding protein specific for TAS, and have begun to study the conformation of the *mom* TIR in response to Com binding (F. G. Wulczyn and R. Kahman, in preparation).

There is little precedent for the specific, positive regulation of translation performed by Com. It is interesting to speculate on the advantage for Mu, especially since *com* and *mom* are transcriptionally linked. One clue might lie in the poor expression of *com*. If initial transcription from P_{mom} leads to slow accumulation of Com, activation of *mom* translation would not be immediate. The instability of the *mom* mRNA in the absence of translation would amplify this effect. As Com accumulates, a burst of *mom* expression would occur slightly delayed compared to the onset of transcription, even though *com* and *mom* are co-transcribed. In this view, Com supplements control at the level of transcription and helps to protect the phage from the deleterious effects of premature *mom* expression. Delaying the onset of *mom* translation would allow a more efficient *mom* TIR to be tolerated. The more tightly *mom* is regulated, the more modification sites Mu can afford to have on its DNA. Mu could thus maximize its protection from host restriction while minimizing the adverse effects of Mom modification. Whether or not this scenario is true, the uniqueness of the regulatory mechanisms used by phage Mu, and the combination of controls at the levels of transcription, translation and mRNA stability have made *mom* a rewarding system for study.

Acknowledgements

We thank all members of the Mu group in Berlin for their contributions to this work. Dr John McCarthy kindly provided us plasmid pJLA503. M. Bölker made unpublished data available for our use, and contributed greatly in the formulation of the model of *mom* transcriptional control. F.G.W. thanks C. Koch for good advise on protein purification. We are especially grateful to the organizers of the NATO/EEC meeting for the wonderful job they did, and to the participants for many stimulating discussions. This work was supported by the Deutsche Forschungsgemeinschaft (Ka 411/4-1).

References

Belasco J G, Higgins CF (1988) Mechanisms of mRNA decay in bacteria: a perspective. Gene 72: 15-23

Bölker M, Kahmann R (1989) The *Escherichia coli* protein OxyR discriminates between methylated and unmethylated states of the phage Mu *mom* promoter. EMBO 8: 2403–2410

Bölker M, Wulczyn F G, and Kahmann R (1989) Role of bacteriophage Mu C protein in activation of the *mom* gene promoter. J Bacteriol 171: 2019-2027

Brot N, Caldwell P, Weissbach H (1980) Autogenous control of the *Escherichia coli* ribosomal protein L10 synthesis in vitro. Proc Natl Acad Sci 77: 2592-2595

Cole J R, Nomura M (1986) Changes in the half-life of ribosomal protein messenger RNA caused by translational repression. J Mol Biol 188: 383-392

Fling S, Gregerson D (1986) Peptide and protein molecular weight determination using a high-molarity Tris buffer system without Urea. Anal Biochem 155: 83-88

Gorski K, Roch J-M, Prentki P, and Krisch H M (1985) The stability of bacteriophage T4 gene 32 mRNA: a 5′ leader sequence that can stabilize mRNA transcripts. Cell 43: 461-469

Gralla J, Steitz J, Crothers D (1974) Direct physical evidence for secondary structure in an isolated fragment of R17 bacteriophage mRNA. Nature 248: 204-208

Hattman S, Ives J, (1984) S1 nuclesae mapping of the phage Mu *mom* promoter: a model for the regulation of *mom* expression. Gene 29: 185-198

Hattman S, Ives J, Wall L, and Maric S (1987) The bacteriophage Mu *com* gene appears to specify a translation factor required for *mom* gene expression. Gene 55: 345-351

Kahmann R (1983) Methylation regulates the expression of a DNA-modification function encoded by bacteriophage Mu. Cold Spring Harbor Symp Quant Biol 47: 639-646

Kahmann R and Hattman S (1987) Regulation and expression of the *mom* gene. In: Symonds N et al (eds) Phage Mu.Cold Spring Harbor, N.Y., Cold Spring Harbor Laboratory, pp 93-110

Kahmann R, Seiler A, Wulczyn F G, and Pfaff E (1985) The *mom* gene of bacteriophage Mu: a unique regulatory scheme to control a lethal function. Gene 39: 61-70

Klippel A, Mertens G, Patchinsky T, Kahmann R (1988) The DNA invertase Gin of phage Mu: formation of a covalent complex with DNA via a phosphoserine at amino acid position 9. EMBO J 7: 1229-1237

Nilsson G, Belasco J G, Cohen S N, and von Gabain A (1987) Effect of premature termination of translation on mRNA stability depends on the site of release. Proc Natl Acad Sci USA 84: 4890-4894

Schauder B, Blöcker H, Frank R, McCarthy J (1987) Inducible expression vectors incorporating the *E. coli atpE* translational initiation region. Gene 52: 279-283

Seiler A, Blöcker H, Frank R, and Kahmann R (1986) The *mom* gene of bacteriophage Mu: the mechanism of methylation-dependent expression. EMBO J 5: 2719-2728

Toussaint A (1976) The DNA modification function of temperate phage Mu-1. Virology 70: 17–27

Toussaint A (1977) The modification function of bacteriophage Mu-1 requires both a bacterial and a phage function. J Virol 23: 825-826

Wulczyn F G, and Kahmann R (1987) Post-transcriptional regulation of the bacteriophage Mu mom gene by the com gene product. Gene 51: 139-147

Wyckoff E, Sampson L, Hayden M, Parr R, Huang W M, and Casjens S (1986) Plasmid vectors useful in the study of translation initiation signals. Gene 43: 281-286

MECHANISMS OF RIBOSOMAL PROTEIN TRANSLATIONAL AUTOREGULATION

David E. Draper
Department of Chemistry
Johns Hopkins University
Baltimore, MD 21218
USA

Introduction

About 10 years ago several laboratories showed that expression of many *E. coli* ribosomal proteins in is translationally regulated (Lindahl & Zengel, 1979; Dennis & Fiil, 1979; Yates et al., 1980; Brot et al., 1980). These observations were eventually summarized in the autoregulation hypothesis (Nomura et al., 1984), which stated that

1) Each ribosomal protein operon contains one ribosomal protein which can bind a target site in the mRNA to repress translational of all ribosomal proteins in the operon;

2) Ribosome assembly competes with the repressor function of a regulatory ribosomal protein, so that repression only takes place when a free pool of repressor protein starts to accumulate; and

3) competition is a consequence of the repressor protein using the same binding site to recognize both ribosomal RNA and the messenger RNA target.

To date all known r-proteins which have repressor activity are also known to bind directly and independently to large or small subunit rRNAs, consistent with the last point. It therefore seems likely that the repression mechanisms arose because the mRNAs evolved structures which take advantage of RNA binding sites pre-existing in some of the ribosomal proteins, rather than because the proteins evolved new RNA-binding functions. This raises the interesting question of how the same protein binding site may interact with two different RNAs to accomplish two very different functions.

About eight of these mRNA - ribosomal protein interactions have now been studied in different laboratories, and some themes have become apparent. In about half the cases it seems evident that the protein

NATO ASI Series, Vol. H 49
Post-Transcriptional Control of Gene Expression
Edited by J. E. G. McCarthy and M. F. Tuite

recognizes a limited RNA structure containing the ribosome binding site and probably competes directly with the ribosome for mRNA binding. The RNA structure in each of these cases bears a strong resemblance to the ribosomal RNA structure. In the remaining systems the mRNA structures involved are more complex, and the translational repression mechanism may involve a more indirect effect of the bound protein on initiating ribosomes. In this review I will summarize the information available for two of the best studied systems, L1 and S4, which represent these two different classes of repression systems, and refer to similarities with other systems.

L1 recognizes a simple mRNA structure with strong similarities to the rRNA recognition site

The L11 operon contains genes for two ribosomal proteins, L1 and L11; translation of both is repressed by L1 *in vivo* (Dean & Nomura, 1980) and *in vitro* (Yates et al., 1980). The mRNA target site for L1 has been defined by deletion analysis (Baughman & Nomura, 1983), site-directed mutagenesis (Baughman & Nomura, 1984), and selection of derepressed mutations (Thomas & Nomura, 1987). The secondary structure of the target site, shown in Figure 1, is supported by the site-directed mutagenesis experiments as well as by enzymatic "structure mapping" experiments (Kearney & Nomura, 1987).

In "bind and chew" experiments the L1 protein was found to recognize a fragment of the large subunit rRNA from both *E. coli* and *Dictyostelium discoideum* (Gourse et al., 1981). Overproduction of this RNA fragment *in vivo* causes derepression of L1 and L11 synthesis, as predicted by the autoregulation hypothesis (Said et al., 1988). As seen in Figure 1, there are striking similarities between the primary and secondary structures of the mRNA and rRNA. Mutations made at similar positions in the two RNAs tend to have similar effects on L1 binding *in vivo* (Thomas & Nomura, 1987; Said et al., 1988). It is undoubtedly the case that the same binding site on L1 protein recognizes both the ribosomal and messenger RNAs.

In several other cases there are strong similarities between the messenger and ribosomal RNA binding sites of the repressor protein; these have been reviewed recently (Draper, 1989). The strong similarities argue for evolution of the mRNA to utilize a pre-existing RNA binding site on the repressor protein.

L1 probably represses translation by a direct displacement mechanism

Procaryotic ribosomes examine ~30 nt of the mRNA in selecting an initiation site; the most prominent features within this ribosome binding site (RBS) are the Shine-Dalgarno sequence and the initiation codon (Gold, 1988). Most translational repressors which have been identified recognize either some primary sequence within the RBS (for instance, the phage T4 *regA* protein, Karam et al., 1981) or a secondary structure which sequesters part of the RBS (the R17 coat protein is a well studied example; Carey et al., 1981). The most obvious mechanism for translational repression in these cases is a direct competition between the repressor protein and the ribosome for binding to the messenger RNA, though competition has yet to be directly demonstrated in any system.

An indirect argument for a competition mechanism in the L1 system can be made from data collected on mRNA mutations selected for defective repression (the equivalent of an mRNA constitutive for translation). It is common to report the results of *in vivo* repression experiments in terms of

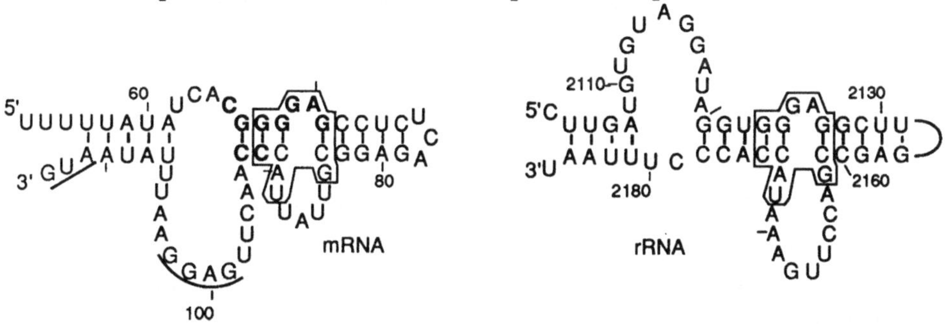

Figure 1. Similarities between the mRNA and rRNA recognition sites of L1. The secondary structure of the rRNA on the right is deduced from phylogenetic comparisons (Gutell & Fox, 1989). The mRNA structure has been drawn in a similar way; only base pairs G49-C76 and G50-C75 have been demonstrated by compensatory base changes. The *box* indicates a region of strong primary and secondary structure similarity, bases in *bold face* type are positions at which derepressed mRNA mutants have been isolated, and *underlines* indicate the initiation codon and Shine/Dalgarno sequence.

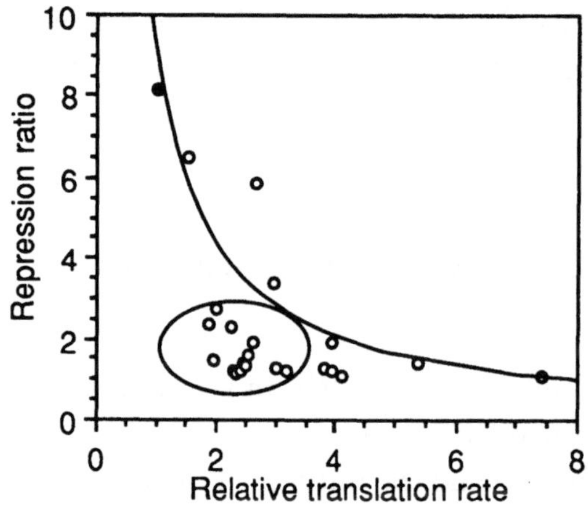

Figure 2. Translational repression as a function of L11 mRNA translation rates for mRNA mutants. Data are taken from Thomas & Nomura (1988). A group of mutants which are probably defective in binding L1 repressor are circled; a smooth curve connects mutants which have elevated translation rates but probably bind L11 normally. Wild type mRNA is the solid dot in the upper left hand corner of the graph.

the "repression ratio", the rate of translation in the absence of repressor divided by the rate in the presence of repressor. In these terms, a competition mechanism predicts that

$$\textbf{repression ratio} = \frac{1+\mathbf{K}_{30\,S}[\mathbf{30S}]}{1+\mathbf{K}_{30\,S}[\mathbf{30S}]+\mathbf{K}_{R}[\mathbf{R}]} \qquad (1)$$

where K_R is the repressor-mRNA binding affinity, K_{30S} the ribosome-mRNA affinity, and [30S] and [R] are the thermodynamically free concentrations of ribosome subunits and repressor, respectively (Draper, 1987). Thus mutations in the mRNA can reduce the repression level by either weakening K_R or strengthening K_{30S}. There appears to be a strong correlation between translation rates and K_{30S}, so the above equation predicts that a plot of repression ratio should hyperbolically approach one as the translation rate (in the absence of repression) increases.

Figure 2 shows such a plot for mutations in the L1 target site of the L11 operon. One group of mutants reduces translational repression without much change in the level of translation; these are likely to be mRNAs which bind the L1 repressor poorly. (Since translation of the L11 operon is repressed by about 50% in normally growing cells, a factor of two increase in expression is expected for mutations which weaken K_R significantly.) Another group of mutants shows decreasing repression with increasing translational efficiency. Although this is qualitatively the behavior predicted by equation (1), the curve is not quantitatively fit by a hyperbola. Since the model used to derive equation (1) does not take into account factors such as changes in mRNA stability with increasing repression, a precise fit is not expected. These data strongly support a competition mechanism for translational repression.

S4 recognizes a complex mRNA pseudoknot structure

The S4 protein regulates translation of the α operon by binding to a ~110 nt sequence near the mRNA 5' terminus (Thomas & Nomura, 1987; Deckman et al., 1987). Measurement of both S4 binding and translational repression with an extensive set of site-directed mutations in the mRNA has established the secondary structure shown in Figure 3 (Tang & Draper, 1989, 1990). Formation of this structure is required for S4 to bind and for regulation to take place, but is not necessarily the most stable or the only structure in solution. Obviously the coding regions must unfold for ribosomes to translate.

S4 recognizes a complex, 460 nt domain of the 16S rRNA (Vartikar & Draper, 1989). S4 clearly cannot interact with all of this RNA, and it is likely that the protein recognizes some portion of this structure which is formed from segments widely separated in the primary sequence. In support of this, we find that some small deletions can be made within the domain without affecting S4 recognition, but it appears that three or four parts of the secondary structure, spread throughout the 460 nt sequence, are required to fold the RNA into the structure recognized by S4 (A. Sapag, J. Vartikar, & D.E.D., unpublished results). Since there are no obvious secondary interactions between these required segments, it appears that the RNA has some specific tertiary structure which is recognized by S4. In contrast to the L1 repression system, the S4 recognition site in the 16S rRNA bears little resemblance to the α messenger RNA. This also hints that S4 recognizes some tertiary structure which is similar between the two RNAs but formed from different secondary structure elements.

There are several other autoregulation systems for which the mRNA

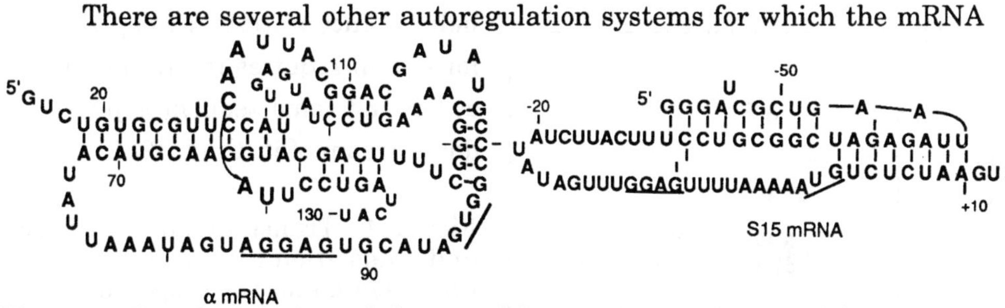

Figure 3. Regulatory sites of the α and S15 mRNAs. The pseudoknot secondary structures are taken from Tang & Draper (1989) and Philippe et al. (1990). The initiation codons and Shine/Dalgarno sequences are underlined.

and rRNA show little similarity. One is the S15 mRNA, regulated by S15; its probable secondary structure is shown in Figure 3 (Philippe et al., 1990). This is also a pseudoknot with some striking similarities to the α mRNA; the implications of this for repression mechanisms is discussed below. Although the S15 protein recognizes a limited region of the 16S rRNA, there is no unambiguous similarity with the mRNA. Again, the pseudoknot folding may achieve a similar three dimensional structure as the 16S rRNA by using a different base pairing strategy.

The S20 protein, which regulates its own synthesis, binds a domain of 16S rRNA only slightly less complex than S4 (Ehresmann et al., 1980), but its mRNA target site is a short sequence with little potential for secondary structure. Parsons et al. (1988) suggested that the lack of similarity between the two RNAs is because S20 represses translation by a fundamentally different mechanism, binding only to the initiation complex, rather than to the mRNA directly. This intriguing proposal was prompted by the observation that S20 - mRNA interactions cannot be detected under the same conditions that allow measurement of the S20 - rRNA binding constant (Donly &Mackie, 1988). As yet there is no direct evidence for S20 interaction with the initiation complex.

S4 binding affects translational initiation by an allosteric mechanism

The regulatory target sites for S4 and S15 are remarkably similar: both contain the RBS on the long, single stranded linker of a pseudoknot, and have the initiation codon situated at the beginning of the pseudoknot helix. Since initiation complex formation requires the initiation codon and some sequences just downstream of it (Gold, 1988;

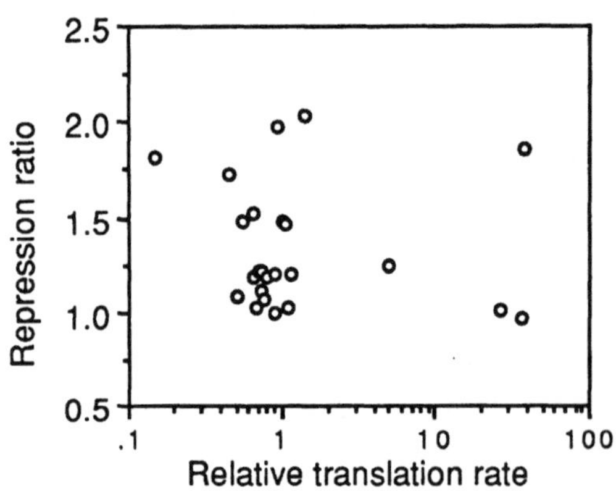

Figure 4. Translational repression and translation rates are uncorrelated for the α operon. Data are plotted as for the L11 operon in Figure 2, only a log scale is used to spread out the larger range of translation rates. Data are taken from Tang & Draper (1990).

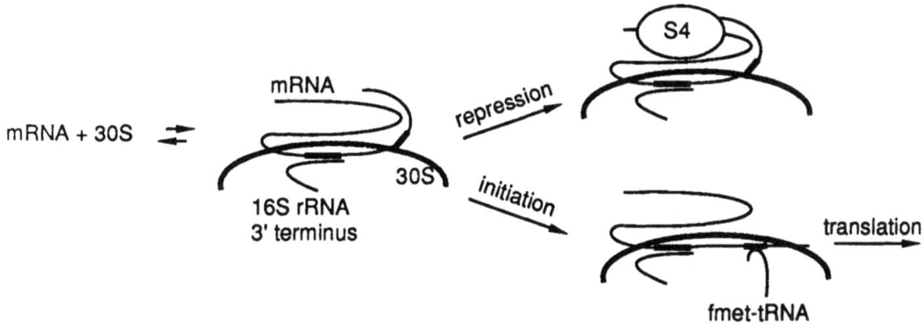

Figure 5. A mechanism for translational repression.in the α operon. 30S ribosomal subunits bind to the α mRNA *via* the Shine-Dalgarno base pairing with the 16S rRNA. The complex can then bind fmet-tRNA to initiation translation, or bind S4 repressor in a dead-end complex.

Looman et al., 1987), it is easy to imagine that repressor binding competes with ribosome initiation by stabilizing the pseudoknot structure. Some of the site-directed mutations made in the α operon have large effects on ribosome translation rates, and therefore should affect the repression ratio measured *in vivo* as predicted by equation (1). This correlation is not seen (Figure 4); in one case even a 40 fold increase in translational efficiency had no effect on repression. It thus seems unlikely that repression of the α operon takes place by the same competition mechanism as the L11 operon.

Another clue to the mechanism of translational repression in the α operon is gained by noting that the level of repression is frequently much less than predicted on the basis of the measured S4-mRNA binding constant (Tang & Draper, 1990). In a couple extreme cases, the binding constant of a mutant mRNA is exactly the same as wild type, but the mRNA is not repressed by S4 *in vivo* (these mRNAs are translated with normal efficiency). An allosteric mechanism can explain these results. We presume that the mRNA contains three functional components: an S4 binding site, a ribosome initiation site which does not overlap with the S4 site, and some structure which allosterically links the two sites, so that bound repressor affects initiation. Mutations may then disrupt the linking structure without affecting either repressor binding or translational efficiency.

A scheme which explains the known features of S4 repression is shown in Figure 5. Ribosome binding to mRNA probably takes place in several steps: a first binding to a "stand-by" site primarily involving the Shine/Dalgarno base pairing, followed by a factor-mediated shift to the true initiation site and eventually the binding of fmet-tRNA (Calogero et al.,

1988; Canonaco et al., 1989). The scheme therefore shows initiation complex formation taking place in two steps. Formation of the Shine-Dalgarno base pairing takes place first, in an mRNA structure with the pseudoknot formed. The second step is binding of the fmet-tRNA, which does disrupt the pseudoknot. It is at this point that bound S4 can slow the initiation rate by stabilizing the pseudoknot structure. Mutations in the mRNA could alter K_{30S} (and the initiation rate) by changing the accessibility of the Shine-Dalgarno sequence to ribosomes, but such mutations would not necessarily affect the ability of S4 to repress translation. A prediction of this mechanism is that a ternary S4-ribosome-mRNA complex should be formed.

Whether the same mechanism can account for the S15 mRNA regulation is open to question. A potentially key difference between them is that the Shine/Dalgarno sequence is quite accessible to chemical reagents in the α mRNA (G. Spedding & D.E.D., unpublished observations), while the similar S15 mRNA sequence is rather protected and may be involved in some additional structure (Philippe et al., 1990). S15 binding may therefore affect a different step in initiation complex formation than S4 does.

It should be clear from the discussion of these autoregulatory systems that different repressors may inhibit translation by acting at different steps in the assembly of a ribosome initiation complex. Analysis of these repression systems may be a useful way to define the detailed pathways of translational initiation in prokaryotes.

Acknowledgments

Work from the author's laboratory was supported by NIH grant GM29048, and the author has been supported by a Research Career Development Award (CA01081).

References

Baughman G, Nomura M (1983) Localization of the Target Site for Translational Regulation of the L11 Operon and Direct Evidence for Translational Coupling in Escherichia coli. Cell 34:979-988

Baughman G, Nomura M (1984) Translational regulation of the L11 ribosomal protein operon of *Escherichia coli*: Analysis of the mRNA target site using oligonucleotide-directed mutagenesis. Proc Natl Acad Sci USA 81:5389-5393

Brot N, Caldwell P, Weissbach H (1980) Autogeneous control of *Escherichia coli* ribosomal protein L10 synthesis *in vitro*. Proc Natl Acad Sci USA 77:2592-2595

Calogero RA, Pon CL, Canonaco MA, Gualerzi CO (1988) Selection of the mRNA translation initiation region by *Escherichia coli* ribosomes. Proc Natl Acad Sci USA 85:6427-6431

Canonaco MA, Gualerzi CO, Pon CL (1989) Alternative occupancy of a dual ribosome binding site by mRNA affected by translation factors. Eur J Biochem 182:501-506

Carey J, Cameron V, de Haseth PL, Uhlenbeck OC (1983) Sequence-specific interaction of R17 coat protein with its ribonucleic acid binding site. Biochemistry 22:2601-2610

Dean D, Nomura M (1980) Feedback regulation of ribosomal protein gene expression in *Escherichia coli*. Proc Natl Acad Sci USA 77:3590-3594

Deckman IC, Draper DE, Thomas MS (1987) S4 - α mRNA translation regulation complex I. Thermodynamics of formation. J Mol Biol 196:313 - 322

Dennis PP, Fiil NP (1979) Transcriptional and post-transcriptional control of RNA polymerase and ribosomal protein genes cloned on composite ColE1 plasmids in the bacterium *Escherichia. coli*. J Biol Chem 254:7540-7547

Donly BC, Mackie GA (1988) Affinities of ribosomal protein S20 and C-terminal deletion mutants for 16S rRNA and S20 mRNA. Nucleic Acids Res 16:997-1010

Draper DE (1987) Translational Regulation of Ribosomal Protein Synthesis in *Escherichia coli*: Molecular Mechanisms. In: Ilan J (ed) Translational Regulation of Gene Expression, Plenum Press, New York London, p 1

Draper DE (1989) How do proteins recognize specific RNA sites? New clues from autogenously regulated ribosomal proteins. Trends Biochem Sci 14:335-338

Ehresmann C, Stiegler P, Carbon P, Ungewickell E, Garrett RA (1980) The topography of the 5' end of 16-S RNA in the presence and absence of ribosomal proteins S4 and S20. Eur J Biochem 103:439-446

Gold L (1988) Posttranscriptional Regulatory Mechanisms in *Escherichia coli*. Ann Rev Biochem 57:199-234

Gourse RL, Thurlow DL, Gerbi SA, Zimmermann RA (1981) Specific binding of a prokaryotic protein to a eukaryotic ribosomal RNA: Implications for evolution and autoregulation. Proc Natl Acad Sci USA 78:2722-2726

Gutell RR, Fox GE (1989) Compilation of large subunit RNA sequences presented in a structural format. Nucleic Acids Res16:r175-r269

Karam J, Gold L, Singer BS,Dawson M (1981) Translational regulation: identification of the site on bacteriophage T4 rIIB mRNA recognized by the regA gene function. Proc Natl Acad Sci 78:4669-4673

Kearney KR, Nomura M (1987) Secondary structure of the autoregulatory mRNA binding site of ribosomal protein L1. Mol Gen Genet 210:60-68

Looman, A. C., Bodlaender, J., Comstock, L. J., Eaton, D., Jhurani, P., de Boer, H. A., & van Knippenberg, P. H. Influence of the codon following the AUG initiation codon on the expression of a modified *lacZ* gene in *Escherichia coli.* (1987) EMBO J 6:2489-2492

Lindahl L., Zengel, JM (1979) Operon-specific regulation of ribosomal protein synthesis in *Escherichia. coli.* Proc Natl Acad Sci USA 76:6542-6546

Nomura M, Gourse R, Baughman G (1984) Regulation of the Synthesis of Ribosomes and Ribosomal Components. Ann Rev Biochem 53:75-118

Parsons GD, Donly BC, Mackie GA (1988) Mutations in the leader sequence and initiation codon of the gene for ribosomal protein S20 (rpsT) affect both translational efficiency and autoregulation. J Bact 170:2485-2492

Phillipe C, Portier C, Mougel M, Grunberg-Manago M, Ebel JP, Ehresmann B, Ehresmann C (1990) Target Site of *Escherichia coli* Ribosomal Protein S15 on its Messenger RNA. J Mol Biol 211:415-426

Said B, Cole JR, Nomura M (1988) Mutational analysis of the L1 binding site of 23S rRNA in *Escherichia coli.* Nucleic Acids Res 16:10529-10545

Tang CK, Draper DE (1989) An Unusual mRNA Pseudoknot Structure is Recognized by a Protein Translational Repressor. Cell 57:531-536

Tang CK, Draper DE (1990) Evidence for allosteric coupling between the ribosome and repressor binding sites of a translationally regulated mRNA. Biochemistry, in press

Thomas MS, Nomura M (1987) Translational regulation of the L11 ribosomal protein operon of *Escherichia coli*: mutations that define the target site for repression by L1. Nucleic Acids Res 15:3085-3096

Thomas MS, Bedwell DM, Nomura M (1987) Regulation of α Operon Gene Expression in *Escherichia coli.* A Novel Form of Translational Coupling. J Mol Biol 196:333-345

Vartikar JV, Draper DE (1989) S4 - 16 S Ribosomal RNA Complex: Binding Constant Measurements and Specific Recognition of a 460 Nucleotide Region. J Mol Biol 209:221-234

Yates JL, Arfsten AE, Nomura M (1980) *In vitro* expression of *Escherichia. coli* ribosomal protein genes: autogenous inhibition of translation. Proc Natl Acad Sci USA 77:1837-1841

TRANSLATIONAL FEEDBACK CONTROL IN *E.COLI*: THE ROLE OF $tRNA^{Thr}$ AND $tRNA^{Thr}$-LIKE STRUCTURES IN THE OPERATOR OF THE GENE FOR THREONYL-tRNA SYNTHETASE

M. Springer, P. Lesage, M. Graffe, J. Dondon, M. Grunberg-Manago.
Institut de Biologie Physico-Chimique.
13, rue Pierre et Marie Curie, 75005 Paris, France.

and

H. Moine, P. Romby, J.-P. Ebel, C. Ehresmann, B. Ehresmann.
Institut de Biologie Moléculaire et Cellulaire.
15, rue Descartes, 67084 Strasbourg Cedex, France.

INTRODUCTION

In prokaryotes, protein-mediated translational control often takes the form of a negative feedback (Gold, 1988; Lindahl *et al.*, 1986). The translational repressor, as proven in some examples, interacts with a regulatory region of its own mRNA, affecting ribosome binding and thus translation. In some cases in bacteriophage and *E.coli*, a particular feedback was shown to belong to a more general regulatory system. For instance, a particular feedback due to a specific ribosomal regulatory protein is modulated by the cellular concentration of the ribosomal RNA (rRNA) to which it binds (Nomura *et al.*, 1984). If the cellular rRNA concentration increases, the specific regulatory protein binds to the excess of rRNA and not to its own mRNA whose translation is increased. This mRNA-rRNA competition permits the ribosomal protein synthesis to be adapted to the cellular rRNA concentration. As suggested in several cases, the binding site of the ribosomal regulatory protein on both its ligands (mRNA and rRNA) could share some similarity. This hypothesis has been called *molecular mimicry* (Campbell *et al.*, 1983) and implies that there is a common site on the regulatory protein that recognises both nucleic acid ligands. This is a simple strategy for adding a regulatory role to a protein involved in nucleic acid binding without having the necessity for a separate regulatory domain.

The expression of the gene for *E.coli* threonyl-tRNA synthetase (*thrS*) is negatively autoregulated at the translational level *in vivo* (Butler *et al.*, 1986; Springer *et al.*, 1985) and *in vitro* (Lestienne *et al.*, 1984). The site acting in *cis* responsible for that regulation was shown to be located immediately upstream to the translation initiation codon of *thrS* (Springer *et al.*, 1986) in a region that has primary and secondary structural analogies

NATO ASI Series, Vol. H 49
Post-Transcriptional Control of Gene Expression
Edited by J. E. G. McCarthy and M. F. Tuite

with the anticodon arm of several threonine isoacceptor species (Moine *et al.*, 1988). Evidence that the synthetase recognises the similar parts of its two RNA ligands --the anticodon-like arm of the mRNA and the true anticodon arm of the tRNA-- in an analogous way comes from studies involving mutants of the synthetase (Springer *et al.*, 1989). In these studies, mutants (called super-repressors) were isolated that repress the translation of their mRNA in *trans* to an extreme level. Other mutants that are completely unable to perform any repression were also isolated. The super-repressors, which are suspected to bind their mRNA with high affinity, were shown to bind their tRNA with an increased affinity. The non-repressing mutants, which are suspected to have lost their capacity to bind the mRNA, were shown to bind their tRNA with less affinity. The binding properties of the mutant enzymes for the other substrates, ATP and threonine, are unchanged.

The present work reviews recent experiments about the mechanism of autoregulation of the expression of the *thrS* gene and how this specific feedback integrates in a more general regulatory scheme. We also provide evidence that the *thrS* translational operator is larger than suspected in previous studies and that it includes other structural domains. Besides the anticodon-like domain that was previously characterised, another domain is shown to be essential for regulation and to have striking homologies with the aminoacid acceptor arm of the threonine isoacceptor tRNAs.

RESULTS

I - Threonyl-tRNA synthetase represses the translation of its own mRNA by binding to it and competing with the ribosome

Although previous genetic data suggested that threonyl-tRNA synthetase interacts with its mRNA to inhibit its translation, the formal proof of that interaction comes from nitrocellulose retention assays and footprinting experiments (Moine *et al.*, 1990). The latter experiments show that the binding of the synthetase to the wild-type RNA causes significant reactivity changes toward both enzymatic and chemical probes : an extensive reduction of reactivity is observed in all the domains with the exception of domain 3 (Figure 3).

In the ribosome binding site (RBS) of *thrS* (domain 1 in Figure 3) one observes in the presence of the synthetase, in addition to an extensive reduction of susceptibility to RNase V1, a complex pattern of reactivity changes with ENU (ethylnitrosourea), in which reductions and enhancements of reactivity are juxtaposed. A reduced reactivity of several uridines towards CMCT (1-cyclohexyl-3-(2-morpholinoethyl) carbodiimide metho-p-toluene sulfonate) is also observed (Moine *et al.*, 1990). Our results suggest that the protein does shield the Shine-Dalgarno region, but also causes some conformational adjustments in this region.

An important reduction of reactivity is observed in domain 2 (Figure 3), containing the previously genetically defined operator, in the presence of the synthetase. All the bases of the anticodon-like loop become almost unreactive toward CMCT, with the notable exception of G_{-30} whose reactivity is even slightly enhanced. The anticodon-like loop is protected from RNase T1. The binding of the synthetase also causes extensive changes in reactivity in domain 4, with both enzymatic and chemical probes. This last result was unexpected since previous selections did not yield operator constitutive mutants in that region of the *thrS* mRNA.

Since in most cases the sites acting in *cis* in translational control were found to be located near to or in the RBS of the controled gene, the general belief was that the repressor would directly compete with the binding of the ribosome to the controled mRNA. This has now been proven in several cases in bacteriophage T4 (McPheeters *et al.*, 1988; Winter *et al.*, 1987). Another possibility was that the binding of the repressor would trap the ribosome on the RBS of the controled gene in such a way that elongation of the polypeptidic chain is impossible (Draper, 1987). These two possibilities were investigated in the case of *thrS*

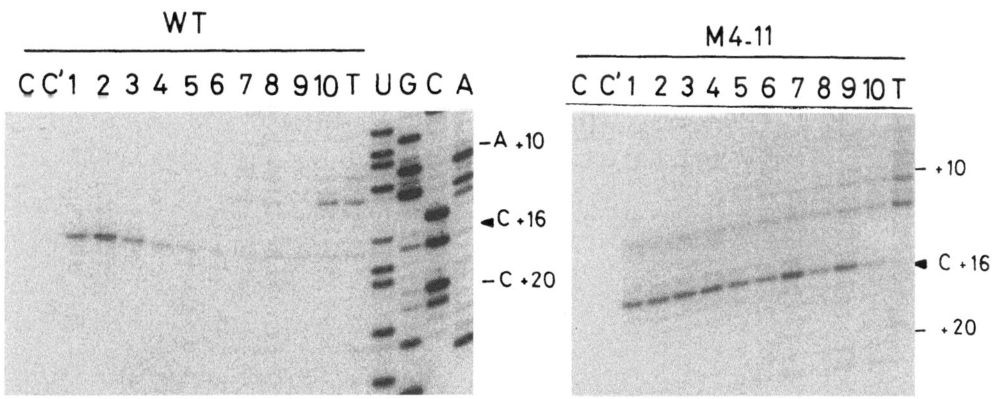

Figure 1 : *Toeprint experiment with wild-type and mutant (M4-11) thrS RNA. The stop of the reverse transcriptase at position +16 (+1 is the A of the AUG initiation codon) is shown with an arrow. The incubation mixture contained RNA, 30S subunits and initiator tRNA. Lane C: control without 30S; lane C': control without initiator* $tRNA_f^{Met}$. *Lane T: control without 30S but with 3x10-5 M threonyl-tRNA synthetase. Incubations with 30S and* $tRNA_f^{Met}$ *but without synthetase are shown in lane 1. In lanes 2 to 10, the concentration of synthetase increases from* $1x10^{-9}$ *M to* $1x10^{-8}$ *M,* $5x10^{-8}$ *M,* $1x10^{-7}$ *M,* $5x10^{-7}$ *M,* $1x10^{-6}$ *M,* $5x10^{-6}$ *M,* $1x10^{-5}$ *and* $3x10^{-5}$ *M respectively. Several unspecific bands, which are also detected in lane T, appear at high synthetase concentration. They most likely arise from degradation due to some RNase contamination of the synthetase. These bands were not found with all enzyme preparations.*

using the reverse transcription inhibition or "toeprint" assay (Hartz *et al.*, 1988). This assay is based on the fact that a 30S-tRNA$_f^{Met}$ complex bound to the RBS blocks the elongation by reverse transcriptase primed with an oligonucleotide complementary to the mRNA 3' from the RBS.

When bound to an *in vitro* synthesised RNA which covers the regulatory regions and the RBS of *thrS*, the 30S-tRNA$_f^{Met}$ complex produces a toeprint at position +16 (Figure 1). In the presence of increasing concentrations of synthetase, the intensity of the toeprint progressively decreases to the point of a complete disappearance of the reverse transcript (Figure 1, left part). The binding of the synthetase alone does not produce any reverse transcriptase stop under the conditions used (Figure 1, left part, lane T). In the presence of an *in vitro* made RNA that carries operator constitutive mutations, a toeprint could also be detected at position +16. However, the addition of increasing amounts of synthetase has no significant effect on the toeprint. Mutant M4-11 (a C_{-19} to U mutation analogous to the mutation (3) in Figure 5) is shown as an example in the right part of Figure 1. These results indicate that the binding of the synthetase is able to inhibit the attachment of the ribosome to the wild type mRNA by competing with rather than trapping the ribosome.

II - The repression caused by the synthetase is relieved by an excess of tRNA^{Thr}

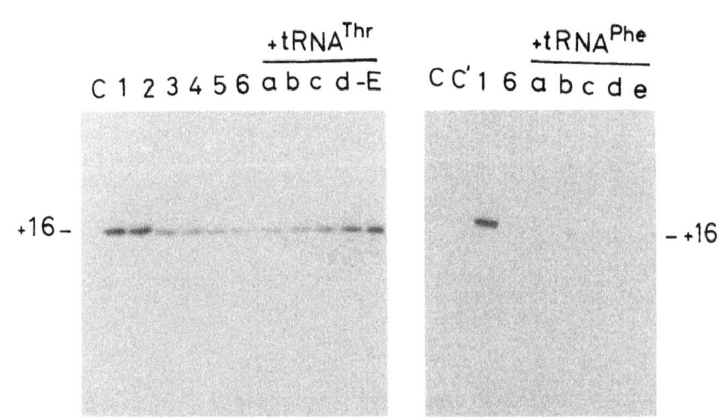

Figure 2: *Effect of tRNA^{Thr} and tRNA^{Phe} on the competition between 30S subunits and threonyl-tRNA synthetase for thrS mRNA binding. The position of the toeprint at -16 is indicated. Lane C: control without 30S, lane C': control without initiator tRNA, lane E: control with $3x10^{-6}$ M tRNA^{Thr} in the absence of synthetase. The synthetase concentration increases in lanes 1 to 6 from 0 to $1x10^{-9}$, $1x10^{-8}$ M, $2.5x10^{-8}$ M, 5 10^{-8} M and 10^{-7} M respectively. The tRNA concentration in lanes a to e raises from $3x10^{-9}$ M to $3x10^{-8}$ M, $3x10^{-7}$ M, $3x10^{-6}$ M and $3x10^{-5}$ M respectively. In the presence of tRNA, the synthetase concentration is$1x10^{-7}$ M.*

The expression of *thrS* is most probably regulated to provide the translation machinery with a sufficiently high concentration of aminoacylated $tRNA^{Thr}$. In other words, the cellular concentration of threonyl-tRNA synthetase should increase if the translation machinery uses more charged $tRNA^{Thr}$. Deacylated $tRNA^{Thr}$ is the obvious signal that should trigger the increase of threonyl-tRNA synthetase expression. The fact that this seems to be the case is indicated by the toeprinting experiment shown in Figure 2. The addition of increasing amounts of $tRNA^{Thr}$ to the [$tRNA_f^{Met}$/30S/wild-type mRNA] complex in the presence of a constant concentration of enzyme (sufficient to inhibit the 30S subunit binding) induces the reappearance of the toeprint (left part of Figure 2). This effect is specific, since the addition of a non-cognate tRNA ($tRNA^{Phe}$) does not produce any effect (right part of Figure 2). This clearly indicates that $tRNA^{Thr}$ is able to suppress the inhibitory effect of the synthetase on the binding of the ribosome to the wild-type mRNA .

III - Comparison of the operator region with threonine isoacceptor tRNAs

Figure 3 shows (lower right side) the composite sequence of the four isoacceptor species of $tRNA^{Thr}$ of *E.coli* in the classical L-shaped representation. An analogous representation is given for the *thrS* mRNA between -130 and +3 (+1 is the A of the ATG initiation codon). This representation is perfectly compatible with the available experimental data concerning the secondary structure of the *thrS* mRNA (Moine *et al.*, 1988). This representation stresses some analogies between the two RNA molecules. A previous theoretical study (McClain *et al.*, 1987) proposed that $tRNA^{Thr}$ identity is defined by nucleotides G_{35} and U_{36} in the anticodon, base pairs C_2-G_{71}, A_5-U_{68} in the acceptor arm and base pair G_{63}-C_{51} in the T stem. The last base pair is replaced by an A-T base pair in $tRNA^{Thr}{}_2$ (Komine *et al.*, 1990) and cannot be involved in the identity of the isoacceptor set. It is striking that all the other bases that could define the $tRNA^{Thr}$ identity (the GC of the anticodon and the two pairs in the acceptor stem that are squared in Figure 3) are found in analogous positions in the mRNA of *thrS*. Moreover, the $tRNA^{Thr}$ isoacceptors all carry an A at the 73 (also called discriminator) position that is also found in the mRNA in an analogous place. Finally, the mRNA also carries a CCA, which is common to all tRNAs, just 3' to the equivalent of the discriminator position (the ACCA sequence is squared in Figure 3 in both the tRNA and the mRNA). The fact that the anticodon of $tRNA^{Thr}$ is involved in identity was suspected for different reasons: an amber derivative of $tRNA^{Thr}{}_4$ was shown to be only very weakly charged by threonyl-tRNA synthetase (Springer *et al.*, 1989) and protection experiments showed that the presence of the synthetase changes the reactivity of the anticodon bases of $tRNA^{Thr}{}_3$ towards chemical and enzymatic structural probes (Théobald *et al.*, 1988). The involvement of the anticodon has recently been proven directly by transplanting a GGU threonine anticodon in a $tRNA^{Met}{}_m$ molecule and showing that this molecule is recognised by threonyl-tRNA synthetase (Schulman *et al.*, 1990). The involvement of the equivalent bases (U_{-31} and G_{-32}) in the mRNA in the control of *thrS* expression was clearly demonstrated by the fact that U_{-31} to A and G_{-32} to A or U changes destroy *thrS*'s negative feedback regulation (Springer *et al.*, 1986). Besides, as stated above, structural mapping of this region shows the existence of a 7 nucleotide long loop in the mRNA as in the tRNA (Moine *et al.*, 1990).

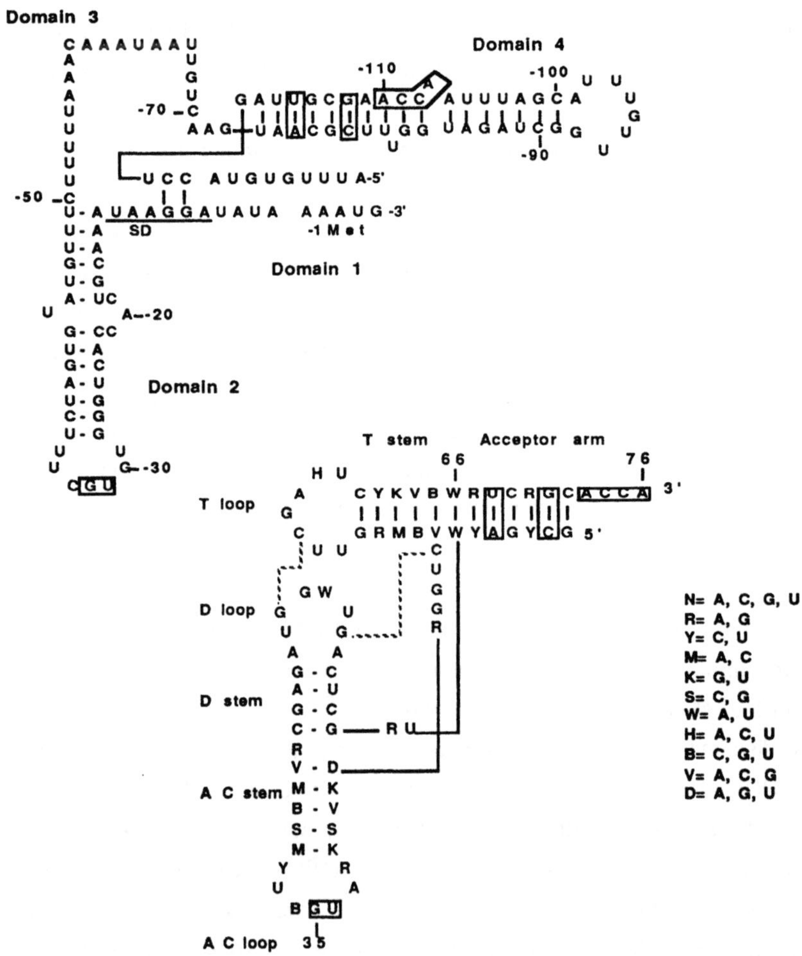

Figure 3 : *L shaped representation of the thrS leader and of the composite sequence of $tRNA^{Thr}$ isoacceptors. The homologous nucleotides are squared in the two molecules. In the case of the thrS leader mRNA only nucleotides -130 to +3 are shown. The Shine-Dalgarno sequence is underlined and shown as SD. The A of the AUG initiation codon is numbered as nucleotide 1.*

IV - The *thrS* operator and its domains

Although previous selections only gave operator constitutive mutants in the domain 2 (the analog of the anticodon arm), the existence of stable secondary structures (domain 4) prompted us to investigate the role of the whole 162 nucleotide long leader in the regulation of the *thrS* mRNA translation.

a) <u>The ribosomal binding site contributes only indirectly to regulation</u>

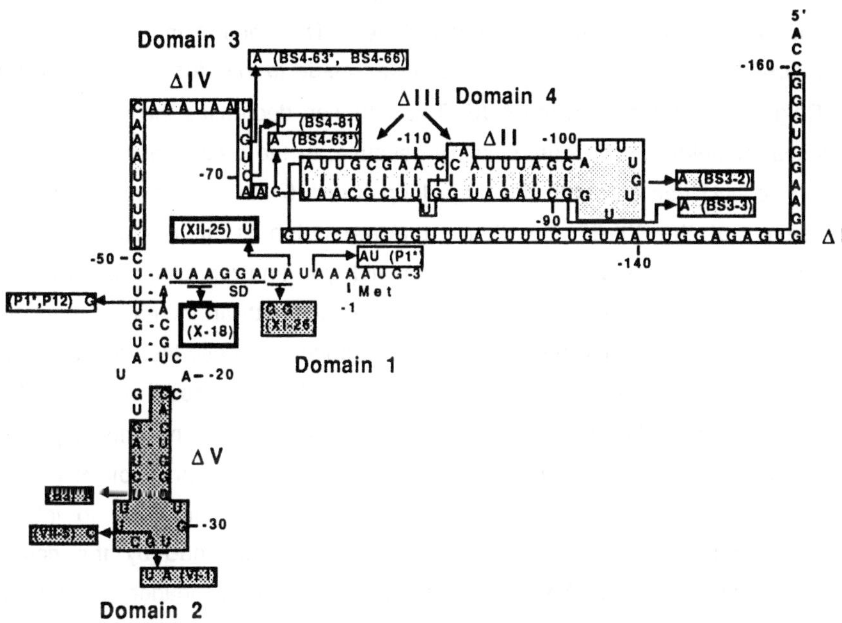

Figure 4 : *In vitro made mutations in the* thrS *leader mRNA. The mutations (deletions or point mutations) that have no effect on regulation are squared with a white background. Those squared with thick lines cause an increase in regulation. Where the background is light grey, the mutation impairs regulation, where the background is dark grey, the mutation eliminates regulation. A star following the name of a mutation indicates that the mutation is found with another change in a non-contiguous nucleotide.*

Four mutations were made by oligonucleotide site directed mutagenesis to study the role of the RBS in the control of the expression of *thrS*. A fifth one (II-12) was found as a revertant as described in the next section. The first mutation (X-18 in Figure 4) is a double change in the second and third base of the Shine-Dalgarno sequence and decreases the complementarity with the 3' end of the 16S rRNA from 6 to 3 base pairs. This change decreases about 50 fold (data not shown) the expression of ß-galactosidase synthesised from a *thrS-lacZ* fusion cloned in a bacteriophage λ vector integrated as single copy in the *E.coli* chromosome. However, the residual ß-galactosidase expression is high enough to allow observation of regulation. In conditions were the cellular level of active threonyl-tRNA synthetase is lowered by the presence of a chromosomal *thrS* mutation, the synthesis of ß-galactosidase from the X-18 RBS mutant is increased 5 times. Under the same conditions, the wild type fusion is derepressed, because of the negative feedback, about 2.5 times, i.e. slightly less. The situation is equivalent when the repression level is measured: in the presence of the RBS mutation X-18, ß-galactosidase levels are slightly more repressed than in the presence of the wild type operator with an excess of wild type threonyl-tRNA synthetase synthesised from a plasmid carrying *thrS*. Thus the X-18 mutation does not lead to loss of control. It rather seems that in the presence of that RBS mutation, *thrS* expression is more sensitive to threonyl-tRNA synthetase levels. The second mutation in this region (XII-25 in Figure 4), changes the A_{-5} to U and was made to introduce a *DraI* site between the Shine-Dalgarno sequence and the AUG. This mutation decreases the expression of *thrS* about 25 times for unknown reasons, possibly because, in the presence of that change, the RBS (from -5 to -11) can pair with the large loop (from -51 to -57) of the domain 3 loop. In the presence of the XII-25 mutation, the regulation is quite clearly increased : derepression is 17 fold instead of 2.5 fold for the wild type and repression is 40 fold instead of 15 fold. Besides these two mutations that are associated with an increased regulation, the other two RBS mutations isolated have no effect on regulation. These last mutations are II-12 which was isolated as a revertant of X-18 (Figure 5) and P1, a double mutant in both the RBS and the begining of the stem of domain 2 (Figure 4), that has no phenotype at all. In conclusion, it seems that changes in the RBS that decrease the expression of *thrS*, do not affect regulation in a negative way but rather increase the sensitivity to intracellular threonyl-tRNA synthetase levels. This is most probably explained by the competition between the synthetase and the ribosome for the binding to the *thrS* leader mRNA which was shown to exist with the toeprint experiments. Thus, if the affinity of the ribosome for the mRNA is lowered, one observes, as expected, that the synthetase has the possibility to modulate translation over a more important range. If this were true, one would expect mutations increasing the affinity of the ribosome for its binding site to partially obliterate the role of the synthetase, i.e. to cause a decreased sensitivity to threonyl-tRNA synthetase levels in the cell. This is exactly what happens in the case of the XI-26 mutation (Figure 4) which is a double change between the Shine-Dalgarno sequence and the AUG that increases the complementarity of the *thrS* RBS and the 3' end of the 16S rRNA from 6 to 9 nucleotides. This mutation increases more than twofold the already high expression of the *thrS-lacZ* hybrid and causes, as expected from the competition model, a strong decrease of the regulation since repression and derepression levels are lowered. We do not believe that this loss of regulation is due to a nucleotide change that directly affects threonyl-tRNA synthetase binding to the operator. The main reason for that belief is that both the XI-26 and XII-25 mutations affect the A_{-5} nucleotide and have opposite effects. One would expect, changing a nucleotide involved in binding the synthetase to have equivalent phenotypes for the two changes.

Taken together the data indicate that mutational changes in the RBS can cause 1) no effect at all; 2) an increased expression of *thrS* (when the Shine-Dalgarno complementarity is increased) and a negative effect on regulation ; 3) a decreased expression of *thrS* and a concommitant increase of regulation. Thus, the effects of the RBS mutations confirm the conclusions drawn from the toeprint experiments that directly prove the competition between ribosome and the synthetase. These effects also indicate that the RBS is not directly involved in binding the synthetase.

b) <u>The anticodon-like arm is essential to regulation</u>

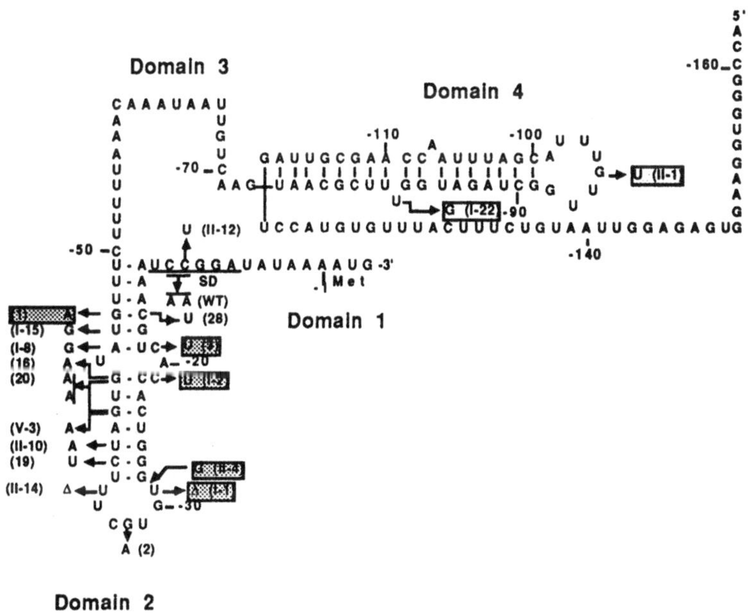

Figure 5 : *In vivo selected mutations in the thrS leader mRNA. See legend to Figure 4 for details.*

A previous selection for increased expression of ß-galactosidase from a *thrS-lacZ* fusion under repressed conditions gave five spontaneous point mutations in that region that includes nucleotides -13 to -49 (Springer *et al.*, 1986). All the mutations caused a complete loss of regulation. We investigated this region more thoroughly by deleting (ΔV in Figure 4) and point mutating (B3, VII-5, VI-1 in Figure 4) the lower part of domain 2 which is analogous to the anticodon arm of $tRNA^{Thr}$. We also isolated 15 transitions, transversions, single nucleotide deletions and insertions in that region (Figure 5). Three among these 15 were already found in the previous selection (Springer *et al.*, 1986). The present mutants were isolated using an in-phase *thrS-lacZα* fusion in M13mp8. With a wild type RBS such phages make deep blue plaques on an adequate indicator strain. If the Shine-Dalgarno X-18 mutation (see previous section) is introduced in the *thrS-lacZα* fusion, the phages make pale blue plaques since *thrS* expression is lowered about 50 times. The pale blue phenotype permitted the screening of mutant phages making dark blue plaques. After mutagenesis of the replicative form of the M13 phage carrying the *thrS-lacZα* fusion and the X-18 mutation with hydroxylamine or growth in a *mutD* strain, deep blue plaques were screened for and the DNA of the corresponding phages sequenced. The physiology of a subset of the isolated mutants (squared in Figure 5) was investigated by reconstitution of the mutated *thrS-lacZ* fusion in λ. Since the double AA to CC change (the X-18 mutation) in the Shine-Dalgarno sequence does not affect regulation in a negative way, the effect on regulation of the secondary mutations in the anticodon-like arm was investigated in the presence of the X-18 mutation. The deletion ΔV and the point mutations that were investigated all cause an increased expression and loss of regulation.

These data and the footprinting experiments indicate that this region is involved in regulation by directly binding the synthetase.

c) The large loop is only indirectly involved in regulation

The role of this region (from -50 to -73) of the *thrS* operator was investigated by isolating a deletion (ΔIV in Figure 4) of the whole domain 3. This deletion was then used to screen for bisulfite induced mutations. We studied the effect of the deletion, two single and a double mutant in that region on *thrS* expression (Figure 4). The deletion causes a strong decrease of the ß-galactosidase level from the same *thrS-lacZ* fusion used throughout this work. The regulation is also affected by the deletion : derepression and repression are lowered when compared to the wild-type operator. In contrast to the deletion, the three point mutation BS4-63, -66 and -81 have no effect whatsoever on the level of expression and on regulation.

Although we isolated only three point mutations, they affect the two Gs and one out of the two Cs of the loop, i.e. the 3/4 of the nucleotides that are in general responsible for stable structures. We thus believe that there is no nucleotide specificity in the loop whose role is to link the two essential arms of the operator: domains 2 and 3 (see discussion).

d) The stable upstream arm is involved in the regulation

The role of the stable upstream arm (domain 4 : from nucleotides -74 to -117) was first investigated with two deletions: ΔII which eliminates only the extremity of the arm and ΔIII which eliminates the entire arm (Figure 4). The deletions were used to isolate two mutations with bisulfite : G_{-91} and G_{-95} were converted independently to As (Figure 4). The screening using a *thrS-lacZα* fusion on M13mp8 and explained in section b) gave also two point mutations in that region affecting G_{-95} and U_{-82} (Figure 5). The effect of the two deletions, the II-1, BS3-2 and BS3-3 point mutations on the expression of the *thrS-lacZ* fusion are very similar: there is an increase of expression and a partial loss of control. The last point mutation, I-22 causes an increased expression of *thrS* without effect on regulation. In other words, most of the mutations in this domain have an effect on expression and regulation, thus domain 4 resembles in that respect the anticodon-like arm of the operator (domain 2). The main difference is that the mutations in domain 4 have a weaker effect than those in the anticodon-like arm on the regulation of *thrS* expression.

e) The 5' region of the thrS leader is not essential to regulation

The 5' end nucleotide of the *thrS* mRNA is either an A at -162 or another A at -163 as shown by primer extention analysis (M.Springer unpublished results). The 5' region of the *thrS* leader was deleted from nucleotides -159 to -118 (ΔI in Figure 4) leaving only three or four nucleotides in front of the stable arm of domain 4. The deletion slightly lowers the expression and affects regulation only very marginally.

DISCUSSION

All the present data on *thrS* expresion and regulation can be explained with a modol, described here, that takes into account the resemblance between domains of the operator and the threonine isoacceptor tRNAs. The structural probing of the *thrS* mRNA in solution shows that there are only two stable arms in the molecule : the anticodon-like arm of domain 2 (which is of intermediate stability) and the very stable arm of domain 4 (Moine *et al.*, 1988). These two arms are found to correspond through partial homology with the anticodon and acceptor arms in the tRNA molecule as shown in Figure 3. Consistent with this homology, recent experiments show that $tRNA^{Thr}$ is able to displace the anticodon-like arm and also the loop of domain 4 mRNA from a mRNA-synthetase complex (Moine *et al.*, 1990). It seems likely that, in solution, the free mRNA forms a molecule where the two structured arms are able to move using the single stranded region of domain 3 as an articulation. An open conformation would permit the ribosome to bind the mRNA at the accessible RBS. It is even possible, to increase accessibility, that the ribosome melts the upper part of the anticodon-like arm (consisting only of A-T base pairs). In the presence of the synthetase, we make the hypothesis that the mRNA is stabilised in an L-like structure similar to that of the tRNA molecule as shown in Figure 3. The domain 2 and 4 of the mRNA would respectively occupy the place of the anticodon and acceptor arm of the tRNA. The presence of the synthetase would fix the mRNA in a rigid structure that is incompatible with ribosome binding at the RBS.

This model explains the mutational data which indicates that only domains 2 and 4 are directly involved in regulation. The footprinting data also clearly indicates that these two domains bind the synthetase. The different effect of the deletion and the point mutations in domain 3 is also well explained : the model assumes that this single stranded region is not recognised by the synthetase but should permit the two structured arms to move sufficiently freely from one another to permit either the interaction with the synthetase or the binding of the ribosome. If this is true, nucleotide substitutions should have no effect on expression or regulation but a deletion of domain 3 should hinder a correct synthetase attachment since the mRNA would not be able to take the L form that we suppose to be essential for regulation by the synthetase. The deletion of domain 3 should cause the stacking of helices of domains 2 and 4 on top of one another. This now very rigid structure would inhibit binding of the ribosome at the RBS. As predicted, the deletion of domain 3 affects regulation and causes a decreased expression of the gene whereas point mutations in the domain have no effect. The non-involvement of domain 3 in the interaction with the synthetase is also perfectly compatible with the footprinting data that indicate no shielding due to the protein in that region. The regions upstream to domain 4 which have absolutely no equivalent in the tRNA molecule are shown by deletion to have no role in translational regulation. Finally we wish to make the point that the RBS region also without equivalent in the tRNA molecule is only indirectly involved in regulation : all the mutations isolated in this region have phenotypes that can be interpreted in terms of the ribosome-synthetase competition that is proven by the toeprinting experiments. We believe that the synthetase does not recognise specifically this region that is involved mainly in binding the ribosome. The footprint data show that the synthetase shields this zone and possibly induces conformational changes. This means that, although the synthetase has no specificity contacts with the RBS, it is in very close proximity to it since it interacts with the two arms of domain 2 and 4. This proximity is essential to explain regulation : it is only because the synthetase is near to the RBS that repression occurs. It is known that a duplication which increases the distance between the Shine-Dalgarno sequence and the start of the arm of domain 2 from 0 to 9 nucleotides causes a complete loss of regulation (Springer *et al.*, 1986). We expect this mutated mRNA to bind the synthetase normally but, because of the increase in distance between the binding site of the enzyme and that of the ribosome, the regulation is lost.

The mutational and footprint data indicate that the regions of the leader mRNA directly involved in regulation are domains 2 and 4, exactly the two same domains that have their equivalents in the tRNA molecule. Even though homologies exist, the two domains are quite different in the mRNA and the tRNA. Some of these differences might not be important if the nucleotides really involved in the specificity of the interaction with the synthetase are shared by the two molecules. A striking difference remains : domain 4 is closed at its end and the acceptor arm of the tRNA molecule is open. The closed extremity of domain 4, which has no obvious counterpart in the tRNA, is involved in regulation most probably by binding the synthetase. This indicates that there are probably regions of the synthetase that are involved in recognising the operator and not the tRNA. Mutations in such regions of the synthetase should affect regulation but not aminoacylation. We intend to specifically look for such mutants although, up to the present time, all the mutant threonyl-tRNA synthetases that were selected as changed in regulation were shown to be also changed in aminoacylation (Springer *et al.*, 1989).

The operator being recognised as a tRNA-like structure makes sense in terms of the physiology of the cell. As stated earlier, the obvious use of the the feedback of *thrS* expression is to provide an adequate quantity of charged $tRNA^{Thr}$ to the translation machinery. If the machinery consumes too much aminoacylated $tRNA^{Thr}$, uncharged $tRNA^{Thr}$ will accumulate and because it competes with the mRNA for synthetase binding, as shown by the toeprint experiments, the enzyme will bind to it and not to the *thrS* mRNA. The free mRNA is thus going to be translated, cause an increase in the concentration of threonyl-tRNA synthetase and as a consequence an increase of aminoacylated $tRNA^{Thr}$. It is interesting to note that $tRNA^{Thr}$ is also the effector for the threonine biosynthetic pathway, mainly encoded by the *thrABC* operon, which is controled by an attenuation mechanism (Gardner, 1979). In other words, $tRNA^{Thr}$ is involved in regulating both the biosynthetic pathway and the synthesis level of the corresponding aminoacyl-tRNA synthetase. This situation is reminiscent of another aminoacid, namely phenylalanine, whose biosynthesis is, through attenuation of the *pheA* gene, also controlled by tRNA aminoacylation levels (Zurawsky *et al.*, 1978). The corresponding aminoacyl-tRNA synthetase is also controled by attenuation (Springer *et al.*, 1985). In both the threonine and the phenylalanine case there is a response to tRNA aminoacylation levels although the corresponding synthetases are regulated with completely different mechanisms (translational autoregulation and attenuation). The reason for these very different mechanisms is yet unknown. In this respect, it is interesting to note that Ames (Ames *et al.*, 1983) proposed that the mRNA of the *his* operon attenuator can be drawn as a tRNA-like structure. It is possible that such structures in front of many genes evolved to mechanisms so radically different as translational control and attenuation but keeping the same effector : the tRNA.

AKNOWLEDGEMENTS

We thank E. Westhof for fruitful discussions, F. Eyermann for skilful technical assistance and D. Popham for reading the manuscript. This work was supported by grants from the CNRS, from the Ministère de la Recherche et de l'Education Nationale, from the "Association pour la Recherche sur le Cancer" (to M. G.-M.), from the CEE (contract SCI*/0194-C(AM) to M. G.-M.), from INSERM (grant 891017 to M.S.) and from E.I. du Pont de Nemours.

REFERENCES

Ames B N, Tsang T H, Buck M, Christman M (1983) Proc. Natl. Acad. Sci. USA 80:5240-5242.

Butler J S, Springer M, Dondon J, Grunberg-Manago M (1986) Posttranscriptional autoregulation of *Escherichia coli* threonyl-tRNA synthetase expression *in vivo*. J.Bacteriol. 165:198-203

Campbell K M, Stormo G D, Gold L (1983) Protein-mediated translational repression; In: J. Beckwith, J. Davies and J. Gallant (eds) Gene function in prokaryotes. Cold Spring Harbor Laboratory, Cold Spring Harbor, N.-Y., p 185-210

Draper D E (1987) Translational regulation of ribosomal proteins in *E.coli*; In: J. Ilan (eds) Translational regulation of gene expression. Plenum publishing corporation, New-York, p 1-25

Gardner J F (1979) Regulation of threonine operon: tandem threonine and isoleucine codons in the control region and translational control of transcription termination. Proc. Natl. Acad. Sci. USA 76:1706-1710

Gold L (1988) Postranscriptional regulatory mechanisms in *Escherichia coli*. Ann. Rev. Biochem. 57:199-233

Hartz D, McPheeters D S, Traut R, Gold L (1988) Extension inhibition analysis of translation initiation complexes; In: H. F. Noller and K. Moldave (eds) Methods in Enzymology. Academic Pres, New-York, p 419-425

Komine Y, Adachi T, Inokuchi H, Ozeki H (1990) Genomic organisation and physical mapping of the Transfer RNA genes in *E.coli*. Submitted.

Lestienne P, Plumbridge J A, Grunberg-Manago M, Blanquet S (1984) Autogenous repression of *E.coli* threonyl-tRNA synthetase expression *in vitro*. J. Biol. Chem. 259:5232-5237

Lindahl L, Zengel J M (1986) Ribosomal genes in *Escherichia coli*. Ann. Rev. Genet. 20:297-326

McClain W H, Nicholas Jr H B (1987) Differences between transfer RNA molecules. J.Molec.Biol. 194:635-642

McPheeters D S, Stormo G D, Gold L (1988) Autogenous regulatory site on the bacteriophage T4 gene 32 messenger RNA. J.Molec.Biol. 201:517-535

Moine H, Romby P, Springer M, Grunberg-Manago M, Ebel J P, Ehresmann C, Ehresmann B (1988) Messenger RNA structure and gene regulation at the translational level in *Esherichia coli*: the case of threonine:$tRNA^{Thr}$ ligase. Proc. Natl. Acad. Sci. USA 85:7892-7896

Moine H, Romby P, Springer M, Grunberg-Manago M, Ebel J P, Ehresmann B, Ehresmann C (1990) *E.coli* threonyl-tRNA synthetase and $tRNA^{Thr}$ modulate the binding of the ribosome to the translation initiation site of the *thrS* mRNA. Submitted:

Nomura M, Gourse R, Baughman G (1984) Regulation of the synthesis of ribosomes and ribosomal components. Ann. Rev. Biochem. 53:73-117

Schulman L H, Pelka H (1990) An anticodon change switches the identity of *E.coli* $tRNA^{Met}{}_m$ from methionine to threonine. Nucleic Acids Res. 18:285-289

Springer M, Graffe M, Butler J S, Grunberg-Manago M (1986) Genetic definition of the translational operator of the threonine tRNA ligase gene in *Escherichia coli*. Proc. Natl. Acad. Sci. USA 83:4384-4388

Springer M, Graffe M, Dondon J, Grunberg-Manago M (1989) tRNA-like structures and gene regulation at the translational level: a case of molecular mimicry in E.coli. EMBO J. 8:2417-2424

Springer M, Mayaux J F, Fayat G, Plumbridge J A, Graffe M, Blanquet S, Grunberg-Manago M (1985) Attenuation control of the *Escherichia coli* phenylalanyl-tRNA synthetase operon. J.Molec.Biol. 181:467-478.

Springer M, Plumbridge J A, Butler J S, Graffe M, Dondon J, Mayaux J F, Fayat G, Lestienne P, Blanquet S, Grunberg-Manago M (1985) Autogenous control of *Escherichia coli* threonyl-tRNA synthetase expression *in vivo*. J. Molec. Biol. 185:93-104

Théobald A, Springer M, Grunberg-Manago M, Ebel J P, Giégé R (1988) Tertiary structure of *E.coli* $tRNA^{Thr}{}_3$ in solution and interaction of this tRNA with the cognate threonyl-tRNA synthetase. Eur. J. Biochem. 175:511-524

Winter R B, Morrissey L, Gauss P, Gold L, Hsu T, Karam J (1987) Bacteriophage T4 regA proteins binds to mRNAs and prevents translation initiation. Proc. Natl. Acad. Sci. USA 84:7822-7826

Zurawsky G, Brown D, Killingly D, Yanofsky C (1978) Nucleotide sequence of the leader region of the phenylalanine operon of *E.coli*. Proc. Natl. Acad. Sci. USA 75:4271-4275

TRANSLATIONAL CONTROL OF THE TRANSCRIPTIONAL ACTIVATOR GCN4 INVOLVES UPSTREAM OPEN READING FRAMES, A GENERAL INITIATION FACTOR AND A PROTEIN KINASE

A.G. Hinnebusch, J.-P. Abastado, E. M. Hannig, B. M. Jackson, P. F. Miller, M. Ramirez, R. C. Wek, and N. P. Williams.
Laboratory of Molecular Genetics, NICHD, National Institutes of Health, Bethesda, MD 20892.

Translational control of GCN4 expression.

The GCN4 protein of S. cerevisiae is a positive regulator of 30-40 unlinked genes encoding enzymes in eleven different amino acid biosynthetic pathways. Transcriptional activation of these genes by GCN4 increases in response to amino acid starvation because synthesis of GCN4 itself is stimulated under these conditions. Regulation of GCN4 expression by amino acid availability occurs primarily at the translational level and requires short open reading frames present in the leader of GCN4 mRNA (reviewed in Hinnebusch 1988).

Upstream open reading frames (uORFs) do not occur in most eukaryotic transcripts (Kozak 1987), being present in only a few percent of all known S. cerevisiae mRNAs (Cigan and Donahue 1987). In addition, introduction of uORFs into eukaryotic mRNAs generally inhibits translation of the downstream protein-coding sequences (Sherman and Stewart 1982, Kozak 1984; Johansen et al. 1984; Liu et al. 1984). This inhibitory effect has been explained by proposing that ribosomes (or 40S subunits) must begin at the 5′ end of the mRNA and traverse the entire leader to reach the AUG start codon (the scanning model). When an uORF is present, initiation occurs preferentially at that site and precludes recognition of downstream start codons, presumably because reinitiation is an inefficient process (Kozak 1983; Kozak 1984).

Removal of the uORFs from the GCN4 transcript by point mutations in the four ATG start codons results in constitutively derepressed GCN4 expression, independent of amino acid availability (Mueller and Hinnebusch 1986). These mutations have little effect on the steady-state abundance of GCN4 mRNA, consistent with the idea that the uORFs repress GCN4 expression at the translational level. In addition, insertion of the four uORFs into the leader of a heterologous yeast transcript causes expression of its protein product to be regulated in the same fashion observed for GCN4 (Mueller et al. 1987). Thus, the four uORFs constitute a translational control element that allows

NATO ASI Series, Vol. H 49
Post-Transcriptional Control of Gene Expression
Edited by J. E. G. McCarthy and M. F. Tuite

ribosomes to reach an AUG start codon downstream in the mRNA only when cells are starved for an amino acid.

The first and fourth uORFs have different functions in translational control.

A combination of uORFs 1 and 4 (numbering from the 5' end) is sufficient for nearly wild-type regulation of GCN4 expression (Fig. 1, top). These two uORFs have very different effects on initiation at the GCN4 start codon when each is present singly in the mRNA leader. Solitary uORF4 constitutes an efficient translational barrier, reducing GCN4 expression to only a few percent of that seen when no uORFs are present. By contrast, uORF1 is a leaky translational barrier when present alone, reducing GCN4 expression by only 50%. In addition, uORF1 functions as a positive element when situated upstream from uORF4, reducing the inhibitory effect of the latter and stimulating GCN4 expression under starvation conditions (Mueller and Hinnebusch 1986; Tzamarias et al. 1986) (Fig. 1, top). Apparently, under starvation conditions a significant fraction of ribosomes that traverse uORF1 can either bypass the uORF4 AUG codon or reinitiate translation more efficiently at the GCN4 start codon following translation of uORF4. uORF3 has the same negative function as uORF4, whereas uORF2 acts as a weak positive element similar to uORF1 (Mueller and Hinnebusch 1986).

Nucleotides surrounding the stop codons of uORFs 1 and 4 determine their distinct roles in translational control.

Point mutations in its stop codon that lengthen or shorten uORF1 by a single codon increase its inhibitory effect on GCN4 expression when present as a solitary uORF. Therefore, either its length or the sequence context of its stop codon is important for uORF1 to act as a weak translational barrier compared to uORF4. Insertions of various single codons between the second and third codons of uORF1 also make it a more inhibitory element, supporting the idea that a three-codon length is necessary for uORF1 to allow efficient initiation downstream. However, the effects of single codon insertions on uORF1 function are not as great as that which occurs when uORF1 is lengthened to four codons by a point mutation in its stop codon. Therefore, sequences 3' to the uORF1 stop codon are also important for its ability to allow efficient initiation at the GCN4 start codon (Miller and Hinnebusch 1989).

Further evidence for this idea came from the fact that replacement of the 10 nucleotides immediately following uORF1 with the corresponding sequence from uORF4 significantly lowered GCN4 expression when uORF1 was present alone. Additionally replacing the third codon of uORF1 with the rare proline codon normally found at that position in uORF4 made uORF1 indistinguishable from uORF4 as a negative element. By contrast, replacing 25 nucleotides upstream from uORF1 with the corresponding nucleotides from uORF4 had no effect on GCN4 expression (Miller and Hinnebusch 1989) (Fig. 1, bottom). The latter suggests that uORFs 1 and 4 have very similar initiation efficiencies, a conclusion reached independently by measuring the synthesis rates of uORF1-lacZ and uORF4-lacZ fusions present as the 5'-proximal coding sequences on GCN4 mRNA (Mueller et al. 1988; Tzamarias and Thireos 1988). The fact that introducing sequences from around the stop codon of uORF4 at the corresponding position in uORF1 makes the latter a more inhibitory element suggests that translation terminates differently at these two uORFs, and that termination at uORF4 is

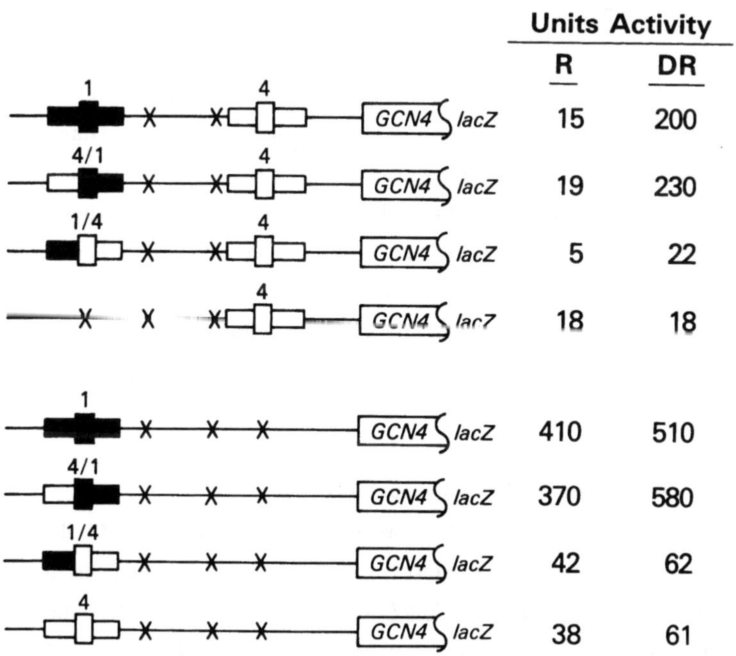

Fig. 1. Sequences surrounding the stop codons of uORFs 1 and 4 distinguish the functions of these elements in translational control. GCN4-lacZ constructs containing point mutations in the ATG codons of uORFs 2 and 3 ("X"s) are shown schematically, with intact uORFs 1 and 4 and segments of 16 and 25 nucleotides upstream and downstream from these uORFs indicated by boxes. Sequences at uORF1 (filled boxes) were replaced with the corresponding sequences from uORF4 (open boxes) in hybrid constructs. GCN4-lacZ enzyme activity was measured in a constitutively repressed gcn2-1 mutant (R) and in a constitutively derepressed gcd1-101 mutant (DR).

incompatible with reinitiation downstream. By contrast, translation of wild-type uORF1 permits considerable reinitiation at the GCN4 start codon.

Introduction of the uORF4 termination sequences at uORF1 also impairs its ability to stimulate GCN4 expression under starvation conditions when uORF4 is present in the leader. By contrast, introduction of the sequences from upstream of uORF4 had no effect on the positive regulatory function of uORF1 (Fig. 1, top) (Miller and Hinnebusch 1989). The fact that reducing the reinitiation potential of uORF1 destroys its ability to overcome the inhibitory effect of uORF4 suggests that ribosomes must first translate uORF1 and resume scanning in order to traverse uORF4. For example, ribosomes might emerge from translation of uORF1 with a different complement of factors than were present during initiation at uORF1, and this altered configuration would allow these reinitiating ribosomes to traverse uORF4 and reach the GCN4 start codon in amino acid-starved cells (Fig. 2). The deleterious effects of the mutations that move the uORF1 stop codon upstream or downstream might result from an altered termination context that favors ribosome dissociation versus reinitiation. The negative effects of the single-codon insertions in uORF1 could be explained by proposing that ribosomes persist in a configuration compatible with reinitiation after forming the first few peptide bonds, but are less likely to reinitiate after completing additional elongation cycles. Of interest in this connection is the fact that certain antibiotics (eg. verrucarin A) inhibit elongation only during formation of the first few peptide bonds, suggesting a mechanistic difference between early and advanced elongation steps (Vazquez 1979).

Trans-acting regulators of GCN4 expression.

The presence of uORF1 in the mRNA leader is not sufficient for increased GCN4 expression in amino acid-starved cells. Also required are *trans*-acting positive regulators encoded by GCN2 and GCN3, as shown by the fact that mutations in these genes impair derepression of GCN4 expression under starvation conditions (Hinnebusch 1984; Hinnebusch 1985). Deletions of GCN2 and GCN3 have no known phenotype except for reduced GCN4 expression (Roussou et al. 1988; Hannig and Hinnebusch 1988), indicating that these factors are specifically required to regulate GCN4 expression.

Mutations in GCD genes lead to constitutive derepression of GCN4 (Hinnebusch 1985; Harashima and Hinnebusch 1986), indicating that GCD factors function as repressors. Because *gcd* mutations have little effect on GCN4 expression when uORF4 is present alone in the leader (Hinnebusch 1985; Mueller et al. 1987), it follows that these factors are not required for the inhibitory effect of uORF4 on initiation at the GCN4 start codon. Instead, GCD factors seem to repress GCN4 primarily by antagonizing the positive function of uORF1 under nonstarvation conditions. Because gcd mutations overcome the low-level GCN4 expression normally seen in gcn2 and gcn3 mutants (Hinnebusch 1985; Harashima and Hinnebusch 1986), GCN2 and GCN3 are thought to act indirectly by antagonism or repression of GCD factors (Fig. 3). Unlike GCN genes, the known gcd deletions are unconditionally lethal (Hill and Struhl 1988; Hannig and Hinnebusch 1988; Paddon and Hinnebusch 1989) and nearly all gcd point mutations lead to temperature-sensitive lethality or unconditional slow growth (Wolfner et al. 1975; Niederberger et al. 1986; Harashima and

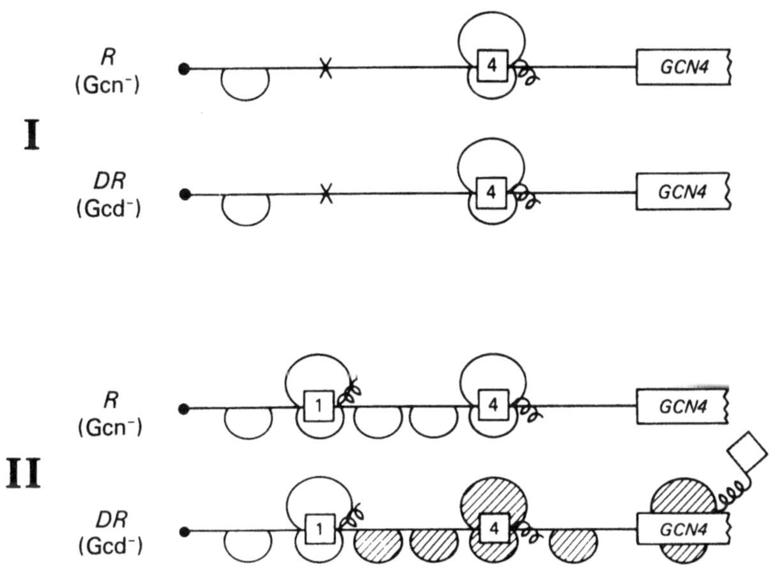

Fig. 2. Hypothetical mechanism for the interactions between uORFs 1 and 4 in regulating initiation at the GCN4 start codon. GCN4 mRNA is shown with uORFs 1 and 4 present, bound to 40S subunits and 80S ribosomes. When the uORF1 AUG codon is missing (first two cases), all 40S subunits scan to uORF4 and initiate translation. Following termination at uORF4, no reinitiation occurs at the GCN4 start codon in repressing (R, gcn2) or derepressing (DR, gcd1) conditions. When uORF1 is present (third and fourth cases), most ribosomes translate uORF1 and scanning resumes following termination at this site. Under repressing conditions, reinitiation occurs at uORF4, but no reinitiation at the GCN4 AUG codon follows. Under derepressing conditions, the behavior of 40S subunits that engaged in prior translation of uORF1 (hatched subunits) is altered due to reduced GCD activity in the cell, making them able to traverse uORF4 and reinitiate at the GCN4 start site by skipping over the uORF4 AUG codon or by more efficient reinitiation following uORF4 translation.

Hinnebusch 1986). These secondary phenotypes indicate that GCD factors carry out essential functions in addition to their roles in regulating GCN4 expression. Perhaps these essential functions are concerned with general protein synthesis.

In the context of the reinitiation model described above, GCD factors would prevent reinitiating ribosomes generated by uORF1 translation from traversing uORF4 and reinitiating at the GCN4 start codon. In this view, an additional modification of ribosome function is required in conjuction with that occurring as a consequence of prior uORF1 translation in order for ribosomes to reach the GCN4 start codon. These additional modifications would arise from partial inactivation of GCD factors by GCN2 and GCN3 under starvation conditions.

Mutations in the structural genes for initiation factor eIF-2 impair GCN4 translational control.

The above hypothesis implies that protein synthesis is altered in amino acid-starved cells in a way that specifically affects translation of GCN4 mRNA. Of interest in this connection is the fact that mutations in the structural genes for two subunits of initiation factor -2 (eIF-2) have the same phenotype as gcd mutations. These two genes, SUI2 and SUI3, were identified by mutations that restore expression from a his4 allele lacking its normal ATG start codon, allowing an in-frame TTG codon at HIS4 to be used as the start site. The deduced amino acid sequences of SUI2 and SUI3 show striking homology to the α and β subunits of human eIF-2 respectively. This initiation factor forms a ternary complex with GTP and Met-$tRNA_i^{Met}$ and associates with the small ribosomal subunit during the initiation process. Ternary complex formation is defective in extracts prepared from SUI2 and SUI3 mutants, supporting the idea that SUI2 and SUI3 are subunits of eIF-2 in *S. cerevisiae* (Donahue et al. 1988b; Cigan et al. 1989). The phenotype of the SUI mutations suggests that eIF-2 plays an important role in AUG recognition during the scanning process.

In addition to their effects on translational start site selection, the SUI mutations lead to elevated HIS4 transcription (Donahue et al. 1988a). This unexpected phenotype was explained by showing that GCN4 expression is constitutively elevated in the SUI mutants. Importantly, this derepression requires multiple uORFs in the GCN4 mRNA leader, since alleles containing uORF4 alone or no uORFs at all are nearly insensitive to the SUI mutations.

In addition, SUI2 and SUI3 mutations overcome the requirement for the positive regulators GCN2 and GCN3 for high-level GCN4 expression (Williams et al. 1989). All of these phenotypes are exhibited by known gcd mutants, suggesting that eIF-2 activity is altered by GCN2 and GCN3 in amino acid-starved cells to allow increased translation of GCN4 mRNA (Fig. 3). In a related development, the binding of charged $tRNA_i^{Met}$ to the small ribosomal subunit (a function requiring eIF-2) was found to be defective in a temperature-sensitive gcd1 mutant after incubation at the non-permissive temperature (Tzamarias et al. 1989), suggesting that GCD1 has a general function in translation initiation involving eIF-2. These results raise the interesting possibility that reduced initiation factor activity under starvation conditions enables reinitiating ribosomes emerging from uORF1 translation to pass through uORF4 and initiate at GCN4.

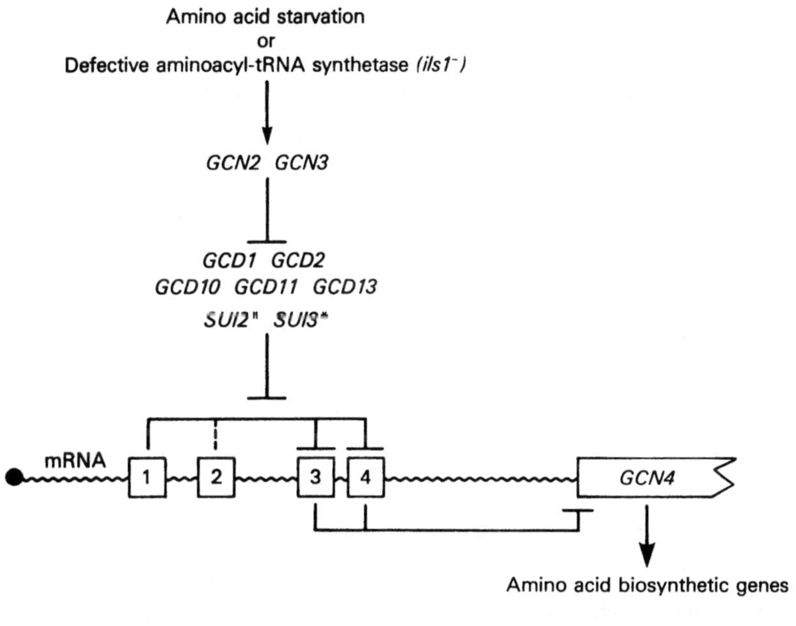

Figure 3. Pathway of trans-acting regulatory factors involved in translational control of GCN4 expression. Arrows indicate stimulatory interactions; bars depict inhibition or repression. The four uORFs in the GCN4 mRNA leader are shown as numbered boxes. uORFs 3 and 4 function as the major translational barriers to GCN4 expression under nonstarvation conditions. The inhibitory effect of these sequences is reduced under starvation conditions by uORFs 1 and 2, the greater effect being exerted by uORF1. The antagonistic interaction between the uORFs is modulated by trans-acting positive (GCN) and negative (GCD and SUI) factors in response to the abundance of aminoacylated tRNA.

Another interesting similarity between SUI and gcd mutations concerns their interactions with the positive regulator GCN3. Deletions of GCN3 have no effect on viability under nonstarvation conditions in otherwise wild-type cells. However, a gcn3 deletion is unconditionally lethal in a gcd1-101 mutant (Hannig and Hinnebusch 1988) and increases the severity of the growth defect associated with other gcd1 mutations (Harashima et al. 1987). Deletion of GCN3 is also lethal in sui2-1 strains (Williams et al. 1989) and certain GCD2 mutations (known as gcd12 alleles) lead to constitutive derepression of GCN4 expression and temperature-sensitivity for growth only in the absence of GCN3 function (Harashima et al. 1987). These genetic interactions suggest that GCN3 is functionally related to GCD1, GCD2 and SUI2, and contributes to the essential activities of these proteins under non-starvation conditions. In fact, there is significant amino acid sequence similarity between GCN3 and the carboxyl-terminus of the GCD2 protein (Paddon et al. 1989). Based on this similarity and the genetic interactions just described, it was suggested that GCN3 competes with GCD2 for binding sites in a complex having an important function in translation initiation, and that the activity of this complex is altered in amino acid-starved cells as the result of post-translational regulation of GCN3 (Paddon et al. 1989). Presumably, this complex would also interact with eIF-2 during the initiation process.

The role of GCN2 protein kinase in recognition of the amino acid starvation signal.

Sequence analysis reveals a strong similarity between a portion of the GCN2 protein-coding sequences and the catalytic domain of eukaryotic protein kinases (Roussou et al. 1988; Wek et al. 1989; Hanks et al. 1988). One of the best characterized of the conserved subdomains in these kinases includes the sequence GlyXGlyXXGlyXVal, followed 13 to 18 residues downstream by AlaXLys. The lysine residue in this motif is believed to function in the phosphotransfer reaction, and amino acid substitutions at this position invariably abolish protein kinase activity (Hanks et al. 1988; Edelman et al. 1987). Substitution of the corresponding lysine residue in GCN2 (Lys-559) with arginine or valine abolished GCN2 positive regulatory function, whereas substitution of an adjacent lysine residue with valine had no effect on GCN4 expression (Wek et al. 1989). These results suggest that GCN2 stimulates GCN4 expression by acting as a protein kinase.

The deduced amino acid sequence of the carboxyl-terminus of GCN2 may provide an important clue about how its protein kinase activity is regulated by amino acid availability: 530 residues in the carboxyl-terminal domain of GCN2 are homologous to histidyl-tRNA synthetases (HisRS) from S. cerevisiae, humans and E. coli. Several two-codon insertions in the HisRS-related coding sequences inactivate GCN2 positive regulatory function, suggesting that this region is required for protein kinase activity under starvation conditions (Wek et al. 1989). Given that aminoacyl-tRNA synthetases bind uncharged tRNA as a substrate, and that accumulation of uncharged tRNA triggers the derepression of GCN4 expression (Messenguy and Delforge 1976), it was proposed that the GCN2 HisRS-related domain monitors the concentration of uncharged tRNA in the cell and activates the adjacent protein kinase moiety under starvation conditions when uncharged tRNA accumulates. Because GCN4 expression increases in response to starvation for any one of at least ten amino acids, perhaps GCN2 has diverged sufficiently from HisRS that it now lacks the ability to discriminate between different uncharged tRNA species (Wek et al. 1989).

No substrates for GCN2 kinase activity have been identified; however, the fact that gcd and sui mutations restore derepression of GCN4 in gcn2 mutants makes eIF-2 and the GCD factors good candidates. Phosphorylation by GCN2 would partially inactivate one or more of these factors, altering the process of translation initiation in a way that allows more ribosomes to traverse the 3'-proximal uORFs and initiate at the GCN4 start codon. This hypothesis is attractive given that eIF-2 activity in mammalian cells is regulated by phosphorylation of the α subunit in stress conditions that include amino acid starvation (Hershey et al. 1986; Clemens et al. 1987). Alternatively, GCN3 (or GCN1, another positive factor implicated in GCN4 control) may be phosphorylated by GCN2 and thereby converted to an antagonist of the GCD proteins.

References

Cigan AM, Donahue TF (1987) Sequence and structural features associated with translational initiator regions in yease - a review. Gene **59**:1-13

Cigan AM, Pabich EK, Feng L, Donahue TF (1989) Yeast translation initiation suppressor *sui2* encodes the alpha subunit of eukaryotic initiation factor 2 and shares identity with the human alpha subunit. Proc Natl Acad Sci USA **86**: 2784-2788

Clemens MJ, Galpine A, Austin SA, Panniers R, Henshaw EC, Duncan R, Hershey JW, Pollard JW (1987) Regulation of polypeptide chain initiation in Chinese hamster ovary cells with a temperature-sensitive leucyl-tRNA synthetase. Changes in phosphorylation of initiation factor eIF-2 and in the activity of the guanine nucleotide exchange factor GEF. J Biol Chem **262**:767-771

Donahue TF, Cigan AM, De Castilho BA, Yoon H (1988a) Translation initiation in yeast: a genetic and mutational analysis. In: "Genetics of Translation", M Tuite, ed, Springer-Verlag, Berlin. pp 361-372

Donahue TF, Cigan AM, Pabich EK, Castilho-Valavicius B (1988b) Mutations at a Zn(11) finger motif in the yeast eIF-2b gene alter ribosomal start-site selection during the scanning process. Cell **54**:621-632

Edelman AM, Blumenthal DK, Krebs EG (1987) Protein serine/threonine kinases. Ann Rev Biochem **56**:567-613

Hanks SK, Quinn AM, Hunter T (1988) The protein kinase family: conserved features and deduced phylogeny of the catalytic domains. Science **241**:41-52

Hannig EM, Hinnebusch AG (1988) Molecular analysis of *GCN3*, a translational activator of *GCN4*: evidence for posttranslational control of *GCN3* regulatory function. Mol Cell Biol **8**:4808-4820

Harashima S, Hannig EM, Hinnebusch AG (1987) Interactions between positive and negative regulators of *GCN4* controlling gene expression and entry into the yeast cell cycle. Genetics **117**:409-419

Harashima S, Hinnebusch AG (1986) Multiple *GCD* genes required for repression of *GCN4*, a transcriptional activator of amino acid biosynthetic genes in *Saccharomyces cerevisiae*. Mol Cell Biol **6**:3990-3998

Hershey JWB, Duncan R, Mathews MB (1986) Introduction: mechanisms of translational control. In: "Current communications in molecular biology translational control", Mathews, M B, ed, Cold Spring Harbor Laboratory, Cold Spring Harbor, NY. pp 1-18

Hill DE, Struhl K (1988) Molecular characterization of *GCD1*, a yeast gene required for general control of amino acid biosynthesis and cell-cycle initiation. Nucleic Acids Res **16**:9253-9265

Hinnebusch AG (1984) Evidence for translational regulation of the activator of general amino acid control in yeast. Proc Natl Acad Sci USA **81**:6442-6446

Hinnebusch AG (1985) A hierarchy of *trans*-acting factors modulate translation of an activator of amino acid biosynthetic genes in yeast. Mol Cell Biol 5:2349-2360

Hinnebusch AG (1988) Mechanisms of gene regulation in the general control of amino acid biosynthesis in *Saccharomyces cerevisiae*. Microbiol Rev **52**:248-273

Johansen H, Schumperli D, Rosenberg M (1984) Affecting gene expression by altering the length and sequence of the 5′ leader. Proc Natl Acad Sci USA **81**:7698-7702

Kozak M (1983) Comparison of initiation of protein synthesis in procaryotes, eucaryotes, and organelles. Microbiol Rev **47**:1-45

Kozak M (1984) Selection of initiation sites by eucaryotic ribosomes: effect of inserting AUG triplets upstream from the coding sequence for preproinsulin. Nucleic Acids Res **12**:3873-3893

Kozak M (1987) An analysis of 5′-noncoding sequences from 699 vertebrate messenger RNAs. Nucleic Acids Res **15**:8125-8149

Liu C, Simonsen CC, Levinson AD (1984) Initiation of translation at internal AUG codons in mammalian cells. Nature (London) **309**:82-85

Messenguy F, Delforge J (1976) Role of transfer ribonucleic acids in the regulation of several biosyntheses in *Saccharomyces cerevisiae*. Eur J Biochem **67**:335-339

Miller PF, Hinnebusch AG (1989) Sequences that surround the stop codons of upstream open reading frames in *GCN4* mRNA determine their distinct functions in translational control. Genes and Development **3**:1217-1225

Mueller PP, Harashima S, Hinnebusch AG (1987) A segment of *GCN4* mRNA containing the upstream AUG codons confers translational control upon a heterologous yeast transcript. Proc Natl Acad Sci USA **84**:2863-2867

Mueller PP, Hinnebusch AG (1986) Multiple upstream AUG codons mediate translational control of *GCN4*. Cell **45**:201-207

Mueller PP, Jackson BM, Miller PF, Hinnebusch AG (1988) The first and fourth upstream open reading frames in *GCN4* mRNA have similar initiation efficiencies but respond differently in translational control to changes in length and sequence. Mol Cell Biol **8**:5439-5447

Niederberger P, Aebi M, Huetter R (1986) Identification and characterization of four new *GCD* genes in *Saccharomyces cerevisiae*. Curr Genet **10**:657-664

Paddon CJ, Hannig EM, Hinnebusch AG (1989) Amino acid sequence similarity between GCN3 and GCD2, positive and negative translational regulators of *GCN4*: evidence for antagonism by competition. Genetics **122**: 551-559

Paddon CJ, Hinnebusch AG (1989) *gcd12* mutations are *gcn3*-dependent alleles of *GCD2*, a negative regulator of *GCN4* in the general amino acid control of *Saccharomyces cerevisiae*). Genetics **122**: 543-550

Roussou I, Thireos G, Hauge BM (1988) Transcriptional-translational regulatory circuit in *Saccharomyces cerevisiae* which involves the *GCN4* transcriptional activator and the GCN2 protein kinase. Mol Cell Biol **8**:2132-2139

Sherman F, Stewart JW (1982) Mutations altering initiation of translation of yeast iso-1-cytochrome *c*; contrasts between the eukaryotic and prokaryotic initiation process. In: "The molecular biology of the yeast Saccharomyces metabolism and gene expression", Strathern, J N, Jones, EW, and Broach, J R, eds, Cold Spring Harbor Laboratory, Cold Spring Harbor, NY. pp 301-334

Thireos G, Driscoll-Penn M, Greer H (1984) 5′ untranslated sequences are required for the translational control of a yeast regulatory gene. Proc Natl Acad Sci USA **81**:5096-5100

Tzamarias D, Alexandraki D, Thireos G (1986) Multiple *cis*-acting elements modulate the translational efficiency of GCN4 mRNA in yeast. Proc Natl Acad Sci USA **83**:4849-4853

Tzamarias D, Roussou I, Thireos G (1989) Coupling of GCN4 mRNA translational activation with decreased rates of polypeptide chain initiation. Cell **57**:947-954

Tzamarias D, Thireos G (1988) Evidence that the *GCN2* protein kinase regulates reinitiation by yeast ribosomes. EMBO J **7**:3547-3551

Vazquez D (1979) Inhibitors of protein synthesis. Springer-Verlag, New York.

Wek RC, Jackson BM, Hinnebusch AG (1989) Juxtaposition of domains homologous to protein kinases and histidyl-tRNA synthetases in GCN2 protein suggests a mechanism for coupling GCN4 expression to amino acid availability. Proc Natl Acad Sci USA **86**: 4579-4583

Williams NP, Hinnebusch AG, Donahue TF (1989) Mutations in the structural genes for eukaryotic initiation factors 2alpha and 2beta of *Saccharomyces cerevisiae* disrupt translational control of *GCN4* mRNA. Proc Natl Acad Sci USA **86**: 7515-7519

Wolfner M, Yep D, Messenguy F, Fink GR (1975) Integration of amino acid biosynthesis into the cell cycle of *Saccharomyces cerevisiae*. J Mol Biol **96**:273-290

TRANSLATIONAL CONTROL BY ARGININE OF YEAST GENE *CPA1*

M. Werner[1], A. Feller, P. Delbecq and A. Piérard
Laboratoire de Microbiologie
Université Libre de Bruxelles and Institut de Recherches du CERIA
Avenue Emile Gryzon, 1
1070 Bruxelles
Belgium

Introduction

Carbamoylphosphate is a common intermediate of the arginine and pyrimidine biosynthetic pathways. In *Saccharomyces cerevisiae*, two independently regulated CPSases produce carbamoylphosphate for the two biosyntheses (Lacroute *et al.*, 1965). One synthetase CPSase P, is repressed and feedback inhibited by the pyrimidines. This nuclear enzyme, encoded by the *URA2* gene, is part of a multienzymatic protein comprising also the aspartate transcarbamylase activity which catalyses the second step of the pyrimidine biosynthesis. The other synthetase, CPSase A is cytosolic and constituted of two nonidentical subunits encoded by the unlinked genes *CPA1* and *CPA2*. The small subunit, the product of gene *CPA1*, binds glutamine, the physiological nitrogen donor of the enzyme, and transfers its amide nitrogen group to the larger subunit, encoded by *CPA2*, that catalyses the synthesis of carbamoylphosphate from ammonia (Piérard and Schröter, 1978). Both *CPA1* and *CPA2* genes are subject to the general control of amino acid biosynthesis. In contrast, the specific repression by arginine acts only upon the expression of gene *CPA1*. This difference in the regulation of the two subunits leads, under condition of repression by arginine, to a situation where the large subunit is over-produced by a five fold excess over the small one (Piérard et al., 1979).

Genetic analysis of the arginine specific repression

The carbamoylphosphate produced by either enzyme is freely available for the other biosynthetic pathway. Indeed, *cpa1* or *cpa2* mutants are able to grow on minimal medium. However they become sensitive to the addition of uracil which inhibits and represses the

[1]Present address: Service de Biochimie, Bât. 142, CEN Saclay, 91191 Gif-sur-Yvette CEDEX, France

NATO ASI Series, Vol. H 49
Post-Transcriptional Control of Gene Expression
Edited by J. E. G. McCarthy and M. F. Tuite

synthesis of carbamoylphosphate by CPSase P, thus making this intermediate unavailable for the biosynthesis of arginine. Conversely *ura2C* mutants (lacking CPSase P) are sensitive to the addition of arginine, due to the repression of gene *CPA1* by arginine.

Advantage has been taken of this situation to select for *ura2C* mutants resistant to the addition of arginine (Thuriaux *et al.*, 1972). Two classes of mutations have been found. The first comprises the *CPA1-0* mutations tightly linked to *CPA1* and cis-dominant. The second is composed of the *cpaR* mutations which are recessive and unlinked to *CPA1* or *CPA2*. The *cpaR* mutations do not affect the synthesis of the other arginine biosynthetic genes. Assays of CPSase A have shown in the case of the *CPA1-O* mutants that the arginine specific repression is almost totally abolished whereas a significant repression is still present in the *cpaR* mutants. In a more recent mutant hunt using the same selection scheme we were unable to find other mutant classes (M.Werner and T. Coulon, unpublished results). These various observations were taken initially to suggest a classical repressor-operator type of regulation acting at the transcriptional level on the expression of gene *CPA1*.

Molecular analysis of the *CPA1* locus

The gene coding for *CPA1* has been cloned by complementation of the *cpa1* mutations and used to measure by Northern blotting the steady state levels of mRNA under various conditions of growth. It has been found that, under condition of arginine repression, the level of glutaminase is repressed 5 to 7 times, whereas the level of *CPA1* mRNA diminishes only of about 30% indicating that the arginine regulation acts at a post-transcriptional level (Messenguy *et al.*, 1983). More recently, pulse-chase experiments (Crabeel *et al.*, 1990) have suggested that this gene might also partly be regulated at the transcriptional level. Crabeel *et al.* also showed that the half-life of *CPA1* mRNA is decreased two-fold under condition of repression, a feature that could be inherent in the mechanism of translational regulation of gene *CPA1*. The general control of amino acid biosynthesis, on the contrary, acts on *CPA1* at a transcriptional level since its mRNA level is correlated with the levels of protein when the arginine specific control is abolished (as is the case in a *CPA1-O* mutant; Messenguy *et al.*, 1983).

Gene *CPA1* has been sequenced and the structure of its mRNA has been investigated by S1 nuclease mapping (Nyunoya and Lusty, 1984; Werner *et al.*, 1985). Interestingly the 5' end of the mRNA comprises a 250 nucleotide long leader upstream of the AUG beginning the open reading frame ORF encoding the glutaminase (small) subunit of the CPSase A. The authenticity of this ORF has been proved by fusing it with *lacZ* and sequencing the N-terminus of the fusion protein. Moreover, S1 mapping experiments with various probes have shown that the mRNA is unspliced (Werner *et al.*, 1985; A. Feller and M. Werner, unpublished results). A striking feature of the mRNA leader sequence is the presence an ORF located upstream (uORF) to the one encoding the small subunit of CPSase A. It begins with an AUG at -134 and ends with a UAA at -59 relative to the glutaminase ORF, potentially encoding a 25 amino acid peptide. This peptide does not bear any similarity to any other protein known to date.

In vitro mutagenesis of the uORF

In order to assess the role of the uORF in the arginine repression, we have introduced various mutations by oligonucleotide directed mutagenesis and evaluated their effect by reintroducing the mutated allele at the *CPA1* locus. Removing the uORF by changing its AUG initiator codon leads to the constitutive expression of *CPA1* at a level which is equivalent to that of the constitutive *CPA1-O* mutants and that of the wild-type under derepressing condition, showing that the uORF is required for proper regulation and that the upstream AUG is not inhibitory for the translation of the glutaminase ORF. Moreover, the creation of a new uORF of 30 amino acid by frameshifting the initiator codon two nucleotides toward the 5' end also leads to an unregulated behaviour. However the level at which the glutaminase subunit is expressed depends on the nucleotidic context around the upstream AUG. These observations tend to show that the sequence of the peptide encoded by the uORF (hereafter called the repressor peptide) is essential for proper repression of *CPA1* to take place. To confirm this hypothesis we have constructed a non-sense mutation which changes the fourth uORF codon to a stop codon. As expected this mutant is also constitutive.

Sequence of the *CPA1-O* mutants

We have sequenced 8 of the *CPA1-O* mutations that were obtained *in vivo* by selecting *ura2C* strains resistant to arginine (Thuriaux *et al.*, 1972; M. Werner and P. Delbecq, unpublished observations). These fall into three categories:

-The first one is made of one mutant where the initiator codon of uORF is mutated.

-The second one is made of three mutants, two of which are identical, having non-sense mutations at various locations along the uORF. Removing as little as five amino acid to the repressor peptide C-terminus is sufficient to abolish the arginine repression.

-The third one is made of four missense mutations affecting three different amino acids of the peptide. This last class once again hint that the important feature in the uORF is the sequence of the peptide it encodes.

The fact that we have obtained all the different types of mutations possibly affecting the sequence of the repressor peptide is in full agreement with the results of the *in vitro* mutagenesis experiments. It should be stressed that all the mutations obtained *in vivo* are cis-dominant and thus that the peptide encoded by the wild-type gene is not able to act at distance. It must

	M.am	M.am+Arg	Repression factor
G			
T A A T A A AA			
↑ ↑ ↑ ↑ ↑ ↑ ↑↑			
ATGTTTAGCTTATCGAACTCTCAATACACCTGCCAAGACTACATATCTGACCACATCTGGAAAACTAGCTCCCACTAA	0,195	0,031	6,3
TCTCTCAGTAATAGCCAG	0,158	0,020	7,9
TCACTTAGTAATAGCCAG	0,140	0,034	4,1
TCTCTTAGCAATAGCCAG	0,196	0,036	5,4
ACTTGTCAGGATTAT	0,162	0,031	5,2
AAGACATCTAGTCAT	0,185	0,029	6,4
AAGACATCAAGTCAT	0,213	0,020	10,7
AAGACCTCAAGTCAT	0,170	0,029	5,9
MetPheSerLeuSerAsnSerGlnTyrThrCysGlnAspTyrIleSerAspHisIleTrpLysThrSerSerHisOch			

Figure 1: Effect of the silent mutations in the uORF. The changed codons are indicated under the wild-type sequence. The nucleotides which differ from the wild-type sequence in the silent mutants, are underlined. The nucleotide changes found in the *CPA1-O* mutants are indicated above the wild-type sequence.

	M.am	M.am +Arg	Repression Factor
*** ** -200 -150			
ACTATAACACATAGATACACTCGAACATCTAATTGTTTAAATACTGCAAAGAATACAAGGTAATCGACTCTTCTACATACCCTTTTTGCAGATTTGAAATAAAAAAAACATTAT ATG TTT ..	0,195	0,031	6,3
ACTATAACACATAGATACACTCGAACATCTAATTGTTTAAATACTGCAAAGAATACAAGGTAATCGACTCTTCTACATACCCTTTTTGCAGATTTGAAATAAAGGATCCTTAT ATG TTT ...	0,161	0,026	6,2
ACTATAACACATAGATACACTCGAACATCTAATTGTTTAAATACTGCAAAGAATACAAGGTAATCGACTCTTCTACATACCCTTTTTGCAGATTTG.......CAGATCCTTAT ATG TTT ...	0,175	0,260	6,7
ACTATAACACATAGATACACTCGAACATCTAATTGTTTAAATACTGCAAAGAATACAAGGTAATCGACTCTTCTACATACCCTTTTTG..............CAGATCCTTAT ATG TTT ...	0,144	0,023	6,3
ACTATAACACATAGATACACTCGAACATCTAATTGTTTAAATACTGCAAAGAATACAAGGTAATCGACTCTT.............................CAGATCCTTAT ATG TTT ...	0,138	0,025	5,5
ACTATAACACATAGATACACTCGAACATCTAATTGTTTAAATACTGCA..CAGATCCTTAT ATG TTT ...	0,261	0,143	1,8
ACTATAACACATAGATACACTCGAACATCTAATTGTTTAAATACTG..CAGATCCTTAT ATG TTT ...	0,099	0,031	3,2
ACTATAACACATAGATACACTCGAACATCTAATTGTTT..CAGATCCTTAT ATG TTT ...	0,118	0,035	4,1
ACTATAACACATAGATACACTCGAACATCTAATTG...CAGATCCTTAT ATG TTT ...	0,107	0,020	5,4
ACTATAACACATAGATACACTCGAAC..CAGATCCTTAT ATG TTT ...	0,095	0,019	5,0
ACTATAACACATAG...GGATCCTTAT ATG TTT ...	0,189	0,086	2,2
ACTATAACACATAG...GGATCCTTAT ATG TTT ...	0,240	0,110	2,2
ACT..GGATCCTTAT ATG TTT ...	0,228	0,133	1,7

Figure 2: Deletion analysis of the 5' untranslated sequence of *CPA1* mRNA. The extent of the deletion is represented by the dots. The nucleotides which, due to the construction of the deletions, differ from the wild-type sequence are underlined. The transcription initiation sites are indicated by stars above the wild-type sequence.

accordingly act on the mRNA or on the ribosome on which it has been synthesized (Werner *et al.*, 1987).

As a final proof that within the uORF, only the coding sequence is important for the repression, we have changed 16 of the 25 codons by groups of 5 or 6 simultaneously, keeping the peptide sequence unchanged. Even though in certain mutants 13 nucleotides of the uORF were changed at the same time, all the mutants constructed were still regulated as the wild-type (Figure 1).

The untranslated sequences of *CPA1* mRNA are not required for the translational repression

The potential role of the untranslated sequences upstream and downstream of the uORF has been investigated by deletion analysis. The length of the 5' end of the mRNA, upstream of the uORF, was varied from 13 to 114 nucleotides (wild-type). The mutants retained full repression until the deletion endpoint reached nucleotide -176 where the repression ratio dropped to 1.8 fold instead of 6. However further deletion restored the full level of repression even though the level of expression dropped steadily as the extent of the deletion increased. Finally when the 5' untranslated sequence was reduced to 24 nucleotides the repressibility of *CPA1* was again reduced to 1.8 fold (Figure 2).

Four deletions located at various locations of the mRNA comprised between the two ORFs and covering from 4 to 11 base pairs have been constructed (Figure 3). In a further mutation 10 nucleotides have been replaced by a randomly chosen sequence. In none of these cases was the regulation of *CPA1* affected. Taken together with the fact that all the *in vivo* mutation map within the uORF, these deletion experiments show that no specific sequence are required outside of the uORF for the arginine specific repression of *CPA1*. We interpret the fact that one of the deletion reduces the level of repression as an indication that some sequences generated by the deletions are detrimental to the regulation.

Attempts at conferring the arginine specific repression to other genes

We have tried to confer the arginine specific regulation to other genes than *CPA1*. This was done by placing BamHI sites at either the 5', the 3' or both ends of the uORF by *in vitro* mutagenesis. First we fused the 5' end of *CPA1* including its uORF to the 5' untranslated leader of *ARG4* gene which codes for argininosuccinate synthetase (*ARG4* is regulated by the general control of amino acid biosynthesis but not by the arginine specific repression). Such a construction is repressed about 3 fold by the addition of arginine. This repression is dependent on the presence of the uORF (Figure 4).

Conversely, 5' untranslated leader sequences from genes *GAL10*, *PGK1* and *ARG4* were placed upstream to the uORF of *CPA1* and CPSase A was assayed. Both *GAL10-CPA1* and *PGK1-CPA1* gene fusions were repressed 4 to 5 times by arginine. Moreover in the case

Sequence (-50 … +1)	M.am	M.am +Arg	Repression Factor
TAA TTTCATTGCTTAATAATCAGAAATTCTATCACAAACCACTCCTAAAAATATTTCAA ATG	0,218	0,032	6,8
TAA GAATTCTGCTTAATAATCAGAAATTCTATCACAAACCACTCCTAAAAATATTTCAA ATG	0,191	0,033	5,8
TAA TTTCATTGCT......TCAGAAATTCTATCACAAACCACTCCTAAAAATATTTCAA ATG	0,174	0,030	5,8
TAA TTTCATTGCTTAATAATC...........CACAAACCACTCCTAAAAATATTTCAA ATG	0,130	0,025	5,2
TAA TTTCATTGCTTAATAATCGCCCGGGCCCGCACAAACCACTCCTAAAAATATTTCAA ATG	0,174	0,027	6,4
TAA TTTCATTGCTTAATAATCAGAAATTCTATC....ACCACTCCTAAAAATATTTCAA ATG	0,216	0,028	7,7
TAA TTTCATTGCTTAATAATCAGAAATTCTATCACAAACCACTCC.......ATTTCAA ATG	0,224	0,029	7,7

Figure 3: Deletion and replacement analysis of the untranslated sequence between the two ORF. The nucleotides which differ from the wild-type sequence are underlined. The deleted nucleotides are indicated by dots.

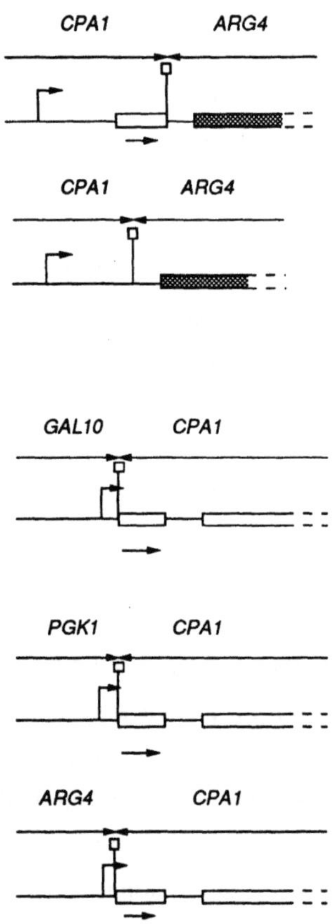

Fusion		Argininosuccinate lyase activity M.am	Argininosuccinate lyase activity M.am +arg.	Ratio
CPA1 – ARG4		2.5	0.8	3,1
CPA1 – ARG4		4,2	3,7	1,1
		Carbamoylphosphate synthetase acitivity M.am	**M.am +arg**	
GAL10 – CPA1	Glucose	0.18	0.05	3,6
	Galactose	2.63	0.52	5,1
	Galactose	11.70	2.80	4,2
PGK1 – CPA1		0.054	0.014	4,2
ARG4 – CPA1		0.173	0.107	1,6

Figure 4: Analysis of the role of the sequences located 3' and 5' to the uORF in the arginine specific repression. The structure of the fusions between *CPA1* and *ARG4*, *GAL10* or *PGK1* is shown on the left. In the case of the constructions with *GAL10*, the carbon source is indicated. +pSJ4 refers to the presence of *GAL4* galactose activator gene on a multicopy plasmid. White squares represent BamHI sites.

of the *GAL10-CPA1* fusion this repression was shown to operate independently of the galactose induction even though the expression of the glutaminase subunit was elevated more than 50 fold relative to its normal level. It should be noted that both the *GAL10* and *PGK1* 5' untranslated leaders do not bear any resemblance to the one of *CPA1*. Moreover, the 5' untranslated leader is 46 nucleotides long in *PGK1* and 32 nucleotides long in *GAL10*, compared to 111 or 100 nucleotides in *CPA1* depending on the initiation site used.

Contrarily to these two cases where the arginine control operates properly, the *ARG4-CPA1* fusion is only repressed 1.6 fold by arginine. One explanation for this behaviour might be that the leader mRNA might be too short (27 nucleotides) for the proper recognition of the upstream AUG. We do not favour this hypothesis since other genes seem to operate normally with even shorter leader sequences. An alternative might be that the secondary structure of the mRNA could interfere with the first AUG recognition process.

We then attempted to confer the arginine specific repression to chimeric constructions using either the *GAL10* or *PGK1* 5' leader, the uORF and the *ARG4* coding sequence as reporter gene (Figure 5). Interestingly a 2 fold repression could be obtained in the *GAL10-CPA1uORF-ARG4* construction. This regulation was also dependent on the presence of the uORF and its orientation. It should be noted that in the inverted orientation a 6 codon ORF also exist and that its presence reduces the level of argininosuccinate lyase activity about 14 fold. On the contrary we were unable to obtain the regulation from a *PGK1-CPA1uORF-ARG4* fusion. This was most unexpected since both *PGK1-CPA1* and *CPA1-ARG4* fusions were repressed 4 fold and 3 fold respectively. However the fact that the *GAL10-CPA1uORF-ARG4* fusion is regulated and that in *CPA1* all the untranslated sequences appear dispensable, strongly argue that the peptide encoded by the uORF is the major element required for the repression.

A model for the translational control

A model of the arginine specific repression must take three facts into account: First, the uORF encodes a 25 amino acid peptide which represses the synthesis of the glutaminase subunit encoded by the second ORF of CPA1 mRNA. Second, the constitutive mutations which change the sequence of the repressor peptide are cis-dominant. Third, the untranslated sequences of *CPA1* mRNA appear to be dispensable for the repression even though some sequence combinations might prevent proper regulation. Moreover, the distances between the 5' end of the mRNA and the uORF, and between the uORF and the glutaminase ORF, can be varied without markedly affecting the control. We propose that the site of action of the repressor peptide is the ribosome which has synthesized it. In the absence of repression by the arginine specific control, most of the ribosomes would not recognize the upstream AUG and initiate translation at the glutaminase ORF. The few ribosomes recognizing the uORF AUG

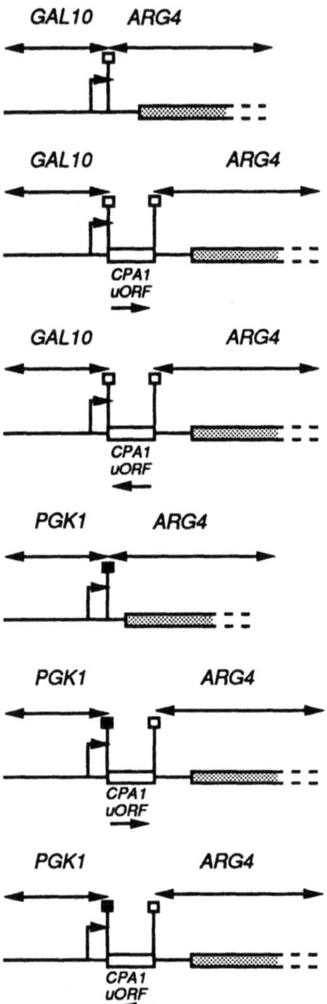

	Argininosuccinate lyase activity		
	M.am	M.am+ Arg	Ratio
Galactose	27,7	28,0	1,0
Galactose	15,3	7,9	1,9
Galactose	2,6	2,3	1,1
Glucose	4,66	4,26	1,1
Glucose	1,07	0,90	1,2
Glucose	1,01	0,90	1,1

Figure 5: Conferring the arginine specific repression with the uORF only. The structure of the constructions is shown on the left. The 5' end of the construction of is either *GAL10* or *PGK1* and *ARG4* is used as reporter gene. The carbon source is indicated in the middle row. White squares represent BamHI sites, black squares represent BglII/BamHI fusion sites.

would synthesize the repressor peptide and release it from the mRNA. In the presence of arginine, a trans-acting factor, possibly the product of gene *CPAR*, would recognize the ribosomes synthesizing the repressor peptide and prevent them from being released from the mRNA. Incoming scanning 40S ribosomal subunits which would not have recognized the uORF would then be blocked by the mRNA-ribosome-repressor peptide-CPAR complex, and would thus be prevented from translating the glutaminase ORF (Werner *et al.*, 1987).

Conclusions and perspectives

The mRNA coding for the glutaminase subunit has an unusual structure for a cellular messenger since it is dicistronic, the first cistron encoding a 25 amino acid repressor peptide which prevents the translation of the second cistron, encoding the glutaminase subunit of CPSase A, under conditions of arginine specific repression. The mRNA of yeast gene *GCN4* also uses the presence of uORFs to repress the synthesis of GCN4 protein. However the sequence of the peptides encoded by these uORFs are unimportant for proper regulation (for a comparative review see Hinnebusch, 1988).

Further work should concentrate on the nature of the trans-acting factors required for the arginine repression. One such factor might be encoded by gene *CPAR*. However, the *CPA1* mRNA levels in a *cpaR* strains are still increased three fold by arginine (Messenguy *et al.*, 1983) and the *cpaR* and *CPA1-O* mutations have additive effects (M. Werner and A. Feller, unpublished observations). Thus the suppression of the arginine sensitivity of *ura2C* mutants by the *cpaR* mutation might be due to an indirect effect i. e. the elevation of the *CPA1* mRNA steady state level.

Our regulatory model also predicts that the repressor peptide should be synthesized in smaller or equal amounts under condition of repression since the ribosome which has just synthesized the peptide would be stalled near the stop codon of the uORF, thus preventing its translation by other ribosomes. The stacking of the ribosomes could be studied by the technique of Wolin and Walter (1988). It should also be noted that, provided the model is correct, the repressor peptide would preferentially be associated with the ribosome under condition of repression.

As noted above, the large subunit of CPSase A is produced in 5 times larger amounts in conditions of repression by arginine. It would be interesting to place gene *CPA2* encoding it under the specific control and ask if this could provide an evolutionary advantage over a wild-type strain.

References

Crabeel, M., Lavallé, R. and Glansdorff, N. (1990). Arginine-specific repression in *Saccharomyces cerevisiae*: kinetic data on *ARG1* and *ARG3* mRNA transcription and stability support a transcriptional control mechanism. Mol. Cell. Biol. *10*, 1226-1233

Hinnebusch, A. G. (1988). Novel mechanisms of translational control in Saccharomyces cerevisiae. Trends Biochem. Sci. *4*, 169-174

Lacroute, F., Piérard, A., Grenson, M. and Wiame, J. M. (1965). The biosynthesis of carbamoylphosphate in *Saccharomyces cerevisiae*. J. Gen. Microbiol. *40*, 127-142

Messenguy, F., Feller, A., Crabeel, M. and Piérard, A. (1983). Control mechanisms acting at the transcriptional and post-transcriptional levels are involved in the synthesis of the arginine pathway carbamoylphosphate synthase of yeast. EMBO J. *2*, 1249-1254

Nyunoya, H. and Lusty, C. J. (1984). Sequence of the small subunit of yeast carbamyl phosphate synthetase and identification of its catalytic domain. J. Biol. Chem. *259*, 9790-9798

Piérard, A. and Schröter, B. (1978). Structure-function relationships in the arginine pathway carbamoylphosphate synthase of *Saccharomyces cerevisiae*. J. Bacteriol. *134*, 167-176

Piérard, A., Messenguy, F., Feller, A. and Hilger, F. (1979). Dual regulation of the synthesis of the arginine pathway carbamoylphosphate synthase of *Saccharomyces cerevisiae* by specific and general controls of amino acid biosynthesis. Mol. Gen. Genet. *174*, 163-171

Thuriaux, P., Ramos, F., Piérard, A., Grenson, M. and Wiame, J. M. (1972). Regulation of the carbamoylphosphate synthetase belonging to the arginine biosynthetic pathway of *Saccharomyces cerevisiae*. J. Mol. Biol. *67*, 277-287

Werner, M., Feller, A. and Piérard, A. (1985). Nucleotide sequence of yeast gene *CPA1* encoding the small subunit of arginine-pathway carbamoyl-phosphate synthetase : Homology of the deduced amino acid sequence to other glutamine amidotransferases. Eur. J. Biochem. *146*, 371-381

Werner, M., Feller, A., Messenguy, F. and Piérard, A. (1987). The leader peptide of yeast gene *CPA1* is essential for the translational repression of its expression. Cell *49*, 805-813

Wolin, S. L. and Walter, P. (1988). Ribosome pausing and stacking during translation of a eukaryotic mRNA. EMBO J. *7*, 3559-3569

Expression from polycistronic cauliflower mosaic virus pregenomic RNA

Johannes Fütterer, Jean-Marc Bonneville, Karl Gordon,
Marc deTapia, Stefan Karlsson and Thomas Hohn
Friedrich-Miescher-Institut
CH-4002 Basel
Switzerland

Cauliflower mosaic virus (CaMV), a plant pararetrovirus, shares many properties with true retroviruses, such as genome replication by transcription/reverse transcription and organisation of the GAG- and POL-genes (Fig.1). It also has a

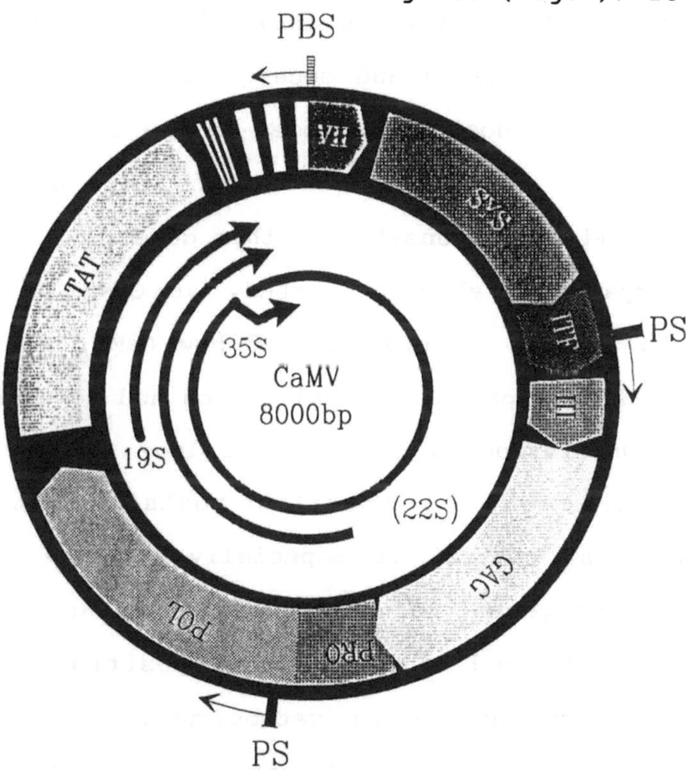

Figure 1. Map of CaMV. The seven major ORFs (functions: VII, unknown; SYS, systemic spreading; ITF, insect transmissibility; III, nucleic acid binding; GAG, structural proteins; PRO, protease, POL, polymerase; TAT, transactivator) and small ORFs in the leader (unnamed) are shown. The inner circles represent the transcripts with their S values. The 22S RNA is putative. PBS (primer binding site) and PS (priming sites) are the origins of minus and plus strand DNA synthesis.

NATO ASI Series, Vol. H 49
Post-Transcriptional Control of Gene Expression
Edited by J. E. G. McCarthy and M. F. Tuite

transactivator gene (TAV; Bonneville et al., 1989; Gowda et al., 1989), like the lentiviruses. Plant viruses spread through their host via plasmodesmata, narrow cytoplasmic connections between cells, and are transmitted by insects, while mammalian retroviruses leave and re-enter cells with the help of envelopes. Reflecting these different mechanisms of virus spread in animals and plants, the CaMV genome includes genes for systemic spreading (SYS) and for insect transmissibility (ITS; Fig. 1) but misses an envelope gene.

CaMV and hepatitis B virus (HBV) genomes, furtheron, do not contain integrase genes and consequently do not integrate obligatorily into the host chromosome. In the case of CaMV the accumulation of several thousand copies of CaMV minichromosomes might compensate for this defect.

These comparisons show that study of CaMV is not only interesting by itself. Comparison of how CaMV, retroviruses and HBV solve their problems of replication and regulation also increase our understanding of retrovirology in general and yield information on the specific hosts. Comparison of posttranscriptional control is especially interesting due to the unusual arrangement of CaMV genes on an apparently polycistronic m-RNA and to the unusual posttranscriptional regulation mechanisms of certain retroviruses (see e.g. Malim et al,; Kingsman et al.; Cao & Hauser; Berkhout & Jeang, all this conference).

The pregenomic CaMV 35S RNA contains eight ORFs larger than 99 codons, at least seven of which are known to be translated and for six of which functions could be assigned

(Fig.1). Only for one of these ORFs a separate mRNA could be found (19S RNA, covering ORF VI) and only for another one suggestive evidence for a separate mRNA exists (22S RNA, covering ORF V [and VI?]; Plant et al., 1985; Hohn et al., 1990; Schultze, PH.D. thesis). The 35S RNA therefore can be considered as mRNA for several geneproducts. ORF VII, the first of the longer CaMV ORFs, is preceded by a 600 nt. long highly structured leader which containins seven small ORFs (sORFs), between two and 30 codons in length (Fütterer et al., 1988). In a eukaryotic system this leader should theoretically pose an obstacle towards the translation of ORF VII, not to speak of translation of ORFs I,II,III and IV, located further downstream. Accordingly, it is not surprising that neither wheat-germ nor reticulocyte *in vitro* translation systems produce substantial amounts of virus proteins from CaMV 35S RNA (Gordon et al., 1988; Fütterer et al., 1988).

Since the CaMV ORFs are apparently translated in infected plants we suspected that host- and/or virus factors are required for translation from 35S RNA and that those factors are missing in the *in vitro* systems. We therefore transfected protoplasts of *Orychophragmus violaceus*, a crucifer and host of CaMV with derivatives of the CaMV genome, each containing the upstream sequences of a specific ORF fused to the chloramphenicol actyltransferase (CAT) or ß-glucuronidase (GUS) reporter gene. In this system the reporter gene fusions to the start codons of ORFs VII and II gave rise to a remarkably high level of reporter activity, i.e. 15% of the respective monocistronic controls (Fig.2). Moreover, this expression was increased to nearly the level of the monocistronic control, if a second plasmid containing the transactivator gene (ORF VI; Bonneville et al., 1989) was provided. In absence of transactivator the reporter gene fusions to ORFs I, III, IV and V, gave not rise to any substantial activity; however, also in these cases cotransfection with the transactivator plasmid raised expression impressively, in some of the cases again to a level approaching the one of the leaderless control (Fig.2; described in detail for ORF I by Bonneville et al., 1989). We conclude that in fact certain viral sequences, certain host factors and the viral transactivator enable and enhance the expression of CaMV ORFs in presence of other ORFs located upstream.

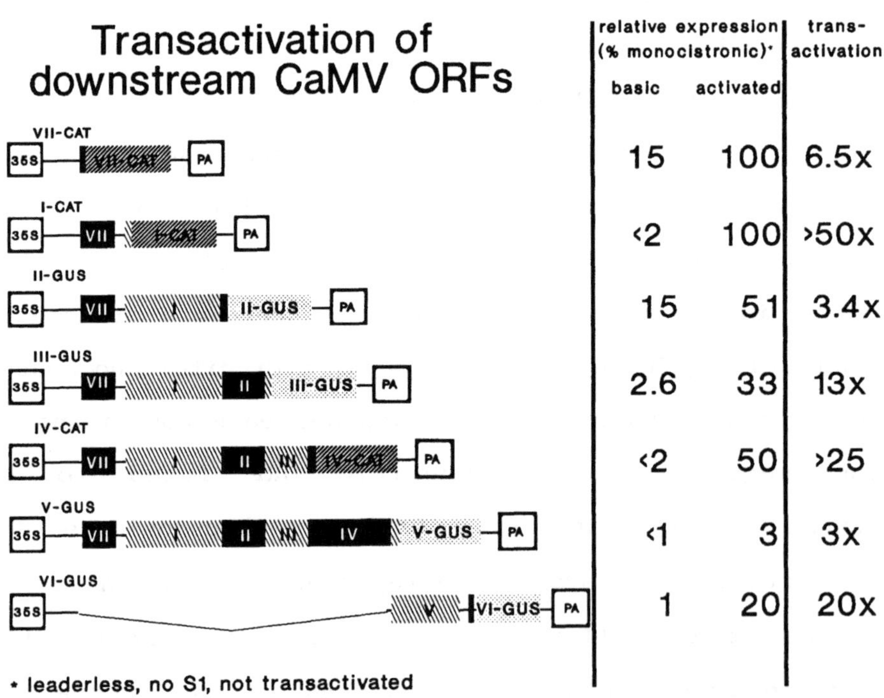

Figure 2

Expression of the ORF VII-CAT fusion in presence of the leader was studied in detail by Fütterer et al. (1990). We found that deletion of certain regions of the CaMV leader increased expression from the downstream ORF and deletion of others diminished expression (Fig.3). As expected from "Kozaks rule" (Kozak, this conference) the sORFs are part of the inhibitory regions (I1 and I2). Region I1, which is only slightly inhibitory, contains the three smallest sORFs (3-6 codons in length including the stop codon), region I2, which is strongly inhibitory when isolated from the rest of the leader,

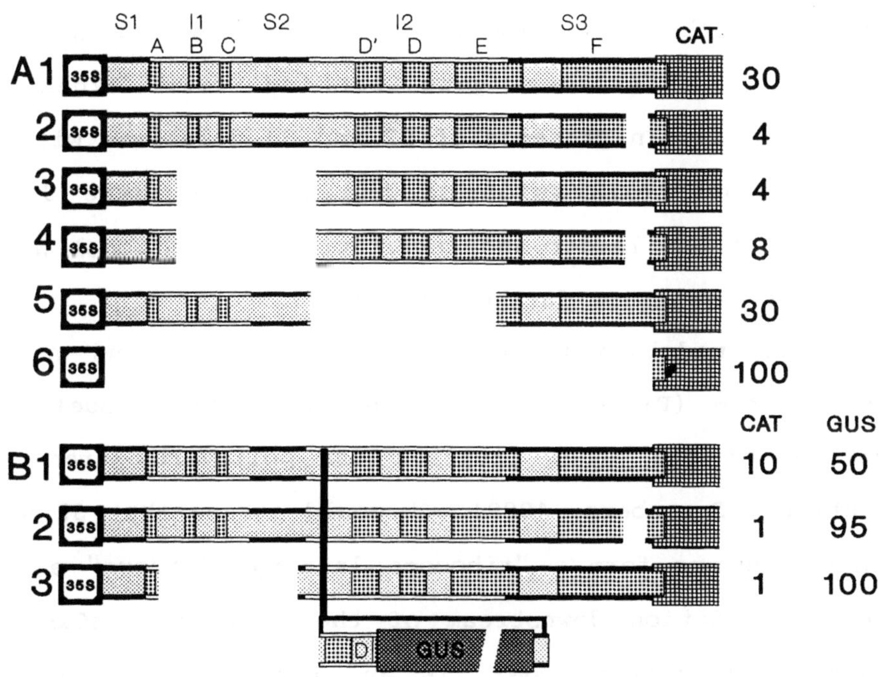

Figure 3. Mutants of the CaMV leader sequence and their effect on downstream gene expression . A: Deletions within S2, S3 and I2 and combinations of them. B: Insertions Of the GUS ORF within I2 as such and in combination of S2 and S3 deletions.

contains the four larger sORFs (10-30 codons in length). Region I2 is flanked by S2 and S3 and both of these stimulatory regions must be present to alleviate its inhibitory effect. In absence of I2 the stimulatory elements S2 and S3 have no effects. Stimulatory sequence S1, located at the 5'-end of the RNA, in contrast is active regardless whether I1 and/or I2 are present or not, perhaps as a general translation enhancer much like the tobacco mosaic virus (TMV) Ω-sequence (Wilson et al.; Gallie and Walbot, both this conference).

Since both of these stimulators must be present to alleviate I2 inhibition, we concluded that the two regions act in concert perhaps by forming a complex together with host factor(s). This idea gets backing from experiments manipulating I2. In presence of S2 and S3 deletions within I2 do not enhance expression of the reporter gene located downstream of the leader and insertions of additional ORFs do not inhibit expression further (Fig.3).

Several models might explain these results and we offer three of them (Fig.4): The **internal-initiation model** is an adaptation of the model proposed for poliovirus translation (Pelletier & Sonenberg, 1989). It would assume that S2 and S3 and host factors form a "ribosome landing platform" allowing internal initiation downstream of the inhibiting sORFs; The **ribosome modification model**, is an adaptation of the model proposed for yeast GCN4 translation (Hinnebusch et al., this conference). It assumes that the set of translation factors bound to the scanning ribosome becomes altered at the S2 region, rendering ribosomes dependent on a cis acting signal present in region S3 for (re)initiation of translation. According to the **bypass model**, finally, a complex between S2, S3 and host factors would allow the scanning ribosomes to derail at S2 and rerail at S3 bypassing the inhibitory region in-between.

Decision between these and perhaps other possibilities has to await further experimentation, however some observations might give us some hints to the mechanism: First, stimulator region S1 can be functionally replaced by the TMV Ω sequence (Fig.5). This suggests that S1 functions similarly than Ω and that ribosomes that finally translate a downstream ORF start

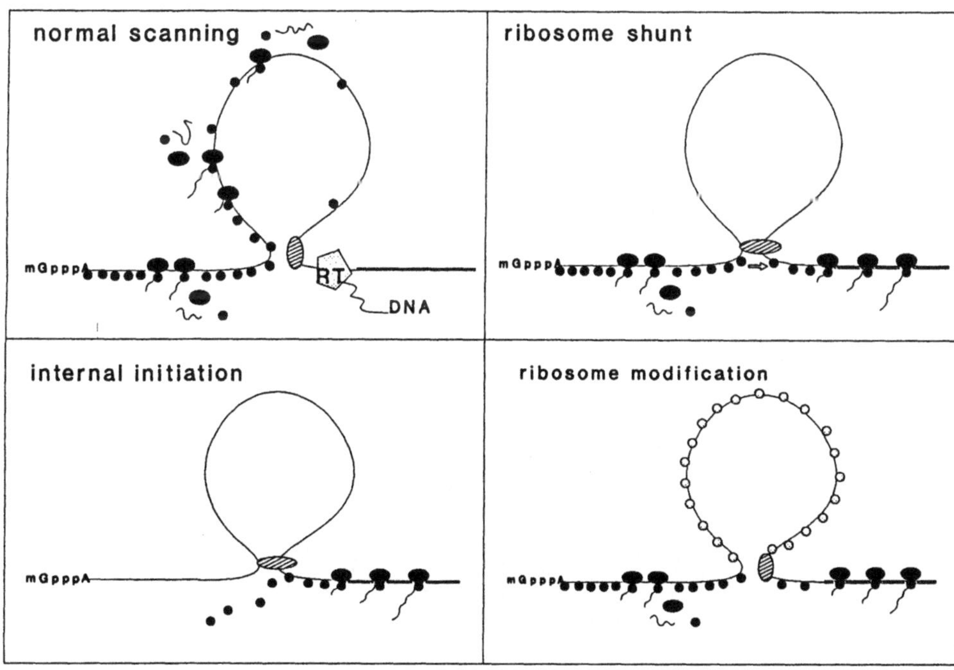

Figure 4. Models of normal scanning and abnormal ways of initiation of transcription in CaMV.

scanning at the cap-site. In accordance with this view, an artificial stem structure involving the immediate 5'end of the RNA inhibits expression strongly. These observations argue against internal initiation. Second, a dicistronic construct containing the GUS reporter gene inserted between S2 and S3, and the CAT reporter gene after S3 expressed both reporters. The expression of CAT in this case is astonishing, since the "leader" sequence preceding the CAT ORF is elongated to 2600 nt, and contains 40 AUG codons. The levels of both reporters depend on the presence of S2 and S3; deletion of either of the stimulating sequences reduced CAT activity to nearly zero and

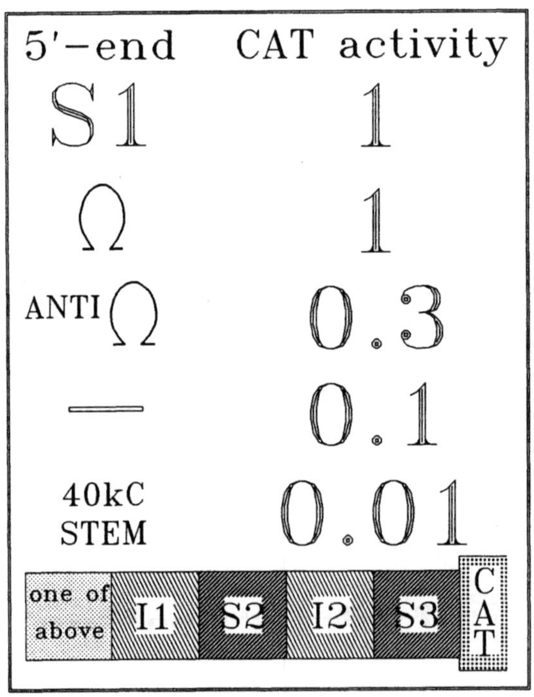

Figure 5. Comparisons of different types of sequences immediately downstream of the CaMV 35S RNA cap.

doubled GUS activity (Fig.3), indicating that the GUS ORF is in fact bypassed in presence of both stimulatory sequences and hence arguing against simple scanning, against modification and for ribosome shunt.

In the dicistronic constructs mentioned above, S3, the region preceding ORF VII, can be replaced by sequences preceding ORFs I ("preI") or IV ("preIV"). Also in these cases the downstream ORF can be expressed, but expression requires absolutely the presence of transactivator and reponds differently to deletions in the upstream leader region, indicating that the mechanisms of ORF VII and ORF I and IV translation are different (Fig.6). This suggests that a

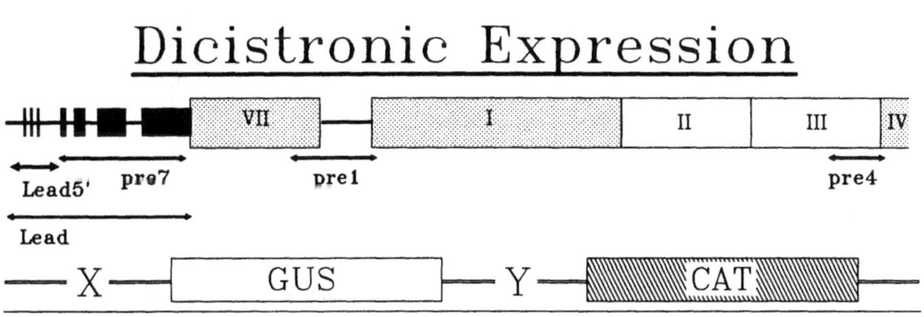

CAT activity

X=	Y=random		Y=pre7		Y=pre1		Y=pre4	
transact	−	+	−	+	−	+	−	+
−	<2	2	<2	2	<2	5	<2	2
lead5'	<2	5	100	300	<2	400	<2	300
lead	<2	2	20	80	<2	300	<2	200

Figure 6. Expression of the downstream coding region in synthetic dicistronic constructs.

function of the leader might be to direct ribosomes to downstream ORFs in a controlled fashion. An interesting common feature of all three regions, S3, preI and preIV, is a stretch of sequence that could potentially base pair with an invariant sequence close to the 3' terminus of 18S ribosomal RNA, thereby opening a stem-loop structure; the replaced arm of the stem could then in turn interact with the 5S RNA of the large ribosomal subunit (Azad & Deacon, 1980). Similar sequences are found at URF4 of the yeast GCN4 gene leader (Miller & Hinnebush, 1989; Fig.7).

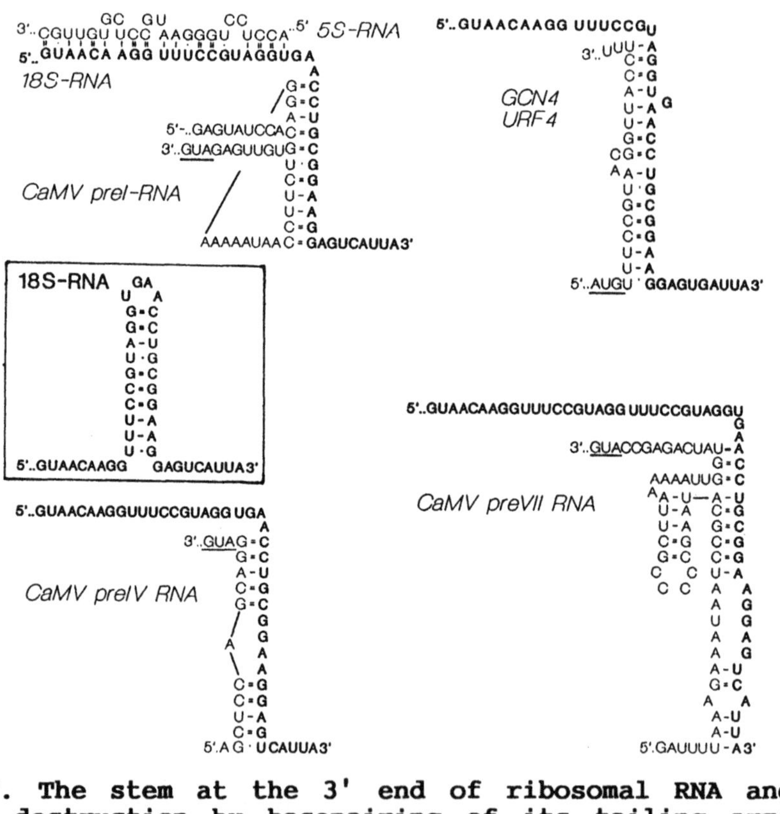

Figure 7. The stem at the 3' end of ribosomal RNA and its possible destruction by basepairing of its tailing arm with various sequences preceeding ORFs of CaMV and yeast GCN4,as indicated. The 18S RNA sequences are drawn in bold. For one of the cases (upper left) also a possible interaction of the leading arm with 5S RNA is shown.

Acknowledgenments. We highly acknowledge the expert technical assistance of Mathias Müller, Gundula Pehling and Hanny Schmid-Grob.

<u>References</u>

Azad AA Deacon N (1980) The 3'-terminal primary structure of five eukaryotic 18S RNAs determined by the direct chemical method of sequencing. Nucleic Acids Res. 8:4365-4376

Bonneville JM Sanfaçon H Fütterer J and Hohn T (1989) Posttranscriptional transactivation in Cauliflower Mosaic Virus. Cell 59:1135-1143.

Fütterer J Gordon K Pfeiffer P Sanfaçon H Pisan B Bonneville JM and Hohn T (1988) Differential inhibition of downstream gene expression by the cauliflower mosaic virus 35S RNA leader. Virus Genes 3:45-55.

Fütterer J Gordon K Sanfaçon H Bonneville JM and Hohn T (1990) A ribosome shunt model for translation of cauliflower mosaic virus pregenomic 35S RNA. EMBO J. in press.

Gordon K Pfeiffer P Fütterer J and Hohn T (1988) <u>In vitro</u> expression of cauliflower mosaic virus genes. EMBO J. 7:309-317.

Gowda S Wu FC Scholthof HB and Shepherd RJ (1989) Gene VI of figwort mosaic virus (caulimo virus group) functions in posttranscriptional expression of genes on the full-length RNA transcript. Proc.Ntl.Acad.Sci.USA 86:9203-9207.

Hohn T Bonneville JM Fütterer J Gordon K Jiricny J Karlsson S Sanfaçon H Schultze M and deTapia M (1988) The use of 35S RNA as either messenger or replicative intermediate might control the cauliflower mosaic virus replication cycle. In "Viral Genes and Plant Pathogenesis" (Pirone, T, Shaw, J, eds.), Springer Verlag, in press.

Miller PF and Hinnebush AG (1989) Sequences that surround the stop codons of upstream open reading frames in GCN4 mRNA determine their distinct functions in translational control. Genes&Development 3:1217-1225.

Pelletier J and Sonenberg N (1989) Internal binding of eukaryotic ribosomes on poliovirus RNA: Translation in HeLa cell extracts. J.Virol. 63:441-443.

Plant AL Covey SN and Grierson D (1985) Detection of a subgenomic mRNA for gene V, the putative reverse transcriptase gene of cauliflower mosaic virus. Nucl. Acids Research 13:8305-8321.

POLYADENYLATION OF CAULIFLOWER MOSAIC VIRUS RNA IS CONTROLLED BY PROMOTER PROXIMITY

Hélène Sanfaçon, Peter Brodman and Thomas Hohn
Friedrich-Miescher-Institut
CH-4002 Basel
Switzerland

Abstract. **Sequences required for efficient cauliflower mosaic virus RNA polyadenylation and dependence of polyadenylation on promoter distance were determined. A comparison of polyadenylation mechanisms for two retro- and two pararetroviruses is given.**

Production of terminally redundant RNA from DNA with long terminal repeats requires selective inhibition of the promoter proximal polyadenylation signal with subsequent use of the distal one. A related mechanism of selective bypass must occur if the terminally redundant RNA is produced from circular, non redundant DNA as in para-retroviruses (Fig.1). To study this phenomenon in cauliflower mosaic virus (CaMV; see contribution by Fütterer et al., this volume), we analyzed polyadenylation of RNA produced from plasmid constructs in which the relevant sequences were placed behind a reporter gene (Fig.2) or directly after the CaMV 35S promoter (Fig.3). In order to rescue all transcripts that could not be processed under the direction of the CaMV sequences, a second polyadenylator (derived from the Agrobacterium tumefaciens nopaline synthase gene [nos]), was installed downstream of the CaMV sequences. Constructs were transfected into plant protoplasts. After overnight incubation CAT activity was assayed and RNA was

NATO ASI Series, Vol. H 49
Post-Transcriptional Control of Gene Expression
Edited by J. E. G. McCarthy and M. F. Tuite

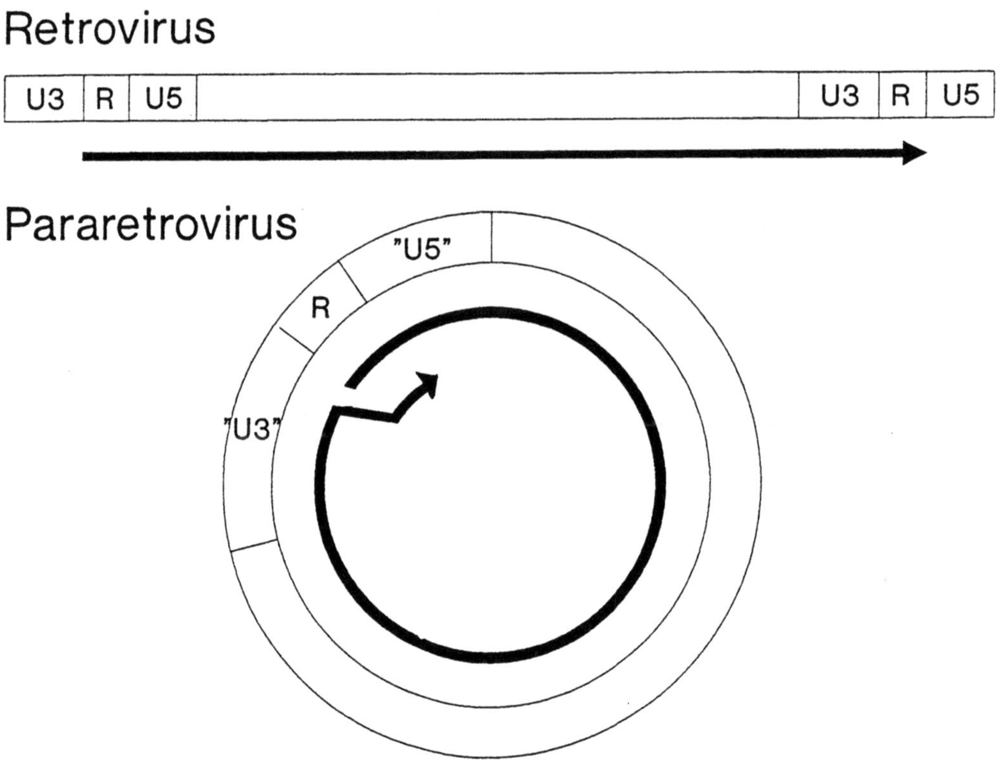

Figure 1. RNA with terminal repeats is produced from integrated retrovirus RNA and from circular pararetrovirus (CaMV,HBV) RNA.

analyzed by RNAse protection assays using the appropriate antisense probes.

All signals necessary for CaMV RNA polyadenylation are contained within the R region, the 180 bp DNA sequence that give rise to the RNA terminal repeat. These results contrast with what is known about Rous sarcoma virus (RSV), hepadnavirus and human immunodeficiency virus (HIV1) pre mRNA processing. Additional sequences within the U3 region are necessary for RSV (Stoltzfus, 1988) and hepadnavirus (Russnak & Ganem, 1990) polyadenylation, additional sequences within the U5 region for efficient 3' end formation of HIV1 (Proudfoot, this conference) and hepadnavirus RNAs (Fig.4).

Increasing deletions of R sequences <u>upstream</u> of the CaMV AATAAA signal progressively diminished polyadenylation (Fig.2). These deleted sequences are rich in GT motifs, similar to those required for mammalian RNA 3'-end formation, although the mammalian sequences are usually located <u>downstream</u> of the

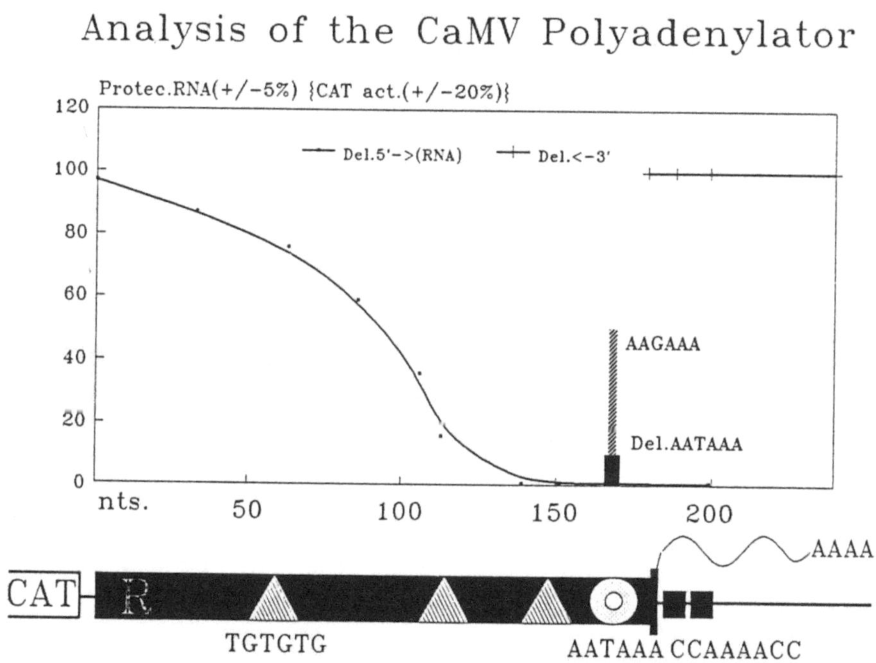

Figure 2. Mutation Analysis The CaMV RNA repeat (R) was placed behind the CAT gene under control of the CaMV 35S promoter. The original construct and progressing deletions starting from the 5' and 3' termini of R (curves from left and right) were analyzed by RNAse protection assays of lysed transfected plant protoplasts (Sanfaçon et al., in prep.). A deletion of the AATAAA motif and T->G replacement mutant of that motif are shown at their position (169 nt. from the cap-site). A map of R (black bar) and neighboring sequences is shown colinearily to the X-axis of the graph. Locations of the AATAAA motif is shown by a circle, of GT rich sequences by triangles and of CCAAAACC motifs by squares. The latter had no effect on polyadenylation.

processing site. Experiments are in progress to confirm that the GT motifs are the active sequences and to determine whether these sequences also function if placed downstream of the AATAAA signal.

In addition, the CaMV polyadenylator differed from the standard mammalian polyadenylator by stringency of usage of the AATAAA consensus sequence. Replacement of the central "T" of

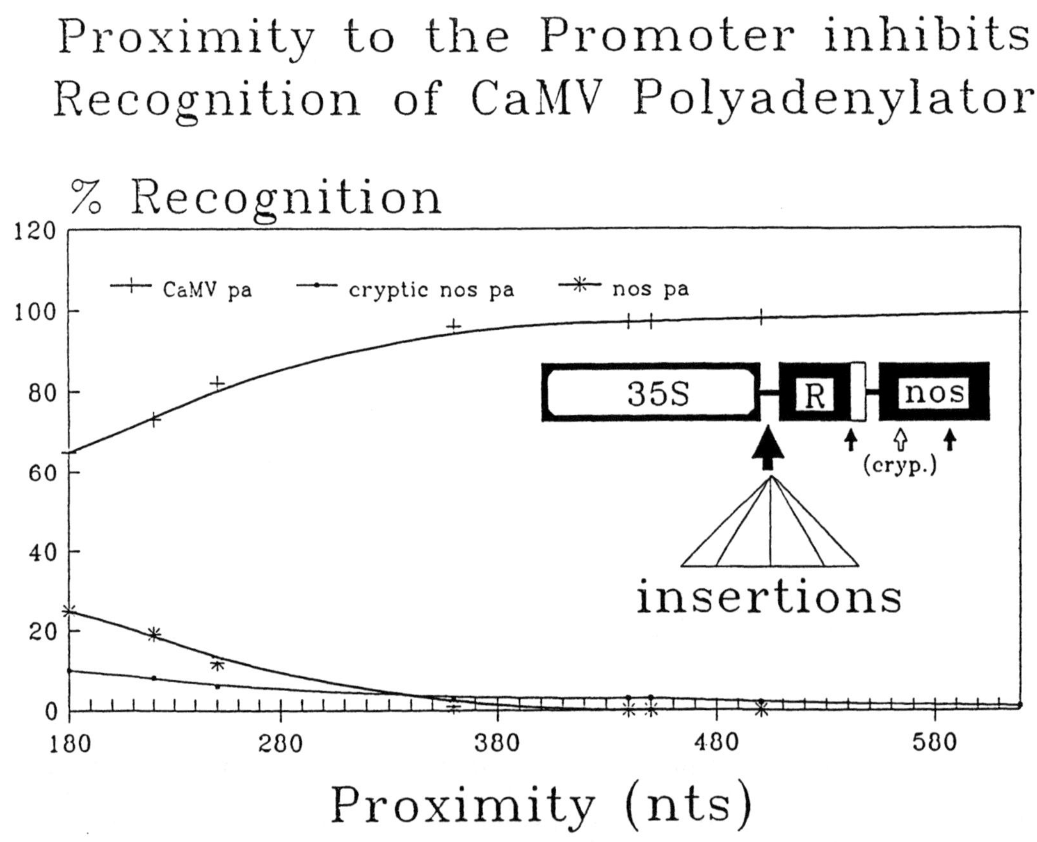

Figure 3. Analysis of polyadenylation directed by sequences within the CaMV RNA repeat (R) placed directly behind the promoter and followed by the <u>nos</u> polyadenylation signal. Polyadenylation at CaMV and <u>nos</u> sites (small filled and open arrows below the inserted map) was assayed by RNAse mapping of transfected plant protoplast lysates as in Fig.2. The original construct and random insertions between the promoter and R (large arrow) were tested.

this motif with a "G" reduced 3'end formation only 50%. Furthermore, deletion of the entire AATAAA sequence allowed residual processing to 10% of wild type (Fig.2).

To study the bypass of the promoter proximal polyadenylator, R was placed directly behind the CaMV promoter. In this case 65% of the transcripts were polyadenylated at the CaMV signal which gave rise to a 180 nt. "short-stop" RNA (Fig.3; Sanfaçon and Hohn, 1990) in comparison to 100%, if R was placed behind the CAT reporter ORF. The other 35%, however, had bypassed the CaMV signal and was processed at a downstream cryptic polyadenylation site and further downstream at the wild type site of the inserted <u>nos</u> sequence. Thus, the promoter proximal polyadenylation sequence was partially repressed.

In order to test whether the proximity of the promoter caused this effect random DNA sequences were inserted between promoter and R. These led sequentially to increased recognition of the CaMV signal. When the AATAAA signal was spaced at least 400 bp downstream of the cap-site, practically all of the transcripts were polyadenylated at the CaMV site (Fig.3). We conclude that the polyadenylation is inhibited by the proximity to the promoter but the polyadenylation site can be fully activated when located approximately 400 bp downstream from the promoter.

It is interesting to note, that CaMV uses a different mechanism for regulated polyadenylation than some other pararetro- and retroviruses. For example, RSV contains a very short R region with the AATAAA signal located upstream of the cap-site. Consequently only the poly(A) site within the downstream LTR is available for use (Fig.4; Stoltzfus,1988). Hepadnaviruses, on the other hand, use TATAAA sequence to direct polyadenylation rather than the ubiquitous AATAAA consensus sequence, and therefore contain a weak polyadenylation signal (Russnak & Ganem, 1990). This defect, however, seems to be compensated for by other AT-rich sequences located in U3, such that effective polyadenylation occurs only

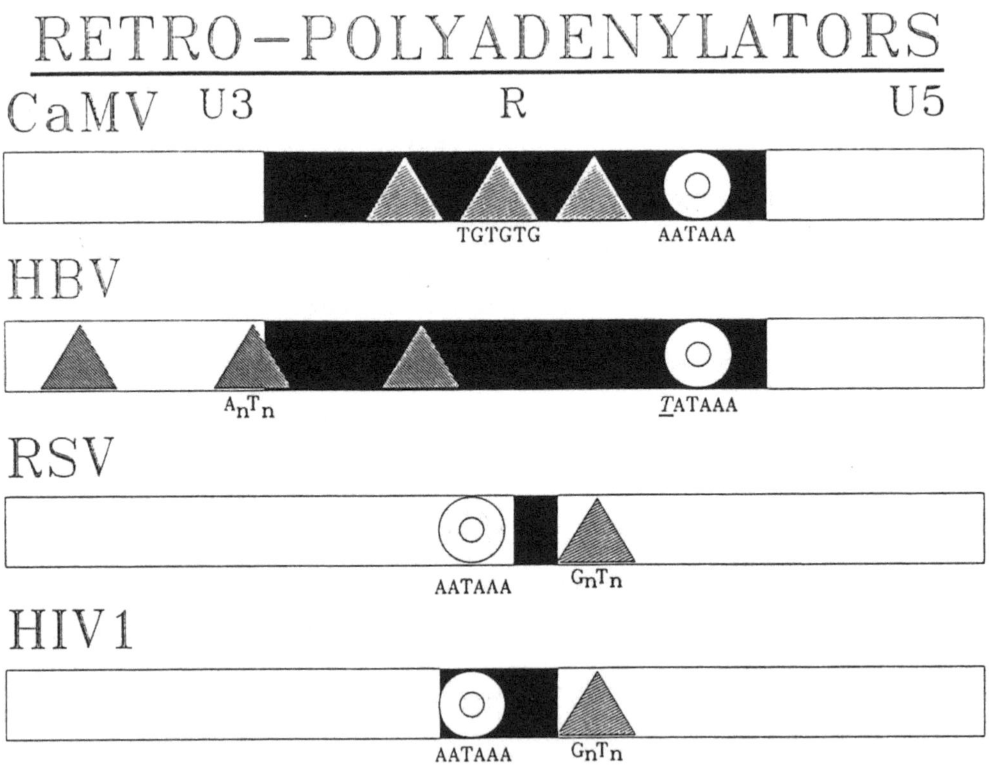

Figure 4. Comparison of "retro-polyadenylators". Polyadenylator and neighbouring sequences of four retro- and pararetroviruses are shown. The black box represents the RNA repeat, circles the main polyadenylation signal and triangles other sequences required for efficient polyadenylation.

after the complete viral RNA is produced. In contrast, HIV-1 contains an intact AATAAA signal located within the short R sequence. This signal is occluded by the proximity of the HIV-1 promoter in the upstream LTR and only used efficiently in the downstream LTR (Proudfoot, this conference). It has not been determined whether HIV1 also produces a "short stop" RNA. However an analogous RNA produced in vitro functions as inhibitor of DAI kinase (Mathews, this conference). In the case of CaMV, we have verified the presence of such a "short stop" RNA in vivo by isolating it as poly(A+) RNA from a CaMV infected plant that hybridized to the expected sequences of an antisense RNA probe (Sanfaçon and Hohn, 1990). The function of this short RNA, however remains a mystery.

Acknowledgments. We highly appreciate the help of Sheila Connelly in preparing this manuscript and we thank Gundula Pehling and Matthias Müller for expert technical assistance.

References

Russnak R and Ganem D (1990) Sequences 5' to the polyadenylation signal mediate differential polyA site use in hepatitis B viruses. Genes & Development, in press

Sanfaçon H and Hohn T (1990) Proximity to the promoter inhibits recognition of CaMV polyadenylation signal. Nature, in press

Stoltzfus CM (1988) Synthesis and processing of avian sarcoma retrovirus RNA. Advances in Virus Research 35:1-38

THE HIV-1 REV TRANS-ACTIVATOR IS A SEQUENCE SPECIFIC RNA BINDING PROTEIN

Michael H. Malim, David F. McCarn, Laurence S. Tiley and Bryan R. Cullen
Howard Hughes Medical Institute,
Department of Medicine and
Department of Microbiology and Immunology
Box 3025
Duke University Medical Center
Durham, North Carolina 27710

INTRODUCTION

The pathogenic human retrovirus, human immunodeficiency virus type 1 (HIV-1), is the major etiologic agent of the acquired immunodeficiency syndrome (AIDS) (Fauci, 1988). HIV-1, along with the related primate immunodeficiency viruses, HIV-2 and simian immunodeficiency virus (SIV), is a member of a subfamily of retroviruses known as lentiviruses. Other members of this subfamily include visna virus, caprine arthritis encephalitis virus (CAEV), equine infectious anemia virus (EIAV) and feline immunodeficiency virus (FIV). Lentivirus infection of the susceptible host typically results in a prolonged and chronic disease state affecting the immune and nervous systems as well as cells in a variety of other tissues (reviewed by Haase, 1986). Interestingly, these viruses display complex patterns of gene expression that are very different from those of the extensively studied Type C family of retroviruses. Specifically, proviral activation from a frequently extended period of latency is followed by a pattern of viral gene expression that is markedly temporal. During this time a number of differentially spliced viral transcripts may be observed. This diversity of mRNAs allows lentiviruses to encode and express a number of novel proteins in addition to the Gag, Pol and Env proteins common to all known replication competent retroviruses. The functions of these proteins in the life cycles of these viruses are currently being

NATO ASI Series, Vol. H 49
Post-Transcriptional Control of Gene Expression
Edited by J. E. G. McCarthy and M. F. Tuite

elucidated; some are essential for viral replication while others may play roles in the establishment and maintenance of latency as well as virion maturation, morphogenesis and infectivity.

HIV-1 encodes at least six of these additional proteins (reviewed by Cullen and Greene, 1989). Two of them, termed Tat and Rev, are nuclear trans-activators essential for viral replication in culture. Tat is required for high level expression of all sequences linked to the viral long terminal repeat (LTR) promoter element (Arya et al., 1985 and Sodroski et al., 1985). The trans-activation response (TAR) element for Tat is located within the LTR immediately downstream from the site of transcription initiation. TAR is functionally recognized by Tat as an RNA stem-loop structure (Berkhout et al., 1989), possibly via a direct Tat:TAR binding event (Dingwall et al., 1989). However, the precise mechanism(s) by which this interaction results in the well documented transcriptional and post-transcriptional effects of Tat remains controversial (reviewed by Sharp and Marciniak, 1989). In contrast to Tat, Rev is only required for the expression of a subset of viral gene products, namely the structural proteins Gag, Pol and Env (Feinberg et al., 1986 and Sodroski et al., 1986). These proteins are exclusively synthesized during the late stages of viral gene expression. They are encoded by a population of unspliced and partially spliced viral transcripts that are constitutively expressed in the cell nucleus where, in the absence of Rev, they are retained by factors that are likely to be involved in splicing (Chang and Sharp, 1989). The function of Rev is to alleviate this block such that structural gene expression is activated and virion formation can occur. The precise mechanism by which Rev activates the nuclear export of unspliced viral transcripts is currently unknown, though models that involve either increased access to a cellular RNA transport pathway or an inhibition of splicing have been proposed (Chang and Sharp, 1989; Emerman et al., 1989; Felber et al., 1989; Hammarskjold et al., 1989; Malim et al., 1989b). The viral RNA target sequence for Rev (the Rev

Response Element or RRE) is located within the viral env gene (Rosen et al., 1988; Felber et al., 1989; Malim et al., 1989b). This sequence is predicted to form a highly ordered complex secondary structure (Fig. 1) that is functional in an orientation dependent manner when positioned in both intronic and exonic sites (Malim et al., 1989b).

The purpose of this report is to briefly review our recent data concerning the RRE and its interaction with Rev. These results will then be discussed in relation to an earlier study on mutagenesis of the Rev protein.

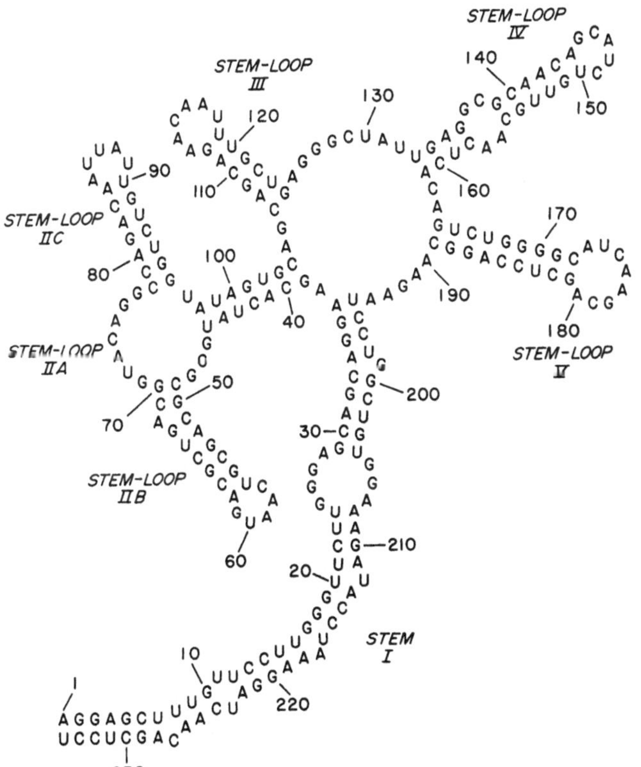

Figure 1. Predicted Secondary Structure of the HIV-1 RRE. The complete RRE structure can be subdivided into five segments, here termed stem I (coordinates 1-37 and 195-234), stem-loop II (39-104), stem-loop III (107-124), stem-loop IV (133-161), stem-loop V (164-189). Stem-loop II can be further subdivided into stem-loops IIA (39-48, 71-77 and 98-104), IIB (49-70), and IIC (78-96).

RESULTS

Mutational Analysis of the RRE. Previous mutagenesis studies have mapped the HIV-1 RRE to a region of the env gene that coincides with the complex RNA secondary structure shown in Fig. 1. We have used extensive site-directed mutagenesis of this structure in an attempt to determine which regions are required for the Rev induced cytoplasmic expression of unspliced viral transcripts. We used two different transient transfection assays in the monkey kidney cell line, COS, to monitor the phenotypes of these mutants. We will only present the findings of these experiments (refer to Table 1) since the assays themselves have been described elsewhere (Malim et al., 1990). Our initial approach was the targeted deletion of specific segments of the RRE. The Rev responsiveness of each deletion (p△) mutant, when compared to the wild type RRE, is indicated in Table 1 along with the precise coordinates of the nucleotides that delineate the mutation and the segment of the RRE that was affected. Mutations that removed stem-loop III (p△106/125), IV (p△130/162) or V (p△163/191) retained greater than 50% activity; indicating that these regions are dispensible for the Rev response. Mutations that affected the long predicted RNA double helix, namely p△1/34 and p△198/234, reduced RRE activity by between 50% and 90%. This suggests that stem I is important, but not essential, for the biological activity of the RRE. In contrast to the above mutations, the deletion of stem-loop II (p△41/105) completely abolished the Rev response. That the deletion of stem-loop IIB (p△50/75) and not IIC (p△73/94) still abrogated RRE function indicates that IIB, and possibly IIA, is required for RRE function in vivo. To more accurately ascribe RRE activity to specific nucleotides within stem-loop II, we constructed a series of scanning substitution (pM) mutations throughout the single- and double-stranded regions of this sequence (Malim et al., 1990). To our surprise, most of these four-to-seven nucleotide changes had minimal (less than four fold) effects on Rev responsiveness. Nevertheless, one mutant, pM63/69,

Table 1. Phenotypic Analysis of HIV-1 RRE Mutants

Clone Tested	Segment Mutated	Phenotype
HIV-1 (wild type)	none	++
Deletions		
p△1/34	I	+
p△41/105	IIA+B+C	-
p△50/75	IIB	-
p△73/94	IIC	+
p△106/125	III	++
p△130/162	IV	++
p△163/191	V	++
p△198/234	I	±
Substitutions		
pM63/69	IIB	±
pM63/69+50/55	IIB	++

The activity of the indicated RRE mutants is expressed relative to the activity of the wild type RRE: ++, 50%-100%; +, 20%-50%; ±, 5%-20%; -, <5%. The mapping of the primary determinant of Rev responsiveness to stem-loop II is in general agreement with the results of Dayton et al. (1989) and Olsen et al. (1990).

did display a greater than 80% inhibition of activity; in this case seven nucleotides that are entirely within the 3' strand of stem IIB were mutated (CGCUGAC → AAUAUUU). To ask whether this dimunition was due to a loss of structural integrity, we introduced a compensatory mutation into the 5' strand of stem IIB (pM63/69+50/55) that was predicted to fully restore secondary structure. This mutant was fully responsive to Rev and suggests that the role of IIB in RRE function is primarily structural. Therefore, although we have shown that Rev responsiveness is conferred by stem-loop II, the precise definition of the critical determinants of RRE function has, to date, remained elusive.

Rev is a sequence specific RNA binding protein. Having defined which sequences within the RRE are required for biological activity, we developed an in vitro RNA binding assay to further examine the Rev:RRE interaction. Binding was assessed by mixing purified Rev protein from E. coli (Daly et al., 1989) with a radiolabeled, in vitro synthesized, 252 nucleotide RRE containing transcript. Rev:RRE complexes were resolved by electrophoresis through native polyacrylamide gels and visualized by autoradiography (Fig. 2, see Malim et al., 1990 for details). At least two such complexes displaying reduced electrophoretic mobility were observed upon addition of Rev to a wild type RRE probe (compare lane 2 to lane 1, Fig. 2). The specificity of this interaction was confirmed by preincubation of Rev with excess unlabeled RNA competitor prior to probe addition. As expected, addition of the wild type RRE to the reaction completely inhibited the formation of Rev:RRE complexes (lane 8) whereas a variety of nonspecific RNA competitors had little or no effect (lanes 3-7). To ask whether the biological activity of selected RRE mutants reflected their Rev binding capabilities in vitro, we performed a series of experiments using mutant RRE transcripts as competitors (Fig. 2, lanes 9-13). It is clear that the nonresponsive mutants △50/75, △41/105 and M63/69 were unable to bind Rev. In contrast, the active mutants △130/162 and △163/191 retained their ability to bind Rev and were therefore effective competitors in this assay. We have recently extended this analysis to a number of radiolabeled probes to show that mutants such as M63/69+50/55 and △198/234 as well as an "excised" stem-loop II are also able to bind Rev (unpublished observations). We therefore conclude that (1) the biological activity of mutant RREs has a close correlation with their ability to bind Rev and (2) stem-loop II contains the major binding site(s) for Rev. Because mutants such as △1/34 and

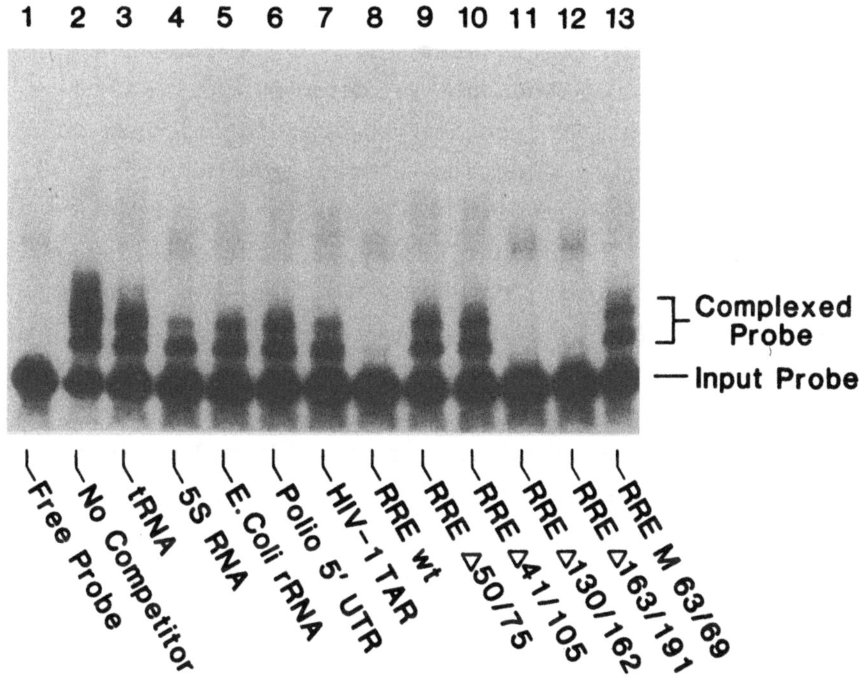

Figure 2. *In vitro* binding of Rev to the RRE correlates with biological activity. Refer to text for details. Polio 5'UTR = poliovirus 5' untranslated region. Similar Rev:RRE binding data has also been obtained by Daly *et al*. (1989), Zapp and Green (1989), Olsen *et al*. (1990) and Heaphy *et al*. (1990).

Δ198/234 exhibit considerable Rev binding *in vitro*, but much reduced activity *in vivo*, it is clear that segments of the RRE, other than stem-loop II, are important for the Rev response *in vivo*. We therefore propose that sequences such as stem I may be required for effective presentation of the Rev binding site to Rev in the cell nucleus.

DISCUSSION

The interaction of the HIV-1 Rev *trans*-activator with the RRE is essential for viral structural gene expression and hence, for viral replication. Here we report that Rev binds specifically to the RRE in a biologically relevant manner. At this point it is interesting to consider a previous mutagenesis study of the Rev protein itself (Malim et al., 1989a). Two distinct regions were identified that are required for in vivo function. Because disruption of these regions generated recessive negative and dominant negative (or transdominant) mutants, it was proposed that these regions are analagous, respectively, to the "binding" and "activation" domains of a variety of transcription factors (reviewed by Ptashne, 1988). Remarkably, two of the mutations within the N-terminal, or "binding", domain affected a sequence that shares considerable identity with the recently described "Arg-rich" RNA binding motif (Lazinski et al., 1989). We therefore suggest that this domain, at least in part, mediates the observed binding of Rev to the RRE. We have recently extended our mutational definition of the C-terminal, or "activation" domain of Rev to demonstrate that the leucine residues at positions 78, 81 and 83 are critical for trans-activation (unpublished observations). The precise way in which this domain interfaces with cellular factors to activate the nuclear export of unspliced viral transcripts is currently unknown.

In conclusion, we believe that the model shown in Fig. 3 represents our current understanding of the domain structure of the HIV-1 Rev protein. The development of binding assays for Rev, in addition to a variety of phenotypically characterized mutant proteins, makes the experimental confirmation of this model readily approachable.

ACKNOWLEDGMENTS

Figures 1 and 2 are reproduced from Malim et al. **Cell** 60: 675-683, 1990 with the permission of Cell Press.

The authors wish to thank Ms. Sharon Goodwin for secretarial assistance.

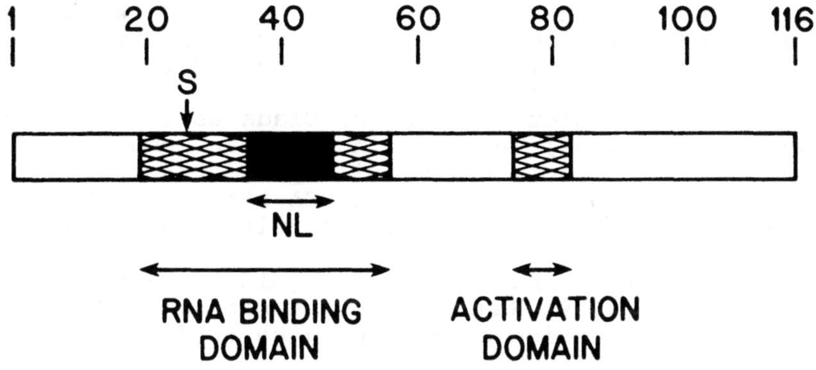

Figure 3. Domain Structure of the HIV-1 Rev Protein. The 116 amino acid full length protein is encoded by two exons separated by an intron largely corresponding to the viral env gene. The "binding" and "activation" domains are shown as hatched boxes encompassing residues 23-56 (approx.) and 78-83, respectively. S = splice junction. NL = highly basic region important for nuclear localization that shares considerable identity with the "Arg-rich" RNA binding motif (see text).

REFERENCES:

Arya SK, Guo C, Josephs SF, Wong-Staal F (1985) Trans-activator gene of human T-lymphotropic virus type III (HTLV-III). Science 229:69-73.

Berkhout B, Silverman RH, Jeang K-T (1989) Tat trans-activates the human immunodeficiency virus through a nascent RNA target. Cell 59:273-282.

Chang DD, Sharp PA (1989) Regulation by HIV Rev depends upon recognition of splice sites. Cell 59:789-795.

Cullen BR, Greene WC (1989) Regulatory pathways governing HIV-1 replication. Cell 58:423-426.

Daly TJ, Cook KS, Gray GS, Maione TE, Rusche JR (1989) Specific binding of HIV-1 recombinant Rev protein to the Rev-responsive element in vitro. Nature 342:816-819.

Dayton ET, Powell DM, Dayton AI (1989) Functional analysis of CAR, the target sequence for the Rev protein of HIV-1. Science 246:1625-1629.

Dingwall C, Ernberg I, Gait MJ, Green SM, Heaphy S, Karn J, Lowe AD, Singh M, Skinner MA, Valerio R (1989) Human immunodeficiency virus 1 Tat protein binds trans-activation-responsive region (TAR) RNA in vitro. Proc Natl Acad Sci USA 86:6925-6929.

Emerman M, Vazeux R, Peden K (1989) The rev gene product of the human immunodeficiency virus affects envelope-specific RNA localization. Cell 57:1155-1165.

Fauci AS (1988) The human immunodeficiency virus: infectivity and mechanisms of pathogenesis. Science 239:617-622.

Feinberg MB, Jarrett RF, Aldovini A, Gallo RC, Wong-Staal F (1986) HTLV-III expression and production involve complex regulation at the levels of splicing and translation of viral RNA. Cell 46:807-817.

Felber BK, Hadzopoulou-Cladaras M, Cladaras C, Copeland T, Pavlakis GN (1989) Rev protein of human immunodeficiency virus type 1 affects the stability and transport of the viral mRNA. Proc Natl Acad Sci USA 86:1495-1499.

Haase AT (1986) Pathogenesis of lentivirus infections. Nature 322:130-136.

Hammarskjold M-L, Heimer J, Hammarskjold B, Sangwan I, Albert L, Rekosh D (1989) Regulation of human immunodeficiency virus env expression by the rev gene product. J Virol 63:1959-1966.

Heaphy S, Dingwall C, Ernberg I, Gait MJ, Green SM, Karn J, Lowe AD, Singh M, Skinner MA (1990) HIV-1 regulator of virion expression (Rev) protein binds to an RNA stem-loop structure located within the Rev response element region. Cell 60:685-693.

Lazinski D, Grzadzielska E, Das A (1989) Sequence-specific recognition of RNA hairpins by bacteriophage antiterminators requires a conserved arginine-rich motif. Cell 59:207-218.

Malim MH, Böhnlein S, Hauber J, Cullen BR (1989a) Functional dissection of the HIV-1 Rev trans-activator—derivation of a trans-dominant repressor of Rev function. Cell 58:205-214.

Malim MH, Hauber J, Le S-Y, Maizel JV, Cullen BR (1989b) The HIV-1 rev trans-activator acts through a structured target sequence to activate nuclear export of unspliced viral mRNA. Nature 338:254-257.

Malim MH, Tiley LS, McCarn DF, Rusche JR, Hauber J, Cullen BR (1990) HIV-1 structural gene expression requires binding of the Rev trans-activator to its RNA target sequence. Cell 60:675-683.

Olsen HS, Nelbock P, Cochrane AW, Rosen CA (1990) Secondary structure is the major determinant for interaction of HIV rev protein with RNA. Science 247:845-848,

Ptashne M (1988) How eukaryotic transcriptional activators work. Nature 335:683-689.

Rosen CA, Terwilliger E, Dayton A, Sodroski JG, Haseltine WA (1988) Intragenic cis-acting art gene-responsive sequences of the human immunodeficiency virus. Proc Natl Acad Sci USA 85:2071-2075.

Sharp PA, Marciniak RA (1989) HIV TAR: an RNA enhancer? Cell 59:229-230.

Sodroski J, Rosen C, Wong-Staal F, Salahuddin SZ, Popovic M, Arya S, Gallo RC, Haseltine WA (1985) Trans-acting transcriptional regulation of human T-cell leukemia virus type III long terminal repeat. Science 227:171-173.

Sodroski J, Goh WC, Rosen C, Dayton A, Terwilliger E, Haseltine W (1986) A second post-transcriptional trans-activator gene required for HTLV-III replication. Nature 321:412-417.

Zapp ML, Green MR (1989) Sequence-specific RNA binding by the HIV-1 Rev protein. Nature 342:714-716.

Control of Protein Synthesis by RNA Regulators

M.B. Mathews, S. Gunnery, L. Manche, K.H. Mellits, and T. Pe'ery
Cold Spring Harbor Laboratory
P.O. Box 100
Cold Spring Harbor, New York 11724

It is increasingly recognized that RNA plays a wide variety of roles in macromolecular synthesis. RNA can serve as the repository of genetic information (as in RNA viral genomes) or as a transient intermediary in the conversion of genetic information from a DNA to a protein form (mRNA). During the act of translation, RNA also serves a structural role (rRNA) and as an adaptor in the decoding operation (tRNA). RNA molecules serve as primers at the origin of DNA replication, as templates for telomere synthesis, as guide sequences during splicing and editing, and doubtless they subserve other functions yet to be described. Perhaps most dramatically, RNA molecules can display catalytic functions previously believed restricted to proteins. Finally, and most relevant here, RNA can play a regulatory role. Central to the present discussion is the observation that RNAs can regulate the efficiency of protein synthesis in the cell and the animal by controlling the activity of a protein kinase. This notion will be elaborated below.

Phosphorylation of eIF-2

In higher cells, protein synthesis can be controlled by changes in the activity of initiation factors (eIFs, eukaryotic initiation factors) (Hershey et al., 1986). In poliovirus-infected cells, eIF-4F, an initiation factor involved in recognition of the 5′ terminal cap structure and mRNA binding, is inactivated by the proteolytic cleavage of one of its subunits, p220. Several initiation factors are substrates for protein kinases, and in many cases the degree of phosphorylation of the factor correlates with the growth state and protein synthetic activity of the cell (reviewed by Hershey, 1989). The best understood of these phosphorylation changes involves eIF-2, the initiation factor responsible for the GTP dependent binding of Met-tRNA$_F$ to the 40S ribosomal subunit. Although eIF-2 can be phosphorylated on both its alpha and beta subunits, only the

NATO ASI Series, Vol. H 49
Post-Transcriptional Control of Gene Expression
Edited by J. E. G. McCarthy and M. F. Tuite

phosphorylation of the alpha subunit has been associated with alterations in protein synthetic activity.

The scheme illustrated in Figure 1 outlines the role of eIF-2 in normal protein synthesis (left side) and its regulation by phosphorylation (right side). In the normal course of protein synthesis, eIF-2 forms a ternary complex with GTP and Met-tRNA$_F$, which in the presence of other initiation factors then binds to the small ribosomal subunit. In the next step, aided by the cap-binding proteins eIF-4A, B, and F, mRNA associates with the 40S subunit complex, followed by the joining of the large ribosomal subunit. At this stage, the initiation factors are released from the ribosome. In the case of eIF-2, the associated nucleotide has been hydrolyzed to GDP and must be replaced with GTP in order to permit a subsequent initiation event. The exchange reaction occurs only slowly at physiological magnesium concentrations and is catalyzed by another initiation factor, the guanine nucleotide exchange factor (GEF or eIF-2B).

While eIF-2 consists of three non-identical polypeptide chains, with an aggregate molecular weight of approximately 125,000, GEF comprises 5 polypeptides with an aggregate molecular weight of approximately 270,000. The reasons for this complexity are far from clear, but it seems reasonable to suppose that it is related to GEF's regulatory role. This was initially discovered in reticulocytes and is depicted on the right side of Figure 1. Phosphorylation of the alpha subunit of eIF-2 on serine-51 (Pathak et al., 1988) prevents the GTP/GDP exchange reaction, probably because phosphorylated eIF-2 traps GEF in a non-productive complex. GEF is thereby sequestered and is prevented from fulfilling its catalytic role. At least in reticulocytes, only approximately 30% of the alpha subunit need be phosphorylated to block protein synthesis initiation completely, presumably because this level of phosphorylation suffices to trap all of the GEF (Ochoa, 1983).

eIF-2 kinases

Two protein kinases are known that are able to phosphorylate the alpha subunit of eIF-2 on serine 51 (Ochoa, 1983). Their chief properties are summarized in Table 1. The first of these, usually known as HCR (the heme controlled regulator) is largely, if not exclusively, found in reticulocytes where its function is to couple the synthesis of globin chains to the availability of iron and of the heme prosthetic group. The starvation of reticulocytes for iron, or the absence of heme from reticulocyte lysates, results in activation of HCR by a mechanism that is unknown in detail, and the consequent inhibition of protein synthesis as illustrated in Figure 1. Through this regulatory mechanism, reticulocytes avoid accumulating globin chains which cannot be assembled into hemoglobin tetramers. The synthesis of most, if not all, other

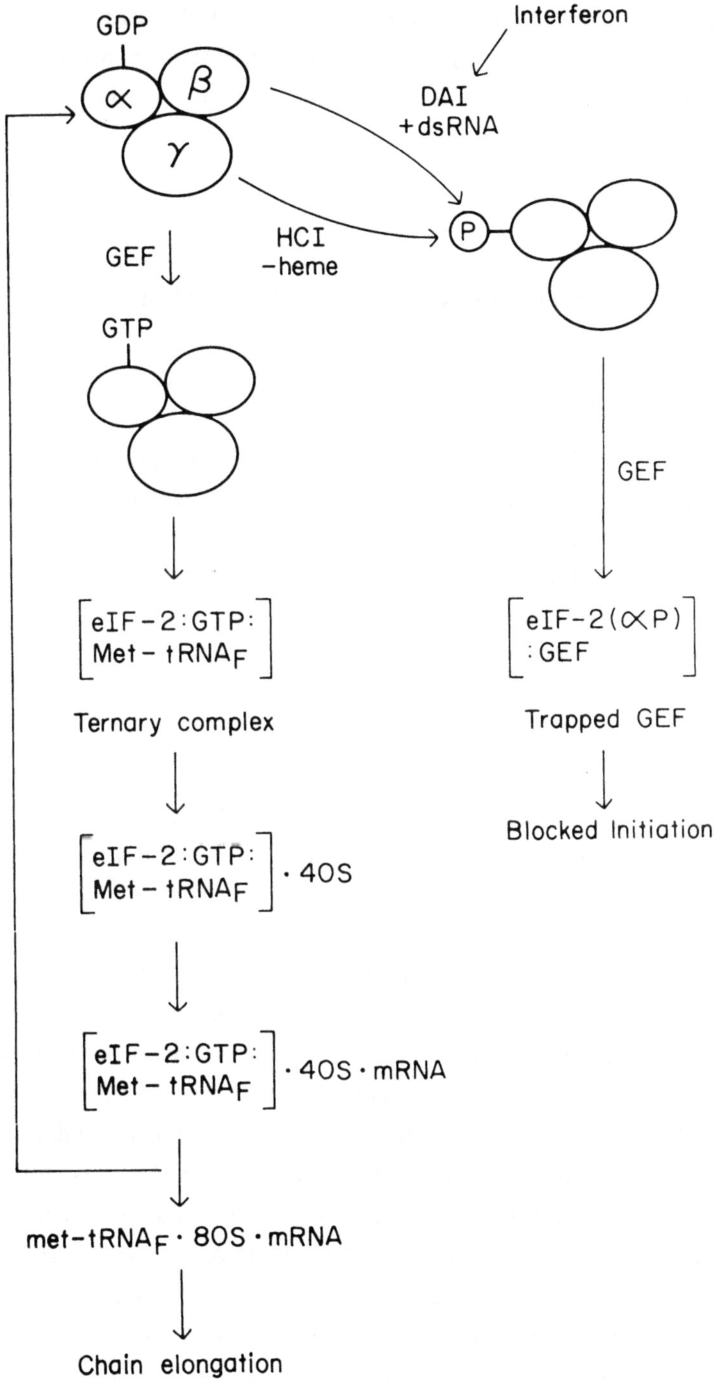

Figure 1. The function and regulation of eIF-2. The structure of eIF-2 is depicted as three ellipses, and its function in the initiation of protein synthesis is shown on the left side of the scheme. Regulation by phosphorylation is illustrated on the right side (HCR=HCI). Reprinted from Nature 313:196, 1985.

reticulocyte proteins is blocked concomitantly (Mathews, et al., 1973). A wide range of other agents, including oxidizing agents and high hydrostatic pressure, can also activate HCR, suggesting that it may regulate protein synthesis under other circumstances.

Table I. Properties of the two eIF-2 alpha kinases

	HCR	DAI
Distribution	Reticulocytes (mainly)	Widespread
Location	Soluble	Ribosome-associated
Synthesis	?	Interferon-induced
Activators	Lack of Fe^{+2}, or hemin; several other stimuli	dsRNA; polyanions
Substrates	eIF-2α	eIF-2α, histones

The second eIF-2 alpha kinase is known as DAI (the double stranded RNA activated inhibitor) and also goes by a number of alternative names (e.g., dsI, p68, PK_{ds}). This enzyme is found in reticulocytes and a very wide range of other tissues (Ochoa, 1983; Hovanessian, 1989; Kostura & Mathews, 1989). Unlike HCR, which appears in the post-ribosomal supernatant, much of the DAI is associated with ribosomes, from which it is released by high salt concentrations. The cellular concentration of DAI can be induced by interferon suggesting an involvement in the interferon-induced anti-viral response. In normal conditions, the enzyme occurs in an inactive or latent state; as its name implies, during infection or under some other conditions such as plasmid transfection, it is activated by the presence of double-stranded (ds) RNA. *In vitro* the enzyme can also be activated by heparin, a sulfated mucopolysaccaride which presumably resembles dsRNA in charge and structure. Activation appears to involve auto-phosphorylation of the enzyme and requires magnesium or manganese ions and ATP. Activation seems to be accompanied by a switch in substrate specificity as the activated enzyme is apparently unable to phosphorylate the latent form of the enzyme: conceivably, phosphorylation of the enzyme changes the conformation of its substrate binding site so that it can accommodate the initiation factor. Activated DAI is able to phosphorylate histones as well as eIF-2, so its range, though narrow, is somewhat broader than that of HCR.

Role of DAI

As far as is known, the only physiologically relevant substrate for DAI is the alpha subunit of eIF-2, implying that DAI's role lies in the regulation of protein synthesis. Perhaps the best example of the operation of this mechanism is afforded by adenovirus-infected cells (summarized in O'Malley et al., 1989). Adenovirus encodes two small RNAs, VA RNA_I and VA RNA_{II}, which serve to counter the action of DAI. The adenovirus mutant, Ad5 dl331, fails to produce VA RNA_I and grows poorly. Unlike the wild-type virus, the mutant is sensitive to interferon suggesting that VA RNA_I also fulfills the role of a viral defense against an aspect of the interferon-induced antiviral response. In the natural host, this may be its chief raison d'être.

The role of VA RNA_I has been studied intensively in cells that have not been treated with interferon. These cells contain latent DAI, albeit at lower levels than in interferon-treated cells. At late times of infection, protein synthesis in cells infected with the VA RNA_I-minus mutant is reduced by 90% compared to cells infected with the wild-type virus. The defect lies at the level of polypeptide chain initiation, and analysis conducted *in vivo* and *in vitro* showed that eIF-2 phosphorylation plays a central part in this process. In a cell-free system, addition of either eIF-2 or GEF remedied the protein synthesis defect. Furthermore, dl331-infected cells exhibit an increase in the activity of DAI, elevated phosphorylation of the alpha subunit of eIF-2, and a reduction in the activity of GEF. dsRNA apparently resulting from symmetrical transcription of the viral genome and capable of activating DAI *in vitro* has been isolated from infected cells. Most importantly, the ability of VA RNA to prevent the dsRNA-mediated activation of DAI has been demonstrated *in vitro*. Finally, cell lines containing a mutant alpha subunit, in which serine-51 has been replaced with an alanine residue, permit the mutant to grow as well as wild-type virus (Davies et al., 1989). Thus, there is an overwhelming body of evidence, both genetic and biochemical, linking the absence of VA RNA_I to a reduction in protein synthesis via the mechanism illustrated in Figure 1. What is known about the interaction of DAI with activating dsRNA and inhibitory VA RNA will be reviewed below.

dsRNA seems to be a frequent concomitant of viral infection. It serves to trigger interferon synthesis, and to activate at least two enzymes (DAI and 2-5A synthetase) that are induced by interferon. Like adenovirus, a number of other viruses have elaborated mechanisms to circumvent DAI activation or avert the consequences (reviewed by Sonenberg, 1990). Epstein-Barr virus (EBV) encodes two small RNAs, known as EBERs, with structures reminiscent of the adenovirus VA RNAs. These RNAs partially substitute for the VA RNAs (Bhat and Thimmappaya, 1985) suggesting that they may play an analogous role in EBV infection, although such a role has not been demonstrated. Influenza virus blocks the activation of DAI via an as yet uncharacterized protein factor, and reovirus encodes a protein (sigma 3) that apparently binds and sequesters dsRNA so that it cannot interact with the kinase. Vaccinia virus also encodes a product, probably a protein, that prevents activation of the kinase. Poliovirus adopts a different strategy, which is to degrade the kinase. The human immunodeficiency virus, HIV-1, appears to interact with DAI at two levels: its RNA affects kinase activity (see below) and like polio it also causes degradation of the enzyme.

In most cases it appears that activation of DAI results in an indiscriminate inhibition of protein synthesis, but selective or localized kinase activation has been observed or invoked in a few cases. De Benedetti and Baglioni (1984) found that a double-stranded region associated with the 3′ end of a messenger RNA preferentially slowed its entry into polysomes in a reticulocyte lysate, and they demonstrated that this effect was due to the activation of DAI. Moreover, the synthesis of proteins from expression vectors introduced into mammalian cells by transfection is increased by VA RNA through a mechanism involving DAI (Akusjärvi et al., 1987; Kaufman and Murtha, 1987). Because global protein synthesis is unaffected, it seems that VA RNA acts by antagonizing the localized activation of DAI in the neighborhood of the mRNAs generated from the plasmids. One possibility is that some regions of the plasmids are transcribed symmetrically, giving rise to partly duplex mRNA which can activate nearby DAI. Finally, despite the presence of VA RNA, eIF-2 alpha phosphorylation occurs to an unexpectedly large extent (approximately 30%) in cells infected with wild-type adenovirus (O'Malley et al., 1989). The infected cells actively synthesize viral proteins even though this degree of eIF-2 phosphorylation would be expected to trap all the GEF. To resolve this paradox, it has been hypothesized that in the infected cells eIF-2 phosphorylation occurs solely in the vicinity of cellular messenger RNAs. Preexisting cellular mRNAs are not degraded during adenovirus infection, nor are they efficiently translated. It may be that VA RNA is compartmentalized in proximity with the viral mRNAs, thereby protecting their translation, while the cellular RNAs are exposed to the consequences of DAI activation and eIF-2 phosphorylation. Thus VA RNA could also contribute to the phenomenon of host cell shut off.

DAI activation

The activation of DAI is intimately associated with its autophosphorylation, and these two processes may indeed be mechanistically linked. It has long been recognized that activation of the enzyme exhibits an unusual dependence on dsRNA concentration. Low concentrations (in the ng/ml range) activate DAI while high concentrations (approaching 1 μg/ml) prevent activation, giving a bell-shaped activation curve. Two models have been advanced to account for this unusual concentration dependence, as outlined in Figure 2. The first, initially proposed by T. Hunt and his colleagues (O'Malley et al.,1986), supposes that autophosphorylation occurs when two DAI molecules interact on a single dsRNA molecule. At supra-optimal concentrations of dsRNA, the dimers will be out-titrated by the excess of nucleic acid, giving monomeric complexes between a single DAI molecule and dsRNA which do not permit activation. This model implies that there is a single site for dsRNA on DAI, that DAI can dimerize on dsRNA, and that the autophosphorylation reaction is intermolecular. A variation on this theme (model 1a) was suggested by D. Gillespie (personal communication), and supposes that only one of the DAI molecules is bound to the dsRNA. This variation has similar requirements to model 1.

A radically different alternative has been proposed by Hovanessian and coworkers (Galabru et al., 1989). In this model (model 2, Figure 2), the DAI has two sites that can bind dsRNA. The first site is a high affinity site, and dsRNA bound to this site activates the enzyme. The second site is of lower affinity and binding of dsRNA (or other RNA) here blocks activation. The high affinity site is occupied preferentially at low dsRNA concentrations, leading to activation of the enzyme. At high concentrations of dsRNA, both sites will be occupied and activation will not occur, assuming that the inhibitory site is dominant. Thus, this model supposes the existence of two sites for dsRNA on DAI, and that DAI activates as a monomer through an intramolecular autophosphorylation reaction.

Evidence from our laboratory favors the former model in that we detect only one site for dsRNA on DAI, and the kinetics of activation are second order with respect to DAI (Kostura & Mathews, 1989). Furthermore, model 1 provides a ready explanation for the observation that short or imperfect duplexes which are unable to activate the enzyme are nevertheless able to prevent its activation by long dsRNA duplexes (Baglioni et al., 1980). Presumably, the short duplex regions are able to bind DAI but are unable to accommodate two enzyme molecules in such a way as to permit intermolecular autophosphorylation. The issue is far from settled, however, as Galabru et al. (1989) report a second, low affinity, site; however, it is not clear whether this site has the physiological function of an inhibitory site.

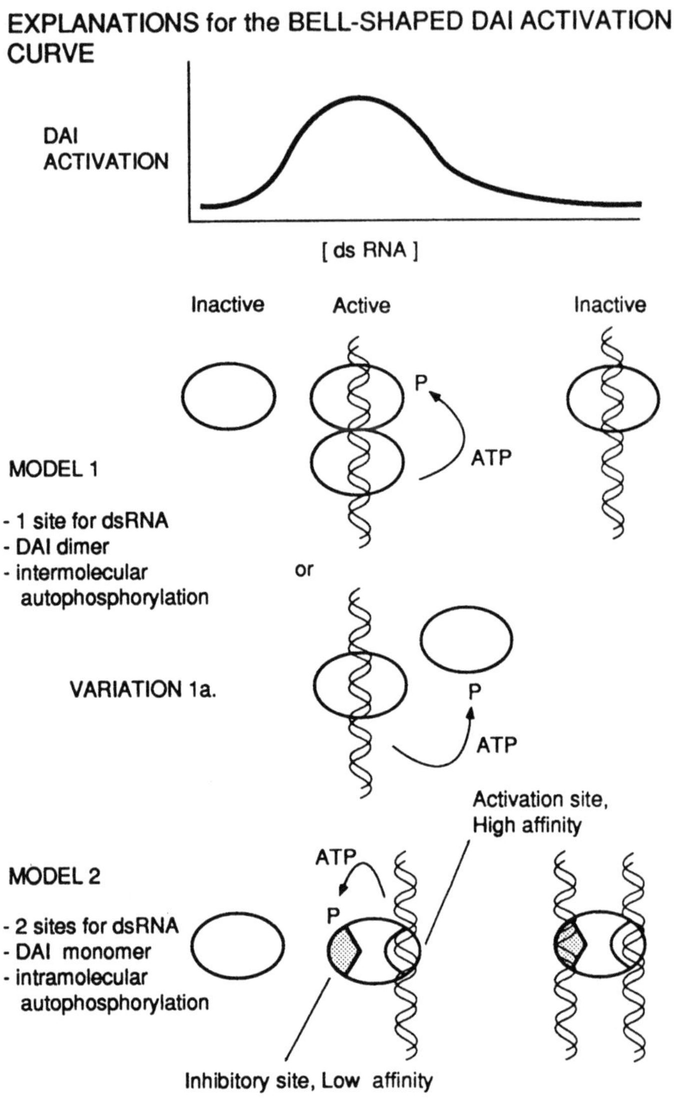

Figure 2. Models for the activation of DAI by dsRNA. Activation involves autophosphorylation of the enzyme and requires suitable, low concentrations of dsRNA. High concentrations block activation (but do not affect the activity of enzyme once it has been activated at low dsRNA concentration). Several models to explain this concentration dependence have been advanced, as illustrated here and discussed in the text.

DAI-RNA interactions

We have recently reinvestigated the influence of dsRNA chain length on interaction with DAI. Our results show that the shortest duplexes that are able to bind are approximately 30 base pairs long, and that binding efficiency increases with chain length up to about 85 base pairs; longer duplexes bind with equal efficiency. Very short duplexes are poor activators of the enzyme but can block activation by longer duplexes. As chain length increases, the balance shifts such that a duplex of 55 base pairs activates reasonably well, although not as well as longer duplexes, and inhibits activation by reovirus dsRNA (with an average chain length of about 1kb) only poorly. TAR RNA, transcribed from the HIV-1 long terminal repeat, forms an imperfect duplex of less than 30 base pairs. In contrast to published reports, but in keeping with the known properties of DAI, we find that purified TAR RNA functions as an inhibitor of DAI autophosphorylation.

Adenovirus VA RNA_I is a highly structured RNA of 160 nucleotides. Its secondary structure has been explored and its interaction with DAI has been studied in some detail. The RNA consists of two stems, 20-25 base pairs in length, a 10 nucleotide bulge, and a region (shown stippled in Figure 3) called the central domain. Although not the most stable structure according to thermodynamic predictions, this structure is supported by direct experimental evidence derived from nuclease sensitivity probing and from mutagenesis (Mellits and Mathews, 1988; Furtado et al., 1989; Mellits et al., 1990). In contrast to early ideas, which anticipated that the duplex regions would play a determinative role in VA RNA function, genetic and biochemical analysis indicate that the central domain is the most important functional region in the molecule. Alterations elsewhere in the molecule are tolerated provided they are not so severe that they destabilize the central domain. Some alterations within the central domain are also compatible with function but others are not, emphasizing the fact that detailed structure-function relations within this region are not yet understood. While the central domain is of paramount importance for blocking DAI activation it does not have the ability to confer high affinity binding upon the RNA; instead, the extended apical duplex appears to be largely responsible for the ability of this molecule to bind to DAI *in vitro*. Thus, it seems that VA RNA is a specialized effector molecule in which a short duplex serves to bring the molecule into close association with its target, and the central domain interferes with DAI function presumably by affecting its enzymatic activity.

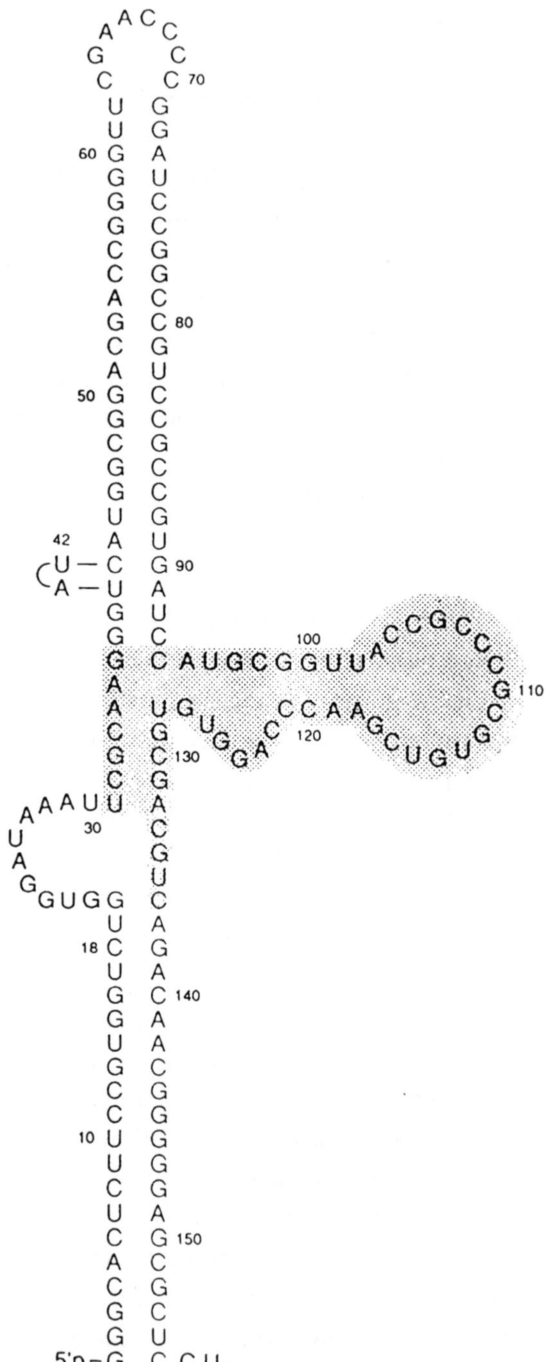

Figure 3. The structure of VA RNA$_I$. Superimposed on the structural model for adenovirus-2 VA RNA$_I$ (redrawn from Mellits & Mathews, 1988) are shown the approximate limits for the central domain (stippled).

Nearly everything that is known about VA RNA_I comes from studies of adenovirus serotype 2 or 5. These serotypes (but not all adenovirus serotypes) also encode a second small RNA species, VA RNA_{II}, which accumulates to much lower levels in virus-infected cells. Like VA RNA_I, this transcript is highly structured, about 160 nucleotides in length, and is produced by RNA polymerase III; unlike the major species, however, VA RNA_{II} is dispensable at least for virus growth in tissue culture. Nonetheless, it is likely that VA RNA_{II} can play the same role as VA RNA_I. A double mutant, lacking both forms, grows more poorly than the VA RNA_I-negative single mutant (Bhat and Thimmappaya, 1984) suggesting that the minor species can partially compensate for the absence of the major species. Possibly VA RNA_{II} represents a redundant version of the gene, which could be effective *in vivo* if it were transcribed to a sufficiently high level; alternatively, there may be a more subtle interplay between the two species. By the same token, it remains to be seen whether cellular regulators of DAI activity exist and what role they play in the regulation of gene expression.

Acknowledgements

This article chiefly reflects studies in the authors' laboratory which were supported by a National Cancer Institute program project grant to MBM (NIH CA13106). To avoid a reference list of burdensome length, the work of others is cited mainly in the form of review articles.

References

Akusjärvi G, Svensson C, Nygård O (1987) A mechanism by which adenovirus virus-associated RNA_I controls translation in a transient expression assay. Mol Cell Biol 7:549-551

Baglioni C, Benvin S, Maroney PA, Minks MA, Nilsen TW, West DK (1980) Interferon-induced enzymes: activation and role in the antiviral state. Ann N Y Acad Sci 350:497-509

Bhat RA, Thimmappaya B (1984) Adenovirus mutants with DNA sequence perturbations in the intragenic promoter of VAI RNA gene allow the enhanced transcription of VAII RNA gene in HeLa cells. Nucl Acids Res 12:7377-7388

Bhat RA, Thimmappaya B (1985) Construction and analysis of additional adenovirus substitution mutants confirm the complementation of VAI RNA function by two small RNAs encoded by Epstein-Barr virus. J Virol 56:750-756

Davies MV, Furtado M, Hershey JWB, Thimmappaya B, Kaufman RJ (1989) Complementation of adenovirus virus-associated RNA I gene deletion by expression of a mutant eukaryotic translation initiation factor. Proc Natl Acad Sci USA 86:9163-9167

De Benedetti A, Baglioni C (1984) Inhibition of mRNA binding to ribosomes by localized activation of dsRNA-dependent protein kinase. Nature 311:79-81

Furtado MR, Subramanian S, Bhat RA, Fowlkes DM, Safer B, Thimmappaya B (1989) Functional dissection of adenovirus VAI RNA. J Virol 63:3423-3434

Galabru J, Katze MG, Robert N, Hovanessian AG (1989) The binding of double-stranded RNA and adenovirus VAI RNA to the interferon-induced protein kinase. Eur J Biochem 178:581-589

Hershey JWB (1989) Protein phosphorylation controls translation rates. J Biol Chem 264:20823-20826

Hershey JWB, Duncan R, Mathews MB (1986) Mechanisms of Translational Control. in Translational Control, pp 1-18 (ed. MB Mathews) Cold Spring Harbor Laboratory Press

Hovanessian AG (1989) The double stranded RNA-activated protein kinase induced by interferon: dsRNA-PK. J Interferon Res 9:641-647

Kaufman RJ, Murtha P (1987) Translational control mediated by eucaryotic initiation factor-2 is restricted to specific mRNAs in transfected cells. Mol Cell Biol 7:1568-1571

Kostura M, Mathews MB (1989) Purification and activation of the double-stranded RNA - dependent eIF-2 kinase DAI. Mol Cell Biol 9:1576-1586

Mathews MB, Hunt T, Brayley A (1973) Specificity of the control of protein synthesis by Haemin. Nature New Biology 243:230-233

Mellits KH, Kostura M, Mathews MB (1990) Interaction of adenovirus VA RNA_I with the protein kinase DAI: Non-equivalence of binding and function. Cell 61:843-852

Mellits KH, Mathews MB (1988) Effects of mutations in stem and loop regions on the structure and function of adenovirus VA RNA_I. EMBO J 7:2849-2859

O'Malley RP, Mariano TM, Siekierka J, Merrick WC, Reichel PA, Mathews MB (1986) The control of protein synthesis by adenovirus VA RNA. Cancer Cells, Vol 4: 291-301 DNA Tumor Viruses.

O'Malley RP, Duncan RF, Hershey JWB, Mathews MB (1989) Modification of protein synthesis initiation factors and shut-off of host protein synthesis in adenovirus-infected cells. Virology 168:112-118

Ochoa S (1983) Regulation of protein synthesis initiation in eucaryotes. Arch Biochem Biophy 223:325-349

Pathak VK, Schindler D, Hershey JWB (1988) Generation of amutant form of protein synthesis initiation factor eIF-2 lacking the site of phosphorylation by eIF-2α kinases. Mol Cell Biol 8:993-995

Sonenberg N (1990) Measures and countermeasures in the modulation of initiation factor activities by viruses. New Biologist 2:1-8

UNTRANSLATED LEADER SEQUENCES AND ENHANCED MESSENGER RNA TRANSLATIONAL EFFICIENCY

Lee Gehrke and Stephen A. Jobling
Division of Health Sciences and Technology
Massachusetts Institute of Technology
Cambridge, MA 02139

The analysis of gene regulation has been facilitated by model experimental systems which represent the extremes of a particular mechanism; that is, by identifying examples of controls which function either with extraordinary efficiency or not at all, investigators are provided with a system which might be manipulated to characterize the underlying control elements. Messenger RNAs which translate with high-level efficiency have been studied both to define fundamental mechanisms of protein synthesis and also to gain a more complete understanding of why some messenger RNAs translate with greater efficiency than others. For some of these efficient mRNAs, experimental evidence has localized control to nucleotides at the 5' terminus of the mRNA preceding the initiation codon--the noncoding or untranslated leader sequence. In this paper, we describe examples of high-level mRNA translational efficiency related to the noncoding 5' leader, summarize current understanding of the mechanism of enhanced translation, and comment upon what has been learned from the study of these mRNAs.

Of the three fundamental steps of protein synthesis (initiation, elongation, and termination), Lodish proposed the initiation step as the most important control point (Lodish, 1976). By inference alone, it seems obvious to correlate the efficiency of mRNA translation with the physical characteristics of the 5' terminus of the mRNA, where initiation of protein synthesis occurs. Analysis of mRNA secondary structure by experimental structure mapping or computer prediction (Gehrke *et al.*, 1983, Iserentant and Fiers, 1979, Pavlakis *et al.*, 1980) established a correlation between translational efficiency and 5'-terminal secondary structure, but direct evidence of the significant role of the 5' untranslated leader sequence in regulating translational efficiency arose from experiments which used chimeric mRNA constructs to either increase (Pelletier and Sonenberg, 1985) or decrease (Jobling and Gehrke, 1987) 5' mRNA secondary structure, resulting in, respectively, inhibition or enhancement of mRNA translation.

The most impressive examples of translationally efficient messenger RNAs are of viral origin (Kozak, 1983). The messenger RNAs which encode the coat proteins of alfalfa, brome and turnip yellow mosaic viruses are among those that translate with extraordinary efficiency (Benicourt and Haenni, 1978, Herson *et al.*, 1979, Shih and Kaesberg, 1973). Other examples of efficient mRNAs include black beetle virus RNA 2 (Friesen and Rueckert, 1981) and encephalomyocarditis virus RNA (Parks *et al.*, 1986).

NATO ASI Series, Vol. H 49
Post-Transcriptional Control of Gene Expression
Edited by J. E. G. McCarthy and M. F. Tuite

Removal of the untranslated leader sequence from inefficient or marginally efficient mRNAs, followed by replacement using the leader from some of the efficient messages leads to a significant increase in the number of protein molecules synthesized per ribosome per unit time (translational efficiency) from the chimeric mRNA. These data demonstrated that the noncoding leader is a *cis-* acting sequence which exercises significant regulatory control over the relative translational efficiency of individual mRNAs. The untranslated leader sequences of alfalfa mosaic virus (Jobling and Gehrke, 1987), tobacco mosaic virus (Gallie *et al.*, 1987a), encephalomyocarditis virus (Parks *et al.*, 1986) and gene 10 of phage T7 (Olins *et al.*, 1988) stimulate the translation of heterologous mRNAs. It has also been demonstrated, however, that "leader-swapping" experiments using noncoding sequences from highly efficient mRNAs does not always result in higher level translation. Although the coat protein mRNAs of brome mosaic virus and turnip yellow mosaic virus support efficient translation, the untranslated leader sequences of these messages do not appear to increase the rates of translation of other mRNAs (Jobling *et al.*, 1988). These results are consistent with the idea that untranslated leader sequence of a messenger RNA may not act in isolation to regulate mRNA translation, but that the characteristics of both the leader and 5'-proximal coding sequences may be important (Browning *et al.*, 1988a, Shakin and Liebhaber, 1986).

MECHANISM OF ENHANCED TRANSLATIONAL EFFICIENCY

The mechanism of increased mRNA translational efficiency mediated through the 5' untranslated leader sequence is not well understood. The noncoding sequences of alfalfa mosaic virus (Gallie *et al.*, 1987b, Jobling and Gehrke, 1987), tobacco mosaic virus (Gallie *et al.*, 1989, Sleat *et al.*, 1987), encephalomyocarditis virus (Parks *et al.*, 1986), and phage T7 gene 10 (Olins *et al.*, 1988) stimulate translation when linked in *cis* to heterologous mRNAs. The length of these leader sequences varies from 35 nucleotides to over 800 nucleotides; therefore, increased translational efficiency does not seem to be directly related to the length of the leader. Of the four sequences cited above, the alfalfa mosaic virus RNA 4 and tobacco mosaic virus untranslated leaders show the greatest sequence homology. The 5' noncoding region of AMV RNA 4 is 35 nucleotides in length, and the nucleotide sequence is remarkably A-U rich (Brederode *et al.*, 1980). The secondary structure of the AMV RNA 4 leader has been analyzed by structure mapping techniques (Gehrke *et al.*, 1983), and these data are evidence that the 5' leader of AMV RNA 4 is principally unstructured in solution. Although a similar analysis has not been reported for the TMV leader sequence, thermodynamic predictions suggest that this leader is also unstructured (L. Gehrke, unpublished results). An alignment of the AMV and

TMV Ω sequences reveals some limited areas of homology; however, deletion data suggest that these sequences are not necessary and sufficient for stimulation of translational efficiency (Gallie *et al.*, 1988). The most striking similarity between the AMV and TMV leader sequences is the absence of internal guanosine residues, and by inference, the inability to form stable intramolecular G:C base pairs.

Translationally efficient messenger RNAs are often also highly competitive messenger RNAs. Co-translation of two mRNAs under conditions of limiting availability of components of the translational apparatus ("mRNA saturation conditions") is accompanied by the greater ability of one mRNA to outpace the other for binding of ribosomes and protein initiation factors. The highly competitive nature of AMV RNA 4 in comparison with satellite tobacco necrosis virus, another efficient mRNA has been described (Herson *et al.*, 1979). The unusual translational characteristics of AMV RNA 4 were further revealed by studies which showed that mRNAs with diminished 5' secondary structure, including AMV RNA 4, can be translated in extracts prepared from poliovirus-infected HeLa cells (Sonenberg *et al.*, 1982), wherein translation of capped cellular mRNAs is blocked. The mechanism of the translational block observed in poliovirus-infected HeLa cells seems to be related to the specific proteolytic cleavage of the p220 component of the cap binding protein complex (Etchison *et al.*, 1982, Lee and Sonenberg, 1982), leading to preferential translation of uncapped viral RNA. These data are consistent with a hypothesis (Sonenberg *et al.*, 1982) stating that the translationally efficient AMV RNA 4, which, by nature of its unstructured 5' leader sequence (Gehrke *et al.*, 1983), will be translated under conditions where the availability of a translational component such as the cap binding protein complex (eIF-4F) is compromised.

Lodish's conclusion (Lodish, 1976) that those mRNAs with the highest rate constant for initiation of protein synthesis would also be the most competitive mRNAs suggests that optimal translation of individual mRNAs may require different levels of protein initiation factors. In other words, the relative translational efficiencies of individual mRNAs would be similar under conditions where initiation factors, ribosomes, tRNAs, enzymes are all in excess; however, as the supply of these components becomes limiting, those mRNAs with the greatest competitive activity would be translated preferentially. One hypothesis to explain the mechanism of the effect is that the competitive mRNA has either 1) a greater affinity for components of the translational machinery, or 2) a diminished requirement for these same components. *In vitro* protein synthesis in a fractionated wheat germ system (Browning *et al.*, 1987, Lax *et al.*, 1986) has been used to directly test the quantitative protein initiation factor requirements of messenger RNAs characterized by structured or unstructured leader sequences (Browning *et al.*, 1988b). The results of the experiments demonstrate that AMV RNA 4 or chimeric mRNAs containing the AMV RNA 4 untranslated leader sequence require addition of significantly less eIF-4F (the cap binding protein complex) in order to achieve half-maximal translation as compared to other mRNAs which have structured untranslated leader sequences. The interpretation of these data is that messenger RNAs containing unstructured leader sequences have a finite but diminished requirement for protein initiation factors required for binding of the 40S ribosomal subunit to messenger RNA. The data are less consistent with the notion that unstructured mRNAs have a higher relative affinity for protein factors.

ENHANCED TRANSLATION, CAP-INDEPENDENT INITIATION AND INTERNAL INITIATION OF PROTEIN SYNTHESIS

Analyses of translation in cells infected by picornaviruses provide evidence that initiation of protein synthesis can occur by a 5' m7GpppG/A cap-independent mechanism (Etchison *et al.*, 1982, Lee and Sonenberg, 1982, Sonenberg *et al.*, 1982); moreover, recent data suggest that ribosomes can initiate protein synthesis by directly binding internal sites in the untranslated leader sequence (Jang *et al.*, 1989, Jang *et al.*, 1988, Pelletier and Sonenberg, 1988). The mechanism of enhanced translation mediated through unstructured 5' untranslated leader sequences may overlap incompletely with the mechanisms of cap-independent and internal initiation of protein synthesis. Infection of HeLa cells by poliovirus leads to a specific proteolytic hydrolysis of the p220 component of the cap binding protein complex (Bonneau and Sonenberg, 1987, Etchison *et al.*, 1982, Lee and Sonenberg, 1982), followed by preferential translation of the uncapped poliovirus RNA. The mechanism of translation of poliovirus RNAs seems to be mediated through the long 5' untranslated leader sequence, for chimeric messenger RNAs containing the poliovirus untranslated leader sequence are also translated virus-infected HeLa cell extracts via a cap-independent internal initiation event (Pelletier *et al.*, 1988, Trono *et al.*, 1988).

Data have been reported which indicate that initiation of protein synthesis directed by the encephalomyocarditis (EMC) viral RNA can also occur by cap-independent and internal initiation mechanisms (Jang *et al.*, 1989, Jang *et al.*, 1988). While the poliovirus untranslated leader sequence is not a particularly strong enhancer of protein synthesis, the EMC RNA leader stimulates translation of heterologous mRNAs (Elroy-Stein *et al.*, 1989, Parks *et al.*, 1986). In this way, translation of the EMC mRNA is cap-independent and seems to involve an internal initiation mechanism on a leader sequence which can enhance translation of other mRNAs. The situation for poliovirus RNA is slightly different in that poliovirus RNA translates by a cap-independent mechanism which includes internal initiation, but the poliovirus leader has not been reported to stimulate the translational efficiency of heterologous mRNAs.

AMV RNA 4 seems to fall somewhere in between the characteristics of poliovirus and EMC RNAs. The activity of the AMV RNA 4 untranslated leader sequence as a translational enhancer was described earlier. Although AMV RNA 4 is a naturally capped messenger RNA

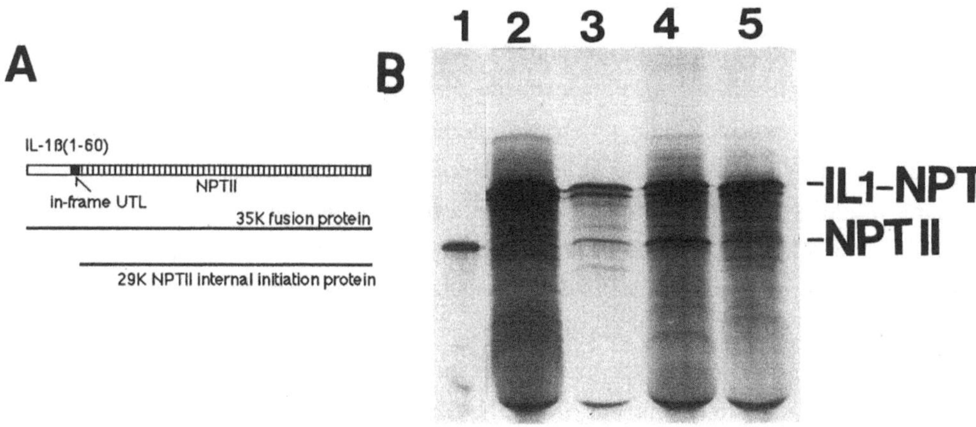

Figure 1. Assay for internal initiation of protein synthesis directed by untranslated leader sequences of AMV RNA 4, rabbit alpha globin, or tobacco mosaic virus. (A) A chimeric interleukin-1ß-neomycin phosphotransferase II (IL1-NPT) mRNA construct was generated by fusing 60 amino-terminal amino acids of the IL-1ß sequence in frame to the NPTII coding region. At the IL-1/NPTII junction, oligonucleotides representing the untranslated leader sequences were inserted so that a continuous open reading frame was maintained. The expected size of the IL1-NPTII fusion protein is 35 kD, while that of the NPTII protein is 29 kD. (B) SDS polyacrylamide gel analysis of proteins synthesized by translation in a rabbit reticulocyte translation system. Lane 1, NPTII marker; lane 2, fusion construct containing the in-frame rabbit alpha globin leader sequence; lane 3, fusion construct containing the in-frame TMV leader sequence; lane 4, fusion construct containing two in-frame head-to-tail repeats of the AMV RNA 4 leader sequence; lane 5, fusion construct containing a single in-frame AMV RNA 4 leader sequence.

(Pinck, 1975), it translates in extracts prepared from poliovirus-infected HeLa cells while translation of other capped mRNAs is blocked (Sonenberg *et al.*, 1982); moreover, translation of this mRNA is relatively insensitive to the addition of cap analogs (Sonenberg *et al.*, 1981). These data are evidence that AMV RNA 4 is capable of translation by a cap-independent mechanism. Unlike the EMC and poliovirus leader sequences, which also translate by cap-independent mechanisms, the AMV RNA 4 leader apparently does not facilitate internal initiation of ribosomes when the 35-nucleotide leader sequence is placed in internal regions of messenger RNAs (Figure 1). To test internal initiation potential, the AMV RNA 4, TMV Ω' or rabbit α-globin untranslated leader sequences were inserted in the translational reading frame between a fragment of the human IL-1β cDNA and the downstream coding region for neomycin phosphotransferase II (Figure 1A). If initiation were to occur at the 5' terminus of the RNA, then a 35 kilodalton fusion protein would be translated; however, if initiation occurred internally through the inserted untranslated leader sequence, then only the 29 kilodalton NPTII product would be generated. The *in vitro* translation data (Figure 1B) demonstrate that the rabbit α-globin leader fusion construct yielded only the full-length IL1-NPTII fusion protein expected in the absence of internal initiation (lane 2). A faint band which comigrates with the NPTII marker can be seen in lane 5, representing translation of mRNA with an in-frame insertion of the AMV RNA 4 leader sequence. These data are consistent with weak internal initiation. The level of apparent internal initiation was increased significantly, however, by inserting a tandem repeat of the AMV RNA 4 leader sequence (lane 4). In this experiment, the overall translation of the mRNA containing the TMV leader was lower than the other mRNAs (lane 3), but the results suggest that the ratio of internal initiation to full-length was similar to that observed with the tandem AMV RNA 4 leader insert (lane 4). Initiation from the 5' terminus was clearly not obviated in the case of the tandem AMV RNA 4 leader repeat (Figure 1B, lane 4), and the possibility of scanning from the 5' terminus as opposed to true internal intiation of protein synthesis cannot be ruled out by these data. As a possible explanation for why the AMV RNA 4 leader does not facilitate internal initiation of protein synthesis, it should be noted that the leader sequence is quite short in length, and inserting a single copy of the leader at an internal region of an mRNA instead of at the 5' terminus may mask it in an inaccessible higher order structure. Tandem leaders might form a larger target for the initiating ribsomes. One hypothesis to explain internal initiation on the poliovirus leader is that the "ribosome landing pad" might represent a locally unstructured area which allows ribosomes to bind internally (Pelletier *et al.*, 1988). The precise character of sequences required to permit internal initiation of protein synthesis is not known, but the current data suggest that the region is significantly larger than the AMV RNA 4 leader sequence (Pelletier *et al.*, 1988).

CONCLUSIONS

Despite the advances that have been made in defining features which characterize translationally efficient and translationally inefficient messenger RNAs, our understanding of protein synthesis is far from complete. The complexity of the initiation step of protein synthesis, involving more that a dozen protein initiation factors (Abramson *et al.*, 1988) in addition to ribosomes, transfer RNAs and messenger RNAs represents a formidable hurdle which stands in the way of a complete elucidation of translational mechanisms. Genetic methods for the analysis of post-transcriptional regulatory mechanisms have already proven to be of great value (Cigan *et al.*, 1988, Donahue *et al.*, 1988, Sachs and Davis, 1989) and the power of these approaches guarantees a role in future work. This article has addressed some of the issues involving the role of 5' untranslated leader sequences in regulating mRNA translational efficiency, but it is unlikely that the leader sequence functions completely in isolation to control how well a messenger RNA is expressed as protein. Complex interactions between noncontiguous regions in messenger RNAs (Sachs and Davis, 1989, Schimmel, 1989) and the functional analysis of proteins which specifically bind messenger RNAs (Bandziulis *et al.*, 1989, Dreyfuss *et al.*, 1988, Grange *et al.*, 1987, Leibold and Munro, 1988, Malter, 1989, Query *et al.*, 1989) are areas for future work.

REFERENCES

Abramson RD, Browning KS, Dever TE, Lawson TG, Thach RE, Ravel JM and Merrick WC (1988) Initiation factors that bind mRNA. J Biol Chem 263:5462-5467

Bandziulis RJ, Swanson MS and Dreyfuss G (1989) RNA binding proteins as developmental regulators. Genes Devel. 3:431-437

Benicourt C and Haenni AL (1978) Differential translation of turnip yellow mosaic virus mRNAs in vitro. Biochem Biophys Res Commun 84:831-839

Bonneau AM and Sonenberg N (1987) Proteolysis of the p220 component of the cap-binding protein complex is not sufficient for complete inhibition of host cell protein synthesis after poliovirus infection. J Virol 61:986-991

Brederode RT, Koper-Zwartoff EC and Bol JF (1980) Complete nucleotide sequence of alfalfa mosaic virus RNA 4. Nucl Acids Res 8:2213-2223

Browning KS, Fletcher L and Ravel JM (1988a) Evidence that the requirements for ATP and wheat germ initiation factors 4A and 4F are affected by a region of satellite tobacco necrosis virus RNA that is 3' to the ribosomal binding site. J Biol Chem 263:8380-8383

Browning KS, Lax SR, Humphreys J, Ravel JM, Jobling SA and Gehrke L (1988b) Evidence that the 5'-untranslated leader of mRNA affects the requirement for wheat germ initiation factors 4A, 4F, and 4G. J Biol Chem 263:9630-9634

Browning KS, Maia DM, Lax SR and Ravel JM (1987) Identification of a new protein synthesis initiation factor from wheat germ. J Biol Chem 262:538-541

Cigan AM, Feng L and Donahue TF (1988) tRNAmet functions in directing the scanning ribosome to the start site of translation. Science 240:93-97

Donahue TF, Cigan AM, Pabich EK and Valavicius BC (1988) Mutations at a Zn(II) finger motif in the yeast eIF-2ß gene alter ribosomal start-site selection during the scanning process. Cell 54:621-632

Dreyfuss G, Swanson MS and Pinol-Roma S (1988) Heterogenous nuclear ribonucleoprotein particles and the pathway of mRNA formation. Tr Biochem Sci 13:86-91

Elroy-Stein O, Fuerst TR and Moss B (1989) Cap-independent translation of mRNA conferred by encephalomyocarditis virus 5' sequence improves the performance of the vaccinia virus/bacteriophage T7 hybrid expression system. Proc Natl Acad Sci USA 86:6126-6130

Etchison D, Milburn SC, Edery I, Sonenberg N and Hershey JWB (1982) Inhibition of HeLa cell protein synthesis following poliovirus infection correlates with the proteolysis of a 220,000-dalton polypeptide associated with eukaryotic initiation factor 3 and a cap binding protein complex. J Biol Chem 257:14806-14810

Friesen PD and Rueckert RR (1981) Synthesis of black beetle virus proteins in cultured Drosophila cells: differential expression of RNAs 1 and 2. J Virol 37:876-886

Gallie DR, Kado CI, Hershey JWB, Wilson MA and Walbot V (1989) Eukaryotic viral 5'-leader sequences act as translational enhancers in eukaryotes and prokaryotes. 237-256 In: Cech, T. ed. Molecular Biology of RNA, Alan R. Liss, Inc., New York.

Gallie DR, Sleat DE, Turner PC and Wilson TMA (1988) Mutational analysis of the tobacco mosaic virus 5' leader for altered ability to enhance translation. Nucl Acids Res 16:883-893

Gallie DR, Sleat DE, Watts JW, Turner PC and Wilson TMA (1987a) The 5'-leader sequence of tobacco mosaic virus RNA enhances the expression of foreign gene transcripts *in vitro* and *in vivo*. Nucl Acids Res 15:3257-3273

Gallie DR, Sleat DE, Watts JW, Turner PC and Wilson TMA (1987b) A comparison of eukaryotic viral 5'-leader sequences as enhancers of mRNA expression *in vivo*. Nucl Acids Res 15:8693-8711

Gehrke L, Auron PE, Quigley GQ, Rich A and Sonenberg N (1983) 5' conformation of capped alfalfa mosaic virus ribonucleic acid may reflect its independence of the cap structure or of cap binding protein for efficient translation. Biochemistry 22:5157-5164

Grange T, Martins de Sa C, Oddos J and Pictet R (1987) Human mRNA polyadenylate binding protein: evolutionary conservation of a nucleic acid binding motif. Nucl Acids Res 15:4771-4787

Herson D, Schmidt A, Seal SN, Marcus A and van Vloten-Doting L (1979) Competitive mRNA translation in an *in vitro* system from wheat germ. J Biol Chem 254:8245-8249

Iserentant D and Fiers W (1979) Secondary structure of the 5' end of bacteriophage MS2 RNA. Eur J Biochem 102:595-604

Jang SK, Davies MV, Kaufman RJ and Wimmer E (1989) Initiation of protein synthesis by internal entry of ribosomes into the 5' nontranslated region of encephalomyocarditis virus RNA in vivo. EMBO J 63:1651-1660

Jang SK, Krausslich HG, Nicklin MJH, Duke GM, Palmenberg AC and Wimmer E (1988) A segment of the 5' nontranslated region of encephalomyocarditis virus RNA directs internal entry of ribosomes during in vitro translation. J Virol 62:2636-2643

Jobling SA, Cuthbert CA, Rogers SG, Fraley RT and Gehrke L (1988) *In vitro* transcription and translational efficiency of chimeric SP6 messenger RNAs devoid of 5' vector nucleotides. Nucl Acids Res 16:4483-4498

Jobling SA and Gehrke L (1987) Enhanced translation of chimeric messenger RNAs containing a plant viral untranslated leader sequence. Nature 325:622-625

Kozak M (1983) Comparison of initiation of protein synthesis in prokaryotes, eukaryotes, and organelles. Microbiol Rev 47:1-45

Lax S, Lauer SJ, Browning KS and Ravel JM (1986) Meth. Enzymol. 118:109-128

Lee KAW and Sonenberg N (1982) Inactivation of cap-binding protein accompanies the shut-off of host protein synthesis by poliovirus. Proc Natl Acad Sci USA 79:3447-3451

Leibold EA and Munro HN (1988) Cytoplasmic protein binds *in vitro* to a highly conserved sequence in the 5' untranslated region of ferritin heavy- and light-subunit mRNAs. Proc Natl Acad Sci USA 85:2171

Lodish HF (1976) Translational control of protein synthesis. Ann Rev Biochem 45:39-72

Malter JS (1989) Identification of an AUUUA-specific messenger RNA binding protein. Science 246:664-666

Olins PO, Devine CS, Rangwala SH and Kavka KS (1988) The T7 phage gene 10 leader RNA, a ribosome-binding site that dramatically enhances the expression of foreign genes in Escherichia coli. Gene 73:227-235

Parks GD, Duke GM and Palmenberg AC (1986) Encephalomyocarditis virus 3C protease: efficient cell-free expression from clones which link viral 5' noncoding sequences to the P3 region. J Virol 60:376-384

Pavlakis GN, Lockard RE, Vamvakopoulos N, Rieser L, RajBhandary UL and Vournakis JN (1980) Secondary structure of mouse and rabbit alpha and beta globin mRNAs: Differential accessibility of alpha and beta initiator codons towards nuclease. Cell 19:91-102

Pelletier J, Kaplan G, Racaniello VR and Sonenberg N (1988) Cap-independent translation of poliovirus mRNA is conferred by sequence elements within the 5' noncoding region. Mol Cell Biol 8:1103-1112

Pelletier J and Sonenberg N (1985) Insertion mutagenesis to increase secondary structure within the 5' noncoding region of a eukaryotic mRNA reduces translational efficiency. Cell 40:515-526

Pelletier J and Sonenberg N (1988) Internal initiation of translation of eukaryotic mRNA directed by a sequence derived from poliovirus RNA. Nature 334:320-325

Pinck L (1975) The 5' end groups of alfalfa mosaic virus are m7GpppG. FEBS Letters 59:24-30

Query CC, Bentley RC and Keene JD (1989) A common RNA recognition motif identified within a defined U1 RNA binding domain of the 70K U1 snRNP protein. Cell 57: 89-101

Sachs AB and Davis RW (1989) The poly(A) binding protein is required for poly(A) shortening and 60S ribosomal subunit-dependent translation initiation. Cell 58:857-867

Schimmel P (1989) RNA pseudoknots that interact with components of the translation apparatus. Cell 58:9-12

Shakin SH and Liebhaber SA (1986) Destabilization of messenger RNA/complementary DNA duplexes by the elongating 80 S ribosome. J Biol Chem 261:16018-16025

Shih DS and Kaesberg P (1973) Translation of brome mosaic viral ribonucleic acid in a cell-free system derived from wheat embryos. Proc Natl Acad Sci USA 70:1799-1803

Sleat DE, Gallie DR, Jefferson RA, Bevan MW, Turner PC and Wilson TMA (1987) Characterization of the 5' leader sequence of tobacco mosaic virus as a general enhancer of translation *in vitro*. Gene 60:217-225

Sonenberg N, Guertin D, Cleveland D and Trachsel H (1981) Probing the function of the eukaryotic 5' cap structure by using a monoclonal antibody directed against cap-binding proteins. Cell 27:563-572

Sonenberg N, Guertin D and Lee KAW (1982) Capped mRNAs with reduced secondary structure can function in extracts from poliovirus-infected cells. Mol Cell Biol 2:1633-1638

Trono D, Pelletier J, Sonenberg N and Baltimore D (1988) Translation in mammalian cells of a gene linked to the poliovirus 5' noncoding region. Science 241:445-448

Coordinate Post-Transcriptional Regulation of Ferritin and Transferrin Receptor Expression: The Role of Regulated RNA-Protein Interaction

Joe B. Harford and Richard D. Klausner
Cell Biology & Metabolism Branch
National Institute of Child Health & Human Development
National Institutes of Health
Bethesda, MD 20892

Two of the best understood examples of post-transcriptional gene regulation in eukaryotic cells have emerged from studies on cellular iron metabolism. Although assimilation and metabolism of iron by higher eukaryotes is a complex process about which many unanswered questions remain (Seligman et al., 1987), some progress has been made in the understanding of iron uptake and sequestration by individual eukaryotic cells (Klausner & Harford, 1990; Theil, 1990; Harford et al., 1990). Iron is delivered to most cells *via* the "transferrin cycle" involving endocytosis of the iron-carrying protein transferrin (Tf). This endocytic pathway is mediated by the transferrin receptor (TfR). From an acidic endosome, iron is released from Tf and transferred by unknown means to the cytoplasm. Once in the cytoplasm, the iron is utilized for the many cellular processes that require iron or it is sequestered into ferritin, a hollow, spherical molecule composed of 24 subunits encoded by two homologous genes (termed H and L) (Munro & Linder, 1978; Theil, 1987). Presumably, this sequestration is largely responsible for the detoxification of unused iron.

The genes which encode ferritin and the TfR are regulated in response to changing levels of iron by alterations in mRNA translation efficiency or mRNA stability, respectively. Limiting iron results in an increase in the expression of the TfR, whereas when iron is plentiful, TfR expression decreases. The expression of ferritin is regulated in the opposite direction, increasing in the presence of iron and decreasing in its absence. TfR regulation allows for modulation of iron uptake, and ferritin regulation allows for adequate sequestration of excess iron and minimizes sequestration when iron is limiting. We are beginning to understand the molecular basis for the coordinate but opposite regulation of these two genes. It had been known since the 1940's that the level of ferritin in tissues varied directly with changes in iron (Granick, 1946). By the 1970's, available evidence indicated that regulation of mammalian ferritin synthesis in response to iron occurred at the level of translation (Munro & Linder, 1978). A similar conclusion was also reached regarding ferritin biosynthetic regulation in amphibia (Shull & Theil, 1982). The molecular cloning of cDNA's corresponding to ferritin subunits (Boyd et al., 1984; Boyd et al., 1985; Costanzo et al., 1986) allowed the direct measurement by hybridization of ferritin mRNA levels confirming the translational control of ferritin (Cairo et al., 1985; Aziz & Munro, 1987; Rouault et al., 1987) and identification of the sequences within the ferritin mRNA responsible for iron regulation (Aziz & Munro, 1987). When cells are exposed to conditions that alter available iron, rapid changes in the rate of synthesis of ferritin are observed. These changes of up to two orders of magnitude in ferritin

NATO ASI Series, Vol. H 49
Post-Transcriptional Control of Gene Expression
Edited by J. E. G. McCarthy and M. F. Tuite

synthesis occurred in the absence of change in the level of cytoplasmic ferritin mRNA. The ferritin mRNA within the cytosol was shown to shift from an mRNP fraction to polysomes as the synthesis of ferritin increased (Rogers & Munro, 1987).

The mRNA for human ferritin H chain (approx 1kb) contains a 212nt 5' untranslated region (UTR). When the 5'UTR was deleted, the ferritin protein was made and assembled normally but all acute iron regulation was lost (Hentze et al., 1987a). Further deletion analysis identified a region approximately 35 bases in length that appeared to contain all the information required for iron-dependent translational control of ferritin synthesis. That this short sequence is all that is required for this translational control was established by cloning the sequence into the 5'UTR of two different reporter genes. The translation of the resultant chimeric mRNA's were iron-regulated (Hentze et al., 1987b; Caughman et al., 1988). Examination of all ferritin genes cloned to date (human, rat, mouse, rabbit, chicken and bullfrog) reveals the presence of this highly conserved sequence in the 5'UTR (Hentze et al., 1988). This sequence of the human ferritin H chain sequence motif (figure 1) as well as all of the corresponding sequence motifs in other species are capable of forming moderately stable stem-loop structures. In all instances, the base pairing of the stem is interrupted by an unpaired C that is found six nucleotides (except one example in which it is five) 5' of a six-membered loop. The sequence of the loop is CAGUGN. Because of its function, we referred to this RNA sequence/structure motif of the ferritin mRNA as an "Iron Responsive Element" or "IRE".

As with ferritin, the rapid regulation of the expression of the TfR is the result of altered rates of protein synthesis. However, in contrast to ferritin, this alteration is directly reflected in changes in the levels of cytoplasmic TfR mRNA (Rao et al., 1986). Initial studies suggested a transcriptional component to this regulation. Indeed the 5' flanking region of the TfR gene can transfer iron-dependent transcriptional regulation to a heterologous reporter gene (Casey et al., 1988a). However, only a 2- to 3-fold effect was observed in contrast to a 20- to 30-fold changes in TfR mRNA levels seen when iron availability is manipulated. The major iron-dependent regulation mapped to within the 3'UTR of the TfR mRNA (Owen & Kühn, 1987; Casey et al., 1988a). That this information was present within the RNA had been suggested by the observation that alternative transcripts of a single gene construct was iron regulated only if it contained a specific region of the 3'UTR (Casey et al., 1988a). The 3'UTR allowed the iron-dependent regulation of the half-life of the TfR mRNA (Müllner & Kühn, 1988). Deletion analysis of the 2.5kb 3'UTR of the TfR mRNA using convenient restriction sites demonstrated that a 680nt fragment contained all of the information for this regulation (Casey et al., 1988b). Iron-dependent control of mRNA half-life can be transferred to a heterologous gene by placing the TfR regulatory region in the 3'UTR of the reporter gene (Müllner & Kühn, 1988; Casey et al., 1988b). Visual examination and computer-assisted analysis suggested that this 680nt region is highly structured. In particular, five stem-loop structures (termed A-E) were found that bore striking resemblance to ferritin IRE's although the sequence of their stems differed (Casey et al., 1988b) (figure 1). It should be noted that the IRE's in the context of the TfR 3'UTR do not confer translational iron regulation upon the TfR mRNA.

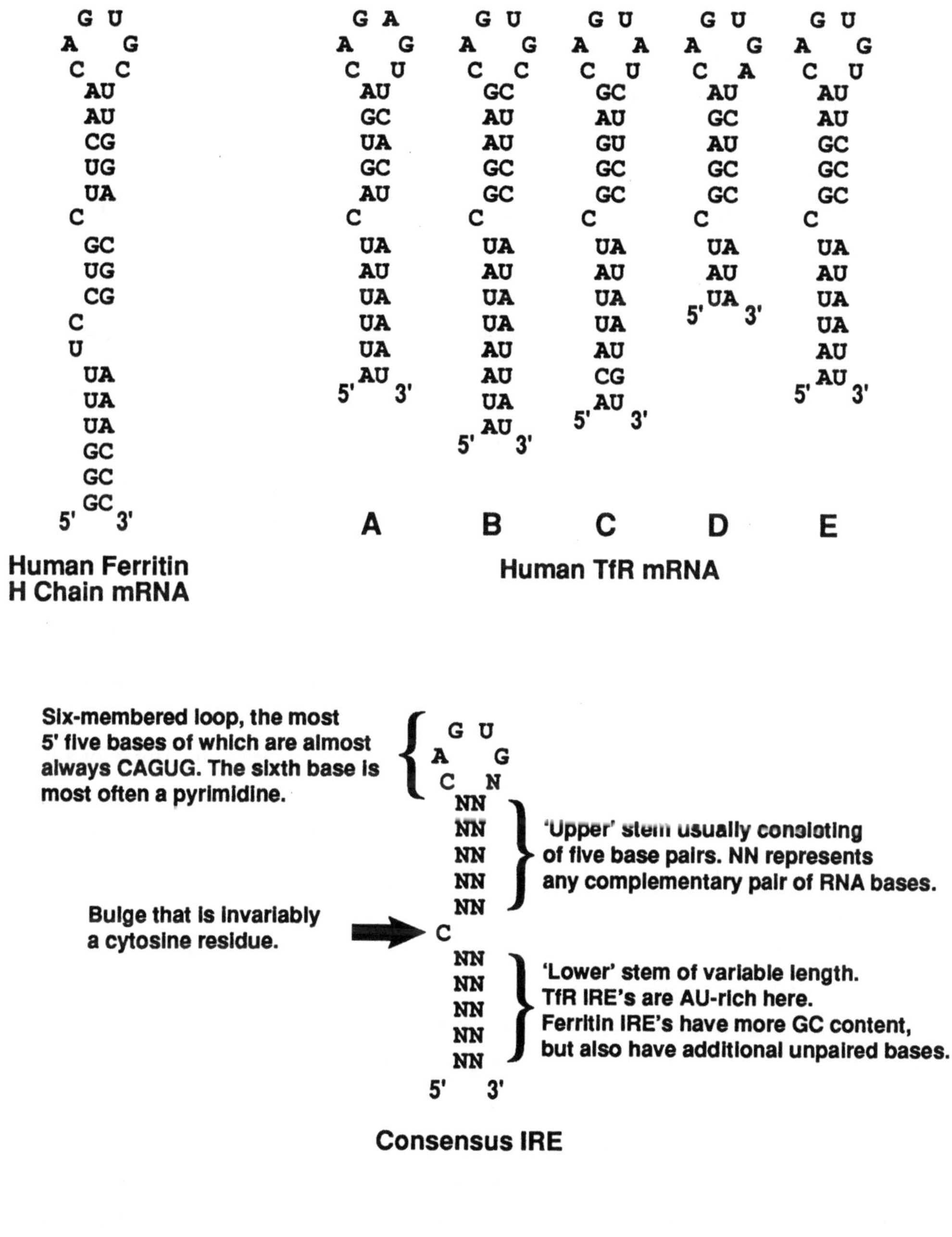

Figure 1. Representative IRE's and a consensus sequence/structure motif. Shown are the single IRE from the 5'UTR of the human ferritin H chain mRNA, the five IRE's from the 3'UTR of the human TfR mRNA and a consensus IRE with its features indicated.

Nonetheless, when we inserted stem-loops B or C from the 3'UTR or the TfR mRNA into the 5'UTR of an indicator gene, each functioned as a ferritin-like translational control element (*i.e.* translational repression in response to iron deprivation).

The location of the IRE in the 5'UTR of ferritin suggested that it may function to control translation initiation. Nucleotides between the IRE and the initiation codon can be deleted without affecting regulation of translation although the distance that the IRE can be displaced from the initiating codon has not been determined. The position dependence of the IRE with respect to the 5' end of the mRNA and the consequences of placing an IRE within the reading frame remain to be examined. Whereas most ferritin IRE's are located within 40nt of the cap site, the sequence of a *Xenopus laevis* ferritin indicates that its IRE is more than 150nt from the 5' end of the mRNA (Moskaitis et al., 1990). A clue as to its mechanism of action came from observations of the effect of the presence or absence of an IRE on the translatability of mRNA (Caughman et al., 1988). Removal of the IRE resulted in a constitutively high level of mRNA translation. Moreover, addition of an IRE to a 5'UTR depressed the rate of translation unless cells were incubated in the presence of iron. These findings suggested that the IRE functions to repress translation if iron is limiting rather than augmenting translation in an iron-dependent fashion, since the presence of an IRE did not yield higher rates of translation than were seen with an identical mRNA lacking an IRE under any iron conditions. The presence of an IRE in the 5'UTR has been shown to repress translation in an *in vitro* translation system (Walden et al., 1988).

One possible mechanism for IRE function was that the IRE serves as the target for a trans-acting repressor molecule. Evidence for an IRE-binding protein (IRE-BP) was provided by an electrophoretic mobility shift assay (Leibold & Munro, 1988; Rouault et al., 1988) analogous to assays that had been employed to identify DNA binding proteins. The human IRE-BP is a 90,000 Dalton cytosolic protein that has been purified to homogeneity from liver by RNA affinity chromatography (Rouault et al., 1989). This polypeptide is the only protein in human liver lysates that is cross-linked to IRE-containing RNA's by ultraviolet (UV) irradiation. Proteins of similar size have been isolated from rabbit liver (Walden et al., 1989) and human placenta (Neupert et al., 1990). IRE-BP activity has been shown to be present in a wide variety of species including mammals, fish, amphibia and insects (Rothenberger et al., 1990). The gene encoding the IRE-BP has been localized to human chromosome 9 (Hentze et al., 1989a) thus removing this protein as a candidate as the primary defect in hereditary hemochromatosis, a human disorder of iron metabolism known to be localized to chromosome 6 (Seligman et al., 1987). The details of the interactions of the IRE-BP with IRE's and its role in regulating the fate of IRE-containing mRNA's will be discussed below. Based on direct binding studies and competition experiments, it was determined that the IRE-BP was capable of interaction with the ferritin IRE and each of the individual IRE motifs of the TfR 3'UTR although the affinities of these IRE's differed over an approximately 10-fold range (Koeller et al., 1989). Partial peptide maps of the rat IRE-BP that had been UV cross-linked to RNA corresponding to either a ferritin or a TfR IRE indicate that a single protein from rat can also interact with either type of IRE (Leibold et al., 1990).

The mechanism by which any regulatory nucleic acid binding protein responds to physiologic signals represents one of the significant open questions in molecular biology. To propose that the IRE-BP is the regulatory protein that allows IRE-containing RNA's to respond to cellular iron status, it was necessary to demonstrate some sensitivity of the IRE-BP activity to iron availability. Initial studies using either gel shift or UV crosslinking demonstrated that the apparent amount of IRE-BP increased in lysates of cells that had been starved of iron and decreased when the cells had been loaded with iron (Leibold & Munro, 1988; Rouault et al., 1988). A more quantitative analysis of the interaction between the IRE-BP and its cognate RNA has helped clarify these observations. Scatchard analysis of this interaction revealed a complex binding curve that could be well fit by a two binding site model suggesting that cytosol contains binding activity with two distinct affinities, one K_D=10-30 pM and the other K_D=2-5 nM (Haile et al., 1989). That these represent the same protein is supported by the following: both gave identical mobilities by gel shift, a process documented to be quite sensitive to the molecular weight of the protein; both affinities were observed using affinity purified IRE-BP; and variant IRE's capable of only binding to the low affinity site could compete (at high concentrations) for both binding sites. The total number of binding sites (high plus low affinity) did not appear to change in response to altered iron status of the cell. What did change was the fraction of total sites displaying 10-30 pM affinity. Less that 1% of the IRE-BP had this affinity in lysates from iron-fed cells, and greater that 50% displayed the higher affinity in iron-starved cells. This shift did not require *de novo* protein synthesis (Hentze et al., 1989b), suggesting that the pre-existent population of IRE-BP had been interconverted by changes in the cell's iron status.

The biochemical basis for the affinity change of the IRE-BP appears to be the reversible oxidation-reduction of one or more disulfides in the protein (Haile et al., 1989). The lower affinity IRE-BP was converted to the higher affinity form *in vitro* by the addition of reducing agents to cell lysates. Conversely, the higher affinity form was switched to the lower affinity state by treatment with agents that catalyze disulfide formation (diamide or copper orthophenanthroline). The oxidized form appears to involve intramolecular disulfide formation as no change in apparent molecular weight accompanies oxidation. In cells in which the IRE-BP had been maximally activated by treatment with an iron chelator, little, if any IRE-BP isolated from the cell was in an oxidized form whereas in cells treated with an iron source, virtually none of the IRE-BP was in a reduced form. Thus, the iron status of the cell appears to set the redox state of the IRE-BP (Haile et al., 1989; Hentze et al., 1989). We have termed this novel biochemical regulatory mechanism a "sulfhydryl switch." It is not known how the cellular iron status throws this switch. We have, so far, been unable to reproduce the switch *in vitro* by the addition or chelation of iron. Interconversion of pools of IRE-BP differing in activity (affinity) and the redox state of the protein has also been reported in a detailed study of the rat IRE-BP (Barton et al., 1990), and the IRE-BP is similarly activated by reduction in a variety of other species (Rothenberger et al., 1990). Recently, from the study of *in vitro* ferritin mRNA translation, it has been proposed that hemin (and by inferrence an intracellular porphyrin) may directly inhibit the binding of the IRE-BP to the IRE and thereby enhance translation (Lin et al., 1990). While there is no debate that hemin can inhibit IRE/IRE-BP interaction *in vitro*, the effects of hemin addition *in vitro* do not appear to be manifestations of changes in the redox state of the IRE-BP. Other evidence points to the fact that the *in vitro* effects of hemin could possibly represents a more general effect of hemin on nucleic acid-protein interactions (Haile et al., 1990). The form of iron that is responsible for the modulation of IRE-BP activity in the living cell's cytoplasm remains an enigma.

In contrast to the simple regulatory structure (a single IRE) that encodes iron-dependent translational control, the regulatory region required for iron-dependent TfR mRNA stability regulation appears to be much more complex. To understand the function of this region demands both a structural and functional analysis of the 3'UTR regulatory element. Direct structural analysis has not yet been accomplished. However, the availability of TfR sequence from two distant species (human and chicken) allows prediction of structure based on phylogenetic similarities. Of 160nt that encode the five IRE's, only six differences exist between the two species (Casey et al., 1988b; Koeller et al., 1989; Chan et al., 1989), and all of these differences are conservative of the IRE sequence/structure motif (figure 1). This analysis also supports the prediction of other structural motifs within the regulatory region including several small non-IRE stem-loop structures and a long base paired structure lying between IRE's B and C (see figure 2). More recently, the sequence of a portion of the 3'UTR of the rat TfR mRNA has been determined (Roberts & Griswold, 1990). It too contains IRE's and, within the previously defined regulatory region, is very similar in sequence (and potential secondary structure) to the human and chicken mRNA's.

Functional dissection of this region has begun to yield information as to how the TfR regulatory element functions (Casey et al., 1989). When the region is split at the central long stem, neither the 5' nor the 3' half of the region is sufficient for regulation. The 680nt segment has been trimmed to about 250nt by removal of the sequences 5' of IRE B, 3' of IRE D and the nucleotides contained within the large central segment which does not contain sequences that are similar between chicken and human. The resultant shorter regulatory fragment, containing only 3 IRE's (B, C & D), cannot be distinguished from the full 680nt region in terms of iron regulation. Two types of deletions within the smaller regulatory fragment can be distinguished: those that specifically affect IRE-BP binding and those that affect non-IRE structures. The alteration that best illustrates an IRE mutant results from the removal of the first C residue from each of loops B, C and D leaving five membered loops. We had previously demonstrated that this altered IRE is incapable of functioning as a translational control element (Hentze et al., 1988) and is incapable of high affinity binding to the IRE-BP (Rouault et al., 1988; Haile et al., 1989). The TfR regulatory fragment with 3 base deletions (from the 250nt) gives no iron-dependent mRNA stability regulation and no longer binds with high affinity to the IRE-BP. A variety of other deletions, particularly of the small non-IRE stem-loops also eliminate regulation but have no effect on the binding of the RNA to the IRE-BP (Casey et al., 1989).

Although both IRE and non-IRE changes can abrogate regulation, a more careful look at their "regulatory phenotypes" yielded revealing differences. These differences most probably reflects two distinct functions of the 3' regulatory domain, both of which are required for iron responsive regulation. Complete removal of the three IRE's from the synthetic element resulted in loss of ability to interact with the IRE-BP and consequent loss of iron regulation. Moreover, deletion of a single cytosine residue from each of the three IRE's in the synthetic regulatory element eliminates high affinity binding to the IRE-BP *in vitro* and results in low levels of iron-independent TfR expression, consistent with production of a constitutively unstable mRNA. These data indicate that the ability of the mRNA to interact with the IRE binding protein is required for regulation of TfR mRNA levels by iron. Certain other deletions within the regulatory region which do not affect the *in vitro* interaction between the TfR mRNA and the IRE-BP result in relatively high levels of iron-independent TfR expression, consistent with production of a constitutively stable mRNA. Collectively, our data support a model for TfR regulation in which the interaction between the IRE-binding protein and the TfR mRNA can protect the transcript from rapid degradation that is mediated by a rapid turnover determinant within the regulatory region.

The above described observations have allowed us to propose a model to explain how the IRE, its binding protein and the sulfhydryl switch can account for the iron-dependent regulation of both translation and mRNA stability (Klausner & Harford, 1990) (figure 2). As cellular iron levels become limiting, a greater fraction of the cell's IRE-BP become recruited into the high affinity binding state through the reduction of one or more critical disulfides in the protein. The high affinity interaction between the protein and an IRE serves to repress translation if an IRE is located within the 5'UTR of an mRNA. Within the regulatory region of the 3'UTR of the TfR mRNA, this same high affinity interaction represses the degradation of the mRNA. In this way the regulated binding of the IRE-BP can coordinately regulate a decrease in the biosynthesis of ferritin (by translational repression) and an increase in the biosynthesis of the TfR (by repression of mRNA degradation). The TfR regulatory region has two distinct functional elements. One is an RNA instability determinant which gives this mRNA a short half life (a requirement for rapid regulation of mRNA levels) and the other is the IRE which can regulate the utilization of this instability element. The instability determinant has not been fully characterized but may contain or overlap with the binding site of the IRE-BP repressor. The IRE-BP may alter the conformation of the RNA such that the target for a nuclease is not formed or may sterically block access to that target. We have recently isolated a cDNA clone of the IRE-BP that has an open reading frame of approximately 87,000 Daltons (Rouault et al., 1990). The encoded protein contains no known consensus sequence elements for RNA or DNA binding proteins but does contain a consensus nucleotide binding site that resembles most closely the NADH binding site of an *E. coli* oxireductase and

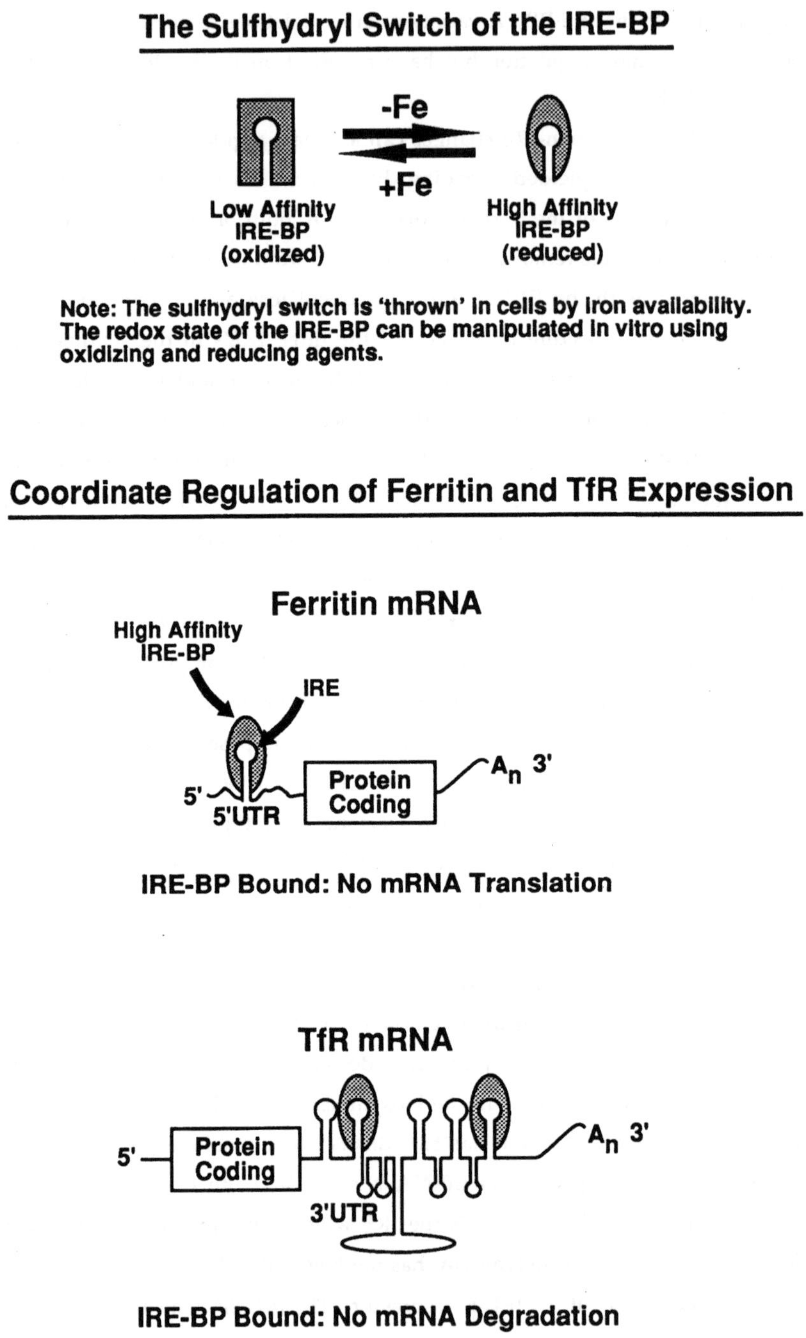

Figure 2. Model for the coordinate regulation of ferritin and the TfR via the sulfhydryl switch of the IRE-BP. When the high affinity (reduced) form of the IRE-BP is bound, ferritin translation and TfR mRNA degradation are suppressed.

cysteine clusters that resemble those found in proteins containing iron-sulfur centers. These features raise the possibility that the IRE-BP binds iron directly and uses NADH (or NADPH) as an electon source to reduce the cysteines that constitute the molecular basis for the sulfhydryl switch.

In the past several years, it has become increasingly clear that eukaryotic gene regulation can occur at points distal to transcription. The regulation of the expression of ferritin and the TfR are two examples. Biosynthesis of these two proteins are regulated by iron in opposite but coordinate fashion. In the case of ferritin, the regulation is translational, and in the case of the TfR, the regulation is at the level of mRNA stability. Nonetheless, similar cis-acting RNA sequence/structure motifs (the IRE's) and a common trans-acting RNA-binding protein (the IRE-BP) serve as key participants in regulation of both genes. To date this is the only example of such overlap in the mechanisms of eukaryotic post-transcriptional gene regulation. It may well be that as more systems are understood in detail and more cis- and trans-acting elements involved in post-transcriptional gene regulation are identified, other examples will emerge.

ACKNOWLEDGEMENT

We wish to thank our past and present colleagues of the Cell Biology & Metabolism Branch whose work cited here is the real foundation of this review.

REFERENCES

Aziz N, Munro HN (1986) Both subunits of rat liver ferritin are regulated at a translational level by iron induction. Nucleic Acids Res 14:915-927.

Aziz N, Munro HN (1987) Iron regulates ferritin mRNA translation through a segment of its 5' untranslated region. Proc Natl Acad Sci U S A84:8478-482.

Barton HA, Eisenstein RS, Bomford A, Munro HN (1990) Determinants of the interaction between the iron-responsive element-binding protein and its binding site in rat L-ferritin mRNA. J Biol Chem 265:7000-7008.

Boyd D, Jain SK, Crampton J, Barrett KJ, Drysdale J. (1984) Isolation and characterization of a cDNA clone for human ferritin heavy chain. Proc Natl Acad Sci U S A 81:4751-4755.

Boyd D, Vecoli C, Belcher DM, Jain SK, Drysdale JW (1985) Structural and functional relationships of human ferritin H and L chains deduced from cDNA clones. J Biol Chem 260:11755-61.

Cairo G, Bardella L, Schiaffonati L, Arosio P, Levi S, Bernelli-Zazzera A. (1985) Multiple mechanisms of iron-induced ferritin synthesis in HeLa cells. Biochem Biophys Res Commun 133:314-321.

Casey JL, Di Jeso B, Rao K, Klausner RD, Harford JB (1988a) Two genetic loci participate in the regulation by iron of the gene for the human transferrin receptor. Proc Natl Acad Sci U S A 85:1787-1791.

Casey JL, Hentze MW, Koeller DM, Caughman SW, Rouault TA, Klausner RD, Harford JB (1988b) Iron-responsive elements: regulatory RNA sequences that control mRNA levels and translation. Science 240:924-928.

Casey JL, Koeller DM, Ramin VC, Klausner RD, Harford JB (1989) Iron regulation of transferrin receptor mRNA levels requires iron-responsive elements and a rapid turnover determinant in the 3' untranslated region of the mRNA. EMBO J 8:3693-3699.

Caughman SW, Hentze MW, Rouault TA, Harford JB, Klausner RD (1988) The iron-responsive element is the single element responsible for iron-dependent translational regulation of ferritin biosynthesis: Evidence for function as the binding site for a translational repressor. J Biol Chem 263:19048-19052.

Chan LN, Grammatikakis N, Banks JM, Gerhardt EM (1989) Chicken transferrin receptor gene: conservation 3' noncoding sequences and expression in erythroid cells. Nucleic Acids Res 17:3763-3771.

Costanzo F, Colombo M, Staempfli S, Santoro C, Marone M, Frank R, Delius H, Cortese R (1986) Structure of gene and pseudogenes of human apoferritin H. Nucleic Acids Res 14:721-736.

Granick S (1946) Ferritin: its properties and significance for iron metabolism Chem Rev 38:379-395.

Haile DJ, Hentze MW, Rouault TA, Harford JB, Klausner RD (1989) Regulation of interaction of the iron-responsive element binding protein with iron-responsive RNA elements. Mol Cell Biol 9:5055-5061.

Haile DJ, Rouault TA, Harford JB, Klausner RD (1990) The inhibtion of the IRE RNA-protein interaction by heme does not mimic *in vivo* iron regulation. J Biol Chem (in press).

Harford JB, Casey JL, Koeller DM, Klausner RD (1990) Structure, function and regulation of the transferrin receptor: insights from molecular biology. In: Steer CJ, Hanover JA, eds. Intracellular trafficking of proteins. Oxford: Oxford Univ Press (in press).

Hentze MW, Rouault TA, Caughman SW, Dancis A, Harford JB, Klausner RD (1987a) A cis-acting element is necessary and sufficient for translational regulation of human ferritin expression in response to iron. Proc Natl Acad Sci U S A 84:6730-6734.

Hentze MW, Caughman SW, Rouault TA, Barriocanal JG, Dancis A, Harford JB, Klausner RD (1987b) Identification of the iron-responsive element for the translational regulation of human ferritin mRNA. Science 238:1570-1573.

Hentze MW, Caughman SW, Casey JL, Koeller DM, Rouault TA, Harford JB, Klausner RD (1988) A model for the structure and functions of iron-responsive elements. Gene 72:201-208.

Hentze MW, Seuanez HN, O'Brien SJ, Harford JB, Klausner RD (1989a) Chromosomal localization of nucleic acid-binding proteins by affinity mapping: assignment of the IRE-binding protein gene to human chromosome 9. Nucleic Acids Res 17:6103-6108.

Hentze MW, Rouault TA, Harford JB, Klausner RD (1989b) Oxidation-reduction and the molecular mechanism of a regulatory RNA-protein interaction. Science 244:357-359.

Klausner RD, Harford JB (1989) Cis-trans models for post-transcriptional gene regulation. Science 246:870-872.

Koeller DM, Casey JL, Hentze MW, Gerhardt EM, Chan LN, Klausner RD, Harford JB (1989) A cytosolic protein binds to structural elements within the iron regulatory region of the transferrin receptor mRNA. Proc Natl Acad Sci U S A 86:3574-3578.

Leibold EA, Munro HN (1988) Cytoplasmic protein binds in vitro to a highly conserved sequence in the 5' untranslated region of ferritin heavy- and light-subunit mRNAs. Proc Natl Acad Sci U S A 85:2171-2175.

Leibold EA, Laudano A, Yu Y (1990) Structural requirements of iron-responsive elements for binding of the protein involved in both transferrin receptor and ferritin mRNA post-transcriptional regulation. Nucleic Acids Research 18:1819-1824.

Lin J-J, Daniels-McQueen S, Patino MM, Gaffield L, Walden WE, Thach RE (1990) Derepression of ferritin messenger RNA translation by hemin *in vitro*. Science 247:74-77.

Moskaitis JE, Pastori RL, Schoenberg, DR (1990) Sequence of Xenopus laevis ferritin. Nucl Acids Res 18:2184.

Müllner EW, Kühn LC (1988) A stem-loop in the 3' untranslated region mediates iron-dependent regulation of transferrin receptor mRNA stability in the cytoplasm. Cell 53:815-825.

Munro HN, Linder MC (1978) Ferritin: structure, biosynthesis, and role in iron metabolism. Physiol Rev 58:317-396.

Neupert B, Thompson NA, Meyer C, Kühn LC (1990) A high yield affinity purification method for specific RNA-binding proteins: isolation of the iron regulatory factor from human placenta. Nucleic Acids Res 18:51-55.

Owen D, Kühn LC (1987) Noncoding 3' sequences of the transferrin receptor gene are required for mRNA regulation by iron. EMBO J 6:1287-1293.

Rao K, Harford JB, Rouault T, McClelland A, Ruddle FH, Klausner RD (1986) Transcriptional regulation by iron of the gene for the transferrin receptor. Mol Cell Biol 6:236-240.

Roberts KP, Griswold MD (1990) Characterization of rat transferrin receptor cDNA: The regulation of transferrin receptor mRNA in testes and in Sertoli cells in culture. Mol Endo 4:531-542.

Rogers J, Munro H (1987) Translation of ferritin light and heavy subunit mRNAs is regulated by intracellular chelatable iron levels in rat hepatoma cells. Proc Natl Acad Sci U S A 84:2277-2281.

Rothenberger S, Müllner EW, Kühn LC (1990) The mRNA-binding protein which controls ferritin and transferrin receptor expression is conserved during evolution. Nucleic Acids Res 18:1175-1179.

Rouault TA, Hentze MW, Dancis A, Caughman W, Harford JB, Klausner RD (1987) Influence of altered transcription on the translational control of human ferritin expression. Proc Natl Acad Sci U S A 84:6335-6339.

Rouault TA, Hentze MW, Caughman SW, Harford JB, Klausner RD (1988) Binding of a cytosolic protein to the iron-responsive element of human ferritin messenger RNA. Science 241:1207-1210.

Rouault TA, Hentze MW, Haile DJ, Harford JB, Klausner RD (1989) The iron-responsive element binding protein: A method for the affinity purification of a regulatory RNA-binding protein. Proc Natl Acad Sci U S A 86:5768-5772.

Rouault TA, Tang CK, Kaptain S, Burgess WH, Haile DJ, Samaniego F, McBride OW, Harford JB, Klausner RD (1990) Cloning of the cDNA encoding an RNA regulatory protein: The human iron responsive element binding protein. Proc Natl Acad Sci U S A (in press).

Seligman PA, Klausner RD, Heubers HA. (1987) Molecular mechanisms of iron metabolism. In: Stamatoyannopoulos G, Nienhaus AW, Leder P, Majerus PW, eds. The molecular basis of blood diseases. Philadelphia: Saunders p. 219-244.

Shull GE, Theil EC (1982) Translational control of ferritin synthesis by iron in embryonic reticulocytes of the bullfrog. J Biol Chem 257:14187-14191.

Theil EC (1987) Ferritin: structure, gene regulation, and cellular function in animals, plants, and microorganisms. Annu Rev Biochem 56:289-315.

Theil EC (1990) Regulation of ferritin and transferrin receptor mRNAs. J Biol Chem 265:4771-4774.

Walden WE, Daniels-McQueen S, Brown PH, Gaffield L, Russell DA, Bielser D, Bailey LC, Thach RE (1988) Translational repression in eukaryotes: partial purification and characterization of a repressor of ferritin mRNA translation. Proc Natl Acad Sci U S A85:9503-9507.

Walden WE, Patino MM, Gaffield L (1989) Purification of a specific repressor of ferritin mRNA translation from rabbit liver. J Biol Chem 264:13765-13769.

TRANSLATION IN YEAST MITOCHONDRIA: A REVIEW OF GENERAL FEATURES AND A CASE OF mRNA-SPECIFIC POSITIVE CONTROL

Thomas D. Fox, Thomas W. McMullin, Pascal Haffter and Linda S. Folley
Section of Genetics and Development
Cornell University
Ithaca, NY 14853-2703
U.S.A.

General features of the mitochondrial translation system

The mitochondrial genetic system has the relatively simple task of supplying seven protein components of the respiratory chain complexes. However, expression of the mitochondrial genes coding these proteins is surprisingly complex (reviewed by Attardi and Schatz 1988; Costanzo and Fox 1990; Grivell 1989). The mitochondrial genes are transcribed and translated by a system whose protein components are, with single known exception (a small subunit ribosomal protein), coded by nuclear genes. Thus, the nucleus must code for about a hundred proteins, most of which function only in mitochondria, to synthesize seven proteins needed for respiration.

No mRNA dependent *in vitro* translation system has ever been derived from mitochondria. However genetic analysis has led to the identification of many components that participate in mitochondrial translation *in vivo*. For example, twelve nuclear genes encoding mitochondrial ribosomal proteins have been identified (Dang et al 1990; Fearon and Mason 1988; Grohmann et al 1989; Matsushita et al 1989; McMullin et al 1990; Myers et al 1987; Partaledis and Mason 1988; Tzagoloff and Dieckmann 1990; K. Fearon & T. L. Mason, personal communication). Surprisingly, only four of them exhibit detectable homology to *E. coli* ribosomal proteins. Ten genes for coding mitochondrial tRNA synthetases have been identified (Gampel

NATO ASI Series, Vol. H 49
Post-Transcriptional Control of Gene Expression
Edited by J. E. G. McCarthy and M. F. Tuite

and Tzagoloff 1989; Herbert et al 1988; Koerner et al 1987; Myers and Tzagoloff 1985; Pape et al 1985; Tzagoloff et al 1988; Tzagoloff and Dieckmann 1990; Tzagoloff et al 1989) including two that code for both the cytoplasmic and mitochondrial enzymes (Chatton et al 1988; Natsoulis et al 1986). Homologs of *E. coli* elongation factors Tu (Nagata et al 1983) and G (Tzagoloff and Dieckmann 1990), that function only in mitochondria, have been identified as well.

Components that play a role in translational fidelity in mitochondria have been identified through the analysis of informational suppressors. Mitochondrial ochre mutations are suppressed by nuclear mutations in at least two genes that probably encode small subunit ribosomal proteins (Boguta et al 1988; Boguta et al 1986) as well as a mutation in the mitochondrial gene encoding the 15S rRNA (Fox and Staempfli 1982; Shen and Fox 1989). A frameshift suppressor has also been found in the 15S rRNA (Weiss-Brummer and Hüttenhofer 1989) as well as in a mitochondrially encoded serine-tRNA (Hüttenhofer et al 1990).

Yeast mitochondrial translation initiation involves mRNA-specific translational activators, as described below. The general features of initiation are poorly understood. Initiation occurs at AUG codons and presumably involves both the mitochondrially encoded initiator tRNA (Canaday et al 1980) and a nuclearly encoded homolog of *E. coli* IF2 (Tzagoloff and Dieckmann 1990). There is no evidence for a defined ribosome binding site located at a fixed distance relative to the initiation codon. Nor is it known whether mitochondrial ribosomes scan for initiation codons. Interestingly, all but one of the major mitochondrial mRNAs have long 5'-leaders with at least one AUG (see Grivell 1989).

A very fruitful genetic approach to the study of translation initiation in the yeast cytoplasm has been the isolation of suppressors of initiation codon mutations (Castilho-Valavicius et al 1990; Cigan et al 1989; Donahue et al 1988). We are attempting to use this approach to study mitochondrial translation initiation. We have taken advantage of the recent development of a yeast mitochondrial transformation system (Fox et al 1988; Johnston et al 1988), and the ability to *replace* wild-type mitochondrial genes with defined mutations (Fox et al 1989), to convert the initiator AUG of the *COX3* gene (coding cytochrome oxidase subunit III (coxIII)) to AUA (L.S.F., unpublished results). This mutation greatly reduced coxIII translation and produced a leaky non-respiratory phenotype. Unlinked suppressors of this mutation have been obtained that map to both the nuclear and mitochondrial genomes. These suppressors are likely to define genes whose products play a role in mitochondrial initiation codon selection.

Specific activators of *COX3* mRNA translation

A especially interesting aspect of yeast mitochondrial translation is the fact that nuclear genes encode translational activators that work on specific mRNAs. Multiple nuclear genes are required for the translation of both the *COB* (Dieckmann and Tzagoloff 1985; Körte et al 1989; Rödel 1986; Rödel and Fox 1987; Rödel et al 1985) and *COX3* mRNAs (see below), and at least one nuclear gene is specifically required for coxII translation (Poutre and Fox 1987; Strick and Fox 1987). It seems likely that specific factors may activate translation of all major mitochondrial mRNAs.

The best understood of these mRNA-specific activators control translation of the *COX3* mRNA. They are encoded by three unlinked nuclear genes, *PET494, PET54, and PET122,* that were identified by recessive mutations specifically blocking accumulation of the COXIII protein (Costanzo et al 1986; Costanzo et al 1988; Ebner et al 1973; Kloeckener-Gruissem et al 1987; Kloeckener-Gruissem et al 1988; Müller et al 1984). Since these activators were identified simply by screening collections of non-respiratory (Pet$^-$) mutants, it is of course possible that there are more such genes awaiting discovery.

The defect in *COX3* gene expression is clearly post-transcriptional, since the *COX3* mRNA is present at wild-type levels in *pet494, pet54* and *pet122* mutant strains (Costanzo et al 1986; Costanzo et al 1988; Kloeckener-Gruissem et al 1988; Müller et al 1984). The mutants block synthesis rather than stability of the coxIII protein, as demonstrated by analysis of bypass-suppressors. Mitochondrial suppressors of null mutations in all three nuclear genes were *COX3* gene rearrangements that led to the production of chimeric *COX3* mRNAs with 5'-leaders from other mitochondrial mRNAs

(Costanzo and Fox 1986; Costanzo et al 1986; Costanzo et al 1988). The fact that these bypass-suppressors allowed *pet494*, *pet54* and *pet122* mutants to respire demonstrated that none of these nuclear genes have other important cellular functions.

The PET494, PET54, and PET122 proteins are all located in mitochondria (Costanzo and Fox 1986; Costanzo et al 1989; Ohmen et al 1988; T.W.M., unpublished results). The site or sites at which they act have been genetically mapped to the 5' two-thirds (about 400 bases) of the COX3 mRNA leader by showing that translation of a chimeric mRNA, bearing this portion of the *COX3* leader fused to the *COB* structural gene, was dependent on *PET494*, *PET54*, and *PET122* function (Costanzo and Fox 1988).

Published data indicate that the synthesis of coxIII may be modulated at the level of translation during release of yeast cells from glucose repression (Falcone et al 1983; Zennaro et al 1985). Expression of both *PET494* (Marykwas and Fox 1989) and *PET54* (Z. Shen and T.D.F., unpublished results) also responds to glucose repression, suggesting that these nuclear genes might play a role in modulating *COX3* expression in mitochondria. However, there is no evidence that the level of either nuclear gene product is rate-limiting for coxIII translation.

Functional interaction between a *COX3*-specific activator and the small ribosomal subunit

The findings described above suggest that one or more of the *COX3*-specific translational activators contact the mRNA leader and mediate an interaction between the leader and a general component of the translation system required for initiation. We are exploring this possibility by looking for allele-specfic suppressors of mutations affecting the mRNA-specific activators. One series of experiments along these lines has demonstrated that the PET122 protein interacts functionally with the small mitochondrial ribosomal subunit.

Deletion mutations that truncate the 254 amino acid long PET122 protein by 67 residues at the C-terminus can be suppressed by unlinked mutations in several genes (Haffter et al 1990a).

Interestingly, these suppressors also suppressed another mutation that truncates PET122 by 24 C-terminal residues, but suppressed neither a missense mutation in the N-terminal portion of the protein nor a complete *pet122* deletion (Haffter et al 1990a). Thus, these suppressors do not bypass *PET122* function but instead enhance residual activity of the truncated proteins.

One of the suppressor mutations by itself caused a heat sensitive non-respiratory phenotype (Pet^{hs}) that resulted from a general blockage of mitochondrial translation at non-permissive temperature. The corresponding gene, termed *PET123*, was cloned and sequenced, but its protein product had no homology with previously described proteins (Haffter et al 1990a). We prepared an anti-PET123 antibody, using an antigen made in *E. coli,* and found that it reacted with a protein located specifically in mitochondria (McMullin et al 1990). Further fractionation revealed that the PET123 protein was a previously undescribed stoichiometric component of the small mitochondrial ribosomal subunit (McMullin et al 1990).

While suppressor mutations affecting a second locus had no effect on mitochondrial translation in otherwise wild-type strains, some of them produced a non-conditional "synthetic defect" in mitochondrial translation when combined with Pet^{hs} alleles of *PET123* (Haffter et al 1990b). This synthetic defect strongly suggested that the second gene coded a protein interacting with PET123, and allowed cloning of the wild-type gene by complementation. This second suppressor gene was sequenced (Haffter et al 1990b) revealing that it was identical to the previously identified gene *MRP1*, which also encodes a small subunit mitochondrial ribosomal protein (Dang et al 1990; Myers et al 1987).

The functional interaction between PET122 and the ribosomal small subunit revealed by these genetic studies could reflect direct contact between PET122 and a site on the ribosome containing PET123 and MRP1. Immunological studies on sub-mitochondrial fractions from strains that overproduce PET122 indicate that PET122 is an integral protein of the inner membrane (T.W.M., unpublished results). Furthermore, PET494 and PET54 appear to be membrane associated as well (T.W.M., unpublished results). Taken together, these findings raise the possibility that the *COX3* mRNA-specific translational activators bring the mRNA together with the ribosome near the inner membrane such that coxIII is synthesized close to the site where it is assembled into cytochrome oxidase.

Acknowledgements

We thank M.C. Costanzo for many helpful discussions. P.H. was supported sequentially by fellowships from EMBO and from the Swiss National Science Foundation. L.S.F. is a Fellow of the Jane Coffin Childs Memorial Fund for Medical Research. Work in the authors' laboratory is supported by a grant (GM29362) from the US NIH.

References

Attardi G, Schatz G (1988) Biogenesis of mitochondria. Ann Rev Cell Biol 4:289-333

Boguta M, Mieszczak M, Zagórski W (1988) Nuclear omnipotent suppressors of premature termination codons in mitochondrial genes affect the 37S mitoribosomal subunit. Curr Genet 13:129-135

Boguta M, Zoladek T, Putrament A (1986) Nuclear suppressors of the mitochondrial mutation *oxi1-V25* in *Saccharomyces cerevisiae*: genetic analysis of the suppressors: absence of complementation between non-allelic mutants. J Gen Microbiol 132:2087-2097

Canaday J, Dirheimer G, Martin RP (1980) Yeast mitochondrial methionine initiator tRNA: characterization and nucleotide sequence. Nucleic Acids Res 8:1445-1457

Castilho-Valavicius B, Yoon H, Donahue TF (1990) Genetic characterization of the *Saccharomyces cerevisiae* translational initiation suppressors *sui1, sui2* ans *SUI3* and their effects on *HIS4* expression. Genetics 124:483-495

Chatton B, Walter P, Ebel J-P, Lacroute F, Fasiolo F (1988) The yeast *VAS1* gene encodes both mitochondrial and cytoplasmic Valyl-tRNA synthetases. J Biol Chem 263:52-57

Cigan AM, Pabich EK, Feng L, Donahue TF (1989) Yeast translation initiation suppressor *sui2* encodes the α subunit of eukaryotic initiation factor 2 and shares sequence identity with the human α subunit. Proc Natl Acad Sci USA 86:2784-2788

Costanzo MC, Fox TD (1986) Product of *Saccharomyces cerevisiae* nuclear gene *PET494* activates translation of a specific mitochondrial mRNA. Mol Cell Biol 6:3694-3703

Costanzo MC, Fox TD (1988) Specific translational activation by nuclear gene products occurs in the 5' untranslated leader of a yeast mitochondrial mRNA. Proc Natl Acad Sci USA 85:2677-2681

Costanzo MC, Fox TD (1990) Control of mitochondrial gene expression in *Saccharomyces cerevisiae.* Ann Rev Genet 24:in press

Costanzo MC, Seaver EC, Fox TD (1986) At least two nuclear gene products are specifically required for translation of a single yeast mitochondrial mRNA. EMBO J 5:3637-3641

Costanzo MC, Seaver EC, Fox TD (1989) The *PET54* gene of *Saccharomyces cerevisiae*: Characterization of a nuclear gene encoding a mitochondrial translational activator and subcellular localization of its product. Genetics 122:297-305

Costanzo MC, Seaver EC, Marykwas DL, Fox TD (1988) Multiple nuclear gene products are required to activate translation of a single yeast mitochondrial mRNA. In: Tuite MF, Picard M, Bolotin-Fukuhara M (eds) Genetics of Translation: New Approaches. Springer-Verlag, Berlin, 373-382

Dang H, Franklin G, Darlak K, Spatola A, Ellis SR (1990) Discoordinate expression of the yeast mitochondrial ribosomal protein MRP1. J Biol Chem in press

Dieckmann CL, Tzagoloff A (1985) Assembly of the mitochondrial membrane system: *CBP6*, a yeast nuclear gene necessary for the synthesis of cytochrome *b*. J Biol Chem 260:1513-1520

Donahue TF, Cigan AM, Pabich EK, Valavicius BC (1988) Mutations at a Zn(II) finger motif in the yeast eIF-2β gene alter ribosomal start-site selection during the scanning process. Cell 54:621-632

Ebner E, Mason TL, Schatz G (1973) Mitochondrial assembly in respiration-deficient mutants of *Saccharomyces cerevisiae*. II. Effect of nuclear and extrachromosomal mutations on the formation of cytochrome *c* oxidase. J Biol Chem 248:5369-5378

Falcone C, Agostinelli M, Frontali L (1983) Mitochondrial translation products during release from glucose repression in *Saccharomyces cerevisiae.* J Bacteriol 153:1125-1132

Fearon K, Mason TL (1988) Structure and regulation of a nuclear gene in *Saccharomyces cerevisiae* that specifies MRP7, a protein of the large subunit of the mitochondrial ribosome. Mol Cell Biol 8:3636-3646

Fox TD, Folley LS, Mulero JJ, McMullin TW, Thorsness PE, Hedin LO, Costanzo MC (1990) Analysis and manipulation of yeast mitochondrial genes. In: Guthrie C, Fink GR (eds) Guide to Yeast

Genetics and Molecular Biology, a volume of Methods in Enzymology. Academic Press, Orlando, in press

Fox TD, Sanford JC, McMullin TW (1988) Plasmids can stably transform yeast mitochondria lacking endogenous mtDNA. Proc Natl Acad Sci USA 85:7288-7292

Fox TD, Staempfli S (1982) Suppressor of yeast mitochondrial ochre mutations that maps in or near the 15S ribosomal RNA gene of mtDNA. Proc Natl Acad Sci USA 79:1583-1587

Gampel A, Tzagoloff A (1989) Homology of aspartyl- and lysyl-tRNA synthetases. Proc Natl Acad Sci USA 86:6023-6027

Grivell LA (1989) Nucleo-mitochondrial interactions in yeast mitochondrial biogenesis. Eur J Biochem 182:477-493

Grohmann L, Graack H-R, Kitakawa M (1989) Molecular cloning of the nuclear gene for mitochondrial ribosomal protein YmL31 from *Saccharomyces cerevisiae*. Eur J Biochem 183:155-160

Haffter P, McMullin TW, Fox TD (1990a) A genetic link between an mRNA-specific translational activator and the translation system in yeast mitochondria. Genetics in press

Haffter P, McMullin TW, Fox TD (1990b) Functional interactions among two yeast mitochondrial ribosomal proteins and an mRNA-specific translational activator. submitted for publication

Herbert CJ, Labouesse M, Dujardin G, Slonimski PP (1988) The NAM2 proteins from *S. cerevisiae* and *S. douglasii* are mitochondrial leucyl-tRNA synthetases, and are involved in mRNA splicing. EMBO J 7:473-483

Hüttenhofer A, Weiss-Brummer B, Dirheimer G, Martin RP (1990) A novel type of +1 frameshift suppressor: a base substitution in the anticodon stem of a yeast mitochondrial serine-tRNA causes frameshift suppression. EMBO J 9:551-558

Johnston SA, Anziano PQ, Shark K, Sanford JC, Butow RA (1988) Mitochondrial transformation in yeast by bombardment with microprojectiles. Science 240:1538-1541

Kloeckener-Gruissem B, McEwen JE, Poyton RO (1987) Nuclear functions required for cytochrome *c* oxidase biogenesis in *Saccharomyces cerevisiae*: multiple trans-acting nuclear genes exert specific effects on the expression of each of the cytochrome *c* oxidase subunits encoded on mitochondrial DNA. Curr Genet 12:311-322

Kloeckener-Gruissem B, McEwen JE, Poyton RO (1988) Identification of a third nuclear protein-coding gene required specifically for posttranscriptional expression of the mitochondrial *COX3* gene in *Saccharomyces cerevisiae*. J Bacteriol 170:1399-1402

Koerner TJ, Myers AM, Lee S, Tzagoloff A (1987) Isolation and characterization of the yeast gene coding for the α subunit of mitochondrial phenylalanyl-tRNA synthetase. J Biol Chem 262:3690-3696

Körte A, Forsbach V, Gottenöf T, Rödel G (1989) In vitro and in vivo studies on the mitochondrial import of CBS1, a translational activator of cytochrome *b* in yeast. Mol Gen Genet 217:162-167

Marykwas DL, Fox TD (1989) Control of the *Saccharomyces cerevisiae* regulatory gene *PET494:* transcriptional repression by glucose and translational induction by oxygen. Mol Cell Biol 9:484-491

Matsushita Y, Kitakawa M, Isono K (1989) Cloning and analysis of the nuclear genes for two mitochondrial ribosomal proteins in yeast. Mol Gen Genet 219:119-124

McMullin TW, Haffter P, Fox TD (1990) A novel small subunit ribosomal protein of yeast mitochondria that interacts functionally with an mRNA-specific translational activator. submitted for publication

Müller PP, Reif MK, Zonghou S, Sengstag C, Mason TL, Fox TD (1984) A nuclear mutation that post-transcriptionally blocks accumulation of a yeast mitochondrial gene product can be suppressed by a mitochondrial gene rearrangement. J Mol Biol 175:431-452

Myers AM, Crivellone MD, Tzagoloff A (1987) Assembly of the mitochondrial membrane system: *MRP1* and *MRP2*, two yeast nuclear genes coding for mitochondrial ribosomal proteins. J Biol Chem 262:3388-3397

Myers AM, Tzagoloff A (1985) *MSW*, a yeast gene coding for mitochondrial tryptophanyl-tRNA synthetase. J Biol Chem 260:15371-15377

Nagata S, Tsunetsugu-Yokota Y, Naito A, Kaziro Y (1983) Molecular cloning and sequence determination of the nuclear gene coding for mitochondrial elongation factor Tu of *S. cerevisiae.* Proc Natl Acad Sci USA 80:6192-6196

Natsoulis G, Hilger F, Fink GR (1986) The *HTS1* gene encodes by the cytoplasmic and mitochondrial histidine tRNA synthetases of *S. cerevisiae.* Cell 46:235-243

Ohmen JD, Kloeckener-Gruissem B, McEwen JE (1988) Molecular cloning and nucleotide sequence of the nuclear *PET122* gene required for expression of the mitochondrial *COX3* gene in *S. cerevisiae.* Nucleic Acids Res 16:10783-10802

Pape LK, Koerner TJ, Tzagoloff A (1985) Characterization of a yeast nuclear gene (*MST1*) coding for the mitochondrial threonyl-tRNA synthetase. J Biol Chem 260:15362-15370

Partaledis JA, Mason TL (1988) Structure and regulation of a nuclear gene in *Saccharomyces cerevisiae* that specifies MRP13, a protein of the small subunit of the mitochondrial ribosome. Mol Cell Biol 8:3647-3660

Poutre CG, Fox TD (1987) *PET111*, a *Saccharomyces cerevisiae* nuclear gene required for translation of the mitochondrial mRNA encoding cytochrome *c* oxidase subunit II. Genetics 115:637-647

Rödel G (1986) Two yeast nuclear genes, *CBS1* and *CBS2*, are required for translation of mitochondrial transcripts bearing the 5'-untranslated *COB* leader. Curr Genet 11:41-45

Rödel G, Fox TD (1987) The yeast nuclear gene *CBS1* is required for translation of mitochondrial mRNAs bearing the *cob* 5'-untranslated leader. Mol Gen Genet 206:45-50

Rödel G, Körte A, Kaudewitz F (1985) Mitochondrial suppression of a yeast nuclear mutation which affects the translation of the mitochondrial apocytochrome *b* transcript. Curr Genet 9:641-648

Shen Z, Fox TD (1989) Substitution of an invariant nucleotide at the base of the highly conserved "530-loop" of 15S rRNA causes suppression of mitochondrial ochre mutations. Nucleic Acids Res 17:4535-4539

Strick CA, Fox TD (1987) *Saccharomyces cerevisiae* positive regulatory gene *PET111* encodes a mitochondrial protein that is translated from an mRNA with a long 5' leader. Mol Cell Biol 7:2728-2734

Tzagoloff A, Akai A, Kurkulos M, Repetto B (1988) Homology of yeast mitochondrial leucyl-tRNA synthetase and isoleucyl- and methionyl-tRNA synthetases of *Escherichia coli*. J Biol Chem 263:850-856

Tzagoloff A, Dieckmann CL (1990) *PET* genes of *Saccharomyces cerevisiae*. Microbiol Rev in press

Tzagoloff A, Vambutas A, Akai A (1989) Characterization of *MSM1*, the structural gene for yeast mitochondrial methionyl-tRNA synthetase. Eur J Biochem 179:365-371

Weiss-Brummer B, Hüttenhofer A (1989) The paromomycin resistance mutation (*par*^{r}-454) in the 15S rRNA gene of the yeast *Saccharomyces cerevisiae* is involved in ribosomal frameshifting. Mol Gen Genet 217:362-369

Zennaro E, Grimaldi L, Baldacci G, Frontali L (1985) Mitochondrial transcription and processing of transcripts during release from glucose repression in "resting cells" of *Saccharomyces cerevisiae*. Eur J Biochem 147:191-196

THE YEAST PYRUVATE KINASE GENE IS REGULATED AT MULTIPLE LEVELS

P A Moore, A J E Bettany* and A J P Brown
Department of Molecular & Cell Biology
University of Aberdeen
Marischal College
Aberdeen AB9 1AS, UK.

INTRODUCTION

Glycolytic enzymes can comprise up to thirty percent of soluble cellular protein under certain conditions in the yeast Saccharomyces cerevisiae (Fraenkel, 1982). In most cases, the synthesis of these enzymes is regulated at the transcriptional level in response to carbon source, increasing ten to twenty-fold when glucose is added to yeast cultures growing on non-fermentable carbon sources (Maitra & Lobo, 1971). Pyruvate kinase occupies a pivotal role in glycolysis, catalysing the last ATP-yielding step of the pathway. It is one of the first glycolytic enzymes to be induced during the switch from non-fermentative to fermentative growth (Maitra & Lobo, 1971). The pyruvate kinase gene (PYK1) has two Upstream Activation Sites (UAS) located between -811 and -634 relative to the initiation codon, one Upstream Repression Site (URS) between -468 and -344 (Nishizawa et al., 1989), and a Downstrean Activation Site (DAS) within the coding region (Purvis et al., 1987), all of which contribute to its transcriptional activity. $UAS_{PYK}1$ (which contains a RAP1/GRF1 binding site; Nishizawa et al., 1989), $UAS_{PYK}2$ and DAS_{PYK} are all required for high level transcription on glucose, but the role of these sequences in carbon source regulation is not clear. The URS_{PYK} partially represses PYK1 transcription in glucose-grown and glycerol/ethanol-grown cultures (Figure 8 in Nishizawa et al., 1989).

Some glycolytic enzymes can be synthesised at extremely high levels in yeast transformants carrying multiple copies of their genes (for example, the HXK, PGI, FBA, TPI, PGK and PGM genes; Alber and Kawasaki, 1982; Kawasaki and Fraenkel, 1982; Mellor et al., 1985; Aguilera and Zimmerman, 1986; Schaaf et al., 1989; Schwelberger et al., 1989). However, we have

---

Footnote: * Current Address: Department of Applied Biology, Massachusetts Institute of Technology, Cambridge 02139, USA.

NATO ASI Series, Vol. H 49
Post-Transcriptional Control of Gene Expression
Edited by J. E. G. McCarthy and M. F. Tuite

demonstrated that only limited over-expression of the pyruvate kinase gene (PYK1) can be achieved under similar conditions (Moore et al., 1990a). Furthermore, in contrast to most other glycolytic genes, a high PYK1 gene copy number has a deleterious effect upon the growth of the host yeast cell (Moore et al., 1990b). We suggest that these effects might be linked by a translational control mechanism which operates on a small subset of glycolytic genes.

RESULTS & DISCUSSION

(1) High levels of expression of the wild-type PYK1 gene have a deleterious effect upon the yeast cell.

The plasmid pLD1(35) comprises a 6.3kb yeast genomic fragment cloned between the HindIII and NdeI sites of the multicopy vector pJDB207 (Bettany et al., 1989). The wild-type PYK1 gene on this plasmid, and the insertion mutants PYK1:1C and PYK1:2D (Figure 1) complement the conditionally-defective pyk1 alleles pyk1-5 and cdc19 at their restrictive temperatures (Bettany et al., 1989). When DBY746 (his3, leu2, trp1, ura3) or X4003-5B (ade1, gal1, his4, leu2, met2, trp5, ura3) is transformed with pLD1(35) (LEU2, PYK1), the transformation frequencies are consistently lower than for the control plasmid pJDB207 (typically about 10^3 transformants per microgram of DNA), and pLD1(35)1C and pLD1(35)2D show the same effect (Moore et al., 1990b).

DBY746:pLD1(35) and X4003-5B:pLD1(35) transformants show a range of colony sizes. Some transformants form strong colonies, but the majority grow weakly on minimal agar (Moore et al., 1990b). Having checked the phenotype of each transformant, RNA was prepared from exponentially growing cultures (on YEPD) inoculated from relatively large or small colonies. The abundance of the PYK1 mRNA was analysed by Northern blotting, and quantitated by dot blotting relative to the actin mRNA (Moore et al., 1990b). Over thirty transformants have been analysed in more than five separate experiments. Small colonies contain significantly elevated levels of the PYK1 mRNA (over 2% of total mRNA) compared with the large colonies (less than 2% of total mRNA) and the untransformed host (0.6% of total mRNA). Therefore, transformants which express the wild-type PYK1 gene at significantly elevated levels grow relatively slowly, and this is particularly obvious on minimal medium. This sets the PYK1 gene apart from most other glycolytic genes which can be expressed at extremely high levels with little apparent effect upon the growth of the host yeast cell (Alber and Kawasaki, 1982; Kawasaki and Fraenkel, 1982; Mellor et al., 1985; Aguilera and Zimmerman, 1986; Schaaf et al., 1989).

(2) A strong selection operates against abnormally high PYK1 copy numbers.

When DBY746:pLD1(35) or X4003-5B:pLD1(35) transformants are subcultured onto fresh minimal medium which selects for maintenance of the plasmid, the growth of the "slow-growing transformants" becomes stronger, the colony sizes approaching that of the "rapidly-growing transformants" (Moore et al., 1990b). RNA was prepared from mid-exponential cultures inoculated with fresh transformants, and from mid-exponential cultures inoculated with the same transformants following several weeks of storage on minimal medium at 4°C. In slow-growing transformants, the abundance of the PYK1 mRNA decreased to 2% of total mRNA or less (relative to the actin mRNA), whereas the PYK1 mRNA level was maintained at 2% of total mRNA or less in the

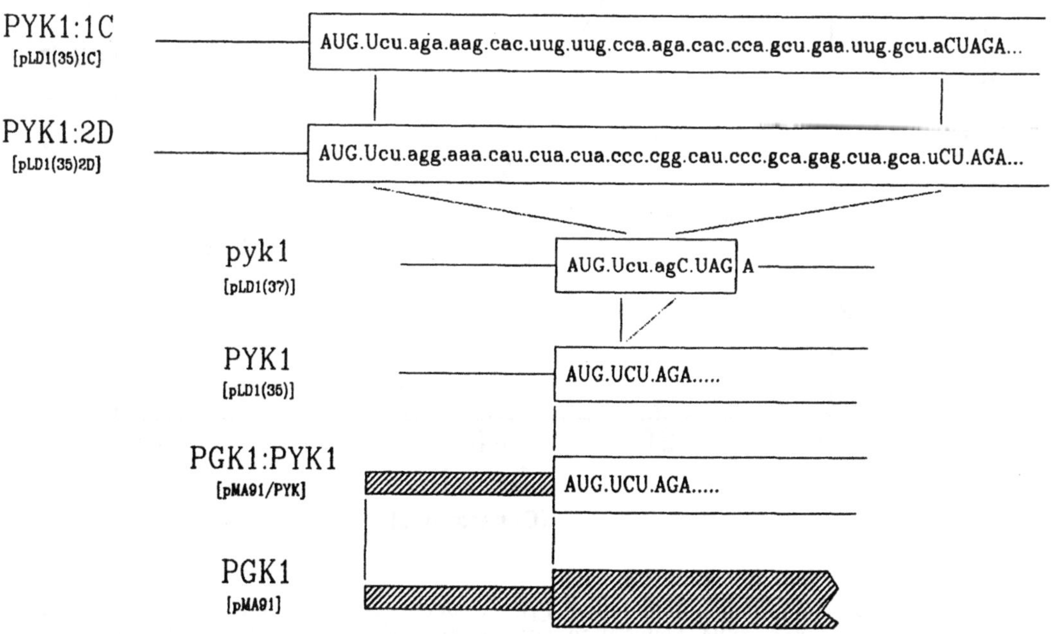

FIGURE 1: PYK1 constructs used in this study.
The plasmids pLD1(35), pLD1(35)1C and pLD1(35)2D (Bettany et al., 1989), and pMA91 (Mellor et al., 1983) have been described previously. PGK1 sequences are hatched, wild-type PYK1 sequences are in upper case, and insertions in lower case. pLD1(35) carries the wild-type PYK1 gene.

rapidly-growing transformants. Therefore, slow-growing transformants which carry significantly elevated levels of the PYK1 mRNA "revert" at high frequency to more rapid growth, and this correlates with decreased levels of PYK1 expression.

This "wind-down" effect was studied further by analysing the behaviour of a slow-growing X4003-5B:pLD1(35) transformant in more detail (Moore et al., 1990b). Minimal medium was inoculated directly from the small colony and serially subcultured into fresh medium as the culture entered stationary phase. The growth of the culture was monitored using the A_{600}, and DNA and RNA were prepared at several timepoints for use in PYK1 copy number and mRNA abundance estimations, respectively (Figure 2). The doubling time of the

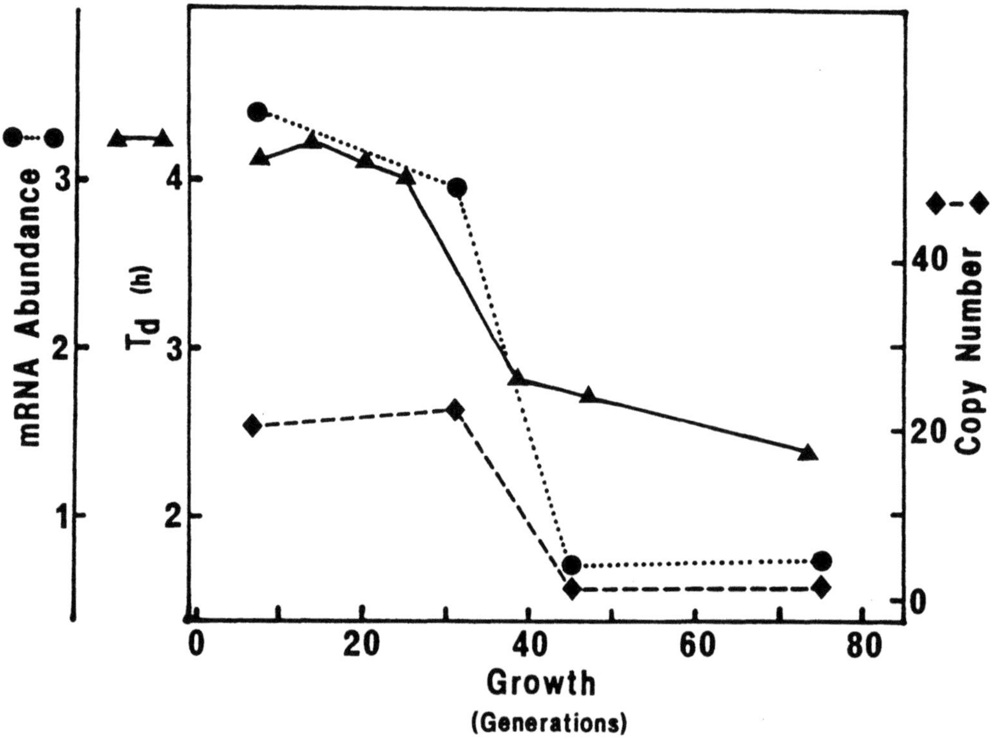

FIGURE 2: Abnormally high levels of PYK1 expression decrease during growth. PYK1 copy number and mRNA abundance were measured during serial batch growth of a slow-growing X4003-5B:pLD1(35) transformant. Fresh cultures were inoculated from cultures that had reached late-exponential growth phase. PYK1 mRNA abundance is expressed as a percentage of total mRNA, and PYK1 copy number is expressed per haploid genome. Doubling times were measured by monitoring the A_{600} of cultures. The dot blotting methods used to measure plasmid copy number and mRNA abundance have been described elsewhere (Santiago et al., 1987; Purvis et al, 1987). Taken from Moore et al., (1990a).

X4003-5B:pLD1(35) transformant which was initially two-fold greater than that of the untransformed host (T_d = 2 hours), decreased by about 40% over 75 generations. This correlated with a 10-fold decrease in PYK1 copy number (measured relative to the single-copy actin gene), and a 4.5-fold reduction in PYK1 mRNA level (relative to the actin mRNA). The inactivation of the PYK1 gene with an N-terminal amber mutation (pLD1(37); Figure 1) disrupts this strong selection for reduced copy number (Moore et al., 1990b).

(3) Over-expression of the PYK1 gene is limited at the level of mRNA abundance.

The abundance of the PYK1 mRNA was measured relative to the copy number of the PYK1 gene in a range of DBY746:pLD1(35) and X4003-5B:pLD1(35) transformants (Moore et al., 1990a). DNA and RNA were prepared from the same mid-exponential cultures to exclude the "wind-down" effects described above. The actin gene and mRNA were used as internal controls for the PYK1 copy number and mRNA abundance measurements, respectively. The data reveal a strong dosage effect (Figure 3). The abundance of the PYK1 mRNA per gene copy decreases 3-fold as the copy number increases 5-fold.

This dosage effect could be mediated either at the transcriptional level, or at a post-transcriptional level, or by a combination of both. Expression of the ADH2 gene is inhibited by the presence in trans of multiple copies of the ADH2 promoter which saturate specific transcriptional activators (Irani et al., 1987). A similar mechanism could account for the reduced levels of PYK1 mRNA per gene copy. However, the effect is unlikely to be mediated by $UAS_{PYK}1$, since RAP1/GRF1 (which activates a large number of yeast genes) is unlikely to become limiting. Alternatively, $UAS_{PYK}2$ may be affected, the transcription of excess PYK1 genes could be repressed via the URS_{PYK} (Nishizawa et al., 1989), or excess PYK1 mRNA may be degraded rapidly.

(4) PYK1 over-expression is limited at the translational level.

Pyruvate kinase activities and PYK1 mRNA abundances were then compared in a range of DBY746:pLD1(35) and X4003-5B:pLD1(35) transformants (Moore et al., 1990a). The specific activity of the enzyme increased about 2-fold (from 6.5 to 11.7 nmoles pyruvate generated.min^{-1}.ug $protein^{-1}$) when the abundance of the PYK1 mRNA increased 2-fold (from 0.6% to 1.1% of total mRNA). However, further increases in PYK1 mRNA abundance (to 3.3% of total mRNA) did not yield equivalent increases in enzyme activity (13.7 nmoles pyruvate generated.min^{-1}.ug $protein^{-1}$). This represents a reduction in the ratio of enzyme/mRNA of 10.9 to 4.2. This dosage effect was observed in a five separate experiments.

The reduction in the ratio of pyruvate kinase/PYK1 mRNA could be due to rapid degradation of excess enzyme, or decreased rates of enzyme synthesis. Therefore, the effect of increased PYK1 mRNA abundance upon its translation was investigated directly using polysome analysis as described previously (Santiago et al., 1987). Sucrose density gradients of yeast polysomes were prepared from DBY746 and a large number of DBY746:pLD1(35) transformants, and the distribution of the PYK1 mRNA across these gradients compared with those of the GAPDH, TCM1 and actin mRNAs (Moore et al., 1990a). Cells from the same cultures were used for measurements of PYK1 mRNA abundance to exclude variation due to the "wind-down" effect described above. The PYK1 and TCM1 mRNA data from DBY746 (PYK1 mRNA abundance = 0.6% of total mRNA)

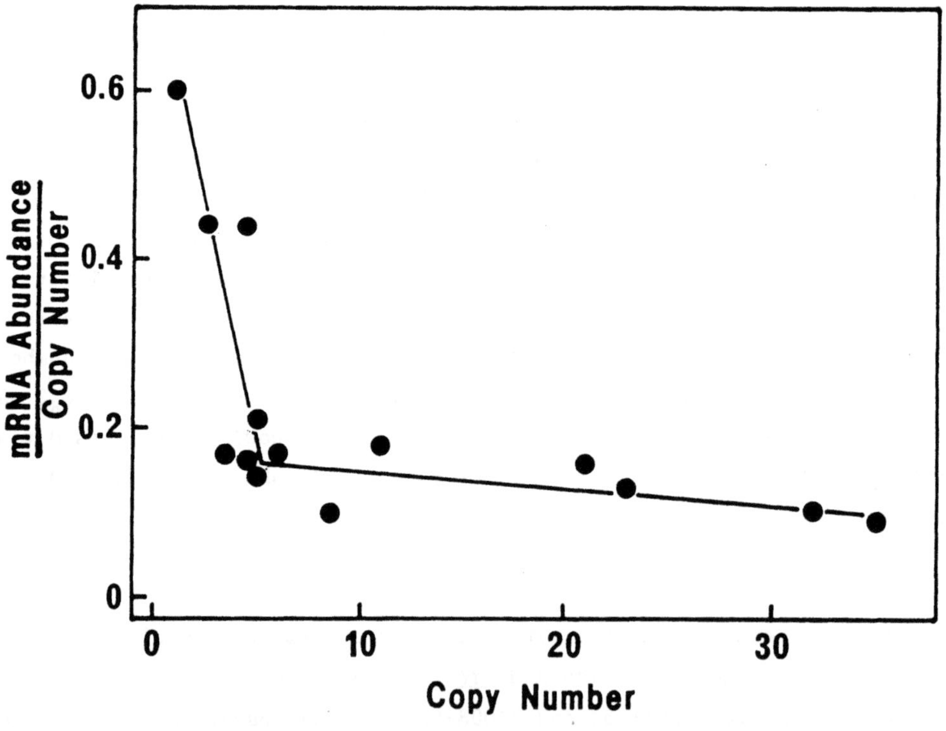

FIGURE 3: Over-expression of PYK1 is limited at the level of mRNA abundance. PYK1 copy number and mRNA abundance were determined in 14 separate pLD1(35) transformants. The amount of PYK1 mRNA per gene copy is plotted against PYK1 copy number for each transformant. Methods for the measurement of PYK1 copy number and mRNA abundance are published (Santiago et al., 1987; Purvis et al, 1987). Taken from Moore et al., (1990a).

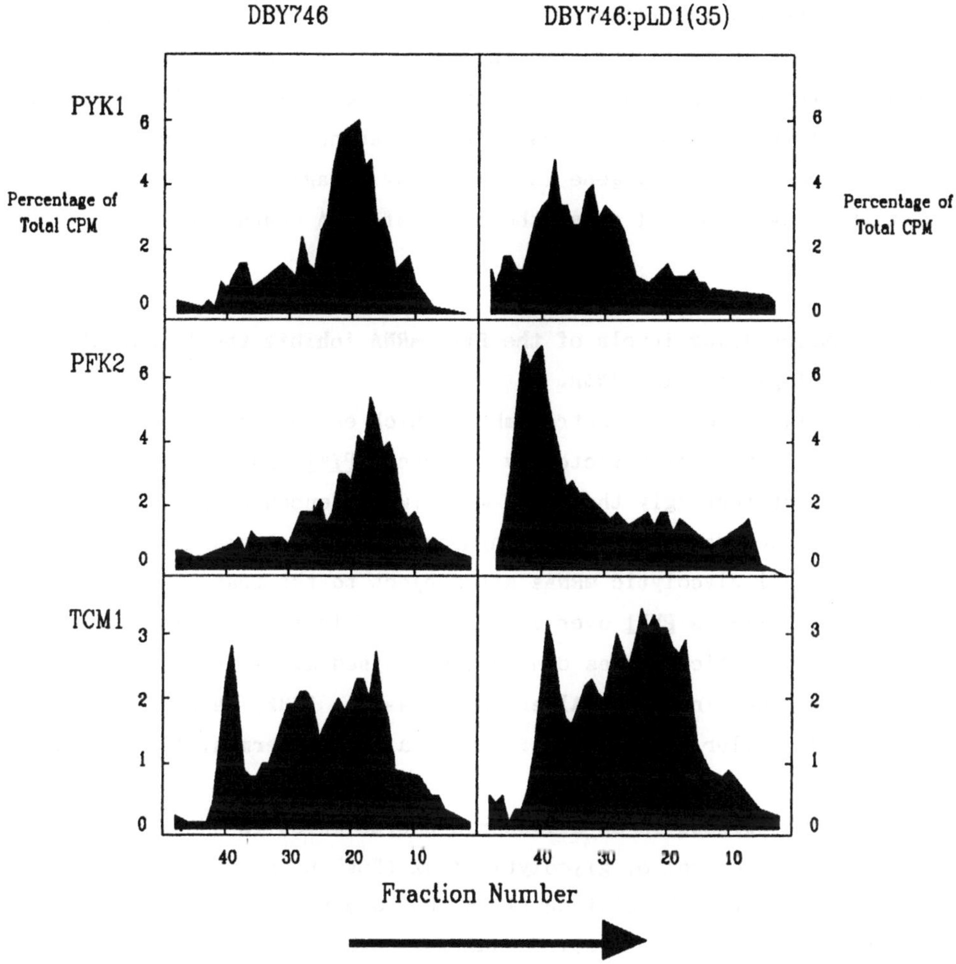

FIGURE 4: Excess PYK1 mRNA inhibits the translation of both the PYK1 and PFK mRNAs.
The translation of the PYK1 and PFK mRNAs are compared with that of the control TCM1 mRNA (which encodes ribosomal protein L3) in a DBY746:pLD1(35) transformant (PYK1 mRNA = 5.9% of total mRNA) and the untransformed host strain (PYK1 mRNA = 0.6% of total mRNA). This was done by measuring the distribution of these mRNAs across sucrose density gradients of yeast polysomes using dot blotting (Santiago et al., 1987). The arrow shows the direction of sedimentation, with the top of the gradient to the left, and heavy polysomes to the right of each box.

and one representative DBY746:pLD1(35) transformant (PYK1 mRNA abundance = 5.9% of total mRNA) are illustrated in Figure 4. As the abundance of the PYK1 mRNA increases its ribosome loading decreases. This confirms that the translation of the PYK1 mRNA decreases as its abundance increases. Hence, over-expression of the PYK1 gene is limited by dosage effects at the translational level as well as at the level of mRNA abundance.

(5) High intracellular levels of the PYK1 mRNA inhibit the translation of a small subset of glycolytic mRNAs.

The translation of the control mRNAs which encode actin and ribosomal protein L3 (TCM1) was not affected by increased PYK1 mRNA levels (Moore et al., 1990a). Interestingly the GAPDH mRNA (which encodes the glycolytic enzyme, glyceraldehyde-3-phosphate dehydrogenase) also remained unaffected. Therefore, not all glycolytic mRNAs are subject to the translational regulation which limits PYK1 over-expression. This is not surprising because most glycolytic enzymes can be synthesised at extremely high levels from multicopy genes in yeast (Alber and Kawasaki, 1982; Kawasaki and Fraenkel, 1982; Mellor et al., 1985; Aguilera and Zimmerman, 1986; Schaaf et al., 1989).

Phosphofructokinase is an allosteric enzyme which is thought by some to contribute to the control of glycolytic flux (Yoshino and Murakami, 1982; Muirhead, 1983), although Heinisch (1986) has demonstrated that it is not the only point of control. Also, Schaaf and coworkers (1989) were unable to over-express the PYK1 and PFK genes in the same yeast cell. We have analysed the effect of increased PYK1 mRNA abundance upon the translation of PFK mRNA using polysome analysis. The translation of the PFK mRNA is severely inhibited in the presence of high PYK1 mRNA levels (Figure 4), demonstrating that a small subset of glycolytic mRNAs is subject to this translational regulation. Furthermore, this shows that the mechanism operates in trans.

(6) The 5'-untranslated region of the PYK1 mRNA is required for the translational regulation.

The plasmid pMA91/PYK (a generous gift from Jane Mellor) carries the PYK1 coding and 3'-untranslated regions fused to the PGK promoter and 5'-leader sequences. The PYK1 sequence in this construct, which starts at -4 with respect to the first nucleotide of the coding region, excludes only the

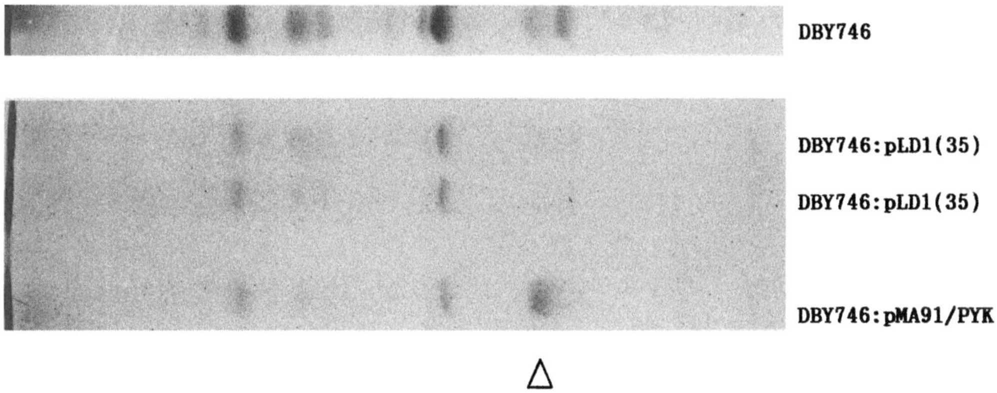

<u>FIGURE</u> <u>5</u>: High levels of pyruvate kinase synthesis are possible using pMA91/PYK.
Total soluble protein extracts from DBY746, two DBY746:pLD1(35) transformants, and a DBY746:pMA91/PYK transformant were electrophoresed on an 8% polyacrylamide/SDS gel and stained with coumassie blue. The abundance of the <u>PYK1</u> mRNA (or the <u>PGK/PYK</u> mRNA in DBY746:pMA91/PYK) are 0.6%, 0.9%, 1.6% and 5.5% of total mRNA, respectively, as measured by dot blotting and confirmed by Northern analysis (not shown). The loading for DBY747 is greater than for the transformants. The arrow indicates the pyruvate kinase band.

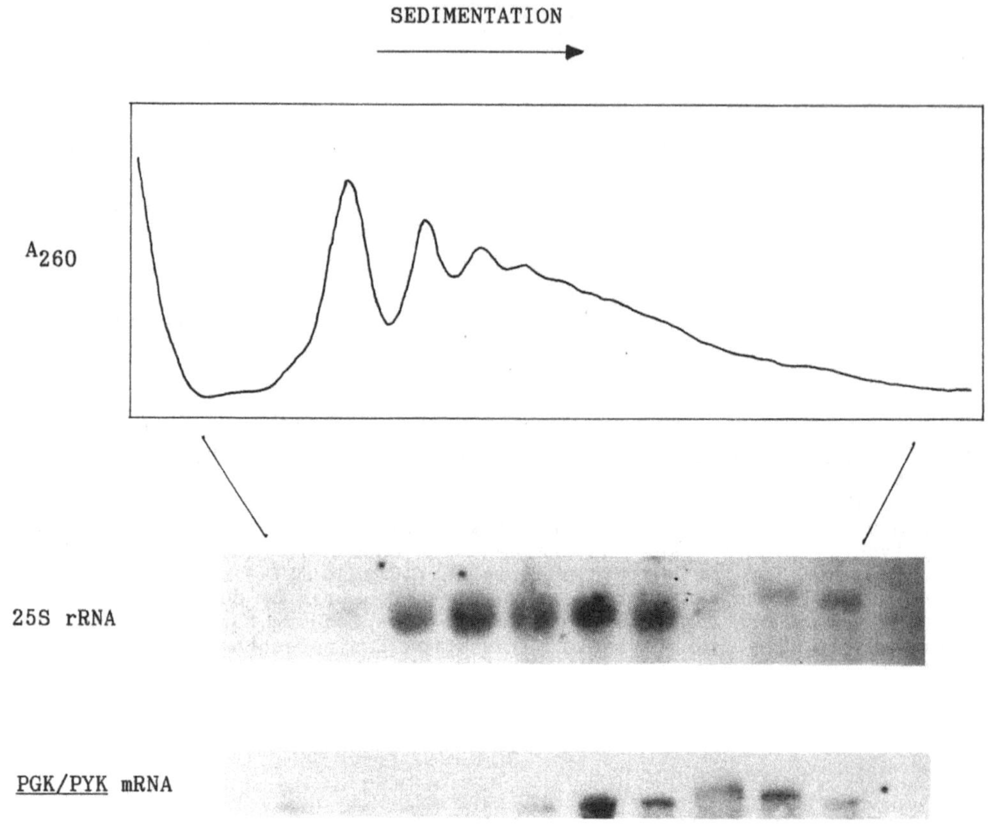

FIGURE 6: Excess PYK/PGK mRNA is translated efficiently.
A polysome gradient was prepared from a DBY746:pMA91/PYK transformant (PGK/PYK mRNA abundance = 4.1% of total RNA) as before (Santiago *et al.*, 1987) except that it was divided into twelve fractions which were collected directly into equal volumes of phenol and chloroform. Fractions were subjected to two phenol-chloroform extractions, followed by two chloroform extractions, and ethanol precipitation. Equal proportions of each fraction were then subjected to Northern blotting as before (Bettany *et al.*, 1989). Filters were probed for PYK, and 25S rRNA sequences. The A_{260} profile of the gradient, and the direction of sedimentation are shown (arrow).

first 32 nucleotides of the 5'-untranslated region on the mRNA. Pyruvate kinase is synthesised at high levels in DBY746:pMA91/PYK transformants (Figure 5) suggesting that this construct is not subject to the dosage effects that limit over-expression of the PYK1 gene. This suggests that the dosage effects that limit increases in PYK1 mRNA abundance are mediated by sequences residing upstream of -4, and hence that they probably operate through transcriptional, rather than post-transcriptional mechanisms. This remains to be confirmed. More importantly, this PGK/PYK fusion mRNA does not appear to be subject to translational regulation. This has been confirmed by analysing the distribution of the PGK/PYK mRNA across polysome gradients (Figure 6). This mRNA is translated efficiently at high abundances (4.1% of total mRNA). Therefore, the 5'-proximal 32 nucleotides of the PYK1 mRNA are required in *cis* for the translational regulation. We are currently testing whether this sequence is *sufficient* for regulation. It has no obvious homology to the 5'-leaders of the PFK mRNAs.

(7) High intracellular levels of pyruvate kinase do not affect cell growth

DBY746:pMA91/PYK transformants form strong colonies, and their doubling times are similar to the untransformed host (T_d = 2 hours). Yet pyruvate kinase is synthesised at extremely high levels in these transformants (Figure 5). Therefore, the growth effects are not caused by excess pyruvate kinase, but by elevated PYK1 gene copy numbers, high levels of PYK1 mRNA, or a combination of both. The deleterious effects upon cell growth might due to the saturation of transcriptional and translational factors required for the expression of other genes, for example PFK.

Why should PYK1 (and probably PFK) be subject to extra levels of regulation compared with other glycolytic genes? Both pyruvate kinase and phosphofructokinase are allosteric enzymes which may play significant roles in controlling glycolytic flux (Yoshino and Murakami, 1982; Muirhead, 1983). Yet, neither appears to catalyse the only control point in the pathway (Schaaf *et al.*, 1989). The regulatory circuits may help to maintain homeostasis during the switches between glycolytic and gluconeogenic growth.

ACKNOWLEDGEMENTS

We are grateful to Jurgen Heinisch for sending us the PFK1 and PFK2 genes, and to Jane Mellor for the plasmid pMA91/PYK. AJEB and PAM were supported by SERC studentships.

REFERENCES

Aguilera, A. and Zimmerman, F.K. (1986) Isolation and molecular analysis of the phosphoglucose isomerase structural gene of Saccharomyces cerevisiae. Mol. Gen.Genet. 202: 83-89.

Alber, T. and Kawasaki, G (1982) Nucleotide sequence of the triose phosphate isomerase gene of Saccharomyces cerevisiae. J.Molec.Appl.Genet. 1: 419-434.

Bettany, A.J.E., Moore, P.A., Cafferkey, R., Bell, L.D., Goodey, A.R., Carter, B.L.A. and Brown, A.J.P. (1989) 5'-secondary structure, in contrast to a short string of non-preferred codons, inhibits the translation of the pyruvate kinase mRNA in yeast. Yeast 5: 187-198.

Fraenkel, D.G. (1982) Carbohydrate metabolism. In: Strathern, J.N., Jones, E.W., Broach, J.R. (eds.) The molecular biology of the yeast Saccharomyces: metabolism and gene expression. Cold Spring Harbor, p1-37.

Heinisch, J. (1986) Isolation and characterisation of the two structural genes coding for phosphofructokinase in yeast. Mol.Gen.Genet. 202: 75-82.

Irani, M., Taylor, W.E. and Young, E.T. (1987) Transcription of the ADH2 gene in Saccharomyces cerevisiae is limited by positive factors that bind competitively to its intact promoter region on multicopy plasmids. Mol.Cell.Biol. 7: 1233-1241.

Kawasaki, G. and Fraenkel, D.G. (1982) Cloning of yeast glycolytic genes by complementation. Biochem.Biophys.Res.Commun. 108: 1107-1112.

Maitra, P.K. and Lobo, Z. (1971) A kinetic study of glycolytic enzyme synthesis in yeast. J.Biol.Chem. 246: 475-488.

Mellor, J., Dobson, M.J., Roberts, N.A., Tuite, M.F., Emtage, J.S, White, S., Lowe, P.A., Patel, T., Kingsman, A.J. and Kingsman, S.M. (1983) Efficient synthesis of enzymatically active calf chymotrypsin in Saccharomyces cerevisiae. Gene 24: 1-14.

Mellor, J., Dobson, M.J., Roberts, N.A., Kingsman, A.J. and Kingsman, S.M. (1985) Factors affecting heterologous gene expression in Saccharomyces cerevisiae. Gene 33: 215-226.

Moore, P.A., Bettany, A.J.E., Brown, A.J.P. (1990a) Expression of a yeast glycolytic gene is subject to dosage limitation. Gene, in press.

Moore, P.A., Bettany, A.J.E., Brown, A.J.P. (1990b) Multiple copies of the pyruvate kinase gene affets yeast cell growth. In preparation.

Muirhead, H. (1983) Triose phosphate isomerase, pyruvate kinase and other alpha/beta-barrel enzymes. Trends Biochem.Sci. 9: 326-330.

Nishizawa, M., Araki, R. and Teranishi, Y. (1989) Identification of an upstream activating sequence and an upstream repressible sequence of the pyruvate kinase gene of the yeast Saccharomyces cerevisiae. Mol.Cell.Biol. 9: 442-451.

Purvis, I.J., Loughlin, L., Bettany, A.J.E. and Brown, A.J.P. (1987) Translation and stability of an Escherichia coli B-galactosidase mRNA expressed under the control of pyruvate kinase sequences in Saccharomyces cerevisiae. Nucleic Acids Res. 15: 7963-7974.

Santiago, T.C., Bettany, A.J.E., Purvis, I.J. and Brown, A.J.P. (1987) Messenger RNA stability in Saccharomyces cerevisiae: the influence of translation and poly(A) tail length. Nucleic Acids Res. 15: 2417-2429.

Schaaf, I. Heinisch, J. and Zimmermann, F.K. (1989) Overproduction of glycolytic enzymes in yeast. Yeast 5: 285-290.

Schwelberger, H.G., Kohlwein, S.D. and Paltauf, F. (1989) Molecular cloning, primary structure and disruption of the structural gene of aldolase from Saccharomyces cerevisiae. Eur.J.Biochem. 180, 301-308.

Yoshino, M. and Murakami, K. (1982) AMP deaminase reaction as a control system of glycolysis in yeast: activation of phosphofructokinase and pyruvate kinase by the AMP deaminase-ammonia system. J.Biol.Chem. 257: 2822-2828.

INITIATION OF PROTEIN SYNTHESIS IN E. COLI: THE TWO CRUCIAL STEPS

L. Gold and D. Hartz
Department of Molecular, Cellular and Developmental Biology
University of Colorado
Campus Box 347
Boulder, CO 80309 USA

Prelude

Our lab has long studied the action of translational repressors (Miller et al., 1985; Andrake et al., 1988; Tuerk et al., 1990; McPheeters et al., 1988) and, more recently, a bacteriophage T4-encoded, Shine and Dalgarno-specific endonuclease (Ruckman et al., 1989). We were concerned with mechanisms for controlling the yield of protein from a given amount of mRNA; repressors provide reversible diminution of yield, while the endonuclease provides irreversible cessation of translation initiation on mRNAs that have been torn from their Shine and Dalgarno. When we invented TOEPRINTING (Hartz et al., 1988), we used that assay to further study repression (Winter et al., 1987; McPheeters et al., 1988; Unnithan et al., submitted). However, we quickly understood that we could contribute to a basic understanding of the unregulated initiation pathway.

The two crucial steps in initiation are selection of mRNA by ribosomes and selection of initiator-tRNA and initiation codon. We will describe initiator tRNA and proper codon selection first because the work is further along and partially published; binary complexes between mRNA and 30S subunits are discussed subsequently.

Selection of Initiator tRNA and Initiation Codon

Toeprinting visualizes ternary complexes between an mRNA, a 30S particle, and a tRNA in the P site; the tRNA is

NATO ASI Series, Vol. H 49
Post-Transcriptional Control of Gene Expression
Edited by J. E. G. McCarthy and M. F. Tuite

basepaired to a cognate codon (Hartz et al., 1989). "Toeprints" are reverse transcriptase stops on an mRNA; toeprints arise when reverse transcriptase collides with the edge of the 30S particle as cDNA is being synthesized on the mRNA template. By changing the availability of tRNAs _in vitro_, the toeprints "move" appropriately; there exists a fixed location in the 30S particle for the P site (and hence the anticodon), while the _sequence_ of the anticodon present in the P site determines which codon is chosen (Hartz et al., 1989). The nucleotides that collectively comprise the ribosome binding site (RBS) of an mRNA limit which nucleotides can be chosen by an anticodon; that is, the "window" for codon selection is defined by, for example, the allowed spacing between the Shine and Dalgarno element and the initiation codon (Hartz et al., 1989). This is all reasonable if 30S particles maintain a fixed distance between the 3′ end of 16S RNA and the P site.

The problem revealed by our very first experiments (Hartz et al., 1989) was that ribosomes themselves don't distinguish elongator tRNAs from the initiator tRNA. Two initiation factors (IF2 and IF3) have clear functions that lead to selection of the initiator tRNA over elongators: IF2 selects for charged tRNAs with a blocked amino group, while IF3 selects for the unique initiator tRNA anticodon stem and loop structure along with a limited set of codons that include AUG, GUG, and UUG (Hartz et al., 1989). IF1 had large effects on proper initiation complex formation only when 70S particles were used (Hartz et al., 1989).

We also studied initiator tRNAs from yeast and wheat germ (Hartz et al., 1990a); these tRNAs have the same features that are found in the _E. coli_ $tRNA^{fmet}$. Interestingly, these initiator tRNAs are selected by _E. coli_ IF3 and 30S subunits over _E. coli_ elongator tRNAs (Hartz et al., 1990a). Thus eukaryotes could have an initiation factor that plays a role analogous to that of IF3. Our guess for the homologue is eIF2-ß, which when mutant will allow UUG to be recognized as an initiation codon in yeast [which normally can use only AUG (Donahue et al., 1988)]. Donahue's work shows clearly that codon:anticodon pairing determines the first amino acid in a protein (Cigan et al., 1988), and that eIF2-ß plays a crucial role in stringent selection of AUG [or UUG in the mutant allele (Donahue et al., 1988)]. This implies a striking mechanistic similarity between E. coli and S. cerevisiae for one of the two crucial steps in translation initiation.

Selection of mRNA by Ribosomes

For several years we held a simple view of the determinants on mRNAs that afforded initiation and, probably, binary complex formation (Gold et al., 1981; Gold, 1988). mRNAs contain information from -20 to +13 surrounding initiating AUGs [where the A of the AUG is position zero (Schneider et al., 1986)], and that information is not present around other AUG's or even around AUGs preceded by Shine and Dalgarno-like sequences (Stormo et al., 1982a,b). Thus primary sequence level "determinants" define RBSs (Schneider et al., 1986).

In addition, secondary structures can hide those determinants from ribosomal inspection (Gold, 1988; Stormo, 1987; Draper et al., 1987). [The most elegant work on this topic, from deSmit and van Duin (1990; and this volume), makes the idea quantitatively simple, as predicted by both David Draper and Gary Stormo.] In the absence of secondary structure, a long Shine and Dalgarno and an AUG with optimal spacing from that Shine and Dalgarno will provide maximal translation (Gold, 1988; Shindeling et al., in preparation). In this view all contacts between 30S particles and mRNA can be RNA/RNA, while the tRNA anticodon selects the AUG through additional base pairing. In our lab we never contemplated ideas about mRNA selection by ribosomal proteins. But a substantial literature existed on the role of S1 during initiation, including recent reviews by Subramanian (Subramanian, 1983; 1984).

We had written an article for Methods in Enzymology about high-efficiency RBS sequences (Gold and Stormo, 1990); in that article some highly efficient RBS were said to be abnormal and unlike other good initiation sites. In Riga, at a meeting on Translation, Irina Boni reported that those very abnormal RBS contained sequences (runs of pyrimidines) that cross-linked to S1 in initiation complexes and could thus be construed as "enhancers." We present here data that are easily interpreted around the positive role of S1 in mRNA binding.

Basically there are two experiments. The first is a modern repeat of an older experiment done in van Duin's lab (van Duin et al., 1980). Using toeprinting rather than full protein synthesis, we have shown (Hartz et al., 1990b) that antibodies prepared against protein S1 block initiation complex formation. Furthermore, initiation complexes can be formed if the 30S particles are mixed with mRNA prior to the addition of the antibodies; no initiator tRNA is needed to provide a complex that is refractive to antibodies against S1. Thus we concluded, as did the earlier authors (Hartz et al., 1990b), that a 30S particle has by itself the capacity to find ribosome binding sites without any tRNA whatsoever.

The second experiment directly measures binary complexes between 30S particles and mRNA, using a variant of the normal toeprinting procedure (Hartz et al., 1990b). When reverse transcriptase is used to make cDNA on an mRNA bound to a 30S particle, the reverse transcriptase yields toeprints right around the Shine and Dalgarno region. Thus binary complexes appear to use the Shine and Dalgarno interaction, as expected (Gold et al., 1981; Gold, 1988). The exact position and strength of these binary complex toeprints <u>depends on protein S1</u>, suggesting that S1 is close to the 3′ end of 16S rRNA and the Shine and Dalgarno portion of the mRNA. Perhaps runs of pyrimidines (Subramanian, 1983;1984) provide extra interactions between mRNA and 30S particles, and in some mRNAs those interactions may even replace the normal Shine and Dalgarno interaction (Boni et al., 1990). As noted by

Boni's lab, sequences identified as "enhancers" by others could function as good S1 binding sites.

Binary complexes represent the other crucial step in translation initiation. Binary complexes occur independently of the addition of initiator tRNA to the ribosomal P site. Binary complexes should be formed in a manner dependent on the length of the Shine and Dalgarno and the amount of secondary structure that occludes it from ribosomal inspection (de Smit and van Duin, 1990). S1 sites are also important; these sites may reside at variable positions with respect to the initiation codon as noted by Boni et al. (1990).

A picture of a ternary complex would show the potential binding contributions from S1, the Shine and Dalgarno, and the codon:anticodon interactions. Initiation of translation may use one, two, or all of those stabilizing interactions. These contributions are different for specific mRNAs. Messages can start with the initiating AUG (Gold et al., 1981) and be translated inefficiently, or mRNAs can have long Shine and Dalgarno regions (without secondary structures) and be translated enormously well (Gold and Stormo, 1990). Runs of pyrimidines may provide supplemental or alternative interactions to afford translation initiation (Boni et al., 1990). Organisms that have no S1 probably utilize longer Shine and Dalgarno regions to make up for the missing binding energy; _B. subtilis_ apparetnly is such an organism. What must be further investigated is the capacity of good S1 sites to promote translation from either poor or occluded Shine and Dalgarno regions.

Eubacteria and Eukaryotes: How Many Biospheres Exist?

Kozak has described a working model for translation of most eukaryotic mRNAs (Kozak, 1989a). In her writing one sees a strong emphasis on the differences between initiation in the two kingdoms. Gold, in reviewing initiation in _E._

coli (1988), noted the likelihood of "conceptual similarities" in all initiation pathways. The selection of an initiation codon in any organism utilizes codon:anticodon pairing. The work of Donahue shows that explicitly in yeast (Cigan et al., 1988), while Hartz et al. (1989) demonstrated that toeprints reflect the cognate codon:anticodon interactions specified by whatever tRNA resides in the P site. S1 binding to mRNAs seems to mirror at least one activity of the CAP-binding complex. Specifically the recognition of the CAP probably serves to narrow the mRNA "space" that is scanned for the initiating AUG. In either case the result of binding would be to increase the local concentration of the initiating AUG in the vicinity of the anticodon of the tRNA locked in the ribosomal P site [by IF2 and IF3 in E. coli, and by (perhaps) eIF2 in eukaryotes].

The major biochemical differences between translation initiation in eubacteria and eukaryotes are two. Eukaryotes utilize ATP and an RNA helicase during the hunt for an initiating AUG codon, whereas the E. coli 30S particle finds the proper AUG (or surrounding codons in a narrow window) without any apparent source of energy. The helicase activity, ATP requirement, and the inhibitory role of extremely long hairpins between the CAP and the initiating AUG (Kozak, 1986; 1989b) are linked phenomena that, when understood, will provide a biochemical picture of the process of "scanning." In E. coli S1 might anchor an mRNA for the hunt for an initiation codon and then use the Shine and Dalgarno to further pin the message around the AUG; in such thinking the Shine and Dalgarno interaction replaces the ATP-powered step that logically utilizes the eIF4A helicase.

The other difference is just the obverse of the first. Most eukaryote mRNAs have no obvious Shine and Dalgarno homologue, nor do the 3′ ends of 18S rRNA contain any nucleotides beyond the universal hairpin stem that is the homologue of the dimethyl-A "kasugamycin" stem in E. coli (Noller and Woese, 1981; Noller, 1984); Giardia lambia is a rare exception to the general eukaryotic case (Sogin et al.,

1989). Probably it is useful to think that when the two kingdoms split, the progenitor of E. coli had not developed the extended 3' end of what is now 16S rRNA and had also not developed base pairing to mRNAs as an initiation component. Selection of that binding capacity during evolution in the eubacterial lineage makes trivial internal translation initiation and polycistronic mRNAs, two hallmarks of eubacterial gene organization and function (Gold et al., 1981; Gold, 1988) that are rather unusual in eukaryotes. In this view a simple difference in the translational machinery between eubacteria and eukaryotes leads to a major difference in the biochemistry of gene expression.

Acknowledgements

This work was sponsored by NIH grant GM28685. The thinking about our data was extended during Irina Boni's recent visit to Boulder.

References

Andrake M, Guild N, Hsu T, Gold L, Tuerk C, Karam, J (1988) The DNA polymerase of bacteriophage T4 is an autogenous translational repressor. Proc Natl Acad Sci USA 85:7942-7946.

Boni IV, Isaeva DM, Musychenko ML, Tzareva NV (1990) Ribosome-messenger recognition: mRNA target sites for ribosomal protein S1. Nucleic Acids Res, submitted

Cigan AM, Feng L, Donahue TF (1988) $tRNA^{met}$ functions in directing the scanning ribosome to the start site of translation. Science 242:93-97

de Smit MH, van Duin J (1990) Control of prokaryotic translational initiation by mRNA secondary structure. Prog Nuc Acid Res and Mol Biol 38:1-35

Donahue, TF, Cigan AM, Pabich EK, Castilho-Valavicius B (1988) Mutations at a Zn(II) finger motif in the yeast eIF-2ß gene alter ribosomal start-site selection during the scanning process. Cell 54:621-632

Draper DE (1987) Translational regulation of ribosomal protein synthesis in E. coli: Molecular mechanisms. In: Translational Regulation of Gene Expression (J. Ilan, eds.), Plenum Publishing, New York, pp. 1-26

Gold L, Pribnow D, Schneider T, Shinedling S, Singer BS, Stormo G (1981) Translational initiation in prokaryotes. Ann Rev Microbiol 35:365-403

Gold L (1988) Post-transcriptional regulatory mechanisms in Escherichia coli. Ann Rev Biochem 57:199-233

Gold L, Stormo GD (1990) How to get high level translation initiation. Meth Enzymol, in press

Hartz D, McPheeters DS, Traut R, Gold L (1988) Extension inhibition analysis of translation initiation complexes. Methods in Enzymol 164:419-425

Hartz D, McPheeters DS, Gold L (1989) Selection of the initiator tRNA by Escherichia coli initiation factors. Genes and Dev 3:1899-1912

Hartz D, Binkley J, Hollingsworth, T, Gold L (1990a) Domains of Initiator tRNA and initiation codon crucial for initiator tRNA selection by Escherichia coli IF3. Genes and Dev, submitted

Hartz D, McPheeters DS, Green L, Gold L (1990b) Detection of Escherichia coli ribosome binding at translation initiation sites in the absence of tRNA. J Mol Biol, submitted

Kozak M (1986) Influences of mRNA secondary structure on initiation by eukaryotic ribosomes. Proc Natl Acad Sci USA 83:2850-2854

Kozak M (1989a) The scanning model for translation: An update. J Cell Biol 108:229-241

Kozak M (1989b) Circumstances and mechanisms of inhibition of translation by secondary structure in eucaryotic mRNAs. Mol Cell Biol 9:5134-5142

McPheeters DS, Stormo GD, Gold L (1988) Autogenous regulatory site on the bacteriophage T4 gene 32 messenger RNA. J Mol Biol 201:517-535.

Miller ES, Winter RB, Campbell KM, Power SD, Gold L (1985) Bacteriophage T4 regA protein: Purification of a translational repressor. J Biol Chem 260:13053-13059

Noller HF, Woese CR (1981) Secondary structure of 16S ribosomal RNA. Science 212:403-411

Noller HF (1984) Structure of ribosomal RNA. Ann Rev Biochem 53:119-162

Ruckman J, Parma D, Tuerk C, Hall DH, Gold L (1989) Identification of a T4 gene required for bacteriophage mRNA processing. New Biologist 1:54-65

Schneider TD, Stormo GD, Gold L, Ehrenfeucht A (1986) The information content of binding sites on nucleotide sequences. J Mol Biol 188:415-431

Sogin ML, Gunderson JH, Elwood HJ, Alonso RA, Peattie DA (1989) Phylogenetic significance of the Kingdom concept: An unusual eukaryotic 16S-like ribosomal RNA from Giardia lambia. Science 243:75

Stormo GD, Schneider T, Gold L (1982) Characterization of translational initiation sites in E. coli Nuc Acids Res 10:2971-2996

Stormo GD, Schneider T, Gold L, Ehrenfeucht (1982) Use of the "perceptron" algorithm to distinguish translational initiation sites in E. coli. Nuc Acids Res 10:2997-3012

Stormo GD (1987) Translation regulation in bacteriophage. In: Translational Regulation of Gene Expression (J. Ilan, eds.), Plenum Publishing, New York, pp. 27-49.

Subramanian AR (1983) Structure and functions of ribosomal protein S1. Progr Nucleic Acid Res Mol Biol 28:101-142

Subramanian, AR (1984) Structure and functions of the largest *Escherichia coli* ribosomal protein. Trends Biochem Sci 9:491-494

Tuerk C, Eddy S, Parma D, Gold, L (1990) The autogenous translational operator recognized by bacteriophage T4 DNA polymerase. J Mol Biol, in press

Unnithan S, Morrissey L, Karam J, Gold L. Binding of the bacteriophage T4 regA protein to mRNA targets: An initiator AUG is required. Nucleic Acids Res, submitted

van Duin J, Overbeek GP, Backendorf C (1980) Functional recognition of phage RNA by 30S ribosomal subunits in the absence of initiator tRNA. Eur J Biochem 110:593-597

Winter RB, Morrissey L, Gauss P, Gold L, Hsu T, Karam J (1987) Proc Natl Acad Sci USA 84:7822-7826

THE FUNCTION OF INITIATION FACTORS IN RELATION TO mRNA-RIBOSOME INTERACTION AND REGULATION OF GENE EXPRESSION

Anna La Teana[1,2], Maurizio Falconi[1], Roman T. Pawlik[2], Roberto Spurio[1], Cynthia L. Pon[1,2] and Claudio O. Gualerzi[1,2]

[1]Laboratory of Genetics, DBC, University of Camerino, 62032 Camerino, Italy and [2]Max-Planck-Institut für Molekulare Genetik, Abt. Wittmann, D-1000 Berlin 33, Germany

Initiation of translation in prokaryotes begins, in ≥90% of the cases, with an AUG triplet which offers the best basepairing with the anticodon CAU of the initiator tRNA; the wobbling triplets GUG, UUG and AUU are found more rarely with the latter being found in a single case (i.e. the infC gene which encodes translation initiation factor IF3) (Gren 1984). The reason for this degeneracy of the initiation triplet and its influence on the level of translational expression remain intriguing. In some cases changing the rare initiation triplet into the more common AUG results in moderate (Reddy et al. 1985; Khudyakov et al. 1988) to large (Brombach and Pon 1987) increases of expression. Inspection of the catalogue of genes starting with the rare triplets, however, seems to argue against the idea that these codons are used to attain a substantial reduction in the level of translation since this catalogue includes some of the most abundantly expressed genes in E. coli such as tufA (EF-Tu), hupB (HU1) as well as some ribosomal protein (r-protein) genes; furthermore, hupB is expressed at approximately the same level as hupA (HU2) and the r-protein genes beginning with GUG or UUG are expressed in amounts stoichiometrically equivalent to those of r-protein genes beginning with AUG.

In the present paper, we examine the influence that the rare initiation triplet AUU might have on gene expression. As mentioned above, the only known case of a natural gene whose coding sequence begins with AUU is that of infC (Sacerdot et al. 1982). In E. coli, infC is located at 38 min in the chromosome map in a region which also carries the genes for threonyl-tRNA synthetase (thrS), r-protein L35 (rpmI), r-protein L20 (rplT), the subunits of phenylalanyl-tRNA synthetase (pheS and pheT) and one of the two subunits of the host integration factor IHF (himA) (Hennecke et al. 1977; Springer et al. 1977; Comer 1981; Wada and Sako

NATO ASI Series, Vol. H 49
Post-Transcriptional Control of Gene Expression
Edited by J. E. G. McCarthy and M. F. Tuite

1987). Since it is known that the IF3/30S ratio is kept constant in the cells (Pon and Gualerzi 1979, Hershey 1987), an important question concerns the mechanism involved in the regulation of infC expression. This could occur at the transcriptional and/or post-transcriptional level.

Owing to the existence of multiple promoters, co-transcription and transcriptional read-through events, the pattern of transcription around the E. coli infC gene is very complex (Wertheimer et al. 1988). Thus, the infC-containing mRNAs are initiated from three separate promoters: two of these are located within the coding region of the thrS gene which is found upstream of infC, while the third promoter is found approximately 170 nucleotides upstream from the translation initiation site of thrS. An additional promoter is found within the infC coding region. Termination of transcription is believed to occur at an inefficient termination signal located in the intergenic spacer between infC and rpmI and at a strong terminator found at the end of rplT. The complexity of this situation and the finding that there is a differential use of the above-mentioned promoters depending on the growth conditions suggest that some type of transcriptional control might be operating. On the other hand, there are both theoretical and experimental grounds to support the premise that infC is under autogenous regulation at the translational level (Gold et al. 1984; Butler et al. 1987). In fact, the initiation region of the E. coli IF3 mRNA was found to break many of the rules followed by most of the known mRNAs and it has been suggested that these features are at the basis of a regulatory mechanism aimed at maintaining the balance between expression of the infC gene and the amount of ribosomes in the cell (Gold et al. 1984).

To determine whether the above peculiarities of infC transcription and of IF3 mRNA are conserved in other bacterial species, we have recently identified and cloned a segment of DNA from Bacillus stearothermophilus which carries the infC gene. As in E. coli, this gene is followed by rpmI and rplT specifying r-proteins L35 and L20, respectively (Pon et al. 1989). Based on the nucleotide sequences, the primary structures of IF3, L35 and L20 of the two organisms share respectively 50%, 48% and 62% identical residues. All three genes are preceded by strong, appropriately spaced, Shine-Dalgarno sequences and have UAA as translation termination codon (Fig. 1). The initiation codon is AUG for both r-protein genes, while, as in E. coli, infC begins with the rare AUU initiation triplet. However, the other peculiar features which make IF3 mRNA substantially different from all other E. coli mRNAs known so far (Gold et al. 1984) seem to be absent in B. stearothermophilus infC (see legend to Fig. 1).

```
   1 GTCAACTGAAAATTTTCTCTTGCTAACGAAGTGAGAATCCATTATAATAAAGGACGTTGTGTTTATCAAGAGCCGCTCCTTGACATGCCATCGCCAAGGT  100
                 transcription initiation
                 ------------->
 101 GGTATAGTGTATAAAGTGGAATGATTGCTTGTTTGTAACGAAAGCAGAAGCGCCCGCTTCTCACCTGACCGACGCTAAGAAGGCTGTTGGCAGGTTGTTA  200

 201 CGGTTCGCCAACTTTTTGTCCGAACGTCGTGCGCGGGTGCTTTGTGCATCCGCGCTTTTTCGTTGAGAGCGGGCTCTGCCGGCAAACAAGCCAGTTGGAA  300

 301 AAAACCTTGGAGGTGGCTCATTATTAGCAAA.........AACAAGTAAGGAGGAAACTGTGATGCCAAAA.........TTGAAATAACGACCGACTAT  1066
                          M S K  .  .  . N K *              M P K  .  .  . L K *
                          IF3 -->                           L35 -->

1067 AGGAGGGAACGATTATGCCACGC........AATAAATAACAACGATGGCCATCCTTGATCAGGATGGCCATTTTCTTATCAAGGAAAGGGATCGAT    1497
                  M P R  .  .  . N K *       transcription termination
                  L20 -->
```

Fig. 1. Initiation and termination signals of transcription and translation in B. stearothermophilus infC operon. The relevant transcription and translation signals are written in bold letters and underlined. The promoter structure identified by computer search but not used in vivo is written in normal script and under- and overlined. Transcription startpoint and direction are indicated by the long arrow. The translational initiation domain of E. coli IF3 mRNA was found to contain two potential Shine-Dalgarno sequences, a unique AUU initiation triplet, a guanosine-rich "structurogenic" sequence just 3' to the AUU initiation triplet and a potential 10 bp base complementarity (from five bases upstream to two bases downstream from the AUU) with 16S rRNA (between G462 and G474) (Gold et al. 1984). Of these unusual features, the only one clearly conserved in B. stearothermophilus IF3 mRNA is the initiation triplet AUU. This is actually preceded by another in-phase AUU triplet which probably is not used as initiation start since only serine was identified as the N-terminal amino acid of B. stearothermophilus IF3 (Kimura et al. 1983). The spacer is of standard length and does not contain additional SD sequences, and the GC-rich region downstream from the initiation triplet is virtually absent. Finally, since the nucleotide sequence of B. stearothermophilus 16S rRNA is not available (Dams et al. 1988), we cannot evaluate whether any additional region of potential complementarity exists between B. stearothermophilus IF3 mRNA and internal regions of the 16S rRNA.

In addition, unlike in *E. coli*, where only three nucleotides separate the stop codon of *thrS* and the beginning of *infC*, and the intercistronic regions separating *infC*, *rpmI* and *rplT* are long, in *B. stearothermophilus*, no *thrS* gene was found within 350 nucleotides upstream of *infC* and the intercistronic regions are extremely short (Fig. 1).

The search for sequences of potential relevance for initiation and termination of transcription in *B. stearothermophilus* revealed two potential promoters approximately 200 bp upstream from the initiation triplet of *infC* and a typical rho-independent termination sequence just downstream from the 3' terminus of *rplT*. Primer extension analysis (Fig. 2) revealed that the first of these potential promoters is used in vivo and that transcription begins at an A 208 nucleotides upstream of the *infC* initiation codon (Fig. 1). Northern blot analysis of the in vivo transcript using as probes the restriction fragments *Cla*I-*Sac*II, *Sac*II-*Hin*dIII and *Hin*dIII-*Cla*I specific for IF3, L35 and L20 coding sequences, respectively, revealed that all three probes hybridize with the same transcript (Fig. 3); this indicates that the *infC*-*rpmI*-*rplT* gene cluster is expressed as a single, polycistronic mRNA. Separate experiments also revealed that the transcript has a half-life of approximately one minute and is degraded uniformly without any indication of a differential stability of any portion of the polycistronic message (not shown). These facts indicate that different patterns of tran-

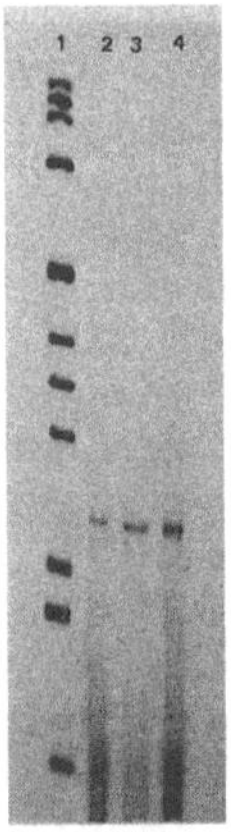

Fig. 2. Primer extension analysis of the in vivo transcriptional start of *B. stearothermophilus infC*. The experiment was carried out essentially as described (La Teana et al. 1989). Lane 1 - molecular weight ladder; Lanes 2, 3, 4 - product of primer extension using increasing amounts of total cellular RNA. The primer used in the extension reaction was a 58 nucleotides long fragment spanning from bases 302 to 361.

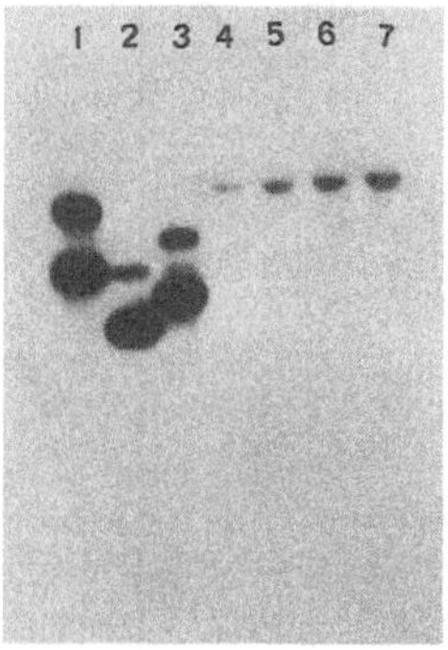

Fig. 3. Northern blot analysis. The restriction fragments ClaI-SacII (Lane 1), SacII-HindIII (Lane 2) and HindIII-ClaI (Lane 3) specific for IF3, L35 and L20, respectively, and increasing amounts of total RNA in Lanes 4, 5, 6 were subjected to electrophoresis, followed by blotting on Hybond-N (Amersham) and hybridization with a mixture of equal amounts of the three nick-translated probes.

scription and possibly of transcriptional regulation are operating in E. coli and B. stearothermophilus and that in the latter organism the differential expression (~1:6:6) of infC, rpmI and rplT must be obtained at the post-transcriptional level. Furthermore, since the main (if not the only) common element between the translational initiation regions of the IF3 mRNAs of the two bacteria is the AUU initiation triplet, the likelihood that a translational regulation mechanism might be linked to the presence of this unusual initiation codon (Gold et al. 1984) seems stronger. Accordingly, it was found that when this AUU is changed to AUG, in vivo expression of infC of both E. coli (Brombach and Pon 1987) and B. stearothermophilus (our unpublished observations) is increased about 40-fold and translational autorepression by IF3 is lost (Butler et al. 1987). To understand better the mechanism by which the initiation triplet AUU may affect the level of translation and IF3 autorepression, we constructed two model mRNAs differing only in their initiation triplet (i.e. AUG or AUU) and compared their activity in various partial reactions of translation initiation.

The properties of the first of these mRNAs (i.e. 002 mRNA-AUG) has been described in detail elsewhere (Calogero et al. 1988); starting from the gene encoding this mRNA, we constructed 002 mRNA-AUU. The second set of model mRNAs (i.e. 022 mRNA-AUG and 022 mRNA-AUU) will be described elsewhere. The coding sequence of the 022 mRNAs is identical to that of the 002 mRNAs but the sequence upstream from the initiation triplet is partly different, the most relevant differences being a shorter potential SD sequence (4 vs 9 bp) and a longer spacer (9 vs 5 nts).

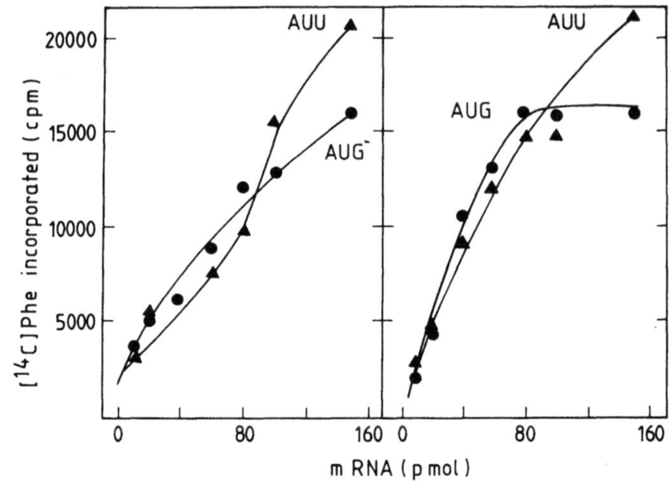

Fig. 4. Incorporation of phenylalanine in response to increasing amounts of 002 mRNA (left) and of 022 mRNA (right) with (●) AUG or (▲) AUU codons. The experimental conditions for the in vitro translation are essentially as described (Calogero et al. 1988).

Having established that the nature of the start codon does not influence the affinity of these mRNAs for the 30S subunit (not shown), we tested the in vitro template activity of these mRNAs and found that this also was hardly influenced by the nature of the initiation codon (Fig. 4). This result, which was unexpected in light of the above-mentioned 40-fold stimulation of IF3 production resulting from the AUU → AUG mutagenesis, prompted an investigation of the effect that IF3 might have on the translation of these mRNAs. Thus it was shown that IF3 at a ~1:1 stoichiometric ratio with ribosomes stimulates the template activity of the AUG-mRNA and, to a slightly lesser extent, of the AUU-mRNA (Fig. 5); when the amount of IF3 is increased above this stoichiometric ratio, however, amino acid incorporation remains fairly constant with the AUG-mRNA but drops to <25% with the AUU-mRNA when IF3 is present in a ten-fold excess over the ribosomes.

Since in the above experiments the translation of each mRNA is unchallenged by other mRNAs, in another set of experiments,

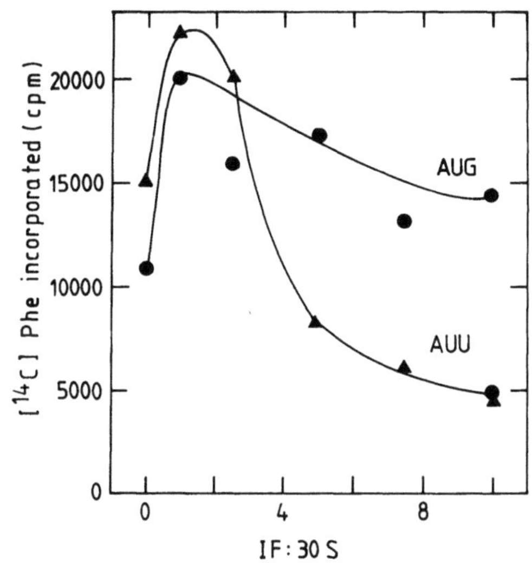

Fig. 5. Incorporation of phenylalanine in response to 022 mRNA with (●) AUG or (▲) AUU startcodon as a function of the indicated IF3:30S stoichiometric ratios in the presence of a 1:1:1 stoichiometric amount of IF1, IF2 and ribosomes. The other conditions were essentially as described (Calogero et al. 1988).

we determined the capacity of the AUG- or AUU-containing 022 mRNAs to compete with another mRNA for initiating ribosomes. Thus, the MS2 RNA-dependent incorporation of valine was measured in the presence of increasing amounts of 022 mRNAs. Since, unlike the MS2 RNA, the 022 mRNAs do not encode this amino acid, competition between mRNAs can be monitored by a decrease of the amount of valine incorporated. The results show that competition indeed takes place and that this occurs over comparable ranges of MS2 RNA and 022 mRNA concentrations indicating that these templates are accepted by the ribosomes with comparable efficiency. In the absence of IF3 as well as at an IF3:ribosome ratio of 1:1, both AUG- and AUU-containing 022 mRNAs compete with the MS2 RNA; the 022 mRNA-AUU is a slightly better competitor in the first (Fig. 6A) and 022 mRNA-AUG in the second case (Fig. 6B). When the IF3:ribosome ratio is increased to 6:1, however, the competition by 022 mRNA-AUU is nearly abolished while that by 022 mRNA-AUG remains completely unaffected (Fig. 6C). These results clearly demonstrate that an excessive amount of IF3 can selectively shut down the protein synthesis of an mRNA

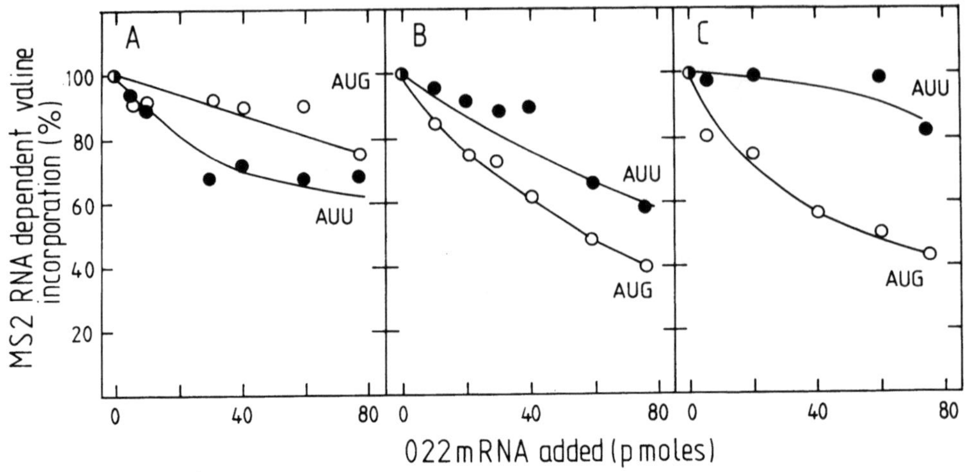

Fig. 6. Inhibition of MS2-dependent valine incorporation as a function of the addition of increasing amounts of 022 mRNA having (o) AUG or (•) AUU startcodons. The ratio between IF1, IF2, IF3 and ribosomes was as follows: A, 1:1:0:1; B, 1:1:1:1; C, 1:1:6:1.

having the AUU initiation triplet, especially in the presence of competing mRNAs.

The next question concerns the mechanism by which this IF3 autorepression operates. As seen in the following experiments, the 30S initiation complexes formed in the presence of IF1 and IF2 but without IF3 are fairly resistant to dissociation induced by dilution and almost equally stable regardless of whether the fMet-tRNA was bound in response to 022 mRNA-AUG or 022 mRNA-AUU (Fig. 7A). In the presence of IF3, however, both complexes dissociate more readily following dilution but the 30S initiation complex formed with 022 mRNA-AUU is much less stable than that obtained with 022 mRNA-AUG (Fig. 7B). Similar results were also obtained when the rate of exchange between pre-bound radioactive and free non-radioactive fMet-tRNA was measured (not shown). These results indicate that the discrimination by IF3 against initiation complexes containing AUU does not stem from an intrinsically lower stability of these complexes but is rather the result of a direct IF3-induced destabilization. This finding is in full agreement with the well established capacity of IF3 to induce a conformational change of the 30S subunit which select-

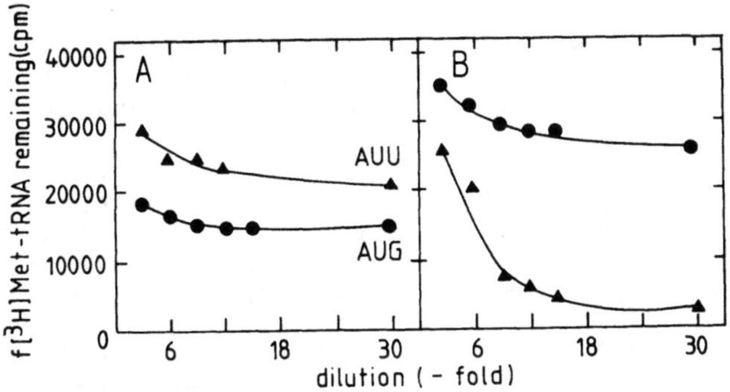

Fig. 7. Effect of IF3 on the stability of 30S initiation complexes formed with 022 mRNA with (●) AUG or (▲) AUU initiation codons. The initiation complexes were made in the presence of IF1 and IF2 (A) or IF1, IF2 and IF3 (B). Each factor was present in stoichiometric equivalence with the 30S subunits. The preformed complexes were diluted as indicated and incubated at 37°C for 15 min before filtering through nitrocellulose discs. The experimental procedure is essentially identical to that previously described (Pon and Gualerzi 1974; Risuelo et al. 1976).

ively affects the dissociation rate of ternary complexes consisting of 30S, aminoacyl-tRNAs and template (Pon and Gualerzi 1974; Risuleo et al. 1976).

To determine whether IF3 autorepression can also occur after the formation of the 70S initiation complex, experiments similar to those described above were performed starting with 70S initiation complexes. No effect of even large excesses of IF3 was detected in these experiments on either the dissociation of 70S complexes or on the rate of exchange between free and 70S-bound fMet-tRNA indicating that once formed the 70S initiation complexes are irreversibly committed to enter the elongation cycle.

In summary, our results offer a physical basis to explain how translational autoregulation by IF3 is lost following the AUU→AUG modification of the infC gene (Butler et al. 1987) by showing that the autoregulation occurs at the level of the 30S subunit and, more precisely, through the influence that IF3 exerts on the dissociation rate of the 30S initiation complexes. The present results also fit the more general model, schematically illustrated in Fig. 8, in which IF3 functions as a fidelity factor in translation initiation by inspecting the correctness of the codon-anticodon basepairing at the P-site and of the

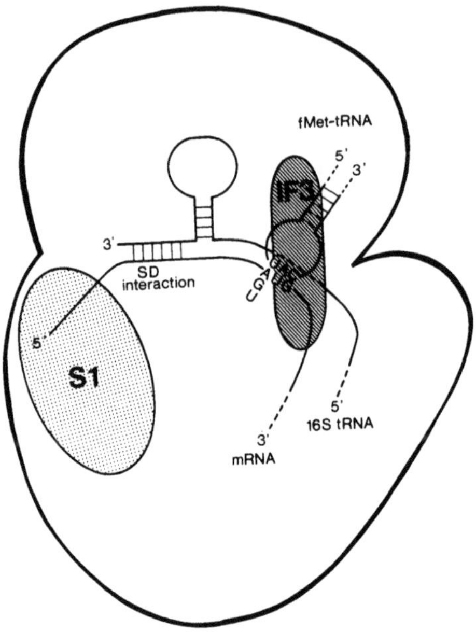

Fig. 8. mRNA-30S interaction at initiation and role of IF3 as initiation fidelity factor. This diagram highlights the three main mRNA-ribosome interactions (the SD - antiSD and the codon-anticodon basepairings and the mRNA - S1 interaction) which dictate translational efficiency. The figure also presents a visual summary of the results obtained in the laboratory of L. Gold concerning the structural elements of initiator tRNA and the portion of the codon-anticodon interaction at the P-site "inspected" by IF3 to ensure fidelity of translation initiation (from Gualerzi and Pon 1990).

stem-loop of the 30S-bound initiator tRNA (Risuleo et al. 1976; Hartz et al. 1989).

Acknowledgments. This work was supported in part by grants from the Italian Ministry of Public Education and National Research Council (CNR) to C.O.G.

References

Brombach M, Pon CL (1987) The unusual translational initiation codon AUU limits the expression of the *infC* (initiation factor IF3) gene of *Escherichia coli*. Mol Gen Genet 208:94-100

Butler JS, Springer M, Grunberg-Manago M (1987) AUU-to-AUG mutation in the initiator codon of the translation initiation factor IF3 abolishes translational autocontrol of its own gene (*infC*) in vivo. Proc Natl Acad Sci USA 84:4022-4025

Calogero RA, Pon CL, Canonaco MA, Gualerzi CO (1988) Selection of the mRNA translation initiation region by *Escherichia coli* ribosomes. Proc Natl Acad Sci USA 85:6427-6431

Comer MM (1981) Gene organization around the phenylalanyl-transfer ribonucleic acid synthetase locus in *Escherichia coli*. J Bacteriol 146:269-274

Dams E, Hendriks L, Van de Peer Y, Neefs J-M, Smits G, Vandenbempt I, De Wachter R (1988) Compilation of small ribosomal subunit RNA sequences. Nucleic Acids Res 16:r87-r173

Gold L, Stormo G, Saunders R (1984) *Escherichia coli* translation initiation factor IF3: a unique case of translational regu-

lation. Proc Natl Acad Sci USA 81:7061-7065
Gren EJ (1984) Recognition of messenger RNA during translational initiation in Escherichia coli. Biochimie 66:1-29
Gualerzi CO, Calogero RA, Canonaco MA, Brombach M, Pon CL (1988) Selection of mRNA by ribosomes during prokaryotic translational initiation. In: Tuite M, Picard M, Bolotin-Fukuhara M (eds) Genetics of Translation - New Approaches, Springer, Berlin, Heidelberg, pp 317-330
Gualerzi CO, Pon CL (1990) Initiation of mRNA translation in prokaryotes. Biochemistry, in press
Hartz, D., McPheeters, D.S. & Gold, L. (1989) Selection of the initiator tRNA by Escherichia coli initiation factors. Genes & Devel 3:1899-1912.
Hennecke H, Springer M, Böck A (1977) A specialized transducing lambda phage carrying the Escherichia coli genes for phenylalanyl-tRNA synthetase. Mol Gen Genet 152:205-210
Hershey JWB (1987) Protein synthesis. In: Neidhardt FC (ed) Escherichia coli and Salmonella typhimurium-cellular and molecular biology. ASM, Washington, D.C., pp. 613-647
Khudyakov YE, Neplyueva VS, Kalinina TI, Smirnov VD (1988) Effect of structure of the initiator codon on traslation in E. coli. FEBS Lett 232:369-371
La Teana A, Falconi M, Scarlato V, Lammi M, Pon CL (1989) Characterization of the structural genes for the DNA-binding protein H-NS in Enterobacteriaceae. FEBS Lett 244:34-38
Pon CL, Gualerzi C (1974) Effect of initiation factor 3 binding on the 30S ribosomal subunits. Proc Natl Acad Sci USA 71:4950-4954
Pon CL, Gualerzi C (1979) Qualitative and semiquantitative assay of Escherichia coli translational initiation factor IF3. Meth Enzymol 60:230-239
Pon CL, Brombach M, Thamm S, Gualerzi CO (1989) Cloning and characterization of a gene cluster from Bacillus stearothermophilus comprising infC, rpmI and rplT. Mol Gen Genet 218:355-357
Reddy P, Peterkofsky A, McKenney K (1985) Translational efficiency of the Escherichia coli adenylate cyclase gene: mutating the UUG initiation codon to GUG or AUG results in increased gene expression. Proc Natl Acad Sci USA 82:5656-5660
Risuleo G, Gualerzi C, Pon C (1976) Specificity and properties of the destabilization, induced by initiation factor IF3, of ternary complexes of the 30S ribosomal subunit, aminoacyl-tRNA and polynucleotides. Eur J Biochem 67:603-613
Sacerdot C, Fayat G, Dessen P, Springer M, Plumbridge JA, Grunberg-Manago M, Blanquet S (1982) Sequence of a 1.26-kb DNA fragment containing the structural gene for E. coli initiation factor IF3: presence of an AUU initiator codon. EMBO J 1:311-315
Springer M, Graffe M, Hennecke H (1977) Specialized transducing phage for the initiation factor 3 gene in Escherichia coli. Proc Nat Acad Sci USA 74:3970-3974
Wada A, Sako T (1987) Primary structures of and genes for new ribosomal proteins A and B in Escherichia coli. J Biochem 101:817-820
Wertheimer SJ, Klotsky R-A, Schwartz I (1988) Transcriptional patterns for the thrS-infC-rplT operon of Escherichia coli. Gene 63:309-320

ALTERNATIVE TRANSLATION AND FUNCTIONAL DIVERSITY OF RELEASE FACTOR 2 AND LYSYL-tRNA SYNTHETASE

Y. Nakamura, K. Kawakami, and O. Mikuni

Department of Tumor Biology
The Institute of Medical Science
The University of Tokyo
P.O. Takanawa, Tokyo 108, Japan

The *prfB* gene encodes peptide-chain-release factor 2 which catalyzes translation termination at nonsense codons UGA and UAA in bacteria. Mutations in *prfB* cause misreading of UGA, *i.e.*, increased frameshift or suppression. These translational errors were due to inefficient translation termination at UGA. As one extreme case, an opal (UGA) mutation in *prfB* is autogenously suppressed. The *prfB* gene is in the same operon with *herC*, a gene defined by a suppressor mutation of a replication-defective ColE1 plasmid. The *herC* gene coincides with *lysS*, a gene encoding a major species of lysyl-tRNA synthetase. Thus, the genetic organization of *prfB* and *lysS* in the same operon may suggest structural, functional and evolutional relevance of these gene products.

I. Release Factor 2 Gene

Polypeptide chain termination requires participation of two peptide-chain-release factors that recognize specific termination codons. Release factor 1 (RF1) catalyzes termination at UAG and UAA, and release factor 2 (RF2) catalyzes termination at UGA and UAA (Scolnick *et al.*, 1968). The RF1 gene has been cloned on the basis of competition between a nonsense suppressor tRNA and a translational release factor (Weiss *et al.*, 1984). The gene encoding RF1 has been named *prfA* and is located at 27 min on the *Escherichia coli* chromosome (Ryden *et al.*, 1986). The RF2 gene has been isolated from the Clarke-Carbon *E. coli* plasmid bank on the basis of RF2 overproduction detected by an anti-RF2 antibody (Caskey *et al.*, 1984). However, the chromosomal location of the RF2 gene had not been reported.

NATO ASI Series, Vol. H 49
Post-Transcriptional Control of Gene Expression
Edited by J. E. G. McCarthy and M. F. Tuite

A. E. coli RF2 Operon. The chromosomal location of the *prfB* gene was found by cloning and sequencing the *E. coli* DNA fragment carrying the *herC* gene mapped at the 62 min region (Kawakami *et al.*, 1988a; 1989). The *herC* gene has been defined by a suppressor mutation of a replication-deficient ColE1 plasmid. The deduced *herC* sequence encodes a 57,603-Da protein composed of 505 amino acids. It is overlapped with an open reading frame encoding a 54,626-Da protein (471 amino acids) in opposite orientation, whose gene product is not identified.

Nine bases upstream of the coding region of *herC*, we found another open reading frame which coincides with COOH-terminal part of the *prfB* gene (Kawakami, *et al.*, 1988a) except one guanine residue which had been lost in the published sequence (Craigen *et al.*, 1985). The complete *prfB* DNA was cloned in an intermediate-copy number plasmid pACYC184. This plasmid (pKK951) is able to complement *prfB* mutations (see below). However, we failed to reclone *prfB* into a high-copy number plasmid, such as pBR322 or pUC119. This is consistent with the previous observation that the pRF2 plasmid, an original isolate in the Clarke-Carbon colony bank (ColE1 replicon), exhibited a low activity of RF2 (Kawakami *et al.*, 1988a). We assume that a leakiness of termination at some UGA (and/or UAA) codons is required for cell growth for natural frameshift or suppression, and excess RF2 activity may be toxic.

The *prfB* and *herC* genes are cotranscribed in a 2,800-base transcript (Kawakami *et al.*, 1988a). The 5' end maps ~40 bp upstream of the initiator AUG codon of *prfB*. The location is consistent with the promoter site predicted by Craigen *et al.* (1985). The 3' end extends beyond the *herC* gene and maps ~40-50 bp downstream of the UAA stop codon of *herC*, where a potential ρ-independent termination signal is located. These results led us to conclude that *prfB* and *herC* genes are in a single operon.

B. *Salmonella typhimurium* RF2 Gene. *S. typhimurium prfB* DNA segments were cloned from λ phage libraries using the *E. coli prfB* probe. The entirety of the *prfB* gene was subcloned in pACYC184. The resulting *S. typhimurium prfB* plasmid (pSRF2) is able to complement both *E. coli* and *S. typhimurium* RF2 mutations. The deduced DNA sequence of *S. typhimurium prfB* was highly homologous to the *E. coli prfB* sequence; 95.6% identical in the amino acid sequence (349/365) and 87.0% identical in the nucleotide sequence (956/1099) (Kawakami & Nakamura, submitted). Both *prfB* genes contain an in-frame premature UGA codon at position 26, indicating that the mature *S. typhimurium* RF2 protein is synthesized by a +1 frameshift as has been shown in *E. coli prfB* (Craigen *et al.*, 1985). The *S. typhimurium prfB* gene is followed by a DNA sequence highly homologous to *E. coli herC*. Therefore, we inferred that the *prfB-herC* operon is conserved in *S. typhimurium*.

II. Release Factor 2 Mutations

Near 27 min on the *E. coli* genetic map, *uar* and *sueB* mutations which affect RF1 have been located (Ryden *et al.*, 1986). These mutants misread UAA and/or UAG and increase the efficiency of suppression of these nonsense codons. On the other hand, no RF2 mutations had been isolated.

A. <u>*E. coli* RF2 Mutations and Misreading UGA</u>. Two rationales of selection were used to isolate RF2 mutations using localized mutagenesis of the *prfB* region. One was to directly select a suppressor mutation of a *lacZ* UGA allele (*lacZ659*), some of which may have a defect in the peptide chain releasing activity of RF2 at UGA. The other was to isolate a temperature-sensitive lethal mutant because RF2 function may be essential to *E. coli* growth. The first selection scheme yielded three RF2 mutations, *prfB1*, *prfB2* and *prfB3*, while the second selection scheme yielded one mutation, *prfB286* (Kawakami *et al.*, 1988b). Of the former mutants, the *prfB2* was also temperature-sensitive in growth. Likewise, the latter mutant *prfB286* was able to suppress UGA. The β-galactosidase levels in the cells harboring *lacZ659*(UGA) and *prfB* mutations were ~7% of the wild-type *lacZ* level and reduced to the background level in the presence of the plasmid pKK951 (*prfB*+). These results indicated that these RF2 mutations are recessive UGA suppressors and the RF2 function is essential to *E. coli* growth.

These RF2 mutations did not suppress UAG and UAA when tested by using *lacI-lacZ* fusion plasmids containing nonsense mutations in *lacI* (Mikuni, unpub.). Although UAA is recognized both by RF1 and RF2 *in vitro*, it may be preferentially recognized *in vivo* by RF1 because the RF1 mutant exhibits a UAA-suppressor activity while the RF2 mutant does not. The efficiency of UGA suppression in *prfB* mutants varied depending on the allele of the UGA mutation; for instance, UGA positioned at 280 in *lacI* was not suppressed. This indicates that UGA suppression is affected by the flanking RNA context (*i.e.*, tight RNA context; Mikuni, unpub.).

Another effect of the RF2 mutation on gene expression is on the RF2 gene itself. Translation of the *prfB* mRNA is autogenously controlled by a +1 frameshift at the premature UGA codon at position 26 (Craigen *et al.*, 1985). Taking advantage of the *prfB* mutant, we observed that a reduced level of RF2 activity leads to an increase in the +1 frameshift (Mikuni, unpub.). These results taken together with similar observations with the *S. typhimurium supK* RF2 mutant described below may yield the first genetic evidence for the autogenous control of RF2 synthesis.

It is not known how RF2 mutations generate UGA suppression. The only experimental relevance was shown in the basal level control of the tryptophan operon (Roesser *et al.*, 1989). The *prfB1* and *prfB3* mutations increased transcription termination two-fold at the *trp* operon attenuator. This was interpreted as indicating that a reduced level of RF2 activity leads to a stall of ribosome at an authentic UGA stop codon of the *trp* leader peptide, giving rise to an increase in formation of a terminating structure of the leader RNA. If the leader peptide contains an engineered UAG or UAA codon instead of UGA, the frequency of attenuation was not affected by *prfB* mutations. We assume that ribosomal pausing may be a general basis for UGA suppression and other translational errors in the mutant.

Mutational base changes in the *E. coli prfB* mutants were defined by cloning and sequencing the mutant RF2 genes or by direct sequencing of amplified DNAs by polymerase chain reaction. Two temperature-sensitive mutations, *prfB2* and *prfB286*, which have been isolated independently, coincidently caused identical amino acid substitutions at position 328. The *prfB1* and *prfB3* mutations caused a single amino acid change at position 89 and 143, respectively, both of which are to reduce acidic charge. These sequence studies aimed to investigate the structural and functional organization of RF2 will be published elsewhere (Mikuni, unpub.).

B. *S. typhimurium supK* Mutations. In *S. typhimurium*, a recessive UGA suppressor mutation, *supK*, had been isolated and mapped at the 62 min region of the chromosome (Reeves & Roth, 1971). The same authors have observed reduced levels of tRNA methyltransferase activities in several *supK* strains and have suggested that an unmodified tRNA causes UGA suppression (Reeves & Roth, 1975). However, we assumed, on the basis of high homology between the *E. coli* and *S. typhimurium* genomes, that the *supK* mutation may affect the RF2 protein (Kawakami *et al.*, 1988a). Two lines of genetic evidence have been consistent with this assumption. First, the *supK* mutation of *S. typhimurium* is complemented by the *E. coli prfB* gene (Kawakami *et al.*, 1988a). Second, RF2 mutants of *E. coli* harbor a recessive UGA suppressor activity (Kawakami *et al.*, 1988b).

The above assumption was directly proven by cloning and sequencing the *S. typhimurium* RF2 gene from the *supK584* mutant. The nucleotide sequence of the mutant *prfB* gene contained a single base substitution of A for G at position 433. No other alteration was found within the coding sequence of *prfB*. This substitution generates a UGA stop codon for a UGG tryptophan codon at amino acid position 144. The pSRF2 plasmid encoding the wild-type RF2 protein eliminated the UGA-suppression activity of both *E. coli prfB* mutants and the *S. typhimurium supK584* mutant. On the other hand, the pSUPK plasmid carrying the above UGG-to-UGA change failed to complement these mutations. Of 10 temperature-resistant colonies isolated from the *supK584* strain, which is temperature-sensitive in growth, four revertants changed the UGA codon to UGG (*i.e.*, true reversion). These results led us to conclude that the *supK* gene encodes *S. typhimurium* RF2 and that the opal *supK584* substitution in the RF2 gene is solely responsible for the UGA suppression and the temperature-sensitivity (Kawakami & Nakamura, submitted).

C. Autogenous Suppression of UGA. It seems extraordinary that an opal mutation in the *prfB* gene generates an opal suppressor activity. The efficiency of termination at this mutational UGA codon was tested by using the *prfB-lacZ* fusion plasmid carrying the *supK584* opal substitution. The synthesis of the opal fusion protein, which is less than 4% of the wild-type fusion protein in the *supK*$^+$ strain, increased three-fold in the *supK584* strain (Kawakami & Nakamura, submitted). The +1 frameshift at the 26th UGA codon was also increased in *supK584*. These results can be interpreted as indicating that suppression is caused by a reduced cellular level of the mature RF2 protein in the opal mutant, presumably due to inefficient termination at UGA. Thus, the opal RF2 mutation is autogenously suppressed under these conditions, generating another feedback regulatory loop for the synthesis of RF2 in the mutant.

III. *herC* Mutation and Lysyl-tRNA Synthetase

E. coli has two forms of lysyl-tRNA synthetases (LysRS). A major form of LysRS is encoded by a gene named *lysS* (Emmerich & Hirshfield, 1987) and a minor form is encoded by *lysU* (Van Bogelen *et al.*, 1983). These genes are located on the distinct chromosomal loci; *lysU* at min 92 and *lysS* at min 62. However, the structural genes had not been isolated. In this article, we describe that the *herC* gene which is in the same operon with *prfB* encodes LysRS (*lysS*).

A. herC Mutation. The *herC180* mutation has been defined in a host mutant which restored maintenance of a replication-defective ColE1 plasmid carrying a primer RNA mutation named *cer114* (Kawakami *et al.*, 1989). Two modes of ColE1 DNA replication are known, one dependent on RNase H, and the other independent on RNase H (Dasgupta, 1987). The *cer114* mutant replicon is defective in both modes of replication and carries a single base pair alteration 95 bp upstream of the replication origin. It substitutes a G for an A at the 3'-terminus position of the AAA triplet in the structure IX loop of primer RNA. One of the revertants of *cer114* inserted an A to this region and regenerated an AAA triplet, thereby suggesting a crucial role of this AAA triplet in ColE1 DNA replication. An *E. coli* mutant which restored maintenance of the *cer114* replicon carried double mutations, one in the RNase H gene and the other in the *herC* gene. Complementation and reconstruction experiments revealed that the *herC180* mutation is recessive to its wild-type allele and supports maintenance of the mutant replicon in the absence of RNase H. The wild-type ColE1 replicon is also maintained in the double mutant. These data suggested that the wild-type *herC* protein (*i.e.*, LysRS as shown below) is not a factor essential for ColE1 DNA replication but a factor which eliminates *cer114* replication or prevents stable maintenance of the *cer114* replicon. The *herC180* mutation alone conferred cold-sensitivity in growth, suggesting that the *herC* function is essential to *E. coli* growth (Kawakami *et al.*, 1989). The DNA sequence analysis disclosed a single base substitution in *herC180* which causes a glycine-to-aspartate change at amino acid position 426 (Mikuni, unpub.).

B. Lysyl-tRNA Synthetase. We found that the *herC* protein shares a significant homology with *S. cerevisiae* cytoplasmic aspartyl-tRNA synthetase. Among several aminoacyl-tRNA synthetase activities tested, the LysRS activity in crude cell lysates increased three-fold in the presence of pKK945 carrying *herC* and further increased in the presence of pKK990 where *herC* is fused to the overexpressing *tac* promoter (Kawakami, unpub.). The *herC* protein labeled with [^{35}S]methionine in the *in vitro* coupled transcription-translation system exactly comigrated with the purified LysRS in the O'Farrell 2D gel electrophoresis.

The overexpressed *herC* protein was purified from the pKK990-bearing cells (*lon$^{\Delta}$100, htpR16*) after induction by IPTG. The fractions containing the *herC* protein were monitored by staining proteins after SDS-PAGE electrophoresis and compared with those of the LysRS activity. Throughout the purification procedures which employed ammonium sulfate fractionation and several column chromatographies, the oversynthesized *herC* protein was exactly copurified with the LysRS activity (Nakamura, unpub.). These results led us to conclude that *herC* encodes LysRS.

IV. Summary and Perspectives

A. Comparative Aspects of Aminoacyl-tRNA Synthetases and Release Factors. We have proven that the RF2 and LysRS proteins are encoded in the same operon in *E. coli* and *S. typhimurium*. In addition to this chromosomal organization, LysRS and RF2 have several similarities in genetic and biochemical properties. First, *prfB* and *lysS* have structural and functional homologs, *prfA* and *lysU*, respectively. Second, they are essential factors in translation and both recognize RNA triplets, *i.e.*, codon UGA for RF2 and anticodon UUU or CUU for LysRS. Finally, LysRS and RF2 catalyze reverse reactions, formation of aminoacyl-tRNAs or hydrolysis of peptidyl-tRNAs. These genetic and functional relevance might suggest that they have evolved from a related family of proteins. From the point of view of UGA recognition and its evolution, it would be interesting to investigate the structural and functional organization of release factors and tryptophanyl-tRNA synthetase of *Mycoplsma capricolum*, in which UGA is used as a sense codon for tryptophan (Yamao *et al.*, 1985).

B. UGA Codon Recognition and Release Factor 2. The UGA codon is a leaky codon and is naturally misread at higher frequency than UAG and UAA. This leakiness is due to misdecoding by tryptophanyl-tRNA and also to limiting intracellular concentration of RF2. In fact, the leakiness is eliminated upon introduction of the plasmid pKK951 which gives rise to an increase in the *prfB* gene dosage (Kawakami *et al.*, 1988b).

The UGA codon plays significant variability in translation. It has been known that several host mutations generate UGA misreading. In addition to this mutational variability, living organisms possess programmed alternatives in reading the UGA codon. In several instances, it is recognized as a signal for frameshift and suppression, or decoded as selenocysteine (reviewed by Parker, 1989). As a common feature, these programmed alternatives have a trick to stall ribosome before UGA and to eliminate proper functioning of the release factor at the specific UGA codon. In this article, we described that a reduced level of the RF2 activity or its cellular amount leads to an increase in translational errors such as suppression and frameshift. In addition, an increase in the *prfB* gene dosage reduces the efficiency of decoding of UGA to selenocysteine in the *E. coli fdhF* gene which encodes formate dehydrogenase (Nakamura & Böck, unpub.). All these results demonstrate that RF2 is directly involved in these alternative translation at UGA.

It is coming obvious that the RNA context plays a crucial role in ribosomal stalling. In the natural +1 frameshift in *prfB*, the in-frame UGA stop codon is preceded by leucine tRNA codon CUU (a *shifty* codon) and a Shine-Dalgarno like sequence AGGGGG three bases upstream of CUU. Weiss *et al.* (1988) have demonstrated that base-pairing between the 3' end of 16S rRNA and this upstream element is required for frameshift, presumably causing ribosomal stalling. Spontaneous high-frequency frameshift has been widely observed in eukaryotic organisms, such as retroviruses, coronaviruses and some yeast transposons (reviewed by Parker, 1989). These systems employ proper RNA contexts which produce RNA stem-loop and RNA pseudoknot structures, or encode minor codons, which cause ribosomal pausing. Development of the *in vitro* system to assay ribosomal stalling in dynamic reactions will certainly contribute to solve many problems in stop codon recognition.

C. Biological Relevance of Lysyl-tRNA Synthetase. LysRS is an exceptional case that it has two distinct forms of the synthetases encoded by *lysS* and *lysU*. The *lysS* gene is constitutively expressed, while *lysU* is heat-inducible though it remains almost silent under normal growth conditions (Hirshfield *et al.*, 1981). Independent of our studies, Gampel and Tzagoloff (1989) have also observed a similarity between yeast LysRS and *E. coli herC* protein. Recently Lévêque *et al.* (1990) have cloned the *lysS* gene by using oligonucleotide probes directed from the NH_2-terminal peptide sequence and have found the identity of the *lysS* sequence to *herC*. These results, taken together with ours, have firmly established that *herC* encodes LysRS. The latter authors have also cloned the *lysU* gene and found 88% amino-acid sequence homology between *lysS* and *lysU*. Biological significance of two distinct LysRS species remains to be investigated.

A stretch of adenine, which includes the lysine codon, is recognized as a site for -1 frameshift in retroviruses, coronaviruses, *E. coli dnaX* and the transposase gene of insertion element IS*1* (reviewed by Parker, 1989; Tsuchihashi & Kornberg, 1990; Sekine & Ohtsubo, 1989). Most of these AAA lysine codons are followed, at some distance, by the stop codon UGA. Therefore, it is plausible that LysRS and the release factor may participate in regulation of these frameshifting.

It is not yet clarified if the *herC* mutation directly restores ColE1 DNA replication of the *cer114* mutant replicon or some other steps involved in the plasmid maintenance under rnh^- conditions. However, the current studies suggested some relevance of LysRS to the mutant primer RNA. It is intriguing that the *cer114* mutation site is in the AAA triplet, which is a codon for lysine; thereby, one might speculate putative interaction between the anticodon loop of lysyl-tRNA and the primer RNA loop. Another possibility is that the suppression is mediated by 5',5'''-diadenosine tetraphosphate, possibly an important regulatory metabolite in cells synthesized by LysRS (Zamecnik, 1983). Diverse roles of tRNAs or aminoacyl-tRNA synthetases in global control of cell growth and gene expression are interesting topics for future work.

References

Caskey CT, Forrester WC, Tate W, Ward, CD (1984) Cloning of the *Escherichia coli* release factor 2 gene. J Bacteriol 158:365-368

Craigen WJ, Cook RG, Tate WP, Caskey CT (1985) Bacterial peptide chain release factors: conserved primary structure and possible frameshift regulation of release factor 2. Proc Natl Acad Sci USA 82:3616-3620

Dasgupta S, Masukata H, Tomizawa J (1987) Multiple mechanisms for initiation of ColE1 DNA replication: DNA synthesis in the presence and absence of ribonuclease H. Cell 51:1113-1122

Emmerich RV, Hirshfield IN (1987) Mapping of the constitutive lysyl-tRNA synthetase gene of *Escherichia coli* K-12. J Bacteriol 169:5311-5313

Gampel A, Tzagoloff A (1989) Homology of aspartyl- and lysyl-tRNA synthetases. Proc Natl Acad Sci USA 86:6023-6027

Garcia GM, Mar PK, Mullin DA, Walker JR, Prather NE (1986) The *E. coli dnaY* gene encodes an arginine transfer RNA. Cell 45:453-459

Hirshfield IN, Bloch PL, Van Bogelen RA, Neidhardt FC (1981) Multiple forms of lysyl-transfer ribonucleic acid synthetase in *Escherichia coli*. J Bacteriol 146:345-351

Kawakami K, Jönsson YH, Björk GR, Ikeda H, Nakamura Y (1988a) Chromosomal location and structure of the operon encoding peptide-chain-release factor 2 of *Escherichia coli*. Proc Natl Acad Sci USA 85:5620-5624

Kawakami K, Inada T, Nakamura Y (1988b) Conditionally lethal and recessive UGA-suppressor mutations in the *prfB* gene encoding peptide chain release factor 2 of *Escherichia coli*. J Bacteriol 170:5378-5381

Kawakami K, Naito S, Inoue N, Nakamura Y, Ikeda H, Uchida H (1989) Isolation and characterization of *herC*, a mutation of *Escherichia coli* affecting maintenance of ColE1. Mol. Gen. Genet. 219:333-340

Lévêque F, Plateau P, Dessen P, Blanquet S (1990) Homology of *lysS* and *lysU*, the two *Escherichia coli* genes encoding distinct lysyl-tRNA synthetase species. Nucl Acids Res 18:305-312

Parker J (1989) Errors and alternatives in reading the universal genetic code. Microbiol Rev 53:273-298

Reeves RH, Roth JR (1971) A recessive UGA suppressor. J Mol Biol 56:523-533

Reeves RH, Roth JR (1975) Transfer ribonucleic acid methylase deficiency found in UGA suppressor strains. J Bacteriol 124:332-340

Roesser JR, Nakamura Y, Yanofsky C (1989) Regulation of basal level expression of the tryptophan operon of *Escherichia coli*. J Biol Chem 264:12284-12288

Ryden M, Murphy J, Martin R, Isaksson L, Gallant J (1986) Mapping and complementation studies of the gene for release factor 1. J Bacteriol 168:1066-1069

Scolnick E, Tompkins R, Caskey T, Nirenberg M (1968) Release factors differing in specificity for terminator codons. Proc Natl Acad Sci USA 61:768-774

Sekine Y, Ohtsubo E (1989) Frameshifting is required for production of the transposase encoded by insertion sequence *1*. Proc Natl Acad Sci USA 86:4609-4613

Tsuchihashi Z, Kornberg A (1990) Translational frameshifting generates the γ subunit of DNA polymerase III holoenzyme. Proc Natl Acad Sci USA 87:2516-2520

Van Bogelen RA, Vaughn V, Neidhardt FC (1983) Gene for heat-inducible lysyl-tRNA synthetase (*lysU*) maps near *cadA* in *Escherichia coli*. J Bacteriol 153:1066-1068

Weiss RB, Dunn DM, Dahlberg AE, Atkins JF, Gesteland RF (1988) Reading frame switch caused by base-pair formation between the 3' end of 16S rRNA and the mRNA during elongation of protein synthesis in *Escherichia coli*. EMBO J 7:1503-1507

Weiss RB, Murphy JP, Gallant JA (1984) Genetic screen for cloned release factor genes. J Bacteriol 158:362-364

Yamao F, Muto A, Kawauchi Y, Iwami M, Iwagami Y, Azumi Y, Osawa S (1985) UGA is read as tryptophan in *Mycoplasma capricolum*. Proc Natl Acad Sci USA 82:2306-2309

Zamecnik PC (1983) Diadenosine-5',5'''-p^1,p^4-tetraphosphate (Ap4A): its role in cellular metabolism. Anal Biochem 134:1-10

Gene Products that Mediate Translation Initiation in Yeast

Heejeong Yoon[1], A. Mark Cigan[2] and Thomas F. Donahue[1]
[1]Department of Biology, Indiana University, Bloomington, IN 47405 USA
[2] Laboratory of Molecular Genetics, NICHD, National Institutes of Health, Bethesda, MD 20892 USA

INTRODUCTION

The general mechanism proposed for higher eukaryotic translation initiation is illustrated by a scanning model (Kozak 1980) in which the preinitiation complex recognizes the 5' end of capped mRNA, migrates along the 5' leader region until the first AUG initiator codon is encountered, where translation begins. Some of the features associated with the yeast initiator regions from over 130 genes were analyzed (Cigan and Donahue 1987). Yeast leader regions are rich in adenine nucleotides, have an average length of 52 nucleotides, and are void of stable stem-loop structures. The sequence context flanking the start codon of yeast mRNAs (5'-AAAAAUGUCU-3', with Y and U printed below the first and third A) is different from the higher eukaryotic consensus (5'-CACCAUGG-3') with the exception of a preference for an A nucleotide at the -3 position relative to the AUG start codon. 95% of yeast mRNAs utilize the AUG codon nearest the 5' end of message as the start site for translation. Converse genetic experiments at the *CYC1* gene (Sherman and Stewart 1982) and the *HIS4* gene (Donahue and Cigan 1988) in yeast establish the preference for the first AUG codon relative to the 5' end of the message to serve as the start site of translation. The "first AUG" rule observed at *CYC1* and *HIS4* (Cigan *et al.* 1988b), in conjunction with the basic features of the leader region of yeast

NATO ASI Series, Vol. H 49
Post-Transcriptional Control of Gene Expression
Edited by J. E. G. McCarthy and M. F. Tuite

mRNAs, is consistent with a scanning mechanism being utilized for translation initiation in yeast.

Alterations in the the leader length by deletion or duplication of sequences in the *HIS4* gene do not seem to affect the efficiency of translation initiation (Cigan *et al.* 1988b). However, insertion of stem-loop structure in the 5' untranslated region of *HIS4* mRNA upstream of the AUG initiator codon decreases the efficiency of *HIS4* expression (Cigan *et al.* 1988b), similar to that effects observed for these types of mutations in leader regions of higher eukaryotic genes (Kozak 1986, Pelletier and Sonenberg 1985). Secondary structure may prevent ribosomal binding to the 5' end of the *HIS4* mRNA or the migration of the ribosome toward the first AUG initiator codon during scanning process.

Although the scanning model is based on sound experimental evidence, it remains unclear which components of the preinitiation complex are involved in establishing the start site of translation. In prokaryotes a complementary base pair interaction between a ribosomal binding site (Shine-Dalgarno sequence) and 16S ribosomal RNA is important in the 30S ribosomal subunit/mRNA recognition process (Shine and Dalgarno 1974). The lack of such ribosomal binding sites in eukaryotic mRNAs and the use of the first AUG codon as the start site of translation has suggested that a base pair interaction between the anticodon of $tRNA^{met}_i$ and the initiator codon on the mRNA may be an important component of the scanning ribosome. In this model, the first AUG codon encountered by the 40S ribosomal subunit in association with $tRNA^{met}_i$ and initiation factors, serves as the signal for the initiation complex to start translation.

$tRNA^{met}_i$ Functions in Start Site Selection of the Scanning Ribosome

To address the role of $tRNA^{met}_i$ during translation initiation, three of the four $tRNA^{met}_i$ genes in yeast were isolated (Cigan and Donahue 1986), and all nine possible point mutations of the AUG start codon at the *HIS4* locus were constructed (Donahue *et al.* 1988a). A mutated anticodon in one of these tRNA genes was then tested for its ability to mediate ribosomal recognition of the mutated *HIS4* initiator codon *in vivo* (Cigan *et al.* 1988a). We specifically mutated the $tRNA^{met}_i$ anticodon from UAC to UCC. This mutated UCC-$tRNA^{met}_i$ would be the best candidate among mutant tRNAs to be aminoacylated *in vivo*, and also the complementary AGG codon did not appear in the 5' leader region of the *HIS4* message, which if present would preclude initiation at a downstream AGG located at the normal initiator region. The complementary base pair interaction between UCC-$tRNA^{met}_i$ and the mutated AGG initiator codon was monitored on SD-histidine plates. This UCC-$tRNA^{met}_i$ present in a single copy yeast vector YCp50 could not restore *HIS4* expression, unless a selected mutation in the *MES1* (methionyl-tRNA synthetase) gene was also present to afford "charging" of the mutated tRNA. The His+ phenotype associated with the revertant strain suggested that the mutant $tRNA^{met}_i$ participated in ribosomal recognition and initiation at an AGG codon. When UCC-$tRNA^{met}_i$ was present in yeast strains containing non-complementary mutations in the initiator codon, it was incapable of restoring *HIS4* expression. This demonstrated the specificity of the UCC-$tRNA^{met}_i$ for the cognate AGG codon and also showed that a codon other than AUG could serve as the start site of translation, provided a cognate anticodon was contained in an initiator tRNA. The ability of this UCC-$tRNA^{met}_i$ to mediate ribosomal recognition of the AGG codon at *HIS4* was in accordance with the scanning model. This was tested with a *HIS4* allele containing an AGG codon in the *HIS4* leader region upstream and out of frame with the AGG codon at the initiator site. When this construct was present at *HIS4* in the UCC-$tRNA^{met}_i$ yeast strain, no *HIS4* expression was observed (His^-). The simplest explanation of this experiment was that UCC-$tRNA^{met}_i$ caused the ribosome to recognize and initiate at the first AGG codon which was out of frame with the *HIS4* coding region. Therefore, recognition of the first AGG codon precluded the ability of the ribosome to initiate at the AGG located at the downstream initiator region in compliance with "first AUG" codon rules.

Genetic Characterization of Translational Initiation Suppressors

Considering the complexity of eukaryotic translation initiation, it is conceivable that, in addition to $tRNA^{met}_i$, other components of the translation initiation complex may participate in ribosomal recognition of an AUG codon. We initiated a genetic reversion analysis at the *HIS4* locus in yeast in an attempt to identify other components of the initiation complex that may be important for ribosomal recognition of an initiator codon. Two copies of initiator codon mutations were introduced into a haploid yeast strain, one copy at *HIS4* (His^-) and a second copy as part of an in-frame *HIS4-lacZ* fusion (white colony on X-gal plate) which was integrated at the *URA3* locus (Donahue *et al.* 1988). Spontaneous revertants were selected on SD-histidine plates and screened on X-gal plates. External suppressor mutations (His+, blue) encoded *trans*-acting factors involved in ribosomal recognition of a start codon (Castilho-Valavicius *et al.* 1990). Three unlinked suppressor loci, restoring both *HIS4* and *HIS4-lacZ* expression in the absence of the AUG initiator codon, *sui1, sui2* and *SUI3* (suppressors of initiator codon mutations) were identified by this selection method. Each suppressor mutation was unlinked to *HIS4* and segregated as a single nuclear mutation. *sui1* and *sui2* were recessive suppressors and conferred temperature sensitivity for growth at 37°C (Ts-), while *SUI3*, a dominant mutation, was Ts+. The temperature-sensitive phenotype and the ability to restore *HIS4* expression associated with either *sui1* or *sui2* cosegregated in crosses. The *sui1* suppressor locus showed tight linkage to *prt2* on chromosome XIV and was composed of both non-complementing and weakly complementing mutants, indicative of a complex genetic locus. The *sui2* suppressor locus exhibited tight linkage to *cdc35*, this in turn being linked to the centromere of chromosome X. *sui1, sui2* and *SUI3* could all suppress the 9 different point mutations of the initiator codon at *HIS4* or *HIS4-lacZ*, a 2 bp change (ACC) and a 3 bp deletion of the AUG, indicating that the site of suppression resided outside the AUG initiator codon. Suppression was specific for initiator codon mutations, and none of the suppressors could suppress known frameshift, missense, or nonsense mutations. The efficiency of suppression determined by assaying ß-galactosidase activities showed that *sui1, sui2* and *SUI3* suppressor mutations resulted in 14-, 11-, 47-fold increases relative to *HIS4-lacZ* parent strains containing AUU at the initiator site. Similar studies with *sui1, sui2* and *SUI3* strains with an AUG initiator codon in the *HIS4-lacZ* fusion showed that *sui2* and *SUI3* suppressor mutations caused 3- and 4-fold increases, respectively, relative to the wild type control. These increases were presumably caused by perturbation of the general amino acid control mechanism in yeast which causes maximal induction of *HIS4* transcription (Castilho-Valavicius *et al.* 1990, Williams *et al.* 1989).

***SUI3* Suppressor Encodes the ß Subunit of Yeast eIF-2**

The *SUI3* gene was isolated as a 1.8 kb *Hind*III fragment and the complete DNA sequence showed that it had a 285 amino acid open reading frame (Donahue *et al.* 1988). This region was transcribed and the 5' ends mapped upstream of ATG that defined the open reading frame. An in-frame fusion of the proximal DNA fragment to the *lacZ* coding region on a YCp50 vector resulted in significant ß-galactosidase expression in yeast. Gene disruption of one copy of *sui3* in a wild type diploid strain resulted in 2 viable and 2 inviable ascospores upon sporulation, indicating that *sui3* encoded an essential gene product. Therefore, *sui3* was transcribed and translated in yeast and essential for cell viability. The sequence of *sui3* showed 42% amino acid homology to the ß subunit of human eIF-2. Mutations that conferred suppression in *SUI3* were identified by DNA sequencing at a putative zinc finger motif located in the carboxyl end of the protein; a motif, that is also present at a similar location in the human eIF-2ß protein as deduced from a cDNA sequence. This suggested that a nucleic acid binding domain of eIF-2 might play an important role in ribosomal recognition of a start codon.

To address the relationship of *SUI3* to yeast eIF-2ß, eIF-2 was partially purified from

SUI3 wild type and mutant strains and analyzed by Western blots using *SUI3* antibody and assayed for eIF-2 activity (GTP-dependent binding of $tRNA^{met}_i$). *SUI3* protein detected by Western blots had a M_r value of 36 kd, similar to the M_r value calculated from the *SUI3* DNA sequence, and copurified with eIF-2 activity. eIF-2 prepared from mutant *SUI3* strains showed very little $tRNA^{met}_i$ binding activity in contrast to wild type eIF-2. Based on the sequence similarities between *SUI3* and human eIF-2ß and biochemical analysis of the *SUI3* protein, we concluded that the *SUI3* gene encoded the ß subunit of yeast eIF-2.

The ability to identify a component of the initiation complex through our genetic reversion analysis suggested that *SUI3* suppressor mutations alter the start site selection process. To identify the site of translation initiation, we purified the His4-lacZ protein produced from *SUI3* suppressor strains and sequenced the amino terminus of the protein (Donahue *et al.* 1988). N-terminal sequence analysis of His4-ß-galactosidase protein showed that *SUI3* allowed initiation to occur at a UUG codon at amino acid position +3 in the absence of an AUG codon. Therefore, mutations in *SUI3* no longer allowed ribosomes to bypass the mutated *HIS4* initiator region but now recognize and initiate at a UUG codon, presumably by stabilizing a mismatched interaction between the anticodon of initiator $tRNA^{met}_i$ and the UUG codon. Thus by genetic reversion studies of initiator codon mutations at *HIS4*, we were able to demonstrate the importance of eIF-2 in ribosomal recognition during the scanning process.

sui2 Suppressor Encodes the α Subunit of Yeast eIF-2

The DNA sequence of the cloned wild type *SUI2* suppressor identified a 304 amino acid open reading frame with a calculated M_r value of 35 kd which was 58% identical with human eIF-2α (Cigan *et al.* 1989). This region was shown to be transcribed and translated in yeast and encoded an essential gene product. Western blot analysis of crude extracts demonstrated that anti-*sui2* antibodies cross-reacted with a protein which was identified as a phosphoprotein and copurified with eIF-2 activity. eIF-2 prepared from *sui2* suppressor strains showed defects in tRNA binding activity. A polysome preparation from a ts *sui2* mutant showed a characteristic profile for a defect in the initiation step of translation; accumulation of 80S ribosomes and loss of polysomes (Pain 1986) at the restrictive temperature.

To determine whether mutations in the α subunit of eIF-2 restored *HIS4* expression by a similar mechanism as *SUI3*, His4-ß-galactosidase protein produced from a *sui2* suppressor strain was purified and the N-terminal amino acid sequence was determined. The protein sequence started at the third codon (UUG) of the *HIS4* coding region as observed by *SUI3* suppression. Thus genetic reversion analysis applied to initiator codon mutations at *HIS4* has proven to be effective in identifying essential components of the translation initiation complex that interact with mRNA or function in mediating start site selection during the scanning process.

Characterization of *sui1* Suppressor Locus

The wild type *SUI1* gene was cloned on an 8.0 kb DNA fragment by transforming ts *sui1* suppressor strains with genomic clone banks (YCp50, YEp24). The *SUI1* gene was localized to a 1.2 kb *Bgl*II-*Hind*III DNA fragment. The DNA sequence of the wild type *SUI1* gene identified a 108 amino acid open reading frame. The proximal *Bgl*II-*Bam*HI *sui1* DNA fragment was fused in-frame with the *lacZ* coding region and subsequently used to transform a wild type yeast strain. The transformants exhibited significant levels of ß-galactosidase activities (258 units), indicating *sui1* was transcribed and translated in yeast. Immunoprecipitation of crude extracts prepared from this strain, with antibodies raised against ß-galactosidase, identified a protein with a M_r value predicted for the fusion protein. Gene disruption of one copy of *SUI1* in a wild type diploid strain resulted in 2 viable and 2 inviable ascospores, indicating that *SUI1* encoded an essential gene product. The mutant *sui1* alleles were isolated by the gap-duplex method and their DNA sequences showed that mutations changed amino acids within the *sui1* coding region.

The *sui1* gene showed tight linkage to the *prt2*, a mutation that caused a conditional defect in protein synthesis. However, *SUI1* and *PRT2* appeared to be different genes since the two mutants could complement and the cloned wild type *SUI1* gene failed to complement the ts phenotype of the *prt2* mutation. Like the *sui2* and *SUI3* suppressors, the *sui1* suppressor allowed translation initiation to occur at a UUG codon in the absence of the AUG initiator codon at *HIS4*. Thus, *sui1* is functionally related to *sui2* and *SUI3*. However, preliminary studies of *sui1* suggested that the *SUI1* gene product is not structually related to eIF-2. Therefore, the *SUI1* gene may encode a component other than eIF-2 that functions in establishing ribosomal recognition of the AUG initiator codon.

LITERATURE REFERENCES

Castilho-Valavicius B, Yoon H, Donahue TF (1990) Genetic Characterization of the *Sacchromyces cerevisiae* translational initiation suppressors *sui1*, *sui2* and *SUI3* and their effects on *HIS4* expression. Genetics 124: 483-495

Cigan AM, Donahue TF (1986) The methonine initiator tRNA genes in yeast. Gene 41: 343-348

Cigan AM, Donahue TF (1987) Sequence and structural features associated with translational initiator regions in yeast. Gene 59: 1-18

Cigan AM, Feng L, Donahue TF (1988a) $tRNA^{met}_{i}$ fuctions in directing the scanning ribosome to the start site of translation. Science 242: 93-97

Cigan AM, Pabich EK, Donahue TF (1988b) Mutational analysis of the *HIS4* translational initiator region in *Sacchromyces cerevisiae*. Mol Cell Biol 8: 2964-2975

Cigan AM, Pabich EK, Feng L, Donahue TF (1989) Yeast translation initiation suppressor *sui2* encodes the a subunit of eukaryotic initiation factor 2 and shares sequence identity with the human a subunit. Proc Natl Acad Sci USA 86: 2784-2788

Donahue TF, Cigan AM (1988) Genetic selection for mutations that reduce or abolish ribosomal recognition of the *HIS4* translational initiator region. Mol Cell Biol 8: 2955-2963

Donahue TF, Cigan AM, Pabich EK, Castilho-Valavicius B (1988) Mutations at a Zn(II) finger motif in the yeast eIF-2ß gene alters ribosomal start site selection during the scanning process. Cell 54: 621-632

Kozak M (1980) Evaluation of the "scanning model" of initiation of protein synthesis in eucaryotes. Cell 22: 7-8

Kozak M (1986) Influences of mRNA secondary structure on initiation by eukaryotic ribosomes. Proc Natl Acad Sci USA 83: 2850-2854

Pain VM (1986) Initiation of protein synthesis in mammalian cells. Biochem J 235: 625-637

Pelletier J, Sonenberg N (1985) Insertion mutagenesis to increase secondary structure within the 5' non-coding region of a eukaryotic mRNA reduces translational efficiency. Cell 40: 515-526

Sherman F, Stewart JW (1982) Mutations altering initiation of translation of yeast iso-1-cytochrome c: contrasts between eukaryotic and prokaryotic initiation process. Cold Spring Harbor Laboratory pp. 301-333

Shine J, Dalgarno L (1974) The 3' terminal sequence of *E. coli* 16S ribosomal RNA complementary to nonsense triplets and ribosome binding sites. Proc Natl Acad Sci USA 71: 1342-1346

Williams NP, Hinnebusch AG, Donahue TF (1989) Mutations in the structural genes for eukaryotic initiation factors 2a and 2ß of *Saccharomyces cerevisiae* disrupt

STRUCTURE/FUNCTION OF MAMMALIAN INITIATION FACTORS.

John W. B. Hershey, Joachim Schnier, Zeljka Smit-McBride, Susan S. Milburn, Nick Gaspar, Sang-Yun Choi and Markus Hümbelin

Department of Biological Chemistry
School of Medicine
University of California
Davis, CA 95616, U.S.A.

The initiation phase of protein synthesis in eukaryotic cells is promoted by at least 10 different initiation factor proteins, comprising over 25 distinct polypeptides. The complexity of the process and the finding that these proteins are post-translationally modified by phosphorylation *inter al.*, suggest that rates of translation are controlled by regulating initiation factor activities. In order to better understand the functional roles played by the initiation factors, we and others have purified the proteins from mammalian cells and have begun to elucidate their structures by cloning cDNAs encoding the factors. Possession of cDNA clones provides not only the protein's sequence, but also tools for cloning the gene and studying its expression, for overexpressing the protein in cells, and for manipulating the protein's structure. This allows us to study the function of the initiation factors *in vivo*, thereby providing confirmation of results obtained by *in vitro* biochemical experiments.

Cloning of cDNAs encoding initiation factors.

cDNAs encoding seven initiation factor polypeptides from mammalian cells have been cloned to date (see Table 1). A variety of screening approaches has been used to obtain these clones. Hans Trachsel and Peter Nielsen were the first to clone an initiation factor cDNA, namely that for eIF-4A; they used specific antibodies to screen a lambda gt11 expression library of mouse cDNAs (Nielsen, et al. 1985). We employed a similar approach to clone the cDNAs for eIF-2α (Ernst, et al. 1987) and eIF-2β (Pathak, et al. 1988) (the latter in collaboration with Trachsel and Nielsen). The cDNAs for eIF-4B (Milburn et al., manuscript submitted) and eIF-4D (Smit-McBride, et al. 1989) were cloned in our laboratory (in collaboration with Randal Kaufman and Bill Merrick, respectively) by using hybridization with oligonucleotide probes based on partial amino acid sequence information. Bob Rhoads and coworkers also used

NATO ASI Series, Vol. H 49
Post-Transcriptional Control of Gene Expression
Edited by J. E. G. McCarthy and M. F. Tuite

Table 1: Cloned Initiation Factor cDNAs

Initiation Factor	#aa's	mass (Da)	mRNA (kb)	species	reference
eIF-2α	315	36,115	1.6	human	Ernst et al., 1987
	315	36,111	1.6		rat
eIF-2β	333	38,404	1.6	human	Pathak et al., 1988
eIF-2γ			2.2, 2.7	human	Gaspar, unpub.
eIF-4A	408		2.0, 1.6	mouse	Nielsen et al., 1985
eIF-4B	611	69,843	4.4, 2.3	human	Milburn et al., 1990
eIF-4D	154	16,703	1.2	human	Smit-McBride et al., 1989
eIF-4E	217	25,117	1.9-2.2	human	Rychlik et al., 1987

hybridization with oligo-DNAs to identify the clone encoding eIF-4E (Rychlik, et al. 1987). In contrast, our attempts to use either antibodies or oligonucleotide probes (based on amino acid sequences provided by W.C. Merrick) to identify the cDNA for eIF-2γ were unsuccessful. Instead, we employed the PCR procedure to amplify a region of DNA in a cDNA library, using primers based on Merrick's sequences, and obtained a DNA fragment encoding 30 amino acid residues of the factor (Nick Gaspar, unpublished results). This fragment was then used to identify longer cDNA clones encoding eIF-2γ, and these currently are being sequenced. Each of the other clones above has been sequenced, the translated coding region identified, and the mRNA size determined by Northern blotting. Various characteristics of the cDNAs are reported in Table 1.

Protein Structures Derived from cDNA Sequences.

As amino acid sequence information became available, insight into the structure and function of the initiation factors began to emerge. Because cDNAs from more than one species were characterized, the extent of conservation also became apparent. Notable conclusions are sketched for each factor below, and structural features are shown in Figure 1.

eIF-2. The protein primary structure for eIF-2α suggested a protein of considerable secondary structure, but no unusual features. No consensus sequence elements for GTP or ATP binding sites were apparent. However, the location of the site of phosphorylation by the eIF-2 kinases, HRI and DAI, was determined to be Ser^{51} (Pathak, et al. 1988) by site-directed mutagenesis of the cDNA to replace Ser with Ala, thereby expressing a protein that resists phosphorylation. The sequence of eIF-2β was more interesting: three blocks of 6 to 8 Lys residues are present in the N-terminal third of the protein, separated by regions rich in acidic residues. A Zn-finger motif is located in the C-terminal third, although Zn^{2+} was not detected in preparations of the factor. The Lys blocks and Zn-finger motif might be involved in RNA binding, a conclusion supported by the isolation of mutants in *Saccharomyces cerevisiae* which are altered in initiator codon recognition due to changes in the Zn-finger

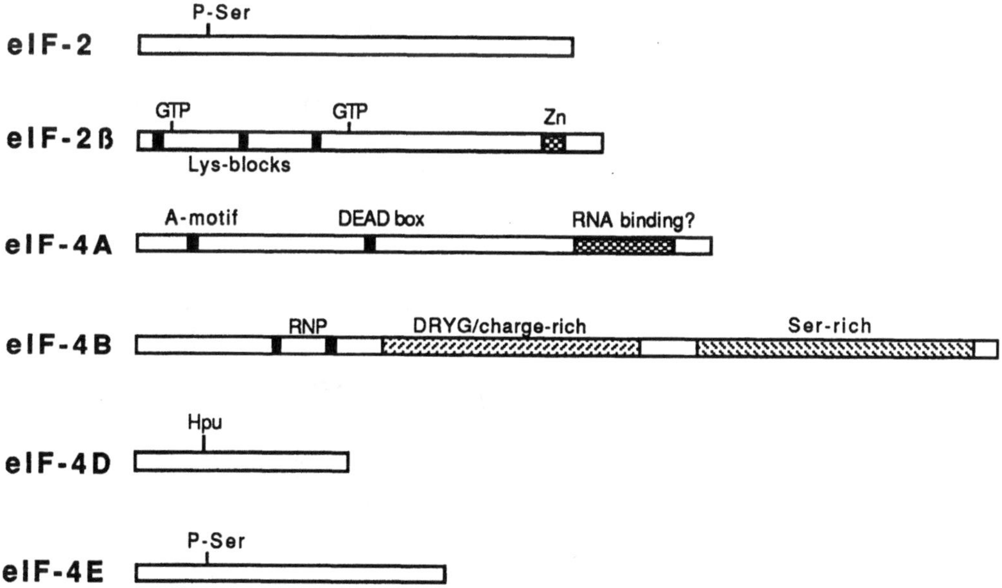

Figure 1. Initiation Factor Structural Elements

motif of eIF-2β. A further feature of the eIF-2β sequence is the occurance of two consensus elements for GTP binding: DEEG (domain 2) and NKKD (domain 3), although the separation of these elements differs from other GTP binding proteins (Dever, et al. 1987). Crosslinking of GTP affinity analogues to eIF-2 subunits results in labeling of both the β- and γ-subunits, suggesting that the GTP binding site might involve both subunits (Bommer and Kurzchalia 1989). One of the partial amino acid sequences from eIF-2γ provided by Merrick's group also contains the NKKD element, suggesting that only one of these elements in the two subunits is the true binding site. A better evaluation of the GTP binding site and other aspects of eIF-2 function should be possible after the full eIF-2γ sequence becomes available.

eIF-4A. eIF-4A is an ATP binding protein with RNA-dependent ATPase activity. It is unique among the initiation factors whose cDNAs are cloned, in that two genes have been identified (Nielsen and Trachsel 1988). The two forms are very similar, with 91% identical amino acid residues and another 7% conserved. eIF-4A is a member of a family of proteins, ranging from prokaryotes to mammals, that share sequence homology and appear to function as RNA helicases (Linder, et al. 1989). Two sequence elements typical of ATP binding proteins are present in nearly all members of the family: DXXXXAXXGXGKT (the A-motif) and VLDEADXMLXXGF (the B-motif, or DEAD box). A third motif, HRIGR, lies in a generally conserved region of 73 amino acid residues, which may be involved in the binding to polynucleotides. Mutation of the K residues of the A-motif to N results in a protein that fails to bind ATP, as assessed by UV-induced crosslinking of [α-^{32}P]dATP (Rozen, et al. 1989), thereby supporting the view that the A-motif is indeed involved in ATP binding.

eIF-4B. eIF-4B is implicated in mRNA binding to ribosomes and appears to promote the functions of eIF-4A and eIF-4F. The primary structure of eIF-4B is unusual in a number of ways. Most interesting is the presence of an RNA binding motif, called the RNP consensus sequence motif (Bandziulis, et al. 1989). A conserved octamer (RNP-1) separated by exactly 33 amino acid residues from an less stringently conserved upstream hexamer (RNP-2) was found in the N-terminal region of the protein (see Figure 1). Proteins that share these elements (*e.g.*, polyA binding protein, RNP proteins involved in splicing) are known to bind RNAs. This is the first report of the presence of this element in an initiation factor, and supports the view that eIF-4B binds to mRNAs (Abramson, et al. 1987). The middle region of the protein is very highly charged, possessing *ca.* 50% charged residues over a stretch of 200 amino acids. The highly charged domain may be responsible for the slow migration rate of eIF-4B upon SDS-PAGE, where the protein usually behaves as an 80 kDa protein rather than the 69.7 kDa size determined by sequencing. A curious feature of this domain is the repeating sequence element, DRYG. Finally, the C-terminal demain is rich in Ser residues which may be targets of protein kinases which phosphorylate the protein on greater than 8 Ser sites (Duncan and Hershey 1985).

eIF-4D. eIF-4D acts late in the initiation pathway, following 80S initiation complex formation, by promoting the formation of the first peptide bond. The primary protein sequence suggests considerable secondary structure, but possesses no striking features. It has enabled us to identify the lysine modificated by the polyamine, spermidine, to create hypusine (*i.e.*, Lys_{50}).

eIF-4E. eIF-4E binds to the 7-methyl-guanosine cap structure of mRNAs and is a component of the cap-binding complex, eIF-4F. The protein sequence contains regions that may be involved in m^7G binding: FKND, SKFD and AKGD resemble the guanine ring consensus element for GTP binding, namely NKXD (Dever, et al. 1987). It also has regions of homology to the influenza polymerase PB2, a cap binding protein. eIF-4E is phosphoyrlated *in vivo* on a single Ser residue, which has been identified in the sequence as Ser_{53} (Rychlik, et al. 1987).

Mammalian Genes and Their Expression.

Most initiation factors are present at about 0.5 copies per ribosome. Exceptions are eIF-4A and eIF-4D, which are *ca.* 6 times more abundant, and eIF-2B and eIF-4E, which appear to be 2 to 5 times less abundant (Duncan and Hershey 1983). In HeLa cells, the rates of synthesis of the initiation factors appear to be mainly responsible for determining their cellular levels, and these rates are regulated in parallel with the rate of total protein synthesis upon serum deprivation or stimulation (Duncan and Hershey 1985). The expression of three representative initiation factor genes, eIF-2α, eIF-4A and eIF-4D, was examined in murine lymphosarcoma cells whose proliferation was inhibited by treatment with dexamethasone (Huang and Hershey 1989). The relative rates of synthesis of the three factors is coordinately down-regulated in parallel with that for ribosomal proteins. However, whereas ribosomal protein synthesis is repressed by inhibition of initiation of protein synthesis, initiation factor synthesis was inhibited by a reduction in the levels of their mRNAs. In the case of eIF-4A, this reduction appears to be due to decreased rates of transcription, as measured by nuclear run-on experiments (S. Huang, unpublished results). Elucidation of the mechanism of coordinate expression awaits a detailed analysis of the structure and expression of initiation factor genes.

The promoter region of the human gene encoding the α-subunit of eIF-2 has been cloned, sequenced and analyzed functionally (Hümbelin, et al. 1989). There appears to be only a single locus encoding eIF-2α, and no pseudogenes have been detected. The gene contains at least 3 exons, but a description of the entire gene is not yet available. The promoter lacks a TATA box, as is typical of housekeeping genes, and multiple start sites over a 30 base region are detectable by S1 mapping and primer extension analyses. Most noteworthy is the large number of putative *cis*-acting elements present in the 5'-flanking region. Safer and coworkers (Jacob, et al. 1989) have purified a 66-68 kDa protein, called α-PAL, which binds to a region (-70 to -10) just upstream from the CAP sites. The α-PAL factor functions as a positive regulator of the eIF-2α gene. It is not related to other known transcriptional factors, although its relatively high abundance suggests that it also may interact with promoters of other genes.

Two intron-containing genes encoding eIF-4A have been cloned from the mouse genome (Nielsen and Trachsel 1988), and as many as 9 to 13 pseudo genes are estimated to be present as well. The second gene, eIF-4AII, is about 8 kb long and comprises 11 exons and 10 introns. Because the 3'-untranslated regions have diverged, probes specific to these regions distinguish the two gene products. Each of the two genes expresses two size-classes of mRNA (1.6 and 2.0 kb), presumably by selection of alternate polyadenylation signals. The relative amounts of the two forms and the relative expression of the two genes vary in different tissues, the latter over a range of 20-fold (*e.g.*, spleen and kidney). The mRNAs have very short 5'-untranslated regions and it has been suggested that their translation might therefore be less dependent on eIF-4A.. Descriptions of the promoters of the two intron-containing genes and how they might be regulated are not yet available.

Applications of Recombinant DNA Techniques to Studies of Initiation Factor Function.

Overexpression. The rather high molar ratio (0.5 to 0.8) of most initiation factors to ribosomes suggests that the initiation factors are present in cells in excess, i.e., that they are not rate-limiting. The less abundant factors, eIF-2B and eIF-4E, may be exceptions, however. In the one case examined thus far, *E. coli* IF-2 was shown to be in 2-fold excess, by down-regulating the expression of its gene in intact cells (Cole, et al. 1987). An approach to determining whether or not a mammalian factor is limiting is to overexpress its cDNA in transiently transfected cells and test the effect of high levels of the protein on protein synthesis. We have used this approach in collaboration with Randal Kaufman (Genetics Institute, Cambridge, MA) to overexpress eIF-2α, eIF-2β, eIF-4B and eIF-4D in COS-1 cells. The factor cDNA was fused to the adenovirus tripartite leader behind the major late promoter and over 10-fold higher rates of synthesis and protein accumulation were obtained. To measure a change in translation rates in the population of cells transfected, a reporter gene, DHFR, was co-transfected and the synthesis of DHFR was monitored. In the case of eIF-4D and the eIF-2 subunits, no effect on DHFR synthesis was seen, indicating that these factors are not limiting. In the case of eIF-4B, the rate of DHFR synthesis (as well as eIF-4B synthesis) was depressed about 5-fold at 48 hours post-transfection when high levels of eIF-4B had accumulated. It appears that excessive levels of eIF-4B inhibit protein synthesis. It will be interesting to see what translational effects may result following eIF-2B and eIF-4E overproduction.

Mechanism of Action. Insight into the function of a protein can be obtained by synthesizing the protein (or a mutant derivative) *in vitro* or *in vivo*, and then testing its activity. Rhoads and coworkers prepared radiolabeled eIF-4E by translating an *in vitro* transcript in a reticulocyte lysate (Hiremath, et al. 1989). We have found that eIF-2α overproduced in COS-1 transfected cells exchanges into the endogenous eIF-2 complex, as well as accumulates as the free subunit (S.-Y. Choi, unpublished results). Sonenberg and coworkers showed that the putative ATP binding site in eIF-4A is in fact involved, by mutating a residue in the site and showing that ATP no longer binds (Rozen, et al. 1989).

Role of Post-translational Modifications. The role of phosphorylation of initiation factors has been studied *in vivo* by employing mutant cDNAs that express proteins that either resist or mimic phosphorylation. eIF-2α was first examined in this way (Kaufman, et al. 1989). When the site of phosphorylation, Ser_{51}, was changed to Asp to mimic phosphoserine, low levels of expression completely inhibited protein synthesis in transfected COS-1 cells. Thus eIF-2 phosphorylation appears to be sufficient to repress translation rates *in vivo*. When Ser_{51} was changed to Ala, the repressive effects of induction of the eIF-2 kinase, DAI, were abrogated (Davies, et al. 1989, Kaufman, et al. 1989). For example, the yields of the mutant adenovirus, dl704 (which lacks the VA RNA genes), were enhanced 100-fold in infected long-term cell lines that overexpress the eIF-2α mutant forms. Rhoads and coworkers have mutated the phosphorylation site of eIF-4E, Ser_{53}, to Ala and shown that the altered factor no longer associates stably with 48S preinitiation complexes (Joshi, et al. 1989).

eIF-4D is modified by reaction of a specific Lys residue with spermidine to generate hypusine. The modification occurs from yeast to man and the eIF-4D protein is highly conserved. To determine whether or not the hypusine modification is essential for the factor's function in translation, we expressed the cDNA in *E. coli*, which lacks the modification enzymes, and purified the precursor form lacking hypusine (Smit-McBride, et al. 1989). The precursor protein fails to stimulate Met-puromycin synthesis, the usual assay for eIF-4D activity, but becomes active following *in vitro* modification to deoxyhypusine (performed by M.H. Park, NIH). This result indicates that reaction with spermidine is essential for the *in vitro* activity of the factor. Attempts to demonstrate the requirement *in vivo* by transfection of the wild-type cDNA or a mutant form where the target of modification was changed from Lys to Arg, did not generate useful results. We then cloned two eIF-4D genes in the yeast, *Saccharomyces cerevisiae*, have constructed knockouts in both genes, and have shown that at least one of the genes is essential for cell growth and viability (J. Schnier, unpublished results). Experiments are in progress to determine if eIF-4D functions in protein synthesis and whether or not the Arg mutant form will support cell growth.

Acknowledgements.

We thank our collaborators, William C. Merrick, Randal J. Kaufman and Myung Hee Park, for their contributions to the research described here. The work was supported by NIH grant GM22135.

References.

Abramson, R. D., T. E. Dever, T. G. Lawson, B. K. Ray, R. E. Thach and W. C. Merrick. (1987). "The ATP-dependent Interaction of Eukaryotic Initiation Factors with mRNA." J. Biol. Chem. **262**: 3826-3832.

Bandziulis, R. J., M. S. Swanson and G. Dreyfuss. (1989). "RNA-binding proteins as developmental regulators." Genes Develop. **3**: 431-437.

Bommer, U.-A. and T. V. Kurzchalia. (1989). "GTP interacts through its ribose and phosphate moieties with different subunits of the eukaryotic initiation factor eIF-2." FEBS Lett. **244**: 323-327.

Cole, J. R., C. L. Olsson, J. W. B. Hershey and M. Nomura. (1987). "Feedback Regulation of rRNA Synthesis in Escherichia coli: Requirement for Initiation Factor IF2." J. Mol. Biol. **198**: 383-392.

Davies, M. V., M. Furtado, J. W. B. Hershey, B. Thimmappaya and R. J. Kaufman. (1989). "Complementation of adenovirus virus-associated RNA I gene deletion by expression of a mutant eukaryotic translation initiation factor." Proc. Natl. Acad. Sci. USA. **86**(23): 9163-9167.

Dever, T. E., M. J. Glynias and W. C. Merrick. (1987). "GTP-binding domain: Three consensus sequence elements with distinct spacing." Proc. Natl. Acad. Sci. USA. **84**: 1814-1818.

Duncan, R. and J. W. B. Hershey. (1983). "Identification and Quantitation of Levels of Protein Synthesis Initiation Factors in Crude HeLa Cell Lysates by Two-dimensional Polyacrylamide Gel Electrophoresis." J. Biol. Chem. **258**: 7228-7235.

Duncan, R. and J. W. B. Hershey. (1985a). "Regulation of Initiation Factors During Translational Repression Caused by Serum Deprivation." J. Biol. Chem. **260**: 5493-5497.

Duncan, R. and J. W. B. Hershey. (1985b). "Regulation of Initiation Factors During Trtanslation Repression Caused by Serum Deprivation: Abundance, Synthesis and Turnover Rates." J. Biol. Chem. **260**: 5486-5492.

Ernst, H., R. F. Duncan and J. W. B. Hershey. (1987). "Cloning and sequencing of complementary DNAs encoding the alpha-subunit of translational initiation factor eIF-2. Characterization of the protein and its messenger RNA." J. Biol. Chem. **262**(3): 1206-1212.

Hiremath, L. S., S. T. Hiremath, W. Rychlik, S. Joshi, L. L. Domier and R. E. Rhoads. (1989). "In vitro Synthesis, Phosphorylation, and Localization on 48S Initiation Complexes of Human Protein Synthesis Initiation Factor 4E." J. Biol. Chem. **264**: 1132-1128.

Huang, S. and J. W. B. Hershey. (1989). "Translational Initiation Factor Expression and Ribosomal Protein Gene Expression Are Repressed Coordinately but by Different Mechanisms in Murine Lymphosarcoma Cells Treated with Glucocorticoids." Mol. Cell. Biol. **9**: 3679-3684.

Hümbelin, M., B. Safer, J. A. Chiorini, J. W. B. Hershey and R. B. Cohen. (1989). "Isolation and Characterization of the Promoter and Flanking Regions of the Gene Encoding the Human Protein Synthesis Initiation Factor 2a." Gene. **81**: 315-324.

Jacob, W. F., T. A. Silverman, R. B. Cohen and B. Safer. (1989). "Identification and Characterization of a Novel Transcriptional Factor Participating in the Expression of Eucaryotic Initiation Factor 2a." J. Biol. Chem. **264**: 20372-20384.

Joshi-Barve, S., W. Rychlik and R. E. Rhoads. (1990). Alteration of the Major Phosphorylation Site of Eukaryotic Protein Synthesis Initiation Factor 4E Prevents Its Association with the 48S Initiation Complex."J. Biol. Chem. **265:** 2979-2983.

Kaufman, R. J., M. V. Davies, V. K. Pathak and J. W. B. Hershey. (1989). "The phosphorylation state of eucaryotic initiation factor 2 alters translational efficiency of specific mRNAs." Mol. Cell. Biol. **9**(3): 946-958.

Linder, P., P. F. Lasko, P. LeRoy, P. J. Nielsen, K. Nishi, J. Schnier and P. P. Slonimski. (1989). "Birth of the D-E-A-D box." Nature. **337**: 121-122.

Nielsen, P. J., G. K. McMaster and H. Trachsel. (1985). "Cloning of eukaryotic protein synthesis initiation factor genes: isolation and characterization of cDNA clones encoding factor eIF-4A." Nucl. Acids Res. **13**: 6867-6880.

Nielsen, P. J. and H. Trachsel. (1988). "The mouse protein synthesis initiation factor 4A gene family includes two related functional genes which are differentially expressed." EMBO J. **7**: 2097-2105.

Pathak, V., D. Schindler and J. Hershey. (1988). "Generation of a Mutant Form of Protein Synthesis Initiation Factor eIF-2 Lacking the Site of Phosphorylation by eIF-2 Kinases." Mol. Cell. Biol. **8**: 993-995.

Pathak, V. K., P. J. Nielsen, H. Trachsel and J. W. B. Hershey. (1988). "Structure of the beta subunit of translational initiation factor eIF-2." Cell. **54**(5): 633-639.

Rozen, F., J. Pelletier, H. Trachsel and N. Sonenberg. (1989). "A Lysine Substitution in the ATP-Binding Site of Eucaryotic Initiation Factor 4A Abrogates Nucleotide-Binding Activity." Mol. Cell. Biol. **9**: 4061-4063.

Rychlik, W., L. L. Domioer, P. R. Gardner, G. M. Hellmann and R. E. Rhoads. (1987). "Amino acid sequence of the mRNA cap-binding protein from human tissues." Proc. Natl. Acad. Sci. USA. **84**: 945-949.

Rychlik, W., M. A. Russ and R. E. Rhoads. (1987). "Phosphorylation Site of Eucaryotic Initiation Factor 4E." J. Biol. Chem. **262**: 10434-10437.

Smit-McBride, Z., T. E. Dever, J. W. B. Hershey and W. C. Merrick. (1989). "Sequence determination and cDNA cloning of eukaryotic initiation factor 4D, the hypusine-containing protein." J. Biol. Chem. **264**(3): 1578-1583.

Smit-McBride, Z., J. Schnier, R. J. Kaufman and J. W. B. Hershey. (1989). "Protein Synthesis Initiation Factor eIF-4D: Functional Comparison of Native and Unhypusinated Forms of the Protein." J. Biol. Chem. **264**: 18527-18530.

INITIATION FACTORS INVOLVED IN mRNA BINDING TO RIBOSOMES IN *SACCHAROMYCES CEREVISIAE*

P.P. Müller, S. Blum, M. Altmann, S. Lanker and H. Trachsel
Institut für Biochemie und Molekularbiologie
Universität Bern
Bühlstrasse 28
CH-3012 Bern
Switzerland

1. Summary

The yeast *Saccharomyces cerevisiae* has become a model system for studies of eukaryotic translation. Amino acid similarity between yeast and mammalian translation factors and *in vitro* or *in vivo* functional assays suggest that the overall translation mechanism as well as individual factor polypeptides are evolutionary conserved and that findings in the yeast system are likely to be significant for our understanding of translation in mammalian cells. To investigate the function of translation initiation factors and their subunits we have prepared *in vitro* systems which are dependent on a single factor. We have obtained factor-dependent lysates by two approaches: (i) Controlled *in vivo* expression of a factor proposed to be involved in unwinding mRNA secondary structure, initiation factor 4A (eIF-4A). An eIF-4A-dependent *in vitro* translation system was then prepared from cells in which transcription of the eIF-4A gene was repressed. (ii) A yeast strain was constructed containing a mutant cap-binding factor (eIF-4E). In lysates prepared from such cells translation of capped mRNA was dependent on the addition of purified eIF-4E.

2. Introduction

Current models of translation initiation in eukaryotes are mainly based on data obtained from crude or reconstituted mammalian cell-free translation systems [for reviews, see Edery et al., 1987; Pain, 1986; Kozak, 1983]. Fig. 1 shows a model for the cap-dependent initiation pathway which accounts for the translation of the majority of cellular mRNAs: 80S ribosomes dissociate into 40S and 60S subunits and

NATO ASI Series, Vol. H 49
Post-Transcriptional Control of Gene Expression
Edited by J. E. G. McCarthy and M. F. Tuite

the 40S subunits associate with the eukaryotic initiation factor 3 (eIF-3) and eIF-4C. Initiation factor eIF-2 then carries the initiator methionyl-tRNA (met-tRNA$_i^{met}$) as part of the ternary complex eIF-2-GTP-met-tRNA$_i^{met}$ to the 40S subunit. The resulting complex binds at or near the mRNA cap structure in a reaction dependent on ATP, ATP hydrolysis and the factors eIF-4A, B, E, F. This initiation complex then moves on the mRNA in the 5' to 3' direction and positions itself at the AUG initiator codon. At this point, the large ribosomal subunit joins the complex in an eIF-5-catalyzed reaction whereby GTP is hydrolyzed and initiation factors are released.

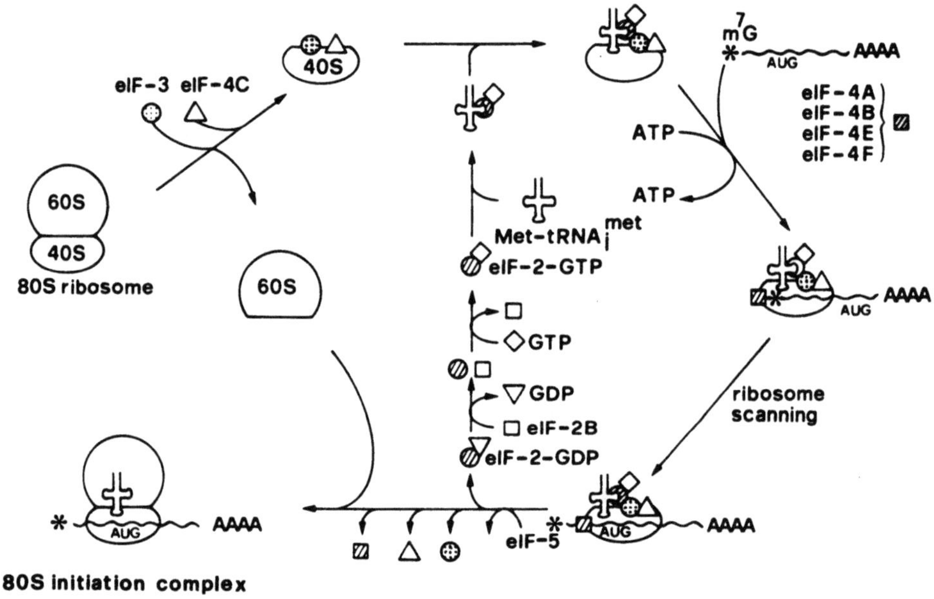

Fig. 1: Scheme of eukaryotic translation initiation. The cycle begins with the 80S ribosome and ends with the 80 initiation complex.
40S, small ribosomal subunit; 60S, large ribosomal subunit; eIF, eukaryotic initiation factor; m^{7}G, mRNA cap structure; Met-tRNA$_i^{met}$, initiator methionyl transfer RNA.

Recently, an alternative pathway was shown to exist in which ribosomes initiate translation of special viral mRNA species in a 5' cap structure independent manner

[Pelletier et al., 1988; Pelletier and Sonenberg, 1988; Bienkowska-Szewczyk and Ehrenfeld, 1988].

Powerful genetic techniques and the development of a cell-free translation system [Gasior et al., 1979], have contributed to establish the yeast *Saccharomyces cerevisiae* system as a model for studies of eukaryotic translation. Today a large body of evidence suggests that the cap-dependent translation initiation pathway in this organism closely resembles the corresponding pathway in higher eukaryotes [for reviews, see Tuite, 1989; Chakraburtty and Kamath, 1988; Müller and Trachsel, 1990]. This system, therefore, appears most suitable to further study the function and regulation of individual translation initiation factors. Towards this end, we have begun to develop initiation factor-dependent *Saccharomyces cerevisiae* cell-free translation systems.

3. Results and Discussion

(a) Initiation factor 4A-dependent system

Initiation factor 4A (eIF-4A) is required for mRNA-binding to ribosomes [Schreier et al., 1977; Trachsel et al., 1977] and appears to melt RNA secondary structure in an ATP-hydrolysis-dependent manner during the scanning process [Ray et al., 1985; Rozen et al., 1990]. The cloning and sequencing of mouse eIF-4A cDNA [Nielsen et al., 1985; Nielsen and Trachsel, 1988] led to the discovery of a number of eIF-4A-related proteins in other species among them TIF (translation initiation factor) from *Saccharomyces cerevisiae* [Linder and Slonimski, 1988; 1989]. To prove that TIF is involved in translation, we produced an anti-TIF antibody using TIF isolated from an overproducing yeast strain as immunogen in mice. This antibody was then affinity-purified by employing TIF protein expressed in *Escherichia coli* [Blum et al., 1989]. We purified TIF from wild-type yeast cells by using chromatographic procedures and the affinity-purified anti-TIF antibody as probe to identify TIF-containing fractions. When yeast cells containing the only *TIF1* gene on a plasmid under the control of the CYC1-GAL10 promoter were grown in medium containing glucose as the carbon source, the production of TIF was shut off leading to growth arrest. Lysates made from these cells were inactive in *in vitro* translation [Fig. 2; Blum et al., 1989]. Addition of TIF to these lysates restored in vitro protein synthesis and thus proved that TIF is a translation factor and therefore the yeast equivalent of mammalian eIF-4A.

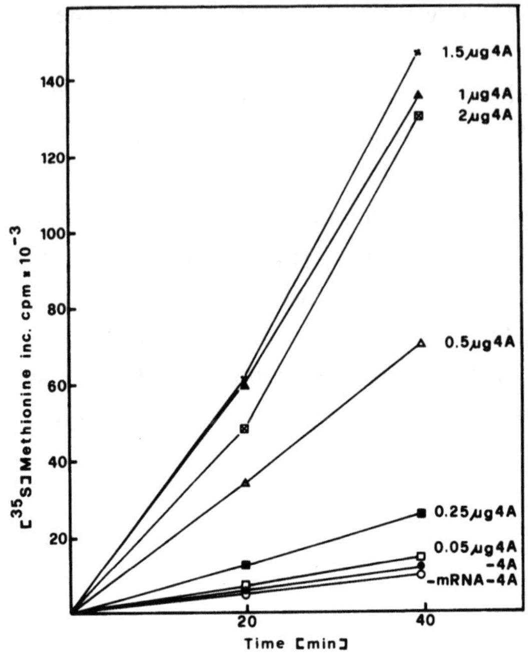

Fig. 2: Initiation factor 4A-dependent system: Titration of TIF. Extract of strain PL49 [Blum et al., 1989] was digested with micrococcal nuclease, reaction mixtures (20 μl) were incubated at 25°C, and 5-μl aliquots were analyzed for [^{35}S]methionine incorporation [Altmann et al., 1985]. o, No addition; o, plus 10 μg of total yeast RNA. The other incubation mixtures contained in addition to 10 μg of total yeast RNA the following components: , 50 ng of TIF; ■, 250 ng of TIF; , 500 ng of TIF; ▲, 1 μg of TIF; x , 1.5 μg of TIF; x , 2 μg of TIF.

This system should be of great value to elucidate the function of eIF-4A, structure-function relationships of eIF-4A and the mRNA structural determinants requiring eIF-4A function for translation.

(b) Initiation factor 4E-dependent system

Initiation factor 4E (eIF-4E) interacts directly with the 5' cap structure of mRNA during initiation and mediates binding of additional factors required for unwinding of RNA secondary structure during scanning [for a review, see Sonenberg, 1988]. It is conceivable that mRNAs devoid of secondary structure in their untranslated leader region might be independent or less dependent on eIF-4E activity for initiation. To test this hypothesis, we constructed an eIF-4E-dependent in vitro translation system from *Saccharomyces cerevisiae* [Altmann et al., 1989]. The gene encoding translation initiation factor eIF-4E from *Saccharomyces cerevisiae* [Altmann et al., 1987; 1989] was mutagenized with hydroxylamine *in vitro*. The mutagenized gene was reintroduced on a plasmid into *Saccharomyces cerevisiae* cells having their only wild-type eIF-4E gene on a plasmid under the control of the regulatable *GAL1* promoter. Transcription from the *GAL1* promoter (and consequently the production of wild-type eIF-4E) was then shut off by plating these cells on glucose-containing medium. Under these conditions, the phenotype conferred upon the cells by the mutated eIF-4E gene became apparent. Temperature-sensitive *Saccharomyces cerevisiae* strains were identified by replica plating (Fig. 3)

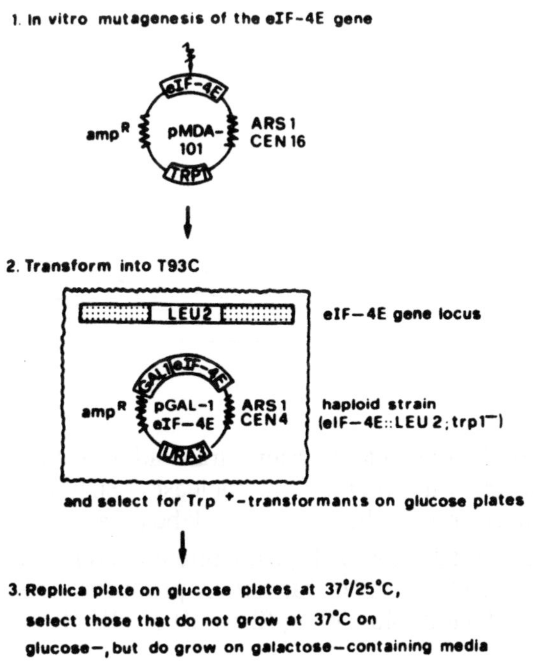

Fig. 3: Strategy to obtain temperature sensitive mutations in the gene coding for eIF-4E.

The properties of one strain, 4-2, were further analyzed. Strain 4-2 has two point mutations in the eIF-4E gene. Upon incubation at 37°C, incorporation of [^{35}S]methionine was reduced to 15 % of the wild-type level [Altmann et al., 1989]. Cell-free translation systems derived from strain 4-2 were dependent on exogenous eIF-4E for efficient translation of total yeast mRNAs (Fig. 4).

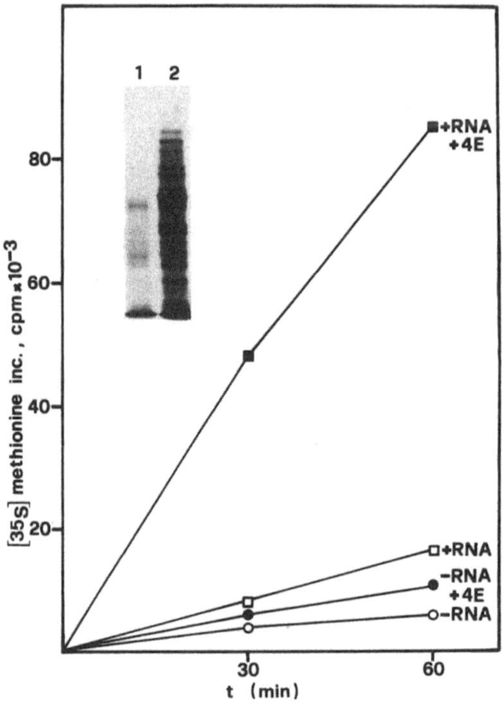

Fig. 4: Initiation factor 4E-dependent system: translation of total yeast RNA. Extract of strain 4-2 was digested with micrococcal nuclease, reaction mixtures (20 µl) were incubated at 37°C for 10 min in the absence of labeled amino acid and mRNA, and translation was started by addition of [^{35}S]methionine and (where indicated) 10 µg of total yeast RNA. (Inset) In vitro translation products after 60-min incubation. Lane 1, Minus eIF-4E; lane 2, plus 50 ng (2 pmol) of eIF-4E.

This eIF-4E-dependent extract allows individual mRNAs to be tested for their eIF-4E requirement for translation. The data in Fig. 5 show that alfalfa mosaic virus

RNA 4 (AMV-4) is only marginally stimulated by added eIF-4E whereas *in vitro* transcribed chloramphenicol acetyltransferase (CAT) mRNA is strongly stimulated.

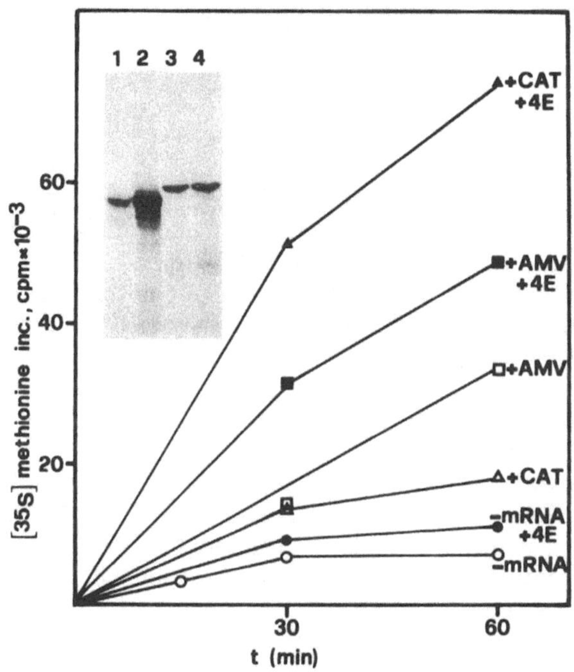

Fig. 5: Initiation factor 4E-dependent system: translation of CAT mRNA and AMV-4 RNA. Extract of strain 4-2 was treated as described in legend to Fig. 4 and reaction mixtures (20 µl) incubated with 0.1 µg of capped CAT mRNA or 0.15 µg of AMV-4 RNA. (Inset) In vitro translation products of CAT mRNA (lanes 1 and 2) and AMV-4 RNA (lanes 3 and 4) after 60-min incubation. Lanes 1 and 3, Minus eIF-4E; lanes 2 and 4, plus 50 ng (2 pmol) of eIF-4E.

Furthermore, insertion of the 5' leader sequence of tobacco mosaic virus (TMV) RNA [Ω', Gallie et al., 1987; Sleat et al., 1987] into CAT mRNA makes translation of this mRNA independent on exogenous eIF-4E (see Table 1, Wilson et al., this volume). Both AMV-4 RNA [Gehrke et al., 1983] and the Ω' sequence of TMV RNA [Sleat et al., 1988] are devoid of RNA secondary structure in agreement with the hypothesis that this feature allows translation under conditions of reduced cap binding activity [Sonenberg, 1987].

These data show that factor-dependent *Saccharomyces cerevisiae* cell-free translation systems should provide powerful tools to analyze translation initiation factor function.

Acknowledgements

This work was supported by grant 31-25565.88 from the Swiss National Foundation to H.T. We thank M. Berger for typing the manuscript.

References

Altmann M, Edery I, Sonenberg N, Trachsel H (1985) Purification and characterization of protein synthesis initiation factor eIF-4E from the yeast Saccharomyces cerevisiae. Biochemistry 24:6085-6089

Altmann M, Handschin C, Trachsel H (1987) mRNA cap-binding protein: Cloning of the gene encoding protein synthesis initiation factor eIF-4E from Saccharomyces cerevisiae. Mol Cell Biol 7:998-1003

Altmann M, Sonenberg N, Trachsel H (1989) Translation in Saccharomyces cerevisiae: Initiation factor 4E-dependent cell-free system. Mol Cell Biol 9:4467-4472

Bienkowska-Szewczyk K, Ehrenfeld E (1988) An internal 5'-noncoding region required for translation of poliovirus RNA in vitro. J Virol 62:3068-3072

Blum S, Mueller M, Schmid SR, Linder P, Trachsel H (1989) Translation in Saccharomyces cerevisiae: Initiation factor 4A-dependent cell-free system. Proc Natl Acad Sci USA 86:6043-6046

Chakraburtty K, Kamath A (1988) Protein synthesis in yeast. Int J Biochem 20:581-590

Edery I, Pelletier J, Sonenberg N (1987) Role of eukaryotic messenger RNA cap-binding protein in regulation of translation. Translational Regulation of Gene Expression, ed Ilan J, pp335-366, Plenum Press

Gallie DR, Sleat DE, Watts JW, Turner PC, Wilson TMA (1987) The 5'-leader sequence of tobacco mosaic virus RNA enhances the expression of foreign gene transcipts in vitro and in vivo. Nucl Acids Res 15:3257-3273

Gasior E, Herrera F, Sadnik I, McLaughlin CS, Moldave K (1979) The preparation and characterization of a cell-free system from Saccharomyces cerevisiae that translates natural messenger ribonucleic acid. J Biol Chem 254:3965-3969

Gehrke L, Auron PE, Quigley GJ, Rich A, Sonenberg N (1983) 5'-Conformation of capped alfalfa mosaic virus ribonucleic acid 4 may reflect its independence of the cap structure or of cap-binding protein for efficient translation. Biochemistry 22:5157-5164

Kozak M (1983) Comparison of initiation of protein synthesis in procaryotes, eucaryotes, and organelles. Microbiol Rev 47:1-45

Linder P, Slonimski PP (1988) Sequence of the genes TIF1 and TIF2 from Saccharomyces cerevisiae coding for a translation initiation factor. Nucl Acids Res 16:10359

Linder P, Slonimski PP (1989) An essential yeast protein, encoded by duplicated genes TIF1 and TIF2 and homologous to the mammalian translation initiation factor eIF-4A, can suppress a mitochondrial missense mutation. Proc Natl Acad Sci USA 86:2286-2290

Müller PP, Trachsel H (1990) Translation and regulation of translation in the yeast *Saccharomyces cerevisiae*. Eur J Biochem, in press

Nielsen PJ, McMaster GK, Trachsel H (1985) Cloning of eukaryotic protein synthesis initiation factor genes: Isolation and characterization of cDNA clones encoding factor eIF-4A. Nucl Acids Res 13:6867-6880

Nielsen PJ, Trachsel H (1988) The mouse protein synthesis initiation factor 4A gene familiy includes two related functional genes which are differentially expressed. EMBO J 7:2097-2105

Pain VM (1986) Initiation of protein synthesis in mammalian cells. Biochem J 235:625-637

Pelletier J, Sonenberg N (1988) Internal initiation of translation of eukaryotic mRNA directed by a sequence derived from poliovirus RNA. Nature 334:320-325

Pelletier J, Kaplan G, Racaniello VR, Sonenberg N (1988) Cap-independent translation of poliovirus mRNA is conferred by sequence elements within the 5' noncoding region. Mol Cell Biol 8:1103-1112

Ray BK, Lawson TG, Kramer JC, Cladaras MH, Grifo JA, Abramson RD, Merrick WC, Thach RE (1985) ATP-dependent unwinding of messenger RNA structure by eukaryotic initiation factors. J Biol Chem 260:7657-7658

Rozen F, Edery I, Meerovitch K, Dever TE, Merrick WC, Sonenberg N (1990) Bidirectional RNA helicases activity of eucaryotic translation initiation factors 4A and 4F. Mol Cell Biol 10:1134-1144

Schreier MH, Erni B, Staehelin T (1979) Initiation of mammalian protein synthesis. I. Purification and characterization of seven initiation factors. J Mol Biol 116:727-753

Sleat DE, Gallie DR, Jefferson RA, Bevan MW, Turner PC, Wilson TMA (1987) Characterization of the 5'-leader sequence of tobacco mosaic virus RNA as a general enhancer of translation in vitro. Gene 60:217-225

Sleat DE, Hull R, Turner PC, Wilson TMA (1988) Studies on the mechansim of translational enhancement by the 5'-leader sequence of tobacco mosaic virus RNA. Eur J Biochem 175:75-86

Sonenberg N (1987) Regulation of translation by poliovirus. Advances in Virus Research 33:175-204

Sonenberg N (1988) Cap-binding proteins of eukaryotic messenger RNA: functions in initiation and control of translation. Prog Nucl Acid Res Mol Biol 35:173-207

Trachsel H, Erni B, Schreier MH, Staehelin T (1977) Initiation of mammalian protein synthesis. II. The assembly of the initiation complex with purified initiation factors. J Mol Biol 116: 755-767

Tuite MF (1989) Protein synthesis. The Yeasts Vol 3 second edition pp161-204 Academic Press

THE EUKAROYOTIC mRNA CAP BINDING PROTEIN (eIF-4E): PHOSPHORYLATION AND REGULATION OF CELL GROWTH

R. Frederickson, A. Lazaris-Karatzas, N. Sonenberg

Department of Biochemistry
McGill University
3655 Drummond St.
Montreal, Quebec
Canada
H3G 1Y6

INTRODUCTION

The binding of eukaryotic mRNA to ribosomes requires the participation of at least three initiation factors: eIF-4A, eIF-4B and eIF-4F, and the hydrolysis of ATP (for reviews see Rhoads, 1988; Sonenberg, 1988). A key functional element of the eukaryotic mRNA in this process is the 5' cap structure, m^7GpppN (where N is any nucleotide). This structure facilitates ribosome binding through its interaction with the 24 kDa cap binding protein, termed eIF-4E (Sonenberg et al, 1978). eIF-4E is present in the cell in a free form and in a complex (termed eIF-4F) with two other polypeptides (Tahara et al, 1981; Grifo et al, 1983; Edery et al, 1983). One of these has been identified as eIF-4A, a 50 kDa polypeptide that possesses ATP binding and single-stranded RNA dependent ATPase activities (Sarkar et al, 1985; Abramson et al, 1987), as well as helicase activity in conjunction with eIF-4B (Ray et al, 1985; Rozen et al, 1990). The second of the polypeptides in eIF-4F is a high molecular weight subunit of 220 kDa, termed p220, whose integrity is essential for eIF-4F function (Sonenberg, 1987). eIF-4F stimulates mRNA binding to the 40S ribosomal subunit by unwinding the secondary structure in the mRNA 5' non-coding region (Ray et al, 1985; Rozen et al, 1990). Accordingly, mRNAs with reduced secondary structure in

NATO ASI Series, Vol. H 49
Post-Transcriptional Control of Gene Expression
Edited by J. E. G. McCarthy and M. F. Tuite

the 5' non-coding region are less dependent on the cap and eIF-4F for translation (Gehrke et al, 1983; Browning et al, 1988).

eIF-4E is the limiting component of the eIF-4F complex, present at 0.02-0.25 molecules per ribosome, compared to 1-3 molecules per ribosome for the other initiation factors (Hiremath et al, 1985; Duncan et al, 1987). This is consistent with mRNA ribosome binding as the rate limiting step in translation initiation (Jagus et al, 1981). Indeed, the cap binding protein complex behaves as a discriminatory factor in in vitro translation systems (Ray et al, 1983; Sarkar et al, 1984), and addition of eIF-4E to a HeLa translation system stimulates translation of capped mRNAs (Sonenberg et al, 1980). The above findings suggest a role for eIF-4E in the control of mRNA translation rates.

REGULATION OF EIF-4E PHOSPHORYLATION

Reversible protein phosphorylation is a fundamental means of regulating the activity of a variety of metabolic prccesses (Krebs and Beavo, 1979). Increased protein synthesis is obligatory for quiescent cells to re-enter the cell cycle (Brooks, 1977). Changes in the phosphorylation state of key translation factors are believed to play a role in this enhancement of mRNA translation following treatment of cells with growth promoting agents (for review see Hershey, 1989). The involvement of the ribosomal protein S6 in this process has received a great deal of attention. There is a temporal correlation between S6 phosphorylation and initiation of protein synthesis when quiescent cells are treated with a variety of agents that promote cell growth (for review see Kozma et al, 1989). However, the functional significance of S6 phosphorylation is not clear as its precise role in protein synthesis remains elusive.

In contrast, translation initiation factor eIF-4E has a relatively well-defined role in translation initiation (as discussed above), and an effect of phosphorylation on the enhancement of its activity has been demonstrated. Metabolic labelling experiments have shown that dephosphorylation of eIF-4E correlates with the decrease in mRNA translation during heat shock (Duncan et al, 1987) and mitosis (Bonneau and Sonenberg, 1987), and enhanced phosphorylation of eIF-4E has been observed in cells stimulated with TPA and insulin (Morley and Traugh, 1989,1990), as well as with tumor necrosis factor (TNF) (Marino et al, 1989). A single major phosphorylation site was mapped to Ser^{53} (Rychlik et al, 1987). Mutation of this serine to alanine prevents the association of eIF-4E with the 48S initiation complex (Joshi-Barve et al, 1990). In summary, these data suggest that regulated phosphorylation of eIF-4E plays a key role in the alteration of translation efficiency in response to growth modulating conditions.

We have extended the correlation between the rapid phosphorylation of eIF-4E and the stimulation of cell growth. Using a rabbit anti-eIF-4E peptide polyclonal antisera we immunoprecipitated eIF-4E from ^{32}P-labelled NIH 3T3 cells treated with a variety of mitogens and growth factors. There was a significant decrease in the degree of eIF-4E phosphorylation upon serum deprivation of these cells which was reversed upon addition of 10% dialyzed fetal bovine serum (data not shown). A time course of stimulation by serum, as well as by PMA and PDGF, is shown in Figure 1A; measurable increases in phosphorylation were consistently detectable within 5-10 minutes after treatment of cells with these agents, peaking at 30 min for PDGF, and 2-4 hours for PMA and serum. A similar stimulation was observed with orthovanadate, an inhibitor of phosphotyrosine phosphatases (data not shown), which suggested that an eIF-4E kinase could be a downstream target of tyrosine kinase activity.

Cells expressing transforming gene products, such as viral tyrosine kinase oncoproteins, continue to grow and divide upon serum deprivation, in contrast to parental lines. The level of phosphorylated eIF-4E is elevated in cells expressing tyrosine specific kinases src and fps. As shown in Fig. 1B, immunoprecipitation of eIF-4E from ^{32}P-labelled,

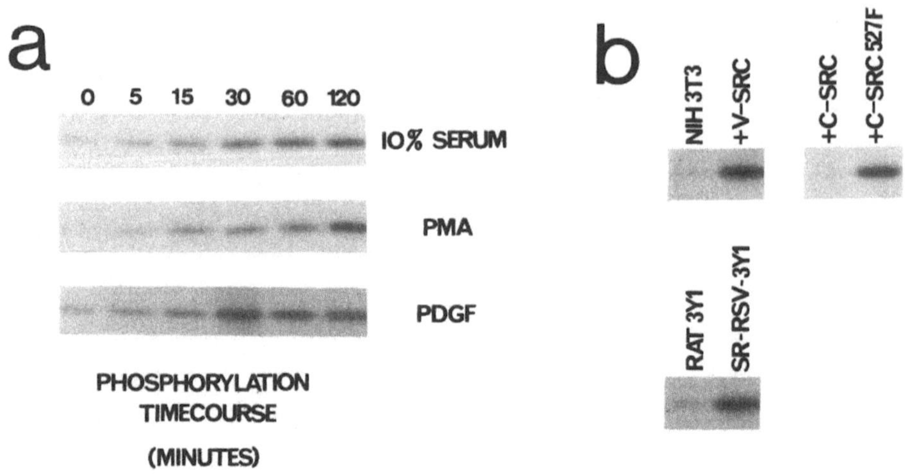

Fig.1 eIF-4E phosphorylation. a) Effect of mitogens and PMA on eIF-4E phosphorylation. Serum-deprived NIH 3T3 cells were treated with 10% dialyzed fetal calf serum, 100ng/ml phorbol-12-myristate-13-acetate (PMA), or 15.5 ng/ml platelet-derived growth factor (PDGF). Cells were labelled in PO_4-free DMEM + ^{32}P-orthophosphate for 1-2 hours. Growth promoting agents were added and incubation continued for the incubated times. Cells were lysed and eIF-4E was immunoprecipitated and electrophoresed on 12.5% polyacrylamide/SDS gels which were dried and autoradiographed. b) Effect of src expression on eIF-4E phosphorylation. Analysis of eIF-4E from serum-deprived cell lines expressing Src oncoproteins: v-src, c-src527F (Kmiecik and Shalloway, 1987), and SR-RSV-3Y1 (Dutta et al, 1990), and the respective parental (or control) lines: NIH 3T3, c-src, Rat 3Y1.

serum-deprived cell lines expressing either $pp60^{v-src}$ (v-SRC, SR-RSV-3Y1), or a transforming derivative of $pp60^{c-src}$ (c-SRC 527F), revealed a 5-10 fold greater extent of phosphorylation relative to the parental cells (NIH 3T3, Rat 3Y1, and c-SRC, respectively). Steady state levels of eIF-4E were unchanged in these cells, as determined by metabolic labelling with ^{35}S-methionine (data not shown). Phosphoamino acid and phosphopeptide analyses demonstrated that the increase in eIF-4E phosphorylation in response to viral tyrosine kinase oncoprotein expression is not the result of a novel site of serine phosphorylation in eIF-4E (Figure 2).

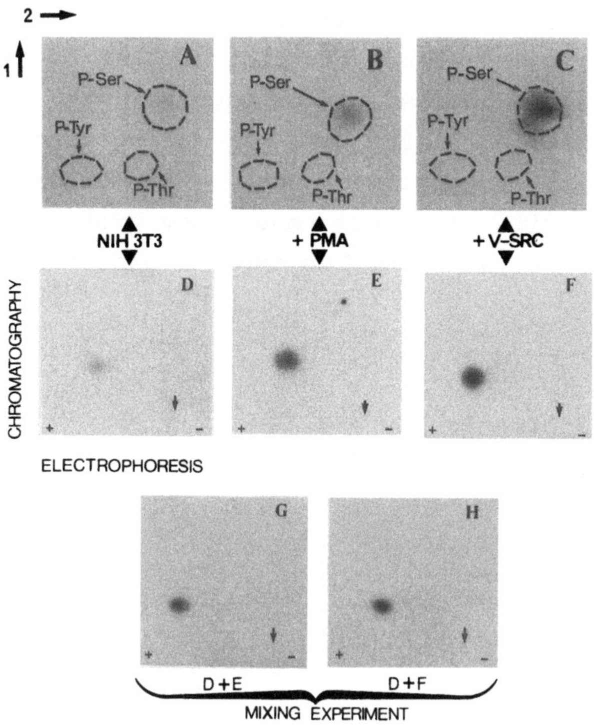

Fig. 2. Analysis of phosphorylated eIF-4E. Phosphoamino acid (A-C) and phosphopeptide (D-H) analyses were performed after treatment of ^{32}P-labelled serum deprived NIH 3T3 cells (A and D) with TPA (B and E) or expression of pp60^{v-src} (C and F). Phosphoamino acid and tryptic peptide analyses were as described in Veillette et al, 1988 a, and b, respectively. Panel G and H: mixing of phosphopeptides from TPA-stimulated NIH 3T3 cells (G) or those expressing pp60^{v-src} (H), with equivalent counts of peptides from untreated serum deprived NIH 3T3, respectively.

These results suggest that activation of a putative eIF-4E serine kinase, and/or conceivably the down-regulation of an eIF-4E phosphatase activity, is dependent on the phosphorylation of an important cellular target of the tyrosine kinase activity associated with transforming Src proteins. Similar kinase activities are associated with receptors for various growth factors, such as PDGF, EGF, insulin, and insulin-like growth factor, implicating regulated tyrosine phosphorylation in the control of normal cellular proliferation and differentiation, as well as by certain transforming gene products. Taken together, it seems likely that eIF-4E functions in the cell signalling cascade, as a key downstream target of tyrosine kinase action, whose activation, in coordination with other key regulatory proteins, is necessary for stimulation of the overall growth rate.

Malignant transformation by eIF-4E

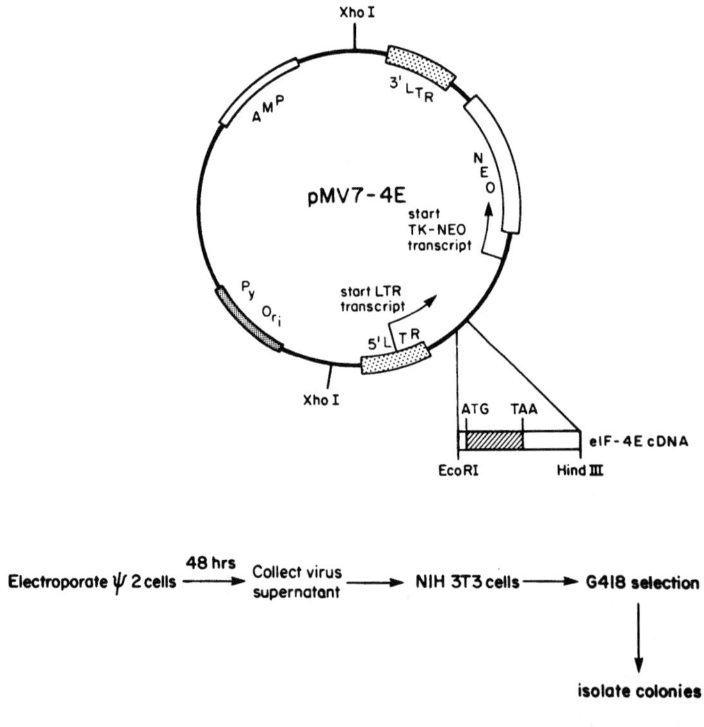

Fig. 3. eIF-4E expression vector. A murine eIF-4E cDNA (Altmann et al, 1989), was subcloned into the retroviral expression vector pMV7 (Kirschmeier et al, 1988), downstream of the Moloney MuLV long terminal repeat. The herpes simplex virus thymidine kinase promoter drives the expression of the bacterial neomycin resistance gene (neomycin phosphotransferase).

As eIF-4E is phosphorylated in response to different growth factors and regulates translation rates, it is likely to play a critical role in control of cell growth. To examine this hypothesis, we overexpressed the murine eIF-4E cDNA in two different immortalized cell lines: NIH 3T3 and Rat 2 fibroblasts. Cells were infected with a retrovirus containing the mouse eIF-4E sequence, pMV7-4E (Fig. 3). In control experiments, cells were infected either with pMV7 virus lacking the eIF-4E cDNA, or containing a mutated cDNA specifying a serine[53] change to alanine [pMV7-4E(Ala)].

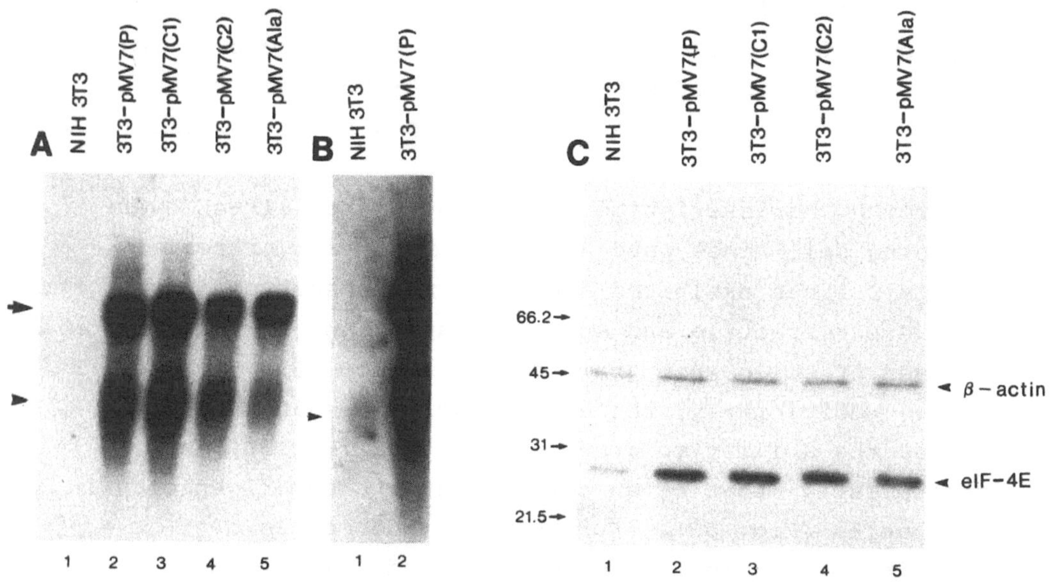

Fig. 4 Overexpression of eIF-4E in NIH 3T3 cells. A) Northern analysis. Poly A^+ RNA was isolated from the different cell lines, subjected to electrophoresis, blotted and hybridized to a α-^{32}P-dATP random primed probe, containing the entire coding region of the eIF-4E cDNA. B) A longer exposure (4 days) of lanes 1 and 2 of panel B. C) Immunoblot (Western) analysis. Cells were lysed, the lysate was centrifuged at 10,000 x g for 20 min and an aliquot was subjected to SDS-gel electrophoresis, blotted to nitrocellulose and reacted with a polyclonal rabbit antibody against a mouse eIF-4E synthetic peptide and a monoclonal mouse β-actin antibody, as a control for amount of protein. Immune complexes were visualized by autoradiography after reaction with ^{125}I-conjugated secondary antibodies. Molecular weights are shown as $M_r \times 10^{-3}$. (For details see Lazaris-Karatzas et al, 1990).

Virus-infected cells were selected for neomycin-resistance, and several colonies were chosen for further characterization, including the two clones: pMV7-4E (C1), and (C2). The remaining colonies were pooled and grown into mass culture [pMV7-4E(P)]. Overproduction of eIF-4E was analyzed by Northern and Western blot analysis. Fig. 4A shows that the pMV7-4E infected cells, including the mutant Ser^{53}→Ala, express vastly increased amounts of the eIF-4E mRNA (~1.6 kb) and a higher molecular weight species of ~5 kb (this RNA species most probably results from termination in the 3' Moloney leukemia virus (MuLV) LTR sequences; See Fig. 3). Protein levels of eIF-4E were also increased relative to endogenous eIF-4E (5-8 fold; Fig.4C). The difference in the level of expression of eIF-4E mRNA (~50-100 fold) versus protein (5-8 fold) is presumably due to a post-transcriptional control mechanism.

Growth characteristics and morphology of eIF-4E over-expressing cell lines were assessed by several criteria. These cell lines exhibited a transformed phenotype in that the cells were refractile and spindle-shaped, and formed foci on a monolayer of cells when grown to confluence (Fig.5; Panels A and B). Additionally, the eIF-4E overexpressing cells acquired the ability to grow in soft agar (Fig.5C), a characteristic that often correlates with tumorigenicity in rodent cells (Fig. 5C). Furthermore, when cells over-expressing eIF-4E were injected into nude mice, tumors developed within a relatively short latent period (10-15 days). Parental NIH 3T3 cells, or cells expressing either the vector alone or the serine mutant eIF-4E, did not show any transformation properties (Fig. 5).

The mechanism by which eIF-4E transforms cells is not clear. One hypothesis is that increasing the concentration of eIF-4E relieves the translational repression of certain mRNAs. Several proto-oncogenes are translationally regulated in vivo and in vitro. These include c-sis (Rao et al, 1986), lck (Marth et al, 1988) and possibly c-myc (Godeau et al, 1986); the mRNAs of these and many other proto-oncogenes differ from the majority of cellular mRNAs in that they contain a long 5' noncoding region. Translation of many of these mRNAs is

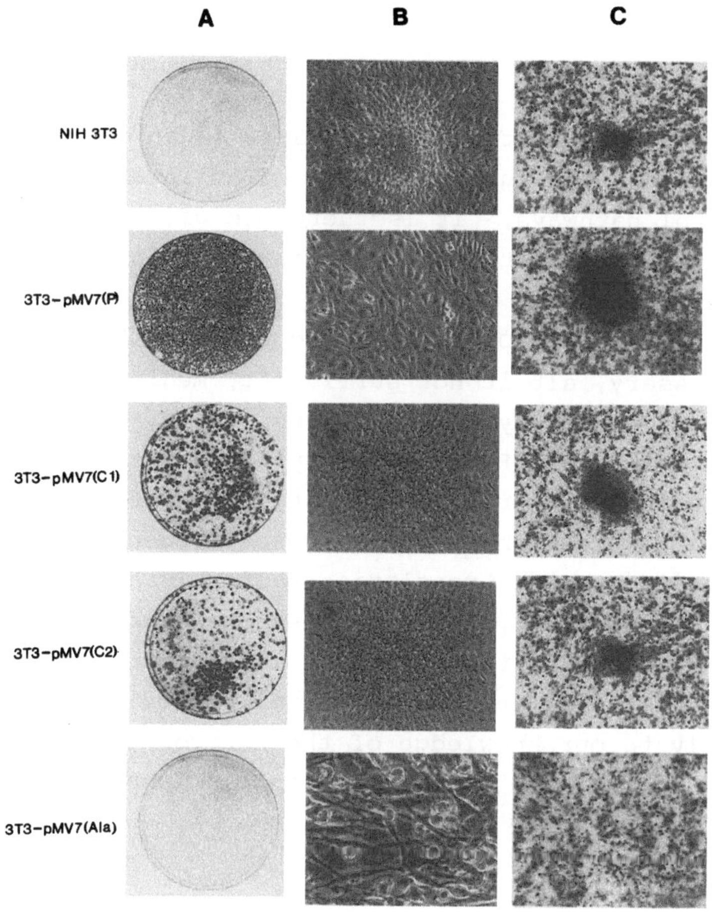

Fig. 5. Characteristics of eIF-4E overexpressing cell lines in monolayer and soft agar. Control NIH 3T3, 3T3-pMV7-4E: (P), (C1), (C2), and (Ala) cell lines were plated at 1×10^5 cells and grown to confluence in 100 mm dishes and maintained for 7 days in DMEM plus 10% FCS, with the addition of fresh medium every 2 days and photographed. A: Methylene blue stained cells. Cells were fixed in 10% formaldehyde for 2-3 hrs, rinsed and stained with 0.1% methylene blue. B: 160X magnification, C: Anchorage-independence growth assay. 160X magnification of colonies in soft agar. (For details see Lazaris-Karatzas et al, 1990).

inefficient in reticulocyte lysate <u>in vitro</u>, but is enhanced by the truncation of the long 5' noncoding region (e.g. Darveau et al, 1985). Thus, a selective mechanism exists for increasing the translational efficiency of inefficiently initiated mRNAs, such as proto-oncogene mRNAs.

Summary

Our results demonstrate that eIF-4E is an intracellular transducer of extracellular signals and an intermediary in growth control pathways. It is therefore of great interest that eIF-4E is the substrate for a kinase(s) in the phosphorylation cascade triggered by diverse mitogenic agents and tyrosine kinases. One intriguing possibility is that eIF-4E is a necessary, albeit not sufficient, mediator of the mitogenic response by tyrosine kinases. Thus, tyrosine kinases might induce eIF-4E phosphorylation through one or more intermediates, thereby activating eIF-4E for the enhanced translational expression of other proto-oncogenes.

The finding that a translation factor is a proto-oncogene (Lazaris-Karatzas et al, 1990) demonstrates that regulation of gene expression at the translational level is critical for cell growth. Elucidating these mechanisms would contribute significantly to our knowledge of the control of cell growth during differentiation, development and in tumorigenesis.

References

Abramson RD, Dever TE, Lawson TG, Ray BK, Thach RE, Merrick WC (1987) The ATP-dependent interaction of eukaryotic initiation factors with mRNA. J Biol Chem 262:3826-3832

Altmann M, Edery I, Trachsel H, Sonenberg N (1988) Site-directed mutagenesis of the tryptophan residues in yeast eukaryotic initiation factor 4E. J Biol Chem 263:17229-17232

Bonneau AM, Sonenberg N (1987) Involvement of the 24-kDa cap binding protein in regulation of protein synthesis in mitosis. J Biol Chem 262:11134-11139

Brooks RF (1977) Continuous protein synthesis is required to maintain the probability of entry into S phase. Cell 12:311-317

Browning KS, Lax SR, Humphreys J, Ravel JM, Jobling SA, Gehrke L (1988) Evidence that the 5' untranslated leader of mRNA affects the requirement for wheat germ initiation factors 4A, 4F and 4G. J Biol Chem 263:9630-9637

Darveau A, Pelletier J, Sonenberg N (1985) Differential

efficiencies of in vitro translation of mouse c-myc transcripts differing in the 5' untranslated region. Proc Natl Acad Sci USA 82:2315-2319

Duncan R, Milburn SC, Hershey JWB (1987) Regulated phosphorylation and low abundance of HeLa cell initiation factor eIF-4F suggest a role in translational control: heat shock effects on eIF-4F. J Biol Chem 262:380-388

Dutta A, Stoeckle MY, Hanafusa H (1990) Serum and v-src increase the level of a CCAAT-binding factor required for transcription from a retroviral long terminal repeat. Genes Dev 4:243-254

Edery I, Humbelin M, Darveau A, Lee KAW, Milburn S, Hershey JWB, Trachsel H, Sonenberg N (1983) Involvement of eukaryotic initiation factor 4A in the cap recognition process. J Biol Chem 258:11398-11403

Gehrke L, Auron PE, Quigley GJ, Rich A, Sonenberg N (1983) 5' Conformation of capped alfalfa mosaic virus ribonucleic acid 4 may reflect its independence of the cap structure or of cap-binding protein for efficient translation. Biochemistry 22:5157-5164

Godeau F, Persson H, Gray HE, Pardee AB (1986) c-myc expression is dissociated from DNA synthesis and cell division in Xenopus oocyte and early embryonic development. EMBO J 5:3571-3577

Grifo JA, Tahara SM, Morgan MA, Shatkin AJ, Merrick WC (1983). New initiation factor activity required for globin mRNA translation. J Biol Chem 258:5804-5810

Hershey JWB (1989) Protein phosphorylation controls translation rates. J Biol Chem 264:20823-20826

Hershko A, Mammont P, Shields R, Tomkins GM (1971) "Pleiotropic response". Nature New Biol 232:206-211

Hiremath LS, Webb NR, Rhoads RE (1985) Immunological detection of the messenger RNA cap-binding protein. J Biol Chem 260:7843-7849

Jagus R, Anderson WF, Safer B (1981) Initiation of mammalian protein biosynthesis. Prog Nucleic Acid Res Mol Biol 25:127-185

Joshi-Barve S, Rychlik W, Rhoads RE (1990) Alteration of the major phosphorylation site of eukaryotic protein synthesis initiation factor 4E prevents its association with the 48S initiation complex. J Biol Chem 265:2979-2983

Kirschmeier PT, Housey GM, Johnson MD, Perkins AS, Weinstein B (1988) Construction and characterization of retroviral vector demonstrating efficient expression of cloned cDNA sequences. DNA 7:219-225

Kmiecik TE, Shalloway D (1987) Activation and supression of $pp60^{c-src}$ transforming ability by mutation of its primary sites of tyrosine phosphorylation. Cell 49:65-73

Kozma SC, Ferrari S, Thomas G (1989) Unmasking a growth factor/oncogene-activated S6 phosphorylation cascade. Cellular Signalling 1:219-225

Krebs EG, Beavo JA (1979) Phosphorylation and dephosphorylation of enzymes. Ann Rev Biochem 48:923-959

Lazaris-Karatzas A, Montine KS, Sonenberg N (1990) Malignant transformation by a eukaryotic initiation factor subunit that binds to mRNA 5' cap. Nature In press

Marino MW, Pfeffer LM, Guidon PT, Donner DB (1989) Tumor necrosis factor induces phosphorylation of a 28 kDa mRNA

cap-binding protein in human cervical carcinoma cells. Proc Natl Acad Sci USA 86:8417-8421
Marth JD, Overell RW, Meier KE, Krebs EG, Perlmutter RM (1988) Translational activation of the lck proto-oncogene. Nature 332:171-173
Morley SJ, Traugh JA (1989) Phorbol esters stimulate phosphorylation of eukaryotic initiation factors 3,4B and 4F. J Biol Chem 264:2401-2404
Morley SJ, Traugh JA (1990) Differential stimulation of phosphorylation of initiation factors eIF-4F, eIF-4B, eIF-3 and ribosomal protein S6 by insulin and phorbol esters. J Biol Chem In press
Rao CD, Pech M, Robbins KC, Aaronson SA (1986) The 5' untranslated sequence of the c-sis/platelet-derived growth factor 2 transcript is a potent translational inhibitor. Mol Cell Biol 8:284-292
Ray BK, Lawson TG, Kramer JC, Cladaras MH, Grifo JA, Abramson RD, Merrick WC, Thach RE (1985) ATP-dependent unwinding of messenger RNA structure by eukaryotic initiation factors. J Biol Chem 260:7651-6758
Rhoads RE (1988) Cap recognition and the entry of mRNA into the protein synthesis initiation cycle. Trends Biochem Sci 13:52-56
Rozen F, Edery I, Meerovitch K, Dever TE, Merrick WC, Sonenberg N (1990) Bidirectional RNA helicase activity of eucaryotic translation initiation factors 4A and 4F. Mol Cell Biol 10:1134-1144
Rychlik W, Gardner PR, Vanaman TC, Rhoads RC (1986) Structural analysis of the messenger RNA cap-binding Protein: presence of phosphate, sulfhydryl, and disulfide groups. J Biol Chem 261:71-75
Rychlik W, Russ MA, Rhoads RE (1987) Phosphorylation site of eukaryotic initiation factor 4E. J Biol Chem 262:10434-10437
Sarkar G, Edery I, Gallo R, Sonenberg N (1984). Preferential stimulation of rabbit α globin mRNA translation by a cap binding protein complex. Biochim Biophys Acta 783:122-129
Sarkar G, Edery I, Sonenberg N (1985) Photoaffinity labeling of the cap-binding protein complex with ATP/dATP. J Biol Chem 260:13831-13837
Sonenberg N (1987) Regulation of translation by poliovirus. Adv Virus Res 33:175-204
Sonenberg N (1988) Cap binding proteins of eukaryotic messenger RNA: Functions in initiation and control of translation. Prog Nucl Acid Res Mol Biol 35:174-207
Sonenberg N, Morgan MA, Merrick WC, Shatkin AJ (1978) A polypeptide in eukaryotic initiation factors that crosslinks specifically to the 5'-terminal cap in mRNA. Proc Natl Acad Sci USA 75:4843-4847
Sonenberg N, Trachsel H, Hecht S, Shatkin AJ (1980) Differential stimulation of capped mRNA translation in vitro by cap binding protein. Nature 285:331-333
Tahara S, Morgan MA, Shatkin AJ (1981) Two forms of purified m^7G-cap binding proteins with different effects on capped mRNA translation in extracts of uninfected and poliovirus-infected HeLa cells. J Biol Chem 256:7691-7694

Veillette A, Horak ID, Bolen JB (1988) Post-translational alterations of the tyrosine kinase $p56^{lck}$ in response to activators of protein kinase C. Oncogene Res 2:385-401
Veillette A, Horak ID, Horak EM, Bookman MA, Bolen JB (1988) Alterations of the lymphocyte-specific protein tyrosine kinase during T-cell activation. Mol Cell Biol 8:4353-4361

PEPTIDE CHAIN INITIATION IN ANIMAL CELLS: MECHANISM AND REGULATION

N.K. Gupta, B.Datta, A.L. Roy and M.K. Ray
Department of Chemistry
University of Nebraska
Lincoln, NE 68588 USA

Mechanism

The detailed mechanism of peptide chain initiation in animal cells is not clearly understood. Significantly different models differing in the basic steps and factor requirements for peptide chain initiation in animal cells have been proposed. In this section, we will briefly review our recently proposed model (Gupta, 1987: Gupta et al, 1987 and Roy et al, 1988) and a widely used model proposed several years ago by Staehelin, Anderson, Hershey and their co-workers (Schreier et al, 1977; Jagus et al, 1981; and Benne and Hershey, 1978). We will then present our views regarding the differences between these two models.

Our Proposed Model

In 1988, we reported that three extensively purified factor preparations from rabbit reticulocytes, namely eIF-2, Co-eIF-2A and Co-eIF-2C promoted formation of stable ternary (Met-tRNA$_f$·eIF-2·GTP) and Met-tRNA$_f$·40S·mRNA complexes. Based on the results of these studies, we have proposed the following model for the early steps in peptide chain initiation.

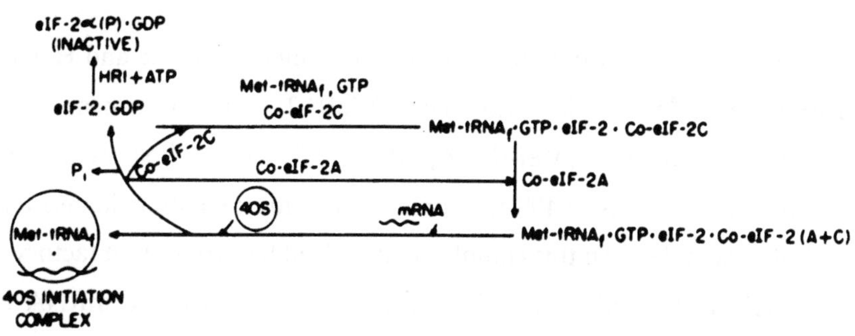

Fig. 1 A proposed model for the early steps of peptide chain initiation.

NATO ASI Series, Vol. H 49
Post-Transcriptional Control of Gene Expression
Edited by J. E. G. McCarthy and M. F. Tuite

A brief description of the proposed model and the characteristics of the protein factors is as follows: Partially purified eIF-2 preparation, used in this work, contained a 67 kDa polypeptide besides the three standard subunits (α, β and γ). Co-eIF-2A was a 94 kDa polypeptide. Co-eIF-2C was similar to eIF-3 preparation and contained apparently similar eight polypeptides (Mr; 180, 110, 65, 63, 53, 50, 43 and 40 kDa). However, this Co-eIF-2C preparation also contained eIF-2B activity and this eIF-2B activity was absent in eIF-3 preparation. Thus Co-eIF-2C activity may be considered as a mixture of eIF-2B + eIF-3 activities. The results obtained for ternary complex formation was essentially similar to those described in the previous chapter (Gupta et al, 1990). eIF-2 alone did not form significant amounts of ternary complex. Co-eIF-2C (eIF-2B + eIF-3) stimulated ternary complex formation in the presence of Mg^{2+} presumably as the eIF-2B activity in Co-eIF-2C promoted GDP-displacement from eIF-2·GDP and the eIF-3 activity in this factor preparation stimulated ternary complex formation. However, as reported (Gupta, 1990), the complex formed in the presence of Co-eIF-2C (eIF-2B + eIF-3) was unstable in the presence of a natural mRNA and Co-eIF-2A activity was required for further stimulation of ternary complex formation and also stabilization of the complex in the presence of a natural mRNA. The ternary complex formed in the presence of eIF-2 + Co-eIF-2C (eIF-2B + eIF-3) efficiently transferred Met-tRNA$_f$ to 40S ribosomes in the presence of AUG codon. However, as expected, the additional protein factor Co-eIF-2A, besides eIF-2 + Co-eIF-2C (eIF-2B + eIF-3) was necessary for a similar transfer of Met-tRNA$_f$ to 40S ribosomes in the presence of a natural mRNA. This Met-tRNA$_f$ transfer to 40S ribosomes was accompanied by hydrolysis of GTP in the ternary complex as release of eIF-2 and eIF-2·GDP.

Staehelin-Anderson-Hershey (S-A-H) Model

According to one version of the originally proposed model (Benne and Hershey, 1978): The eukaryotic peptide chain initiation factor 2 (eIF-2) alone forms near stoichiometric amount of ternary complex (Met-tRNA$_f$·eIF-2·GTP) as the first step. The ternary complex subsequently binds to 40S ribosomes and forms near stoichiometric amount of Met-tRNA$_f$·40S complex and this complex is stabilized by two protein factors, eIF-3 and eIF-4C. Protein factors, eIF-4A, eIF-4B and eIF-1 then promote mRNA binding to Met-tRNA$_f$·40S complex and this reaction is accompanied by ATP hydrolysis.

However, after the discovery of eIF-2 kinases, Trachsel and Staehelin (1978) observed that an eIF-2 kinase, namely dsI, did not inhibit eIF-2 promoted Met-tRNA$_f$·40S complex formation under their assay condition. Similarly, after the discovery of eIF-2B, Konieczny and Safer (1983) noted that this protein factor was not required for near stoichiometric amount of ternary complex formation with either eIF-2 or eIF-2α(P)(pre-phosphorylated using eIF-2 kinases) in the presence of Mg^{2+}.

To explain the above observations, it was later postulated that eIF-2 was purified as free protein (GDP-free) and both eIF-2 and eIF-2α(P) were fully active during one cycle of peptide chain initiation. After the first cycle, eIF-2 was released as eIF-2·GDP from ribosomes and required eIF-2B for a second cycle of ternary complex formation. However, eIF-2α(P)·GDP, similarly released after one cycle was inactive as eIF-2B formed an abortive complex with eIF-2α(P)·GDP.

The above modification has now been incorporated in an updated model (Merrick, 1987).

Our views regarding the differences between the above two models

In a TIBS article published in 1987, it was suggested (Gupta, 1987) that Mg^{2+}-insensitive eIF-2 activity and also eIF-2 -ancillary protein factor(s)-independent ternary and Met-tRNA$_f$·40S complex formation as reported by Staehelin, Anderson, Hershey and their co-workers and also described in the S-A-H Model, were due to the use of extremely high Met-tRNA$_f$ concentrations by these workers in their assays for ternary and Met-tRNA$_f$·40S complex formation; excess nucleotides in added tRNA bound Mg^{2+} in the reaction mixture and this promoted non-enzymatic GDP displacement and subsequent ternary complex formation by both eIF-2·GDP and eIF-2·α(P)·GDP. In support of this suggestion, we have provided experimental evidence that addition of excess tRNA in the concentration range of 5 to 10A_{260} unit per ml reaction promoted non-enzymatic GDP-displacement from eIF-2·GDP and also subsequent ternary complex formation by both phosphorylated and non-phosphorylated eIF-2 in the presence of Mg^{2+} (Roy et al, 1987). Also, our recent collaborative work with the Hershey and Merrick Laboratories clearly demonstrates that eIF-2 prepared in all three laboratories does not by itself form ternary complex in the presence of Mg^{2+} and requires three additional protein factors, eIF-2B, eIF-3 and Co-eIF-2A for such complex formation. (Gupta et al, 1990).

The above results thus show that the factor independent ternary and Met-tRNA$_f$·40S complex formation as described in the S-A-H Model cannot now be experimentally reproduced in the presence of physiological concentrations of Mg^{2+} and natural mRNA. It should be apparent, therefore, that the subsequent reaction, namely binding of mRNA to Met-tRNA$_f$·40S complex in the presence of several eIF-4 group proteins, is also not feasible in the absence of a substrate (Met-tRNA$_f$·40S complex) and an essential factor (Co-eIF-2A). These results, therefore, indicate that, at the present, we have no direct evidence that several protein factors, specifically the eIF-4 group proteins, are required for Met-tRNA$_f$·40S·mRNA complex formation as described in the S-A-H Model.

However, we recognize that numerous reports from many laboratories have indicated that several of the eIF-4 group proteins: (i) are required for *in vitro* protein synthesis (Griffo et al, 1983, Merrick et al, 1987), (ii) bind to natural mRNAs accompanying ATP hydrolysis (Griffo et al, 1983, Merrick et al, 1987) and (iii) cross-link specifically to the cap-site in mRNA (Sonenberg, 1987). Also, we recognize that our Co-eIF-2C preparation is a high molecular weight protein complex and contains eIF-4A polypeptide and also possibly other eIF-4 group proteins in trace amounts. It will now be necessary to extensively purify different protein factors and establish their requirements using our easily reproducible assay method from Met-tRNA$_f$·40S·mRNA complex formation.

Regulation

Animal cells contain one or more protein synthesis inhibitors, such as HRI (heme-regulated inhibitor) and dsI (double-stranded RNA activated inhibitor). These inhibitors are also eIF-2 kinases and phosphorylate specifically the eIF-2 α-subunit. Protein synthesis becomes inhibited as phosphorylated eIF-2 (eIF-2 α(P)·GDP) does not from ternary complex. In reticulocyte lysates, these inhibitors remain in inactive forms. One of these inhibitors, HRI, becomes activated in the absence of hemin and the other inhibitor, dsI is activated in the presence of double stranded RNA and ATP. Numerous reports have now indicated that animal cells widely use eIF-2 α-subunit phosphorylation to regulate protein synthesis under different physiological conditions which include nutritional deprivation, heat shock and virus infection (for a review, see Gupta et al, 1987). There are also indications that protein synthesis during several virus infections may also be regulated both by eIF-2 kinases and eIF-2 kinase inhibitors. Three different eIF-2 kinase inhibitors have been characterized: (1) High level of a small RNA transcript called VAI RNA inhibits activation of eIF-2 kinase in adenovirus infected cells (Schneider et al, 1985; Siekierka et al, 1985); (2) A protein factor synthesized in vaccinia virus infected cell, inhibits activation of eIF-2 kinase (Rice and Kerr, 1984; Akkaraju et al, 1989); (3) A protein factor synthesized in polio virus infected cells inhibits activity of already activated eIF-2 kinase (Ransoni and Dasgupta, 1988).

Consistent with the above reports, we have now identified and characterized an eIF-2 kinase inhibitor in reticulocyte lysate (Datta et al, 1988 a). The background and recent developments in this study are described below.

Several laboratories have previously reported that protein synthesis inhibition in heme-deficient reticulocyte lysate can be reversed by addition of either eIF-2 or a reticulocyte cell supernatant factor which we term RF (reversal factor). Several laboratories have also claimed that the eIF-2B activity in the RF preparation is responsible for this reversal activity. For a recent review, see Gupta et al, 1987.

We have reported (Datta et al, 1988 a) that a partially purified eIF-2 preparation which reverses protein synthesis inhibition in heme-deficient reticulocyte lysate contains an extra 67 kDa polypeptide (p^{67}) besides the usual three eIF-2 subunits (α,β and γ). Similarly, an active RF preparation contains p^{67} and also free eIF-2 α-subunit. We have provided evidence that this p^{67} and also the eIF-2 α-subunit are necessary for protein synthesis inhibition reversal and p^{67} in both eIF-2 and RF preparations, protects eIF-2 α-subunit from eIF-2 kinase phosphorylation.

In this section we will briefly describe a few significant results related to the roles of p^{67} in regulation of protein synthesis in animal cells.

Figure 2 describes the SDS-Page of two eIF-2 preparations (eIF-2 IV and V) and isolated p^{67}. Partially purified eIF-2 (IV) contains, besides the three standard subunits (α, β and γ) a 67 kDa polypeptide (lane 1). Upon further purification, eIF-2 containing three subunits (eIF-2 V)(lane 2) and "free" p^{67} were isolated (lane 3).

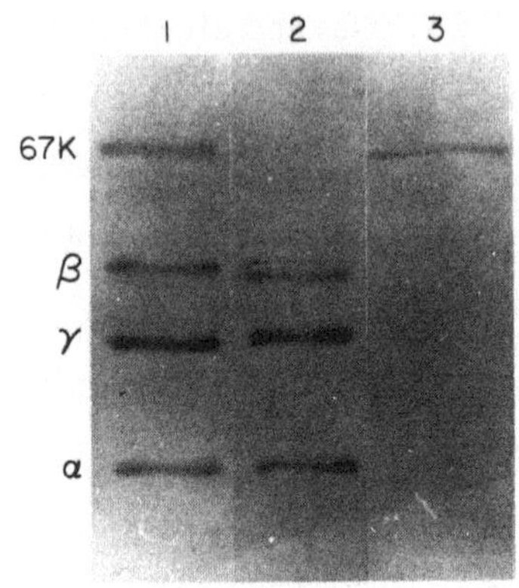

Fig. 2 SDS-Page of different protein factors. Data obtained from Datta et al (1988 a).

Figure 3 shows the activities of these three preparations to reverse protein synthesis inhibition in heme-deficient reticulocyte lysates.

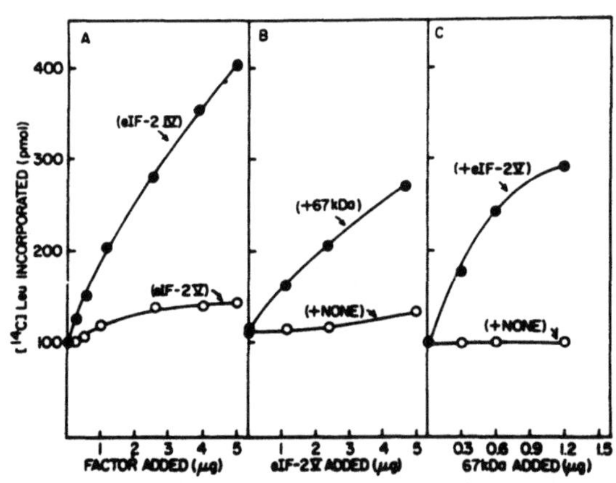

Fig. 3 Protein synthesis inhibition reversal activity of Fraction IV eIF-2, Fraction V, eIF-2 and p^{67}. Data obtained from Datta et al (1988 a).

Fraction IV eIF-2 efficiently reversed protein synthesis inhibition in heme-deficient reticulocyte lysates and Fraction V. eIF-2 preparation was completely inactive. Also, p^{67} alone was almost completely inactive. The reversal activity could be reconstituted, however, by combining Fraction V eIF-2 and the isolated p^{67} indicating that both the three subunits eIF-2 and p^{67} are essential for reversal action.

The results presented in Figure 4 shows that p^{67} protects eIF-2 α-subunit from HRI-catalyzed phosphorylation.

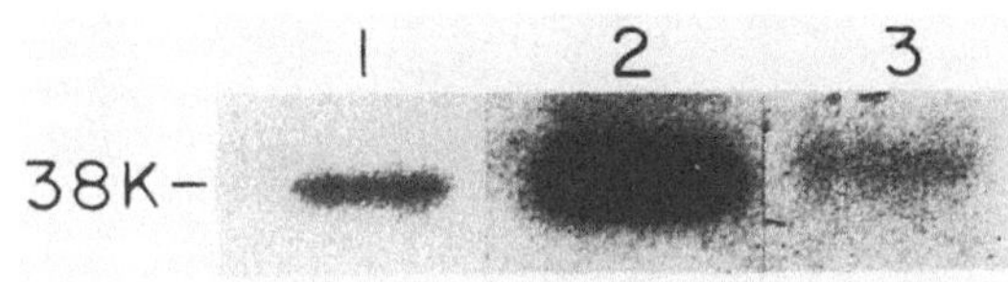

Fig. 4 Phosphorylation of eIF-2 α-subunit (38 kDa) with ATP and HRI in the presence and absence of p^{67}. Data obtained from Datta et al, 1988 b).

In this experiment, eIF-2 (IV), eIF-2 (V) and eIF-2 V + equimolar amount of p^{67} was incubated in the presence of HRI and γ p^{32}-ATP and the reaction products were analyzed by SDS-PAGE followed by autoradiography. As shown in Figure 4, the α-subunit in the 3-subunit eIF-2 was extensively phosphorylated by HRI (lane 2). Under similar conditions, the α-subunit in eIF-2 containing p^{67} was very poorly phosphorylated (lane 1). Also, the addition of p^{67} to the 3-subunit eIF-2 strongly inhibited eIF-2 α-subunit phosphorylation (lane 3). Results of several control experiments have indicated (Datta et al, 1988 a) that p^{67} is not an eIF-2 phosphatase and also does not inactivate eIF-2 kinase activity in HRI. These results thus suggest that p^{67} protects the eIF-2 α-subunit from eIF-2 kinase catalyzed phosphorylation.

Recently, we have reported that p^{67} contains multiple (≈12) O-linked N-Acetyl Glucosamine (GlcNAc) residues and that these GlcNAc residues on p^{67} may be necessary for p^{67} activity to protect eIF-2 α-subunit from eIF-2 kinase phosphorylation (Datta et al, 1989).

In other studies, we have examined the roles of eIF-2 kinases and p^{67} in regulating protein synthesis in animal cells *in culture* (Datta et al, 1988 b). These studies have indicated that protein synthesis is regulated by eIF-2 kinases and also by the availability of p^{67}. The results of a recent experiment describing the changes in p^{67} level in quiescent tumor hepatoma cells before and after TPA (12-O-tetradecanoyl-phorbol-13 acetate) addition are shown in Figure 5. It has been previously suggested that TPA addition enhanced protein synthesis rate in these quiescent cells (Trevllyan et al, 1984).

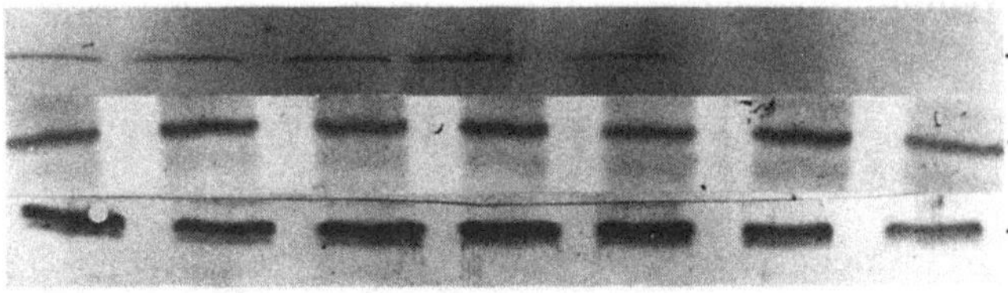

Fig. 5. Immuno blot analysis of eIF-2 α-subunit dsI and p^{67} before and after TPA addition to queiscent tumor hepatoma cells (KRC-7).

The cloned cell line (KRC-7) derived from Reuber H35 rat hepatoma cells was obtained from John Koontz (Univ. Tennessee, Knoxville, TN). The experimental procedures were essentially similar to those described by Trevillyn et al (1984). The serum starved cells were stimulated by addition of TPA. Cells were harvested at different times and proteins were analyzed by SDS-PAGE followed by Immuno-blotting using polyclonal antibodies against eIF-2 α-subunit and dsI and a monoclonal antibody against p^{67} following the previously described procedures (Datta et al, 1988 b).

As shown in Fig. 5, levels of eIF-2 α-subunit and more importantly the level of dsI remained essentially unchanged in quiescent cells before (0 hr) and after TPA addition. On the other hand p^{67} was almost non-existent in quiescent cells but became prominent after TPA addition.

Based on our studies on p^{67} action, we have made the following postulations:

(i) Protein synthesis in animal cells is regulated by eIF-2 kinases and also by the level of the eIF-2 associated p^{67}. In the presence of excess p^{67}, eIF-2·p^{67} thus formed becomes resistant to eIF-2 kinase phosphorylation and protein synthesis continues. On the other hand, as p^{67} level decreases, eIF-2 kinase(s) phosphorylate free three-subunit eIF-2 and this inhibits protein synthesis. Similarly, eIF-2 kinases when formed in excess, such as during heme-deficiency in reticulocyte lysate, may more actively phosphorylate eIF-2 α-subunit and shift the reaction in favor of eIF-2 α-subunit phosphorylation.

(ii) p^{67} is an essential component of functioning eIF-2 molecules. Several reports have indicated that many highly efficient *in vitro* protein synthesizing systems including heme-supplemented reticulocyte lysates (Leroux and London, 1982) contain variable amounts of phosphorylated eIF-2, presumably due to the presence of eIF-2 kinases (albeit at a very low level in some cases) in these preparations. It is expected that p^{67} will be necessary to protect eIF-2 α-subunit from eIF-2 kinase phosphorylation and thus to promote protein synthesis.

(iii) p^{67} activity may be regulated both transcriptionally and also at the post-transcription level. The post-transcriptional regulation will involve glycosylation of p^{67} polypeptide.

(iv) The "shut-off" of host protein synthesis during some viral infection may be accomplished by differential expression of eIF-2 kinases and also p^{67} activity. During early phase of viral infection, activation of eIF-2 kinases will lead to inhibition of host protein synthesis and untranslated mRNAs will decay. During a later stage of infection, induced synthesis of active p^{67} will promote translation of viral mRNAs available at that stage.

Acknowledgement: This work was supported by NIGMS Grant GM22079.

References

Akkaraju GR, Whitaker-Dowling P, Younger JS, Jagus R (1989) *J Biol Chem.* **264**:10321-10325.

Benne R, Hershey JWB (1978) *J Biol Chem.* **253**:3078.

Datta B, Chakrabarti D, Roy AL, Gupta NK (1988a) *Proc Natl Acad Sci USA.* **85**:3324-3328.

Datta B, Ray MK, Chakrabarti D, Gupta NK (1988b) *Indian J Biochem.* **25**:478-482.

Datta B, Ray MK, Chakrabarti D, Wylie D, Gupta NK (1989) *J Biol Chem. 264:20620-20624.*

Gupta NK (1987) *Trends Biochem Sci.* **11**:15-18.

Gupta NK, Ahmad MF, Chakrabarti D, Nasrin N (1987) In Translation Regulation in Gene Expression, Ilan J (ed) Plenum Press p 287-334.

Gupta NK, Roy AL, Nag MK, Kinzy TG, MacMillan S, Hileman RE, Devon TE, Wu S, Merrick WC, Hershey JWB (1990) this volume.

Jagus R, Anderson WF, Safer B (1981) *Proc Nucl Acid Res Mol Biol.* **25**:127-183.
Leroux A, London IM (1982) *Proc Natl Acad Sci USA.* **79**:2147-2151.
Konieczny A, Safer B (1983) *J Biol Chem.* **258**:3402-3408.
Merrick WC (1987) Molecular Biology of the Gene. In: Benjamin Cumming Publishing Co (eds) Watson JD, Hopkins NH, Roberts JW, Steitz JA, Weiner AM, p 724.
Merrick WC, Abramson RD, Anthony DD, Dever TE, Caliendo AM (1987) Translation Regulation of Gene Expression (ed) Ilan J, Plenum Press, p 265-285.
Ransoni LJ, DasGupta A (1988) *J Virol.* **62**:3551-3558.
Rice AP, Kerr IM (1984) *J Virol.* **50**:229-236.
Roy AL, Chakrabarti D, Datta B, Hileman RE, Gupta NK (1988) *Biochemistry.* **27**:8203-8209.
Schneider RJ, Safer B, Munemitsu SM, Samuel CE, Shenk T (1985) *Proc Natl Acad Sci USA.* **82**:4321-4325.
Schreier MH, Erni B, Staehelin T (1977) *J Mol Biol.* **116**:727-753.
Sierkierka J, Mariano T, Reichel PA, Matheses MB (1985) *Proc Natl Acad Sci USA.* **82**:1259-63.
Sonenberg N (1987) *Adv Virus Res.* **33**:175-204.
Trachsel H, Staehelin T (1978) *Proc Natl Acad Sci USA.* **75**:204-208.
Trevillyan JM, Kulkarni RK, Byers CV (1984) *J Biol Chem.* **259**:897-902.

NEW INSIGHTS INTO AN OLD PROBLEM: TERNARY COMPLEX (Met-tRNA$_f$·eIF-2·GTP) FORMATION IN ANIMAL CELLS

N.K. Gupta,[1] A.L. Roy,[1] M. K. Nag,[1] T.G. Kinzy,[2] S. MacMillan,[3] R.E. Hileman,[1] T.E. Dever,[2] S. Wu,[1] W.C. Merrick[2] and J.W.B. Hershey[3]
[1]*Department of Chemistry, University of Nebraska, Lincoln, NE 68588 USA.*
[2]*Department of Biochemistry, Case Western Reserve University, Cleveland, OH 44106 USA.*
[3]*Department of Biological Chemistry, University of California, Davis, CA 95616 USA.*

An early step in the initiation of protein synthesis in eukaryotic cells is the formation of a ternary complex between initiation Factor 2, Met-tRNA$_f$ and GTP; Met-tRNA$_f$·eIF-2·GTP. This complex is an obligate intermediate in the binding of Met-tRNA$_f$ to 40S ribosomes and thus is important for translation of all mRNAs. Several reports have indicated that formation and/or stability of the ternary complex *in vitro* may require a number of additional protein factors (for a review, see Gupta et al, 1987). For example, it has been reported that the bulk of reticulocyte eIF-2 is purified as eIF-2·GDP and a protein factor eIF-2B is required to displace GDP from eIF-2·GDP and subsequently to form ternary complex in the presence of Mg^{2+}. Recently, Roy et al (1988) reported that two additional protein factor preparations, Co-eIF-2A and Co-eIF-2C are required for the formation of a stable ternary complex in the presence of natural mRNAs. Co-eIF-2A and Co-eIF-2C are two highly purified protein preparations that possess eIF-2 stimulatory activities described in a number of papers from Gupta and co-workers (see Gupta et al, 1987, Roy et al, 1988). Co-eIF-2A possesses a mass of 94 kDa and appears to be unrelated to any of the other known initiation factors. On the other hand, Co-eIF-2C resembles the previously described initiation factor eIF-3. Because of the considerable controversy surrounding the roles of these auxiliary factors in ternary complex formation, our three laboratory groups collaborated to compare the structures of the relevant initiation factors prepared independently and to re-examine their roles in ternary complex formation.

Materials and Methods

Initiation factors prepared in different laboratories were made available for independent

NATO ASI Series, Vol. H 49
Post-Transcriptional Control of Gene Expression
Edited by J. E. G. McCarthy and M. F. Tuite

studies in all three laboratories. All three laboratories purified eIF-2 according to previously described procedures (Benne et al, 1979; Merrick, 1979; Roy et al, 1988). eIF-3 was purified in Hershey's Laboratory (Benne et al, 1979) and also in Merrick's Laboratory (1979). Co-eIF-2A and eIF-2B were purified in Gupta's Laboratory following the previously described procedures (Roy et al, 1988; Grace et al, 1984). Antibodies against isolated 180 kDa polypeptide component in eIF-3 was prepared following the procedure described previously (Datta et al, 1988). Standard Millipore filtration method was used for assays of ternary complex formation (Roy et al, 1988).

Results

As mentioned earlier, all three laboratories prepared initiation factors and made them available to the others. The results obtained using similar preparations from different laboratories were essentially the same. Figure 1 shows the SDS-Page of the eIF-2, Co-eIF-2A and eIF-3 preparations used in this work.

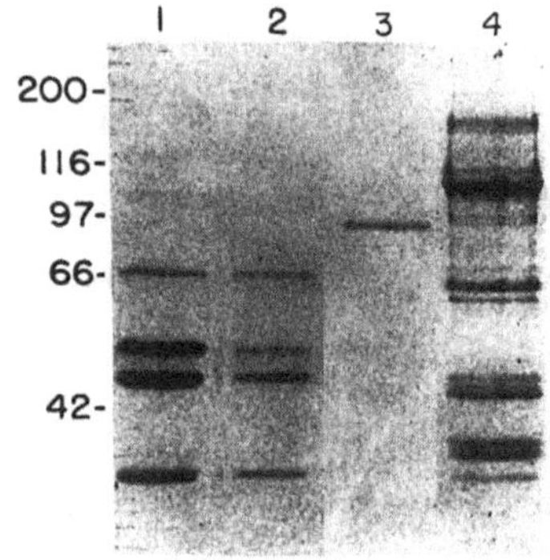

Fig. 1. SDS-polyacrylamide gel electrophoresis of different factor preparations. Lane 1, eIF-2 from Merrick; lane 2, eIF-2 from Gupta; lane 3, Co-eIF-2A from Gupta; lane 4, eIF-3 from Merrick.

As shown in Fig. 1, both eIF-2 preparations contained, in addition to the usual three subunits, the 67 kDa polypeptide (p^{67}). Co-eIF-2A gave a single polypeptide band (94 kDa) and eIF-3 gave multiple polypeptide bands including the 180 kDa polypeptide (p^{180}).

Figure 2 describes the characteristics of ternary complex formation at varying eIF-2 concentrations and the effects of addition of Mg^{2+} and natural mRNA.

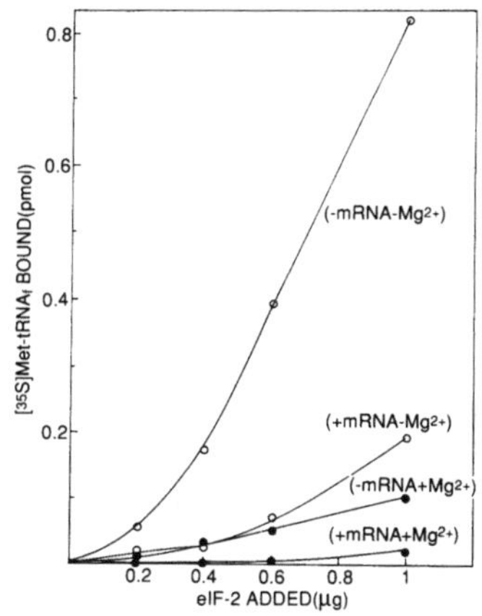

Fig. 2. Characteristics of ternary complex formation by eIF-2. Where indicated 1 mM Mg^{2+} and 0.5 μg globin mRNA were added. eIF-2 from Merrick was used.

At the lowest concentration used (0.2 μg; approximately 1 pmol assuming a molecular weight of 200 kDa for eIF-2 containing p^{67}), eIF-2 formed negligible ternary complex. In the absence of Mg^{2+}, ternary complex formation by eIF-2 followed a sigmoidal curve. At higher concentration, 5 pmol input eIF-2 (1μg) bound approximately 0.8 pmol Met-$tRNA_f$ i,e 16% of the input eIF-2 molecules formed ternary complex. However, as reported previously (Gupta et al, 1987; Roy et al, 1988), ternary complex formation by eIF-2 alone was almost completely inhibited by addition of globin mRNA and Mg^{2+}.

The results described in Figure 3 show that both Co-eIF-2A and eIF-3 strongly stimulated ternary complex formation by eIF-2. In the presence of either excess Co-eIF-2A or eIF-3, 0.2 μg (1 pmol) eIF-2 formed approximately 0.7 pmol ternary complex i.e 70% of the input eIF-2 molecules bound Met-$tRNA_f$. However, as reported previously (Gupta et al, 1987; Roy et al, 1988), the ternary complex formed in the presence of Co-eIF-2A was mostly stable in the presence of mRNA, whereas, the ternary complex formed in the presence of eIF-3 was unstable to mRNA.

In the presence of Mg^{2+}, eIF-2 becomes inactive and dependent on eIF-2B for ternary complex formation. The results presented in Figure 4 describes ternary complex formation by eIF-2 in the presence of 1 mM Mg^{2+} and additional factors. As shown in Fig. 4, eIF-2 alone or eIF-2 in the presence of either Co-eIF-2A or eIF-3 formed very little ternary complex in the presence of Mg^{2+}. At the catalytic concentration used, eIF-2B alone did not stimulate ternary complex formation by eIF-2 but when added in the presence of either Co-eIF-2A or eIF-3 significantly stimulated ternary complex formation by eIF-2. As before, the ternary complex formed in the presence of eIF-2B + Co-eIF-2A was stable to natural mRNA and a similar complex formed in the presence of eIF-2B + eIF-3 was unstable to mRNA (Data not shown).

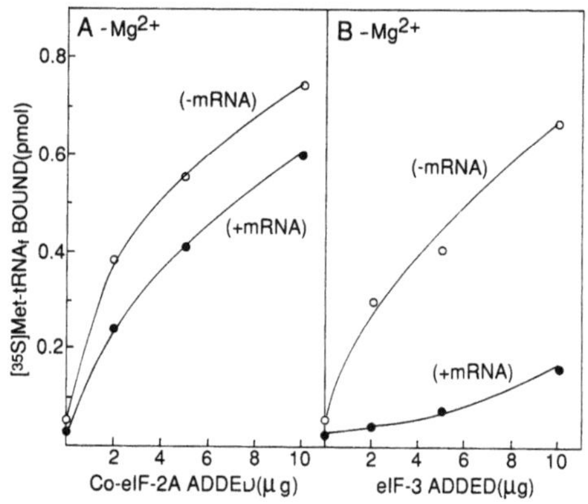

Fig. 3. Ternary complex formation by eIF-2 in the presence of Co-eIF-2A and eIF-3. Ternary complex formation was analyzed in the absence of Mg^{2+}. Where indicated Co-eIF-2A (panel A) and eIF-3 (panel B) were added. eIF-2 from Merrick was used.

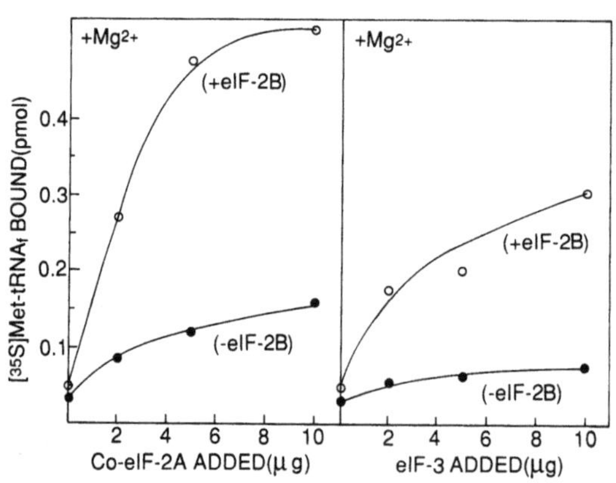

Fig. 4. Ternary complex formation by eIF-2 in the presence of eIF-2B, Co-eIF-2A and eIF-3. Ternary complex formation was analyzed in the presence of 1 mM Mg^{2+}. Where indicated 1 μg eIF-2B, Co-eIF-2A and eIF-3 were added. eIF-2 from Gupta was used.

We have further investigated the mechanism of eIF-3 stimulation of ternary complex formation. First, the stimulation of ternary complex formation was strongly inhibited by antibodies against the p^{180} subunit, whereas other antibodies had no significant effect.

Second, preparations of eIF-3 deficient in p^{180} (as a consequence of chromatography) was much less effective in stimulating ternary complex formation. These observations suggest that the p^{180} component in eIF-3 is involved in eIF-3 stimulation of ternary complex formation.

Discussion

The results presented in this paper, demonstrate that three distinct protein factors, eIF-2B, eIF-3 and Co-eIF-2A are required for ternary complex formation by eIF-2 in the presence of Mg^{2+} and natural mRNA. A brief description of these factor activities are as follows:

(1) eIF-2B: This activity is presumably required for GDP-displacement from eIF-2·GDP during ternary complex formation in the presence of Mg^{2+}. It has been previously reported that approximately 40-60% of the isolated eIF-2 molecules contain bound GDP (Gupta et al, 1987). However, as shown in Figures 1 & 3, almost all the eIF-2 molecules (with or without bound GDP) remained in an inactive form in the presence of 1mM Mg^{2+} and ternary complex formation was totally dependent on the presence of eIF-2B. These results suggest that all the eIF-2 molecules assume an inactive conformation in the presence of Mg^{2+} and eIF-2B restores the active conformation of the eIF-2 molecules. In this active conformation, GDP from eIF-2·GDP can be displaced and ternary complex can be formed in the presence of eIF-3 and/or Co-eIF-2A.

(2) eIF-3: The precise mechanism of eIF-3 stimulation of eIF-2 activity is not clear. Previous work suggested that eIF-3 stabilizes the ternary complex bound on 40S ribosomes (Peterson et al, 1979). The current work indicates that eIF-3 may stabilize the ternary complex even in the absence of 40S ribosomes and thus shift the equilibrium from free components to complexed components.

(3) Co-eIF-2A: Ternary complex formed with eIF-2B + eIF-3 was unstable in the presence of natural mRNA. Addition of Co-eIF-2A caused further stimulation of ternary complex formation and the complex was stable to physiological concentrations of a natural mRNA. Co-eIF-2A is thus required for stabilization of the ternary complex in the presence of natural mRNAs.

Based on the results presented in this paper, we propose the following tentative model for ternary complex formation by eIF-2 in the presence of eIF-2B, eIF-3 and Co-eIF-2A.

eIF-2·GDP + Met-tRNA$_f$+GTP --eIF-2B--> 3° complex (unstable) --eIF-3--> 3° complex·eIF-3

(unstable + mRNA) --Co-eIF-2A--> 3° complex·eIF-3·Co-eIF-2A (stable + mRNA)

Figure 5: A tentative proposed model for ternary complex formation by eIF-2.

In the presence of Mg^{2+}, eIF-2B at catalytic concentrations promotes GDP displacement from eIF-2·GDP and facilitates ternary complex formation. This ternary complex formation is significantly stimulated by eIF-3. However, the ternary complex formed in the presence of eIF-2B + eIF-3 is unstable to natural mRNA. Co-eIF-2A is required to further stimulate ternary complex formation by eIF-2 and subsequently to stabilize the complex in the presence of natural mRNA.

Acknowledgements: This work was supported by NIH Grants GM 22079 (to NG), GM 26796 (to WM) and GM22135 (to J.H.).

References

Benne R, Brown-Luedi ML, Hershey JWB (1979) *Methods Enzymol.* **60**:15-35.
Datta B, Chakrabarti D, Roy AL, Gupta NK (1988) *Proc Natl Acad Sci USA.* **85**:3324-3328.
Grace M, Bagchi M, Ahmad F, Yeager T, Olson C, Chakravarty I, Nasrin N, Gupta NK (1984) *Proc Natl Acad Sci USA.* **81**:5379-5383.
Gupta NK, Ahmad MF, Chakrabarti D, Nasrin N (1987) In Translation Regulation in Gene Expression. In: Ilan J (ed) Plenum Press, p 287-334.
Merrick WC (1979) *Methods Enzymol.* **60**:101-108.
Peterson DT, Merrick WC, Safer B (1979) *J Biol Chem.* **254**:2509-2516.
Roy AL, Chakrabarti D, Datta B, Hileman RE, Gupta NK (1988) *Biochemistry.* **27**:3203-3209.

PHOSPHORYLATION OF INITIATION AND ELONGATION FACTORS AND THE CONTROL OF TRANSLATION

Christopher G. Proud, Nicholas T. Redpath and Nigel T. Price

Department of Biochemistry, School of Medical Sciences, University of Bristol, Bristol, BS8 1TD, United Kingdom.

Introduction

Several protein components of the translational machinery of eukaryotic cells are subject to phosphorylation *in vitro* and/or *in vivo*. They include initiation and elongation factors, and the ribosomal protein S6. Changes in the phosphorylation states of these proteins accompany alterations in rates of translation under a variety of conditions, suggesting that their phosphorylation provides a regulatory mechanism. However, in only a few cases is there clear evidence that phosphorylation actually affects translational activity (reviewed by Hershey [1]), i.e. initiation factor eIF-2 and elongation factor EF-2, for which effects of phosphorylation on activity are now well established, and the cap-binding initiation factor eIF-4E and ribosomal protein S6, for which circumstantial evidence for regulation by phosphorylation is available. The work described here concerns the phosphorylation of eIF-2 and EF-2, and in particular the identification of the phosphorylation sites in these proteins, the characterisation of the kinases which phosphorylate them and the identification of the protein phosphatases responsible for their dephosphorylation.

Initiation factor eIF-2

eIF-2 is composed of subunits termed α (36kDa), β (38kDa) and γ (52kDa). It mediates the binding of the initiator tRNA (Met-$tRNA_i^{met}$) to the 40S ribosomal subunit as a ternary complex with GTP [eIF-2·GTP·Met-$tRNA_i^{met}$]. The GTP is hydrolysed during the subsequent formation of the complete 80S initiation complex and the eIF-2 is released as an inactive complex with GDP. Exchange of the GDP for GTP to regenerate the active [eIF-2·GTP] species requires another multimeric factor, GEF (= guanine nucleotide exchange factor). This 'recycling' step is an important control point for peptide-chain initiation.

Phosphorylation of eIF-2

Phosphorylation of the α-subunit (eIF-2α): Phosphorylation of eIF-2α is associated with inhibition of translation. The best-characterised examples of this are, in reticulocyte lysates, haem-deprivation or addition of low concentrations of double-stranded RNA (dsRNA), which lead to activation of distinct eIF-2α-specific protein kinases termed HCR (= haem-controlled repressor) and dsI (= dsRNA-activated inhibitor), respectively. Phosphorylation of eIF-2α by these kinases inhibits translation because eIF-2 phosphorylated in its α-subunit is a potent competitive inhibitor of GEF and prevents recycling of eIF-2, which then accumulates as inactive [eIF-2·GDP] complexes (see Hershey, 1989 for a review).

Increased phosphorylation of eIF-2α also occurs in other cell-types under conditions where peptide-chain initiation is impaired, e.g. due to serum-deprivation, heat-shock or absence of an essential amino acid (Duncan & Hershey, 1985; Montine & Henshaw, 1989; Rowlands *et al.*, 1988). The

NATO ASI Series, Vol. H 49
Post-Transcriptional Control of Gene Expression
Edited by J. E. G. McCarthy and M. F. Tuite

mechanism by which phosphorylation of eIF-2α is increased is unclear: it is not known for example whether the same site in eIF-2α is phosphorylated here as in reticulocyte lysates or which kinase is responsible for phosphorylation of eIF-2α under these conditions.

Phosphorylation site in eIF-2α: We have shown earlier that serine-51 is phosphorylated by HCR and by dsI *in vitro* (Colthurst *et al.*, 1987). Limited tryptic cleavage of native eIF-2 yields a fragment of approximately 6kDa corresponding to residues 1-53 of eIF-2α which contains all the radioactivity incorporated by HCR (Fig. 1A) or dsI (not shown). Digestion of this fragment with *Staphylococcus aureus* V8 proteinase gives a single phosphopeptide corresponding to residues 50-53 and containing only one serine (serine-51, Fig. 1B). This demonstrates that only serine-51 is phosphorylated by HCR or dsI *in vitro*, and that the adjacent serine-48 is not phosphorylated by these kinases (Price and Proud, 1990), in contradiction to an earlier report by Wettenhall and co-workers (1986). The sequence around the phosphorylation site in eIF-2α is:

Met-Ile-Leu-Leu-^{48}Ser-Glu-Leu-^{51}Ser(P)-Arg-Arg-Arg-Ile-Leu

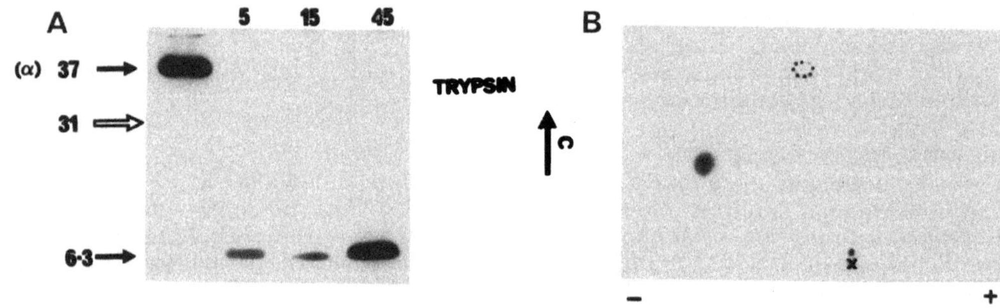

Fig. 1. Proteolytic digestion of eIF-2(α^{32}P): eIF-2 was phosphorylated by HCR using [γ-^{32}P]ATP and subjected to limited tryptic digestion (Colthurst *et al.*, 1987) and the products were analysed on Tricine gels (Price & Proud, 1990). An autoradiograph of the dried gel is shown (Panel A). The positions of eIF-2α and the 6.3kDa fragment (residues 1-53) are shown. The labelled fragment was excised, eluted and digested with V8 proteinase, and then analysed by electrophoresis at pH 3.6 (polarity shown) and chromatography (Clark *et al.*, 1989). An autoradiograph of the peptide map is shown (Panel B).

Furthermore, we have shown that serine-51 is the sole site of phosphorylation in eIF-2α in reticulocyte lysates deprived of haem or supplemented with dsRNA, where either HCR or dsI are activated (Fig. 2, Price & Proud, 1990). The techniques developed for this are applicable to the identification of phosphorylation sites in eIF-2α in intact cells. Pathak *et al.* (1988) have used a quite different approach, involving site-directed mutagenesis of either serine-48 or serine-51 to alanine, to show that serine-51 and not -48 is phosphorylated in reticulocyte lysates (Wettenhall *et al.*, 1986). In contrast to some other reports we have so far been unable to detect a second phosphorylation site in eIF-2α (Maurides *et al.*, 1989).

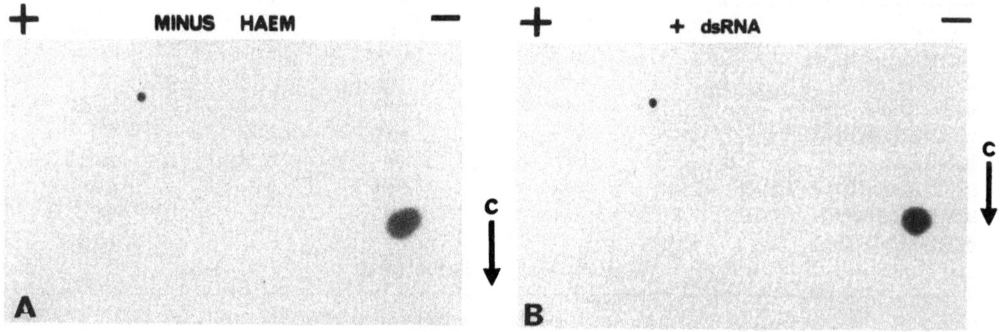

Fig. 2. Peptide maps derived from eIF-2α labelled in reticulocyte lysates incubated in the absence of haem (panel A) or the presence of dsRNA (panel B): eIF-2 excised from gel chips was digested with trypsin and V8 and the resulting peptides were analysed by two-dimensional mapping (see legend to Fig. 1).

Specificity of eIF-2α kinases: Four arginine residues lie immediately C-terminal to serine-51 in eIF-2α, suggesting that basic residues are important in the specificity of eIF-2α kinases as for several other protein kinases (although usually located in N-terminal positions). Together with Professor L. Pinna (Padua) we have tested several proteins and synthetic peptides containing clusters of basic residues as substrates for HCR and dsI, including one corresponding to residues 45-56 of eIF-2α (Table 1).

Peptides with basic residues C-terminal to the phosphorylatable residue are phosphorylated by dsI and HCR, although three adjacent basic residues are apparently insufficient. In contrast, peptides containing basic residues only in N-terminal positions relative to the serine or threonine residue are not phosphorylated at all (HCR) or only very poorly (dsI). Against protamines, HCR and dsI show differing specificities (Table 2).

Both kinases phosphorylate the synthetic peptide [eIF-2α(45-56)], although it is notable that both enzymes show a much higher K_M for this peptide than for eIF-2 itself (Table 1). This is the case for other peptide substrates of HCR and dsI (Table 1) and suggests that features other than just the primary structure are important in the recognition of substrates by these kinases. The sequence around the phosphorylation site is conserved in yeast (*Saccharomyces cerevisiae*) eIF-2α (Cigan *et al.*, 1989), suggesting yeast eIF-2 may be subject to phosphorylation (Romero & Dahlberg, 1986).

Table 1. Activities of HCR and dsI on synthetic peptides.

Substrate Peptide	Efficacy as Substrate for HCR	Efficacy as Substrate for dsI	K_M HCR(μM)	K_M dsI(μM)
native eIF-2			1	0.9
ILLSELSRRRIR (eIF-2α)	++++++	++++++	150	60
RRSTVA	-	(+)	-	n.d.
LRRASLG	-	(+)	-	n.d.
GSRRR	-	-		
GSRRRRRY	++	+++	n.d.	n.d.
RRRRYGSRRRRRRY	++	+++	100	90
RRLSLRA	(+)	+	n.d.	n.d.
RRSTVARRRRRVVRRRR	+++	+++	50	43
PRRRRRSSRPVR	++	++	75	180
RRRRRYRRSTVA	-	+	-	n.d.

n.d. = not determined;

Table 2. Phosphorylation of protamines by HCR and dsI.

Name:	CLUPEINE Y1	
Sequence:	ARRRR SSSRP IRRRR PRRRT TRRRR AGRRR R 'Site-1' 'Site-2'	
K_M(HCR):	10μM	(phosphorylates only site-1)
K_M(dsI):	11μM	(phosphorylates sites-1 and -2)
Name:	SALMINE A1	
Sequence:	PRRRR SSSRP VRRRR PRVSR RRRRR GGRRR R 'Site-1' 'Site-2'	
K_M (HCR):	12μM	(phosphorylates only site-1)
K_M (dsI):	12μM	(phosphorylates sites-1 and -2)

Phosphorylation of the ß-subunit of eIF-2 (eIF-2ß): eIF-2ß is phosphorylated on several sites in intact cells, and its phosphorylation changes under conditions where peptide chain initiation is modulated (Duncan & Hershey, 1984, 1985). *In vitro*, eIF-2ß can be phosphorylated by several protein kinases including casein kinase-2, protein kinase C and cAMP-dependent protein kinase (Tuazon *et al.*, 1980; Schatzman *et al.*, 1983; Alcazar *et al.*, 1988). We have recently shown that casein kinase-2 phosphorylates serine-2 and protein kinase C acts on serine-13 in the N-terminal section of eIF-2ß (Clark *et al.*, Price *et al.*, Pathak *et al.*, 1988a). The sequence of the N-terminal portion of eIF-2ß containing these sites is:

X-Met-^{2}Ser(P)-Gly-Asp-Glu-Met-Ile-Phe-Asp-Pro-Thr-Met-^{13}Ser(P)-$(Lys)_8$

Casein kinase-2 also phosphorylates at least one other residue which is located in the largest CNBr fragment (residues 205-333). cAMP-dependent protein kinase phosphorylates a serine, probably serine-218, which has two adjacent N-terminal lysine residues, the recognition feature for this kinase (unpublished data) (Fig. 3).. We are currently studying which sites are phosphorylated in intact cells. The effect of phosphorylation of eIF-2ß on eIF-2 activity is unclear, as indeed is the function of eIF-2ß itself. Preparations lacking eIF-2ß are functional in initiation complex formation

(Colthurst & Proud, 1986 and references therein). eIF-2ß appears to be involved in binding guanine nucleotides (Bommer & Kurzchalia, 1989). Recent work suggests that eIF-2ß plays a role in interaction with mRNA and initiation site selection (Donahue *et al.*, 1988; Dasso *et al.*, 1990), and it will be important to test whether phosphorylation of eIF-2ß affects this.

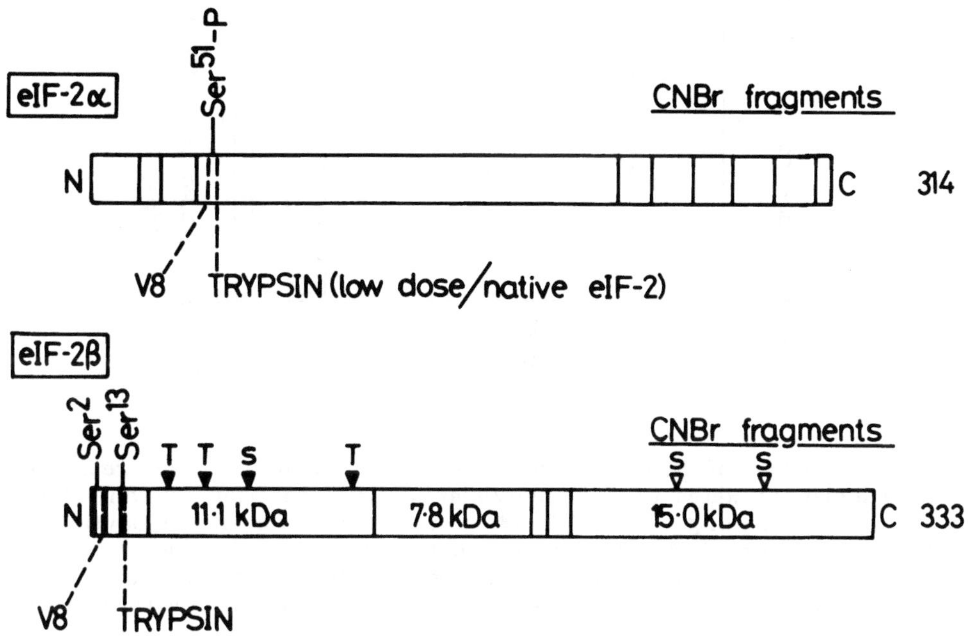

Fig. 3. **Location of Phosphorylation Sites in eIF-2α and eIF-2ß:** The N- and C-termini of each chain, and their total numbers of amino acids are indicated. Unmarked vertical lines show CNBr cleavage points, and dotted lines show selected cleavage sites for V8 or trypsin. Serine-51 of eIF-2α, phosphorylated by HCR and dsI, is marked, as are serine-2 and serine-13 of eIF-2ß, the sites of phosphorylation for casein kinase-2 and protein kinase C. Triangles indicate possible additional phosphorylation sites for casein kinase-2 in eIF-2ß (S = serine; T = threonine).

Phosphorylation of elongation factor-2

Phosphorylation of EF-2 inhibits elongation: Elongation factor EF-2 is a monomeric protein (M_r 100 000) which mediates the translocation step of peptide chain elongation on the ribosome. This process requires GTP-hydrolysis and EF-2 binds this nucleotide. EF-2 is phosphorylated, on threonine residues, only by a specific Ca^{2+}/calmodulin-dependent protein kinase (now termed EF-2 kinase) (Nairn & Palfrey, 1987; Ryazanov *et al.*, 1988). Phosphorylation of EF-2 impairs its ability to support translation of poly(U) and to mediate the translocation step. Recently we have shown that phosphorylation of EF-2 inhibits the translation of natural mRNA.

Okadaic acid is a tumour promoting agent which is a specific inhibitor of protein phosphatases (see below). It potently inhibits translation in the

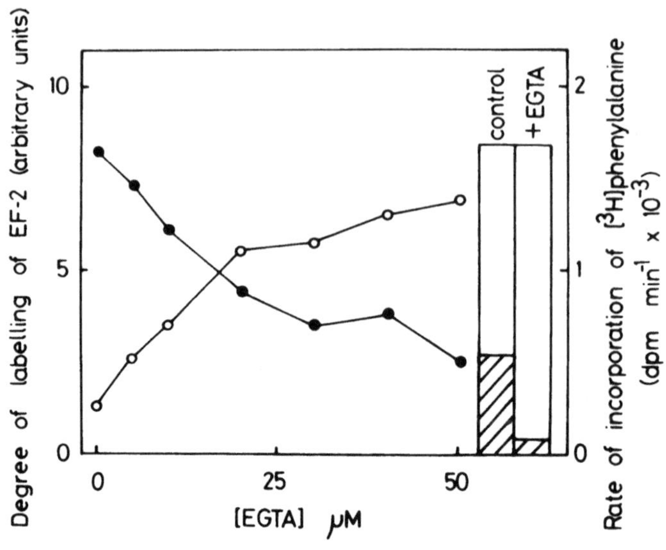

Fig. 4. Influence of EGTA on the effects of okadaic acid on reticulocyte lysate translation: Rates of [^{3}H]phenylalanine incorporation (O) and relative levels of [^{32}P]labelling of EF-2 (●) are shown for reticulocyte lysate incubations containing 50nM okadaic acid and indicated concentrations of EGTA. Open bars show [^{3}H]phenylalanine incorporation and hatched bars EF-2 labelling in incubations in the absence of okadaic acid with or without 50μM EGTA. Taken from Redpath and Proud, 1989.

in the reticulocyte lysate cell-free system (Redpath & Proud, 1989). At 20-50nM, okadaic acid causes aggregation of polysomes (not shown), characteristic of inhibition of elongation rather than initiation, and causes enhanced phosphorylation of a prominent protein of M_r 100kDa. Both effects are antagonised by chelation of Ca^{2+}-ions (using EGTA, Fig. 4) and by calmodulin antagonists (not shown). These data demonstrate that phosphorylation of EF-2 inhibits the translation of natural mRNA in an authentic translation system. The mechanism by which phosphorylation of EF-2 inhibits elongation is unclear. Since the EF-2 kinase depends on Ca^{2+}/calmodulin, this mechanism could provide a link between Ca^{2+}-ion concentrations and the control of translation, although its physiological significance remains unknown (particularly since effects of Ca^{2+} on translation appear to be exerted mainly at initiation rather than elongation (Brostrom *et al.*, 1989; Kumar *et al.*, 1989)).

Okadaic acid also inhibits protein synthesis in intact cells, although higher concentrations are required (0.5μM). This effect appears to result from inhibition of initiation as well as elongation (Redpath & Proud, unpublished).

Phosphorylation sites in EF-2: Phosphorylation of EF-2 by EF-2 kinase occurs exclusively on threonine residues. Incorporation of up to three phosphates occurs *in vitro*, the first and second being incorporated much faster than the third, based on isoelectric focusing analysis (Fig. 5). Earlier work suggested that the phosphorylated residues were located in the tryptic fragment containing residues 51-60 (Nairn & Palfrey, 1987):

Ala-Gly-Glu-^{54}Thr-Arg-Phe-^{57}Thr-Asp-^{59}Thr-Arg.

Our data confirm that phosphorylation occurs only within the cyanogen bromide fragment corresponding to residues 23-156 (Price & Proud, data not shown). Digestion of this fragment with trypsin yields three major and one minor radiolabelled peptide as analysed by mapping (not shown) or reversed-phase chromatography (Fig. 6). An additional peak, eluting just after peak 2, is the main species in monophosphorylated EF-2. Work is in hand (in collaboration with Dr. D. Campbell and Professor P. Cohen, Dundee) to identify the phosphopeptides and the sites of phosphorylation.

Protein phosphatases acting on eIF-2 and EF-2

Eukaryotic cells contain just four types of protein phosphatase which act on phosphoserine or phosphothreonine residues and are termed protein phosphatases-1, -2A, -2B and -2C (Cohen & Cohen, 1989). Protein phosphatases-1 and -2A account for almost all the cellular protein phosphatase activity against most phosphoproteins. The different classes of protein phosphatase can be distinguished by their differing susceptibilities to specific inhibitors and/or dependence on divalent cations. Type-1 protein phosphatase is inhibited by two specific inhibitor proteins (inhibitors-1 and -2) which allows it to be distinguished from the other protein phosphatases. Type-2A phosphatase is inhibited by nanomolar concentrations of okadaic acid (see above), which only inhibits protein phosphatases-1 and -2B at much higher concentrations. Protein phosphatases-2B and -2C can be identified by their requirements for Ca^{2+} and Mg^{2+}-ions, respectively (Cohen *et al.*, 1989). We have used these properties to identify the protein phosphatases acting on EF-2 and eIF-2α in reticulocyte lysates and extracts of other cells.

Activity of purified protein phosphatases against EF-2 and eIF-2: The two major cellular protein phosphatases were tested against EF-2 and eIF-2(αP), and the rates compared with those against a standard, common substrate for both enzymes, phosphorylase *a*, arbitrarily assigned as 100. EF-2 is a very poor substrate for phosphatase-1 (4±1% of rate with phosphorylase *a*) but a very good one for phosphatase-2A (150±3% of rate with phosphorylase *a*), while eIF-2(αP) is a moderate substrate for both enzymes (18±2% and 24±2% of rates with phosphorylase *a* for phosphatases-1 and -2A respectively, Redpath & Proud, unpublished observations). Using a different approach, Gschwendt *et al.* (1990) showed that EF-2 was dephosphorylated by phosphatase-2A.

Protein phosphatase activities in cell extracts against EF-2 and eIF-2: The activities of phosphatases-1 and -2A against phosphorylase *a*, EF-2 and eIF-2(αP) were assessed using the technique of Cohen *et al.* (1989). The extent of inhibition by inhibitor-1 or 1nM okadaic acid indicates the contributions due to phosphatases-1 and -2A. Both together should give complete inhibition as the protein phosphatases not inhibited under these conditions are dependent on Ca^{2+} (2B) or Mg^{2+}. Addition of these cations then gives the activity of these two protein phosphatases. The results for reticulocyte lysates are shown in Fig. 7.

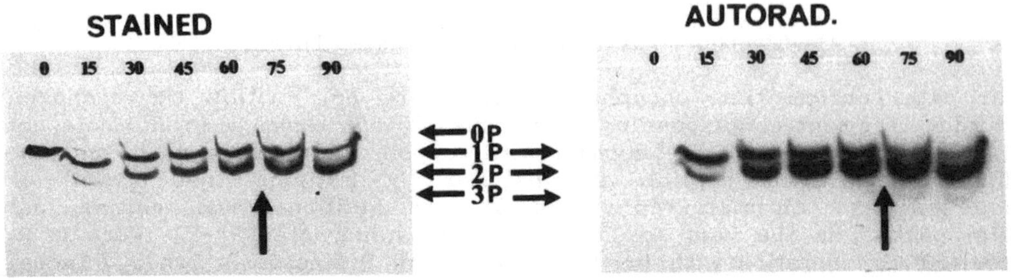

Fig. 5. Time course of phosphorylation of EF-2: EF-2 was incubated with EF-2 kinase and [γ-^{32}P]ATP/$MgCl_2$ (0.1mM/1mM). Samples were removed and analysed by slab-gel isoelectric focusing. The stained gel and corresponding autoradiograph are shown. Positions of species of differing pI apparently corresponding to unphosphorylated and mono-, di- and tri-phosphorylated EF-2 are shown by labelled arrows. After 60 minutes more kinase (three times the initial concentration) was added and the incubation continued (vertical arrow). The anode was at the bottom.

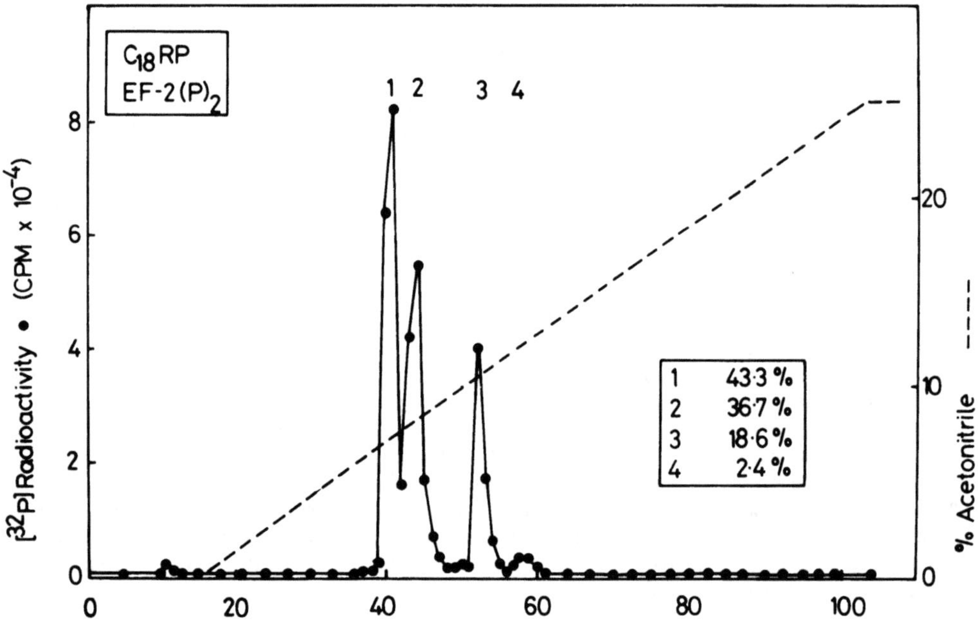

Fig. 6. Analysis of tryptic fragments of diphosphorylated EF-2 by reverse-phase chromatography: Diphosphorylated EF-2 was digested with CNBr, carboxymethylated and digested with trypsin. The digest was analysed on a C_{18} reverse-phase column in 0.1% TFA with increasing acetonitrile. Numbers indicate radioactive peaks referred to in the text.

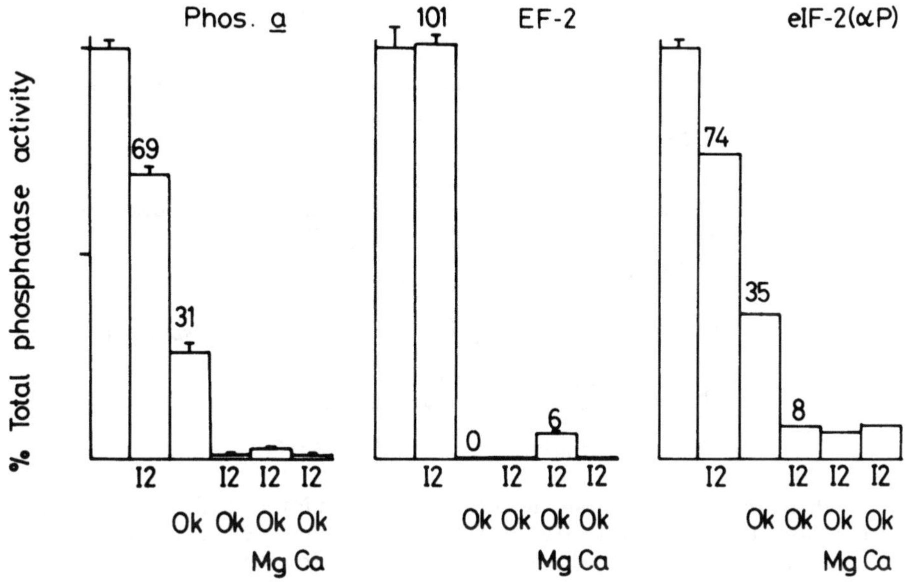

Fig. 7. Protein phosphatase activities in reticulocyte lysates against phosphorylase *a*, EF-2 and eIF-2(αP): Dephosphorylation of [^{32}P]labelled substrates by reticulocyte lysate samples was measured in the absence and presence of inhibitor-protein-2 (0.2µM), okadaic acid (2nM), both, and both plus $MgCl_2$ (20mM) or $CaCl_2$ (20mM) as shown. Results are given as % of the activity measured with each substrate in the absence of effectors.

Phosphatase activity against EF-2 is almost entirely due to phosphatase-2A (with some activity of protein phosphatase-2C). For eIF-2(αP), the contribution of phosphatase-2A is about twice that of phosphatase-1, similar to the relative activities against phosphorylase *a*. Protein phosphatase-2A is also the major EF-2 phosphatase in extracts of Swiss 3T3 fibroblasts and hepatocytes, while phosphatase-1 makes a significant contribution to eIF-2(αP) phosphatase activity (Fig. 8) in both cases (Redpath & Proud, unpublished observations. It is notable that phosphatase-2C provides significant activity against EF-2 in hepatocyte extracts under these assay conditions (20mM $MgCl_2$), although the contribution at physiological Mg^{2+} concentrations is likely to be much lower.

The activity of protein phosphatase-1 can be regulated by several mechanisms involving phosphorylation of inhibitor proteins 1 and 2, and of 'targetting' proteins (Cohen and Cohen, 1989). Therefore the level of phosphorylation of eIF-2α may be controlled by alterations in protein phosphatase as well as eIF-2α kinase activities.

Acknowledgements: This work was supported by Grants to CGP from the Medical Research Council, the Science and Engineering Research Council, the Nuffield Foundation and the British Diabetic Association. We are grateful to Professor Philip Cohen (Dundee) for his invaluable and generous help with our investigations of protein phosphatases.

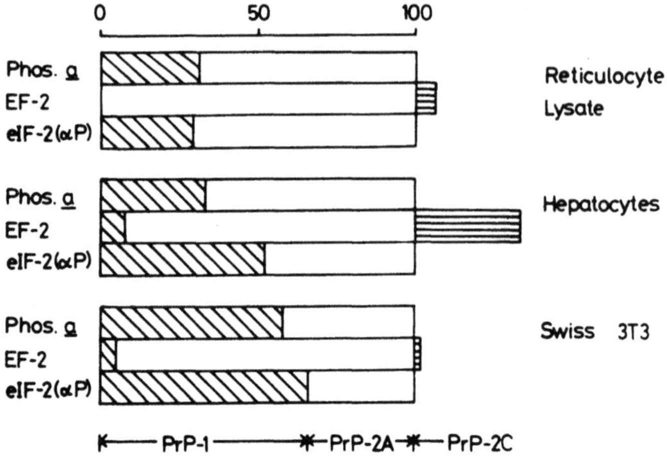

Fig. 8. Protein phosphatase activities in various cell extracts against phosphorylase *a*, EF-2 and eIF-2(αP): Activities are expressed as % of phosphatase activity measured without effectors for each substrate.

Alcazar A, Mendez E, Fando J, Salinas M (1988) Specific phosphorylation of the ß subunit of eIF-2 factor from brain by three different protein kinases. Biochem Biophys Res Commun 153: 313-320.

Bommer U-A, Kurzchalia TV (1989) GTP interacts through its ribose and phosphate moieties with different subunits of the eukaryotic initiation factor eIF-2. FEBS Lett 244: 323-327.

Brostrom CO, Chin KV, Wong WL, Cade C, Brostrom MA (1989) Inhibition of translation initiation in eukaryotic cells by calcium ionophore. J Biol Chem 264: 1644-1649.

Cigan AM, Pabich EK, Feng L, Donahue TF (1989) Yeast translational suppressor *SUI2* encodes the alpha subunit of eukaryotic initiation factor 2 and shares sequence identity with the human alpha subunit. Proc Natl Acad Sci USA 86: 2784-2788.

Clark SJ, Colthurst DR, Proud CG (1988) Structure and phosphorylation of eukaryotic initiation factor-2. Casein kinase-2 and protein kinase C phosphorylate distinct but adjacent sites in the ß-subunit. Biochim Biophys Acta 968: 211-219.

Clark SJ, Ashford AJ, Price NT, Proud CG (1989) Casein kinase-2 phosphorylates serine-2 in the ß-subunit of initiation factor-2. Biochim Biophys Acta 1010: 377-380.

Cohen P, Cohen PTW (1989) Protein phosphatases come of age. J Biol Chem 264: 21435-21438.

Cohen P, Klumpp S & Schelling DL (1989) An improved procedure for identifying and quantitating protein phosphatases in mammalian tissues. FEBS Lett 250: 596-600.

Colthurst DR, Proud, CG (1986) Eukaryotic initiation factor-2 from rat liver: no apparent function for the ß-subunit in the formation of initiation complexes. Biochim Biophys Acta 868: 77-86.

Colthurst DR, Campbell DG, Proud CG (1987) Structure and regulation of eukaryotic initiation factor eIF-2. Sequence of the site in the α-subunit phosphorylated by the haem-controlled repressor and by the double stranded-RNA activated inhibitor. Eur J Biochem 166: 357-363.

Dasso MC, Milburn SC, Hershey JWB, Jackson RJ (1990) Selection of the 5'-proximal translation initiation site is influenced by mRNA and eIF-2 concentrations. Eur J Biochem 187: 361-371.

Donahue TF, Cigan AM, Pabich EK, Valavicius BC (1988) Mutations at a Zn(II) finger motif in the yeast eIF-2ß gene alter ribosomal start-site selection during the scanning process. Cell 54: 621-632.

Duncan R, Hershey JWB (1985) Regulation of initiation factors during translational repression caused by serum depletion. Covalent modification. J Biol Chem 260: 5493-5497.

Gschwendt M, Kittstein W, Mieskes G, Marks F A type-2A phosphatase dephosphorylates the elongation factor 2 and is stimulated by the phorbol ester TPA in mouse epidermis *in vivo*. FEBS Lett. 257: 357-360.

Hershey JWB (1989) Protein phosphorylation controls translation rates. J Biol Chem 264: 20823-20826.

Kumar RV, Wolfman A, Panniers R, Henshaw EC (1989) Mechanism of inhibition of polypeptide chain initiation in calcium-depleted Ehrlich ascites tumour cells. J Cell Biol 108: 2107-2115.

Maurides PA, Akkaraju GR, Jagus R (1989) Evaluation of protein phosphorylation state by a combination of vertical slab gel isoelectric focusing and immunoblotting. Anal Biochem 183: 144-151.

Montine KS, Henshaw EC (1989) Serum growth factors cause raoid activation of protein synthesis and dephosphorylation of eIF-2 in serum-deprived Ehrlich cells. Biochim Biophys Acta 1014: 282-288.

Nairn AC, Palfrey HC (1987) Identification of the major M_r 100 000 substrate for calmodulin-dependent protein kinase III in mammalian cells as elongation factor-2. J Biol Chem 262: 17299-17303.

Pathak VK, Nielsen PJ, Trachsel H, Hershey JWB (1988a) Structure of the ß-subunit of translational initiation factor eIF-2. Cell 54: 633-639.

Pathak VK, Schindler D, Hershey JWB (1988b) Generation of mutant form of protein synthesis initiation factor eIF-2 lacking the site of phosphorylation by eIF-2 kinases. Mol Cell Biol 8: 993-995.

Price NT, Proud CG (1990) Phosphorylation of protein synthesis initiation factor-2. Identification of the site in the α-subunit phosphorylated in reticulocyte lysates. Biochim Biophys Acta, In Press.

Redpath NT, Proud CG (1989) The tumour promoter okadaic acid inhibits reticulocyte lysate protein synthesis by increasing the net phosphorylation of elongation factor-2. Biochem J 262: 69-75.

Romero DP, Dahlberg AE (1986) The alpha subunit of initiation factor-2 is phosphorylated *in vivo* in the yeast *Saccharomyces cerevisiae*. Mol Cell Biol 6: 1044-1049.

Rowlands AG, Montine KM, Henshaw EC, Panniers R (1988) Physiological stresses inhibit guanine-nucleotide-exchange factor in Ehrlich cells. Eur J Biochem 175: 92-99.

Ryazanov AG, Shestakova EA, Natapov PG (1988) Phosphorylation of elongation factor-2 by EF-2 kinase affects rate of translation. Nature 334: 170-173.

Schatzman R C, Grifo JA, Merrick WC, Kuo JF (1983) Phospholipid-sensitive, Ca^{2+}-dependent protein kinase phosphorylates the ß-subunit of eIF-2. FEBS Lett 159: 167-170.

Tuazon PT, Merrick WC, Traugh JA (1980) Site-specific phosphorylation of initiation factor-2 by three cyclic nucleotide-independent protein kinases. J Biol Chem 255: 10954-10958.

Wettenhall REH, Kudlicki W, Kramer G, Hardesty B (1986) The NH_2-terminal sequence of the α and γ subunits of eukaryotic initiation factor-2 and the phosphorylation site for the haem-regulated eIF-2α kinase. J Biol Chem 261: 12444-12447.

REGULATION OF PROTEIN SYNTHESIS IN MAMMALIAN SYSTEMS BY THE GUANINE NUCLEOTIDE EXCHANGE FACTOR

Jaydev N. Dholakia and Albert J. Wahba
Department of Biochemistry
The University of Mississippi Medical Center
2500 North State Street
Jackson, MS 39216
U.S.A.

In mammalian cells, the guanine nucleotide exchange factor (GEF) plays a major role in regulating protein synthesis. The first step in polypeptide chain initiation is the formation of a ternary (eIF-2·GTP·Met-tRNA$_f$) complex where eIF-2 is eukaryotic initiation factor (eIF) 2. This is followed by the transfer of this complex to a 40 S ribosomal subunit (Wahba and Woodley, 1984; Pain, 1986). In the presence of mRNA, other initiation factors and the 60S ribosomal subunit, an 80S initiation complex is formed, thereby setting the stage for polypeptide chain elongation. Upon formation of the 80S initiation complex, GTP is hydrolyzed and eIF-2 is released as a binary complex of eIF-2·GDP. The binary complex is stable in the presence of Mg^{2+} and is functionally inactive. Regeneration of the ternary (eIF-2·GTP·Met-tRNA$_f$) complex for a new cycle of initiation requires GEF which catalyzes the exchange of eIF-2-bound GDP for GTP (Dholakia and Wahba, 1989). It is at this point in the eIF-2 cycle that the initiation process is regulated (Fig. 1). Phosphorylation of the α-subunit of eIF-2 by either the heme-controlled repressor or the double-stranded RNA induced kinase is associated with the cessation of protein synthesis and is due to the inability of GEF to catalyze the GDP/GTP exchange from eIF-2(α-P)·GDP (Pain, 1986). Our studies demonstrate that GEF activity may also be regulated directly by phosphorylation of one of its subunits and the redox state of the cell (Dholakia *et al.*, 1986; Dholakia and Wahba, 1988). GEF is a five subunit protein which is isolated with bound NADPH (Dholakia *et al.*, 1986). Recent experiments illustrate that the GEF-catalyzed reaction follows a sequential enzyme mechanism. The addition of GEF alone does not cause the displacement of eIF-2-bound GDP. The release of bound GDP is dependent on the presence of both GTP and GEF. An important observation is the binding of GTP to GEF. Guanine binding domains on GEF and eIF-2 may be identified by covalent modification of these proteins with the photoreactive 8-azido analog of GTP, [γ-^{32}P]8N$_3$GTP.

NATO ASI Series, Vol. H 49
Post-Transcriptional Control of Gene Expression
Edited by J. E. G. McCarthy and M. F. Tuite

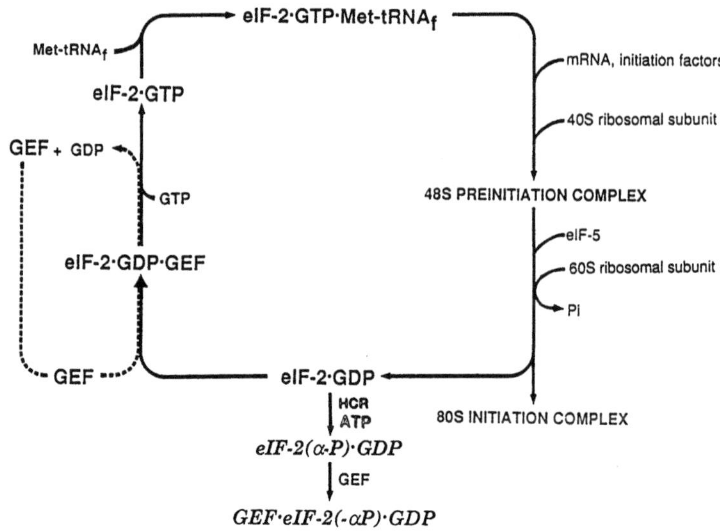

FIG 1. **eIF-2 cycle during polypeptide chain initiation.**

CHARACTERIZATION OF GEF AND eIF-2

Rabbit reticulocyte GEF and eIF-2 may be purified to apparent homogeneity from the postribosomal supernatant and the 0.5 M KCl wash of ribosomes, respectively (Dholakia and Wahba, 1987, 1988). GEF consists of five polypeptides of M_r 82,000, 65,000, 55,000, 40,000 and 34,000 and eIF-2 is composed of three non-identical subunits with apparent M_r 52,000-γ, 50,000-β and 38,000-α (Fig. 2). Until the last step in the purification procedure, GEF is isolated as a 1:1 complex with eIF-2. Chromatography on a FPLC Mono Q column results in a GEF preparation which is free of eIF-2.

REGULATION OF GEF BY THE REDOX STATE OF THE CELL

Fluorescence spectroscopy reveals a fluorescence maximum near 435 nm when GEF is excited at 335 nm (Fig. 3). The fluorescence spectrum is characteristic of protein-bound reduced pyridine dinucleotides and suggests the presence of NADH or NADPH bound to GEF. Two independent experiments confirm the presence of NADPH as the reduced pyridine dinucleotide associated with GEF (Dholakia *et al.*, 1986): (i) The addition of NADPH to GEF previously depleted of reduced pyridine dinucleotide results in quenching the tryptophan fluorescence of the protein by 60%,

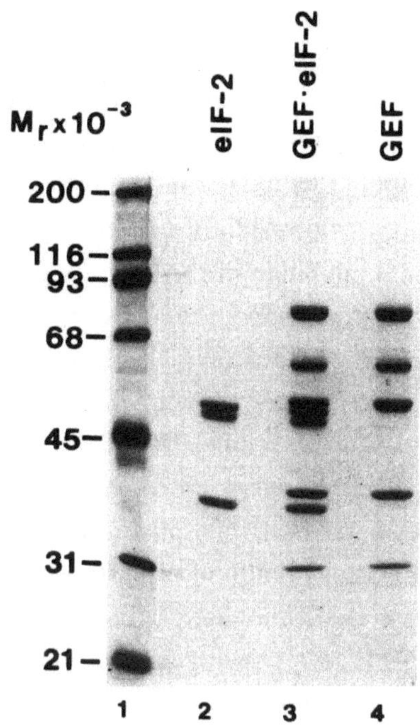

FIG. 2. **Gel electrophoretic analysis of the factors.** The polypeptide composition of the purified preparations (2μg each) of rabbit reticulocyte eIF-2 (*lane 2*), eIF-2 and GEF (*lane 3*), and GEF (*lane 4*) are analyzed by sodium dodecyl sulfate polyacrylamide gel electrophoresis followed by silver staining. *Lane 1* contains standard molecular weight markers.

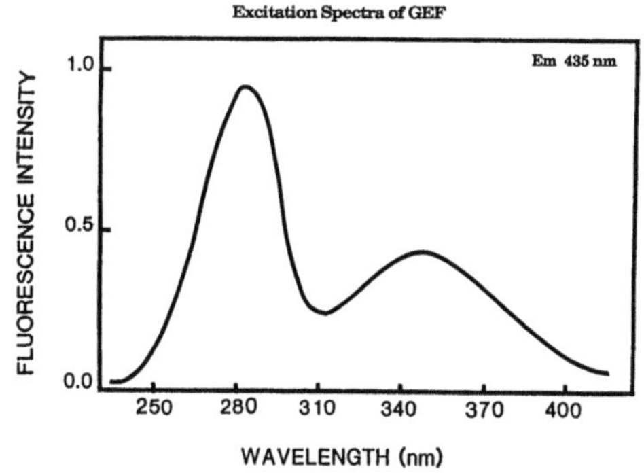

FIG. 3. **Excitation spectrum of GEF.** Emission is monitored at 435 nm.

whereas the same level of NADH results in only 10% quenching of tryptophan fluorescence; (ii) NADPH may be identified by extraction from GEF and subsequent chromatography on reverse-phase or ion-exchange columns, or by a $NADP^+$(H)-specific enzyme assay (Dholakia *et al.*, 1986). In order to determine the effect of pyridine dinucleotides on its activity, GEF may be assayed in the presence of oxidized or reduced pyridine dinucleotides (Table I). The addition of 0.5-0.75 mM NAD^+ or $NADP^+$ results in the complete inhibition of the GEF-catalyzed exchange of GTP for eIF-2-bound [^{3}H]GDP. On the other hand, the addition of NADH or NADPH does not affect the exchange reaction. These results, therefore, provide evidence that oxidized and reduced pyridine dinucleotides may directly interact with GEF and consequently establish a link between the redox state of the cell and the regulation of polypeptide chain initiation. They may also provide a basis for many earlier observations on the effects of NADPH and sugar phosphates on protein synthesis in cell-free extracts and the requirement of glycerol and dithiothreitol for the maintenance of GEF activity. The 20-fold increase in the rate of protein synthesis which is paralleled by a 3-5-fold increase in NADPH levels during fertilization of sea urchin eggs may be due to the activation of GEF (Colin *et al.*, 1987).

Table I

Effect of pyridine dinucleotides on GEF-dependent release of [^{3}H]GDP from the eIF-2 · GDP binary complex

Concentration, mM	[^{3}H]GDP released, cpm			
	NAD^+	NADH	$NADP^+$	NADPH
0	4050	4060	4100	4080
0.25	4050	4070	2340	4385
0.50	2210	4100	1810	4365
0.75	0	4200	110	4400
1.00	0	4250	0	4300
2.00	0	4090	0	4040

Each reaction mixture contains in 75 μl: 20 mM Tris · HCl (pH 7.8), 1 mM Mg^{2+}, 1 mM dithiothreitol, 100 mM KCl, 0.2 mM GTP, 10 μg of bovine serum albumin, 0.25 pmol of GEF and 1.0 pmol of the isolated eIF-2 · [^{3}H]GDP (5000 cpm/pmol). The reactions are initiated by the addition of the preformed binary complex. In the absence of GEF, the pyridine dinucleotides have no effect on the nucleotide exchange reaction.

PHOSPHORYLATION OF GEF REGULATES ITS ACTIVITY

The guanine nucleotide exchange reaction may also be regulated by phosphorylation/dephosphorylation of GEF (Dholakia and Wahba, 1988). The M_r 82,000 subunit of GEF is phosphorylated *in vitro* by ATP and casein kinase II (Fig. 4). The phosphorylation of the isolated protein results in a 2.4-fold increase in its activity (Table II). Treatment with alkaline phosphatase of phosphorylated GEF decreases its activity approximately 5-fold, and rephosphorylation restores it to the level of the phosphorylated factor. Casein kinase II is active in all cell types examined and is regulated by direct interaction with specific intracellular acidic and basic compounds (Hathaway and Traugh, 1981). The enzyme is stimulated by monovalent cations and polyamines

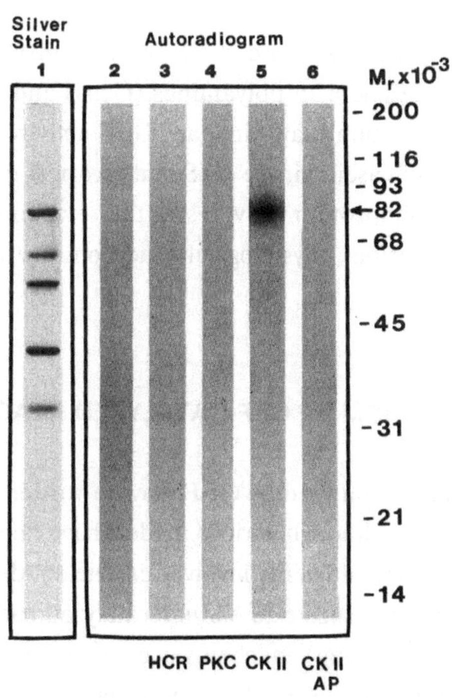

FIG. 4. **Phosphorylation of GEF by various kinases.** GEF (2 μg) is phosphorylated in the presence of [γ-^{32}P]ATP with either the heme-controlled repressor (HCR), protein kinase C (PKC) or casein kinase II (CKII) as indicated. The reaction mixture for lane 6 contains GEF phosphorylated by CKII which is further incubated with alkaline phosphatase.

Table II

Effect of phosphorylation of GEF on the release of [^{3}H]GDP from the eIF-2·GDP binary complex

Addition(s)	[^{3}H]GDP released	Relative activity
	mol/mol of GEF	*-fold*
Isolated GEF	4.6	2.0
Phosphorylated GEF	11.1	4.8
Dephosphorylated GEF	2.3	1.0
Rephosphorylated GEF	11.1	4.8

GEF is phosphorylated or dephosphorylated as described in Fig. 4 and isolated by chromatography on a Mono S column.

and inhibited by heparin, and a decrease in GEF activity may be observed when rabbit reticulocyte lysates are deficient in polyamines (Gross and Rubino, 1989). Recent studies also indicate that insulin may stimulate GEF activity (Kimball and Jefferson, 1988) through activation of casein kinase II (Sommercorn *et al.*, 1987). The characterization of GEF as a phosphoprotein provides further evidence that protein synthesis may be regulated, under various physiological conditions, by the phosphorylation state of this factor.

MECHANISM OF GEF-CATALYZED REACTION

The overall reaction catalyzed by GEF may be written as: eIF-2·GDP + GTP $\rightleftharpoons$ eIF-2·GTP + GDP. Although various models have been proposed for its mode of action, the exact sequence of events involved in nucleotide exchange is not clearly established (Pain, 1986). Two substrate enzyme-catalyzed pathways may follow either a substituted enzyme (ping-pong) mechanism or a sequential enzyme mechanism (Fig. 5). We studied the guanine nucleotide exchange reaction by three different techniques: (a) membrane filtration assays to measure the release of [^{3}H]GDP from the eIF-2·[^{3}H]GDP binary complex (Table III), (b) changes in the steady-state polarization of fluorescamine-GDP during the exchange reaction (Table IV), and (c) sucrose gradient analysis of the total reaction (Dholakia and Wahba, 1989). In the presence of GTP the smallest amount of GEF (2.5 nM) is sufficient to catalyze the

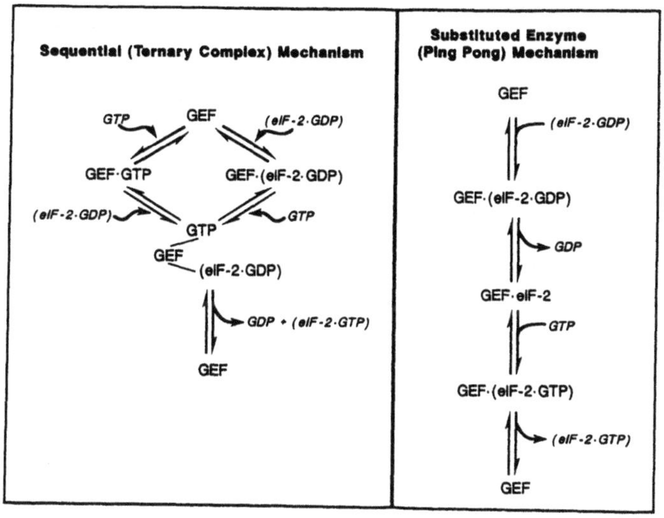

FIG. 5. **Possible mechanisms for the GEF-catalyzed guanine nucleotide exchange reaction.**

Table III

Release of [³H]GDP from eIF-2·[³H]GDP binary complex

Additions		[³H]GDP·eIF-2	
GEF	GTP	5 nM	15 nM
nM	*20 μM*	*cpm*	
0	-	3000	9010
0	+	3010	9005
2.5	+	301	490
2.5	-	3005	9005
5.0	-	2980	8990
10.0	-	2990	9050
20.0	-	2890	8880
40.0	-	2870	8850

-, no GTP added; +, GTP added.

complete release of eIF-2-bound [³H]GDP from 15 nM eIF-2·[³H]GDP binary complex. These results clearly indicate that GEF functions catalytically in promoting the release of eIF-2-bound GDP. However, in the absence of added GTP, a 5-8-fold excess of GEF over the eIF-2·[³H]GDP binary complex does not displace the eIF-2-bound GDP. Similar results are obtained when fluorescence polarization is used to monitor the binding and release of the fluorescent nucleotide, fluorescamine-GDP (Table IV). Addition of either GEF or GTP alone to the eIF-2·fluorescamine-GDP complex has no effect on the polarization value, but the presence of both GTP and GEF results in the decrease in polarization from 0.3 to 0.06, a value obtained with the free nucleotide. Therefore, the results obtained do not support the reaction as written: eIF-2·GDP + GEF $\rightleftharpoons$ eIF-2·GEF + GDP. The release of eIF-2-bound GDP is dependent on the presence of both GTP and GEF, and this argues against the possibility of a ping-pong mechanism.

Enzyme kinetic analysis of the GEF-catalyzed reaction was performed by assaying the release of eIF-2-bound GDP at several constant concentrations of one substrate (GTP or eIF-2·GDP) while varying the second substrate concentration (Dholakia and Wahba, 1989). The patterns of the resulting double-reciprocal plots strongly suggest that this reaction may follow a sequential mechanism via the formation of a [GTP·GEF·(eIF-2·GDP)] ternary complex (Fig. 5). This implies that in addition to having a site for eIF-2·GDP, GEF should also have a specific GTP binding site. Changes in the intrinsic tryptophan fluorescence of GEF may be used to monitor GTP binding to GEF. As shown in Fig. 6, the addition of GTP to GEF causes a decrease in the intensity of the tryptophan fluorescence. A K_d of 4.1 μM for GTP binding to GEF is estimated when the decrease in fluorescence is plotted against the corresponding GTP concentration. A similar K_d (3.9 μM) for GTP binding to GEF may also be calculated when [^{3}H]GTP binding to GEF is determined by nitrocellulose membrane filtration assays (Dholakia and Wahba, 1989). In summary, our results which indicate GTP and eIF-2·GDP binding to GEF, the requirement of both substrates for the release of

Table IV

Changes in the fluorescence properties of fluorescamine-GDP

Additions	Fluorescence polarization
F-GDP	0.05
F-GDP + eIF-2	0.31
F-GDP + eIF-2 + Mg^{2+}	0.30
F-GDP + eIF-2 + Mg^{2+} + GEF	0.31
F-GDP + eIF-2 + Mg^{2+} + GTP	0.30
F-GDP + eIF-2 + Mg^{2+} + GEF + GTP	0.06

F-GDP denotes fluorescamine-GDP.

eIF-2-bound GDP and the pattern of the double-reciprocal plots, strongly suggest that the guanine nucleotide exchange reaction follows a sequential enzyme mechanism. Therefore, the GEF-catalyzed reaction is different from the nucleotide exchange reaction involved during polypeptide chain elongation (EF-Tu-EF-Ts) but may be considered analogous to the eukaryotic receptor G-protein system which proceeds via a hormone-receptor-(G·GDP) ternary complex.

CHARACTERIZATION OF GEF AS A GTP BINDING PROTEIN

The characteristics of GEF, a multisubunit protein, suggest that this factor may be composed of several distinct structural and functional domains. It is important to identify the site of some of these domains in order to understand the catalytic and

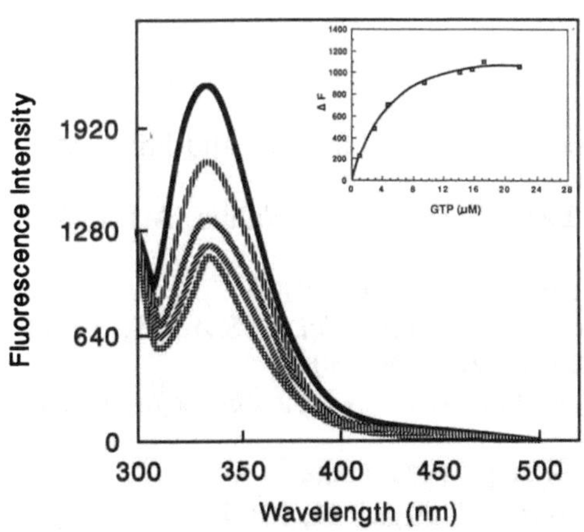

FIG. 6. **Effect of GTP on the tryptophan emission spectrum of GEF.** GEF (0.1 mg/ml) is excited at 286 nm. The emission spectrum is measured in the absence (—) or the presence of 3, 9, 15 and 21 μM GTP. *Inset*, the decrease in the fluorescence intensity (F) is plotted against the corresponding GTP concentration.

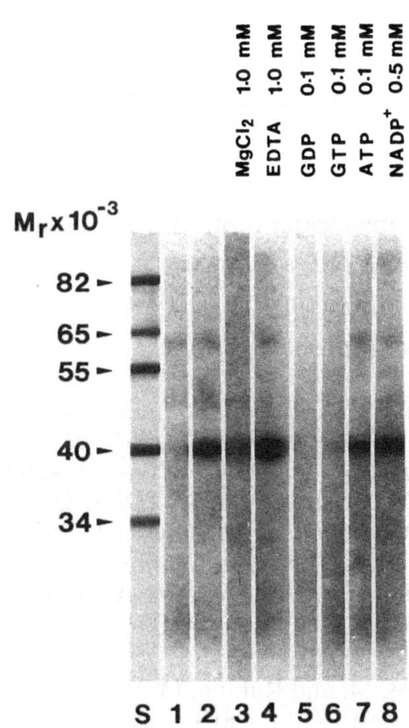

FIG. 7. **Photolabeling of GEF with [γ-^{32}P]8N$_3$-GTP.** Photoaffinity labeling of GEF with [γ-^{32}P]-8N$_3$GTP is carried out either with (*lanes 3-8*) or without (*lanes 1 and 2*) further additions as indicated. Molecular weights of the GEF subunits are indicated on the stained gel (*lane S*). *Lane 1*, no UV irradiation; *lanes 1-8*, autoradiogram.

regulatory functions of the different subunits. As a first step towards understanding the molecular details of GEF, the protein may be covalently modified with the photoreactive 8-azido analog of GTP, [γ-^{32}P]8N_3GTP. UV irradiation results in the photoinsertion of 8-N_3GTP in the M_r 40,000 subunit of GEF and demonstrates that the M_r 40,000 subunit of the factor constitutes the GTP binding domain (Fig. 7). The specificity of the interaction of [γ-^{32}P]8N_3GTP with GEF is examined by measuring the extent of photoinsertion in the presence of excess GDP, GTP, ATP, and NADP$^+$. As shown in Fig. 7, only GDP and GTP inhibit the photoinsertion of [γ-^{32}P]8N_3GTP to GEF, demonstrating that nucleotide binding to the M_r 40,000 subunit is specific for the guanine-binding domain. Photoinsertion of the 8-azido analog of GTP to GEF is inhibited by Mg^{2+} and stimulated by EDTA. The characteristics of [^{3}H]GDP binding to GEF as determined by membrane filtration are identical to those observed in the photoaffinity labeling experiments (Dholakia *et al.*, 1989). Of the five subunits of GEF, the M_r 55,000 and 65,000 polypeptides bind [γ-^{32}P]8N_3ATP. Both 8-N_3GTP and 8-N_3ATP specifically label the β-subunit of eIF-2 (Dholakia *et al.*, 1989). The significance of ATP binding to either GEF or eIF-2 is at present not known. In enzymatic assays, 8-N_3GTP supports the activity of GEF and eIF-2 (Dholakia *et al.*, 1989), indicating that its interaction with these proteins is catalytically relevant.

Acknowledgement: This work was supported in part by United States Public Health Services Grant GM 25451.

References:

Colin, A.M., Brown, B.D., Dholakia, J.N., Woodley, C.L., Wahba, A.J. and Hille, M.B. (1987) ***Dev. Biol.* 123**, 354-363.

Dholakia, J.N., Mueser, T.C., Woodley, C.L., Parkhurst, L.J. and Wahba, A.J. (1986) ***Proc. Natl. Acad. Sci. U.S.A.* 83** 6746-6750.

Dholakia, J.N. and Wahba, A.J. (1987) ***J. Biol. Chem.* 262**, 10164-10170.

Dholakia, J.N. and Wahba, A.J. (1988) ***Proc. Natl. Acad. Sci. U.S.A.* 85**, 51-54.

Dholakia, J.N. and Wahba, A.J. (1989) ***J. Biol. Chem.* 264**, 546-550.

Dholakia, J.N., Francis, B.R., Haley, B.E. and Wahba, A.J. (1989) ***J. Biol. Chem.*. 264**, 20638-20642.

Gross, M. and Rubino, M.S. (1989) ***J. Biol. Chem.* 264**, 21879-21884.

Hathaway, G.M. and Traugh, J.A. (1981) ***Curr. Top. Cell. Regul.* 21**, 101-112.

Kimball, S.R. and Jefferson, L.S. (1988) ***Biochem. Biophys. Res. Comm.* 156**, 706-711.

Pain, V.M. (1986) ***Biochem. J.* 235**, 625-637.

Sommercorn, J., Mulligan, J.A., Lozeman, F.J. and Krebs, E.G. (1987) ***Proc. Natl. Acad. Sci. U.S.A.* 84**, 8834-8838.

Wahba, A.J. and Woodley, C.L. (1984) ***Prog. Nucleic Acid Res. Mol. Biol.* 31**, 221-265.

ISOLATION OF THE *lsd* MUTATIONS IN *SACCHAROMYCES CEREVISIAE*

Alan B. Sachs
The Whitehead Institute for Biomedical Research
9 Cambridge Center
Cambridge, MA 02142
USA

ABSTRACT

The poly(A)-binding protein (PAB) is required for translation initiation and poly(A) tail shortening in *Saccharomyces cerevisiae* (Sachs and Davis, 1989). Extragenic suppressors of a null mutation in *PAB1* also affect the 60S ribosomal subunit. One of these *spb* genes (suppressor of *PAB1*) encodes ribosomal protein L46 (*SPB2*), and another encodes a putative rRNA helicase (*SPB4*) (Sachs and Davis, 1990). Mutagenesis of an *spb4-1* strain revealed that many lethal mutations in genes other than *PAB1* are suppressed by the *spb4-1* mutation. Furthermore, approximately 75% of these were also suppressed by deleting *RPL46*. These data suggest that some essential genes in yeast are made dispensable by altering the translational apparatus.

INTRODUCTION

The regulation of an mRNA's expression in the cytoplasm can be controlled by both its degradation and translation initiation rates. A potentially important element on mRNA for the control of these reactions is the poly(A) tail. In the cell, poly(A) is bound by the poly(A)-binding protein (PAB), and it is likely that this complex is the functional form of poly(A) in vivo (Baer and Kornberg, 1983; Blobel, 1973; Kwan and Brawerman, 1972;

NATO ASI Series, Vol. H 49
Post-Transcriptional Control of Gene Expression
Edited by J. E. G. McCarthy and M. F. Tuite

Sachs and Kornberg, 1985; Setyono and Greenberg, 1981). In order to understand the regulation of translation initiation and mRNA degradation, a detailed analysis of PAB's roles in these processes has been undertaken.

That PAB is required for these reactions has been documented in studies examining the effects of a conditional mutation in *PAB1* on translation and poly(A) tail shortening (Sachs et al., 1987, Sachs and Davis, 1989). Depletion of PAB by promoter inactivation or utilization of a temperature-sensitive mutation results in the inhibition of translation initiation and poly(A) tail shortening. Whether depletion of PAB affects mRNA stability remains to be determined.

PAB is essential for viability in *S. cerevisiae* in the presence of a wild-type 60S ribosomal subunit (Sachs and Davis, 1989). The genetic interaction of the 60S subunit with a protein bound to the poly(A) tail raises the possibility that this subunit "checks" an mRNA for the presence of the 3' end (i.e. PAB and a poly(A) tail) before it initiates translation. Such a mechanism defines the function of poly(A) as marking an mRNA as being intact, thereby allowing its efficient translation. This poly(A) requirement would prevent the efficient translation of trunctated mRNAs, which is both a waste of cellular energy and a potential source of toxic proteins. Poly(A) deficient mRNAs, while not completely untranslated, would be initiated at a basal poly(A)-independent rate.

Alterations in the 60S ribosomal subunit allows cell growth in the absence of PAB. These alterations are created by mutations in the *spb* genes, which were isolated as suppressors of a temperature sensitive mutation in *PAB1* (Sachs and Davis, 1989). These mutations include a null mutation in the ribosomal protein L46 gene (*RPL46=SPB2*) or a modification of a putative rRNA helicase (*SPB4*) involved in the maturation of 25S rRNA (Sachs and Davis, 1990).

The work described here was designed to test the hypothesis that the *spb* mutations could also suppress mutations in genes that are required for PAB activity. For instance, the potential initiation factors needed to mediate the requirement for PAB may be dispensable in the *spb* backgrounds. Another potential class of dispensable gene products are those involved in the synthesis and degradation of poly(A). This possibility would be realized if the essential function of poly(A) was to bind PAB.

Analysis of $5x10^5$ EMS mutagenized *spb4-1* cells revealed that lethal mutations in at least 75 independent complementation groups are suppressed by *spb4-1* but not *SPB4*. Approximately 75% of these mutations are also suppressed by deleting *RPL46*, indicating that suppression of these mutations almost certainly occurs through alterations in the 60S ribosomal subunit. This suggests that suppression of a *PAB1* deletion by the *spb* mutations is not a unique event, and it indicates that many essential genes and processes may be made dispensable in an *spb* background.

RESULTS

Mutations in genes that depend on wild type 60S ribosomal subunits for their essential phenotype (*lsd*=large subunit dependent) were identified by looking for cell death in the presence of a wild-type but not a mutated subunit (Figure 1). Yeast strains YAS535 and YAS536 contain the *spb4-1* mutation as well as a galactose inducible *SPB4* gene. These strains have mutated 60S subunits when grown in glucose medium and wild-type subunits when grown in galactose medium. EMS mutagenesis of YAS535 and YAS536, followed by plating on glucose and then galactose medium revealed that approximately 0.3% of the cells were only able to grow on glucose. Over 90% of these *lsd* mutants fell into the *lsd1* complementation group. The *lsd1* mutation also arises spontaneously, suggesting it confers a growth advantage to the *spb4-1* strain.

Complementation analysis of the remaining isolates revealed that approximately 66% of them fell into one of 17 complementation groups (Table 1). There were also 57 individual isolates that could not be placed into a complementation group. Interestingly, none of the *lsd* mutants contained a mutated *PAB1* .

As an independent measure of how close the screen came to identifying all of the *lsd* mutations, those cells that were unable to utilize galactose as a carbon source were analyzed. These gal$^-$ mutants appeared at a frequency of $4x10^{-5}$. They fell into five complementation groups (Table 2) that have been arbitrarily labeled *galA* through *galE*. The presence of multiple alleles in some of these groups indicates that the screen was technically successful. However, the absence of multiple members in other groups, together with the results of the *lsd* complementation analysis, suggests that some potential *lsd* mutations have not been isolated.

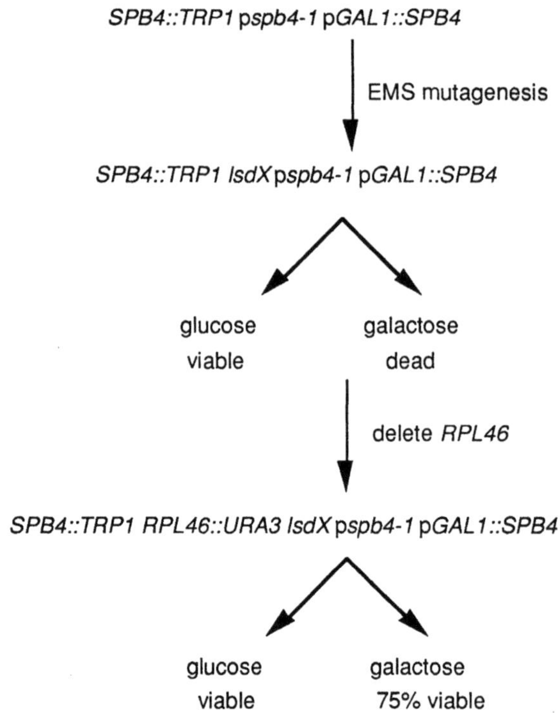

Figure 1: Strategy for creating and catagorizing the *lsd* mutations into *SPB4* specific and 60S ribosomal subunit specific suppression. Stationary phase YAS535 (MAT**a** *SPB4::TRP1 ade2 leu2 his3 ura3 trp1 pspb4-1 pGAL1::SPB4*) or YAS536 (MATα *SPB4::TRP1 ade2 leu2 his3 ura3 trp1 pspb4-1 pGAL1::SPB4*) cells were mutagenized with EMS (Sherman et al., 1986) until approximately 30% of the cells were killed, and plated onto minimal medium containing glucose. Single colonies were then replica plated onto minimal medium containing galactose or glucose. Cells unable to grow on the galactose medium were divided into *lsd* mutants and gal$^-$ mutants. After complementation analysis, a member of each complementation group had its single copy *RPL46* gene (Leer et al., 1985) deleted by transplacement with a linear *RPL46::URA3* construction (Sachs and Davis, 1990). Cell growth on galactose after this step indicated that the *spb* mutation was suppressed by the *RPL46* deletion in the presence of a wild type SPB4 protein. p*spb4-1* is a LEU2 centromeric plasmid containing *spb4-1*, and p*GAL1::SPB4* is either a URA3 (YAS535) or HIS3 (YAS536) centromeric plasmid containing the *GAL1* promoter fused to the *SPB4* gene.

To confirm that the lethality of the *lsd* mutants on galactose medium was due to the presence of a wild-type 60S subunit, a member from each of the complementation groups and approximately 30 individual isolates had their single copy of the *RPL46* gene deleted. Following this, cells were scored for their ability to grow on galactose. Approximately 75% of the *lsd* mutants were able to grow on galactose, indicating that suppression of their *lsd* phenotype could also occur by removing RPL46 from the 60S subunit (Table 1).

TABLE 1: Complementation analysis of *lsd* mutants. *Lsd* isolates from either YAS535 or YAS536 were mated and then scored for growth on galactose medium. Diploids unable to grow on galactose were placed into the same complementation group. The ability of a member from each group to grow on galactose in the presence of *SPB4* and an *RPL46* deletion is indicated.

Complementation Group	# of Alleles	Rescue by *RPL46* Deletion
lsd2	2	Y
lsd3	7	N
lsd4	2	Y
lsd5	15	N
lsd6	5	Y
lsd7	6	Y
LSD8	1	Y
lsd9	5	N
lsd10	2	Y
lsd11	2	Y
lsd12	5	Y
lsd13	2	Y
lsd14	3	Y
lsd15	3	N
lsd16	3	Y
lsd17	3	Y

TABLE 2: Complementation analysis of gal⁻ mutants. Isolates unable to grow on galactose in the absence of the plasmid p*GAL::SPB4* were mated and then scored for their ability to grow on galactose medium. Diploids unable to grow on galactose were placed into the same complementation group. The *GALC* mutant displayed a dominant gal⁻ phenotype.

Complementation Group	# of Alleles
galA	5
galB	12
GALC (DOMINANT)	1
galD	1
galE	6

CONCLUSION

The experiments described here confirm the hypothesis that genes other than *PAB1* can have their lethal mutations suppressed by a modified 60S ribosomal subunit. This group of *lsd* mutations could include those genes whose products are required for PAB-dependent translation initiation or poly(A) metabolism. The inability to place all of the *lsd* mutants into complementation groups, as well as the failure to isolate a mutation in *PAB1*, suggests that not all of the *lsd* mutations have been identified.

Some of the *lsd* mutants could not have their lethal phenotype rescued by a deletion of *RPL46*. This suggests that suppression of these *lsd* mutations could be specific for *spb4-1*, or that these mutations confer a lethal phenotype when *SPB4* is overexpressed. This subgroup of *LSD* genes may have some common physiological function. These could include being part of a pre-rRNA processing complex that may also contain SPB4, or being part of the 60S ribosomal subunit that is made dispensable in a *spb4-1* background but not in a *RPL46* deletion background.

The finding that over 70 different lethal mutations can be suppressed by mutations affecting the 60S ribosomal subunit indicates that the ribosome is responsible for determining the essential nature of many different genes. This suppression effect does not appear to be due to translation through stop codons since the *spb4-1* strain does not suppress nonsense mutations in either highly suppressible auxotrophic markers or in a synthetic β-galactosidase gene containing an amber or ochre mutation (data not shown). Instead, the defect in *spb4-1* strains appears to be at the level of initiation since their polyribosomes are smaller and contain an abundance of halfmers (data not shown). Furthermore, analysis of extracts from *spb4-1* and *SPB4* strains by two-dimensional gel electrophoresis indicates large differences in the amounts of certain proteins. As a result, a tentative hypothesis is that alterations in the relative initiation efficiencies of mRNA in *S. cerevisiae* due to a ribosomal subunit mutation results in an altered cellular metabolism that makes normally essential genes dispensable. Future work will focus on identifying some of these genes and determining if any of them are involved in PAB-dependent translation initiation or poly(A) metabolism.

Supported by grants from the NIH GM43164 and the Lucille P. Markey Charitable Trust. I thank M. Tuite for the mutated β–galactosidase gene, and the technical assistance of Fiona Griffin is acknowledged.

REFERENCES

Baer B, Kornberg RD (1983) The protein responsible for the repeating structure of cytoplasmic poly(A) ribonucleoprotein. J Cell Biol 96:717-721

Blobel G (1973) A protein of molecular weight 78,000 bound to the polyadenylate region of eukaryotic messenger RNAs. Proc Natl Acad Sci USA 70:924-928

Kwan SW, Brawerman G (1972) A particle associated with the polyadenylate segment in mammalian messenger RNA. Proc Natl Acad Sci USA 69:3247-3250

Leer JR, Raamsdonk-Duin MC, Kraakman P, Mager WH, Planta R (1985) The genes for ribosomal proteins S24 and L46 are adjacent and divergently transcribed. Nucl Acids Res 13:701-709

Sachs AB, Kornberg RD (1985) Nuclear polyadenylate binding protein. Mol Cell Biol 5:1193-1996

Sachs AB, Bond MW, Kornberg RD (1986) A single gene from yeast for both nuclear and cytoplasmic polyadenylate-binding protein: Domain structure and expression. Cell 45:827-835

Sachs AB, Davis RW, Kornberg RD (1987) A single domain of yeast poly(A)-binding protein in necessary and sufficient for RNA binding and cell viability. Mol Cell Biol 7:3268-3276

Sachs AB, Davis RW (1989) The poly(A)-binding protein is required for poly(A) shortening and 60S ribosomal subunit-dependent translation initiation. Cell 58:857-867

Sachs AB, Davis RW (1990) Translation initiation and ribosomal biogenesis: Involvement of a putative rRNA helicase and RPL46. Science 247:1077-1079

Setyono B, Greenberg JR (1981) Proteins associated with poly(A) and other regions of mRNA and hnRNA molecules investigated by crosslinking. Cell 24:775-783

Sherman F, Fink GR, Lawrence C (1986) Methods in Yeast Genetics. (Cold Spring Harbor, New York: Cold Spring Harbor Laboratory).

FUNCTIONAL ROLE AND BIOCHEMICAL PROPERTIES OF YEAST PEPTIDE ELONGATION FACTOR 3 (EF-3)

Masazumi Miyazaki, Masahiro Uritani, Yoshihisa Kitaoka, Kazuko Ogawa and Hideto Kagiyama

Department of Molecular Biology
School of Science
Nagoya University
Nagoya 464-01
Japan

INTRODUCTION

The eukaryotic peptide elongation cycle is well known to be driven by the two complementary factors EF-1α and EF-2, functionally analogous to the bacterial EF-Tu and EF-G, respectively, and the two GTP hydrolysis steps catalyzed by those factors have been considered to be essential for the cycle to run (Kaziro, 1978; Moldave, 1985). On yeast ribosomes, however, the elongation process additionally requires the third soluble factor, which was found by Skogerson and collaborators (1976, 1977) using poly-(U)-dependent protein synthesis systems and designated as EF-3. Inhibition experiments with monoclonal (Hutchison et al., 1984) and polyclonal (Dasmahapatra and Chakraburtty, 1981; Miyazaki and Kagiyama, to be published) antibody raised against EF-3 demonstrated that the factor was essential for the elongation phase in the translation of natural mRNA as well as poly(U). A temperature-sensitive yeast mutant producing a thermolabile EF-3 was blocked in the elongation cycle at a non-permissive temperature, indicating the factor indispensable for the *in vivo* translation (Herrera et al., 1984; Kamath and Chakraburtty, 1986b; Qin et al., 1987).

Purification and biochemical characterization have revealed that EF-3 consists of a single polypeptide chain with a high molecular weight of 120,000 to 125,000 and exhibits strong ATPase and GTPase activities that are dependent on yeast but not on *Tetrahymena* ribosomes (Skogerson and Wakatama, 1976; Dasmahapatra and Chakraburtty, 1981; Uritani and Miyazaki, 1988a). This unique factor occurs widely in yeasts and fungi species but neither in

NATO ASI Series, Vol. H 49
Post-Transcriptional Control of Gene Expression
Edited by J. E. G. McCarthy and M. F. Tuite

other lower eukaryotes such as *Tetrahymena* and *Chlamydomonas* nor higher eukaryotes (Skogerson and Engelhardt, 1977; Miyazaki and Uritani, 1985; Chakraburtty and Kamath, 1988). Ribosomes from rat liver (Skogerson and Engelhardt, 1977) and *Tetrahymena* (Miyazaki and Uritani, 1985; Miyazaki and Kagiyama, to be published) carry strong ATPase and GTPase activities even after being extensively washed and did not require yeast EF-3 for the elongation cycle, suggesting that EF-3 is a solublized form of ribosomal component(s) responsible for the nucleotidase activity. ATP hydrolysis is prerequisite for EF-3 to function in the yeast elongation cycle, in addition to the GTP hydrolysis by EF-1α and EF-2 (Uritani and Miyazaki, 1988a). Elucidation of the role of EF-3 ATPase is expected to disclose a new mechnism in the elongation phase of translation of yeast and moreover, of eukaryotes.

ESSENTIAL REQUIREMENTS OF EF-3 AND ATP HYDROLYSIS FOR POLYMERIZATION AND BIOCHEMICAL PROPERTIES OF THE FACTOR

The first observation of the ATP stimulation of poly(U)-dependent polyphenylalanine synthesis on yeast ribosomes was reported by Skogerson (1979). We have confirmed the results in our yeast systems reconstituted with the three homogeneous factors and well-washed ribosomes free from ATPase activity (Uritani and Miyazaki, 1985 and 1988a). This ATP stimulation is most likely promoted by EF-3 ATPase since it is the sole factor having affinity to ATP in the reaction system. While the GTPase activity of EF-1α and EF-2 was strictly directed at guanine nucleotide but not at ATP, the EF-3 ATPase showed its affinity to both purine and pyrimidine nucleotides (XTPs) irrespective of 2'-hydroxylation of the sugar moiety. The highest affinity was to ATP among the tested eight XTPs (Km values for ATP and GTP were 0.12 mM and 0.20 mM, respectively). The wide substrate specificity of the ATPase was reflected on the different stimulation of polyphenylalanine synthesis by various XTPs as well as ATP (see Table I, 2nd column).

The ATP effect on polyphenylalanine synthesis varied as a function of GTP concentration in the reaction systems (Fig. 1). At lower concentrations of GTP (less than 10 μM), the reconstituted

Table I. Stimulation by Various Nucleoside Triphosphates (XTPs) of Polyphenylalanine Synthesis, Phe-tRNA Binding and N-AcPhe-Puromycin Formation Reactions Dependent on Poly(U).

Nucleotides added (0.3 mM)	[^{14}C]Phe-Polymerization[a)] (pmol)	[^{14}C]Phe-tRNA Binding[b)] (pmol)	N-Ac[^{14}C]Phe-Puromycin Formation[c)] (pmol)
None	21.5	1.87	0.18
ATP	46.3	7.64	0.43
GTP	36.1	6.10	0.41
UTP	44.5	5.64	0.42
CTP	24.2	2.98	0.27
dATP	42.2	5.60	0.34
dGTP	28.2	2.27	0.42
dTTP	42.3	4.61	0.36
dCTP	24.5	1.60	0.26

a) The reaction system includes 100 μM GTP, EF-1α, EF-2, EF-3 and yeast ribosomes. From Uritani and Miyazaki (1988a).
b) The rection system includes 10 μM GTP, EF-1α, EF-3 and yeast ribosomes. From Uritani and Miyazaki (1988b).
c) The reaction system includes 10 μM GTP, EF-2, EF-3 and yeast ribosomes. Uritani and Miyazaki, unpublished.

elongation system could no longer synthesize polyphenylalanine without ATP (open circles). The addition of ATP (0.3 mM) was needed for the polymerization (closed circles). Taking account of a high Km value of EF-3 for GTP (0.20 mM), 10 μM or less of GTP should be too low for EF-3 to function, while the other two factors, EF-1α and EF-2, are expected to operate at least partially at those concentrations of GTP since their Km values for GTP (EF-1α = 8.5 μM and EF-2 = 30 μM) are low enough. Getting higher concentrations of GTP (more than 30 μM, which could at least partially satisfy the demand for EF-3 to function), the system became capable of synthesizing polyphenylalanine (open circles) and so, the stimulation of the polymerization was observed by the addition of ATP (closed circles) by which EF-1α and EF-2 have not been activated. Therefore, the ATP effect emerges as an essential requirement at lower concentrations of GTP and as a stimulation at higher concentrations of GTP. It is natural that no polymerization occurred at zero concentration of GTP even by the addition of ATP since EF-1α and EF-2 can not work. Neither ATP stimu-

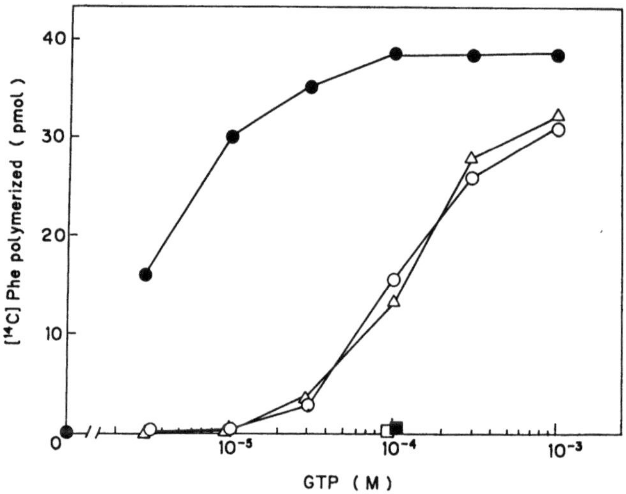

Fig. 1. Changes in the ATP effect on polyphenylalanine synthesis as a function of GTP concentration. Polyphenylalanine synthesis was measured with (●) or without (○) ATP (0.3 mM) in the presence of EF-1α, EF-2 and EF-3 on yeast ribosomes. In (△), ATP was replaced by AMPPNP. No activity was observed without EF-3 either in the presence (■) or in the absence (□) of ATP. From Uritani and Miyazaki (1988a).

lation nor polymerization was observed in the absence of EF-3 (Fig. 1., squares). ATP hydrolysis was required for the ATP effect. ATP was unable to be replaced by unhydrolyzable AMPPNP or by the products of the EF-3 ATPase, ADP and Pi (Uritani and Miyazaki, 1988a). Further evidence to show the ATPase activity essential for the EF-3 function was provided from inactivation experiments of EF-3 by N-ethylmaleimide, which reacts with thiol groups of cysteine, (Uritani and Miyazaki, 1988a) and with anti-EF-3 antibody (Dasmahapatra and Chakraburtty, 1981; Miyazaki and Kagiyama, to be published), and from heat inactivation of thermolabile EF-3 (Kamath and Chakraburtty, 1986a). In those experiments, the ATPase and polyphenylalanine activities were demonstrated to decline along nearly the same inactivation curves.

Very recently, Qin et al. (1990) reported the total sequence of yeast EF-3 gene in which an open reading frame codes a deduced protein of 1,044 amino acids including the internal repeats (A and B) of a region with about 200 amino acids. The repeats harbor the sequence motifs for ATP-binding. This essential gene was located on chromosome XII. According to this sequence and N-terminal sequence of a tryptic fragment (Miyazaki et al., 1988), it was noticed that the tryptic cleavage between 774Arg and 775 Gln located in the repeat B (C-terminal side) was strongly protected by the presence of ATP and at the same time exerted a fatal effect on ATPase, Phe-tRNA binding and polyphenylalanine synthesis activities of EF-3.

Several lines of evidence mentioned above lead us to conclude that ATP hydrolysis by the ATPase of EF-3 is an essential step to drive the elongation cycle on yeast ribosomes (Uritani and Miyazaki, 1988a).

Both the catalytic and binding sites of the ATPase were demonstrated to reside intrinsically in the elongation factor itself but not on ribosomes (Miyazaki et al., 1988). The intrinsic activity of the ATPase was stimulated up to two orders of magnitude by the presence of yeast ribosmes which are fully active in polyphenylalanine synthesis. The small subunits of yeast ribosomes were responsible for the stimulation. In this connection, it was shown that polyphenylalanine synthesis by ribosomes reconstituted with yeast 40S and wheat germ 60S is significantly stimulated by EF-3 while ribosomes reconstituted with yeast 60S and wheat germ 40S were fully functional without added EF-3 or with anti-EF-3 (Chakraburtty and Kamath, 1988). This indicates that ribosomal component(s) responsible for EF-3 function (probably ATPase activity) is also located on the small subunits of wheat germ ribosomes. The ribosome-enhanced activity of the EF-3 ATPase was attained by enhancing the catalytic activity (k_{cat}) to much greater extent (about 30-fold stimulated) than the binding affinity (Km: the affinity reduced to two-fifths). Divalent (but not monovalent) cations such as Mg^{++} are required for the enzymic catalysis (Miyazaki et al., 1988). In contrast to the GTPase of EF-1α and EF-2, the ATPase activity of EF-3 was inhibited by vanadate, suggesting that a phosphoenzyme intermediate might be involved in a elongation step like the cation-pumping P-ATPase.

The ribosome-enhanced activity of the ATPase was remarkably inhibited by the essential components for the elongation cycle, poly(U) and tRNA (AA-tRNA) but the intrinsic activity was not affected by them, indicating that those polynucleotides suppressively influence the ATPase activity through their interaction with ribosomes but have no direct interaction with EF-3 (Uritani and Miyazaki, 1988a; Miyazaki et al., 1988). According to a quantitative radioimmunoassay, it was estimated that approximately 50 % of the polysomal ribosomes and 2 % of the 80S ribosomes contained EF-3. However, the subunits (40S and 60S) was unreactive to the anti-EF-3 antibody. EF-3 is expected to cycle in the elongation phase between the particles and the soluble fraction of the cytoplasm (Hutchison et al., 1984).

ROLE OF EF-3 AT THE AA-TRNA BINDING STEP

In order to define the exact function of EF-3 in the peptide

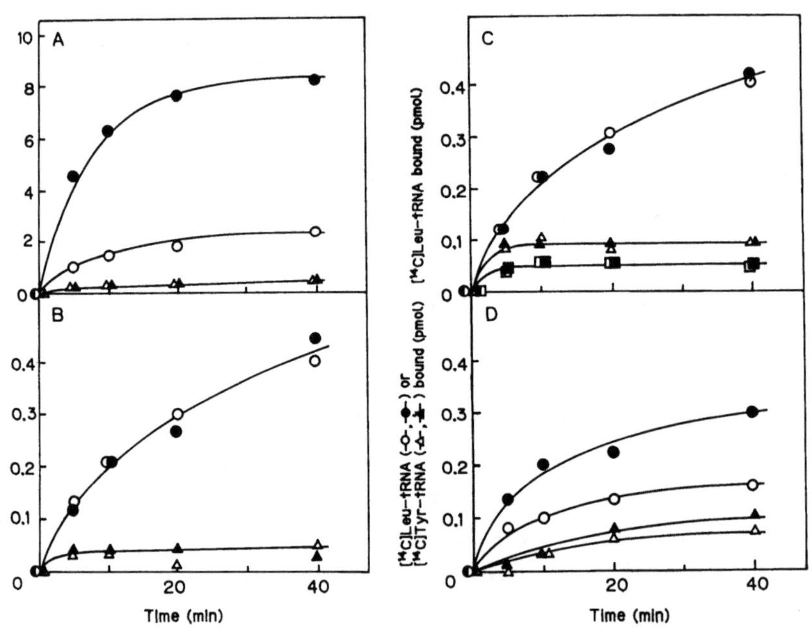

Fig. 2. Comparison of the binding stimulation by EF-3 and ATP between cognate and non-cognate AA-tRNA. A and B, poly-(U)-dependent binding; C, poly(U)- (○, ●), poly(A)- (△, ▲) and poly(C)- (□, ■) dependent binding; D, endogenous mRNA-directed binding. Binding was measured with (●, ▲, ■) or without (○, △, □) EF-3 and ATP. From Uritani and Miyazaki (1988b)

elongation cycle of yeast, studies were required on the role of EF-3 in the partial reactions comprizing the cycle. The most prominent effect of EF-3 was on the AA-tRNA binding to ribosomes and was observed as stimulation rather than requirement (Fig. 2A). On yeast ribosomes, EF-1α apparently functions in a stoichiometric manner in the binding reaction. The addition of EF-3 and ATP to this binding system strikingly stimulated the reaction, which proceeded catalytically with respect to both EF-1α and EF-3, accompanied by ATP hydrolysis (Uritani and Miyazaki, 1988b; Kamath and Chakraburtty, 1989).

Two ribosomal complexes binding AA-tRNA are assumed to be functionally distict between the one formed with EF-1α alone and the other formed with EF-3 and ATP added, since AA-tRNA on the latter complex was eligible for translocation or for transpeptidation by the addition of EF-2 but AA-tRNA on the former complex was not (Uritani and Miyazaki, 1988b). The two binding states of AA-tRNA were revealed by the measurement of the stability of AA-tRNA on

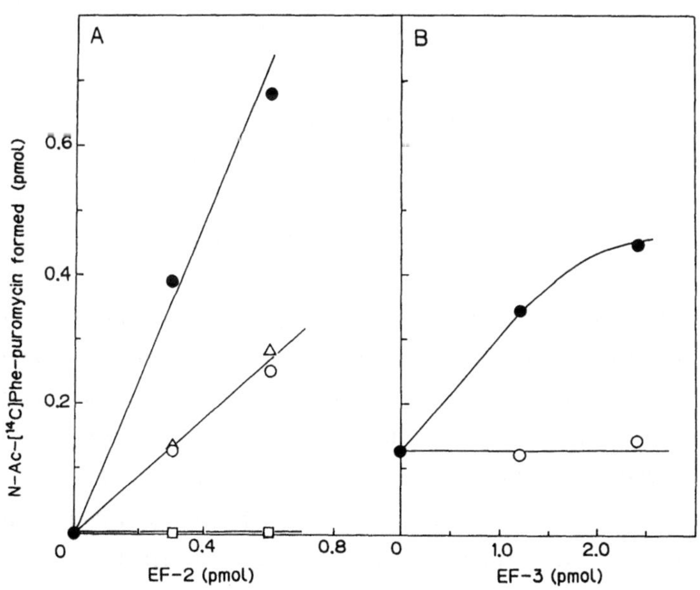

Fig. 3. Effect of increasing amounts of EF-2 (A) and EF-3 (B) on the poly(U)-programmed N-Ac[^{14}C]Phe-Puromycin formation (EF-2-dependent) on yeast ribosomes. The translocation reaction was measured by the addition of ATP (0.3 mM) (●) or AMPPNP (0.3 mM) (△) to or the omission (□) of GTP from the reaction system (○) including 10 μM GTP and EF-3 (A) or EF-2 (B). Uritani and Miyazaki, unpublished.

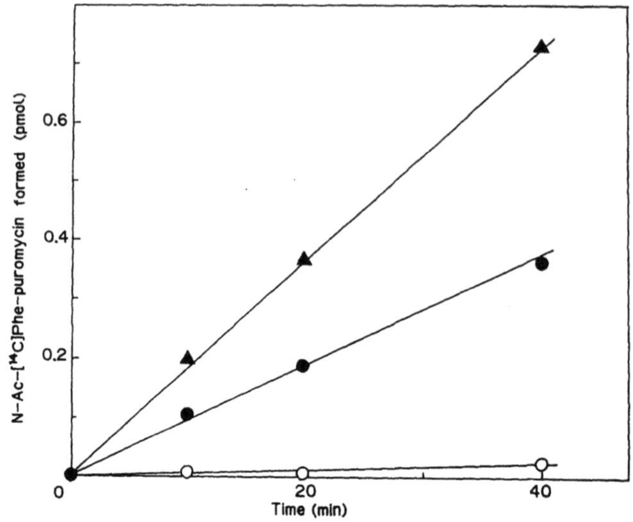

Fig. 4. Effect of EF-3 and ATP on the EF-2-dependent N-Ac[^{14}C]-Phe-Puromycin formation (translocation) on yeast ribosomes in the presence of GDP (10μM) substituted for GTP. The reactions were carried out with (●, ▲) or without (○) ATP (0.3 mM) in the presence of GDP (●, ○) or GTP (▲) in the system driven with EF-2 and EF-3 on yeast ribosomes. From Uritani and Miyazaki, unpublished.

the ribosomal complexes formed with EF-1α alone and with EF-1α and EF-3 with ATP. To examine the involvement of EF-3 and ATP in the decoding process (translational accuracy), the binding stimulation by EF-3 and ATP was compared between cognate and non-cognate AA-tRNA (Fig. 3). In contrast to the great enhancement of cogante Phe-tRNA binding dependent on poly(U), no enhancement of non-cognate Leu- and Tyr-tRNA binding was demonstrated by EF-3 and ATP (Fig. 3A, B and C). EF-3 was unable to promote the AA-tRNA binding to ribosomes and to stimulate a ternary complex formation between EF-1α GTP AA-tRNA (Uritani and Miyazaki, 1988b; Kamath and Chakraburtty, 1989). We concluded that at first EF-1α carries AA-tRNA in the ternary complex to the ribosomal introducing site (I-site with a little binding of noncognate AA-tRNA, and then EF-3 select strictly cognate AA-tRNA and transfer from the I- to the A-site, by changing its binding state, accompanied by ATP hydrolysis (Uritani and Miyazaki, 1988b). The transferring step from the I- to the A-site is likely a basic essential process in the elongation cycle catalyzed by EF-3.

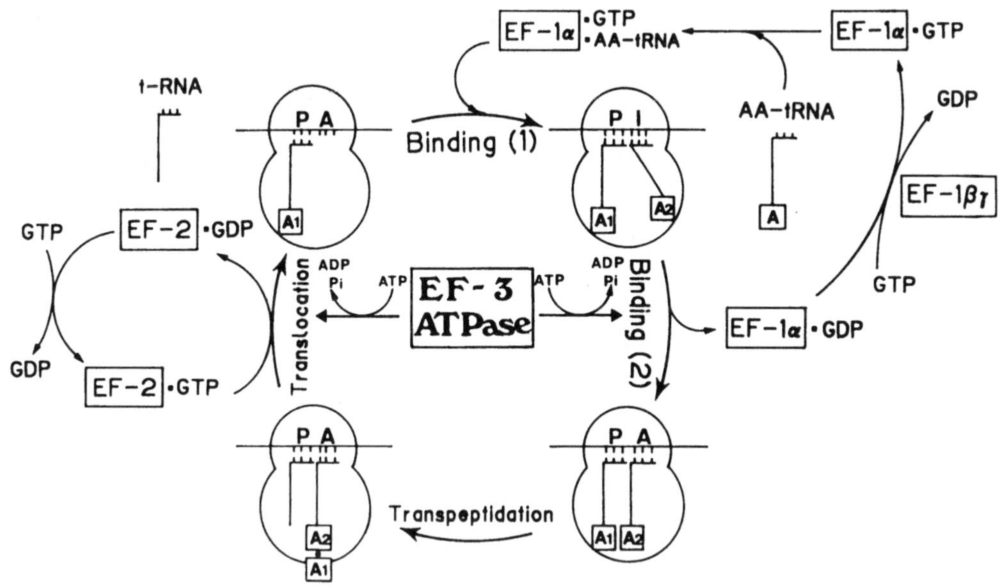

Fig. 5. A proposed yeast peptide elongation cycle.

A STIMULATION BY EF-3 AND ATP OF TRANSLOCATION BUT NOT OF TRANSPEPTIDATION

EF-3 and ATP exerted no effect on the transpeptidation reaction by analyzing N-AcPhe-puromycin formation between high salt-translocated N-AcPhe-tRNA and puromycin. However, EF-2-dependent N-AcPhe-puromycin formation was remarkably enhanced by EF-3 and ATP or other XTPs (Table I and Fig. 3). Again ATP can not be replaced by AMPPNP. Even in the presence of GDP for GTP, EF-3 and ATP enabled the system to operate the translocation reaction (Fig. 4) as observed in polyphenylalanine synthesis and AA-tRNA binding. Here, we propose a revised elongation cycle in yeast (Fig. 5).

REFERENCES

Chakraburtty K, Kamath A(1988) Protein synthesis in yeast. Int J Biochem vol 120:581-590

Dasmahapatra B, Chakraburtty K(1981) Protein synthesis in yeast. I. Purification and properties of elongation factor 3 from *Saccharomyces cerevisiae*. J Biol Chem 256:9999-10004

Herrera F, Martinez JA, Moreno N, Sadnik I, McLaughlin CS, Feinberg B, Moldave K(1984) Identification of an altered elongation factor in temperature-sensitive mutants ts7'-14 of *Saccharomyces cerevisiae*. J Biol Chem 259:14347-14349

Hutchison JS, Feinberg B, Rothwell TC, Moldave K(1984) Monoclonal antibody specific for yeast elongation factor 3. Biochemistry 23:3055-3063

Kamath A, Chakraburtty K(1986a) Purification of elongation factor 3 from temperature-sensitive mutant 13-06 of the yeast *Saccharomyces cerevisiae*. J Biol Chem 261:12596-12598

Kamath A, Chakraburtty K(1986b) Identification of an altered elongation in the thermolabile mutants of the yeast *Saccharomyces cerevisiae*. J Biol Chem 261:12593-12595

Kaziro Y (1978) The role of guanosine 5'-triphosphate in polypeptide chain elongation. Biochim Biophys Acta 505:95-127

Miyazaki M, Uritani M(1985) Studies on the occurrence of elongation factor 3 in lower eukaryotes other than yeasts. 13th Int Congr Biochem Abstracts p60

Miyazaki M, Uritani M, Kagiyama H(1988) The yeast peptide elongation factor 3 (EF-3) carries an active site for ATP hydrolysis which can interact with various nucleoside triphosphates in the absence of ribosomes. J Biochem (Tokyo) 104:445-450

Miyazaki M, Kagiyama H (to be published) Soluble factor requirements for the *Tetrahymena* ribosomes and the ribosomal ATPase as an counterpart of yeast elongation factor 3 (EF-3). J Biochem (Tokyo)

Moldave K(1985) Eukaryotic protein synthesis. Ann Rev Biochem 54: 1109-1149

Qin S, Moldave K, McLaughlin CS (1987) Isolation of the yeast gene encoding elongation factor 3 for protein synthesis. J Biol Chem 262:7802-7807

Qin S, Xie A, Christina M, Bonato M, McLaughlin CS (1990) Sequence analysis of the translational elongation factor 3 from *Saccharomyces cerevisiae*. J Biol Chem 265:1903-1912

Skogerson L, Wakatama E (1976) A ribosome-dependent GTPase from yeast distinct from elongation factor 2. Proc Natl Acad Sci US 73:73-76

Skogerson L, Engelhardt D (1977) Dissimilarity in protein chain elongation factor requirements between yeast and rat liver ribosomes. J Biol Chem 252:1471-1475

Skogerson L (1979) Separation and characterization of yeast elongation factors. In; Moldave K and Grossman L(ed) Methods in enzymology vol LX Academic Press, New York, p676

Uritani M, Miyazaki M (1985) Interaction of yeast polypeptide chain elongation factor 3 (EF-3) with different nucleotides. Nucleic Acids Res Symp Ser No 16:229-232

Uritani M, Miyazaki M (1988a) Characterization of the ATPase and GTPase activities of elongation factor 3 (EF-3) purified from yeasts. J Biochem (Tokyo) 103:522-530

Uritani M, Miyazaki M (1988b) Role of yeast peptide elongation factor 3 (EF-3) at the AA-tRNA binding step. J Biochem 104: 118-126

TRANSLATIONAL MOVEMENTS

C.G. Kurland
Dept. of Molecular Biology
Biomedical Center, Box 590
S-751 24 UPPSALA

Introduction

Gene expression involves the successive synthesis of two polymers that must be elaborated in a strictly processive manner. This means that the coherence of gene expression rests on the precision of the molecular movements that animate RNA and protein synthesis. We understand very little about these movements.

The availability of antibiotics and mutants with suitable disruptive properties has been useful in studying the mechanisms of tRNA selection by mRNA-programmed ribosomes (Murgola,1985; Buckingham and Grosjean,1986; Kurland and Ehrenberg, 1987). Likewise, we expect this to be the case for the analysis of translational movements. In addition, new assays are required not the least because we have discovered that the most widely used assays for defective translation movements, namely, measurements of reading frameshifts, only score a small fraction of the movement errors that normally disrupt translation (Jörgensen and Kurland,1990).

NATO ASI Series, Vol. H 49
Post-Transcriptional Control of Gene Expression
Edited by J. E. G. McCarthy and M. F. Tuite

Finally, we need some ideas. Here, it is worth recalling that important progress was made in understanding the mechanism of tRNA selection when the model of kinetic proofreading was introduced (Hopfield,1974; Ninio,1975). This was because the proofreading model focused attention on the dynamics of codon function and in this way lifted our attention beyond phenomenolgy. Although it is difficult to predict when new ideas will emerge, it seems for the moment that we have to concentrate on enriching the phenomenology of the processivity problem.

Reading frameshifts

Because of its relative ease of application the most widespread assay for processivity errors is the measurement of frameshift suppression. Here, a typical experiment consists of measuring the frequency with which a frameshift mutation in an indicator gene such as *lacZ* is spontaneously suppressed by a compensatory reading frame error. Nonallelic mutations, typically in genes coding for translational components, can be classified according to their impact on such suppression frequencies. In this way mutations affecting *ribosomes* (Atkins et al 1972; Kruzewskia and Slonimski 1984; Weiss-Brummer et al 1987; Murgola et al 1988), *tRNA isoacceptors* (Roth1974; Kohno and Roth 1978; Bossi and Roth 1981; Curran and Yarus 1987; O'Mahony et al 1989; Tucker et al 1989), and *elongation factor Tu* (Hughes et al 1987) can be scored as enhancers or restrictors of reading frame errors.

Several useful generalizations have emerged from these studies of suppressors and antisuppressors of frameshifting: First, the analysis of ribosome mutants and antibiotics such as streptomycin indicate that there is a positive correlation between the missensse error rates and the reading frame error rates during translation (Atkins et al 1972). Second, this and many other observations (Weissand Gallant,1983; Murgola,1985; Buckingham and Grosjean,1986; Curran and Yarus,1987; Murgola, 1989; Tucker et al 1989) support the interpretation that the movement of mRNA is dependent on the character of the codon-anticodon interaction as well as other positions in the tRNA structure (Kurland,1979). Similarly, interactions between the tRNA and the ribosome at sites distinct from the anticodon modulate simultaneously the accuracy of both tRNA selection and mRNA movement (Kurland and Ehrenberg, 1985;1987). Finally, the movements of the mRNA-tRNA complexes through the successive ribosome states is most easily understood as an incremental rather than as a single step process (Kurland and Ehrenberg, 1985;1987; Moazed and Noller,1989).

Evidently, the 10Å movement of the mRNA with respect to the ribosomal translation sites is an elaborate business. We naturally wonder what is driving all of this machinary, and this brings us just as naturally to the elongation factors and their GTP reactions.

Elongation factor G

Elongation factor G (EF-G) is involved in the translocation of peptidyl-tRNA on the ribosome (Nishizuka and Lipmann, 1966;

Tanaka et al 1968; Okura et al 1970) In addition, data cited in the foregoing section suggest that at a minimum the codon-anticodon interaction couples mRNA movements to the translocation of peptidyl-tRNA. Accordingly, we have searched for an experimental connection between the functions of EF-G and the movement of mRNA during translation.

Our approach has been to identify mutants of EF-G by virtue of their resistance to fusidic acid, and to screen these by conventional methods for their enhancement or restriction of reading frame errors *in vivo.* This has been done both in *Salmonella typhimurium* (D.Hughes, unpublished) and *Escherichia coli* (Richter and Kurland, 1990a) with similar results. So far only two mutant EF-G phenotypes have been identified in these screenings: conventional fusidic acid resistant mutants with normal reading frame accuracy, and fusidic resistant mutants that restrict +1 and -1 reading frame errors in certain contexts by up to a factor of four.

Biochemical comparisons of the mutant EF-G 's with wild type factor reveal provocative differences between the variants (Richter and Kurland,1990a). The conventional fusidic acid resistant variant turns out to have a partially cryptic phenotype. Thus, it has a defective binding interaction with GTP in vitro that is masked by the relatively high *in vivo* concentrations of GTP, but its interactions with the ribosome are kinetically normal. In contrast, the frameshift-restrictive variant shows a very clear perturbation of its interaction with the ribosome as well as a marked decrease of its maximum turnover rate in translation. All of this is consistent with the interpretation that the restrictive EF-G variant is abnormal in its functional contribution to the mRNA movement mechanism.

In order to determine whether or not the restrictive phenotype is associated with significant changes in the GTP flows during translocation we have also measured the stoichiometry of the EF-G dependent GTP dissipation during steady state elongation with wild type and mutant factors. The results are consistent even if they are not inspiring: The wild type, the conventional fusidic acid resistant and the restrictive variants of EF-G function with a stoichiometry very close to one GTP dissipated per peptide bond (Richter and Kurland,1990b).

Finally, we have studied mutants of the ribosome that influence the EF-G functions. These mutants are spectinomycin resistant mutants that are expressed either as changes in the nucleotide at position 1192 in 16S RNA or in the structure of ribosomal protein S5 (Bilgin et al 1990). Our data show that the antibiotic is a specific inhibitor of EF-G function and that the mutant ribosomes retain to varying degrees their normal functions with EF-G. Comparison of the responses of the mutant and wild type translation systems to spectinomycin *in vitro* and *in vivo* suggest that the concentrations of EF-G in the bacteria are not sufficient to saturate the ribosomes.

In summary, we have identified an antibiotic that very specifically inhibits EF-G-ribosome functions, spectinomycin, and that can be used to select useful ribosome mutants. In addition, we have used the antibiotic fusidic acid to identify a mutant phenotype of EF-G that reflects a direct role for the factor in determining the precision of mRNA movements on the translating ribosome. We next try to set the frameshift error in perspective with the other errors of processivity.

Processivity

There are, as far as we know, three sorts of accidents besides frameshifting that can disrupt the normal processing of polypeptides: One is false termination i.e. termination at a sense codon; this sort of event to our knowledge remains to be identified. Another is polypeptidyl-tRNA drop off, which has been described by Menninger (1976,1978). Finally, there are all of the accidents that produce truncated mRNA such as abortive termination of transcription and degradation by nucleases (Adhya and Gottesman, 1978; Friedman et al 1987).

Previously, Manley (1978) showed that 23% of *lacZ* products could be recovered as truncated polypeptides, which suggests that close to 31% of all translation starts on this mRNA were aborted for one reason or another. While our studies were under way, Tsung et al (1989) measured the decrease in the amount of polypeptide produced from a series of tandemly duplicated genes. Their results indicate that up to 50% of all polypeptide starts are abortively terminated. However, their measurements were made with ribosome mutants, which as we have shown (see below) may process polypeptides at efficiecies very different from wild type.

We have tried to estimate the global rate of all abortive events by a method that does not depend on the recovery or identification of the aborted polypeptide products. We have used a strategy similar to that used by Tsung et al (1989), but we have confined ourselves to comparing the yields of polypeptide produced from monomer and dimer constructions of *lacZ* (Jörgensen and Kurland,1990). Here, the difference in yield of the two separable products is a measure of the relative frequency with which a ribosome that completes the monomer failes to complete the dimer. In other words, this difference corresponds to the sum of all processing errors that can occur in attempting to translate a monomer product of *lacZ*.

Our data suggest that wild type ribosomes have a 0.25 probability of failing to process a full beta-galactosidase chain (Jörgensen and Kurland, 1990). A small fraction of this attrition, at most one third, is due to abortive termination of mRNA transcription. Most important is the observation of an augmented processivity error rate for ribosome mutants that are more accurate with respect to missense errors than wild type. In particular, a partially steptomycin dependent mutant (Ruusala at al 1984) has a probability close to 0.73 of failing to complete the translation of a *lacZ* polypeptide. When streptomycin is present, the probability of aborting translation in the mutant is reduced to 0.5 even though the missense error rate is enhanced by the antibiotic. In other words, there is a reciprocal relationship between the missense error rate and the total processivity error rate for this mutant.

Partitioning Processivity

Atkins et al (1972) demonstrated a positive correlation between the missense error rate and the reading frame error rate for a series of ribosome mutants. In contrast, we observe that the processivity error rates and the missense frequencies are reciprocally related for the one mutant that we have studied so far (Jörgensen and Kurland,1990). Since this mutant has properties that fall into the general pattern of restrictive S12 mutants (Ruusala et al 1984; Andersson et al 1986), we suspect that the reciprocal correlation also will be general. We explain the superficial contradiction between these two tendencies by suggesting that the total processivity error rate is much larger than the reading frame error rate (Jörgensen and Kurland,1990).

The data of Atkins et al (1972) has been used to calculate an average reading frameshift frequency that is close to 3×10^{-5} (Kurland, 1979). This is almost certainly an overestimate because termination codons seem to be the dominant sites for frameshifting within the suppression windows normally studied (Weiss et al 1987). Indeed, it has been shown recently that frameshift error frequencies can be reduced by increasing the concentrations of release factors present in the bacteria (Ryden and Hughes, 1990). Nevertheless, even the overestimation of the frameshift frequency yields a number that is much smaller than the average processivity error rate per codon, which corresponds to roughly 2.5×10^{-4} (Jörgensen and Kurland, 1990). The question then is what event is dominating the processivity figures?

We are not yet certain. However, it does seem likely that drop-off events make up the majority of processivity errors. Thus, the average rate of drop-off per codon is estimated (very roughly) to be 4×10^{-4} (Menninger, 1976). This is close enough to the estimate of the global abortion rate to qualify it as the dominant processivity error if we make ample allowance for strain differences and the like.

Indeed, we have made independent observations that clearly show that frameshift events make a negligible contribution to the total processivity error rate. Thus, we have measured the change in the efficiency of processing polypeptides when mutant elongation factors are present *in vivo.* In particular, when either a frameshift enhancing EF-Tu mutant or a frameshift restrictive EF-G mutant is present, the efficiency of processing is influenced by less than 20% (Hughes, unpublished; Richter, unpublished).

A Little Mechanism

We have previously suggested that there are at least two separable structural interactions that influence the accuracy of the movements of peptidyl-tRNA-mRNA complex during translation (Kurland and Ehrenberg,1985; 1987). One of these would be the codon-anticodon interaction, the quality of which has been identified as an important factor influencing the frameshift error frequency (Kurland,1979; Weiss and Gallant, 1983). The second is the quality of the tRNA-ribosome interaction which would influence the stability of binding of the peptidyl-tRNA-mRNA complex to the ribosome, and therefore it would tend among other things to influence the drop off frequencies. Furthermore, each of the two sorts of interactions could be preferentially influencing a different partial step in the movement process.

Thus, it has been suggested that the movements of the peptidyl-tRNA-mRNA complex are incremental, and that for example, the coupled codon-anticodon domains might be moved in a step separate from that for the remaining domains of the peptidyl-tRNA (Kurland and Ehrenberg, 1985;1987). This view is consistent with the observations made more recently by Moazed and Noller (1989). Thus, it seems likely that the opposing correlations of missense error rates on the one hand with the reading frame error rate and on the other hand with the processivity (presumably drop off) error rate reflect selective influences of these two functional interactions on two separate partial movements.

In summary, there is a pattern emerging from the data concerning translational movements. It remains to be seen if the partitioning of the translation movements into incremental steps of the sort described here is correct. Nevertheless, it does seems as though we are beginning to acumulate sufficiently sharp tools to find out.

Acknowledgements

I thank F. Jörgensen and Diarmaid Hughes for useful discussions and for the sharing of unpublished results, for which A. Richter also is acknowledged. This work is supported by the Swedish Cancer Society and Natural Sciences Research Council.

References

Adhya, S. & Gottesman, M. (1978). Ann. Rev. Biochem. 47, 967-996.

Andersson, D.I., van Verseveld, H.W., Stouthammer, A.H. & Kurland, C.G. (1986). Arch. Microbiol. 144, 96-101.

Atkins, J.F., Elseviers, D.& Gorini, L. (1972). Proc. Natl. Acad. Sci. USA 69, 1192-1195.

Bilgin, N., Richter, A.A., Ehrenberg, M., Dahlberg, A.E. & Kurland, C.G. (1990). Submitted.

Bossi, L. & Roth, J.R. (1981). Cell 25, 489-496.

Buckingham, R.H. & Grosjean, H. (1986). In Accuracy in Molecular Processes, eds. Kirkwood, Rosenberger, Galas, London, Chapman and Hall.

Curran, J.F. & Yarus, M. (1987). Science 238, 1545-1550

Friedman, D.I., Imperiale, M.J. & Adhya, S.L. (1987). Ann. Rev. Genet. 21, 453-488.

Hopfield, J.J. (1974). Proc. Natl. Acad. Sci. USA 71, 4135-4139.

Hughes, D., Atkins, J.F. & Thompson, S. (1987). EMBO J. 6, 4235-4239.

Jörgensen, F. & Kurland, C.G. (1990). Submitted.
Kohno, T. & Roth, J.R. (1978). J. Mol. Biol. 126, 37-52.
Kruzewskia, A. & Slonimski, P.P. (1084). Curr. Genet 9, 11-19.
Kurland, C.G. (1979). In Nonsense mutations and tRNA-suppressors. (Celis, J. E. Smith, J. D. eds.) 97-108, Academic Press, London.
Kurland, C.G. & Ehrenberg, M. (1985). Quart. Rev. Biophys. 18, 423-450.
Kurland, C.G. & Ehrenberg, M. (1987). Ann. Rev. Biophys. 16, 291-317.
Manley, J.L. (1978). J. Mol. Biol. 125, 407-432.
Menninger, J.R. (1976). J. Biol. Chem. 251, 3392-3398.
Menninger, J.R. (1978). J. Biol. Chem. 253, 6808-6813
Moazed, D. & Noller, H.F. (1989). Nature, in press.
Murgola, E.J. (1985). Ann. Rev. Genet. 19, 57-80
Murgola, E.J. (1989). In Transfer RNAs and other soluble RNAs. Cherayil J. D. (ed.), CRC Press.
Murgola, E.J., Hijazi, K.A., Göringer, H.U. & Dahlberg, A.E. (1988). Proc. Natl. Acad. Sci. USA 85, 4162-4165.
Ninio, J. (1975). Biochimie 57, 587-595.
Nishizuka, Y. & Lipmann, F. (1966). Proc. Natl. Acad. Sci. USA 55, 212-219.
Okura, A., Kinoshita, T. & Tanaka, N. (1970). Biochem. Biophys. Res. Comm. 41, 1545-1550.
O'Mahony, D.J., Mims, B.H., Thompson, S., Murgola, E.J. & Atkins, J.F. (1989). Proc. Natl. Acad. Sci. USA 86, 7979-7983.
Richter, A.A. & Kurland, C.G. (1990a). Submitted.
Richter, A.A. & Kurland, C.G. (1990b). Submitted.
Roth, J.R. (1974). Ann. Rev. Genet. 8, 319-346.
Ruusala, T., Andersson, D.I., Ehrenberg, M. & Kurland, C.G. (1984). EMBO J. 3, 2575-2580.
Rydén-Aulin, M. & Hughes, D. (1990). Submitted.
Tanaka, N., Kinoshita, T. & Masukawa, H. (1968). Biochem. Biophys. Res. Comm. 30, 278-283.
Tsung, K., Inouye, S., Inouye, M. (1989). J. Biol. Chem. 264, 4428-4433.
Tucker, S.D., Murgola, E.J., Pagel, F.T. (1989) Biochimie 71, 729-739
Weiss, R. & Gallant, J. (1983). Nature 302, 389-393.
Weiss, R.B., Dunn, D.M., Atkins, J.F. & Gesteland, R.F. (1987). CSHSQB vol 52, 687-693.
Weiss-Brummer, B., Sakai, H. & Kaudewitz, F. (1987). Curr. Genet. 11, 295-301.

RIBOSOMAL FRAMESHIFT AND FRAME-JUMP SITES AS CONTROL POINTS DURING ELONGATION

Robert B. Weiss
The Howard Hughes Medical Institute
Department of Human Genetics
The University of Utah
Salt Lake City, Utah 84132

Introduction

As the details of the interactions amongst tRNA, elongation factors, ribosomes and mRNA during the elongation phase of protein synthesis come into increasing focus, several new avenues for translational control have emerged. The recent advent of the high-level ribosomal frameshift site has opened up one such avenue for alternative events during the elongation and termination phases. These extreme cases of sequence-dependent variation in reading frame step size have also exposed a new lead into unsuspected interactions that occur during decoding and translocation. Analysis of ribosomal frameshifting is thus informative both from the standpoint of novel genetic control mechanisms, as well as illuminating new, and complex, macromolecular interactions occurring in protein synthesis.

At first glance, the capacity for altering the progress or reading frame of an elongating ribosome would not seem to unveil enormous vistas of genetic regulation. Indeed, the increasingly common recurrence of the retroviral gag-pol type -1 frameshift class in diverse eukaryotic and prokaryotic genes argues for a limited set of mechanisms and circumstances where efficient ribosomal frameshifting makes sense.

Currently, there are three examples of high-level ribosomal frameshifting where the site of the reading frame change has been pinpointed by protein sequencing. This article will briefly consider the 'how' of these various frameshifts before engaging in a more speculative treatment of the 'why'.

NATO ASI Series, Vol. H 49
Post-Transcriptional Control of Gene Expression
Edited by J. E. G. McCarthy and M. F. Tuite

High-level ribosomal frameshift sites comprise multiple sequence features

Historically, translational frameshifting was studied from either a genetic approach, such as extragenic frameshift suppressors (a veritable plethora of alterations in both tRNAs, elongation factors and rRNA have been discovered by this approach; for a recent synopsis see O'Connor et al., 1989; and Culbertson, et al., 1989) or a physiological approach that investigated the relationship between aminoacyl-tRNA selection and reading frame maintenance (for an excellent review of all these subjects see Parker, 1989). Many of these studies emphasized the importance of interactions between the ribosome and ternary complex necessary for correct substrate selection, as well as the details of codon:anticodon interaction that gives the major definition to the ribosome's reading frame. What was not well understood in these early studies was the contribution of mRNA sequence to reading frame maintenance, although variation in the leakiness of some frameshift mutations presaged this topic (Fox and Weiss-Brummer, 1980; Atkins et al., 1983).

Appreciation of the mRNA's contribution to reading frame maintenance for the normal translation apparatus began to grow as the list of naturally-occurring high-level ribosomal frameshift events in certain genes began to expand (Dunn and Studier, 1983; Jacks and Varmus, 1985; Craigen et al., 1985; Clare and Farabaugh, 1985; Mellor et al., 1985; Moore et al., 1987, Brierly et al.; 1987, Huang et al., 1988; Clare et al., 1988; Sekine and Ohtsubo, 1989). This remarkable class of translational events range in rates from 5% to nearly 100% of ribosomes frameshifting at a specific sequence, and in altered step size of -1 or +1 nucleotides for the frameshift class, up to +50 nucleotides for the frame-jump class.

A unifying rule for the examples where the frameshift site has been localized by protein sequencing is the potential for stable anticodon pairing of the frameshifting tRNA (or tRNAs) with a codon outside the zero frame. Frameshifting on strings of repetitive nucleotides obey this rule, as seen with the +1 frameshift in the *E. coli* release factor 2 gene (Craigen et al.,

1985) and with the -1 frameshift in the retroviral gag-pol-type class (Jacks et al., 1988). Another sequence combination conforming to this rule is the passage of tRNA from its initial zero frame codon to a non-overlapping synonymous codon. This type of event, termed a tRNA hop (Weiss et al., 1987), has been observed with normal tRNAs and suppressor tRNAs (Falahee et al., 1988; O'Connor et al., 1989), and has been shown to be one component of the 50 nt. frame-jump across the coding gap of gene 60 of bacteriophage T4 (Huang et al., 1988; Weiss et al., 1990a).

The -1 frameshifts found in retroviral gene overlaps and the +1 frameshift found in the E.coli RF2 gene both utilize extensive flanking mRNA sequence context to increase the amount of frameshifting at a specific site. Investigation of these new examples has already revealed unexpected interplay between mRNA, ribosomes, and tRNA inside and outside of the decoding sites (Weiss et al., 1988; Jacks et al., 1988b; Brierly et al., 1989).

Figure 1 depicts the mRNA sequence features which contribute to maximal +1 frameshifting at the RF2 frameshift site. The amount of +1 frameshifting is between 30 and 50%, with the CUU-decoding leucyl-tRNA slipping +1, both in the natural gene (Craigen et al., 1985) and in a synthetically-derived lacZ fusion (Weiss et al., 1987). When the CUU-U shift site is examined in a heterologous mRNA context, the detectable level of +1 frameshifting is < 0.2%, indicating that the mRNA context plays a major role (Weiss et al., 1987).

Genetic dissection of the flanking mRNA sequence has revealed two distinct elements. The 3' element (sequence feature 2b, Figure 1) is a simple stop codon located next to the codon at which the frameshifting tRNA initially decodes. In E.coli, any of the three stop codons can stimulate frame 'slipping' of a tRNA decoding strings of repetitive nucleotides (Weiss et al., 1987). This property may indicate a relatively slow step in the termination pathway prior to release factor binding (Curran and Yarus, 1989; Weiss et al., 1990b).

The second flanking element was shown to be an RNA:RNA hybrid (Weiss et al., 1988) between the AGGGGG mRNA sequence 4 nts. 5' of the frameshift site and nucleotides 1535-1540 of 16S

Sequence Features: 1 single slippery codon
2a 5' mRNA:rRNA hybrid, critical spacing
2b 3' stop codon

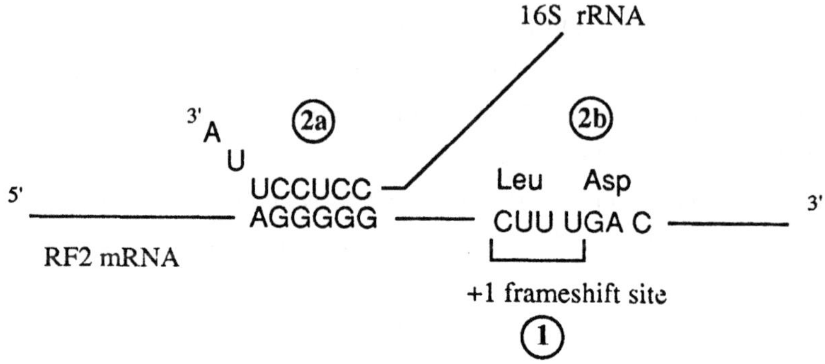

Figure 1. Multiple sequence features that contribute to the level of site-specific +1 frameshifting in the release factor 2 gene of E. coli.

Three elements determine the amount of +1 frameshifting at the RF2 frameshift site: a slippery codon, a 3' flanking stop codon and a 5' flanking mRNA:rRNA transient hybrid formation. Disruptions of the slippery codon (i.e., negated the ability of a well-decoded tRNA to base-pair effectively in an overlapping reading frame) reduce frameshifting by > 100-fold, while stop to sense alterations and removal or spacing changes of 5' Shine-Dalgarno sequence cause 10-20 fold reductions (Weiss et al., 1987).

rRNA ; essentially a Shine-Dalgarno interaction during the elongation phase. This observation implies a continuous cycle of hybridization and melting of the 3' end of 16S rRNA with the mRNA as the ribosome translocates along the message. The function of this cycle is unknown, but it can serve to dampen, as well as stimulate certain frameshift sites (Weiss et al., 1990b).

The -1 frameshifts first observed in retroviral gag-pol overlaps (Jacks and Varmus, 1985) constitute a widely distributed class of events. They are comprised of two major sequence features (Figure 2): a tandem set of 'slippery' codons at the frameshift site and a flanking 3' stimulating element. The hallmark of this class of -1 frameshifts is that effective offset base-pairing potential is required for the set of tandem codons and the frameshift occurs after the second codon has been correctly decoded in the zero frame.

Sequence Features: 1 tandem slippery codons
2 3' mRNA structure (stem-loop or pseudoknot)

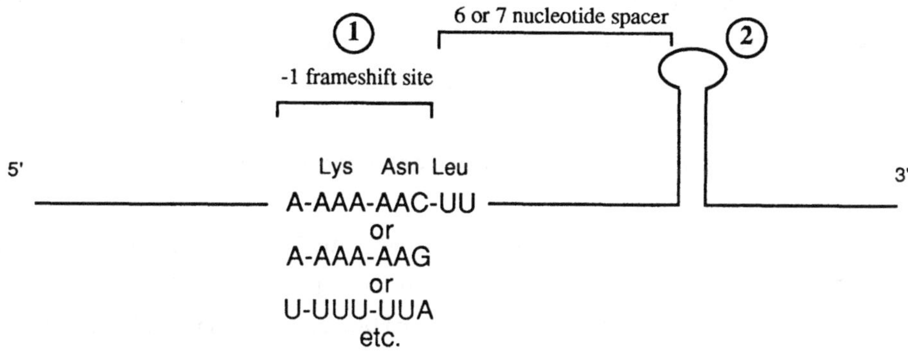

Figure 2. mRNA sequence features utilized by retroviral gag-pol overlaps and analogous high-level -1 frameshift sites.

Two essential elements are found in these -1 frameshift sites: a tandem slippery codon at the frameshift site and a flanking 3' mRNA structure. Both eukaryotic and prokaryotic ribosomes are able to respond to these elements.

The second sequence element found associated to various extent with these -1 frameshift sites are a 3' flanking element. The most impressive element found to date is an RNA pseudoknot, located 6 nucleotides 3' of the putative frameshift site in the F1/F2 overlap of the coronavirus IBV (Brierly et al., 1989). Potential stem-loop structures are located approximately 7 nucleotides from the putative frameshift sites in most viral examples, and compensating base-pair substitutions have implicated these directly in the Rous sarcoma virus gag-pol shift (Jacks et al., 1988).

When the eukaryotic viral sequences are translated in vivo by E.coli ribosomes, they are still capable of eliciting the same type of frameshift (Weiss et al., 1989). A simple C to G substitution at the MMTV gag-pro frameshift site (A-AAA-AAC to A-AAA-AAG) causes a 50% level of -1 frameshifting at this sequence by E.coli ribosomes, indicating a basic conservation of the tRNA:mRNA:ribosome interactions. The frameshifting rate of E.coli ribosomes also responds to 3' stem-loop structures, and one can speculate that the ribosomal process responsible for melting mRNA secondary structure of bacterial attenuators (Landick and Yanofsky,1988) and translational control regions of antibiotic resistance genes (Alexieva et al., 1988), is interacting with this 3' flanking element. Detailed interpretations of why the melting of mRNA secondary structure located at least 6 nts. 3' of the decoding sites should influence the position of the tandem tRNAs in the decoding sites have been proposed (Jacks et al., 1988; Weiss et al., 1989).

An enigma: the gene 60 50 nucleotide frame-jump

The discovery of an internal 50 nucleotide nontranslated coding gap within the coding region of gene 60 mRNA (Huang et al., 1988), revealed a new class of ribosomal frameshifts, which may more properly be termed 'frame-jumps'. The 50 nt. coding gap separates the first 46 codons from the last 114 codons of gene 60; these 160 discontinuous codons encode the 18 kd subunit of the bacteriophage T4 DNA topoisomerase. The ability of translating ribosomes to bypass the gap can be re-constructed in E.coli with gene 60-lacZ fusions, where the gap bypass efficiency in vivo approaches 100% (Weiss et al. 1990a). A mutational analysis of these gene 60-lacZ fusions has defined 5 key sequence features that contribute to high-level bypass (Figure 3). These features are: a segment of the nascent peptide translated from the sequence preceding the gap, a stop codon at the 5' junction, a short stem at the beginning of the gap, a duplicated codon flanking the gap and the distance between these two codons all contribute to gap bypass (Weiss et al., 1990a).

The sequence features in the coding gap region (features 2 through 5, Figure 3) suggest that the bypass mechanism may be akin to tRNA hopping. That is, the gly-tRNA initially decoding GGA^{46} locates the 3' junction by dissociating from mRNA, passing over the gap and binding at the GGA codon situated 50 nts. away. The stop codon and short stem-loop may facilitate the initial dissociation from the 5' GGA codon. As yet unexplained are the optimal 50 nt. spacing between these two codons necessary for maximal bypass and the mechanism which supplies the impetus to the ribosome so that it may track 50 nts. along the mRNA.

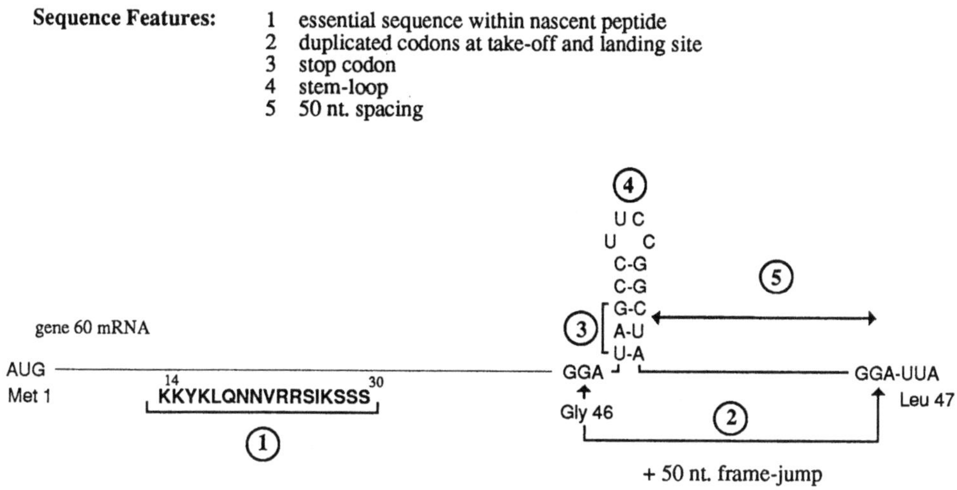

Figure 3. Multiple sequence features contributing to the high-level ribosomal frame-jump observed in gene 60.

Five sequence elements necessary for maximal bypass of the 50 nt. coding gap listed. The order (1 through 5) indicates the relatively severity of inactivating mutations in the elements, with #1 being the most severely affected. This interpretation of coding gap bypass is based on a mutational analysis of gene 60-lacZ fusion expressed in E. coli (Weiss et al., 1990a).

The most unexpected sequence feature that has been found within the gene 60 sequence is the requirement for a cis-acting nascent peptide sequence centrally located in the 46 amino acid 5' coding region (Weiss et al., 1990a). Elimination of this sequence by phasing the ribosome across this region in an alternate frame decreases the bypass efficiency by nearly three orders of magnitude. The loss of gap bypass seen with the phase variants implicates amino acid residues 17 to 32 as activators of gap bypass, but does not indicate which features of the KKYKLQNNVRRSIKSSS sequence are important or indicate by what possible mechanism such a sequence could elicit the jumping over a 50 nt. section of mRNA.

Figure 4 lists the scores from a FASTA sequence similarity search of the 46 amino acid 5' peptide versus the NBRF protein sequence database, and Figure 5 displays a few chosen alignments.

(Peptide) FASTA of: Wai5seq.Pep from: 1 to: 46 March 24, 1990 22:15
5' end of gene60
TO: NBRF:* Sequences: 12,476 Symbols: 3,406,022 Word Size: 2

The best scores are:

		init1	initn	opt..
New:A24897	DNA topoisomerase II - Yeast (Schizosaccharom...	38	55	42
Protein:Mfnzb3	Matrix protein - Bovine parainfluenza vir...	51	51	54
Protein:Qqvz25	Hypothetical protein D-551 - Vaccinia vir...	46	46	53
New:S00111	Ribosomal protein L5 - Rat	45	45	52
New:A27380	Ribosomal protein L5 - Rat	45	45	52
Protein:Isectb	DNA topoisomerase (ATP-hydrolyzing) (EC 5...	33	44	38
New:A23807	30S ribosomal protein S13 - Escherichia coli	39	39	45
Protein:R3ec13	30S ribosomal protein S13 - Escherichia coli	39	39	45
Protein:Wmbe71	57K protein - Herpes simplex virus (type 1)	38	38	46
Protein:Sybycs	Carbamoyl-phosphate synthetase (EC 6.3.5....	38	38	42
New:A26596	Cell division control protein 25 - Yeast (Sac...	37	38	37
Protein:R3ec18	30S ribosomal protein S18 - Escherichia coli	38	38	39
Protein:Fovwvl	gag polyprotein - AIDS virus LV (lymphade...	38	38	46
Protein:Gnljvs	pol polyprotein - Visna lentivirus (strai...	38	38	40
Protein:Fovwa2	gag polyprotein - AIDS virus ARV-2 (AIDS-...	38	38	48
Protein:Fovwh3	gag polyprotein - AIDS virus HTLV-III (T-...	38	38	46
Protein:Fovwlv	gag polyprotein - AIDS virus LAV-1a (lymp...	38	38	46
New:A26480	Knob protein - Plasmodium falciparum (fragment)	37	37	42
New:A29454	Knob-associated histidine-rich protein precur...	37	37	42
New:A05205	Hypothetical protein 1708 - Common tobacco ch...	30	37	39
Protein:Mfnzsv	Matrix protein - Parainfluenza 1 virus (s...	37	37	42
New:A25630	DNA topoisomerase II - Yeast (Saccharomyces c...	37	37	63
Protein:Mfnzs	Matrix protein - Parainfluenza 1 virus	37	37	42
New:B28821	Phospholipase C (EC 3.1.4.3) III - Rat	31	36	32
Protein:Xqbs	Amidophosphoribosyltransferase (EC 2.4.2.14...	36	36	37
Protein:Rgeciy	Regulatory protein ilvY - Escherichia coli	36	36	41
Protein:R3nt15	Ribosomal protein S15 - Common tobacco ch...	36	36	37
Protein:P2wl11	Probable L2 protein - Papillomavirus (typ...	36	36	36
Protein:Tvbyq4	Transforming protein (ras) - Yeast (Sacch...	36	36	36
New:D24444	Hypothetical protein 4 (16S-23S rRNA spacer r...	36	36	36
New:A27787	Thyroid peroxidase - Human	35	35	37
New:A27416	Thyroid peroxidase (EC 1.11.1.8) - Pig	34	35	36
Protein:Ntsreb	Neurotoxin XI - Scorpion (Buthus occitanus)	35	35	35
Protein:R5bs2f	50S ribosomal protein L2 - Bacillus stear...	35	35	40
New:A29496	Cytochrome f precursor - Rice chloroplast	35	35	43

Figure 4. Protein sequence similarity search of the 46 amino acid 5' peptide from gene <u>60</u> versus the NBRF data base.

The score table from a FASTA search (Pearson and Lipman, 1988) of the National Biomedical Research Foundation (NBRF) protein sequence database (Release 21.0, 6/89) is shown. The PAM250 matrix was used in the second step of the comparison. The search was run on a VAX 8300 using the UWGCG software package (Devereux et al., 1984).

```
Ribosomal protein L5 - Rat
Gaps: 0  Quality: 21.40  Ratio: 0.465  Score: 21.40
                   .         .         .         .
       1 MKFVKIDSSSVDMKKYKLQNNVRRSIKSSSMNYANVAIMTDADHDG 46
         | |||: .... :|:|.:. . ||. |..   .  .:.| .....:.
       1 MGFVKVVKNKAYFKRYQVRFRRRREGKTDYYARKRLVIQDKNKYNT 46


30S ribosomal protein S13 - Escherichia coli
Gaps: 0  Quality: 22.80  Ratio: 0.518  Score: 22.80
                   .         .         .         .
       1 MKFVKIDSSSVDMKKYKLQNNVRRSIKSSSMNYANVAIMTDADH 44
         :. ..||. . :: |: ::.::||.|. | ... :::.  :  |
      47 LSEGQIDTLRDEVAKFVVEGDLRREISMSIKRLMDLGCYRGLRH 90


30S ribosomal protein S18 - Escherichia coli
Gaps: 0  Quality: 20.50  Ratio: 0.456  Score: 20.50
                   .         .         .         .
       1 MKFVKIDSSSVDMKKYKLQNNVRRSIKSSSMNYANVAIMTDADHD 45
         .||.::.. :|:  .||  ..::. |..|:.  :.  . | |..:
       7 RKFCRFTAQGVQEIDYKDIATLKNYITESGKIVPSRITGTRAKYQ 51


core polyprotein (gag) p17-Human immunodeficiency virus I
Gaps: 0  Quality: 23.20  Ratio: 0.516  Score: 23.20
                   .         .         .         .
       2 KFVKIDSSSVDMKKYKLQNNVRRSIKSSSMNYANVAIMTDADHDG 46
         |: ||  ...:.|||||.: |: | . ..:. . . : |..:  .
      15 KWEKIRLRPGGKKKYKLKHIVWASRELERFAVNPGLLETSEGCRQ 59
```

Figure 5. Pair-wise alignment of the 46 amino acid 5'peptide with three ribosomal proteins and the gag-encoded core polypeptide of HIV-1.

The PROFILE programs (UWGCG software package) were used to align the gene 60 5' peptide with a few selected high-scoring matches found with the FASTA search.

Besides the similarity to other type II topoisomerases, intriguing weak similarities are found with a number of ribosomal proteins (Rat L5, E.coli S13 and S18; Figure 5), and to retroviral gag-encoded core proteins. Such similarities and the net + charge of this segment (isoelectric pH=11.82) may indicate that this segment is an RNA binding module. The exact nature of the activation of gap bypass by this nascent chain remains to be determined.

Frameshift and frame-jump sites as control points

The apparent design of both the RF2 and retroviral frameshift sites seems sensible. The RF2 level in the cell can feedback on RF2 synthesis by competing with the frameshift at the stop codon (Craigen and Caskey, 1986; Weiss et al., 1990b). Most retroviruses gauge the ratio of gag and gag-pol polyproteins with the -1 frameshift events, and thus balance the production of gag-derived structural proteins and pol-derived enzymes found within the viral particle. The retroviral-type of -1 frameshift observed in some insertion elements may balance the ratio of transposase to transposition inhibitor (Sekine and Ohtsubo, 1989). Since slippery strings in these special contexts seem to be the culprits in the majority of leaky frameshifts, careful pruning of these sequences from open reading frames would serve to minimize loss of normally elongating ribosomes.

The puzzle of why gene 60 carries a high-level ribosome jump is however quite unclear. There is no genetic evidence to implicate a regulatory function to the jump. Ribosomes do not normally skip over 50 nts. in the middle of translating messenger RNAs. The event observed in gene 60 is the first of its kind, although a potential 12 nt. in-frame jump has recently been postulated to occur from codons 4 to 8 within the carA mRNA from P.aeruginosa (Wong and Abdelal, 1990). Thus, other genes may also be taking advantage of the possibilities of high-level ribosome jumping for reasons yet unknown.

Acknowledgements

I thank my colleagues Ray Gesteland, John Atkins and Diane Dunn for the many discussions regarding ribosomal frameshifting.

References

Alexieva Z, Duvall EJ, Ambulos NP, Kim UJ, Lovett PS (1988): Chloramphenicol induction of cat-86 requires ribosome stalling at a specific site in the leader. Proc. Natl. Acad. Sci. U.S.A. 85:3057-3060.

Atkins JF, Nichols BP, Thompson S (1983): The nucleotide sequence of the first externally suppressible -1 frameshift mutant, and of some nearby leaky frameshift mutants. EMBO J. 2:1345-1350.

Brierly I, Boursnell MEG, Binns MM, Bilimoria B, Blok VC, Brown TDK, Inglis SC (1987): An efficient ribosomal frameshifting signal in the polymerase-encoding region of the coronavirus IBV. EMBO J. 6:3779-3785.

Brierly I, Digard P, Inglis SC (1989): Ribosomal frameshifting signal: requirement for an RNA pseudoknot. Cell 57:537-547.

Clare JJ, Farabaugh P (1985): Nucleotide sequence of a yeast Ty element: evidence for an unusual mechanism of gene expression. Proc.Natl.Acad.Sci. USA 82:2829-2833.

Clare JJ, Belcourt M, Farabaugh PJ (1988): Efficient translation frameshifting occurs within a conserved sequence of the overlap between the two genes of yeast Ty1 transposon. Proc. Natl. Acad. Sci. USA 85:6816-6820.

Craigen WJ, Cook RG, Tate WP, Caskey CT (1985): Bacterial peptide chain release factors: conserved primary structure and possible frameshift regulation of release factor 2. Proc.Natl.Acad.Sci. USA 82:3616-3620.

Craigen WJ and Caskey, CT (1986): Expression of peptide chain release factor 2 requires high-efficiency frameshift site. Nature 322:273-275.

Culbertson MR, Leeds P, Sandbaken MG, Wilson PG in "Ribosomes:

Structure and Function" (W.Hill, P. Moore, R. Garrett, J. Warner, A. Dahlberg and D. Schlessinger, eds) in press, ASM publications, 1990.

Curran, J.F. and Yarus, M. (1989). Rates of aminoacyl-tRNA selection at 29 sense codons in vivo. J.Mol.Biol. 209, 65-78.

Devereux J, Haeberli P, and Smithies O (1984) A comprehensive set of sequence analysis programs for the VAX. Nuc.Acids.Res. 12:387-395.

Dunn JJ, Studier FW (1983): Complete nucleotide sequence of bacteriophage T7 DNA and the locations of T7 genetic elements. J.Mol.Biol. 166:477-535.

Falahee MB, Weiss RB, O'Connor M, Doonan S, Gesteland RF, Atkins JF (1988): Mutants of translational components that alter reading frame by two steps forward or one step back. J.Biol.Chem. 263(34):18099-18103.

Fox TD, Weiss-Brummer B (1980): Leaky +1 and -1 frameshift mutations at the same site in a yeast mitochondrial gene. Nature 288:60-63.

Huang WM, Ao SZ, Casjens S, Orlandi R, Zeikus R, Weiss R, Winge D, Fang M (1988): A persistant untranslated sequence within bacteriophage T4 DNA topoisomerase gene 60. Science 239:1005-1012.

Jacks T, Varmus HE (1985): Expression of Rous sarcoma virus pol gene by ribosomal frameshifting. Science 230:1237-1242.

Jacks T, Madhani HD, Masiarz FR, Varmus HE (1988): Signals for ribosomal frameshifting in the Rous Sarcoma Virus gag-pol region. Cell 55:447-458.

Landick R, Yanofsky C: in Neidhardt FC(ed) Escherchia coli and Salmonella typhimurium, Cellular and Molecular Biology, Washington DC, American Society of Microbiology, 1988, p 1276.

Mellor J, Fulton SM, Dobson MJ, Wilson W, Kingsman SM, Kingsman AJ (1985): A retrovirus-like strategy for expression of a fusion protein encoded by yeast transposon Ty1. Nature 313:243-246.

Moore R, Dixon M, Smith R, Peters G, Dickson C (1987): Complete nucleotide sequence of a milk-transmitted mouse mammary tumor virus: two frameshift suppression events are required for translation of gag and pol. J.of Virology 61(2):480-490.

O'Connor, M., Gesteland, R.F., and Atkins, J.F. (1989). tRNA hopping: enhancement by an expanded anticodon. EMBO J. 13(8), 4315-4323.

Parker, J (1989): Errors and alternatives in reading the universal genetic code. Microbiological Reviews 53:273-298.

Pearson WR, Lipman DJ (1988) Improved tools for biological sequence comparison. Proc. Natl. Acad. Sci. USA 85:2444-2448.

Sekine Y, Ohtsubo E (1989): Frameshifting is required for production of the transposase encoded by insertion sequence 1. Proc. Natl. Acad. Sci. USA 86:4609-4613.

Weiss RB, Dunn DM, Atkins JF, Gesteland RF (1987): Slippery runs, shifty stops, backward steps and forward hops: -2, -1, +5 and +6 ribosomal frameshifting. Cold Spring Harbor Symp. Quant. Biol. 52:687-693.

Weiss RB, Dunn DM, Dahlberg AE, Atkins JF, Gesteland RF (1988): Reading frame switch caused by base-pair formation between the 3' end of 16S rRNA and the mRNA during elongation of protein synthesis. EMBO J.:1503-1507.

Weiss RB, Dunn DM, Shuh M, Atkins JF, Gesteland RF (1989) E. coli ribosomes re-phase on retroviral frameshift signals at rates ranging from 2 to 50 percent. The New Biologist 1:159-169.

Weiss RB, Huang WM, Dunn DM (1990a) Mutational analysis of the coding gap in gene 60 of bacteriophage T4. Cell, in press.

Weiss RB, Dunn DM, Atkins JF, Gesteland RF (1990b) Ribosomal Frameshifting from -2 to +50 Nucleotides. Progress in Nucleic Acid Research and Molecular Biology, Vol. 39, edited by W.E. Cohn and K. Moldave.

Wong SC. Abdelal AT (1990) Unorthodox expression of an enzyme: evidence for an untranslated region within carA from Pseudomonas aeruginosa. J. Bacteriol. 172:630-642.

FRAMESHIFTING IN THE EXPRESSION OF THE <u>trp</u>R GENE OF <u>ESCHERICHIA COLI</u>

I. Benhar, C. Miller and H. Engelberg-Kulka
Department of Molecular Biology
Hebrew University - Hadassah Medical School
Jerusalem, Israel

Introduction

The genetic code is read, three bases at a time, from a fixed point of reference on messenger RNA (mRNA). The choice of the reading frame is believed to be determined by the proper positioning of the ribosome relative to the initiation site. It is well known that point mutations can cause the translation apparatus to be shifted into an improper reading frame. Two types of mutations have been described: a) frameshift mutations, in which nucleotide pairs are added or removed from the coding sequence of a gene (Roth 1974) and b) frameshift suppressor mutations in which a tRNA is the altered component so that it occasionally induces a shift in reading frames in either direction (Murgola 1985; Roth 1981). We call a single forward shift +1 and a single backward shift -1. Recently, a third kind of frameshift has been revealed in which neither the mRNA nor the tRNA are mutated. A normal tRNA molecule may sometimes read a normal mRNA molecule in a frame shifted +1 or -1 from the normal one (Craigen & Caskey 1987a; Parker 1989; Varmus 1988). We call such an event normal frameshifting. Normal frameshifting seems to be programmed by the sequence of the mRNA and sometimes also by its structure (Parker 1989). Normal frameshifting provides a mechanism of gene expression that permits the synthesis of two different proteins from two separate reading frames of a single sequence of an mRNA molecule. Most examples of normal frameshifting have been found in prokaryotic and eukaryotic viral genes (Parker 1989; Varmus 1988; and see Discussion). Until now, the only example in the literature of a cellular gene expressed by such a mechanism has been the <u>Escherichia coli</u> (<u>E</u>. <u>coli</u>) gene coding for the protein release factor 2 (RF2) (Craigen and Caskey 1986, 1987b; Craigen <u>et al</u> 1985.)

NATO ASI Series, Vol. H 49
Post-Transcriptional Control of Gene Expression
Edited by J. E. G. McCarthy and M. F. Tuite

Here we describe a second cellular gene whose expression involves a normal +1 frameshifting event: The E. coli trpR gene which codes for the trpR repressor protein. We shall compare the characteristics of the trpR frameshifting process with those of other genes that can be expressed in this way.

Figure 1: Schematic diagram of the trpR gene open reading frame, the corresponding frame 0 product, and the putative +1 frameshift product.

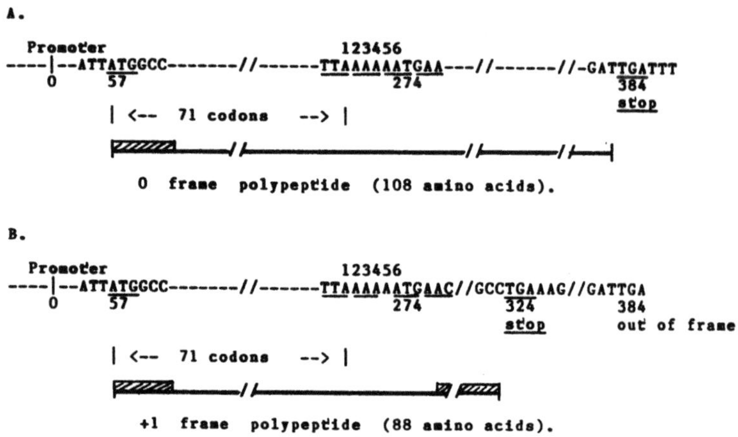

1A: trpR and the 0 reading frame polypeptide (Gunsalus and Yanofsky 1980). The hatched regions here and in Fig. 1B represent the polypeptide region against which antibodies were prepared. The numbers below the sequence here and in 1B represent the position of the nucleotide in trpR; the first digits of the numbers are below the corresponding nucleotide. 1B: trpR and the putative +1 frameshift polypeptide as suggested in this paper.

The E.coli trp repressor has been extensively studied. It regulates the transcription initiation of three operons involved in the biosynthesis of tryptophan: trpEDCBA (Rose *et al* 1973), aroH (Zurawski *et al* 1981), and trpR (Gunsalus and Yanofsky 1980). The repressor polypeptide is 108 amino acids long and is encoded by an open reading frame of the trpR gene. This open reading frame is 324 nucleotides long, spanning from positions 57 to position 381 of the gene (Fig 1A and Gunsalus and Yanofsky 1980). In our laboratory, we noticed that there are six consecutive adenine (A) residues between positions 269 and 274 of the trpR gene (Fig. 1). Based on early reports of Atkins *et al* (1979) and Bermand and Blumenthal (1979), we hypothesized that such a site might permit a +1 frameshift resulting in the synthesis of a +1 frameshift product (Fig. 1B). The product of the trpR frame 0 which ends at the UGA codon at position 384 is 108 amino acids long. In contrast, the putative +1 product is likely to be shorter: we expect it to be only 88 amino acids long, and to end at the UGA codon at position 324. Both polypeptides should start at the same initiation site, position 57, and should be identical for the length of the first 71 codons. Their C-termini should differ (Fig. 1).

Here we report that a +1 frameshifting mechanism really is involved in the expression of the trpR gene. The +1 frameshifting event results in the synthesis of a polypeptide whose N-terminus is identical with that of the known frame 0 product of the gene, but whose length and C-terminus are different.

Results

Use of trpR-lac'Z fusions for studies of trpR frameshifting.

To test whether frameshifting is realy involved in trpR gene expression, we constructed an experimental system in which we used the E. coli lac'Z gene as a reporter gene for trpR frameshifting. In each of its -1, 0, and +1 reading frames, we fused the trpR gene to the eighth codon of the lac'Z gene (Fig. 2). The lac'Z gene was fused at position 294, 307, or 309 of the trpR gene (Fig. 4). These are respectively 20, 33, and 35 base pairs downstream from the adenine hexa-homopolymer (6 A's) of trpR, a site which we suspected is involved in trpR frameshifting (Fig. 1). We called the trpR-lac'Z fusions having 0, +1, and -1 reading frames $trpR_{0}$-lac'Z, $trpR_{+1}$-lac'Z, and $trpR_{-1}$-lac'Z, respectively. The β-galactosidase activity of each fusion was taken as a measure of the rate of the translation activity of the trpR gene in the respective reading frame. The level of gene expression due to frameshifting was determined by comparing the level of β-galactosidase activity of the fusion product in the appropriate frame to that the fusion product in the 0 reading frame of trpR. The gene expression of these fusions was tested when under the control of either the trpR promotor or the lamda (λ) p_L promotor. Experiments with λ p_L containing plasmids were carried out in E. coli strains lysogenic for λ cI857; λp_L derepression

Figure 2: Sequences of trpR-lac'Z fusions in reading frames -1, 0, and +1 of the trpR gene.

```
Frame 0:                                                                   lacZ
-----------------------123456--- trpR------------------------------------>|8th codon
..ATGACGCAGGCTGAGTTAAAAAATGAACTCGGCGCAGGCATCGCGAATACGCGCGCGGATCCC|GTC
   ---..---..---..---..---..---..---..---..---..---..---

Frame +1:
-----------------------123456--- trpR----------------------------------->
                                   ---..---..---..---..---..---..---|lacZ 8th codon
..ATGACGCAGGCTGAGTTAAAAAATGAACTCGGCGCAGGCATCGCGAATACGCGCGGATCCC|GTC
   ---..---..---..---..---..---..---..---..---..---..-

Frame -1:
-----------------------123456--- trpR------------------------------->|lacZ 8th codon
..ATGACGCAGGCTGAGTTAAAAAATGAACTCGGCGCAGGCATCGCGAATACGCGGATCCC|GTC
   ---..---..---..---..---..---..---..---..---..---..
```

The fusion point is at position 294 of the trpR gene. ___...___ below the sequence represents the normal open reading frame 0 of trpR. ___...___ above the sequence represents the open reading frame of trpR resulting by the postulated frame-shifting, either +1 or -1. The six consecutive adenine (A) residues in the trpR gene are represented by the numbers 1-6. Termination codons in the described region of the trpR gene +1 reading frame are emphasized by bold type.

was achieved by thermal induction. Table 1 shows the level of β-galactosidase activity resulting from translation in each of the three frames of trpR, in a pair of $trpR^{+}$ (NK5301) and $trpR^{-}$ (SP361) isogenic *E. coli* strains. For both strains, the $trpR_{0}$-lac'Z fusion product had the highest level of activity. No activity was detected for the $trpR_{-1}$-lac'Z fusion product. However, some β-galactosidase activity was detected for the $trpR_{+1}$-lac'Z fusion product. The percentage of β-galactosidase activity in the +1 frame relative to that of the 0 frame was in the range of 2.6% - 6.7%. The experiments reported in Table 1 were carried out with the trpR-lac'Z fusions illustrated in Fig. 2, where the trpR fusion point is at position 294 of the gene. Similar results were obtained with other trpR-lac'Z fusion constructs where the fusion point was at trpR positions 307 or 309 (data not shown). All these strains are frameshift suppressorfree. These results suggest that +1 frameshifting is involved in the expression of the trpR gene.

Identification of the trpR gene translation products.

The trpR frame 0 product is 108 amino acids long (Fig. 1) and migrates as a polypeptide of 12,000 daltons in SDS-polyacrylamide gels (Gunsalus and Yanofsky 1980; Paluh and Yanofsky

Table 1: Expression of trpR-lac'Z fusions in trpR gene reading frames -1, 0, and +1.

	β-galactosidase activity (units) in strain							
	NK5031 ($trpR^+$)				SP361 ($trpR^-$)			
	trpR reading frames							
	-1	0	+1	%*	-1	0	+1	%*
Promotor λp_L	0	5867	151.3	2.58	0	4606	134.4	2.58
PtrpR	0	4865	204.6	4.2	0	3667	224.7	6.67

*: % is percent activity of $trpR_{+1}$-lac'Z fusion product relative to activity in $trpR_0$-lac'Z fusion product.

1986). On the other hand, the product of the putative +1 frameshift of the trpR gene is expected to be only 88 amino acids long (Fig. 1B) and should migrate as a polypeptide of about 10,000 daltons. Both polypeptides should be identical in their N-termini but not in their C-termini. Therefore, we would expect these polypeptides to be separated by gel migration and identified by their immunological specificity. To test this hypothesis, we used antibodies prepared against two different domains of the polypeptides: a) the first 14 amino acids, a common domain of the two polypeptides; and b) the last 16 amino acids of the +1 frame which are not common to the two polypeptides (see Fig. 1). Thus, the first kind of antibody should react with both 0 and +1 frame products, while the second kind with the +1 frame product only.

We studied trpR translation in an *E. coli in vitro* transcription-translation system. The template for the trpR gene was provided by adding plasmid pHEK-trpR in which we had placed the trpR gene under the control of the strong λp_L promotor. The polypeptides were labled with [^{35}S]-methionine for 60 minutes at 37°C, and immunoprecipitated (Oliver and Beckwith 1982) with each of the two described antibodies. Aliquots were then applied for electrophoresis to urea-SDS-polyacrylamide gels and autoradiographed (Dekel-Gorodetsky *et al* 1986). The migration pattern of the labeled polypeptides in the gels can be seen in Fig. 3. From the sample immunoprecipitated with an-

Figure 3: in vitro translation products of the trpR gene.

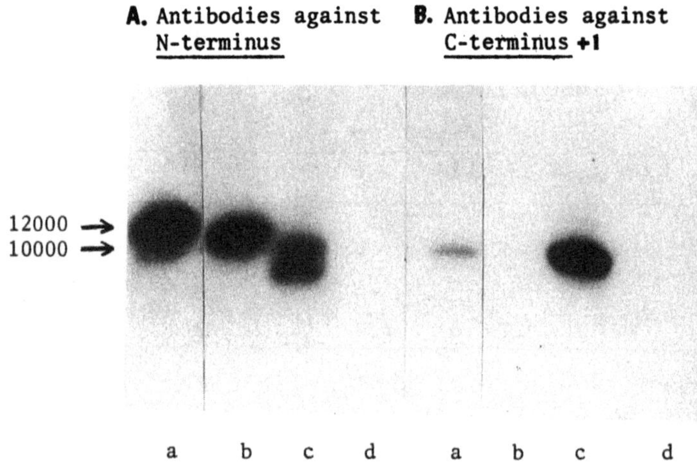

Plasmids pHEK-trpR and its mutant derivatives were added to an E.coli in vitro transcription translation system. Aliquots were immunoporecipitated with antibodies against the first 14 amino acids of the trpR product (Fig.3A) and with antibodies against the last 16 amino acids of frame +1 product (Fig. 3B). a) pHEK-trpR carrying the wild type trpR gene; b) pHEK-trpR$_{UGG}$ and c) pHEK-trpR$_{5A's}$ carrying mutated genes; d) pKC30 carrying no trpR gene.

tibodies against the common N-termini of the polypeptides there are two bands: a major band of 12,000 daltons, and a minor band of 10,000 (Fig. 3Aa). On the other hand, from the sample immunoprecipitated with antibodies against the C-terminus of the +1 frameshift polypeptide, only a single 10,000 dalton band appeared in the gel (Fig. 3Ba). In addition, as seen in Fig.s 3Ad and 3Bd, there are no equivalent radioactive bands in the gels when we added plasmid pKC30 instead of plasmid pHEK-trpR to the extracts. Plasmid pHEK-trpR carries the whole trpR gene, while plasmid pKC30 does not.

The results of these experiments suggest that the trpR gene directs the synthesis of two products in the in vitro S-30 system of E. coli: a) the confirmed 12,000 dalton trpR frame 0 product; and b) the additional 10,000 dalton polypeptide which migrates as would the predicted +1 frameshift product. It appears to have the same amino acid composition at its N-terminus as does the frame 0 product, but to differ from it in its carboxy terminus and length. These characteristics, as well as the immunological specificity to the antibodies used, were further confirmed by experiments for which we pre-pared two mutated trpR genes to be used in place of a wild type template. These mutated trpR genes were prepared on plasmid pHEK-trpR. From the first, pHEK-trpR$_{5'As}$, we deleted an A residue from the run of six A's at positions 260 - 274 of trpR. We predicted that as a major product, trpR$_{5'As}$ should encode a polypeptide migrating as a major band at 10,000 daltons, and reacting with both of the two kinds of antibodies prepared for these experiments. In Fig.s 3Ac and 3Bc we see that this is the case. However, there is an unexpected additional labelled band at about 8,000 daltons reacting only with antibodies against the N-terminus (Fig. 3Ac) and not with those against the C-terminus of the +1 product (Fig. 3Bc). We assume that this shorter 8,000 dalton polypeptide is produced either by cleavage of the C-terminus of the trpR$_{5A's}$ gene directed 10,000 dalton polypeptide, or by a +1 frameshifting event during the expression of this gene. We are presently investigating these two possibilities.

We called the second mutant pHEK-trpR$_{UGG}$. In this case, the UGA codon at position 324 of the +1 frame of the gene was mutated to UGG. We found that such a change inhibits the translation of the trpR +1 reading frame, but does not affect the 0 reading frame. Here we confirm this result using an *in vitro* *E. coli* transcription-translation system (Fig. 3). In the following section we confirm it using an *in vivo* trpR-lac'Z fusion system (Table 2). *In vitro*, we found that pHEK-trpR$_{UGG}$ directs only the synthesis of the 12,000 dalton major product of frame 0 which reacts with the antibodies against the N-terminus (Fig. 3Ab). Plasmid pHEK-trpR$_{UGG}$ does not direct the synthesis of the 10,000 dalton polypeptide that reacts with the C-terminus of the +1 frame (Fig. 3Bb).

The effects of point mutations in trpR on its expression by frameshifting.

To further characterize the frameshifting event as involved in trpR gene expression, we also studied the *in vivo* effects of several point mutations in trpR on the expression of our trpR-lac'Z fusions. The mutations were produced by oligonucleotide-directed site-specific mutagenesis. Then we compared the level of β-galactosidase activity of the mutated trpR-lac'Z fusion product relative to that of the wild type fusion product. The effects of the mutations on the translation of trpR were studied in reading frames 0 and +1 in each of the trpR-lac'Z fusions (Table 2). We first investigated the effect of a change from ATG to ATC in the initiation codon at positions 57 to 59 of the trpR gene. As shown in Table 2, this change completely inhibits the translation in both the 0 and +1 reading frames of trpR.

We conclude that the products of trpR in frame 0 and in frame +1 are translated *in vivo* from the same initiation site. We considered the possibility that the +1 product of trpR results an internal initiation at an ATG codon located at positions 274-276 in the +1 frame of the gene. This is not the case, as is shown in Table 2. Changing this internal ATG codon to either CTG or ATC does not affect the level of translation of trpR in either the 0 or the +1 reading frames of the gene. In addition, our experiments show that a change of the termination codon from TGA to TGG completely inhibits the translation

Table 2: Effects of point mutations in the trpR gene on the translation of the trpR-lac'Z fusion products.

trpR gene nucleotide*		trpR-lac'Z expression** in frame	
Wild Type	Mutation	0	+1
ATG^{59}	ATC^{59}	0	0
^{274}ATG	^{274}CTG	100	100
ATG^{276}	ATC^{276}	100	100
TGA^{324}	TGG^{324}	100	0

* The number near the nucleotide represents its position in the trpR gene. ** The number represents the percentage of β-galactosidase activity of the point-mutated trpR-lac'Z fusion relative to the wild type.

of frame +1, but does not affect that of frame 0 (Table 2). These *in vivo* results agree with our *in vitro* experiments (see previous section). They suggest that the TGA termination codon of frame +1 has a role in the frameshifting event of trpR gene translation.

Discussion

The *trp* repressor protein is 108 amino acids long and weighs 12,000 daltons. It is encoded by the *E. coli trp*R gene in a 324 nucleotide open reading frame, from positions 57 to 384 of the *trp*R gene (Gunsalus and Yanofsky 1980). Here we confirm both *in vivo* and *in vitro* that a second +1 reading frame is translated. As as result, an additional 10,000 dalton +1 frame polypeptide is synthesized. This +1 frame product is identical in its N-terminus with the known frame 0 product, but differs from it in the amino acid composition of its C-terminus.

Recent research has disclosed numerous examples of gene expression occuring by a natural +1 or -1 frameshifting mechanism taking place on the ribosomal level. Until now, most of the examples of this kind of control in higher eukaryotes were found in retroviruses and retrotransposons where frameshifting appears to be a mechanism for the regulation of the expression of reverse transcriptase (Craigen & Caskey 1987a; Parker 1989; Varmus 1988). The phenomenon of frameshifting as a regulating mechanism of gene expression in a non-retroviral system of higher eukaryotes was only recently described in the avian coronavirus infectious bronchitis virus (IBV) (Brierley *et al* 1987, 1989).

Thus, the frameshifting mechanisms of eukaryotic ribosomes are now established for viral genes. However, no eukaryotic cellular gene has yet been found that requires the ribosomes' natural frameshift potential. There are two prokaryotic genes described whose expression involves natural frameshifting. The first is the *E. coli* gene RF2 which codes for the peptide chain release factor two (RF2) required for termination of protein synthesis.

Here we report on the second prokaryotic cellular gene in which +1 frameshifting is involved: the *E. coli trp*R gene. We believe that this frameshfting event also occurs during translation by a normal tRNA. We are now identifying the site(s) involved in *trp*R frameshifting. Having originally suspected that the six A hexa-homopolymer is involved in the +1 frameshifting, we initially based our assumption on early reports suggesting that such a site is involved in frameshifting of E. coli viruses (Atkins et al 1979; Beremand & Blumenthal 1979), and on later reports that such a site is involved in the frameshifting of several retroviruses, including mouse mammary tumor

Figure 4: The region of the trpR gene where the putative +1 frameshifting is expected to take place.

Fig 4. The seqeunce of the end of trpR.

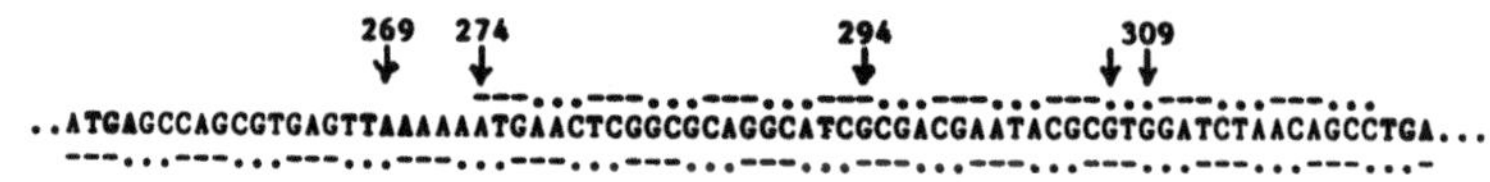

___...___below the sequence represents the trpR open reading frame 0. ___...___ above the sequence represents the region of the open reading frame of the +1 frameshift to which we prepared antibodies against the corresponding peptide. The numbered nucleotides emphasized by arrows are of special interest for this work.

virus (MMTV), bovine leukemia virus (BLV), and human T cell leukemia virus II (HTLV II) (Jacks et al 1988). Our inspection of the trpR sequence has also revealed that the +1 frame of gene contains multiple termination codons the last two of which are shown in Fig. 4. Thus there is no +1 open reading frame up to position 269 of the gene so that no +1 frameshifting can occur upstream from the six A hexa-homopolymer. Furthermore, as seen in Fig. 3Ba, the 10,000 dalton product of in vitro +1 frameshift translation immunoprecipitates with antibodies made against the last 16 amino acids of the C-terminus of the +1 frame product. We therefore suggest that the frameshift event takes place either at a site located between positions 269 and 274, or not far downstream from position 274 (Fig. 4). We further suggest that additional sequences located outside the frameshift site are also involved in the frameshift event because: a) a change of the UGA codon of frame +1 to UGG inhibits the +1 frameshifting process (Fig. 3 and Table 2); b) deletion of 500 base pairs beginning at 10 nucleotides downstream from the the trpR stop codon prevents in vitro +1 translation (data not shown). Experiments to identify precisely the frameshift site and to determine whether other RNA sequences and/or structures are involved in the frameshift mechanisms of trpR are now in progress.

We believe that in at least one way trpR frameshifting is unique among the examples of natural frameshifting described so far. For both prokaryotes and eukaryotes, natural frameshifting suppresses a termination codon located in-frame with the initiation codon: instead of polypeptide chain termination there is a -1 or a +1 frameshift allowing the continuation of protein synthesis until the next termination codon (Craigen & Caskey 1987; Parker 1989; Varmus 1988). This cannot occur in the trpR gene because there is no such in-frame termination codon (Fig. 1). Thus trpR natural frameshifting represents a regulatory mechanism of gene expression unlike the known mechanism of frameshift suppression: it does not involve suppression of a termination codon. Natural frameshifting as described here may also participate in the regulation of some other genes. Such a mechanism should lead to the synthesis of two polypeptides from one gene: a major frame 0 product and a minor frameshift product. The two products will have a common N-terminus and divergent C-termini. Because their C-termini differ, the two products may have different functions. Natural frameshifting may control these functions by regulating the ratio of the synthesis of the two products.

References

Atkins JF, Gesteland RF, Reid BR and Anderson CW (1979) Normal tRNAs promote ribosomal frameshifting. Cell 18:1199-1131.

Beremand MN and Blumenthal T.(1979) An overlapping gene in RNA phage for a protein implicated in lysis. Cell 18:257-286.

Brierley I, Boursnell MEG, Binns MM, Bilimoria B, Blok VC, Brown TDK and Inglis SC. (1987) An efficient ribosomal frameshifting signal in the polymerase-encoding signal of the coronavirus IBV. EMBO J 6:3779-3785.

Brierly I, Digard P and Inglis SC. (1989) Characterization of an efficient coronavirus ribosomal frameshifting requirment for an RNA pseudoknot. Cell 57:537-547.

Craigen WJ and Caskey CT. (1986) Expression of peptide chain release factor 2 requires high-efficiency frameshift. Nature (London) 322:273-275.

Craigen WJ and Caskey CT. (1987a) Translational frameshifting: where will it stop? Cell 50:1-2.

Craigen WJ and Caskey CT. (1987b) The function, structure and regulation of E. coli peptide chain release factors. Biochimie 69:1031-1041.

Craigen WJ, Cook RG, Tate WP and Caskey CT. (1985) Bacterial peptide chain release factors: conserved primary structure and possible frameshift regulation of release factor 2. Proc Natl Acad Sci USA 82:3616-3620.

Dekel-Gorodetsky L, Schoulaker-Schwarz R and Engelberg-Kulka H. (1986) Escherichia coli tryptophan operon directs the in vivo synthesis of a leader peptide. J Bact 1965:1046-1048.

Gunsalus RP and Yanofsky C. (1980) Nucleotide sequence and expression of Escherichia coli trpR, the structural gene for trp aporepressor. PNAS USA 77:7117-7121.

Jacks T, Madhani HD, Masiarz FR and Varmus HE. (1988) Signals for ribosomal frameshifting in the Rous sarcoma virus gag-pol region. Cell 55:446-458.

Murgola EJ. (1985) tRNA suppression and the code. Ann Rev Genet 19:57-80.

Oliver DB and Beckwith J. (1982) Regulation of a membrane component required for protein secretion in Escherichia coli. Cell 30:311-319.

Paluh JL and Yanofsky C. (1986) High level production and rapid purification of the E. coli trp repressor. Nuc Acids Res 14:7851-7860.

Parker J. (1989) Errors and alternatives in reading the universal genetic code. Microbiol Rev 53:273-298.

Rose, JK, Squires CL, Yanofsky C, Yang HL and Zubay G. (1973) Regulation of in vitro transcription of the tryptophan operon by purified RNA polymerase in the presence of partially purified repressor of tryptophan. Nature (London) New Biol 245:133-137.

Roth JR. (1974) Frameshift mutations. Ann Rev Genet 8:319-346.

Roth JR. (1981) Frameshift supression. Cell 24:601-602.

Weiss RB, Dunn DM, Dahlberg AE, Atkins JF and Gesteland RF. (1988) Reading frame switch caused by base-pair formation between the 3′ end of 16s rRNA and the mRNA during elongation of protein synthesis in Escherichia coli. EMBO J. 7:1503-1507.

Varmus HE (1988) Retroviruses. Science 240:1427-1435.

Zurawski G, Gunsalus RP, Brown KD and Yanofsky C.(1981) Structure and regulationof aroH, the structural gene for the tryptophan-repressible-3-Deoxy-D-arabino-haptulosonic acid -7-phosphate synthetase of Escherichia coli.JMB 145:47-73.

Acknowledgments

We thank Hanna Alexander and Dr. Haughton (Scrips Clinic, La Jolla, California USA) for preparing for us the antibodies against chemically snythesized peptides. We thank FR Warshaw Dadon for a critical reading of the manuscript. This research was supported by the endowment fund for Basic Research Foundation in Life Sciences: Charles H. Revson Foundation administered by the Israel Academy of Sciences and Humanities. I. Benhar is supported by a Levy Eshkol fellowship from the Ministry of Sciences and the National Council for Research and Development.

THE RIBOSOMAL FRAME-SHIFT SIGNAL OF INFECTIOUS BRONCHITIS VIRUS

S.C. Inglis, N. Rolley and I. Brierley
Division of Virology,
Department of Pathology,
University of Cambridge,
Tennis Court Road,
Cambridge, U.K.

Introduction

We recently described the first non-retroviral example of ribosomal frame-shifting in higher eukaryotes (Brierley et al., 1987). The shift occurs during translation of the genomic RNA of the coronavirus infectious bronchitis virus (IBV), just at the end of the F1 open reading frame (located at the 5' end of the genome); its consequence is that a proportion of ribosomes reading the F1 frame fail to terminate at the F1 stop codon, and instead begin reading the F2 open reading frame (ORF), which overlaps the end of F1, leading to the production of an F1-F2 fusion protein. This "-1" frame-shift is highly efficient (about 30%) and can be reproduced *in vitro* by cloning a short sequence from the junction of the F1/F2 ORFs into a suitable reporter gene (Figure 1). The recombinant gene may then be transcribed using the phage T7 RNA polymerase, and the resulting mRNA translated in the rabbit reticulocyte lysate cell-free system.

Figure 1 - Organisation of plasmid pFS8

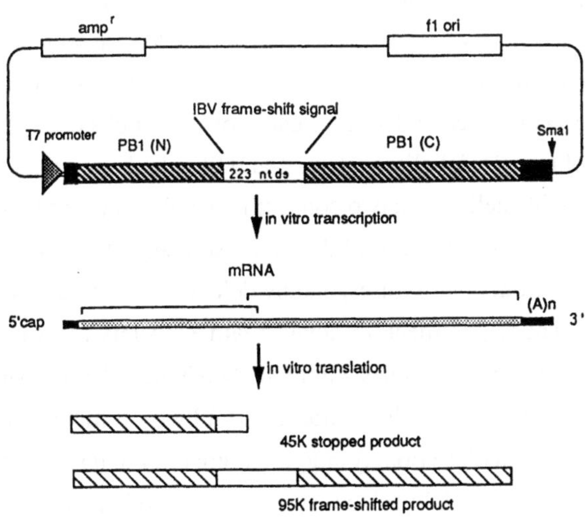

NATO ASI Series, Vol. H 49
Post-Transcriptional Control of Gene Expression
Edited by J. E. G. McCarthy and M. F. Tuite

Figure 2 - Structure of the IBV frame-shift signal

(a) - Primary sequence

5' UUUAAACGGGUACGGGGUAGCAGUGAGGCUCGGCUGAUACCCCUUGCUAGUGGAUGUGAUCCUGAUGUUGUAAAGCGAGCCUUU3'

(b) - Proposed tertiary structure

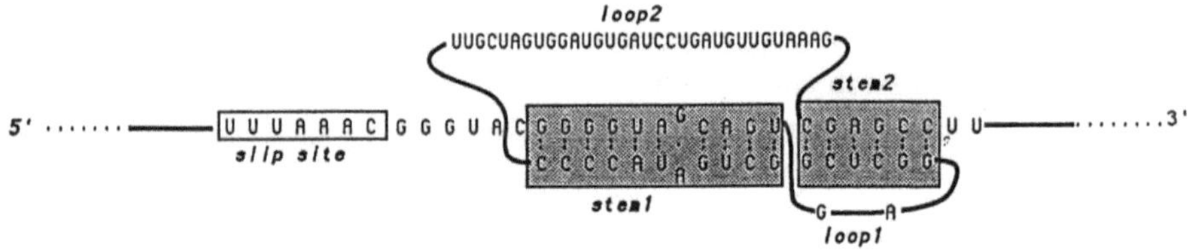

Frame-shifting can therefore be monitored readily through production of a 'read-through' product .

We have previously analysed in detail the sequence requirements for the IBV frameshift signal (Brierley et al., 1989) by deletion analysis and site-directed mutagenesis on the plasmid pFS8 (Figure1). These studies indicated that the signal can be narrowed down to a stretch of 86 nucleotides (ntds), which is in itself sufficient to direct efficient ribosomal frame-shifting in a heterologous genetic context (Figure 2). This sequence appeared to consist of two separate essential elements: a "slippery sequence", located at the 5'end at which the ribosome was believed actually to change frame, followed at a critical distance by a kind of tertiary RNA structure which has been called an RNA pseudoknot (Studnicka et al., 1978; Pleij et al., 1985). This is composed of two RNA helices (stems 1 and 2), which are co-axially-stacked to form a quasi-continuous double helix, with two connecting single-stranded loops (loops 1 and 2). The location of a slippery site at the 5' end of the frame-shift signal was suggested by comparison with known slip-sites from other viruses, and by deletion analysis and site-directed mutagenesis within this region. The requirement for downstream RNA tertiary structure was suggested by the observation that nucleotide changes within the two helices 1 and 2 were highly inhibitory to frame-shifting, but that these could be compensated fully by complementary changes on the opposite strand. These studies strongly suggested not only that such a structure was likely to be part of the frame-shift signal, but also that it was absolutely required for ribosomal slippage to occur with high efficiency. However as yet the mechanism by which ribosomal slippage occurs, and the precise structure and contribution of the pseudoknot remains unclear.

Here we describe further mutagenic studies on the frame-shift signal which seek to elucidate more precisely its structure and the mechanism by which it direct efficient ribosomal slippage.

Analysis of the "Slippery Site"

Based on the work of Jacks and colleagues (1985, 1988a, 1988b), who identified the precise location at which ribosomal frame-shifting occurred during translation of the gag-pol suggested that the heptanucleotide sequence UUUAAAC was likely to be the point at which region of Rous sarcoma virus (RSV) and Human immunodeficiency virus (HIV) we previously ribosomes change frame within the IBV sequence (Figure 2). This possibility was supported, but not formally proven by the deletion analysis described above (Brierley et al., 1989). We therefore introduced two new termination codons around the putative slip site (Figure 3) by site-directed mutagenesis, as described before (Brierley et al., 1989). The stop codon introduced downstream of the slip site is in the F1 reading frame, and therefore serves as a terminator for ribosomes which do not change frame. The stop codon introduced upstream of the slip site is in the "-1" frame (i.e. the F2 reading frame) and consequently the upstream and downstream open reading frames now overlap solely by the seven nucleotides which constitute the putative slip site UUUAAAC. For ribosomal frame-shifting to be observed in this mutant, ribosomes would have to slip within the heptanucleotide sequence. Translation of mRNA bearing this altered signal indicated that frame-shifting occurs with wild type efficiency (Figure 3)*, demonstrating that ribosomal slippage does indeed occur at this position.

Figure 3 - Definition of the IBV frame-shift site

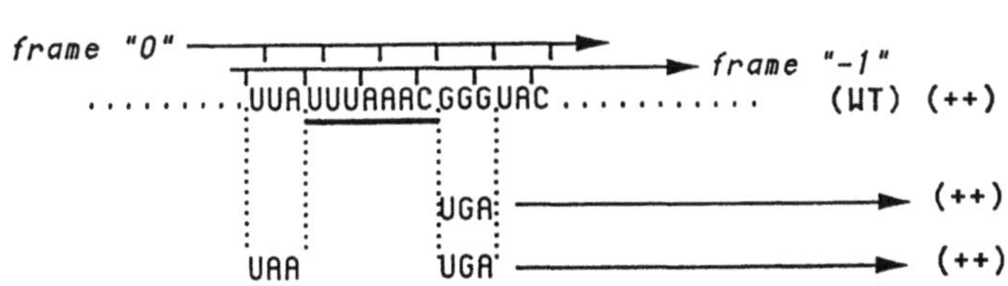

*In this, and all subsequent figures, the symbol (++) represents the aproximate level of wild type frame-shifting (about 30%), (+) denotes frame-shifting between 10% and 20%, (+/-) - 5-10%, and (-) - less than 2%.

Figure 4 - Double-slippage model for -1 frame-shifting

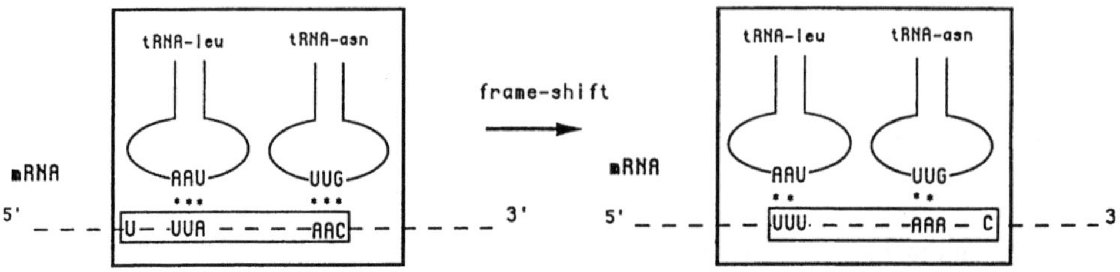

Jacks et al. (1988) proposed a model for "-1" frame-shifting in retroviruses which involved simultaneous slippage on the messenger RNA of tRNAs occupying both the A and P site of the the translating ribosome (figure 4). The model suggests that such frame-shifts would only be allowed if 2 out of 3 base pair contacts are retained between the mRNA and each tRNA after slippage (i.e. 4 out of 6 contacts in total). Thus seven nucleotide sequences which begin with two repeated nucleotide triplets (XXXYYY) would be potentially slippery, and indeed almost all the sequences known, or thought, to direct -1 frame-shifting in retroviruses conform to this pattern, as does the IBV frame-shift site (UUUAAAC). One might expect therefore that alterations made to the slippery site which decreased potential pairing after slippage would be inhibitory to frame-shifting, and this is indeed the case. Mutation of the UUUAAAC sequence to either UUUAUAC or UUUACAC completely abolished frame-shifting, and a change to CUUAAAC was highly inhibitory (Brierley et al., 1989). If however the potential for formation of post-slippage pairs is all that is required, it might also be expected

Figure 5 - Mutational analysis of the IBV slip site

	P site A site	post-slip contacts	Frame-shifting	%
WT	U UUA AAC	4/6	++	30
	U UUA AAG	4/6	+/-	1
	U UUA AAU	4/6	++	30
	U UUU UUC	5/6	++	30
	U UUG GGC	4/6	+/-	1

that any sequence which can still allow potential 4/6 pairing (or greater) after slippage would be slippery. Consistent with this, the sequence UUUUUUC which should allow 5/6 pairs to form after slippage is at least as efficient as the WT IBV sequence (Figure 5). However the sequence UUUGGGC, though still active, was much less efficient, even though it also has the potential to form 4/6 post-slippage contacts. One possible explanation for this result is that heptanucleotide sequences involving several G:C pairs may be less slippery due to the extra energy required to break these contacts. However when we changed only the last nucleotide of the IBV slippery sequence, from a C to a G residue (which should alter neither the potential for pairing post-slippage, nor the overall energy of the tRNA:mRNA interactions) we found once again a marked reduction in frame-shift efficiency. This suggests that the "slipperiness" of a particular sequence is not simply due to the number and type of pairs which can form post-slippage, but in addition depends on the particular tRNAs with which it interacts. Thus the lysine-tRNA (which decodes AAG) appears less prone to slippage than the asparagine tRNA which decodes AAC. In this respect our results differ somewhat from those of Jacks et al., (1988b) who found that any nucleotide substitution could be tolerated equally well at the last position within the RSV frame-shift site. However in this case the sequence of the wild type slip site is quite different (AAAUUUA), and it could be that the tRNAs which decode the final UUA (leu) and third position variants thereof (UUU-phe, UUG-val and UUC-leu) may all be inherently slippery. In vitro translation studies in cell-free systems supplemented with particular kinds of tRNAs may help to resolve this question.

Analysis of pseudoknot structure

Our previous results (Brierley et al., 1989) indicated that formation of a pseudoknot downstream of the slippery site was required for efficient ribosomal slippage, and we proposed a likely model of the structure based on our initial mutagenesis data and on nucleotide sequence analysis. However the precise configuration of the pseudoknot remained uncertain, and so we set out to examine in detail the structure through site-directed mutagenesis, on the premise that nucleotide changes which destabilise the structure should be inhibitory to frame-shifting.

Our analysis of the nucleotides proposed to be part of stems 1 and 2 is summarised in Figure 6. In general the results bear out the model displayed in Figure 2, in that individual nucleotide changes within stems 1 and 2 are inhibitory to frame-shifting. However the results show that the inhibitory effect is much less for nucleotides located at the extremities of the helices than for those in the middle. This would be consistent with the idea that the overall stability of the structure is related to its ability to promote frame-shifting; thus the slight destabilisation of the structure produced by mismatching at the ends, would have a less dramatic

Figure 6 - Mutational analysis of the pseudoknot stems

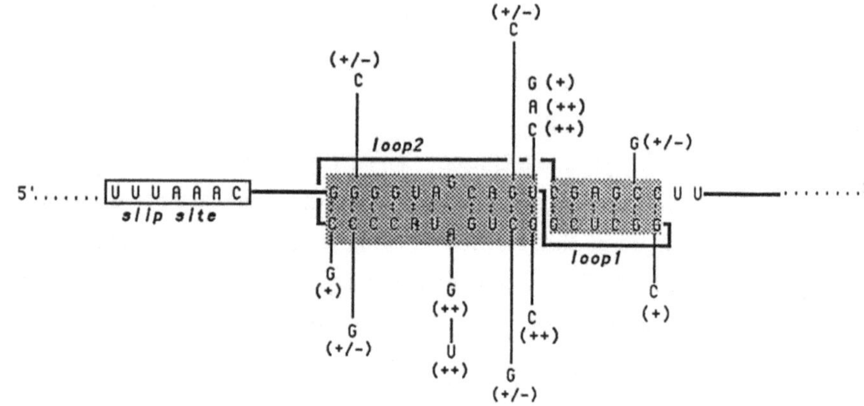

Figure 7 - Mutational analysis of the pseudoknot loops

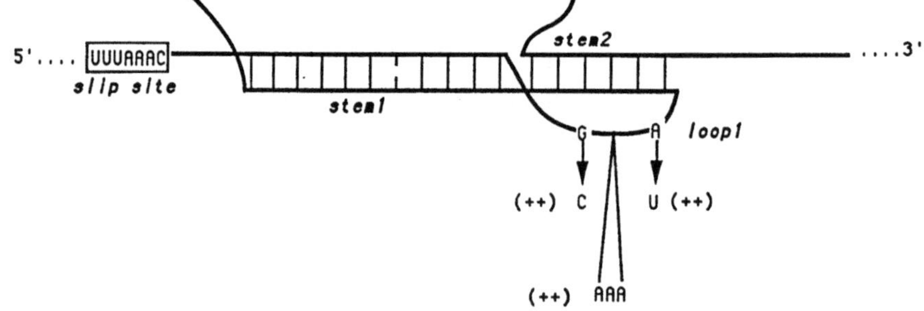

effect than more highly destabilising mismatches introduced within the stems. One exception to this general finding was observed however. Mutations affecting the G:U pair proposed to form at the top of stem1 were not found to affect frame-shifting efficiency greatly. This was a somewhat surprising result since one of the central features of the pseudoknot building principle (Pleij et al., 1985) is that the stems contributing to the tertiary structure should stack co-axially form at the top of stem1 were not found to affect frame-shifting efficiency greatly. This was a somewhat surprising result since one of the central features of the pseudoknot building principle (Pleij et al., 1985) is that the stems contributing to the tertiary structure should stack co-axially to form a quasi-continuous double helix, and that these stacking forces will help to stabilise the structure. Thus we expected that introduction of a mismatched nucleotide pair at the top of stem1 would destabilise the pseudoknot considerably and hence inhibit frame-shifting. That this did not happen suggests either that direct stacking is not required and that a bulge can be tolerated at this position, or perhaps that a mismatched nucleotide pair can be "held" in position by virtue of the stacking forces on either side. Direct structural analysis of the pseudoknot will be required to resolve this point.

Our model for the structure of the pseudoknot proposes that the two RNA helices are connected by two single-stranded loops of 2 and 32 nucleotides (Figure 2) which span the 6 base pairs of stem2 (loop1) and the 11 base pairs of stem1 (loop2) respectively. Theoretical considerations (Pleij et al., 1985) suggest that for the structure to form, loop1 would have to be a minimum of 2 nucleotides long, while loop2 would have to be 7 nucleotides or more. However, provided the loops are long enough, their precise nucleotide sequence shoud not be important. Thus if frame-shifting requires only the formation of the correct tertiary structure, and does not depend on the presence of particular nucleotides within the loops, then alteration of the sequence and length of the loops within the constraint of the required length should not have a dramatic effect. The results of this kind of analysis, summarised in Figure 7, show that this is indeed the case. We found that either of the nucleotides proposed to comprise loop1 could be changed without affecting frame-shifting, and that insertion of an extra 3 nucleotides in the loop could also be tolerated. Loop2 may be reduced in length to 8 nucleotides, and furthermore each of these 8 nucleotides can be substituted without affecting frame-shifting. However deletion of a further 3 nucleotides (leaving only 5 nucleotides) was highly inhibitory. Since we predicted that loop2 would have to be a minimum of 7 nucleotides in order for the pseudoknot structure to form, this result was not unexpected and increases our confidence that the proposed structure is largely correct. However direct structural analysis will be once again required to confirm the hypothesis.

Brierley I, Digard P, Inglis SC (1989) Characterisation of an efficient coronavirus ribosomal frameshifting signal: Requirement for an RNA pseudoknot. Cell 57: 537-547.

Jacks T, Varmus HE (1985) Expression of the Rous sarcoma virus *pol* gene by ribosomal frameshifting. Science 230: 1237-1242.

Jacks T, Madhani HD, Masiarz FR, Varmus HE (1988a) Signals for ribosomal frameshifting in the Rous sarcoma virus *gag-pol* region. Cell 55: 447-458.

Jacks T, Power MD, Masiarz FR, Luciw PA, Barr PJ, Varmus HE (1988b) Characterisation of ribosomal frameshifting in HIV-1 *gag-pol* expression. Nature 331: 280-283.

Pleij CWA, Rietveld K, Bosch L. (1985) A new principle of RNA folding based on pseudoknotting Nucl. Acids Res. 13: 1717-1731.

Studnicka GM, Rahn GM, Cummings IW, Salser WA (1978) Computer method for predicting the secondary structure of single-stranded RNA. Nucleic Acids Res 5: 3365-3387.

Tang CK, Draper DE (1989) Unusual mRNA pseudoknot structure is recognised by a protein translational repressor. Cell 57: 531-536.

CONTROL OF TRANSLATIONAL ACCURACY IN YEAST: THE ROLE OF THE Sal4 (Sup45) PROTEIN

Mick F. Tuite, Akhmaloka, Mandy Firoozan, Julio A.B. Duarte and Chris M. Grant

Biological Laboratory,
University of Kent,
Canterbury,
Kent,
CT2 7NJ,
England.

1. Introduction

It is clearly important for an organism to ensure that translational errors are kept to a minimum. The isolation of genetic modifiers of nonsense suppressors from both *E. coli* and yeast has provided a powerful genetic screen for mutants with defects in the control of translational accuracy (Eggertsson and Soll, 1988; Sherman, 1982). That such mutants are defective in the control of translational accuracy can be readily confirmed by demonstrating their hypersensitivity to antibiotics and other agents known to perturb translational accuracy. In *E. coli* this approach has uncovered a number of key components of the translational machinery important for maintaining the accuracy level; for example the *ram* (ribosomal ambiguity) mutants which increase translational misreading have demonstrated important roles for the ribosomal proteins S4 and S5 (Gorini, 1974; Cabezon *et al*, 1976). We are taking a similar approach in the yeast *Saccharomyces cerevisiae* in an attempt to uncover analogous translational components in this simple eukaryote.

Nonsense suppressors in *S. cerevisiae* fall into two basic classes; tRNA-mediated, codon-specific suppressors (termed tRNA suppressors) and non-tRNA-mediated, codon-non-specific suppressors (termed omnipotent suppressors). Yeast tRNA suppressors generally arise as a consequence of anticodon mutations in either $tRNA^{Tyr}$, $tRNA^{Ser}$ or $tRNA^{Leu}$-encoding genes (Sherman, 1982). In the case of the omnipotent suppressors, at least 11 loci have been identified (Ono *et al*., 1989; Wakem and Sherman, 1990; All-Robyn *et al*, 1990) although in only two cases have the gene products

NATO ASI Series, Vol. H 49
Post-Transcriptional Control of Gene Expression
Edited by J. E. G. McCarthy and M. F. Tuite

been characterised, namely those of the SUP35 (SUP2) and SUP45 (SUP1) genes (Table 1). Mutations at these two loci generate phenotypes that are indicative of a loss of translational accuracy; omnipotent suppression, hypersensitivity to error-inducing antibiotics such as paromomycin, and osmotic-sensitivity (Masurekar et al, 1981; Eustice et al, 1986; Crouzet et al, 1988; Wakem and Sherman, 1990). In neither case, however, is the product of the gene a bona fide ribosomal protein (Table 1).

Table 1

Omnipotent suppressor genes SUP35 (SUP2) and SUP45 (SUP1): allelism, phenotypes and encoded gene products

Gene	Alleles	Phenotypes	Gene Product	Refs
SUP35 (SUP2)		omnipotent suppression	polypeptide; 78,000 M_r;	1,2
	SUF12	frameshift suppression	GTP-binding; EF-1α-related	3
	SAL3	allosuppression		4,5
	GST1	cell-cycle defect		6
SUP45 (SUP1)		omnipotent suppression	polypeptide; 49,000 M_r;	1,2
	SAL4	allosuppression	ATP-binding (?); ribosome-	4,7
	NOV1	novobiocin resistance	associated	8
	MOS1	ochre tRNA misreading		9

References: 1, Hawthorne and Leupold (1974); 2, Inge-Vechtomov and Adrianova (1970);3, Wilson and Culbertson (1988); 4, Cox (1977); 5, Crouzet and Tuite (1987); 6, Kikuchi et al. (1988); 7, Crouzet et al. (1988); 8, E. Orr, personal communication; 9, Gelugne and Bell (1988).

A large number of extragenic modifiers of yeast tRNA suppressors have also been described which fall into one of two classes: suppressor-enhancing or allosuppressor (sal) mutants (Cox, 1977), and suppressor-reducing or antisuppressor (asu) mutants (McCready and Cox, 1973). In addition, the efficiency of certain yeast tRNA suppressors (e.g. the $tRNA^{Ser}$ encoded by the SUQ5 locus) is dramatically affected by an extrachromosomal determinant, [psi] (Cox et al, 1988). Genetic modifiers of omnipotent suppressors have also been described which belong to either the allosuppressor (Song and Liebman, 1987) or antisuppressor (Liebman and Cavenagh, 1980; Liebman et al, 1980; Ishiguro, 1981) class.

The efficiency of omnipotent suppressors is also affected by the extrachromosomal determinants [psi] and [eta] (Wakem and Sherman, 1990; All-Robyn et al, 1990).

Phenotypic analysis of both antisuppressor (asu) and allosuppressor (sal) mutants also suggests that they have lost the ability to maintain translational accuracy. Thus, a potentially large number of loci have been identified in S. cerevisiae that encode components of the translational machinery important for its accuracy although allelism studies between omnipotent suppressor and allosuppressors and antisuppressors is far from complete.

We have focussed our attention on the role of the product of the allosuppressor gene SAL4.

2. The SAL4 Gene and Its Product

Mutations at the SAL4 locus can generate a variety of allele-specific phenotypes in addition to allosuppression against the tRNA suppressor SUQ5 (Cox, 1977); these include omnipotent suppression, hypersensitivity to the antibiotics paromomycin, hygromycin B and G418, and temperature-sensitivity (Cox, 1977; Crouzet et al, 1988). Furthermore, we have isolated and sequenced the SAL4 gene (Crouzet et al, 1988; Akhmaloka and M.F. Tuite, unpublished) confirming our earlier deductions that the SAL4 gene is identical to the previously isolated and sequenced SUP1 (SUP45) gene (Breining and Piepersberg, 1986) with the exception of a single silent amino acid replacement; Gln^{41} in Sal4 for Leu^{41} in Sup45. The predicted amino acid sequence of 438 amino acids shows no significant overall homology with any proteins in the NBRF data base (release 19) although there are small regions of homology with aminoacyl-tRNA synthetases (Breining and Piepersberg, 1986) and chloroplast ATPases (Grant, 1990). As shown in Figure 1, there is also strong homology, starting at amino acid 150, with the consensus ATP-binding site described for protein kinases (Patterson et al, 1986). These homologies strongly suggest that Sal4 may bind a nucleotide, most probably ATP.

A polyclonal antibody, generated against a Sal4-β-galactosidase fusion protein (Grant, 1990), was used to probe the cellular location of the Sal4 polypeptide. Using an affinity purified antibody, in conjunction

Homology 1

Consensus	G . G . F G . V	(8-16)	V A . K	
Sal4	G Q G T L F G S V	(12)	V D L P K	
	150			

Homology 2

Consensus	G . G . F G . V	(8-16)	V A . K
Sal4	G Q G T L F G S V	(6)	V L H K
	150		

Figure 1. Putative ATP-binding site in the Sal4 polypeptide. Comparison with the consensus sequence for an ATP-binding site found in protein kinases (Patterson et al., 1986). Two potential homologies are shown each starting at amino acid position 150.

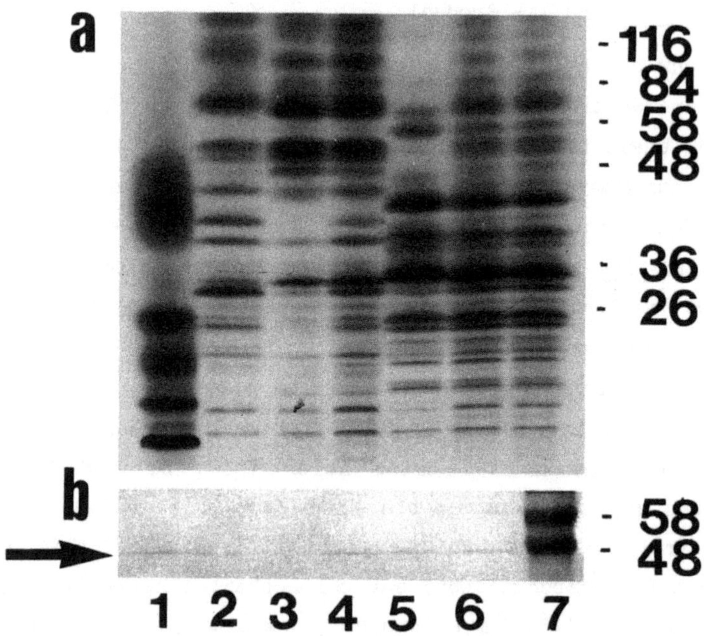

Figure 2. Subcellular distribution of the Sal4 protein. (a) SDS-PAGE of: lane 1, total lysate; lane 2, S30 lysate; lane 3, post-ribosomal supernatant; lane 4, ribosome fraction; lane 5, salt-washed ribosome fraction; lane 6, salt-wash fraction; lane 7, molecular weight markers. (b) Western blot analysis of samples shown in (a) blotted with affinity-purified, anti-Sal4 polyclonal antibody. The Sal4 protein is indicated by the arrow.

with subcellular fractionation and Western blot analysis, the Sal4 protein is found to be tightly bound to the 80S ribosome. As shown in Figure 2, it cannot be detected in a post-ribosomal supernatant fraction (lane 3) and is only partially removed from the ribosome by a high salt (0.5M KCl) wash (lane 6).

The analyses described above thus suggest that the product of the SAL4 gene plays an important role in maintaining translational accuracy in S. cerevisiae. Furthermore, the encoded polypeptide is not a ribosomal protein, but rather may be an nucleotide-binding translation factor that is tightly associated with the ribosome. In an attempt to further probe the role of the Sal4 polypeptide in controlling the fidelity of protein synthesis, we have characterised a series of both conditional-lethal and normal sal4 mutants.

3. Molecular Analysis of sal4 Mutants

A conditional-lethal allele of SAL4, designated sal4-2 and originally isolated by Cox (1977), is an allosuppressor at 30°C but fails to grow at 37°C. The mutant shows a decline in both protein synthesis and the proportion of ribosomes on polysomes following a shift from the permissive (30°C) to the non-permissive (37°C) temperature over a period of 60 min. (Figure 3). There is no decline in RNA synthesis over this period (data not shown) strongly suggesting a temperature-sensitive defect in protein synthesis probably at the level of elongation, although a "leaky" effect on initiation cannot be ruled out from these data.

The sal4-2 gene has been isolated using allele-rescue methods, and its nucleotide sequence determined. A single amino acid change (Ile to Ser) at amino acid position 222 is the only detectable alteration in the Sal4 protein encoded by this allele. From the predicted hydropathicity plot of the Sal4 polypeptide (Figure 4) it can be seen that this change occurs within one of the most hydrophobic (and therefore buried) regions of the molecule. A ts sup45 (sup1) allele has previously been shown to arise from a Leu to Ser change at amino acid position 34 (Breining and Piepersberg, 1986), which again is within a hydrophobic region of the molecule (Figure 4). In both cases thermostability is presumably caused by a disruption of these hydrophobic regions by the serine residue.

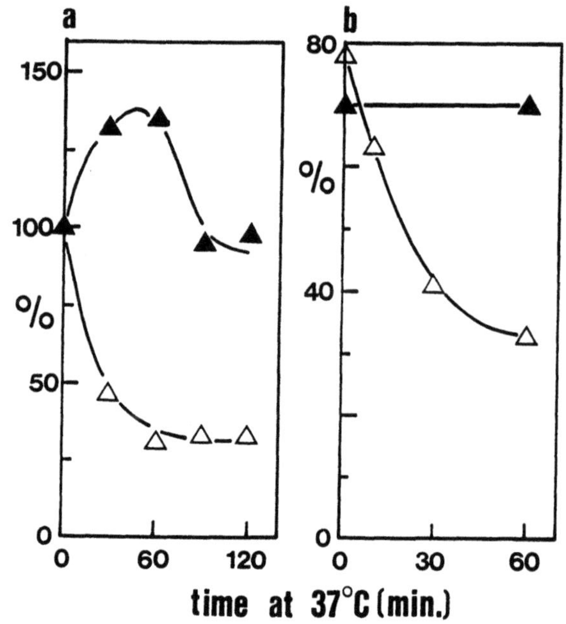

Figure 3. The sal4-2 mutant shows a temperature-sensitive defect in protein synthesis at 37°C. (a) the rate of protein synthesis over 0-120 min. after a shift to 37°C from 28°C. (b) the proportion of ribosomes as polysomes over 0-60 min. after a shift to 37°C from 28°C. ▲ , $SAL4^+$ strain; △ , sal4-2 strain.

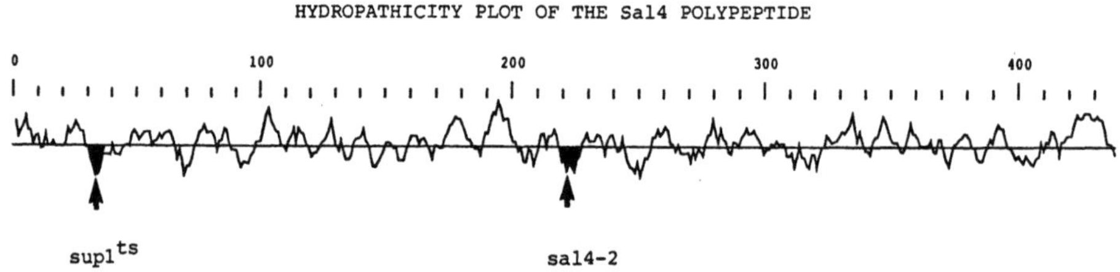

Figure 4. Hydropathicity plot of the Sal4 polypeptide. The location of the mutations in the $sup1^{ts}$ and sal4-2 alleles are indicated.

We have also studied four non-ts sal4 alleles isolated either spontaneously or following EMS mutagenesis, in the strain BSC483/1a (α SUQ5 ade2-1 his5-2 can1-100 lys1-1 PNM1 ura3-1 [psi$^-$]). Two of the alleles (sal4-18, sal4-22) have a paromomycin-sensitive (Pm^s) phenotype as compared with the SAL$^+$ parent strain (BSC483/1a) and the other two sal4 alleles (sal4-28, sal4-42) studied (Crouzet et al., 1988). As shown in Table 2, these four alleles show different levels of ochre suppressor efficiency, as determined by a novel quantitative assay for ochre suppression based on readthrough of a plasmid-based PGK1-UAA-lacZ gene fusion (Firoozan et al, 1990). The two Pm^s alleles show much higher levels of ochre suppression than the Pm^r alleles and over 30-fold higher than the SUQ5 SAL$^+$ [psi$^-$] parent strain. The sal4-2 ts allele is the weakest allosuppressor as defined by this assay.

Using allele-rescue methods all four non-ts sal4 alleles have been isolated and their nucleotide sequences determined. Remarkably, in each case the mutational defect in the sal4 gene is the occurrence of the UAA nonsense codon within the SAL4 open reading frame i.e. all four non-ts sal4 mutants, that enhance the efficiency of the ochre suppressor tRNA

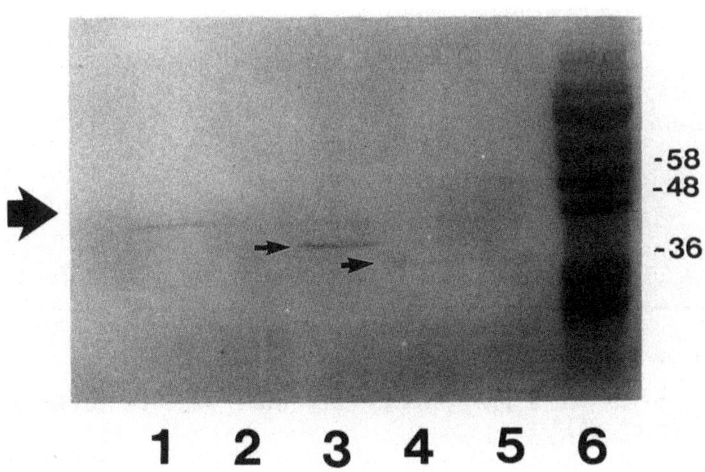

Figure 5. Western blot analysis of various sal4 alleles using an anti-Sal4 polyclonal antibody. lane 1, SAL4$^+$; lane 2, sal4-22 lane 3, sal4-42 lane 4, sal4-18 lane 5, sal4-28 Lane 6, molecular weight markers;. The position of the full length and truncated Sal4 polypeptides are indicated by arrows.

SUQ5 are themselves ochre mutants. The location of the ochre mutations is summarised in Table 2. In each case Western blot analysis has confirmed the predictions of the DNA sequence analysis (Figure 5). A common feature in all four sal4 strains is a very low level of the full length Sal4 polypeptide that is also detected in the Western blot analysis (Figure 5).

4. The Role of the Sal4 Polypeptide in Translation

The product of the SAL4 allosuppressor gene plays an important role in maintaining the accuracy of translation in S. cerevisiae and represents one of at least a dozen gene products implicated in such a role by genetic and/or biochemical studies.

Table 2

Ochre suppressor efficiency, sensitivity to paromomycin and mutational defect in five sal4 mutants

Allele[1]	Ochre Suppressor Efficiency[2]	Sensitivity[3] to Pm	Mutation
$SAL4^+$	6.31 ± 0.09	r	-
sal4-2	57.21 ± 2.09	r	Ile^{222} → ser
sal4-18	175.57 ± 5.96	s	Gln^{359} → UAA
sal4-22	196.15 ± 2.15	s	Gln^{46} → UAA
sal4-28	111.23 ± 1.22	r	Glu^{360} → UAA
sal4-42	83.72 ± 2.06	r	Lys^{411} → UAA

Footnotes

1. All mutations were analysed in the same genetic background with the exception of sal4-2.

2. Ochre suppressor efficiency was determined by the indicated levels of β-galactosidase (nmol/min/mg) in the strains transformed with the single copy plasmid pUKC817, carrying a PGK1-UAA-lacZ gene fusion (Firoozan et al., 1990).

3. r, resistant to concentrations ≥ 1mg/ml Pm; s, sensitive to ≤ 500μg/ml Pm.

The existence of sal4 alleles that are temperature-sensitive for growth together with gene disruption experiments (Himmelfarb et al., 1985), clearly demonstrate that the product of the SAL4 gene (Sal4) is essential for viability. Sequence analysis and regulation studies strongly suggest that Sal4 is not a ribosomal protein (Himmelfarb et al., 1985; Breining and Piepersberg, 1986) although our cellular fractionation studies demonstrate that Sal4 is tightly associated with the ribosome (Figure 2). Based on these facts together with the phenotypes associated with the Sal4 mutation suggests two possible models for the role of Sal4 in translation; either it is a termination release factor (eRF) or it is a translation factor that proofreads the codon-anticodon interactions at the ribosomal A site.

Model 1: Sal4 as a release factor. There is presumably in yeast, as in E. coli, competition between RFs and the suppressor tRNA for their "cognate" stop codon at the ribosomal A site. The allosuppression phenotype could therefore arise by shifting this competition in favour of the suppressor tRNA via impaired RF function or by a reduction in the cellular levels of RF. The four non-ts sal4 alleles we have characterised all show significantly reduced levels of the Sal4 polypeptide strongly suggesting that the allosuppressor phenotype is indeed due to lowering the cellular levels of Sal4. Is the Sal4 polypeptide therefore an S. cerevisiae eRF? The eRF (or eRFs) of S. cerevisiae have yet to be described. However, comparisons we have carried out between the amino acid sequence of Sal4 and that of the E. coli RF's (Craigen et al., 1985) and the recently published rat eRF (Lee et al., 1990), show no detectable regions of homology. There is also no homology between the E. coli RFs and the rat eRF (Lee et al., 1990). If Sal4 were an eRF it would be difficult to reconcile its essential role in translation with its central role in controlling both missense and nonsense fidelity.

Model 2: Sal4 as a proofreading factor. In E. coli the accuracy of the codon:anticodon interaction, upon binding of aminoacyl-tRNA to the ribosomal A site, is maintained by a GTP-dependent ribosomal proofreading mechanism (Thompson and Stone, 1977). Impaired proofreading can come from either ribosomal protein or EF-Tu mutations (Thompson, 1988). The Sal4 protein may therefore function in a similar proofreading role at the yeast ribosomal A site ensuring that the correct codon:anticodon interactions occur subsequent to EF-1-mediated binding of the aminoacyl-tRNA to the ribosome. In each of the sal4 mutants we have studied a full length but mutant Sal4 protein is made; for sal4-2 the mutant Sal4 protein contains an Ile to Ser change, whereas for the four sal4 nonsense alleles the full length protein is only generated by translation of the UAA codon by the SUQ5-encoded $tRNA^{Ser}$. These mutant Sal4 polypeptides may therefore be impaired in their proofreading functions which may in turn effect eRF recognition of a stop codon (in this case UAA) at the ribosomal A-site.

A definitive distinction between these two models will require detailed biochemical studies using purified wild type and mutant proteins. These can now be obtained by overexpression in either *E. coli* or yeast using the cloned genes described here.

Acknowledgement **This work was supported in part by a project grant from the SERC Biotechnology Directorate, and by studentships from NATO (JABD) and the World Bank (A).**

5. References

All-Robyn, J.A., Kelly-Geraghty, D., Griffin, E., Brown, N. and Liebman, S.W. (1990) Isolation of omnipotent suppressors in an [eta^+] yeast strain. Genetics 124: 505-514.

Breining, P. and Piepersberg, W. (1986) Yeast omnipotent suppressor SUP1 (SUP45): nucleotide sequence of the wild type and a mutant gene. Nucleic Acids Res. 14: 5187-5197.

Cabezon, T., Herzog, A., DeWilde, M., Villarroel, M. and Bollen, A. (1976) Cooperative control of translational fidelity by ribosomal proteins in *Escherichia coli*. III A ram mutation in the structural gene for protein S5 (rpxE). Molec. gen. Genet. 144: 59-62.

Cox, B.S. (1977) Allosuppressors in yeast. Genet. Res. 30: 187-205.

Cox, B.S., Tuite, M.F. and McLaughlin, C.S. (1988) The ψ factor of yeast: a problem of inheritance. Yeast 4: 159-178.

Craigen, W.J., Cook, R.G., Tate, W.P. and Caskey, C.T. (1985) Bacterial peptide chain release factors; conserved primary structure and possible frameshift regulation of release factor 2. Proc. Natl. Acad. Sci. USA 82: 3616-3620.

Crouzet, M. and Tuite, M.F. (1987) Genetic control of translational fidelity in yeast: molecular cloning and analysis of the allosuppressor gene SAL3. Mol. gen. Genet. 210: 581-583.

Crouzet, M., Izgu, F., Grant, C.M. and Tuite, M.F. (1988) The allosuppressor gene SAL4 encodes a protein important for maintaining translational fidelity in *Saccharomyces cerevisiae*. Current Genet. 14: 537-543.

Eggertsson, G. and Soll, D. (1988) Transfer RNA-mediated suppression of termination codons in *Escherichia coli*. Microbiol. Rev. 52: 354-374.

Eustice, D.C., Wakem, P., Wilhelm, J.M. and Sherman, F. (1986) Altered 40S ribosomal subunits in omnipotent suppressors of yeast. J. Mol. Biol. 188: 207-214.

Firoozan, M., Grant, C.M., Duarte, J.A.B. and Tuite, M.F. (1990) Quantitation of readthrough of termination codons in yeast using a novel gene fusion assay. Submitted for publication.

Gelugne, J-P. and Bell, J.B. (1988) Modifiers of ochre suppressors in Saccharomyces cerevisiae that exhibit ochre suppressor-dependent amber suppression. Current Genet. 14: 345-354.

Gorini, L. (1974) Streptomycin and misreading of the genetic code. In 'Ribosomes' (M. Nomura, A. Tissieres and P. Lengyel, eds.) pp. 791-803. Cold Spring Harbor Laboratory, New York.

Grant,C.M. (1990) Control of translational fidelity in yeast. Ph.D. Thesis. University of Kent at Canterbury.

Hawthorne, D.C. and Leupold, U. (1974) Suppressor mutations in yeast. Curr. Top. Microbiol. Immunol. 64: 1-47.

Himmelfarb, H.J., Maicas, E. and Friesen, J.D. (1985) Isolation of the SUP45 omnipotent suppressor gene of Saccharomyces cerevisiae and characterisation of its gene product. Mol. Cell. Biol. 5: 816-822.

Inge-Vechtomov, S.G. and Andrianova, V.M. (1970) Recessive super-suppressors in yeast. Genetika 6: 103-115.

Ishiguro, J. (1981) Genetic and biochemical characterisation of antisuppressor mutants in the yeast Saccharomyces cerevisiae. Current Genet. 4: 197-204.

Kikuchi, Y., Shimatake, H. and Kikuchi, A. (1988) A yeast gene required for the G_1-to-S transition encodes a protein containing an A-kinase target site and GTPase domain. EMBO J. 7: 1175-1182.

Lee, C.C., Craigen, W.J., Muzny, D.M., Harlow, E. and Caskey, C.T. (1990) Cloning and expression of a mammalian peptide chain release factor with sequence similarity to tryptophanyl-tRNA synthetases. Proc. Natl. Acad. Sci. USA 87: 3508-3512.

Liebman, S.W. and Cavenagh, M. (1980) An antisuppressor that acts on omnipotent suppressors in yeast. Genetics 95: 49-61.

Liebman, S.W., Cavenagh, M. and Bennett, L.N. (1980) Isolation and properties of an antisuppressor in Saccharomyces cerevisiae specific for an omnipotent suppressor. J. Bacteriol. 143: 1527-1529.

Masurekar, M., Palmer, E., Ono, B., Wilhelm, J.M. and Sherman, F. (1981) Misreading of the ribosomal suppressor SUP46 due to an altered 40S subunit in yeast. J. Mol. Biol. 147: 381-390.

McCready, S.J. and Cox, B.S. (1973) Antisuppressors in yeast. Mol. gen. Genet. 124: 305-320.

Ono, B., Tanaka, M., Awano, I., Okamoto, F., Satoh, R., Yamagishi, N. and Ishino-Arao, Y. (1989) Two new loci that give rise to dominant omnipotent suppressors in Saccharomyces cerevisiae. Current Genet. 16: 323-330.

Patterson, M., Sclafani, R.A., Fangman, W.L. and Rosamond, J. (1986) Molecular characterization of cell cycle gene CDC7 from Saccharomyces cerevisiae. Mol. Cell. Biol. 6: 1590-1598.

Sherman, F. (1982) Suppression in the yeast Saccharomyces cerevisiae. In: 'Molecular Biology of the Yeast Saccharomyces: Volume II, Metabolism and Gene Expression' (J.N. Strathern, E.W. Jones and J.R. Broach, eds.) pp. 463-486. Cold Spring Harbor Laboratory, New York.

Song, J.M. and Liebman, S.W. (1987) Allosuppressors that enhance the efficiency of omnipotent suppressors in Saccharomyces cerevisiae. Genetics 115: 451-460.

Thompson, R.C. (1988) EF-Tu provides an internal kinetic standard for translational accuracy. Trends Biochem. Sci. 13: 91-93.

Thompson, R.C. and Stone, P.J. (1977) Proofreading of the codon-anticodon interaction on ribosomes. Proc. Natl. Acad. Sci. USA 74: 198-202.

Wakem, L.P. and Sherman, F. (1990) Isolation and characterisation of omnipotent suppressors in the yeast Saccharomyces cerevisiae. Genetics 124: 515-522.

Wilson, P.G. and Culbertson, M.R. (1988) SUF12 suppressor protein of yeast: a fusion protein related to the EF-1 family of elongation factors. J. Mol. Biol. 199: 559-573.

HIV *pol* EXPRESSION VIA A RIBOSOMAL FRAMESHIFT

Alan J. Kingsman, Wilma Wilson and Susan M. Kingsman

Department of Biochemistry
South Parks Road
Oxford
OX1 3QU
U.K.

INTRODUCTION

The genetic relationships of the *gag* and *pol* genes of all retroviruses are approximately the same and the strategy for expression of the protein products of these genes is also strongly conserved (Weiss *et al*., 1982) (e.g. Figure 1). The *gag* and *pol* genes are adjacent and in many cases the 3' end of *gag* and the 5' end of *pol* overlap by up to a few hundred nucleotides. Where there is an overlap *pol* is generally in the -1 translational phase with respect to *gag*. Both genes are expressed from the full length genomic RNA to produce two primary translation products, a GAG precursor protein and a GAG:POL fusion precursor protein. The production of the fusion protein is achieved by the *gag* and *pol* reading frames being brought into translational phase. For several years it was assumed that this translational shift was mediated by a splice and the absence of any evidence for this was explained by proposing that the splice was small and therefore hard to detect (Weiss *et al*., 1982). However, in 1985 two pieces of data suggested that the splicing hypothesis was wrong. First, in the retrovirus-like yeast transposon Ty it was shown that frameshifting between the *TYA* gene, a *gag* analogue, and the *TYB* gene, a *pol* analogue, was not due to splicing (Mellor *et al*., 1985; Clare and Farabaugh, 1985). Secondly, Jacks and Varmus (1985) showed that RSV frameshifting could be achieved when an RNA synthesised *in vitro* was used in an *in vitro* translation system. This clearly excluded the possibility of a splice and suggested that the frameshift was due to some event at the ribosome. The phenomenon is now refered to as ribosomal frameshifting.

Some retroviruses fuse the products of *gag* and *pol* by a different mechanism. For example in MLV the *gag* and *pol* genes are adjacent and in phase but separated by a UAG termination codon.

NATO ASI Series, Vol. H 49
Post-Transcriptional Control of Gene Expression
Edited by J. E. G. McCarthy and M. F. Tuite

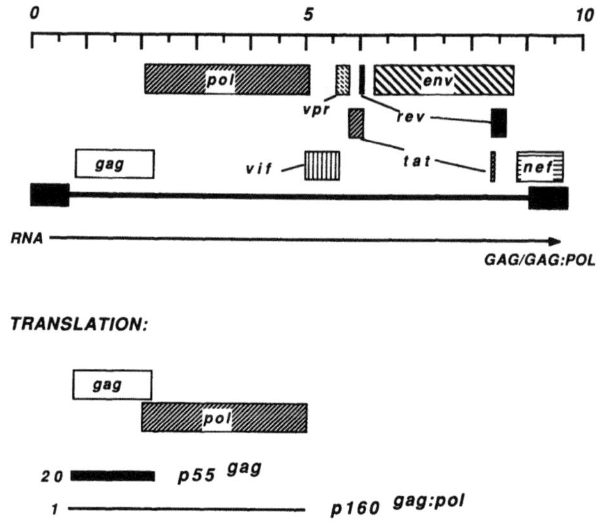

Figure 1 The genetic organisation of HIV-1. The upper figure shows the HIV map and the lower figure shows the simple translation product of the full length RNA, p55 and the frameshift product p160.

Production of a GAG:POL fusion protein is achieved by suppression of termination by a glutamine-tRNA (Yoshinaka et al., 1985).

Clearly in retroviruses with gag and pol out of phase frameshifting is essential for the expression of the enzyme activities of the virus, protease, reverse transcriptase, RNaseH and integrase. Also the frequency of shifting, generally about 5%, determines the relative levels of GAG and POL proteins in the cell. In addition, the attachment of the POL proteins to the GAG proteins either via the shift, or by termination suppression, not only achieves the genetic economy commonly seen in viruses but also ensures that the enzyme activities are packaged into the virus, as it is the GAG proteins that assemble into the viral core.

In HIV-1, gag and pol overlap by 241 nucleotides with pol in the -1 phase with respect to gag (Figures 1 and 2) (Ratner et

```
      GCAGGGCCCCTAGGAAAAAGGGCTGTTGGAAATGTGGAAAGGAAGGACACCAAATGAAAG
 0     R  A  P  R  K  K  G  C  W  K  C  G  K  E  G  H  Q  M  K
-1
-2                                                         ***

                        p7 | pol->
      ATTGTACTGAGAGACAGGCTAATTTTTTAGGGAAGATCTGGCCTTCCTACAAGGGAAGGC
 0    D  C  T  E  R  Q  A  N| F  L  G  K  I  W  P  S  Y  K  G  R
-1                          *** F  F  R  E  D  L  A  F  L  Q  G  K  A
-2                                 ***

               | p6
      CAGGGAATTTTCTTCAGAGCAGACCAGAGCCAACAGCCCCACCATTTCTTCAGAGCAGAC
 0    P  G  N  F| L  Q  S  R  P  E  P  T  A  P  P  F  L  Q  S  R
-1     R  E  F  S  S  E  Q  T  R  A  N  S  P  T  I  S  S  E  Q  T
-2

      CAGAGCCAACAGCCCCACCAGAAGAGAGCTTCAGGTCTGGGGTAGAGACAACAACTCCCC
 0    P  E  P  T  A  P  P  E  E  S  F  R  S  G  V  E  T  T  T  P
-1     R  A  N  S  P  T  R  R  E  L  Q  V  W  G  R  D  N  N  S  P
-2                                               ***

      CTCAGAAGCAGGAGCCGATAGACAAGGAACTGTATCCTTTAACTTCCCTCAGATCACTCT
 0    P  Q  K  Q  E  P  I  D  K  E  L  Y  P  L  T  S  L  R  S  L
-1     S  E  A  G  A  D  R  Q  G  T  V  S  F  N  F  P==Q==I==T==L
-2                      ***                   ***
                       <-gag
      TTGGCAACGACCCCTCGTCACAATAAAGATAGGGGGGCAACTAAAGGAAGCTCTATTAGA
 0    F  G  N  D  P  S  S  Q ***   ***           ***           ***
-1    ==W==Q==R==P==L==V==T==I==K==I==G==G==Q==L==K==E==A==L==L==D
-2

      TACAGGAGCAGATGATACAGTATTAGAAGAAATGAGTTTGCCAGGAAGATGGAAACCAAA
 0                            ***      ***
-1    ==T==G==A==D==D==T==V==L==E==E==M==S==L==P==G==R==W==K==P==K
-2                  ***

      AATGATAGGGGGAATTGGAGGTTTTATCAAAGTAAGACAGTATGATCAGATACTCATAGA
 0      ******                         ***                      ***
-1    ==M==I==G==G==I==G==G==F==I==K==V==R==Q==Y==D==Q==I==L==I==E
-2                                                 ***

      AATCTGTGGACATAAAGCTATAGGTACAGTATTAGTAGGACCTACACCTGTCAACATAAT
 0                         ***           ******                 ***
-1    ==I==C==G==H==K==A==I==G==T==V==L==V==G==P==T==P==V==N==I==I
-2                  ***

      TGGAAGAAATCTGTTGACTCAGATTGGTTGCACTTTAAATTTTCCCATTAGCCCTATTGA
 0                  ***                     ***
-1    ==G==R==N==L==L==T==Q==I==G==C==T==L==N==F  P  I  S  P  I  E
-2                                                   ***      ***
```

Figure 2. The nucleotide sequence of the gag:pol overlap region of HIV. The sequence is derived from the IIIB isolate reported by Ratner et al (1985). The frameshift site and the potential stem- loop region are underlines. 0, -1 and -2 denote translational reading phases where gag is arbitrarily given 0. The limits of the gag and pol open reading frames are marked as are cleavage sites that give rise to the p7 and p6 mature GAG proteins. The amino acid sequence of the viral protease is marked by '='. Stop codons are maked by ***.

al., 1985; Wain-Hobson et al., 1985; Sanchez-Pescador et al., 1985). The GAG precursor protein, $Pr55^{gag}$, is the primary product of simple translation of full length genomic viral RNA. The GAG:POL fusion protein, $Pr160^{gag:pol}$, also a precursor, is the product of frameshifted translation of the same full length RNA. The frequency of the shift is about 5% and therefore the relative abundance of the two precursors is about 20:1 respectively, although this has only been determined in vitro (Jacks et al., 1988a; Wilson et al., 1988).

The HIV overlap begins almost exactly at the 3' end of the p7 coding region. Protein p6 is encoded entirely by the overlap in the GAG phase and the protease coding sequence in the POL phase overlaps p6 by 12 codons (Figure 2).

SEQUENCE REQUIREMENTS FOR THE GAG:POL RIBOSOMAL FRAMESHIFT

The standard assay for frameshifting uses an in vitro translation reaction to translate an SP6 generated RNA (Jacks and Varmus, 1985). The RNA is constructed with a preshift sequence,

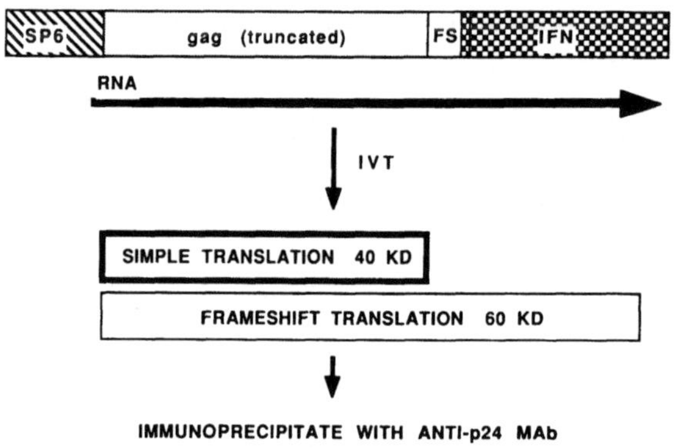

Figure 3. An SP6 frameshift assay system. IVT = in vitro translation. In this case the truncated gag is a pre-shift sequence, FS marks the position of a candidate shifting sequence and IFN (interferon) is a post-shift sequence that adds about 20 kD onto the 40kd GAG protein. This system is taken from Wilson et al. (1988).

a candidate shift site and then a post-shift sequence (Figure 3). Shifting is detected either by the presence of a second higher molecular weight band in an immunoprecipitation with an antibody directed against the protein encoded by the preshift sequence or by expression of a reporter gene that is dependent on the shift.

Frameshifting must occur in the *gag:pol* overlap region so that the shift into the -1, *pol* phase is achieved before the ribosome reaches the stop codon of the gag open reading frame. Jacks et al. (1988a) showed that shifting occured at codon 3 of the pol open reading frame with the sequence TTT.TTA.GGG being read as PHE.LEU.GLY in GAG and as PHE.LEU.ARG in the GAG:POL fusion. However, the data reveal that there is also substantial (at least 30%) phenylalanine in the second position of the shifted product giving an alternative shift site sequence of PHE.PHE.ARG (see later). These data were in agreement with the observation of Wilson et al. (1988) who showed that the sequences required for HIV frameshifting were located within the first 16 nucleotides of the overlap region.

The observation that such a short stretch of the overlap region was required for shifting was surprising in the context of what was known about other retroviral requirements for frameshifting (Jacks et al., 1987; Jacks et al., 1988b). In almost every case where a virus makes use of frameshifting as a gene expression strategy the shift site or putative shift site is followed within a very short distance, usually less that 10 nucleotides, by a region of secondary structure (Table 1 and Figure 4)). The secondary structure may be a simple stem-loop or a pseudoknot in which the loop sequence is capable of base-pairing with another region downstream to form a complex structure (Jacks et al., 1988b; Brierley et al., 1989). Stem-loop structures that may form downstream of a selection of retroviral shift sites, including those of the HIV/SIV family are shown in Figure 4. It is possible that these secondary structures are recognised by soluble "shifting factors' or by ribosomal components that mediate the shift. A simpler suggestion is that these structures slow the ribosome making it more likely that it will 'slip' back at the shift site. In RSV there is good evidence (Jacks et al., 1988b) that the downstream secondary structure is required from mutational studies that show that destabilisation of the stem-loop substantially decreases frameshifting efficiency and unpublished data (cited in Jacks et al. 1988b) indicates that ribosomal pausing may occur at the stem loop. In HIV, however, although a stem-loop structure exists (Figure 4) it does not lie within the first 16 nucleotides of the overlap region. It would seem therefore that HIV frameshifting does not require ribosomal stalling and is therefore not mediated by the same mechanism as is used in RSV (Wilson et al., 1988). We will return to this later.

Table 1. Frameshift classes.

RETROVIRUS/ RETROELEMENT	OVERLAP	FRAMESHIFT SEQUENCE	DISTANCE TO SS.
Class I		X XXY YY	
		A AAU UU	
RSV	gag/pol	ACA AAU UUA UAG	7
SRV-1	pro/pol	GGA AAU UUU UAA	8
MPMV	pro/pol	GGA AAU UUU UAA	8
		G GGU UU	
Mouse IAP	gag/pol	CUG GGU UUU CCU	6
		G GGA AA	
SRV-1	gag/pro	CAG GGA AAC GAC	8
MPMV	gag/pro	CAG GGA AAC GGG	8
Visna	gag/pol	CAG GGA AAC AAC	7
		U UUA AA	
BLV	pro/pol	CCU UUA AAC UAG	7
HTLV-1	pro/pol	CCU UUA AAC CAG	7
HTLV-2	pro/pol	CCU UUA AAC CUG	7
Class II		X XXX XX	
		U UUU UU	
HIV-1	gag/pol	AAU UUU UUA GGG	8
HIV-2	gag/pol	GGU UUU UUA GGA	5
SIV	gag/pol	GGU UUU UUA GGC	4
gypsy	gag/pol	AAU UUU UUA GGG	8
		A AAA AA	
MMTV	gag/pro	UCA AAA AAC UUG	8
BLV	gag/pro	UCA AAA AAC UAA	8
HTLV-1	gag/pro	CCA AAA AAC UCC	7
HTLV-2	gag/pro	GGA AAA AAC UCC	8
EIAV	gag/pol	CCA AAA AAC GGG	10
EXCEPTION			
MMTV	pro/pol	CAG GAU UUA UGA	5
Ty1-15	TYA/TYB	CAU CUU AGG CCA GAA	

Table 1. Two classes of retroviral and retroelement frameshifting Examples of class I and II demonstrated or suspected shift sites are listed. The sequences are grouped to show gag or gag-equivalent codons. The last column shows the distance from the actual or purtative frameshift sequence to a region of downstream secondary structure. References for nucleotide sequences: RSV, Rous sarcoma virus (Schwartz et al., 1983); SRV-1, simian retrovirus type 1 (Power et al., 1986); MPMV, Mason-Pfizer monkey virus (Sonigo et al., 1986); 17.6 (Saigo et al., 1984); Visna virus (Sonigo et al., 1985); Mouse IAP, mouse intracisternal A particle (Meitz et al., 1987); BLV, bovine leukemia virus (Sagata et al., 1985; Rice et al., 1985); HTLV-1, human T cell leukemia virus type 1 (Hiramatsu et al., 1987); HTLV-2, human T cell leukemia virus type 2 (Shimotohno et al., 1985; Mador, Panet and Honigman, 1989); HIV-1 (Ratner et al., 1985); HIV-2 (Guyader et al., 1987); SIV, simian immunodeficiency virus (Chakrabarti et al., 1987); gypsy (Marlor et al., 1986); MMTV, mouse mammary tumor virus (Jacks et al., 1987; Moore et al., 1987); EIAV, equine infectious anemia virus (Stephens et al., 1986); Ty1-15 (Mellor et al., 1985; Wilson et al., 1986; Clare, Belcourt and Farabaugh, 1988).

Shift sites are generally regarded as being heptanucleotide sequences. Examination of these sequences reveals that they fall into two broad groups (Table 1). In what we will call Class I shift sites the first six nucleotides are of the general sequence X.XXY.YY where the stops represent GAG codons. In Class II shift

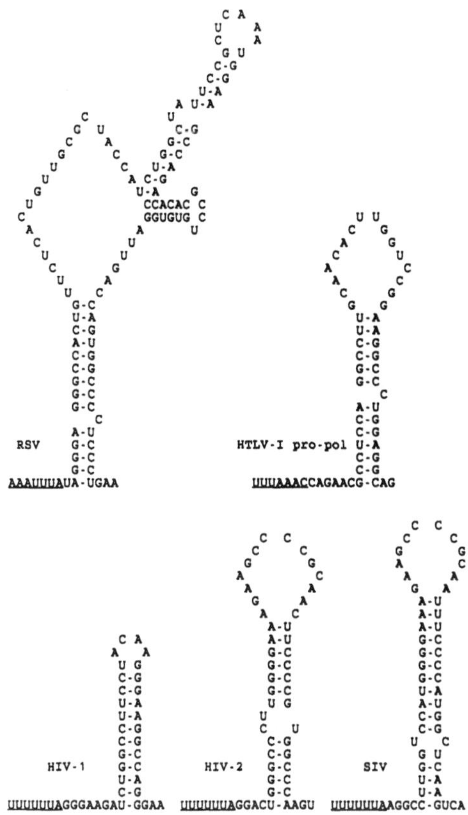

Figure 4. A selection of potential secondary structures found associated with retroviral frameshift sites. In each case the heptanucleotide shift site is underlined. Sequence information is the same as for Table 1.

sites the general sequence is X.XXX.XX (Table 1). HIV has a heptanucleotide sequence of U.UUU.UUA. Jacks et al (1988a) have shown that mutating the terminal UUUA to CUUA or UCUA dramatically reduces frameshifting efficiency. On the basis of this and the fact that UUUA appears in many shifty sequences it has been suggested that this tetranucleotide is a key signal in determining shifting. However, Wilson et al (1988) have shown that deletion of three of the six T/Us in the heptanucleotide also disrupts shifting even though reading frame and the UUUA sequence are preserved. More subtle changes in the 6T/U sequence also disrupt shifting. For example mutation of the third T/U or the sixth T/U to any of the other three nucleotides substantially reduces the shift. However, mutation of the A at position 7 has no effect except in the case of a change of A to T/U when frameshifting increases. This, and other studies (Wilson et al., unpublished data), suggest that the signal for shifting in HIV is the 6T/U sequence rather than UUUA.

THE MECHANISM OF RIBOSOMAL FRAMESHIFTING IN RETROVIRUSES

Most studies of viral frameshifting have been carried out on viruses with sequences of the type that we have called Class I, such as RSV (Jacks et al., 1988b) and IBV (Brierley et al., 1989). In both of these, downstream secondary structure, in the form of a stem-loop or a pseudoknot, is required for efficient shifting. The only virus with a Class II sequence that has been analysed is HIV and this does not require a region of secondary structure just downstream of the shift site, even though one exists (Wilson et al., 1988; Madhani et al., 1988). It is tempting, and perhaps useful, to look for an explanation for this in the potential shiftiness of the shift sites alone. In Class I, such as the RSV gag:pol frameshift site, there are two short, adjacent, homopolymeric runs of three nucleotides of the general structure X XXY YY (Table 1). The phase relationship of these two triplets to the gag and pol reading frames is always the same irrespective of sequence composition. In RSV the heptanucleotide sequence AAAUUUA is the shift site and is thought to mediate shifting through a -1 slip of codon:anticodon interactions at both the A and P sites (Figure 5, Class I) (Jacks et al., 1988b). Following the slip the $tRNA^{Leu}$ (UUA) and the $tRNA^{Asn}$ (AAU) would be held on the RNA by 2 out of 3 base pairs each. Normal translocation would then take place and translation would proceed, in phase with pol, to the end of pol. Such a mechanism can be brought about by any adjacent, homopolymeric triplets as long as the distribution of the triplets with respect to the gag and pol reading phases is the same as in RSV (Table 1). This mechanism does not allow a shift to the +1/-2 phase. In Class II, such as the HIV gag:pol frameshift site, there is a single long homopolymeric run of six nucleotides of the general structure X XXX XX (Table 1). Like Class I the relationship of these six nucleotides to the gag and pol reading phases is conserved and is the same as the relationship of the two triplets in Class I (Table 1). In HIV the heptanucleotide shift sequence is UUUUUUA. Shifting to -1 can be achieved by a mechanism almost identical to that proposed for Class I (Figure 5, Class IIa). In this case the tRNA^{Phe} (UUU) and tRNA^{Leu} (UUA) would both slip back one nucleotide in the A and P sites. The only difference between Class I and Class II would be that after the slip one of the tRNAs, tRNA^{Phe}, would be held on the RNA by three out of three base pairs. tRNA^{Leu} would be held, as in Class I, by a 'two-out-

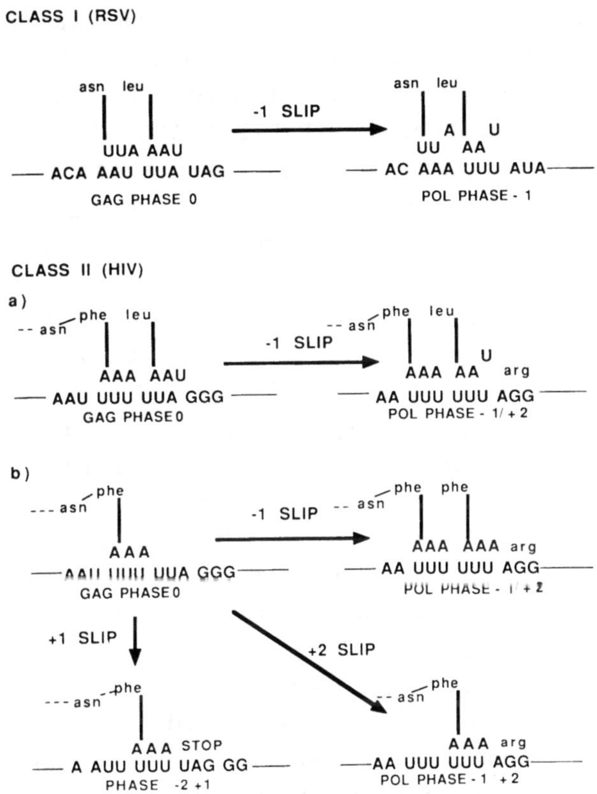

Figure 5. Class I and class II shifting. See text for explanation.

of-three' association. In Class II, therefore, the slip is maintained by five base pairs as opposed to four in Class I. Consequently, the slip may be more stable. The amino acid sequence over the frameshift region that would be predicted by this mechanism is ASN-PHE-LEU-ARG (Figure 5), the sequence determined for this region by Jacks et al. (1988a).

Slippage at the A and P sites is not the only mechanism open to a Class II shift sequence. It is possible that in HIV, for example, the $tRNA^{Phe}$ (UUU) slips during translocation (Figure 5, Class IIb) prior to the $tRNA^{Leu}$ (UUA) entering the A site. In this case the $tRNA^{Phe}$ slips to -1, maintaining three out of three base pairing, and exposing a free UUU codon in the A site. Rather than $tRNA^{Leu}$ (UUA) entering the A site, a second $tRNA^{Phe}$ (UUU) enters and the -1 shift is completed. The slip is maintained by six out of six base pairs and is likely, therefore, to be quite stable. This mechanism predicts that the amino acid sequence over the HIV frameshift region would be ASN-PHE-PHE-ARG. Both A and P site shifting and translocation shifting seem equally plausible for HIV. If both occured then a mixture of two gag:pol products would be produced differing at one amino acid position at the frameshift site. Close examination of the sequence data of Jacks et al (1988a) shows exactly that. At least 30% of the HIV-1 shifts produced the sequence ASN-PHE-PHE-ARG. We would suggest, therefore, that the frameshift in HIV might produce microheterogeneity in the gag:pol fusion protein.

Translocation slippage of the $tRNA^{Phe}$ could also produce a +2 slip (Figure 5, Class IIb) resulting in the amino acid sequence ASN-PHE-ARG. This apparently is not seen (Jacks et al., 1988a). Furthermore the HIV frameshift sequence could mediate shifting into the +1/-2 phase as well as into the -1/+2 phase. The translocation slip of $tRNA^{Phe}$ (UUU) could accommodate this via a +1 slip (Figure 5, Class IIb). As there is a UAG termination codon in the +1/-2 phase immediately after the shift site in HIV this observation predicts that there should be a truncated gag precursor protein of about 48 kD present at about 5% of the level of the authentic p55 GAG precursor protein. Clearly this is testable.

Our scheme suggests that Class II sequences are much shiftier than Class I and this may be reflected in differences in their respective requirements for downstream secondary structures. The requirements for HIV frameshifting appear to be simple. The virus makes use of a generally shifty sequence, T/U_6, to express its pol gene and this is all that is necessary. In contrast, RSV not only requires a shifty sequence, A_3T/U_3, but also a region of RNA secondary structure immediately downstream of the shift site (Jacks et al., 1988b). We would suggest that this is not a fundamental difference. Jacks et al (1988b) have shown that in RSV the sequence of the region that forms the secondary structure is not critical to shifting efficiency as long as the potential for secondary structure is maintained. This argues against sequence specific cellular factors being involved in a complex mechanism for RSV frameshifting. Class I frameshift sites are likely to be less efficient than Class II sites. There are fewer opportunities for shifting and the slip may be less stable in Class I sites. Viruses with class I, low efficiency, shifty sites may have therefore evolved to use a downstream secondary and possibly tertiary RNA structure (Jacks et al., 1988b; Brierley et al., 1989) to act as a ribosomal stalling site. During the stall there is an increased probability of a shift at the low efficiency site. In contrast, viruses with a class II, high efficiency site do not require a ribosomal stalling site to achieve shifting at the frequency required to deliver the correct ratio of GAG and GAG:POL fusion.

The notion that there are two classes of retroviral and retroelement frameshift sites is easily tested but it also needs to be qualified. There are shift sites that do not conform to either Class I or Class II. The MMTV -1 pro:pol shift site is one and may represent a member of a third class. The retrotransposon, Ty, also does not fit into the scheme. Its TYA and TYB gene products are fused by frameshifting occuring in the sequence, CATCTTAGGCCAGAA (in Ty1-15), (Wilson et al., 1986; Clare, Belcourt and Farabaugh, 1988). This mediates a +1 shift and may, therefore, resemble the +1 shifts observed in some prokaryotic systems, such as the E.coli RF2 gene (Craigen and Caskey, 1986; Curran and Yarus, 1988).

The number of shift sites that have been analysed is remarkably small, Ty, RSV, MMTV, IBV and HIV. A broader study will reveal whether the division of shifting into these two Classes is appropriate or whether the story is more complex.

ACKNOWLEDGEMENTS

We would like to thank our colleagues in the Virus Molecular Biology Group for comments on the manuscript. The MRC supports our work on HIV frameshifting.

REFERENCES

Brierley,I., Digard,P. and Inglis,S.C. (1989). Characterization of an efficient coronavirus ribosomal frameshifting signal: requirement for an RNA pseudoknot. Cell 57, 537-547.

Chakrabarti,L., Guyader,M., Alizon,M., Daniel,M.D., Desrosiers,R.C., Tiollais,P. and Sonigo,P. (1987). Sequence of simian immunodeficiency virus from macaque and its relationship to other human and simian retroviruses. Nature 328, 543-547.

Clare, J. and Farabaugh, P. (1985) Nucleotide sequence of a yeast Ty element: evidence for an unusual mechanism of gene expression. Proc. Natl. Acad. Sci. USA 82, 2829-2833.

Clare,J.J., Belcourt,M. and Farabaugh,P.J. (1988). Efficient translational frameshifting occurs within a conserved sequence of the overlap between the two genes of a yeast Ty1 transposon. Proc. Natl. Acad. Sci. USA 85, 6816-6820.

Craigen,W.J. and Caskey,C.T. (1986). Expression of peptide chain release factor 2 requires high-efficiency frameshift. Nature 332, 273-275.

Curran,J.F. and Yarus,M. (1988). Use of tRNA suppressors to probe regulation of Escherichia coli release factor 2. J. Mol. Biol. 203, 75-83.

Guyader,,M., Emerman,M., Sonigo,P., Claver,F., Montagnier,L. and Alizon,M. (1987). Genome organization and transcription of the human immunodeficiency virus type 2. Nature 326, 662-669.

Hiramatsu,K., Nishida,J., Naito,A. and Koshikura,H. (1987). Molecular cloning of the closed circular provirus of human T cell leukemia virus type 1: a new open reading frame in the gag-pol region. J.Gen. Virol. 68, 213-218.

Jacks,T. and Varmus,H.E. (1985). Expression of the Rous sarcoma virus pol gene by ribosomal frameshifting. Science 230, 1237-1242.

Jacks,T., Townsley,K., Varmus,H.E. and Majors,J. (1987). Two efficient ribosomal frameshifting events are required for synthesis of mouse mammary tumor virus gag-related polyproteins. Proc. Natl. Acad. Sci. USA 84,4398-4302.

Jacks,T., Power,M.D., Masiarz,F.R., Luciw,P.A., Barr,P.J. and Varmus,H.E. (1988a). Characterization of ribosomal frameshifting in HIV-1 gag-pol expression. Nature 331, 280-283.

Jacks,T., Madhani,H.D., Masiarz,F.R. and Varmus,H.E. (1988b). Signals for ribosomal frameshifting in the Rous sarcoma virus gag-pol region. Cell 55, 447-458.

Madhani,H.D., Jacks,T. and Varmus,H.E. (1988). Signals for the expression of the HIV pol gene by ribosomal frameshifting. In The Control of HIV pol Gene Expression, R.Franza, B.Cullen and F.Wong-Staal,eds. (Cold Spring Harbor, New York: Cold Spring Harbor Laboratory), pp. 119-125.

Mador,N., Panet,A and Honigman,A. (1989). Translation of gag, pro, and pol gene products of human T-cell leukemia virus type 2. J.Virol. 63, 2400-2404.

Marlor,R.L. Parkhurst,S.M. and Corces,V.G. (1986). The Drosophila melanogaster gypsy transposable element encodes putative gene products homologous to retroviral proteins. Mol. Cell. Biol. 6, 1129-1134.

Meitz,J.A., Grossman,Z., Leuders,K.K. and Kuff,E.L. (1987). Nucleotide sequence of a complete mouse intracisternal A-particle genome: no relationship to known aspects of particle assembly and function. J. Virol. 61, 3020-3029.

Mellor,J., Fulton,A.M., Dobson,M.J., Wilson,W., Kingsman,S.M. and Kingsman,A.J. (1985). A retrovirus-like strategy for expression of a fusion protein encoded by the yeast transposon, Ty1. Nature 313, 243-246.

Moore,R., Dixon,M. Smith,R., Peters,G. and Dickson,C. (1987). Complete nucleotide sequence of a milk-transmitted mouse mammary tumor virus: two frameshift suppression events are required for translation of gag and pol. J. Virol. 61, 480-490.

Power,M.D. Marx,P.A. Bryant,M.L., Gardner,M.D., Barr,P.J. and Luciw,P.A. (1986). Nucleotide sequence of SRV-1, a type D simian acquired immune deficiency syndrome retrovirus. Science 231, 1567-1572.

Ratner,L., Haseltine,W., Patarca,R., Livak,K.J., Starcich,B., Josephs,S.F., Doran,E.R., Rafalski,J.A., Whitehorn,E.A., Baumeister,K., Ivanoff,L., Petteway,Jr.,S.R. Pearson,M.L., Lautenberger,J.A., Papas,T.S., Ghrayeb,J., Chang,N.T., Gallo,R.C. and Wong-Staal,F. (1985). Complete nucleotide sequence of the AIDS virus, HTLV-III. Nature 313, 277-284.

Rice,C.M. and Strauss,J.H. (1981). Nucleotide sequence of the 26S mRNA of sindbis virus and deduced sequence of the encoded virus structural proteins Proc. Natl. Acad. Sci. USA 78, 1062-1066.

Sagata,N., Yasunaga,T., Tsuzuku-Kawamura,J., Ohishi,K., Ogawa,Y. and Ikawa,Y. (1985). Complete nucleotide sequence of the genome of bovine leukemia virus: its evolutionary relationship to other retroviruses. Proc. Natl. Acad. Sci. USA 82, 677-681.

Saigo,K., Kugimiya,W., Matsuo,Y., Inouye,S., Koshioka,K. and Yuki,S. (1984). Identification of the coding sequence for a reverse transcriptase-like enzyme in a transposable genetic element in *Drosophila melanogaster*. Nature 312, 659-661.

Sanchez-Pescador,R., Power,M.D., Barr,P.J., Steimer, K.S., Stempien,M.M., Brown-Shimmer,S.L., Gee,W.W., Renard,A., Randolph,A., Levy,J.A., Dina,D. and Luciw,P.A. (1985). Nucleotide sequence and expression of an AIDS-associated retrovirus (ARV-2). Science 227,484-492.

Schwartz,D.E., Tizard,R. and Gilbert,W. (1983). Nucleotide sequence sequence of Rous sarcoma virus. Cell 32, 853- 869.

Shimotohno,K., Takahashi,Y., Shimizu,N., Gojobori,T., Golde,D.W., Chen,I.S.Y., Miwa,M. and Sugimura,T. (1985). Complete nucleotide sequence of an infectious clone of human T-cell leukemia virus type II: an open reading frame for the protease gene. Proc. Natl. Acad. Sci. USA 82, 3101-3105.

Sonigo,P., Alizon,M., Staskus,K., Klatzmann,D., Cole,S., Danos,O., Retzel,E., Tiollais,P., Haase,A. and Wain-Hobson,S. (1985). Nucleotide sequence of the visna lentivirus: relationship to the AIDS virus. Cell 42, 369-382.

Sonigo,P., Barker,C., Hunter,E. and Wain-Hobson,S. (1986). Nucleotide sequence of Mason Pfizer monkey virus: an immunosuppressive D-type retrovirus. Cell, 45, 375-385.

Stephens,R,M., Casey,J.W. and Rice,N.R. (1986). Equine infectious anemia virus gag and pol genes: relatedness to visna and AIDS virus. Science 231, 589-594.

Wain-Hobson,S., Sonigo,P., Danos,O., Cole,S. and Alizon,M.

(1985). Nucleotide sequence of the AIDS virus, LAV. Cell 40, 9-17.
Weiss,R., Teich,N., Varmus,H. and Coffin,J. (1982). Molecular Biology of Tumor Viruses. (Cold Spring Harbor, New York: Cold Spring Harbor Laboratory).
Wilson,W., Malim,M.H., Mellor,J., Kingsman,A.J. and Kingsman,S.M. (1986). Expression strategies of the yeast transposon Ty: a short sequence directs ribosomal frameshifting. Nucl. Acids Res. 14, 7001-7015.
Wilson,W., Braddock,M., Adams,S.E., Rathjen,P.D., Kingsman,S.M. and Kingsman,A.J. (1988). HIV expression strategies: ribosomal frameshifting is directed by a short sequence in both mammalian and yeast systems. Cell 55, 1159-1169.
Yoshinaka, Y., Katoh, I., Copeland, T.D. and Oroszlan, S. (1985) Murine leukaemia virus protease is encoded by the gag-pol gene and is synthesised through suppression of an amber termination codon. Proc. Natl. Acad. Sci. USA 82, 1618-1622.

Protein Modifications and Mitochondrial Import of Yeast Cytochrome *c*: An Overview

Fred Sherman
Departments of Biochemistry and Biophysics
University of Rochester Medical School
Rochester, NY 14642, U.S.A.

Introduction

In addition to transcription and translation, the biosynthesis of mitochondrial cytochrome *c* involves the following co-translational and post-translational steps for formation and localization of the final product: excision of the amino-terminal methionine residue and amino-terminal acetylation of certain cytochromes *c*; trimethylation of specific lysine residues in fungal and plant cytochromes *c*; and the covalent attachment of protoheme to two cysteinyl residues, concomitant with transport into the mitochondrial membrane. Subsequently, mature cytochrome *c* interacts with a variety of other heme proteins to carry out various redox reactions. These processes, schematically outlined in Fig. 1, have been systematically investigated with the yeast *Saccharomyces cerevisiae*.

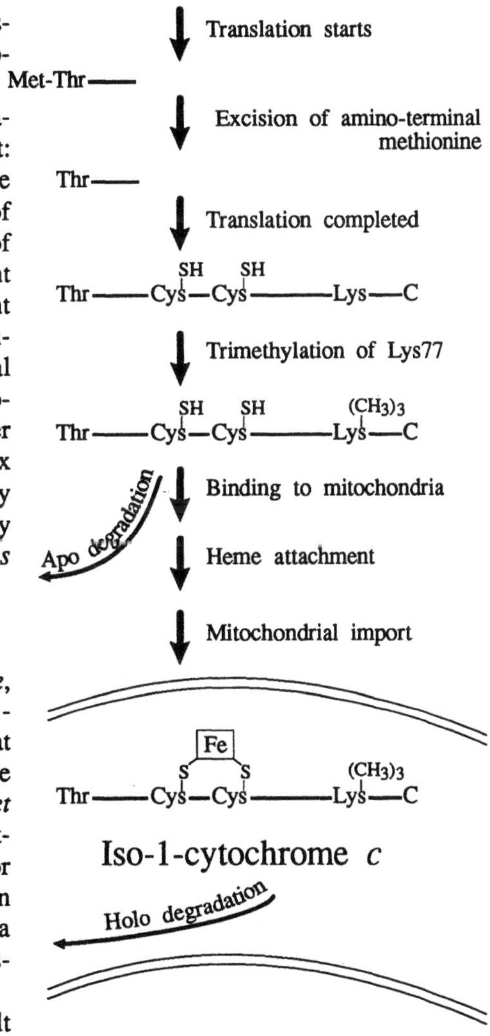

Fig. 1 Co-translational and post-translational processing and mitochondrial import of yeast iso-1-cytochrome *c*.

Iso-cytochromes *c* System

The yeast *Saccharomyces cerevisiae*, contains two forms of cytochrome *c*, iso-1-cytochrome *c* and iso-2-cytochrome *c* that constitute, respectively, 95% and 5% of the total cytochrome *c* complement (Sherman *et al.*, 1965). Mutations that directly or indirectly affect the structure or levels of just one or both of these iso-cytochromes *c* have been isolated and characterized in order to pursue a genetic approach for investigating the expression and maturation of these proteins.

Cytochrome *c* deficiencies can result from a variety of distinct classes of mutations that act generally or specifically; the *hem* mutants, defective in porphyrin or

NATO ASI Series, Vol. H 49
Post-Transcriptional Control of Gene Expression
Edited by J. E. G. McCarthy and M. F. Tuite

Table 1. Loci specificly and solely affecting cytochrome *c*

Locus	Gene product	Deficiencies in mutants		Reference
		Iso-1	Iso-2	
CYC1	Iso-1-cytochrome *c*	0	+	Sherman *et al.*, 1966
CYC7	Iso-2-cytochrome *c*	+	0	Downie *et al.*, 1977
CYC3	Heme lyase	0	0	Dumont *et al.*, 1987
CYC2	Import factor	±	±	J. B. Schlichter, M. E. Dumont and F. Sherman, unpublished

heme synthesis, lack cytochrome *c* and all other heme proteins as well. (Gollut *et al.*, 1977). Similarly, mutants particularly sensitive to catabolite repression have low levels of cytochrome *c* as well as low levels of other mitochondrial cytochromes (Parker and Mattoon, 1969). Also the *hap* mutations, which generally control transcription of certain heme proteins, either have little effect on the levels of iso-1-cytochrome *c* (*hap1*), (Verdière *et al.*, 1986) or are pleiotropic, reducing the levels of various different cytochromes (*hap2* and *hap3*) (Pinkham and Guarente, 1985). However, there are genes which specifically and solely effect iso-1-cytochrome *c* and iso-2-cytochrome *c*. The genetic symbol *cyc* is reserved for such genes, and these major loci are listed in Table 1. Transcription, translation, co-translational and post-translational processing, mito-chondrial import and protein stability have been examined with numerous alterations of the *CYC1* gene (Sherman, 1990). Furthermore, mitochondrial import has been investigated with mutations of the *CYC2* and *CYC3* genes. These studies have revealed detailed features not only of cytochrome *c* expression but also of general protein processing, such as excision of amino-terminal methionine residues and amino-terminal acetylation.

METHIONINE AMINOPEPTIDASE

The biosynthesis of all proteins from all prokaryotes and eukaryotes begins with methionine. During and after translation, the amino terminal regions of most proteins are subjected to a variety of modification and processing events. The N-formyl group is removed from proteins in prokaryotes by a deformylase enzyme, leaving a methionine residue at amino termini. The amino-terminal methionine subsequently can be cleaved from prokaryotic as well as eukaryotic proteins by a general methionine aminopeptidase. Early work with cell-free systems revealed enzymatic activities that remove amino-terminal residues of methionine from certain proteins. This excision is believed to occur before the completion of nascent polypeptide chains and before the occurrence of other amino-terminal processing events such as amino-terminal acetylation. Further amino-terminal maturation can take place if the protein is secreted, usually by removal of the signal sequence which is generally 15 to 30 residues in length. Because the amino-terminal region is dispensible and not required for mitochondrial import, altered forms of iso-1-cytochromes *c* are ideally suited for investigating the specificity of methionine aminopeptidase and amino-terminal acetylation. A large number of altered iso-1-cytochromes *c* have been uncovered by analyzing the series of mutations of the type: $CYC1^{+} \rightarrow cyc1\text{-}x \rightarrow CYC1\text{-}x\text{-}y$, where $CYC1^{+}$ denotes the wild-type gene that determines iso-1-cytochrome *c*, *cyc1-x* denoted mutations that cause deficiency or nonfunction of iso-1-cytochrome *c*, and *CYC1-x-y* denotes intra-genic reversions that restore at least partial activity and that give rise to either the normal or

Fig. 2 The percent cleavage of the amino-terminal methionine residue from two series of iso-1-cytochromes *c* having the sequences:

Met-XXX-Phe-Leu- ◎ and Met-XXX-Pro-Leu- ✦

(Moerschell *et al.*, 1990).

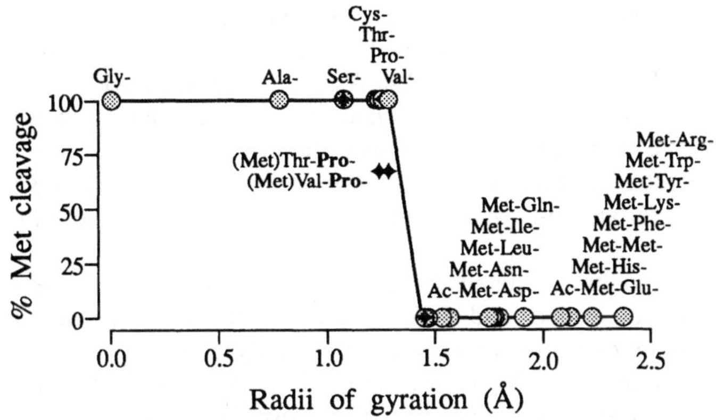

altered iso-1-cytochrome *c*. Over 500 *cyc1-x* mutants have been isolated and characterized and over 100 different iso-1-cytochrome *c* sequences have been obtained from *CYC1-x-y* revertants (Hampsey *et al.*, 1988). Some of these contain alterations in the amino-terminal region and have been the subject of a study on amino-terminal processing *in vivo* (Tunasawa *et al.*, 1985). Furthermore, transformation of a *cyc1* mutant directly with synthetic oligonucleotides (Moerschell *et al.*, 1988) has been used to systematically generate series of iso-1-cytochromes *c* with alterations in the amino-terminal region (Moerschell *et al.*, 1990). These iso-1-cytochromes *c* generated "randomly" and by synthetic oligonucleotides have revealed that cleavage of methionine is dependent primarily on the penultimate amino-terminal residue and modified by proline at position 3. The early results (Tunasawa *et al.*, 1985) was the bases for our hypothesis (Sherman *et al.*, 1985) that methionine is completely removed from penultimate residues having radii of gyration of 1.29 Å or less (glycine, alanine, serine, cysteine, threonine, proline, and valine). This hypothesis was confirmed by our systematic study (Moerschell *et al.*, 1990; Fig. 2), as well as other studies (Huang *et al.*, 1987; Hirel *et al.*, 1989). In addition, only partial cleavage occurred in the sequences Met-Thr-Pro-Leu- and Met-Val-Pro-Leu-, demonstating that proline at the third position inhibits methionine cleavage when the penultimate residue has an intermediate radius of gyration (Moerschell *et al.*, 1990; Fig. 2). Apparently, the retention of methionine is critical for preventing degradation of proteins with large residues at amino termini (Finley *et al.*, 1985).

AMINO-TERMINAL ACETYLATION

Amino-terminal acetylation is one of the most common protein modifications, present on 50 to 90% of all eukaryotic proteins. Amino-terminal acetylation occurs co-translationally in *in vitro* translation systems, usually when there are between 20 and 50 residues extruding from the ribosome. Acetylation may occur on the initiator methionine residue, but often the methionine is cleaved by a specific aminopeptidase and the newly exposed residue is then acetylated. Although the residues most frequently found to be acetylated are serine, alanine and methionine, other residues may also be substrates for this modification.

In normal strains, the amino-terminal methionine of iso-1-cytochrome *c* is cleaved and the newly exposed threonine is not acetylated (Fig. 1). However, mutant forms having certain alterations in the amino-terminal region are acetylated (Tsunasawa *et al.*, 1985; Moerschell *et al.*, 1988; 1990). These altered forms are being used to deduce the consensus sequences required for acetylation. In some sequences, only two or three specific residues specify acetylation. Acetylated forms included Ac-Met-Ile-Arg-, Ac-Met-Ile-Lys-, Ac-Met-Met-Asn-, Ac-Met-Asn-Asn-, Ac-Met-Glu-Phe-, Ac-Met-Asp-Phe-, and Ac-Ser-Glu-Phe-. Although the consenus for acetylation of the retained amino-terminal methionine is not completely known, these results and the results of published sequences indicated that Ac-Met-Glu- and Ac-Met-Asp- (methonine followed by an acidic residue) is sufficient for amino-terminal acetylation in eukaryotes but not in prokaryotes (Moerschell *et al.*, 1990).

In addition, altered forms of iso-1-cytochromes *c* are being used to define the specificity of distinct and different amino-terminal acetyltransferases. *NAT1* and *ARD1* encode proteins that function together to catalyze the amino-terminal acetylation of a subset of yeast proteins (Mullen *et al.*, 1989). The *CYC1-793* gene codes for an iso-1-cytochrome *c* that undergoes methionine cleavage followed by acetylation of the newly exposed amino-terminal serine: Ac-Ser-Glu-Phe-. However, iso-1-cytochrome *c* in a *CYC1-793 nat1*$^-$ strain is not acetylated, demonstrating that the NAT1-ARD1 amino-terminal acetyltransferase acts on Ser-Glu- proteins.

TRIMETHYLLYSINE 77

Mitochondrial cytochrome *c* from fungi and higher plants contain trimethyllysine; iso-1-cytochrome *c* and iso-2-cytochrome *c* from the yeast *S. cerevisiae*, cytochrome *c* from *Neurospora crassa*, as well as from other fungi, have a trimethyllysine residue confined to a single position corresponding to the amino acid position 72 of vertebrate cytochromes *c*, whereas cytochrome *c* from wheat germ and other higher plants have two trimethyllysine residues at positions 72 and 86 (Frost and Paik, 1990). In contrast, animal cytochromes *c* do not contain any trimethyllysine. A specific cytochrome *c* *S*-adenosylmethionine: protein-lysine *N*-methyltransferase, purified from the fungi *S. cerevisiae* and *N. crassa*, and from wheat germ, methylates only lysine 72 of the naturally unmethylated horse heart cytochrome *c* (Frost and Paik, 1990). Because apo-cytochrome *c* is preferred over holo-cytochrome *c* as a substrate, this methylase appears to act before heme attachment and mitochondrial import (DiMaria *et al.*, 1979). The specific activity of the methylase is lower in extracts of yeast grown under conditions of catabolite repression or anaerobiosis, the same conditions where cytochrome *c* is low. Furthermore, during anaerobic to aerobic adaptation, the methylase is induced parallel with cytochrome *c*, suggesting that the synthesis of cytochrome *c* and the methylase are at least partially coordinately regulated (Liao and Sherman, 1979).

In order to investigate the biological function and importance of cytochrome *c* methylation in yeast, we have used oligonucleotide-directed mutagenesis to construct strains having a single copy of the gene encoding iso-1-cytochrome *c* with an Arg77 replacing Lys77 (or position 72 of vertebrate cytochrome *c*) (Holzschu *et al.*, 1987). Although arginine residues share certain properties of lysine residues, it is not methylated. Low-temperature (-196°C) spectroscopic examination of intact cells indicated that the Arg77 protein was synthesized at the normal rate, or at least within 20% of the normal rate. Furthermore, growth curves in lactate medium indicated that the activity *in vivo* was normal or near normal; no differences in the growth pattern curves was observed with normal and Arg77 strains, although as little as a 10% reduction was detected in other studies with strains having other alterations. Enzymatic studies of the iso-1-cytochromes *c* with their normal physiological redox partners, cytochrome *c* peroxidase cytochrome b_2 and cytochrome *c* oxidase, indicated that the Arg77 iso-1-cytochrome *c* had normal or even enhanced activity *in vitro* (Holzschu *et al.*, 1987). Furthermore, normal or near normal levels and activities *in vivo* were observed with an iso-1-cytochrome *c* having an Asp77 replacement (F. Sherman and T. Cardillo, unpublished results). Because of the normal or near normal activity and levels of these altered iso-1-cytochromes *c in vivo*, no function could be assigned to the trimethylated Lys77, Also, these results with the Arg77 and Asp77 iso-1-cytochromes *c* throws doubt on the suggested role of trimethyllysine on import, deduced from *in vitro* experiments (Parks *et al.*, 1987). Nevertheless, the disadvantage of having replacements for Lys77, however slight, is sufficient to not only evolutionarily preserve this residue but also to retain the specific methylating enzyme.

CYTOCHROME *C* HEME LYASE AND MITOCHONDRIAL IMPORT

Cytochrome *c* is a mitochondrial protein located on the outer surface of the inner mitochondrial membrane where it participates in electron transport. Like most mitochondrial proteins, cytochrome *c* is encoded in the nucleus, synthesized in the cytoplasm, and than translocated into mitochondria. Like all known proteins that are imported into mitochondria, it is capable of being imported post-translationally. However, in contrast to most other imported mitochondrial proteins, cytochorme *c* does not require a membrane potential across the inner membrane (Zimmerman *et al.*, 1981), the precursor is not cleaved during import (see Hartl *et al.*, 1989, for review) and the amino-terminal region is not required for proper localization (Sherman and Stewart, 1973). Nevertheless, cytochrome *c* does undergo a major post-translational modification: attachment of heme to the apoprotein to form holo-cytochrome *c*. This modification is catalyzed by the enzyme cytochrome *c* heme lyase, which attaches the cysteine residues 19 and22 in the apoprotein to the propionic side chains.

Dumont *et al.* (1987) demonstrated that cytochrome *c* heme lyase was encoded in yeast by the nuclear gene *CYC3*. Strains carrying *cyc3*$^-$ null mutations completely lacked both iso-1-cytochrome *c* and iso-2-cytochrome *c* (Table 1). Enzyme assays of attachment of heme to apo-cytochrome *c* by mitochondrial extracts from yeast bearing different copy numbers of the functional *CYC3* gene suggested that *CYC3* encodes cytochrome *c* heme lyase. Mitochondrial extracts prepared from *cyc3*$^-$ strains exhibited greatly reduced heme lyase activity in the assay while extracts from strains bearing multiple copies of *CYC3* were greatly enriched in the activity (Dumont *et al.*, 1988). Furthermore, heme lyase activity was detected in an *Escherichia coli* expression system containing the yeast *CYC3* gene (Dumont *et al.*, 1988). Because of the important role of the heme in cytochrome *c* folding and evidence that an analog of protohemin could block uptake of apocytochrome *c* into mitochondria isolated from *Neurospora* (Hennig and Neupert, 1981), Dumont *et al.* (1988) examined the relationship between heme attachment and import of cytochrome *c* into mitochondria in an *in vitro* system. Apo-cytochrome *c* transcribed and translated *in vitro* could be imported with high efficiency into mitochondria isolated from normal yeast strains. However, no import of apo-cytochrome *c* occurred with mitochondria isolated from *cyc3*$^-$ strains, which lack cytochrome *c* heme lyase. In addition, amino acid substitutions in apo-cytchrome *c* at either of the 2 cysteine residues that are the sites of the thioether linkages to heme, or at an immediately adjacent histidine that serves as a ligand of the heme iron, resulted in a substantial reduction in the ability of the precursor to be translocated into mitochondria. In contrast, replacement of the methionine serving as the other heme ligand had no detectable effect on import of apo-cytochrome *c* in this system. Thus, covalent heme attachment is a required step for import of cytochrome *c* into mitochondria. These results suggest that protein folding initiated by heme attachment to apo-cytochrome *c* is required for import into mitochondria.

The results obtained *in vitro* are consistent with results obtained *in vivo*. Of the 30 single amino acid substitutions in iso-1-cytochrome *c* examined *in vivo* by Hampsey *et al.*

(1988), only replacements at residues Cys19, Cys22 and His23, were found to result in a complete deficiency of holo-iso-1-cytochrome *c*. The 2 cysteine residues are the sites of the thioether bond linkages in holo-cytochrome *c*. The histidine immediately adjacent to these cysteines is one of the ligands of the heme iron and may play a role in recognition by the heme lyase. The lack of cytochrome *c in vivo* apparently results from a failure to have heme attached, causing a blockage in import.

APO-CYTOCHORME *c* DEGRADATION AND REGULATION

Because heme attachment is required of cytochrome *c* import, a block of heme attachment *in vivo* would be expected to lead to an accumulation of apo-cytochrome *c* in the cytoplasm. Indeed, *cyc3*$^-$ mutants lack both holo-iso-1-cytochrome *c* and holo-iso-2-cytochrome *c*. Surprisingly, apo-iso-1-cytochrome *c* is absent in *cyc3*$^-$ strains, although apo-iso-2-cytochrome *c* is present at approximately the same level at which holo-iso-2-cytochrome *c* is found in *CYC3*$^+$ strains (Dumont *et al.*, 1990). The lack of apo-iso-1-cytochrome *c* is not due to a deficiency of either transcription of translation, but to rapid degradation of the protein. Apo-cytochromes *c* encoded by composite cytochrome *c* genes composed of the central portion of iso-2-cytochrome *c* flanked by amino and carboxyl regions of iso-1-cytochrome *c* exhibit increased stability compared with apo-iso-1-cytochrome *c*. A region encompassing no more than four amino acid differences between iso-1-cytochrome *c* and iso-2-cytochromes c is sufficient to partially stabilize the protein. In contrast to what is observed *in vivo* with the apo forms, the holo forms of the composite iso-cytochromes *c* were even less stable to thermal denaturation than iso-1-cytochrome *c* or iso-2-cytochrome *c*. Apparently certain regions or structures of apo-iso-1-cytochrome *c* are susceptible to a cytosolic degradation system (Dumont *et al.*, 1990).

Degradation of apo-iso-1-cytochrome *c* may play a role in regulating levels of the various forms of cytochromes *c* in cells. Blocking heme attachment to apo-cytochrome *c* prevents import into isolated mitochondria *in vitro*. Because apo-iso-cytochrome *c* is stable, the same block in import can be observed *in vivo* with heme lyase-deficient strains that express high levels of *CYC7* (M. E. Dumont and F. Sherman, unpublished results). The differential stability of the apo-iso-cytochromes *c* may be part of a regulatory process that increases the relative amount of iso-2-cytochrome *c*, compared with iso-1-cytochrome *c* under certain physiological conditions. Although holo-iso-1-cytochrome *c* and holo-iso-2-cytochrome *c* are normally present at approximately 95% and 5% relative levels, respectively, in derepressed cells, the absolute amount and relative proportions of the two iso-cytochromes c are strongly dependent on growth conditions. Catabolite (glucose) repression and the degree of aeration, two conditions that affect the absolute amount of the iso-cytochromes *c*, also influence their relative amounts. Under anaerobic conditions, or under aerobic conditions in the presence of high glucose levels, there is generalized repression of the synthesis of a number of mitochondrial and other enzymes, including those involved in the utilization of non-fermentable carbon sources. During the initial phase of catabolite derepression, in cultures that have exhausted fermentable substrates and are just inititating synthesis of these enzymes, holo-iso-2-cytochrome *c* can constitute as much as 25% of the total amount of the iso-cytochromes *c* (Sherman and Stewart, 1971). Similarly, when anaerobically grown yeast are induced with oxygen, the two iso-cytochromes *c* are produced at different rates, depending on the initial physiological state of the cultures, with holo-iso-2-cytochrome *c* sometimes comprising more than half the total complement of cytochrome *c*. Thus, yeast grown under partially catabolite-repressed or partially anaerobic conditions generally contain high proportions of holo-iso-2-cytochrome *c*. On the other hand, yeast grown under certain conditions of extreme aeration can have less than 1% holo-iso-2-cytochrome *c*. Even though mutants lacking iso-2-cytochrome *c* do not exhibit any obvious defect, and the structure and function of the two cytochromes *c* are very similar, higher levels of holo-iso-2-cytochrome *c* may confer some slight advantage in cells that are partially repressed.

Partially repressed yeast may use both differential transcription and differential stability of the apo-proteins as regulatory mechanisms for maintaining elevated proportions of iso-2-cytochrome c. Laz *et al.* (1984) demonstrated that *CYC7* mRNA is present at elevated levels compared with *CYC1* mRNA at the onset of derepression when glucose is exhausted from growth media. Furthermore, *CYC7* mRNA levels do not appear to be regulated by heme (Laz *et al.*, 1984) in spite of the fact that heme strongly affects *CYC1* mRNA levels, (Guarente and Mason, 1983) and appears to be an important general mediator of cellular responses to growth conditions.

Repression may diminish synthesis of heme lyase or other components required for import, including heme, causing a reduction in the efficiency of import. Apo-iso-2-cytochrome *c*, the more stable of the two apo-forms, would then be expected to be enriched in the cytoplasm and would thus constitute a higher fraction of the material that ultimately is imported and converted to holo-cytochrome *c*.

STABILIZATION OF HOLO-CYTOCHROME *C*

Individual amino acid residues each contribute to the overall stability of the protein. Evolutionary selection presumably optimizes protein structure with respect to function and stability so that "random" amino acid replacements generally are neutral or detrimental. Consistent with this view are the numerous replacements of iso-1-cytochrome *c* that resulted in a full range of altered stabilities, from little or no affect to greatly diminished stabilities. However, Das *et al.* (1989) combined genetic selection with oligonucleotide-directed mutagenesis to produce an altered iso-1-cytochrome *c* with an increase in stability *in vitro* and *in vivo*. Reversion of two missense mutants resulted in second-site replacements of Asn57 to Ile57. Introduction of the Ile57 replacement in an otherwise normal sequence caused a 17°C increase in the transition temperature (T_m) corresponding to a greater than two-fold increase in the energy change of thermal unfolding; ΔG^o for the normal iso-1-cytochrome *c* and the Ile57 mutant protein were equal to 5.4 and 12.5 kcal/mol, respectively, at 25°C. Although the Ile57 iso-1-cytochrome *c* is under intensive investigation, the reason for its unusual stability is not presently understood.

Surprisingly, strains containing the Ile57 iso-1-cytochrome *c* contain higher levels and activity *in vivo* (D. R. Hickey and F. Sherman, unpublished results). Low-temperature (-196°C) spectroscopic examination of intact cells revealed an increase of approximately 1.5 fold whereas growth in lactate medium revealed an activity of over 2 fold. Thus, the Ile57 iso-1-cytochrome *c* has both an increased thermodynamic stability *in vitro* and is resistant to a degration system *in vivo*. The proteolytic activities uncovered in mammalian mitochondria (Duque-Magalhães and Ferreira, 1980; Desautels and Goldberg, 1982) are likely candidates for this hypothetical degrading system. It is indeed puzzling why the Ile57 protein was not naturally selected and maintained by evolutionary pressures. It is tempting to speculate that the wild-type protein is evolutionarily maintained at its partially labile because of a regulatory requirement based on its degradation rate.

CONCLUDING REMARKS

Yeast iso-1-cytochrome *c* undergoes a number of co-translation and post-translation steps for synthesis and localization of the final holo-protein as outlined in Fig. 1. We have taken advantage of the genetic and biochemical techniques developed for the iso-cytochromes *c* system to investigate these processes. As outlined in Table 2, some of the processes are universal, whereas some are restricted to cytochrome *c* from a limited group of organisms. Methionine aminopeptidase appears to act similarly in *E. coli*, yeast and mammals, suggesting that the enzyme has the same or similar action in all organisms (Sherman *et al.*, 1985). Amino-terminal acetylation appears identical or similar in all eukaryotes, but different in prokaryotes (Moerschell. *et al.*, 1990). Because of the near normal amount of cytochrome *c* in yeast strains with genes encoding vertebrate cytochromes *c* (Clements *et al.*, 1989; Hickey *et al.*, 1990), heme attachment and mitochondrial import are generally the same for all eukaryotes. However, trimethylation of lysine in cytochrome *c* residues is restricted to the fungal and plant kingdoms (Frost and Paik, 1990). It remains to be seen if other organisms have degradation systems acting on apo-cytochrome *c* and holo-cytochrome *c* as postulated for yeast.

Table 2. Post-translational modifications:
Equivalent specificities amoung organisms and proteins

Modifications	Organisms and proteins
Methionine cleavage	Prokaryotes and Eukaryotes
Amino-terminal acetylation	Eukaryotes
Heme attachment and import	Eukaryotic Cytochromes *c*
Trimethylation of Lys77	Plant and Fungal Cytochromes *c*

ACKNOWLEDGMENT

The writing of this article was supported by United States Public Health Service Research Grant R01 GM12702 from the National Institutes of Health.

REFERENCES

Clements JM, O'Connell LI, Tsunasawa S, Sherman F (1989) Expression and activity of a gene encoding rat cytochrome *c* in the yeast *Saccharomyces cerevisiae*. Gene **83**: 1-14

Das G, Hickey DR, McLendon D, McLendon G, Sherman F (1989) Dramatic thermostabilization of yeast iso-1-cytochrome *c* by an Asn57 → Ile57 replacement. Proc Natl Acad Sci USA **86**:496-499

Desautels M, Goldberg AL (1982) Demonstration of an ATP-dependent, vanadate-sensitive endoprotease in the matrix of rat liver mitochondria. J Biol Chem **257**: 11673-11679

DiMaria P, Polastro E, DeLange RJ, Kim S, Paik WK (1979) Studies on cytochrome *c* methylation in yeast. J Biol Chem **254**:4645-4652

Downie JA, Stewart JW, Brockman N, Schweingruber AM Sherman F (1977) Structural gene for yeast iso-2-cytochrome *c*. J Mol Biol **113**:369-384

Dumont ME, Ernst JF, Sherman F (1987) Identification and sequence of the gene encoding cytochrome *c* heme lyase in the yeast *Saccharomyces cerevisiae*. EMBO J **6**:235-241

Dumont ME, Ernst JF, Sherman F (1988) Coupling of heme attachment to import of cytochrome *c* into yeast mitochondria: Studies with heme lyase deficient mitochondria and altered apocytochromes *c*. J Biol Chem **263**:15928-15937

Dumont ME, Mathews AJ, Nall BT, Baim SB, Eustice DC, Sherman F (1990) Differential stability of two apo-isocytochromes *c* in the yeast *Saccharomyces cerevisie*. J Biol Chem **265**:2733-2739

Duque-Magalhães MC, Ferreira MMM (1980) Cytochrome *c* degrading activity in rat liver mitochondria. Bichem Biophys Res Comm **93**:106-112

Finley D, Varshavsky A, (1985) In vivo half-life of a protein is a function of its amino-terminal residure. Trends Biochem Sci **10:** 343-347

Frost B, Paik WK (1990) Cytochrome *c* methylation. In: Paik WK, Kim S (eds) Protein Methylation. CRC Press, Boca Raton, Florida, pp. 60-77

Gollub EG, Liu K, Dayan J, Adlersberg M, Sprinson DB (1977) Yeast mutants deficient in heme biosynthesis and a heme mutant additionally blocked in cyclization of 2,3-oxidosqualene. J Biol Chem **252**:2846-2854

Guarente L, Mason T (1983) Heme regulates transcription of the *CYC1* gene of *Saccharomyces cerevisiae* via an upstream activation site. Cell **32**:1279-1286

Hampsey DM, Das G, Sherman F (1988) Yeast iso-1-cytochrome *c*: genetic analysis of structure-function relationships. FEBS Letters **231**:275-283

Hartl F-U, Pfanner N, Nicholson DW, Neupert W (1989) Mitochondrial protein import. Biochim Biophys Acta **998**:1-45

Hennig B, Neupert W (1981) Assembly of cytochrome *c*. Apocytochrome *c* is bound to specific sites on mitochondria before its conversion to holocytochrome *c*. Eur J Biochem **121**:302-312

Hickey, DR, Jayaraman K, Goodhue CT, Shah J, Clements JM, Tsunasawa S, Sherman F (1990) Synthesis and expression of genes encoding tuna, pigeon, and horse cytochromes *c* in the yeast *Saccharomyces cerevisiae.* J Biol Chem (in press)

Holzschu D, Principio L, Taylor K, Hickey DR, Short J, Rao R, McLendon G, Sherman F (1987) Replacement of the invariant lysine 77 by arginine in yeast iso-1-cytochrome *c* results in enhanced and normal activities *in vitro* and *in vivo.* J Biol Chem **262**:7125-7131

Hirel H-P, Schmitter J-M, Dessen P, Fayat G, Blanquet S (1989) Extent of N-terminal methionine excision from *Escherichia coli* proteins is governed by the side-chain length of the penultimate amino acid. Proc Natl Acad Sci USA **86**: 8247-8251

Huang S, Elliott RC, Liu PS, Koduri RK, Weickmann JL, Lee JH, Blair LC, Gosh-Dastidar P, Bradshaw RA, Bryan KM, Einarson B, Kendall RL, Kolacs KH, Saito K. (1987) Specificity of cotranslational amino-terminal processing of proteins in yeast. Biochemistry **26**: 8242-8246

Laz TM, Pietras DF, Sherman F (1984) Differential regulation of the duplicated iso-cytochrome *c* genes in yeast. Proc Natl Acad Sci USA **81**:4475-4479

Liao HN, Sherman F (1979) Yeast cytochrome *c*-specific protein-lysine methyltransferase: coordinate regulation with cytochrome *c* and activities in *cyc* mutants. J Bacteriol **138**:853-860

Moerschell RP, Tsunasawa S, Sherman F (1988) Transformation of yeast with synthetic oligonucleotides. Proc Natl Acad Sci USA **85**:524-528

Moerschell RP, Hosokawa Y, Tsunasawa S, Sherman F (1990) The specificities of yeast methionine aminopeptidase and acetylation of amino-terminal methionine *in vivo*: Processing of altered iso-1-cytochromes *c* created by oligonucleotide transformation. J Biol Chem (in press)

Mullen JR, Kayne PS, Moerschell RP, Tsunasawa S, Gribskov M, Colavito-Shepanski M, Grunstein M, Sherman F, Sternglanz R (1989) Identification and characterization of genes and mutants for an *N*-terminal acetyltransferase from yeast. EMBO J 8:2067-2075

Park KS, Frost B, Tuck M, Ho LL, Kim S, Paik WK (1987) Enzymatic methylation of *in vivo* synthesized apocytochrome *c* enhances its transport into mitochondria. J Biol Chem **262**:14702-14708

Parker JH, Mattoon JR (1969) Mutants of yeast with altered oxidative energy metabolism: Selection and genetic characterization. J Bact **100**:647-657

Pinkham JL, Guarente L (1985) Cloning and molecular analysis of the *HAP2* locus: a global regulator of respiratory genes in *Saccharomyces cerevisiae*. Mol Cell Biol 5:3410-3416

Sherman F (1990) Studies of yeast cytochrome *c*: How and why they started and why they continued. Genetics **124**: (in press)

Sherman F, Stewart JW (1971) Genetics and biosynthesis of cytochrome *c*. Ann Rev Genetics **5**:257-296

Sherman F, Stewart JW (1973) Mutations at the end of the iso-1-cytochrome *c* gene of yeast. In: Pollak JK, Lee JW (eds) The Biochemistry of Gene Expression in Higher Organisms. Australian and New Zealand Book Co PTY LTD, Sydney, pp. 56-86

Sherman F, Taber H, Campbell W (1965) Genetic determination of iso-cytochromes *c* in yeast. J Mol Biol **13**:21-39

Sherman F, Stewart JW, Margoliash E, Parker J, Campbell W (1966) The structural gene for yeast cytochrome *c*. Proc Natl Acad Sci USA **55**:1498-1504

Sherman F, Stewart JW, Tsunasawa S (1985) Methionine or not methionine at the beginning of a protein. BioEssays **3**:27-31

Tsunasawa S, Stewart JW, Sherman F (1985) Amino-terminal processing of mutant forms of yeast iso-1-cytochrome *c*. J Biol Chem **260**:5382-5391

Verdière J, Creusot F, Guarente L, Slonimski PP (1986) The overproducing *CYP1* mutation and the underproducing *hap1* mutations are alleles of the same gene which regulates in *trans* the expression of the structural genes encoding iso-cytochromes *c*. Curr Genet **10**:339-342

Zimmerman R, Hennig B, Neupert W, (1981) Different transport pathways of individual precurssor proteins into mitochondria. Eur J Biochem **116**:455-460

INDEX

Allosuppressors, 612 - 622
α-globin mRNA, 67, 250
Amino-terminal acetylation, 639 - 640
AMV RNA L, 390 - 394
Antibiotics
  chloramphenicol, 35
  sparsomycin, 229, 268
  spectinomycin, 571
  streptomycin, 573
Antisense oligodeoxyribonucleotides, 103 - 112
Antisense RNA, 93 - 102
Apo-cytochrome c
  degradation, 642 - 643
  regulation, 642 - 643
Arginine codons, 217 - 226
  specific repression, 341 - 345
atp operon, 159
atp TIR, 159 - 161, 294
AV-rich element (ARE), 67
Autoregulation, 129, 299 - 308
Autorepression, 450 - 452

Bacillus subtilis
  sdh and odh genes, 31
Bacteriophage f1, 197
Bacteriophage Lambda
  cIII gene, 277
  cII gene, 277
  N gene, 125 - 133
Bacteriophage MS2
  coat gene, 169
Bacteriophage Mu
  mom operon, 285 - 298
  com gene, 285 - 298
Bacteriophage T4
  gene 32, 21
ß-globin, 249 - 260
Binary complex formation, 436 - 437
boxA, 113 - 123, 125

Cap-binding proteins, 233
Cap-binding complex, 497 - 498
Cap-independent initiation, 391 - 393
Carbamoyl phosphate synthetase regulation, 337 - 345
Cauliflower mosaic virus (CaMV)
  35S RNA expression, 348 - 357
  genome organisation, 347 - 349
  leader mutants, 351
  polyadenylation, 359 - 365
Cell-free translation systems
  eIF-4A-dependent, 489 - 490
  eIF-4E-dependent, 490 - 493
Co-eIF-2A, 521 - 526
c-fos gene, 65 - 71
ColE1 DNA replication, 460
Chloroplasts
  mRNA stability, 73 - 82
  atpB gene, 75
Circular RNA templates, 227
Codon bias, 438
Codon usage, 207 - 216, 217 - 226
Coupling
  initiation and elongation, 153
  mRNA degradation/translation, 31 - 44, 52
COX3 gene, 413 - 416
CPA1 gene
  translational control, 337 - 345
  regulatory mutants, 339 - 345
Cytochrome-heme lyase

(<u>CYC3</u>), 641 - 642

Double stranded RNA
activator/inhibitor (DAI/dsI), 380 - 385, 527
activation, 382 - 384
RNA interactions, 385 - 387

<u>Drosophila</u>
heat shock genes, 83 - 91

<u>E.coli</u>
<u>ampC</u> gene, 145
<u>atp</u> operon, 159
<u>galE</u> gene, 146
<u>gnd</u> gene, 145
<u>lamB</u> gene, 146
L11 operon, 300
<u>ompA</u> gene, 31
OxyR protein, 287
<u>rne</u> gene, 28
<u>trp</u> operon, 135

EF-2 kinase, 531 - 533
Elongation factor 2 (EF-2), 531 - 535
EF-3
ATP hydrolysis, 558 - 562
role in elongation, 562 - 566
EF-G, 569 - 572
Endoribonucleases, 31, 21, 1 - 11, 77 - 78
RNase K, 31
RNase E, 21
RNase III, 21, 127, 165, 278, 282
Enhancing sequences, 160
Epsilon sequence, 160, 186
Exoribonucleases, 1 - 11, 36
RNase II, 79

Ferritin
post-transcriptional regulation, 399 - 407
Filamentous phages, 197
Fold-back interaction (fbi), 96
Foreign gene expression, 159, 169
Frame-jumping, 584 - 588
Frame-shifting, 579 - 584, 591 - 601, 603, 623 - 633
RF2, 463, 581 - 583
retroviral, 582 - 583, 623 - 633
<u>trpR</u>, 592 - 601
Frameshift errors, 573 - 575
Frameshift suppressors, 412, 568
Fusidic acid, 570

Gag-pol gene(s), 605, 623 - 633
Gamma-globin mRNA; 249 - 260
<u>GCD</u> factors, 328 - 333
<u>GCN4</u> gene
translational control, 325 - 335, 345, 356
<u>trans</u>-acting regulators, 328 - 333
Gene 60, 584 - 588
General control of amino acid biosynthesis, 325 - 326, 337
Glycolytic genes, 421 - 422, 428
ß-glucuronidase (GUS), 349
Growth rate effects, 33, 145 - 155, 214, 222
Guanidine nucleotide exchange factor (GEF), 378, 525, 539 - 548

Hairpin structures, 15, 169, 241
compensatory mutants in, 171
Heat-shock response, 83
Helicase, 140
Heme-controlled regulator (HCR), 378 - 380, 523 - 531

Heparin, 380
Hereditary hemochromatosis, 402
Histone H4 mRNA, 103 - 112
HIV-1
    frameshifting, 624 - 633
    polyadenylation, 364 - 365
    REV response element (RRE), 369 - 372
    REV trans-activator, 367 - 375
    TAR element, 368, 385
    Tat trans activator, 368
HIV-2, 367
Hsp 70, 83

Initiation codon
    selection in E.coli, 434 - 437
    selection in yeast, 467 - 468
Initiation codon mutations
    HIS4, 467 - 468
    mitochondrial, 412 - 416
    suppressors, 468 - 470
Initiation factors: prokaryotic
    IF1, 434
    IF2, 434
    IF3, 181, 187, 443 - 452
Initiation factors: eukaryotic
    eIF-2, 249 - 260, 330 - 332, 377 - 384, 481, 511 - 515, 521 - 531
    eIF-3, 522 - 525
    eIF-4A, 237, 267
    eIF-4D, 482
    eIF-4E, 267, 490 - 493, 497 - 506
    eIF-4F, 237
Initiation factor genes: prokaryotic
    infC (E.coli), 443 - 447
    infC (B.stearothermophilus), 444 - 447
Initiation factor genes: eukaryotic
    mammalian, 475 - 482
    SUI2 (eIF-2α), 471 - 472
    SUI3 (eIF-2β), 470 - 471
    TIF (eIF-4A), 489 - 490
Initiator tRNA
    start site selection, 467 - 468
Insertion element IS10, 93
    transposase, 93
    'fbi', 96
    'pot', 100
    antisense RNA, 93
Intercistronic regions, 166
Internal initiation, 352, 391 - 393
Inverted repeat (IR) sequences, 74
IR-binding proteins, 80
IRE-binding protein, 402 - 407
Iron responsive element (IRE), 400 - 407
Iso-1-cytochrome c
    mitochondrial import, 642 - 645
    modifications, 638 - 642

Lodish model, 392
lsd mutants, 551 - 554
Lysyl-tRNA synthetase
    gene (herC/lysS), 460 - 463

Malignant transformation, 502 - 505
Metallothionein promoter, 85
Methionine aminopeptidase, 52 - 53
Mitochondrial import, 641 - 642
Mitochondrial translation system, 411 - 413
mRNA
    cap, 227 - 236, 262, 264
    cis-acting determinants, 45 - 54
    chimeric, 49
    degradation, 1 - 92

3′end formation, 135
leader sequence, 229, 240, 241, 269
mutations in, 55 - 63
pseudoknot, 303
recognition sequences, 158 - 160
secondary structure, 6, 130, 147, 237, 240, 265, 278, 291
structure in yeast, 237
tertiary structure, 303
trans-acting factors, 51

Nonsense suppressors, 611 - 613
Nus factors, 125 - 133
nut sequences, 125 - 133

Okadaic acid, 531 - 532
Omega sequence, 160, 189, 261, 352 - 354, 392, 493
Orchophragmus violaceus, 349

p67, 512 - 515
Peptide chain initiation mechanism, 511 - 519
PET mutants, 413 - 416
PET 122/123 proteins, 414 - 416
Phosphofructokinase gene (PFK1), 428
Phosphorothioate linkages, 103, 107
Phosphorylation
EF-2, 531 - 533
eIF-2, 377 - 384, 513 - 515, 527 - 531
eIF-4E, 498 - 502
GEF, 543 - 544
Plant viruses, 262
Plasmid copy number, 423 - 424
P_L-promoter, 126
Polarity, 148
Poly(A)-binding protein (PAB), 61, 549 554
Polyadenylation
AATAAA signal, 361 - 365
CaMV, 359 - 365
control, 363 - 365
signal, 141
tail, 60
Polycistronic operons, 157, 166, 197
Polysome analysis, 426 - 428
Poliovirus RNA, 391
Processivity, 572-575
Proofreading factors, 619
Protein phosphatases, 533 - 535
puf mRNA, 13, 37
Pyruvate kinase gene (PYK1)
overexpression, 422 - 431
regulation, 421 - 431

Reinitiation of translation, 328 - 330
Release factors, 619
Release factor 1 (RF1)
gene (prfA), 455
Release factor 2 (RF2)
autogenous control, 458
codon recognition, 461 - 462
frameshifting, 581 - 582, 587
gene (prfB), 455 - 459
REP sequence, 36, 165
Retroviruses, 360
Rev response element (RRE), 369 - 372
Rho protein, 135
Rhodobacter capsulata, 13
Ribonucleases, 1 - 11
Ribosomes
binding, 169, 279
binding site (RBS), 145, 158, 169, 301
43S complex, 237

frameshifting, 579 - 584, 591 - 601, 623 - 636
landing pad, 394
mRNA decay, 18, 34
scanning, 227 - 236
30S, 157
40S subunits, 227 - 236, 237, 261
subunits: L1, 300
L11, 300
L46, 550 - 554
S1, 161, 436 - 437
S4, S15, S20, 303 - 304
S21, 187
RNA-binding proteins, 372 - 74
RNA end formation (REF) genes, 139
Rous Sarcoma Virus (RSV), 363 - 364
16S rRNA, 158, 185 - 196, 438 - 439
18S rRNA, 356, 438

Sal4 (SUP45) gene
gene product, 613 - 615, 618 - 620
mutants, 615 - 618
Scanning model, 227 - 236, 325, 467 - 468
Shine-Dalgarno sequence, 145, 158, 181, 185, 434 - 436
spb mutants, 550 - 554
Start codon
context, 227 - 236, 237 - 248
selection, 227 - 236
Stem-loop structures, 15, 36 , 74, 130, 169, 229, 291
Super-repressors, 310

Ternary complex, 181, 521 - 526
Threonyl-tRNA synthetase, 309
Tobbacco mosaic virus (TMV), 261
126 kDA polypeptide, 262
Toeprinting, 95, 281, 312, 433 - 437
TPA (12-O-tetradecanoyl-phorbol-13-acetate), 514
Transferrin receptor (TFR)
post-transcriptional regulation, 399 - 406
Translational
activation, 282, 285 - 298
coupling, 197
enhancers, 185 - 196, 266
initiation region (TIR), 157, 294
kinetics, 181, 207 - 216
operator, 309
repression, 299 - 308, 309

Transcriptional
antitermination, 125 - 133
termination, 135, 165
Transposase, 93
Trimethyllysine, 640 - 641
tRNA
concentration in the cell, 210. 217
precursors, 7
processing, 7, 77
trp operon, 135
trp repressor (trpR)
frameshifting, 592 - 601
mutants, 597 - 598
Ty transposon, 623, 633

Untranslated leader sequence
AMV RNA4, 390 - 395
enhancing translational efficiency, 393 - 395
ferritin, 400
HIS4, 465 - 466
PYKI, 428 - 431
Upstream open reading frames (uORF)

CPA1, 339 - 345
GCN4, 325 - 330
3′-UTR, 52, 75, 87, 256

VA RNA, 381 - 387
DAI interactions, 381 - 387
Verrucarin A, 328

Xenopus, 103 - 112

Yeast
ADH1 gene, 57
HIS3 gene, 45 -54
MATα1 gene, 45 - 54
mRNA stability, 45 - 54
PGK gene, 242
REF genes, 139
RNA polymerase II, 45
RNA15 protein, 59
STE3 gene, 54 - 54

NATO ASI Series H

Vol. 1: **Biology and Molecular Biology of Plant-Pathogen Interactions.**
Edited by J.A. Bailey. 415 pages. 1986.

Vol. 2: **Glial-Neuronal Communication in Development and Regeneration.**
Edited by H.H. Althaus and W. Seifert. 865 pages. 1987.

Vol. 3: **Nicotinic Acetylcholine Receptor: Structure and Function.**
Edited by A. Maelicke. 489 pages. 1986.

Vol. 4: **Recognition in Microbe-Plant Symbiotic and Pathogenic Interactions.**
Edited by B. Lugtenberg. 449 pages. 1986.

Vol. 5: **Mesenchymal-Epithelial Interactions in Neural Development.**
Edited by J.R. Wolff, J. Sievers, and M. Berry. 428 pages. 1987.

Vol. 6: **Molecular Mechanisms of Desensitization to Signal Molecules.**
Edited by T.M. Konijn, P.J.M. Van Haastert, H. Van der Starre, H. Van der Wel, and M.D. Houslay. 336 pages. 1987.

Vol. 7: **Gangliosides and Modulation of Neuronal Functions.**
Edited by H. Rahmann. 647 pages. 1987.

Vol. 8: **Molecular and Cellular Aspects of Erythropoietin and Erythropoiesis.**
Edited by I.N. Rich. 460 pages. 1987.

Vol. 9: **Modification of Cell to Cell Signals During Normal and Pathological Aging.**
Edited by S. Govoni and F. Battaini. 297 pages. 1987.

Vol. 10: **Plant Hormone Receptors.** Edited by D. Klämbt. 319 pages. 1987.

Vol. 11: **Host-Parasite Cellular and Molecular Interactions in Protozoal Infections.**
Edited by K.-P. Chang and D. Snary. 425 pages. 1987.

Vol. 12: **The Cell Surface in Signal Transduction.**
Edited by E. Wagner, H. Greppin, and B. Millet. 243 pages. 1987.

Vol. 13: **Toxicology of Pesticides: Experimental, Clinical and Regulatory Perspectives.**
Edited by L.G. Costa, C.L. Galli, and S.D. Murphy. 320 pages. 1987.

Vol. 14: **Genetics of Translation. New Approaches.**
Edited by M.F. Tuite, M. Picard, and M. Bolotin-Fukuhara. 524 pages. 1988.

Vol. 15: **Photosensitisation. Molecular, Cellular and Medical Aspects.**
Edited by G. Moreno, R.H. Pottier, and T.G. Truscott. 521 pages. 1988.

Vol. 16: **Membrane Biogenesis.** Edited by J.A.F. Op den Kamp. 477 pages. 1988.

Vol. 17: **Cell to Cell Signals in Plant, Animal and Microbial Symbiosis.**
Edited by S. Scannerini, D. Smith, P. Bonfante-Fasolo, and V. Gianinazzi-Pearson. 414 pages. 1988.

Vol. 18: **Plant Cell Biotechnology.**
Edited by M.S.S. Pais, F. Mavituna, and J.M. Novais. 500 pages. 1988.

Vol. 19: **Modulation of Synaptic Transmission and Plasticity in Nervous Systems.**
Edited by G. Hertting and H.-C. Spatz. 457 pages. 1988.

Vol. 20: **Amino Acid Availability and Brain Function in Health and Disease.**
Edited by G. Huether. 487 pages. 1988.

NATO ASI Series H

Vol. 21: **Cellular and Molecular Basis of Synaptic Transmission.**
Edited by H. Zimmermann. 547 pages. 1988.

Vol. 22: **Neural Development and Regeneration. Cellular and Molecular Aspects.**
Edited by A. Gorio, J. R. Perez-Polo, J. de Vellis, and B. Haber. 711 pages. 1988.

Vol. 23: **The Semiotics of Cellular Communication in the Immune System.**
Edited by E. E. Sercarz, F. Celada, N. A. Mitchison, and T. Tada. 326 pages. 1988.

Vol. 24: **Bacteria, Complement and the Phagocytic Cell.**
Edited by F. C. Cabello und C. Pruzzo. 372 pages. 1988.

Vol. 25: **Nicotinic Acetylcholine Receptors in the Nervous System.**
Edited by F. Clementi, C. Gotti, and E. Sher. 424 pages. 1988.

Vol. 26: **Cell to Cell Signals in Mammalian Development.**
Edited by S. W. de Laat, J. G. Bluemink, and C. L. Mummery. 322 pages. 1989.

Vol. 27: **Phytotoxins and Plant Pathogenesis.**
Edited by A. Graniti, R. D. Durbin, and A. Ballio. 508 pages. 1989.

Vol. 28: **Vascular Wilt Diseases of Plants. Basic Studies and Control.**
Edited by E. C. Tjamos and C. H. Beckman. 590 pages. 1989.

Vol. 29: **Receptors, Membrane Transport and Signal Transduction.**
Edited by A. E. Evangelopoulos, J. P. Changeux, L. Packer, T. G. Sotiroudis, and K. W. A. Wirtz. 387 pages. 1989.

Vol. 30: **Effects of Mineral Dusts on Cells.**
Edited by B. T. Mossman and R. O. Bégin. 470 pages. 1989.

Vol. 31: **Neurobiology of the Inner Retina.**
Edited by R. Weiler and N. N. Osborne. 529 pages. 1989.

Vol. 32: **Molecular Biology of Neuroreceptors and Ion Channels.**
Edited by A. Maelicke. 675 pages. 1989.

Vol. 33: **Regulatory Mechanisms of Neuron to Vessel Communication in Brain.**
Edited by F. Battaini, S. Govoni, M. S. Magnoni, and M. Trabucchi. 416 pages. 1989.

Vol. 34: **Vectors as Tools for the Study of Normal and Abnormal Growth and Differentiation.**
Edited by H. Lother, R. Dernick, and W. Ostertag. 477 pages. 1989.

Vol. 35: **Cell Separation in Plants: Physiology, Biochemistry and Molecular Biology.**
Edited by D. J. Osborne and M. B. Jackson. 449 pages. 1989.

Vol. 36: **Signal Molecules in Plants and Plant-Microbe Interactions.**
Edited by B. J. J. Lugtenberg. 425 pages. 1989.

Vol. 37: **Tin-Based Antitumour Drugs.** Edited by M. Gielen. 226 pages. 1990.

Vol. 38: **The Molecular Biology of Autoimmune Disease.**
Edited by A. G. Demaine, J-P. Banga, and A. M. McGregor. 404 pages. 1990.

Vol. 39: **Chemosensory Information Processing.** Edited by D. Schild. 403 pages. 1990.

Vol. 40: **Dynamics and Biogenesis of Membranes.**
Edited by J. A. F. Op den Kamp. 367 pages. 1990.

Vol. 41: **Recognition and Response in Plant-Virus Interactions.**
Edited by R. S. S. Fraser. 467 pages. 1990.

NATO ASI Series H

Vol. 42: **Biomechanics of Active Movement and Deformation of Cells.**
Edited by N. Akkaş. 524 pages. 1990.

Vol. 43: **Cellular and Molecular Biology of Myelination.**
Edited by G. Jeserich, H. H. Althaus, and T. V. Waehneldt. 565 pages. 1990.

Vol. 44: **Activation and Desensitization of Transducing Pathways.**
Edited by T. M. Konijn, M. D. Houslay, and P. J. M. Van Haastert. 336 pages. 1990.

Vol. 45: **Mechanism of Fertilization: Plants to Humans.**
Edited by B. Dale. 710 pages. 1990.

Vol. 46: **Parallels in Cell to Cell Junctions in Plants and Animals.**
Edited by A. W. Robards, W. J. Lucas, J. D. Pitts, H. J. Jongsma, and D. C. Spray. 296 pages. 1990.

Vol. 47: **Signal Perception and Transduction in Higher Plants.**
Edited by R. Ranjeva and A. M. Boudet. 357 pages. 1990.

Vol. 48: **Calcium Transport and Intracellular Calcium Homeostasis.**
Edited by D. Pansu and F. Bronner. 456 pages. 1990.

Vol. 49: **Post-Transcriptional Control of Gene Expression.**
Edited by J. E. G. McCarthy and M. F. Tuite. 671 pages. 1990.

MIX
Papier aus verantwortungsvollen Quellen
Paper from responsible sources
FSC® C105338

If you have any concerns about our products,
you can contact us on
ProductSafety@springernature.com

In case Publisher is established outside the EU,
the EU authorized representative is:
Springer Nature Customer Service Center GmbH
Europaplatz 3, 69115 Heidelberg, Germany

Printed by Libri Plureos GmbH
in Hamburg, Germany